Design of Wood Structures— ASD

Donald E. Breyer, P.E.
Professor Emeritus
Department of Engineering Technology
California State Polytechnic University
Pomona, California

Kenneth J. Fridley, Ph.D.
Professor and Head
Department of Civil and Environmental Engineering
University of Alabama
Tuscaloosa, Alabama

David G. Pollock, P.E., Ph.D.
Associate Professor
Department of Civil and Environmental Engineering
Washington State University
Pullman, Washington

Kelly E. Cobeen, S.E.
Principal
Cobeen & Associates Structural Engineering
Lafayette, California

Fifth Edition

McGraw-Hill

New York   Chicago   San Francisco   Lisbon   London   Madrid
Mexico City   Milan   New Delhi   San Juan   Seoul
Singapore   Sydney   Toronto

The McGraw·Hill Companies

Cataloging-in-Publication Data is on file with the Library of Congress

Copyright © 2003, 1998, 1993, 1988, 1980 by The McGraw-Hill Companies, Inc. All rights reserved. Printed in the United States of America. Except as permitted under the United States Copyright Act of 1976, no part of this publication may be reproduced or distributed in any form or by any means, or stored in a data base or retrieval system, without the prior written permission of the publisher.

4 5 6 7 8 9 0   DOC/DOC   0 9 8 7 6 5

ISBN 0-07-137932-0

The sponsoring editor for this book was Larry S. Hager and the production supervisor was Pamela A. Pelton. It was set in Century Schoolbook by Pro-Image Corporation. The art director for the cover was Anthony Landi.

Printed and bound by RR Donnelley.

McGraw-Hill books are available at special quantity discounts to use as premiums and sales promotions, or for use in corporate training programs. For more information, please write to the Director of Special Sales, McGraw-Hill Professional, Two Penn Plaza, New York, NY 10121-2298. Or contact your local bookstore.

Contents

Preface

The purpose of this book is to introduce engineers, technologists, and architects to the design of wood structures. It is designed to serve either as a text for a course in timber design or as a reference for systematic self-study of the subject.

The book will lead the reader through the complete design of a wood structure (except for the foundation). The sequence of the material follows the same general order that it would in actual design:

1. Vertical design loads and lateral forces

2. Design for vertical loads (beams and columns)

3. Design for lateral forces (horizontal diaphragms and shearwalls)

4. Connection design (including the overall tying together of the vertical- and lateral-force-resisting systems)

The need for such an overall approach to the subject became clear from experience gained in teaching timber design at the undergraduate and graduate levels.

This text pulls together the design of the various elements into a single reference. A large number of practical design examples are provided throughout the text. Because of their wide usage, buildings naturally form the basis of the majority of these examples. However, the principles of member design and diaphragm design have application to other structures (such as concrete formwork and falsework).

This book relies on practical, current industry literature as the basis for structural design. This includes publications of the American Forest and Paper Association, the International Code Council, the American Society of Civil Engineers, APA—The Engineered Wood Association, and the American Institute of Timber Construction.

In the writing of this text, an effort has been made to conform to the spirit and intent of the reference documents. The interpretations are those of the authors and are intended to reflect current structural design practice. The material presented is suggested as a guide only, and final design responsibility lies with the structural engineer.

The fifth edition of this book was promoted by three major developments:

1. Publication of new wood design criteria in the *2001 National Design Specification for Wood Construction (NDS)*

2. Publication of the comprehensive *Allowable Stress Design Manual for Engineered Wood Construction.*

3. Publication and increased adoption nationally of the 2003 *International Building Code.*

The National Design Specification (NDS) is published by the American Forest and Paper Association (AF&PA) and represents the latest structural design recommendations by the wood industry. The 2001 NDS contains new chapters covering prefabricated wood I-joists, structural composite lumber, wood structural panels, shearwalls and diaphragms, and fire design. Prior to the 2001 NDS, the design of prefabricated I-joists, structural composite lumber, and wood structural panels were not explicitly included in the NDS. In addition to these new chapters, the 2001 NDS has significantly revised chapters on connection design. While the basis for the design of mechanical fasteners remains largely unchanged, the equations and procedures have been unified.

The 2002 *Allowable Stress Design* (ASD) *Manual for Engineered Wood Construction* includes five product design supplements, four product guidelines, a manual providing additional guidance for the design of some of the most commonly used components of wood-frame buildings, and a separate supplement dealing specifically with special provisions for wind and seismic design. The ASD Manual was first introduced in 1999 for the 1997 NDS and for the first time brought together all necessary information required for the design of wood structures.

The *International Building Code* (IBC) is a product of the International Code Council (ICC). The ICC brought together the three regional model building code organizations to develop and administer a single national building code. The first edition of the IBC was published in 2000, and now nearly all of the states have adopted all or part of the IBC at either the state or local level.

Past performance of wood structures indicates wood to be a safe, durable and economical building material when it is used properly. The 1997 NDS represents the latest in structural design recommendations for wood. While the 1997 NDS does not contain as extensive of changes as the 1991 NDS (see the third edition of this text), a number of new or improved provisions for member and connection design are introduced. For example, more comprehensive provisions are now available for the design of notched beams and for wood-to-concrete connections. New provisions for the design of certain connection types such as wood-to-masonry connections and nailed connections in combined lateral and withdrawal loading have been added to the NDS.

The NDS is based on the principles of what is termed *allowable stress design* (ASD). In ASD *allowable stresses* of a material are compared to calculate *working stresses* resulting from *service loads*. Recently, the wood industry and design community completed the development of a *load and resistance factor design* (LRFD) specification for wood construction. In LRFD, *factored nominal capacities* (resistance) are compared to the effect of *factored loads*. The factors are developed for both resistance and loads such that uncertainty and consequence of failure are explicitly recognized. The LRFD approach to wood design is provided in the LRFD Manual for Engineered Wood Construction, which is published by the American Forest and Paper Association. It is expected that LRFD will eventually replace the ASD approach of the NDS, but currently the NDS is the popular choice among design professionals and is the focus of the fourth edition of this text. The authors are currently preparing an LRFD version of *Design of Wood Structures*. To easily distinguish between the ASD and LRFD versions of the text, ASD was added to the title of the text, and LRFD will be added to the title of the forthcoming LRFD version.

If this book is used as a text for a formal course, an Instructor's Manual is available. Requests on school letterhead should be sent to: Civil Engineering Editor, McGraw-Hill Professional, 2 Penn Plaza, New York, NY 10121-2298.

Questions or comments about the text or examples may be addressed to any of the authors. Direct any correspondence to:

Prof. Donald E. Breyer
Department of Engineering Technology
California State Polytechnic University
3801 West Temple Avenue
Pomona, CA 91768

Prof. David G. Pollock
Department of Civil and Environmental
  Engineering
Washington State University
P.O. Box 642910
Pullman, WA 99164-2910

Prof. Kenneth J. Fridley
Department of Civil and Environmental
  Engineering
University of Alabama
Box 870205
Tuscaloosa, AL 35487-0205

Ms. Kelly E. Cobeen
Cobeen & Associates Structural Engineering
251 Lafayette Circle, Suite 230
Lafayette, CA 94549

Acknowledgment and appreciation for help in writing the fifth edition are given to Philip Line and Bradford Douglas of the American Forest and Paper Association; Jeff Linville of the American Institute of Timber Construction; John Rose, Thomas Skaggs, and Thomas Williamson of APA—The Engineered Wood Association; and Kevin Cheung of the Western Wood Products Association. Numerous other individuals also deserve recognition for their contributions to the previous three editions of the text, including Russell W. Krivchuk, William A. Baker, Michael Caldwell, Thomas P. Cunningham, Jr., Mike Drorbaugh, John R. Tissell, Ken Walters, B. J. Yeh, Thomas E. Brassell, Frank Stewart, Lisa Johnson, Edwin G. Zacher, Edward F. Diekmann, Lawrence A. Soltis, Robert Falk, Don Wood, William R. Bloom, Frederick C. Pneuman, Robert M. Powell, Sherm Nelson, Bill McAlpine, Karen Colonias, and Ronald L.

Carlyle. Suggestions and information were obtained from many other engineers and suppliers, and their help is gratefully recognized.

Dedication

To our families:

Matthew, Kerry, Daniel, and Sarah

Paula, Justin, Connor, and Alison

Lynn, Sarah, and Will

Chris and Matthew

Donald E. Breyer, P.E.
Kenneth J. Fridley, Ph.D.
David G. Pollock, P.E., Ph.D.
Kelly E. Cobeen, S.E.

Nomenclature

Organizations

AF&PA
American Forest and Paper Association
American Wood Council (AWC)
1111 19th Street, NW, Suite 800
Washington, DC 20036
www.afandpa.org
www.awc.org

AITC
American Institute of Timber
  Construction
7012 South Revere Parkway, Suite 140
Englewood, CO 80112
www.aitc-glulam.org

ALSC
American Lumber Standard Committee,
  Inc.
P.O. Box 210
Germantown, MD 20875-0210
www.alsc.org

APA
APA—The Engineered Wood Association
P.O. Box 11700
Tacoma, WA 98411-0700
www.apawood.org

ASCE
American Society of Civil Engineers
1801 Alexander Bell Drive
Reston, VA 20191
www.asce.org

ATC
Applied Technology Council
201 Redwood Shores Parkway,
  Suite 240
Redwood City, CA 94065
www.atcouncil.org

AWPA
American Wood-Preservers' Association
P.O. Box 5690
Granbury, TX 76049
www.awpa.com

AWPI
American Wood Preservers Institute
12100 Sunset Hills Road, Suite 130
Reston, VA 20190
www.preservedwood.com

BSSC
Building Seismic Safety Council
National Institute of Building Sciences
1090 Vermont Avenue, N.W., Suite 700
Washington, DC 20005
http://www.bssconline.org/

CANPLY
Canadian Plywood Association
735 West 15 Street
North Vancouver, British Columbia,
  Canada V7M 1T2
www.canply.org

CWC
Canadian Wood Council
1400 Blair Place, Suite 210
Ottawa, Ontario, Canada K1J 9B8
www.cwc.ca

CPA–CWC
Composite Panel Association
Composite Wood Council
18922 Premiere Court
Gaithersburg, MD 20879-1574
301-670-0604
www.pbmdf.com

FPL
U.S. Forest Products Laboratory
USDA Forest Service
One Gifford Pinchot Drive
Madison, WI 53726-2398
www.fpl.fs.fed.us

ICC
International Code Council
5203 Leesburg Pike
Suite 600
Falls Church, VA 22041
www.iccsafe.org

ISNATA
International Staple, Nail and Tool
 Association
512 West Burlington Avenue, Suite 203
La Grange, IL 60525-2245

MSRLPC
MSR Lumber Producers Council
P.O. Box 6402
Helena, MT 59604
www.msrlumber.org

NFBA
National Frame Builders Association
4840 West 15th Street, Suite 1000
Lawrence, KS 66049-3876
www.postframe.org

NHLA
National Hardwood Lumber Association
P.O. Box 34518
Memphis, TN 38184-0518
www.natlhardwood.org

NLGA
National Lumber Grades Authority
#406 First Capital Place
960 Quayside Drive
New Westminster, British Columbia,
 Canada V3M 6G2
www.nlga.org

NELMA
Northeastern Lumber Manufactures
 Association
272 Tuttle Road
P.O. Box 87A
Cumberland Center, ME 04021
www.nelma.org

NSLB
Northern Softwood Lumber Bureau
272 Tuttle Road
P.O. Box 87A
Cumberland Center, ME 04021
www.nelma.org

NWPA
Northwest Wood Products Association
64672 Cook Avenue, Suite B
Bend, OR 97701
www.nwpa.org

PLIB
Pacific Lumber Inspection Bureau
33442 First Way South, #300
Federal Way, WA 98003-6214
www.plib.org

SEAOC
Structural Engineers Association of
 California
555 University Avenue, Suite 126
Sacramento, CA 95825
www.seaoc.org

SLMA
Southeastern Lumber Manufactures
 Association
P.O. Box 1788
Forest Park, GA 30298-1788
www.slma.org

SFPA
Southern Forest Products Association
P.O. Box 641700
Kenner, LA 70064-1700
Street address: 2900 Indiana Avenue
Kenner, LA 70065
www.sfpa.org
www.southernpine.com

SPIB
Southern Pine Inspection Bureau, Inc.
4709 Scenic Highway
Pensacola, FL 32504-9094
www.spib.org

SBA
Structural Board Association
45 Sheppard Avenue East, #412
Toronto, Ontario, Canada M2N 5W9
www.osbguide.com

TPI
Truss Plate Institute
583 D'Onofrio Drive, Suite 200
Madison, WI 53719
www.tpinst.org

WCLIB
West Coast Lumber Inspection Bureau
P.O. Box 23145
Portland, OR 97281-3145
www.wclib.org

WRCLA
Western Red Cedar Lumber Association
1200-555 Burrard Street
Vancouver, British Columbia, Canada
V7X 1S7
www.wrcla.org

WWPA
Western Wood Products Association
522 Southwest Fifth Avenue, Suite 500
Portland, OR 97204-2122
www.wwpa.org

WIJMA
Wood I-Joist Manufacturing Association
200 East Mallard Drive
Boise, ID 83706
www.i-joist.org

WTCA
Wood Truss Council of America
One WTCA Center
6300 Enterprise Lane
Madison, WI 53719
www.woodtruss.com

Publications

ASCE 7: American Society of Civil Engineers (ASCE). 2002. *Minimum Design Loads for Buildings and Other Structures* (ASCE 7-02), ASCE, New York, NY.

ASD Manual: American Forest and Paper Association (AF&PA). 2001. *Allowable Stress Design Manual for Engineered Wood Construction and Supplements and Guidelines,* 2001 ed., AF&PA, Washington DC.

IBC: International Codes Council (ICC). 2003. *International Building Code,* 2003 ed., ICC, Falls Church, VA.

NDS: American Forest and Paper Association (AF&PA). 2001. *National Design Specification (NDS) for Wood Construction,* 2001 ed., AF&PA, Washington DC.

TCM: American Institute of Timber Construction (AITC). 1994. *Timber Construction Manual,* 4th ed., AITC, Englewood, CO.

Additional publications given at the end of each chapter.

Units

ft	foot, feet	mph	miles per hour
ft^2	square foot, square feet	pcf	pounds per cubic foot (lb/ft^3)
in.	inch, inches	plf	pounds per lineal foot (lb/ft)
in.^{2}	square inch, square inches	psf	pounds per square foot (lb/ft^2)
k	1000 lb (kip, kilopound)	psi	pounds per square inch (lb/in.^{2})
ksi	kips per square inch (k/in.^{2})	sec	second

Abbreviations

Allow.	allowable	DF-L	Douglas Fir-Larch
ASD	allowable stress design	Ecc.	eccentric

B&S	Beams and Stringers		
C.-to-c.	center to center	MSR	machine stress rated lumber
cg	center of gravity		
EMC	equilibrium moisture content	NA	neutral axis
FBD		o.c.	on center
FS	free-body diagram	OM	overturning moment
FSP	factor of safety	OSB	oriented strand board
glulam	fiber saturation point	PL	plate
ht	structural glued laminated timber	P&T	Posts and Timbers
IP	height	PSL	parallel strand lumber
J&P	inflection point (point of reverse curvature and point of zero moment	Q/A	quality assurance
lam	of zero moment	Req'd	required
LF	Joists and Planks	RM	resisting moment
LRFD	lamination	S4S	dressed lumber (surfaced four sides)
LFRS	Light Framing		
LVL	load and resistance factor design	Sel. Str.	Select Structural
max.		SCL	structural composite lumber
MC	lateral-force-resisting system	SJ&P	Structural Joists and Planks
MDO	laminated veneer lumber		
MEL	maximum	SLF	Structural Light Framing
min.	moisture content based on oven-dry weight of wood	Tab.	tabulated
		T&G	tongue and groove
	medium density overlay (plywood)	TL	total load (lb, k, lb/ft, k/ft, psf)
	machine evaluated lumber		
	minimum	trib.	tributary
		TS	top of sheathing
		WSD	working stress design

Symbols

A	area (in.^{2}, ft^2)
a	acceleration
a	length of end zone for wind pressure calculations
A_g	gross cross-sectional area of a tension or compression member (in.^{2})
$A_{group\text{-}net}$	net cross-sectional area between outer rows of fasteners for a wood tension member (in.^{2})
A_h	projected area of hole caused by drilling or routing to accommodate bolts or other fasteners at net section (in.^{2})

A_n	cross-sectional area of member at a notch (in.^{2})
A_n	net cross-sectional area of a tension or compression member at a connection (in.^{2})
a_p	in-structure component amplification factor
A_s	area of reinforcing steel (in.^{2})
A_s	sum of gross cross-sectional areas of side member(s) (in.^{2})
A_t	tributary area for a structural member or connection (ft^2)
A_{web}	cross-sectional area of the web of a steel W-shaped beam or wood I joist (in.^{2})
A_x	diaphragm area immediately above the story being considered (ft^2)
b	length of shearwall parallel to lateral force; distance between chords of shearwall (ft)
b	width of horizontal diaphragm; distance between chords of horizontal diaphragm (ft)
b	width of rectangular beam cross section (in.)
B_t	allowable tension on anchor bolt embedded in concrete or masonry
B_v	allowable shear on anchor bolt embedded in concrete or masonry
C	compression force (lb, k)
c	buckling and crushing interaction factor for columns
c	distance between neutral axis and extreme fiber (in., ft)
C_b	bearing area factor
C_D	load duration factor
C_d	seismic deflection amplification factor
C_{di}	diaphragm factor for nail connections
C_e	exposure factor for snow load
C_{eg}	end grain factor for connections
C_F	size factor for sawn lumber
C_f	form factor for bending stress
C_{fu}	flat use factor for bending stress
C_G	grade and construction factor for wood structural panels
C_g	group action factor for connections
C_i	incising factor for sawn lumber
C_L	beam stability factor
C_M	wet service factor for high-moisture conditions
C_P	column stability factor
C_p	seismic response coefficient for determining force on a portion of a structure
C_r	repetitive-member factor (bending stress) for Dimension lumber
C_s	roof slope factor for snow load
C_s	seismic response coefficient

C_s	panel size factor for wood structural panels
C_{st}	metal side plate factor for 4-in. shear plate connections
C_T	buckling stiffness factor for 2×4 and smaller Dimension lumber in trusses
C_t	seismic coefficient depending on type of LFRS used to calculate period of vibration T
C_t	temperature factor
C_t	thermal factor for snow load
C_{tn}	toenail factor for nail connections
C_V	volume factor for glulam
C_{vx}	seismic vertical distribution factor
C_Δ	geometry factor for connections
D	dead load (lb, k, lb/ft, k/ft, psf)
D	diameter (in.)
d	cross-sectional dimension of rectangular column associated with axis of column buckling (in.)
d	depth of rectangular beam cross section (in.)
d	dimension of wood member for shrinkage calculation (in.)
d	pennyweight of nail or spike
d_1	shank diameter of lag bolt (in.)
d_2	pilot hole diameter for the threader portion of lag bolt (in.)
d_e	effective depth of member at a connection (in.)
d_n	effective depth of member remaining at a notch (in.)
d_x	width of rectangular column parallel to y axis, used to calculate column slenderness ratio about x axis
d_y	width of rectangular column parallel to x axis, used to calculate column slenderness ratio about y axis
E	earthquake force (lb, k)
E	length of tapered tip of lag bolt (in.)
E, E'	tabulated and allowable modulus of elasticity (psi)
e	eccentricity (in., ft)
E_{axial}	modulus of elasticity of glulam for axial deformation calculation (psi)
E_m	modulus of elasticity of main member (psi)
E_s	modulus of elasticity of side member (psi)
E_x	modulus of elasticity about x axis (psi)
E_y	modulus of elasticity about y axis (psi)
F	force or load (lb, k)
F	roof slope in inches of rise per foot of horizontal span
f_1	live load coefficient for special seismic load combinations
F_a	acceleration-based seismic site coefficient at 0.3 second period

f_b	actual (computed) bending stress (psi)
F_b, F_b'	tabulated and allowable bending (psi)
F_b^*	tabulated bending stress multiplied by all applicable adjustment factors except C_L (psi)
F_b^{**}	tabulated bending stress multiplied by all applicable adjustment factors except C_V (psi)
F_{bE}	critical buckling (Euler) value for bending member (psi)
f_{bx}	actual (computed) bending stress about strong (x) axis (psi)
F_{bx}, F_{bx}'	tabulated and allowable bending stress about strong (x) axis (psi)
f_{by}	actual (computed) bending stress about weak (y) axis (psi)
F_{by}, F_{by}'	tabulated and allowable bending stress about weak (y) axis (psi)
f_c	actual (computed) compression stress parallel to grain (psi)
F_c	out-of-plane seismic forces for concrete and masonry walls (lb, plf, k, klf)
F_c, F_c'	tabulated and allowable compression stress parallel to grain (psi)
F_c^*	tabulated compression stress parallel to grain multiplied by all applicable adjustment factors except C_P (psi)
F_{cE}	critical buckling (Euler) value for compression member (psi)
$f_{c\perp}$	actual (computed) compression stress perpendicular to grain (psi)
$F_{c\perp}, F_{c\perp}'$	tabulated and allowable compression stress perpendicular to grain (psi)
$F_{c\perp 0.2}, F_{c\perp 0.2}'$	reduced and allowable compression stress perpendicular to grain at a deformation limit of 0.02 in. (psi)
F_e	dowel bearing strength (psi)
$F_{e\parallel}$	dowel bearing strength parallel to grain for bolt or lag bolt connection (psi)
$F_{e\perp}$	dowel bearing strength perpendicular to grain for bolt or lag bolt connection (psi)
$F_{e\theta}$	dowel bearing strength at angle to grain θ for bolt or lag bolt connection (psi)
F_{em}	dowel bearing strength for main member (psi)
F_{es}	dowel bearing strength for side member (psi)
f_g	actual (computed) bearing stress parallel to grain (psi)
F_g, F_g'	tabulated and allowable bearing stress parallel to grain (psi)
F_p	allowable bearing stress for fastener in steel member (psi, ksi)
F_{px}	seismic story force at level x for designing the horizontal diaphragm (lb, k)
F_θ'	allowable bearing stress at angle to grain θ (psi)
f_s	stress in reinforcing steel (psi, ksi)
f_t	actual (computed) tension stress in a member parallel to grain (psi)
F_t, F_t'	tabulated and allowable tension stress parallel to grain (psi)
F_u	ultimate tensile strength for steel (psi, ksi)

F_v	velocity-based seismic site coefficient at 1.0 second period
F_v, F_v'	tabulated and allowable shear stress parallel to grain (horizontal shear) in a beam (psi)
f_v	actual (computed) shear stress parallel to grain (horizontal shear) in a beam using full design loads (psi)
f_v'	reduced (computed) shear stress parallel to grain (horizontal shear) in a beam obtained by neglecting the loads within distance d of face of support (psi)
F_x	seismic story force at level x for designing vertical elements (shearwalls) in LFRS (lb, k)
F_y	yield strength (psi, ksi)
F_{yb}	bending yield strength of fastener (psi, ksi)
g	acceleration of gravity
G_s	specific gravity of side member
h	building height or height of wind pressure zone (ft)
h	height of shearwall (ft)
h_i, h_x	height above base to level i level x (ft)
h_{mean}	mean roof height above ground (ft)
h_n	height above base to nth or uppermost level in building (ft)
h_x	the elevation at which a component is attached to a structure relative to grade, for design of portions of structures (ft)
I	moment of inertia (in.^{4}, ft^4)
I	importance factor for seismic force
I_p	seismic component importance factor
I_s	importance factor for snow load
I_W	importance factor for wind force
K	Code multiplier for DL for use in beam deflection calculations to account for creep effects.
k	exponent for vertical distribution of seismic forces related to the building period
K_{bE}	Euler buckling coefficient for beams
K_{cE}	Euler buckling coefficient for columns
K_D	diameter coefficient for nail and spike connections
K_e	effective length factor for column end conditions (buckling length coefficient for columns)
K_f	column stability coefficient for bolt and nail built-up columns
K_{LL}	live load element factor for influence area
K_θ	angle to grain coefficient for bolt and lag bolt connections
KS	effective section modulus for plywood (in.^{3})
L	live load (lb, k, lb/ft, k/ft, psf)
L	beam span length (ft)

L	length (ft)
l	length (in.)
l	length of bolt in main or side members (in.)
l	length of fastener (in.)
l	unbraced length of column (in.)
L_0	unreduced floor live load (lb, k, lb/ft, k/ft, psf)
l/D	bolt slenderness ratio
l_b	bearing length (in.)
L_c	cantilever length in cantilever beam system (ft)
l_e	effective unbraced length of column (in.)
l_e/d	slenderness ratio of column
$(l_e/d)_x$	slenderness ratio of column for buckling about strong (x) axis
$(l_e/d)_y$	slenderness ratio of column for buckling about weak (y) axis
l_e	effective unbraced length of compression side of beam (in.)
l_m	dowel bearing length of fastener in main member (in.)
L_r	roof live load (lb, k, lb/ft, k/ft, psf)
l_s	dowel bearing length of fastener in side member(s) (in.)
l_u	laterally unbraced length of compression side of beam (in.)
l_w	length of shearwall for calculation of l (ft)
l_x	unbraced length of column considering buckling about strong (x) axis (in.)
l_y	unbraced length of column considering buckling about weak (y) axis (in.)
M	bending moment (in.-lb, in.-k, ft-lb, ft-k)
M	mass
M_p	plastic moment capacity (in.-lb, in.-k)
M_u	ultimate (factored) bending moment (in.-lb, in.-k, ft-lb, ft-k)
M_y	yield moment (in.-lb, in.-k)
N	normal reaction (lb, k)
N	number of fasteners in connection
N, N'	nominal and allowable lateral design value at angle to grain θ for a single split ring or shear plate connector (lb)
n	number of fasteners in row
n	number of stories (seismic forces)
N_a	seismic near-source factor
n_{row}	number of rows of fasteners in a fastener group
N_v	seismic near-source factor
P	total concentrated load or force (lb, k)
P, P'	nominal and allowable lateral design value parallel to grain for a single split ring or shear plate connector (lb)

p	parallel-to-grain component of lateral force z on one fastener
p	penetration depth of fastener in wood member (in.)
p_g	ground snow load (psf)
p_{net}	net design wind pressure for components and cladding (psf)
p_{net30}	net design wind pressure for components and cladding at a height of 30 ft in Exposure B conditions (psf)
p_s	simplified design wind pressure for main wind force-resisting systems (psf)
p_{s30}	simplified design wind pressure for main wind force-resisting systems at a height of 30 ft in Exposure B conditions (psf)
P_u	collapse load (ultimate load capacity)
P_u	ultimate (factored) concentrated load or force (lb, k)
Q	static moment of an area about the neutral axis (in.^{3})
Q, Q'	nominal and allowable lateral design value perpendicular to grain for a single split ring or shear plate connector (lb)
q	perpendicular-to-grain component of lateral force z on one fastener
q	soil bearing pressure (psf)
q_a	soil bearing pressure under axial loads (psf)
q_b	bending soil bearing pressure caused by overturning moment (psf)
Q_E	effect of horizontal seismic forces (lb, k)
R	nominal calculated resistance of structure (*see* LRFD)
R	rain load (lb, k, plf, klf, psf)
R	reaction (lb, k)
R	seismic response modification factor
r	radius of gyration (in.)
R_1	roof live load reduction factor for large tributary roof areas
R_1	seismic force generated by mass of wall that is parallel to earthquake force being considered
R_2	roof live load reduction factor for sloped roofs
R_B	slenderness ratio of laterally unbraced beam
r_i	portion of story force resisted by a shearwall element.
r_{maxx}	element—story shear ratio
R_p	seismic response modification factor for a portion of a structure
R_u	ultimate (factored) reaction (lb, k)
R_{u1}	ultimate (factored) seismic force generated by mass of wall that is parallel to earthquake force being considered
S	snow load (lb, k, plf, klf, psf)
S	section modulus (in.^{3})
S	shrinkage of wood member (in.)
s	center-to-center spacing between adjacent fasteners in a row (in.)

s	length of unthreaded shank of lag bolt (in.)
S_1	mapped maximum considered earthquake spectral acceleration at 1 one-second period (g)
s_{crit}	critical spacing between fasteners in a row (in.)
S_{D1}	design spectral response acceleration at a one-second period (g)
S_{DS}	design spectral response acceleration at short periods (g)
G	specific gravity
G_m	specific gravity of main member
S_{MS}	maximum considered earthquake spectral response acceleration at short periods (g)
S_{M1}	maximum considered earthquake spectral response acceleration at a one-second period (g)
S_S	mapped maximum considered earthquake spectral acceleration at short periods (g)
SV	shrinkage value for wood due to 1 percent change in moisture content (in./in.)
T	fundamental period of vibration of structure in direction of seismic force under consideration (sec)
T	tension force (lb, k)
t	thickness (in.)
T_a	approximate fundamental building period (sec)
t_m	thickness of main member (in.)
T_o	response spectrum period at which the S_{DS} plateau is reached (sec)
T_s	response spectrum period at which the S_{DS} and S_{D1}/T curves meet (sec)
t_s	thickness of side member (in.)
T_u	ultimate (factored) tension force (lb, k)
t_{washer}	thickness of washer (in.)
U	wind uplift resultant force (lb, k)
V	basic wind speed (mph)
V	seismic base shear (lb, k)
V	shear force in a beam, diaphragm, or shearwall (lb, k)
v	unit shear in horizontal diaphragm or shearwall (lb/ft)
V'	reduced shear in beam determined by neglecting load within d from face of supports (lb, k)
v_2	unit shear in second-floor diaphragm (lb/ft)
v_{12}	unit shear in shearwall between first- and second-floor levels (lb/ft)
v_{2r}	unit shear in shearwall between second-floor and roof levels (lb/ft)
V_{px}	diaphragm forces created by the redistribution of forces between vertical elements
v_r	unit shear in roof diaphragm (lb/ft)

V_{story}	story shear force (lb, k)
v_u	ultimate (factored) unit shear in horizontal diaphragm or shearwall (lb/ft)
v_{u2}	ultimate (factored) unit shear in second-floor diaphragm (lb/ft)
v_{u12}	ultimate (factored) unit shear in shearwall between first- and second-floor levels (lb/ft)
v_{u2r}	ultimate (factored) unit shear in shearwall between second-floor and roof levels (lb/ft)
v_{ur}	ultimate (factored) unit shear in roof diaphragm (lb/ft)
V_{wall}	wall shear force in the wall with the highest unit shear (lb, k)
W	lateral force due to wind (lb, k, lb/ft, psf)
W	weight of structure or total seismic dead load (lb, k)
w	tabulated withdrawal design value for single fastener (lb/in. of penetration)
w	uniformly distributed load or force (lb/ft, k/ft, psf, ksf)
W, W'	nominal and allowable withdrawal design value for single fastener (lb)
W_1	dead load of 1-ft-wide strip tributary to story level in direction of seismic force (lb/ft, k/ft)
W_2	total dead load tributary to second-floor level (lb, k)
W_2'	that portion of W_2 which generates seismic forces in second-floor diaphragm (lb, k)
w_2	uniform load to second-floor horizontal diaphragm (lb/ft, k/ft)
W_D	dead load of structure (lb, k)
W_{foot}	dead load of footing or foundation (lb, k)
w_i, w_x	tributary weight assigned to story level i, level x (lb, k)
W_p	weight of portion of structure (element or component) (lb, k, lb/ft, k/ft, psf)
w_{px}	uniform load to diaphragm at level x (lb/ft, k/ft)
W_r	total dead load tributary to roof level (lb, k)
w_r	uniform load to roof horizontal diaphragm (lb/ft, k/ft)
W_r'	that portion of W_r which generates seismic forces in roof diaphragm (lb, k)
w_u	ultimate (factored) uniformly distributed load or force (lb/ft, k/ft, psf, ksf)
w_{u2}	ultimate (factored) uniform load to second-floor horizontal diaphragm (lb/ft, k/ft)
w_{upx}	ultimate (factored) uniform load to diaphragm at level x (lb/ft, k/ft)
w_{ur}	ultimate (factored) uniform load to roof horizontal diaphragm (lb/ft, k/ft)
x	exponent dependant on structure type used in calculation of the approximate fundamental period
x	width of triangular soil bearing pressure diagram (ft)

Z	plastic section modulus (in.^{3})
z	lateral force on one fastener in wood connection (lb)
Z, Z'	nominal and allowable lateral design value for single fastener in a connection (lb)
Z'_α	allowable resultant design value for lag bolt subjected to combined lateral and withdrawal loading (lb)
Z'_{GT}	allowable connection capacity due to group tear-out failure in a wood member (lb)
Z'_{NT}	allowable connection capacity due to net tension failure in a wood member (lb)
Z'_{RT}	allowable connection capacity due to row tear-out failure in a wood member (lb)
$Z_\parallel$	nominal lateral design value for single bolt or lag bolt in connection with all wood members loaded parallel to grain (lb)
$Z_\perp$	nominal lateral design value for single bolt or lag bolt in wood-to-metal connection with wood member(s) loaded perpendicular to grain (lb)
$Z_{m\perp}$	nominal lateral design value for single bolt or lag bolt in wood-to-wood connection with main member loaded perpendicular to grain and side member loaded parallel to grain (lb)
$Z_{s\perp}$	nominal lateral design value for single bolt or lag bolt in wood-to-wood connection with main member loaded parallel to grain and side member loaded perpendicular to grain (lb)
Δ	deflection (in.)
Δ	design story drift (amplified) at center of mass, $\delta_x - \delta_{x-1}$ (in.)
Δ_a	anchor slip contribution to shearwall deflection (in.)
Δ_b	bending contribution to shearwall deflection (in.)
Δ_D	deflection of diaphragm
Δ_{MC}	change in moisture content of wood member (percent)
Δ_n	nail slip contribution to shearwall deflection (in.)
Δ_s	deflection of shearwall (in.)
Δ_v	sheathing shear deformation contribution to shearwall deflection (in.)
δ_x	amplified deflection at level x, determined at the center of mass at and above level x (in.)
δ_{xe}	deflection at level x, determined at the center of mass at and above level x using elastic analysis (in.)
ϕ	resistance factor
γ	load factor
γ	load/slip modulus for a connection (lb/in.)
λ	height and exposure factor for wind pressure calculations
μ	coefficient of static friction
ρ	redundancy/reliability factor for seismic design
ρ_x	redundancy/reliability factor calculated for story x

θ angle between direction of load and direction of grain (longitudinal axis of member) (degrees)

θ_m angle of load to grain θ for main member (degrees)

θ_s angle of load to grain θ for side member (degrees)

ω wind load coefficient for use in basic ASD load combinations

Ω_o overstrength factor for seismic design

Wood Buildings and Design Criteria

1.1 Introduction

There are probably more buildings constructed with wood than any other structural material. Many of these buildings are single-family residences, but many larger apartment buildings as well as commercial and industrial buildings also use wood framing.

The widespread use of wood in the construction of buildings has both an economic and an aesthetic basis. The ability to construct wood buildings with a minimal amount of specialized equipment has kept the cost of wood-frame buildings competitive with other types of construction. On the other hand, where architectural considerations are important, the beauty and the warmth of exposed wood are difficult to match with other materials.

Wood-frame construction has evolved from a method used in primitive shelters into a major field of structural design. However, in comparison with the time devoted to steel and reinforced-concrete design, timber design is not given sufficient attention in most colleges and universities.

This book is designed to introduce the subject of timber design as applied to wood-frame building construction. Although the discussion centers on building design, the concepts also apply to the design of other types of wood-frame structures. Final responsibility for the design of a building rests with the structural engineer. However, this book is written to introduce the subject to a broad audience. This includes engineers, engineering technologists, architects, and others concerned with building design. A background in statics and strength of materials is required to adequately follow the text. Most wood-frame buildings are highly redundant structures, but for design simplicity are assumed to be made up of statically determinate members. The ability to analyze simple trusses, beams, and frames is also necessary.

1.2 Types of Buildings

There are various types of framing systems that can be used in wood buildings. The most common type of wood-frame construction uses a system of horizontal diaphragms and shearwalls to resist lateral forces, and this book deals specifically with the design of this basic type of building. At one time building codes classified a shearwall building as a *box system,* which was a good physical description of the way in which the structure resists lateral forces. However, building codes have dropped this terminology, and most wood-frame shearwall buildings are now classified as *bearing wall systems.* The distinction between the shearwall and diaphragm system and other systems is explained in Chap. 3.

Other types of wood building systems, such as glulam arches and post-frame (or pole) buildings, are beyond the scope of this book. It is felt that the designer should first have a firm understanding of the behavior of basic shearwall buildings and the design procedures that are applied to them. With a background of this nature, the designer can acquire from currently available sources (e.g., Refs. 1.5 and 1.11) the design techniques for other systems.

The basic bearing wall system can be constructed entirely from wood components. See Fig. 1.1. Here the *roof, floors, and walls* use wood framing. The calculations necessary to design these structural elements are illustrated throughout the text in comprehensive examples.

In addition to buildings that use only wood components, other common types of construction make use of wood components in combination with some other type or types of structural material. Perhaps the most common mix of structural materials is in buildings that use *wood roof and floor systems* and *concrete tilt-up* or *masonry (concrete block or brick) shearwalls.* See Fig. 1.2. This type of construction is very common, especially in one-story commercial and industrial buildings. This construction is economical for small buildings, but its economy increases as the size of the building increases. Trained crews can erect large areas of *panelized* roof systems in short periods of time.

Design procedures for the wood components used in buildings with concrete or masonry walls are also illustrated throughout this book. The connections

Figure 1.1 Two-story wood-frame building. (*Photo by Mike Hausmann.*)

Figure 1.2a *Foreground:* Office portion of wood-frame construction. *Background:* Warehouse with concrete tilt-up walls and wood roof system. (*Photo by Mike Hausmann.*)

Figure 1.2b Building with reinforced-concrete block walls and a wood roof system with plywood sheathing. (*Photo by Mark Williams.*)

between wood and concrete or masonry elements are particularly important and are treated in considerable detail.

This book covers the *complete* design of wood-frame *box*-type buildings from the roof level down to the foundation. In a complete building design, *vertical loads and lateral forces* must be considered, and the design procedures for both are covered in detail.

Wind and seismic (earthquake) are the two lateral forces that are normally taken into account in the design of a building. In recent years the design for lateral forces has become a significant portion of the design effort. The reason for this is an increased awareness of the effects of lateral forces. In addition, the building codes have substantially revised the design requirements for both wind and seismic forces. These changes are the result of extensive research in wind engineering and earthquake-resistant design.

1.3   Required and Recommended References

The fifth edition of this book was prompted by three main developments:

1. Publication of the 2001 *National Design Specification for Wood Construction* (Ref. 1.2)

2. Publication of the comprehensive 2001 *Allowable Stress Design Manual for Engineered Wood Construction* (Ref. 1.3)

3. Publication and increased adoption nationally of the 2003 *International Building Code* (Ref. 1.8)

The *National Design Specification* (NDS) is published by the American Forest and Paper Association (AF&PA) and represents the latest structural design recommendations by the wood industry. The 2001 NDS contains new chapters covering prefabricated wood I-joists, structural composite lumber, wood structural panels, shearwalls and diaphragms, and fire design. Prior to the 2001 NDS, the design of prefabricated I-joists, structural composite lumber, and wood structural panels was not explicitly included in the NDS. Furthermore, the design of shearwalls and diaphragms, while comprising the primary lateral-force-resisting system (LFRS) in most wood structures, was also not explicitly part of the NDS. These additions allow the designer one primary industry document to cover the design of wood structures. In addition to these new chapters, the 2001 NDS has significantly revised chapters on connection design. While the basis for design with mechanical fasteners remains largely unchanged, the equations and procedures have been unified. Previously, the designer was presented with a set of design equations for each connector type (e.g., nails, screws, and bolts). With the latest revision of the NDS, all dowel-type connections are designed using the same set of equations.

The 2001 *Allowable Stress Design* (ASD) *Manual for Engineered Wood Construction* includes five product design supplements, four product guidelines, a manual providing additional guidance for the design of some of the most commonly used components of wood-frame buildings, and a separate supplement dealing specifically with special provisions for wind and seismic design. With the exception of the wind and seismic supplement, these supplements and guidelines are organized according to specific product lines. Each document contains design information and details specific to each product type. The supplements are documents that include complete design information, including design values, which enable the designer to fully design the specific product or system in accordance with the provisions of the NDS. The guidelines cover design using proprietary product lines. As such, the guidelines do not contain specific design values, which must be obtained from the product supplier, but otherwise the guidelines do include information that is required to design with the various products.

The ASD Manual was first introduced in 1999 for the 1997 NDS, and for the first time brought together all necessary information required for the design of wood structures. Previous to this, the designer referred to the NDS for the design of solid sawn lumber and glulam members, as well as the design of many connection details. For the design of other wood components and systems, the designer was required to look elsewhere. For example, the design of shearwalls and diaphragms was not covered in the NDS, but through various other sources including publications by the APA—The Engineered Wood Association. The 2001 ASD Manual represents a major advancement for the design of engineered wood structures by bringing together all of the necessary design information into a single package.

All or part of the design recommendations in the NDS will eventually be incorporated into the wood design portions of most building codes. However, the 2001 NDS was not available until early 2002, and the code change process can take considerable time. This book deals specifically with the design provisions of the 2001 NDS, and the designer should verify local building code acceptance before basing the design of a particular wood structure on this criterion.

Because of the subject matter, the reader must have a copy of the NDS to properly follow this book. The NDS is actually the formal *design section* of what is a series of interrelated design documents. This series of documents comprises a package referred to as the *Allowable Stress Design Manual for Engineered Wood Construction* (Ref. 1.3). The ASD Manual includes what was traditionally considered the two core components of the NDS:

1. The actual *National Design Specification for Engineered Wood Construction,* and

2. The NDS *Supplement* (design values for wood construction)

In addition to this, the ASD Manual includes a collection of additional *supplements* covering:

1. Structural lumber

2. Structural glued laminated timber

3. Timber poles and piles

4. Wood structural panels

5. Wood structural panel shearwalls and diaphragms

A collection of *guidelines* are included in the ASD Manual as well. *Guidelines* differ from *supplements* in that guidelines address the design and use of proprietary products. The following guidelines are included with the ASD Manual:

1. Wood I-joists

2. Structural composite lumber

3. Metal plate connected wood trusses

4. Pre-engineered metal connectors

The ASD package includes what is specifically referred to as the *Manual.* The Manual provides additional guidance for the design of some of the most commonly used components of wood-frame buildings. The last component of the ASD package is a separate supplement dealing with *special provisions for wind and seismic design.*

The numerous tables of member properties, allowable stresses, fastener design values, and shearwall and diaphragms are lengthy. Rather than reproducing these tables in this book, it is felt that the reader is better served to have a copy of the basic documents for wood design. Having a copy of the *Allowable Stress Design Manual for Engineered Wood Construction,* including the NDS and the NDS Supplement, is analogous to having a copy of the AISC *Steel Manual* (Ref. 1.4) in order to be familiar with structural steel design.

This book also concentrates heavily on understanding the loads and forces required in the design of a structure. Emphasis is placed on both gravity loads and lateral forces. Toward this goal, the design loads and forces in this book are taken from the 2003 IBC (Ref. 1.8). The IBC is published by the International Code Council (ICC), and it is highly desirable for the reader to have a copy of the 2003 IBC to follow the discussion in this book. However, the IBC is not used in all areas of the country, and a number of the IBC tables that are important to the understanding of this book are reproduced in Appendix C. If a copy of the IBC is not available, the tables in Appendix C will allow the reader to follow the text.

Frequent references are made in this book to the ASD Manual, the NDS, and the IBC. In addition, a number of cross references are made to discussions or examples in this book that may be directly related to a particular subject. The reader should clearly understand the meaning of the following references:

Example reference	Refers to	Where to look
NDS Sec. 15.1	Section 15.1 in 2001 NDS	2001 NDS (required reference)
NDS Supplement Table 4A	Table 4A in 2001 NDS Supplement	2001 NDS Supplement (comes with NDS)
ASD Shear Wall Supplement Table 4.1A	Table 4.1A in the 2001 Shear Wall and Diaphragm Supplement	Wood Structural Panel Shear Wall and Diaphragm Supplement to the 2001 ASD Manual (required reference)
IBC Chap. 16	Chapter 16 in 2003 IBC	2003 IBC (recommended reference)
IBC Table 1617.6.2	Table 1617.6.2 in 2003 IBC	2003 IBC (recommended reference) or Appendix C of this book
Section 4.15	Section 4.15 of this book	Chapter 4 in this book
Example 9.3	Example 9.3 in this book	Chapter 9 in this book
Figure 5.2	Figure 5.2 in this book	Chapter 5 in this book

Another reference that is often cited in this book is the *Timber Construction Manual* (Ref. 1.5), abbreviated *TCM*. This handbook can be considered the basic reference on structural glued-laminated timber. Although it is a useful reference, it is not necessary to have a copy of the TCM to follow this book.

1.4 Building Codes and Design Criteria

Cities and counties across the United States typically adopt a building code to ensure public welfare and safety. Until recently, most local governments used one of the three *regional model codes* as the basic framework for their local building code. The three major model codes are the

1. *Uniform Building Code* (Ref. 1.9)
2. *The BOCA National Building Code* (Ref. 1.7)
3. *Standard Building Code* (Ref. 1.10)

Generally speaking, the *Uniform Building Code* was used in the western portion of the United States, *The BOCA National Building Code* in the north, and the *Standard Building Code* in the south. The model codes were revised and updated periodically, usually on a 3-year cycle.

While regional code development had been effective, engineering design now transcends local and regional boundaries. The International Code Council (ICC) was created in 1994 to develop a single set of comprehensive and coordinated national model construction codes without regional limitations. The *International Building Code* (IBC) is one of the products of the International Code Council. The ICC includes representation from the three regional model building code organizations: the Building Officials and Code Administrators International, Inc. (BOCA), which maintains the National Building Code; the International Conference of Building Officials (ICBO), which oversees the Uniform Building Code; and the Southern Building Code Congress International, Inc. (SBCCI), which administers the Standard Building Code. The first edition of the IBC was published in 2000. Since then, most states have adopted all or part of the IBC at either the state or local level.

The standard *Minimum Design Loads for Buildings and Other Structures* (Ref. 1.6) is commonly referred to as ASCE 7-02 or simply ASCE 7. It serves as the basis for some of the loading criteria in the IBC and the regional model codes. The IBC directly references ASCE 7, as does this book. Several tables and figures from ASCE 7 are reproduced in Appendix D for assisting the reader in understanding the load criteria.

In writing this design text, it was considered desirable to use one of the model building codes to establish the loading criteria and certain allowable stresses. The *International Building Code* (IBC) is used throughout the text for this purpose. The IBC was selected because it is widely used throughout the United States, and because it represents the latest national consensus with respect to load and force criteria for structural design. Prior editions of this book referenced the UBC for all loading criteria.

Throughout the text reference is made to the *Code* and the *IBC*. As noted in the previous section, when references of this nature are used, the design criteria are taken from the 2003 edition of the *International Building Code*.

Design load and force criteria for this book are taken from the IBC. Users of other codes will be able to verify this by referring to IBC tables reproduced in Appendix C. By comparing the design criteria of another code with the information in Appendix C, the designer will be able to determine quickly whether the two are in agreement. Appendix C will also be a helpful cross-reference in checking future editions of the IBC against the values used in this text.

Although the NDS is used in this book as the basis for determining the allowable loads for wood members and their connections, note that the IBC also has a chapter that deals with these subjects. However, the latest design criteria are typically found in industry-recommended design specifications such as the NDS.

The designer should be aware that the local building code is the legal authority, and the user should verify acceptance by the local code authority before applying new principles. This is consistent with general practice in structural design, which is to follow an approach that is both rational and conservative. The objective is to produce structures which are economical and safe.

1.5 Future Trends in Design Standards

Recently, the wood industry and design community completed the development of *a load and resistance factor design* (LRFD) specification for wood construction. The NDS is based on what is termed *allowable stress design* (ASD), wherein *allowable stresses* of a material are compared to calculated *working stresses* resulting from *service loads*. In LRFD, *factored nominal capacities* (resistance) are compared to the effect of *factored loads*. The factors are developed for both resistance and loads such that uncertainty and consequence of failure are explicitly recognized. The LRFD approach to wood design is provided in the *LRFD Manual for Engineered Wood Construction* (Ref. 1.1), which is published by the American Forest and Paper Association. It is expected that LRFD will eventually replace the ASD approach, but currently ASD is the popular choice among design professionals for wood design.

Currently, the Wood Design Standards Committee (WDSC) of the American Forest and Paper Association (AF&PA), which is responsible for maintaining and revising the NDS, is considering the development and publication of a combined ASD/LRFD version of the NDS for its next revision. If this concept is adopted, then both ASD and LRFD would be published simultaneously in one single document or set of documents. At present, the ASD provisions of the NDS and the new LRFD provisions are overseen by two separate committees. To ensure consistency and compatibility in design, this *dual-format* approach to the NDS has advantages. However, useability and clarity are a concern with the dual format. Whether or not the dual format is adopted for the next revision of the NDS, both ASD and LRFD will be maintained and promoted for use for the foreseeable future in some form.

1.6   Organization of the Text

The text has been organized to present the complete design of a wood-frame building in an orderly manner. The subjects covered are presented roughly in the order that they would be encountered in the design of a building.

In a building design, the first items that need to be determined are the design loads. The Code requirements for vertical loads and lateral forces are reviewed in Chap. 2, and the distribution of these in a building with wood framing is described in Chap. 3.

After the distribution of loads and forces, attention is turned to the design of wood elements. As noted previously, there are basically two systems that must be designed, one for *vertical loads* and one for *lateral forces.*

The vertical-load-carrying system is considered first. In a wood-frame building this system is basically composed of beams and columns. Chapters 4 and 5 cover the characteristics and design properties of these wood members. Chapter 6 then outlines the design procedures for beams, and Chap. 7 treats the design methods for columns and members subjected to combined axial and bending.

As one might expect, some parts of the vertical-load-carrying system are also a part of the lateral-force-resisting system. The sheathing for wood roof and floor systems is one such element. The sheathing distributes the vertical loads to the supporting members, and it also serves as the *skin* or *web* of the horizontal diaphragm for resisting lateral forces. Chapter 8 introduces the grades and properties of wood structural panels and essentially serves as a transition from the vertical-load- to the lateral-force-resisting system. Chapters 9 and 10 deal specifically with the lateral-force-resisting system. In the typical bearing wall type of buildings covered in this text, the lateral-force-resisting system is made up of a diaphragm that spans horizontally between vertical shear-resisting elements known as shearwalls.

After the design of the main elements in the vertical-load- and lateral-force-resisting systems, attention is turned to the design of the connections. The importance of proper connection design cannot be overstated, and design procedures for various types of wood connections are outlined in Chaps. 11 through 14.

Chapter 15 describes the anchorage requirements between horizontal and vertical diaphragms. Basically anchorage ensures that the horizontal and vertical elements in the building are adequately tied together.

The text concludes with a review of building code requirements for seismicly irregular structures. Chapter 16 also expands the coverage of overturning for shearwalls.

1.7   Structural Calculations

Structural design is at least as much an *art* as it is a *science.* This book introduces a number of basic structural design principles. These are demonstrated through a large number of practical numerical examples and sample calculations. These should help the reader understand the technical side of the problem, but the application of these tools in the design of wood structures is an art that is developed with experience.

Equation-solving software or *spreadsheet* application programs on a personal computer can be used to create a *template* which can easily generate the solution of many wood design equations. Using the concept of a template, the design equations need to be entered only once. Then they can be used, time after time, to solve similar problems by changing certain variables.

Equation-solving software and spreadsheet applications relieve the user of many of the tedious programming tasks associated with writing dedicated software. Dedicated computer programs certainly have their place in wood design, just as they do in other areas of structural design. However, equation-solving software and spreadsheets have leveled the playing field considerably. Templates can be very simple, or they can be extremely sophisticated. Regardless of programming experience, it should be understood that a very simple template can make the solution of a set of bolt equations easier than looking up a design value in a table.

It is highly recommended that the reader become familiar with one of the popular equation-solving or spreadsheet application programs. It is further recommended that a number of the sample problems be solved using such applications. With very little practice, it is possible to create templates which will solve problems that are repetitive and tedious on a hand-held calculator.

Even with the assurance given about the relatively painless way to implement the design equations for wood, some people will remain unconvinced. For those who simply refuse to accept or deal with the computer, or for those who have only an occasional need to design a wood structure, the NDS contains tables that cover a number of common applications. The advantage of the equations is that a wider variety of connection problems can be handled, but the NDS tables can accommodate a number of frequently encountered problems.

Although the NDS tables can handle a number of common situations, some problems will require the solution of wood design equations. The form and length of the equations are such that the solution by hand-held calculator may not be convenient, and the recommended approach is to solve the problem once using a spreadsheet or equation-solving application. The "document" thus created can be saved, and it then becomes a *template* for future problems. The template remains intact, and the values for a new problem are input in place of the values for the original problem.

Although the power and convenience of equation-solving and spreadsheet applications should not be overlooked, all the numerical problems and design examples in this book are shown as complete hand solutions. Lap-top computers may eventually replace the hand-held calculator, but the problems in this book are set up for evaluation by calculator.

With this in mind, an expression for a calculation is first given in general terms (i.e., a formula is first stated), then the numerical values are substituted in the expression, and finally the result of the calculation is given. In this way the designer should be able to readily follow the sample calculation.

Note that the conversion from pounds (lb) to kips (k) is often made without a formal notation. This is common practice and should be of no particular concern to the reader. For example, the calculations below illustrate the axial load capacity of a tension member:

$$\text{Allow. } T = F_t' A$$

$$= (1200 \text{ lb/in.}^2)(20 \text{ in.}^2)$$

$$= 24.0 \text{ k}$$

where T = tensile force
F_t' = allowable tensile stress
A = cross-sectional area

The following illustrates the conversion for the above calculations, which is normally done mentally:

$$\text{Allow. } T = F_t' A$$

$$= (1200 \text{ lb/in.}^2)(20 \text{ in.}^2)$$

$$= (24{,}000 \text{ lb}) \left(\frac{1 \text{ k}}{1000 \text{ lb}} \right)$$

$$= 24.0 \text{ k}$$

The appropriate number of significant figures used in calculations should be considered by the designer. When structural calculations are done on a calculator or computer, there is a tendency to present the results with too many significant figures. Variations in loading and material properties make the use of a large number of significant figures inappropriate. A false degree of *accuracy* is implied when the stress in a wood member is recorded in design calculations with an excessive number of significant figures.

As an example, consider the bending stress in a wood beam. If the calculated stress as shown on the calculator is 1278.356 ··· psi, it is reasonable to report 1280 psi in the design calculations. Rather than representing sloppy work, the latter figure is more realistic in presenting the degree of accuracy of the problem.

Although the calculations for problems in this text were performed on a computer or calculator, intermediate and final results are generally presented with three or four significant figures.

An attempt has been made to use a consistent set of symbols and abbreviations throughout the text. Comprehensive lists of symbols and abbreviations, and their definitions, follow the Contents. A number of the symbols and abbreviations are unique to this book, but where possible, they are in agreement with those accepted in the industry. The NDS uses a comprehensive notation system for many of the factors used in the design calculations for wood structures. This notation system is commonly known as the *equation format* for wood design and is introduced in Chap. 4.

The units of measure used in the text are the U.S. Customary System units. The abbreviations for these units are also summarized after the Contents. Factors for converting to SI metric units are included in Appendix E.

1.8 Detailing Conventions

With the large number of examples included in this text, the sketches are necessarily limited in detail. For example, a number of the building plans are shown without doors or windows. However, each sketch is designed to illustrate certain structural design points, and the lack of full details should not detract from the example.

One common practice in drawing wood structural members is to place an ×️ in the cross section of a *continuous* wood member. But a *noncontinuous* wood member is shown with a single diagonal line in cross section. See Fig. 1.3.

1.9 Fire-Resistive Requirements

Building codes place restrictions on the materials of construction based on the occupancy (i.e., what the building will house), area, height, number of occupants, and a number of other factors. The choice of materials affects not only the initial cost of a building, but the recurring cost of fire insurance premiums as well.

The fire-resistive requirements are very important to the building designer. This topic can be a complete subject in itself and is beyond the scope of this

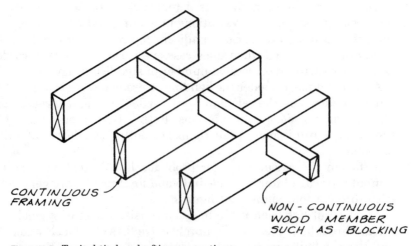

CONTINUOUS FRAMING

NON - CONTINUOUS WOOD MEMBER SUCH AS BLOCKING

Figure 1.3 Typical timber drafting conventions.

book. However, several points that affect the design of wood buildings are mentioned here to alert the designer.

Wood (unlike steel and concrete) is a combustible material, and certain *types of construction* (defined by the Code) do not permit the use of combustible materials. There are arguments for and against this type of restriction, but these limitations do exist.

Generally speaking, the *unrestricted use* of wood is allowed in buildings of limited floor area. In addition, the height of these buildings without automatic fire sprinklers is limited to one, two, or three stories, depending upon the occupancy.

Wood is also used in another type of construction known as *heavy timber*. Experience and fire endurance tests have shown that the tendency of a wood member to ignite in a fire is affected by its cross-sectional dimensions. In a fire, large-size wood members form a protective coating of char which insulates the inner portion of the member. Thus large wood members may continue to support a load in a fire long after an uninsulated steel member has collapsed because of the elevated temperature. This is one of the arguments used against the restrictions placed on "combustible" building materials. (Note that properly insulated steel members can perform adequately in a fire.)

The minimum cross-sectional dimensions required to qualify for the heavy timber fire rating are set forth in building codes. As an example, the IBC states that the minimum cross-sectional dimension for a wood column is 8 in. Different minimum dimensions apply to different types of wood members, and the Code should be consulted for these values. Limits on maximum allowable floor areas are much larger for wood buildings with heavy timber members, compared with buildings without wood members of sufficient size to qualify as heavy timber.

1.10 Industry Organizations

A number of organizations are actively involved in promoting the proper design and use of wood and related products. These include the model building code groups as well as a number of industry-related organizations. The names and addresses of some of these organizations are listed after the Contents. Others are included in the list of references at the end of each chapter.

1.11 References

[1.1]   American Forest and Paper Association (AF&PA). 1996. *Load and Resistance Factor Design Manual for Engineered Wood Construction and Supplements.* 1996 ed., AF&PA, Washington DC.

[1.2]   American Forest and Paper Association (AF&PA). 2001. *National Design Specification for Wood Construction,* 2001 ed., AF&PA, Washington DC.

[1.3]   American Forest and Paper Association (AF&PA). 2001. *Allowable Stress Design Manual for Engineered Wood Construction and Supplements and Guidelines,* 2001 ed., AF&PA, Washington DC.

[1.4]   American Institute of Steel Construction (AISC). 2001. *Manual of Steel Construction—Load and Resistance Factor Design,* 3rd ed., AISC, Chicago, IL.

[1.5]   American Institute of Timber Construction (AITC). 1994. *Timber Construction Manual,* 4th ed., AITC, Englewood, CO.

[1.6]   American Society of Civil Engineers (ASCE). 2003. *Minimum Design Loads for Buildings and Other Structures, ASCE 7-02,* ASCE, New York, NY.

[1.7]   Building Officials and Code Administrators International, Inc. (BOCAI). 1999. *The BOCA National Building Code/1999,* 13 ed., BOCAI, Country Club Hills, IL.

[1.8]   International Codes Council (ICC). 2003. *International Building Code,* 2003 ed., ICC, Falls Church, VA.

[1.9]   International Conference of Building Officials (ICBO). 1997. *Uniform Building Code,* 1997 ed., ICBO, Whittier, CA.

[1.10] Southern Building Code Congress International, Inc. (SBCCI). 1997. *Standard Building Code,* 1997 ed., SBCCI, Birmingham, AL.

[1.11] Walker, J.N., and Woeste, F.E. (eds.) 1992. *Post-Frame Building Design Manual,* ASAE—The Society for Engineering in Agricultural, Food, and Biological Systems, St. Joseph, MI.

Design Loads

2.1 Introduction

The calculation of design loads for buildings is covered in this chapter and Chap. 3. Chapter 2 deals primarily with Code-required design loads and forces and how these are calculated and modified for a specific building design. Chapter 3 is concerned with the distribution of these design loads throughout the structure.

In ordinary building design, one normally distinguishes between two major types of design criteria: (1) *vertical (gravity) loads* and (2) *lateral forces.* Although certain members may function only as vertical-load-carrying members or only as lateral-force-carrying members, often members may be subjected to a combination of vertical loads and lateral forces. For example, a member may function as a beam when subjected to vertical loads and as an axial tension or compression member under lateral forces (or vice versa).

Regardless of how a member functions, it is convenient to classify design criteria into these two main categories. Vertical loads offer a natural starting point. Little introduction to gravity loads is required. "Weight" is something with which most people are familiar, and the design for vertical loads is often accomplished first. The reason for starting here is twofold. First, gravity loading is an ever-present load, and quite naturally it has been the basic, traditional design concern. Second, in the case of lateral seismic forces, it is necessary to know the magnitude of the vertical loads before the earthquake forces can be estimated.

The terms *load* and *force* are often used interchangeably. Both are used to refer to a vector quantity with U.S. Customary System units of pounds (lb) or kips (k). There are no hard-and-fast rules regarding the use of these terms. In order to best mirror traditional earthquake design provisions, this text will use the term *vertical load* to refer to gravity effects (dead load, live load, snow load) and *lateral force* to refer to horizontal wind and seismic effects. The *load* versus *force* terminology used will vary in some cases from what is used in the IBC, however the intent is not affected by use of one term versus the other.

Also note that the design of structural framing members usually follows the reverse order in which they are constructed in the field. That is, design starts with the lightest framing member on the top level and proceeds downward, and construction starts at the bottom with the largest members and proceeds upward.

Design loads are the subject of the *International Building Code* (IBC) Chap. 16 (Ref. 2.9). The 2003 IBC has substantially incorporated structural load provisions from the American Society of Civil Engineers (ASCE) Standard 7-02, *Minimum Design Loads for Buildings and Other Structures* (Ref. 2.5). It is suggested that the reader accompany the remaining portions of this chapter with a review of ASCE 7 and Chap. 16 of the IBC. For convenience, a number of figures and tables from the IBC and ASCE 7 are reproduced in Appendix C and Appendix D.

2.2   Vertical Loads

There are three primary categories of vertical loads: dead loads, live loads, and snow loads. However, other types of vertical loads are defined in the IBC and may need to be considered, including ponding loads and wind uplift. The loading specified in the IBC represents minimum criteria; if the designer has knowledge that the actual load will exceed the Code minimum loads, the higher values must be used for design. In addition, it is required that the structure be designed for loading that can reasonably be anticipated for a given occupancy and structure configuration. If loads which are not addressed by the IBC are anticipated, ASCE 7 and its commentary may provide guidance for the loads and combinations of loads.

2.3   Dead Loads

The notation D is used by the IBC and this book to denote dead loads. Included in dead loads are the weight of all materials which are permanently attached to the structure. In the case of a wood roof or wood floor system, this would include the weight of the roofing or floor covering, sheathing, framing, insulation, ceiling (if any), and any other permanent materials such as piping or automatic fire sprinklers.

The IBC identifies a 20 psf unit floor dead load to account for partition loads in buildings where the locations of partitions may be subject to change. This often occurs in office buildings where different tenants will want different office layouts. The allowance of 20 psf will accommodate many layouts, however the adequacy of framing should be verified for specific layouts when they are known.

Another dead load which must be included, but one that is easily overlooked (especially on a roof), is the mechanical or air-conditioning equipment. Often this type of load is supported by two or three beams or joists side by side which are the same size as the standard roof or floor framing members. See Fig. 2.1. The alternative is to design special larger (and deeper) beams to carry these isolated equipment loads.

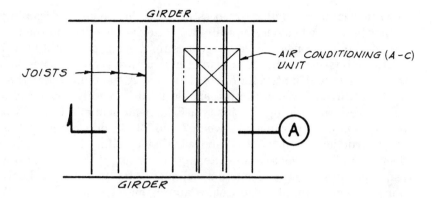

ROOF FRAMING PLAN

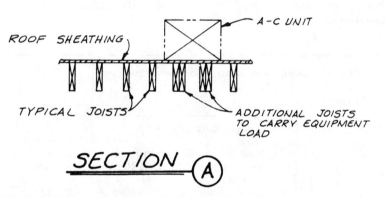

Figure 2.1 Support of equipment loads by additional framing.

The magnitude of dead loads for various construction materials can be found in a number of references. A fairly complete list of weights is given in Appendix B, and additional tables are given in Refs. 2.3 and 2.5.

Because most building dead loads are estimated as uniform loads in terms of pounds per square foot (psf), it is often convenient to convert the weights of framing members to these units. For example, if the weight per lineal foot of a wood framing member is known, and if the center-to-center spacing of parallel members is also known, the dead load in psf can easily be determined by dividing the weight per lineal foot by the center-to-center spacing. For example, if 2×12 beams weighing 4.3 lb/ft are spaced at 16 in. on center (o.c.), the equivalent uniform load is 4.3 lb/ft $\div$ 1.33 ft = 3.2 psf. A table showing these equivalent uniform loads for typical framing sizes and spacings is given in Appendix A.

It should be pointed out that in a wood structure, the dead load of the *framing members* usually represents a fairly minor portion of the total design load.

For this reason a small error in estimating the weights of framing members (either lighter or heavier) typically has a negligible effect on the final member choice. *Slightly* conservative (larger) estimates are preferred for design.

The estimation of the dead load of a structure requires some knowledge of the methods and materials of construction. A "feel" for what the unit dead loads of a wood-frame structure should total is readily developed after exposure to several buildings of this type. The dead load of a typical wood floor or roof system usually ranges between 7 and 20 psf, depending on the materials of construction, span lengths, and whether a ceiling is suspended below the floor or roof. For wood wall systems, values might range between 4 and 20 psf, depending on stud size and spacing and the type of wall sheathings used (for example, ⅜-in. plywood weighs approximately 1 psf whereas ⅞-in. stucco weighs 10 psf of wall surface area). Typical load calculations provide a summary of the makeup of the structure. See Example 2.1.

The dead load of a wood structure that differs substantially from the typical ranges mentioned above should be examined carefully to ensure that the various individual dead load (D) components are in fact correct. It pays in the long run to stand back several times during the design process and ask, "Does this figure seem reasonable compared with typical values for other similar structures?"

EXAMPLE 2.1   Sample Dead Load D Calculation Summary

Roof Dead Loads

Roofing (5-ply with gravel)	=	6.5 psf
Reroofing	=	2.5
½-in. plywood (3 psf × ½ in.)	=	1.5
Framing (estimate 2 × 12 at 16 in. o.c.)	=	2.9
Insulation	=	0.5
Suspended ceiling (acoustical tile)	=	1.0
Roof dead load D	=	14.9
Say Roof D	=	15.0 psf

Floor Dead Loads

Floor covering (lightweight concrete 1½ in. at 100 lb/ft³)	=	12.5
1⅛-in. plywood (3 psf × 1⅛ in.)	=	3.4
Framing (estimate 4 × 12 at 4 ft-0 in. o.c.)	=	2.5
Ceiling supports (2 × 4 at 24 in. o.c.)	=	0.6
Ceiling (½-in. drywall, 5 psf × ½ in.)	=	2.5
Floor dead load D	=	21.5 psf
Partition load*	=	20.0
	=	41.5 psf
Say Floor D	=	42.0 psf

*Uniform partition loads are required when the location of partitions is unknown or subject to change.

In the summary of roof dead loads in Example 2.1, the load titled "reroofing" is sometimes included to account for the weight of roofing that may be added at some future time. Subject to the approval of the local building official, new roofing materials may sometimes be applied without the removal of the old roof covering. Depending on the materials (e.g., built-up, asphalt shingle, wood shingle), one or two overlays may be permitted.

Before moving on to another type of loading, the concepts of the *tributary area* and *influence area* of a member should be explained. The area that is assumed to load a given member is known as the tributary area A_T. For a beam or girder, this area can be calculated by multiplying the *tributary width* times the span of the member. See Example 2.2. The tributary width is generally measured from mid-way between members on one side of the member under consideration to mid-way between members on the other side. For members spaced a uniform distance apart, the tributary width is equivalent to the spacing between members. Since tributary areas for adjacent members do not overlap, all loads are assumed to be supported by the nearest structural member. When the load to a member is uniformly distributed, the load per foot can readily be determined by taking the unit load in psf times the tributary width (lb/ft^2 × ft = lb/ft). The concept of tributary area will play an important role in the calculation of many types of loads.

The area that is assumed to influence the structural performance of a member is known as the *influence area* A_I. The influence area is specified in the IBC as an integer multiple of the tributary area. The integer multiple is called the *live load element factor* K_{LL} in the IBC. The influence area typically includes the full area of all members that are supported by the member under consideration. As discussed previously, the tributary area approach assumes that each load on a structure is supported entirely by the nearest structural member. In contrast, the influence area concept recognizes that the total load experienced by a structural member may be influenced by loads applied outside the tributary area of the member. For example, the influence area concept assumes that any load applied between two beams influences the performance of both beams, even though the load is located within the tributary area of only one of the beams. As provided in IBC Table 1607.9.1, the influence area is defined as four times the tributary area for most columns and twice the tributary area for most beams. Thus, influence areas for adjacent structural members will typically overlap, while tributary areas never overlap. Influence areas are often used to determine live load reductions for large floor areas in structures.

EXAMPLE 2.2   Tributary Areas

In many cases a uniform spacing of members is used throughout the framing plan. This example is designed to illustrate the *concepts* of tributary area and influence area rather than typical framing layouts. See Fig. 2.2.

Tributary Area Calculations

	A_T = trib. width $\times$ span	$A_I = A_T \times K_{LL}$
Joist J1	$A_T = 2 \times 12 = 24\ \text{ft}^2$	$A_I = 24 \times 2 = 48\ \text{ft}^2$
Joist J2	$A_T = 2 \times 14 = 28\ \text{ft}^2$	$A_I = 28 \times 2 = 56\ \text{ft}^2$
Girder G1	$A_T = (^{12}\!/_2 + {}^{14}\!/_2)20 = 260\ \text{ft}^2$	$A_I = 260 \times 2 = 520\ \text{ft}^2$
Girder G2	$A_T = (^{12}\!/_2 + {}^{14}\!/_2)24 = 312\ \text{ft}^2$	$A_I = 312 \times 2 = 624\ \text{ft}^2$
Interior column C1	$A_T = (^{12}\!/_2 + {}^{14}\!/_2)(^{20}\!/_2 + {}^{24}\!/_2) = 286\ \text{ft}^2$	$A_I = 286 \times 4 = 1144\ \text{ft}^2$
Exterior column C2	$A_T = (^{12}\!/_2)(^{20}\!/_2 + {}^{24}\!/_2) = 132\ \text{ft}^2$	$A_I = 132 \times 4 = 528\ \text{ft}^2$
Corner column C3	$A_T = (^{14}\!/_2)(^{20}\!/_2) = 70\ \text{ft}^2$	$A_I = 70 \times 2 = 140\ \text{ft}^2$

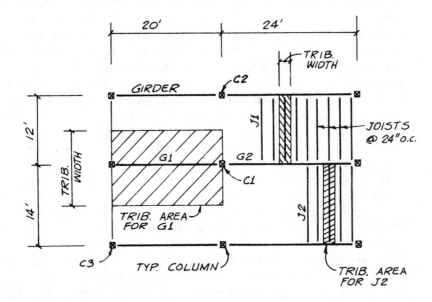

ROOF or FLOOR FRAMING PLAN

Figure 2.2

2.4 Live Loads

The term L_r is used by the IBC and this book to denote *roof live loads*. The symbols L and L_0 are used for live loads other than roof. Included in *live loads* L are loads associated with *use* or *occupancy* of a structure. Roof live loads L_r are generally associated with maintenance of the roof. While dead loads are applied permanently, live loads tend to fluctuate with time. Typically included are people, furniture, contents, and so on. Building codes typically specify the minimum roof live loads L_r and minimum floor live loads L that must be used in the design of a structure. For example, IBC Table 1607.1 specifies unit floor live loads L in psf for use in the design of floor systems. Unit roof live loads L_r are given in IBC Sec. 1607.11. Note that this book uses the italicized terms L, L_1 and L_2 to denote span. The variable L denoting a span will always be shown in italics, while L and L_r denoting live loads will be shown in standard text.

The minimum live loads in the IBC are, with some exceptions, intended to address only the use of the structure in its final and occupied configuration. Construction means and methods, including loading and bracing during construction, are generally not taken into account in the design of the building. This is because these loads can typically only be controlled by the contractor, not the building designer. In wood frame structures, the construction loading can include stockpiling of construction materials on the partially completed structure. It is incumbent on the contractor to ensure that such loading does not exceed the capacity of the structural members.

For reduction of both roof live loads L_r and floor live loads L, the influence area of the member under design consideration is taken into account. The concept that the influence area should be considered in determining the magnitude of the unit uniform live load, not just total load, is as follows:

> If a member has a small influence area, it is likely that a fairly high unit live load will be imposed over that relatively small surface area. On the other hand, as the influence area becomes large, it is less likely that this large area will be uniformly loaded by the same high unit load considered in the design of a member with a small influence area.

Therefore, the consideration of the influence area in determining the unit live load has to do with the probability that high unit loads are likely to occur over small areas, but that these high unit loads will probably not occur over large areas.

It should be pointed out that no reduction is permitted where live loads exceed 100 psf or in areas of public assembly. Reductions are not allowed in these cases because an added measure of safety is desired in these critical structures. In warehouses with high storage loads and in areas of public assembly (especially in emergency situations), it is possible for high unit loads to be distributed over large plan areas. However for the majority of wood-frame structures, reductions in live loads will be allowed.

Floor live loads

As noted earlier, minimum floor unit live loads are specified in IBC Table 1607.1. These loads are based on the occupancy or use of the building. Typical occupancy or use floor live loads range from a minimum of 40 psf for residential structures to values as high as 250 psf for heavy storage facilities. These Code unit live loads are for members supporting small influence areas. A small influence area is defined as 400 ft^2 or less. From previous discussion of influence areas, it will be remembered that the magnitude of the unit live load can be reduced as the size of the influence area increases. For members with an influence area of $A_I \geq 400$ ft^2, the reduced live load L is determined as follows:

$$L = L_0 \left(0.25 + \frac{15}{\sqrt{A_I}} \right)$$

where L_0 = unreduced floor live load specified in IBC Table 1607.1. The reduced live load L is not permitted to be less than $0.5L_0$ for members supporting loads from only one floor of a structure, nor less than $0.4L_0$ for members supporting loads from two or more floors.

The calculation of reduced floor live loads is illustrated in Example 2.3.

EXAMPLE 2.3   Reduction of Floor Live Loads

Determine the total axial force required for the design of the interior column in the floor framing plan shown in Fig. 2.3. The structure is an apartment building with a floor dead load D of 10 psf and, from IBC Table 1607.1, a tabulated floor live load of 40 psf. Assume that roof loads are not part of this problem and the load is received from one level.

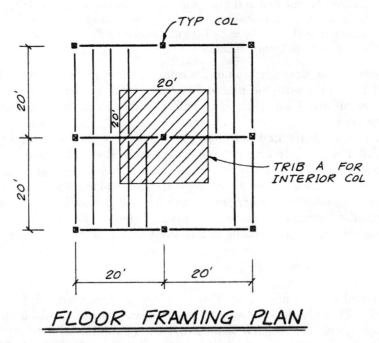

Figure 2.3

Floor Live Load

$$\text{Trib. } A = A_T = 20 \times 20 = 400 \text{ ft}^2$$

$$A_I = K_{LL} A_T = 4(400) = 1600 \text{ ft}^2 > 400 \text{ ft}^2$$

∴ floor live load can be reduced

$$L_0 = 40 \text{ psf}$$

$$L = L_0 \left(0.25 + \frac{15}{\sqrt{A_I}} \right) = 40 \left(0.25 + \frac{15}{\sqrt{1600}} \right) = (40)(0.625) = 25 \text{ psf}$$

Check:

$$0.5L_0 = 0.5(40) = 20 \text{ psf} < 25 \text{ psf} \qquad OK$$

Use L = 25 psf.

Total Load

$$TL = D + L = 10 + 25 = 35 \text{ psf}$$

$$P = 35 \times 400 = 14.0 \text{ k}$$

In addition to basic floor uniform live loads in pounds per square foot, the IBC provides special alternate concentrated floor loads. The type of live load, uniform or concentrated, which produces the more critical condition in the required load combinations (Sec. 2.16) is to be used in sizing the structure.

Concentrated floor live loads other than vehicle wheel loads can be distributed over an area 2½ feet square (2½ ft by 2½ ft). Their purpose is to account for miscellaneous nonstationary equipment loads which may occur. Vehicle loads are required to be distributed over an area of 20.25 in² (4.5 in. by 4.5 in.), which is approximately the contact area between a typical car jack and the supporting floor.

It will be found that the majority of designs will be governed by the uniform live loads. However, both the concentrated loads and the uniform loads should be checked. For certain wood framing systems, NDS Sec. 15.1 (Ref. 2.1) provides a method of distributing concentrated loads to adjacent parallel framing members.

IBC Sec. 1607.9.2 also provides an alternate method of calculating floor live load reductions based on the tributary area of a member. The formula for calculating the live load reduction is different from the formula for the influence area approach.

Roof live loads

The IBC specifies minimum unit live loads that are to be used in the design of a roof system. The live load on a roof is usually applied for a relatively short period of time during the life of a structure. This fact is normally of no concern in the design of structures other than wood. However, as will be shown in subsequent chapters, the length of time for which a load is applied to a *wood structure* does have an effect on the load capacity.

Roof live loads are specified to account for the miscellaneous loads that may occur on a roof. These include loads that are imposed during the roofing process. Roof live loads that may occur after construction include reroofing operations, air-conditioning and mechanical equipment installation and servicing, and, perhaps, loads caused by fire-fighting equipment. Wind forces and snow loads are not normally classified as live loads, and they are covered separately.

Unit roof live loads are calculated based on the provisions of IBC Sec. 1607.11. The standard roof live load for small tributary areas on flat roofs is 20 psf. A reduced roof live load may be determined based on the slope or pitch of the roof and the tributary area of the member being designed. The larger the tributary area, the lower the unit roof live load. As discussed for floor live load reductions, the consideration of tributary area has to do with the reduced probability of high unit loads occurring over large areas. Consideration of roof slope also relates to the probability of loading. On a roof that is relatively flat, fairly high unit live loads are likely to occur. However, on a steeply pitched roof much smaller unit live loads will be probable. Reduced roof live loads may be calculated for tributary areas A_T greater than 200 ft^2 and for roof slopes F steeper than 4 in./ft, as illustrated in the following formula:

$$L_r = 20R_1R_2$$

$$\text{where } R_1 = \begin{cases} 1 & \text{for } A_T \leq 200 \text{ ft}^2 \\ 1.2 - 0.001A_T & \text{for } 200 \text{ ft}^2 < A_T < 600 \text{ ft}^2 \\ 0.6 & \text{for } A_T \geq 600 \text{ ft}^2 \end{cases}$$

$$R_2 = \begin{cases} 1 & \text{for } F \leq 4 \\ 1.2 - 0.05F & \text{for } 4 < F < 12 \\ 0.6 & \text{for } F \geq 12 \end{cases}$$

A_T = tributary area supported by a structural member, ft^2
F = the number of inches of rise per foot for a sloped roof

The smallest roof live load permitted is 12 psf. Example 2.4 illustrates the determination of roof live loads for various structural members based on the tributary area of each member.

EXAMPLE 2.4   Calculation of Roof Live Loads

Determine the uniformly distributed roof loads (including dead load and roof live load) for the purlins and girders in the building shown in Fig. 2.4. Also determine the total load on column C1. Assume that the roof is flat (except for a minimum slope of ¼ in./ft for drainage). Roof dead load D = 8 psf.

Tributary Areas

Purlin P1: $A_T = 4 \times 16 = 64 \text{ ft}^2$

Girder G1: $A_T = 16 \times 20 = 320 \text{ ft}^2$

Column C1: $A_T = 16 \times 20 = 320 \text{ ft}^2$

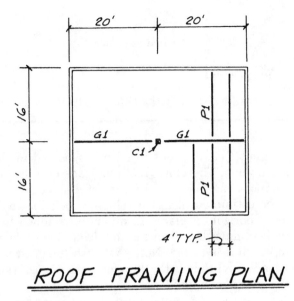

ROOF FRAMING PLAN

Figure 2.4

Roof Loads

Flat roof:

$$\therefore R_2 = 1$$

    a. Purlin

$$A_T = 64 \text{ ft}^2 < 200 \text{ ft}^2$$

$$\therefore R_1 = 1$$

$$\therefore L_r = 20 \text{ psf}$$

$$w = (D + L_r)(\text{trib. width})$$

$$= [(8 + 20) \text{ psf}](4 \text{ ft}) = 112 \text{ lb/ft}$$

    b. Girder

$$200 \text{ ft}^2 < (A_T = 320 \text{ ft}^2) < 600 \text{ ft}^2$$

$$R_1 = 1.2 - 0.001A_T = 1.2 - 0.001(320) = 0.88$$

$$L_r = 20R_1R_2 = (20)(0.88)(1) = 17.6 \text{ psf}$$

$$w = [(8 + 17.6) \text{ psf}](16 \text{ ft}) = 409.6 \text{ lb/ft}$$

c. Column

$$A_T = 320 \text{ ft}^2 \qquad \text{same as girder}$$

$$\therefore L_r = 17.6 \text{ psf}$$

$$P = [(8 + 17.6) \text{ psf}](320 \text{ ft}^2) = 8192 \text{ lb}$$

It should be pointed out that the unit live loads specified in the IBC are applied on a horizontal plane. Therefore, roof live loads on a flat roof can be added directly to the roof dead load. In the case of a sloping roof, the dead load would probably be estimated along the sloping roof; the roof live load, however, would be on a horizontal plane. In order to be added together, the roof dead load or live load must be converted to a load along a length consistent with the load to which it is added. Note that both the dead load and the live load are gravity loads, and they both, therefore, are *vertical* (not inclined) vector resultant forces. See Example 2.5.

In certain framing arrangements, unbalanced live loads (or snow loads) can produce a more critical design situation than loads over the entire span. Should this occur, the IBC requires that unbalanced loads be considered.

EXAMPLE 2.5   Combined D + L$_r$ on Sloping Roof

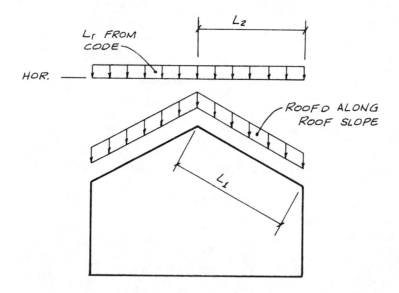

Figure 2.5

The total roof load ($D + L_r$) can be obtained either as a distributed load along the roof slope or as a load on a horizontal plane. The lengths L_1 and L_2 on which the loads are applied must be considered.

Equivalent total roof loads ($D + L_r$):

Load on horizontal plane:

$$w_{TL} = w_D \left(\frac{L_1}{L_2}\right) + w_{L_r}$$

Load along roof slope:

$$w_{TL} = w_D + w_{L_r} \left(\frac{L_2}{L_1}\right)$$

Special live loads

IBC Sec. 1607 also requires design for special loads. Because these loads have to do with the occupancy and use of a structure and tend to fluctuate with time, they are identified as live loads. It should be noted that the direction of these live loads is horizontal in some cases. Examples of special live loads include ceiling vertical live loads and live loads to handrails (which are applied both horizontally and vertically). The notation L is generally used for all live loads other than roof live loads L_r.

2.5   Snow Loads

Snow load is another type of gravity load that primarily affects roof structures. In addition, certain types of floor systems, including balconies and decks, may be subjected to snow loads.

The magnitude of snow loads can vary greatly over a relatively small geographical area. As an example of how snow loads can vary, the design snow load in a certain mountainous area of southern California is 100 psf, but approximately 5 miles away at the same elevation, the snow load is only 50 psf. This emphasizes the need to be aware of local conditions. For this reason, local building officials often specify design snow loads in lieu of calculation procedures and maps provided in the IBC and ASCE 7.

Snow loads can be extremely large. For example, a basic snow load of 240 psf is required in an area near Lake Tahoe. It should be noted that the specified snow loads are on a horizontal plane (similar to roof live loads). Unit snow loads (psf), however, are not subject to the tributary area reductions that can be used for roof live loads.

The exposure of a building to wind, the slope of the roof, and the thermal condition of the roof have a substantial effect on the magnitude of snow accumulation on the roof surface. The IBC and ASCE 7 provide a method by which the design snow load may be determined based on the ground snow load, building exposure conditions, roof slope, and roof thermal conditions.

In order to understand the snow load provisions in the IBC, it is important to introduce the concept of *exposure categories*. Previous editions of ASCE 7 and the IBC defined four exposure categories, which were intended to account for the effects of different types of terrain on both wind load and snow load. Although Exposure A is no longer defined in ASCE 7-02 for wind loads, it is briefly discussed here since the 2003 IBC snow load provisions still reference Exposure A. Exposure A is applicable only for buildings located in large city centers and sheltered by nearby structures having heights in excess of 70 ft. Exposure B includes terrain with buildings, wooded areas, or surface irregularities approximately the height of a single-family dwelling extending 1500 ft or more from the site. Exposure C is characterized by generally flat and open terrain extending ½ mile or more from the site. Exposure D applies for unobstructed flat terrain that faces a large body of water. Exposure D extends inland from the shoreline a distance of 1500 ft or 10 times the building height, whichever is greater. The general effect of exposure categories is to specify higher snow loads for sheltered areas such as Exposures A and B, and lower snow loads for open areas subjected to higher wind such as Exposures C and D.

The IBC requires that a particular building site be analyzed and assigned to one of the exposure categories. The description of Exposure D is quite specific, and the assignment of this exposure should be clear. Exposure C is to be used for open country and grasslands where only scattered obstructions less than 30 ft in height are present. Exposure B applies to most urban and suburban areas or other terrain that has closely spaced obstructions the size of single-family dwellings or larger. Exposure A has only limited applicability for buildings surrounded by tall structures, typically in city centers.

The basic formula for calculating the design snow load in the IBC and ASCE 7 is:

$$S = 0.7 C_e C_t C_s p_g I_S$$

Each of the terms in this expression is defined as follows:

p_g = **ground snow load.**   In the absence of snow loads specified by the local building official, the ground snow load for a particular geographic area can often be read from a map of the United States given in IBC Fig. 1608.2. The snow loads on this map are associated with an annual probability of exceedence of 0.02 (mean recurrence interval of 50 years). Ground snow loads shown on the map for many locations in the western and northeastern regions of the United States are applicable only below specified elevations. Snow loads for higher elevations should be determined based on site-specific data and historical records. In the formula given above for design snow load, ground snow loads are reduced by a factor of 0.7 to account for the fact that snow accumulation is greater on the ground than at the roof level for most structures.

I_S = **importance factor.**   Importance factors are a fairly recent development in the determination of design loads. An importance coefficient was first included in seismic force calculations, and has more recently been incorporated into the calculation of snow and wind design loads. The concept behind the importance factor is that certain structures should be designed for larger loads than ordinary structures. The IBC lists the importance factors for snow, wind, and seismic forces in the same table (Table 1604.5). Except for the default value of 1.0 for standard occupancies, note that the importance factors for snow, wind, and seismic forces are not equal.

Importance factors for snow are intended to ensure that essential facilities and hazardous facilities are designed to support heavier snow loads than other structures. *Essential facilities* are those that must remain safe and usable for emergency purposes. Examples of essential facilities include hospitals, fire and police stations, and communications centers. *Hazardous facilities* encompass buildings whose failure would cause a substantial hazard to the lives of a large number of people. Examples of hazardous facilities include schools, jails, and public buildings where large numbers of people may congregate, as well as buildings that contain toxic or explosive substances in such quantities as to threaten public safety. In the IBC the importance factor for essential facilities is I_S = 1.2, and the importance factor for hazardous facilities is I_S = 1.1. The importance factor of I_S = 1.2 is associated with an approximate conversion of the 50-year mean recurrence interval in IBC Fig. 1608.2 to a 100-year recurrence interval. The importance factor for "standard occupancy" for many wood structures is I_S = 1.0.

C_e = **snow exposure factor.**   As specified in IBC Table 1608.3.1, the snow exposure factor varies based on the exposure category for the terrain at the building site, as well as the roof exposure conditions. Roofs that have no immediate shelter provided by trees, structures, or surrounding terrain are classified as "fully exposed" and have snow exposure factors that range from C_e = 0.9 in Exposure B to C_e = 0.7 in Exposure D. Roofs that are surrounded by tall conifers are classified as "sheltered" and have snow exposure factors that range from C_e = 1.3 in Exposure A to C_e = 1.0 in Exposure D. All other roofs are classified as "partially exposed" and have snow exposure factors that range from C_e = 1.1 in Exposure A to C_e = 0.9 in Exposure D. IBC Table 1608.3.1 also provides lower snow exposure factors for buildings located above the treeline in mountainous areas and in treeless regions of Alaska.

C_t = **thermal factor.**   As the name implies, the thermal factor varies based on the thermal condition of the roof of a structure. Thermal factors are provided in IBC Table 1608.3.2. For unheated structures, or for buildings with well-ventilated roofs that have high thermal resistance (R-values) and will remain relatively cold during winter months, thermal factors greater than unity are specified since heat transfer from inside the structure will not melt much of the snow on the roof. For continuously heated greenhouses with roofs that have low thermal resistance (R-values), a thermal factor of C_t = 0.85 is specified since heat transfer from within the structure will tend to melt substantial amounts of snow on the roof. All other structures are assigned a thermal factor of $C_t = 1.0$.

C_s = **roof slope factor.** The roof slope factor is specified in ASCE 7 and provides reduced snow loads based on roof slope, type of roof surface, and thermal condition of the roof. The roof slope factor is intended to address the likelihood of snow sliding to the ground from a sloped roof. Roof surfaces are classified as either "unobstructed slippery surfaces" (e.g., metal, slate, glass, or membranes with smooth surfaces) that facilitate snow sliding from the roof, or as "all other surfaces" (including asphalt shingles, wood shakes or shingles, and membranes with rough surfaces). The thermal condition of a roof is categorized as either "warm" (roofs with $C_t \leq 1.0$) or "cold" (roofs with $C_t > 1.0$). As provided in ASCE 7 Fig. 7-2, for each category of roof the roof slope factor varies linearly between unity and zero for a specific range of roof slopes. For example, warm roofs that are not slippery or unobstructed (a typical condition for many wood-frame structures) are assigned the following C_s values:

$C_s = 1$ for roof slopes less than 30 degrees (slopes of approximately 7 in./ft or less)

$C_s = 0$ for roof slopes greater than 70 degrees

C_s varies linearly between 1 and 0 for roof slopes between 30 degrees and 70 degrees.

Calculation of the snow load for a typical structure based on IBC and ASCE 7 provisions is illustrated in Example 2.6. This example also illustrates the effects of using a load on a horizontal plane in design calculations.

EXAMPLE 2.6   Snow Loads

Assuming that the basic design snow load is not specified by the local building official, determine the total design dead load plus snow load for the rafters in the building illustrated in Fig. 2.6a. The building is a standard residential occupancy located near Houghton, Michigan in Exposure C terrain, with trees providing shelter on all sides of the structure. The building is heated, the rafters are sloped at 6 in./ft, and the roof covering consists of cement asbestos shingles. Determine the design shear and moment for the rafters if they are spaced 4 ft-0 in. o.c. Roof dead load D has been estimated as 14 psf along the roof surface.

Snow Load

Ground snow load:	$p_g = 80$ psf	from IBC Fig. 1608.2
Importance factor:	$I_S = 1.0$	from IBC Table 1604.5
Snow exposure factor:	$C_e = 1.1$	from IBC Table 1608.3.1 for "sheltered" roof
Thermal factor:	$C_t = 1.0$	from IBC Table 1608.3.2

Roof slope factor: $C_s = 1.0$ from ASCE 7 Fig. 7-2

Design snow load:

$$S = 0.7C_eC_tC_sp_gI_S = (0.7)(1.1)(1.0)(1.0)(80)(1.0) = 61.6 \text{ psf}$$

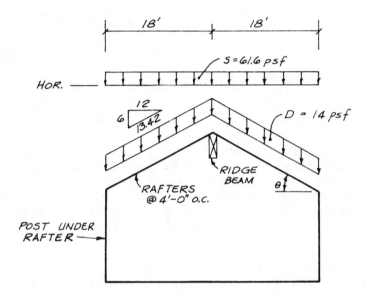

TYPICAL SECTION

Figure 2.6a

Total Loads

In computing the total load to the rafters in the roof, the different lengths of the dead and snow loads must be taken into account. In addition, the shear and moment in the rafters may be analyzed using the sloping beam method or the horizontal plane method. In the *sloping beam method,* the gravity load is resolved into components that are parallel and perpendicular to the member. The values of shear and moment are based on the normal (perpendicular) component of load and a span length equal to the full length of the rafter. In the *horizontal plane method,* the gravity load is applied to a beam with a span that is taken as the horizontal projection of the rafter. Both methods are illustrated, and the maximum values of shear and moment are compared.

$$TL = D + S$$
$$= 14 + 61.6\left(\frac{18}{20.12}\right)$$
$$= 69 \text{ psf}$$

$$TL = D + S$$
$$= 14\left(\frac{20.12}{18}\right) + 61.6$$
$$= 77.2 \text{ psf}$$

$$w = 69 \text{ psf} \times 4 \text{ ft}$$
$$= 276 \text{ lb/ft}$$

$$w = 77.2 \text{ psf} \times 4 \text{ ft}$$
$$\approx 309 \text{ lb/ft}$$

Use load normal to roof and rafter span parallel to roof.

Use total vertical load and projected horizontal span.

$$V = \frac{wL}{2} = \frac{0.247(20.12)}{2}$$
$$= 2.48 \text{ k}$$

$$V = \frac{wL}{2} = \frac{0.309(18)}{2}$$
$$= 2.78 \text{ k} \quad \text{(conservative)}$$

$$M = \frac{wL^2}{8} = \frac{0.247(20.12)^2}{8}$$
$$= 12.5 \text{ ft-k}$$

$$M = \frac{wL^2}{8} = \frac{0.309(18)^2}{8}$$
$$= 12.5 \text{ ft-k} \quad \text{(same)}$$

NOTE: The horizontal plane method is commonly used in practice to calculate design values for inclined beams such as rafters. This approach is convenient and gives equivalent design moments and conservative values for shear compared with the sloping beam analysis. (By definition *shear* is an internal force *perpendicular* to the longitudinal axis of a beam. Therefore, the calculation of shear using the sloping beam method in this example is theoretically correct.)

Sloping beam method
(left rafter illustrated)

Horizontal plane method
(right rafter illustrated)

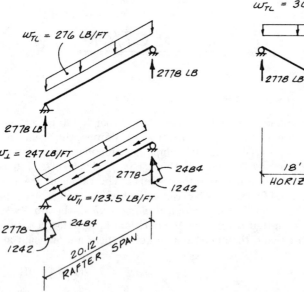
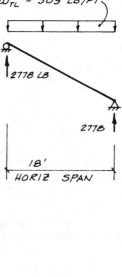

Figure 2.6b Comparison of *sloping beam method* and *horizontal plane method* for determining shears and moments in an inclined beam.

In addition to these basic guidelines for snow loads on flat or sloped roofs, ASCE 7 provides more comprehensive procedures for evaluating snow loads under special conditions. For example, ASCE 7 provisions include consideration of drifting snow and unbalanced snow loads, sliding snow from higher roof surfaces, rain-on-snow surcharge loads for flat roofs, and minimum design snow loads for low-slope roofs (slope ≤ 5 degrees).

2.6 Other Minimum Loads

The IBC contains a series of miscellaneous minimum design loads. In order to use the *basic load combinations* (Sec. 2.16), the type of loading for each of these miscellaneous loads is identified. As an example, the 5 psf horizontal force on partitions is identified as a live load L. It is intended that this live load be combined with other applicable design loads in the basic load combinations.

2.7 Deflection Criteria

The IBC establishes deflection limitations for beams, trusses, and similar members that are not to be exceeded under certain gravity loads. The deflection criteria are given in IBC Table 1604.3 and apply to roof members, floor members, and walls. These deflection limits are intended to ensure user comfort and to prevent excessive cracking of plaster ceilings and other interior finishes.

The question of user comfort is tied directly to the confidence that occupants have regarding the safety of a structure. It is possible for a structure to be very safe with respect to satisfying stress limitations, but it may deflect under load to such an extent as to render it unsatisfactory.

Excessive deflections can occur under a variety of loading conditions. For example, user comfort is essentially related to deflection caused by live loads only. The IBC therefore requires that the deflection under live load be calculated. This deflection should typically be less than or equal to the span length divided by 360 ($\Delta_{\text{L or } L_r} \leq L/360$) for floor members and for roof members that support plaster ceilings.

Another loading condition that relates to the cracking of plaster and the creation of an unpleasant visual situation is that of deflection under dead load plus live load. For this case, the actual deflection is often controlled by the limit of the span divided by 240 ($\Delta_{\text{D}+(\text{L or } L_r)} \leq L/240$). Although the IBC does not explicitly define the factor K, the footnotes to IBC Table 1604.3 explain that for structural wood members the dead load may be multiplied by a factor of either 0.5 or 1.0 when calculating deflection. The magnitude of the factor depends on the moisture conditions of the wood. Additional deflection limits are provided in IBC Table 1604.3 for various types of loads, including snow and wind.

Notice that in the second criterion above, the calculated deflection is to be under K times the dead load D plus the live load L or L_r. This K factor is an attempt to reflect the tendency of wood to creep under sustained load. Recall that when a beam or similar member is subjected to a load, there will be an *instantaneous deflection*. For certain materials and under certain conditions, additional deflection may occur under long-term loading, and this added deflection is known as *creep*. In practice, a portion of the live load on a floor may be a long-term or sustained load, but the IBC essentially treats the dead load as the only long-term load that must be considered.

Some structural materials are known to undergo creep, and others do not. Furthermore, some materials may creep under certain conditions and not under others. The tendency of wood beams to creep is affected by the moisture content (see Chap. 4) of the member. The drier the member, the less the deflection under sustained load. Thus, for seasoned lumber, a K factor of 0.5 is used; for unseasoned wood, K is taken as 1.0. *Seasoned lumber* here is defined as wood having a moisture content of less than 16 percent at the time of construction, and it is further assumed that the wood will be subjected to dry conditions of use (as in most covered structures). Although the K factor of 0.5 is included in the IBC, many designers take a conservative approach and simply use the full dead load (that is, $K = 1.0$) in the check for deflection under D + L or L_r in wood beams. See Example 2.7.

EXAMPLE 2.7   Beam Deflection Limits

The deflection that occurs in a beam can be determined using the principles of strength of materials. For example, the maximum deflection due to bending in a simply supported beam with a uniformly distributed load over the entire span is

$$\Delta = \frac{5wL^4}{384EI}$$

There are several limits on the computed deflection which are not to be exceeded. See Fig. 2.7.

The IBC requires that the deflection of roof beams that support a rigid ceiling material (such as plaster) and floor beams be computed and checked against the following criteria:

1. Deflection under live load only shall not exceed the span length divided by 360:

$$\Delta_{(L \text{ or } L_r)} \leq \frac{L}{360}$$

2. Deflection under K times the dead load plus live load shall not exceed the span length divided by 240:

$$\Delta_{[KD+(L \text{ or } L_r)]} \leq \frac{L}{240}$$

The values of K may be used:

$$K = \begin{cases} 1.0 & \text{for } \textit{unseasoned} \text{ or green wood} \\ 0.5 & \text{for } \textit{seasoned} \text{ or dry wood} \end{cases}$$

As an alternative, the deflection limit for a wood member under total load (that is, $K = 1.0$) may be conservatively used:

$$\Delta_{\text{TL}} \leq \frac{L}{240}$$

For members not covered by criteria given in the IBC, the designer may choose to use the deflection limits given in Fig. 2.8.

Fabricated wood members, such as glulam beams and wood trusses, may have curvature built into the member at the time of manufacture. This built-in curvature is known as *camber*, and it opposes the deflection under gravity loads to provide a more pleasing visual condition. See Example 6.15 in Sec. 6.6 for additional information. Solid sawn wood beams are not cambered.

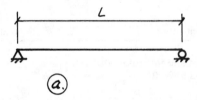

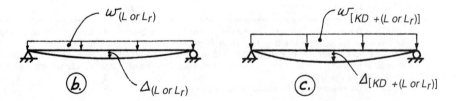

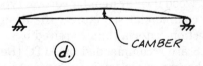

Figure 2.7 *a.* Unloaded beam. *b.* Deflection under live load only. *c.* Deflection under K times dead load plus live load. *d.* Camber is curvature built into fabricated beams that opposes deflection due to gravity loading.

Recommended Deflection Limitations		
Use classification	Applied load only	Applied load + dead load
Roof beams		
Industrial	L/180	L/120
Commercial and institutional		
Without plaster ceiling	L/240	L/180
With plaster ceiling	L/360	L/240
Floor beams		
Ordinary usage*	L/360	L/240
Highway bridge stringers	L/300	
Railway bridge stringers	L/300 to L/400	

*The ordinary usage classification is for floors intended for construction in which walking comfort and minimized plaster cracking are the main considerations. These recommended deflection limits may not eliminate all objections to vibrations such as in long spans approaching the maximum limits or for some office and institutional applications where increased floor stiffness is desired. For these usages the deflection limitations in the following table have been found to provide additional stiffness.

Deflection Limitations for Uses Where Increased Floor Stiffness Is Desired		
Use classification	Applied load only	Applied load + K (dead load)*
Floor beams		
Commercial, office and institutional		
Floor joists, spans to 26 ft†		
L ≤ 60 psf	L/480	L/360
60 psf < L < 80 psf	L/480	L/360
L ≥ 80 psf	L/420	L/300
Girders, spans to 36 ft†		
L ≤ 60 psf	L/480‡	L/360
60 psf < L < 80 psf	L/420‡	L/300
L ≥ 80 psf	L/360‡	L/240

*$K = 1.0$ except for seasoned members where $K = 0.5$. Seasoned members for this usage are defined as having a moisture content of less than 16 percent at the time of installation.
†For girder spans greater than 36 ft and joist spans greater than 26 ft, special design considerations may be required such as more restrictive deflection limits and vibration considerations that include the total mass of the floor.
‡Based on reduction of live load as permitted by the IBC.

Figure 2.8 Recommended beam deflection limitations from the TCM (Ref. 2.4). (*AITC.*)

Experience has shown that the IBC deflection criteria may not provide a sufficiently stiff wood floor system for certain types of buildings. In office buildings and other commercial structures, the designer may choose to use more restrictive deflection criteria than required by the IBC. The deflection criteria given in Fig. 2.8 are recommended by AITC (Ref. 2.4). These criteria include limitations for beams under *ordinary usage* (similar to the IBC criteria) and limitations for beams where *increased floor stiffness* is desired. These latter criteria depend on the type of beam (joist or girder), span length, and magnitude of floor live load. The added floor stiffness will probably result in increased user comfort and acceptance of wood floor systems.

Other deflection recommendations given in Fig. 2.8 can be used for guidance in the design of members not specifically covered by the IBC deflection criteria. In Fig. 2.8, the applied load is live load, snow load, wind load, and so on.

The deflection of members in other possible critical situations should be evaluated by the designer. Members over large glazed areas and members which affect the alignment or operation of special equipment are examples of two such potential problems.

The NDS takes a somewhat different position regarding beam deflection from the IBC and the TCM. The NDS (Ref. 2.1) does not recommend deflection limits for designing beams or other components, and it essentially leaves these *serviceability criteria* to the designer or to the building code. However, the NDS recognizes the tendency of a wood member to creep under sustained loads in NDS Sec. 3.5.2 and NDS Appendix F.

According to the NDS, an unseasoned wood member will creep an amount approximately equal to the deflection under sustained load, and seasoned wood members will creep about half as much. With this approach the total deflection of a wood member including the effects of creep can be computed.

For green lumber, or for seasoned lumber, glulam, and wood structural panels used in wet conditions:

$$\Delta_{\text{Total}} = 2.0(\Delta_{\text{long term}}) + \Delta_{\text{short term}}$$

For seasoned lumber, glulam, I-joists, and structural composite lumber used in dry conditions:

$$\Delta_{\text{Total}} = 1.5(\Delta_{\text{long term}}) + \Delta_{\text{short term}}$$

where $\Delta_{\text{long term}}$ = immediate deflection under long-term load. Long-term load is dead load plus an appropriate (long-term) portion of live load. Knowing the type of structure and nature of the live loads, the designer can estimate what portion of live load (if any) will be a long-term load.

$\Delta_{\text{short term}}$ = deflection under short-term portion of design load

The NDS thus provides a convenient method of estimating total deflection including creep. With this information, the designer can then make a judgment about the stiffness of a member. In other words, if the computed deflection is excessive, the design may be revised by selecting a member with a larger moment of inertia.

In recent years, there has been an increasing concern about the failure of roof systems associated with excessive deflections on *flat* roof structures caused by the entrapment of water. This type of failure is known as *ponding* failure, and it represents a progressive collapse caused by the accumulation of water on a flat roof. The initial beam deflection allows water to become trapped. This trapped water, in turn, causes additional deflection. A vicious cycle is generated which can lead to failure if the roof structure is too flexible.

Ponding failures may be prevented by proper design. The first and simplest method is to provide adequate drainage together with a positive slope (even on essentially flat roofs) so that an initial accumulation of water is simply not possible. IBC Sec. 1611 and ASCE 7 require that a roof have a minimum slope of one unit vertical to 48 units horizontal (¼ inch per foot) unless it is specifically designed for water accumulation. An adequate number and size of roof drains must be provided to carry off this water unless, of course, no obstructions are present. See ASCE 7 for additional requirements for roof drains and loads due to rain.

The second method is used in lieu of providing the minimum ¼-in./ft roof slope. Here ponding can be prevented by designing a sufficiently stiff and strong roof structure so that water cannot accumulate in sufficient quantities to cause a progressive failure. This is accomplished by imposing additional deflection criteria for the framing members in the roof structure and by designing these members for increased stresses and deflections. The increased stresses are obtained by multiplying calculated actual stresses under service loads by a magnification factor. The magnification factor is a number greater than 1.0 and is a measure of the sensitivity of a roof structure to accumulate (pond) water. It is a function of the total design roof load ($D + L_r$) and the weight of ponding water.

Because the first method of preventing ponding is the more direct and less costly method, it is recommended for most typical designs. Where the minimum slope cannot be provided for drainage, the roof structure should be designed as described above for ponding. Because this latter approach is not the more common solution, the specific design criteria are not included here. The designer is referred to Ref. 2.4 for these criteria and a numerical example.

Several methods can be used in obtaining the recommended ¼-in./ft roof slope. The most obvious solution is to place the supports for framing members at different elevations. These support elevations (or the *top-of-sheathing,* abbreviated TS, elevations) should be clearly shown on the roof plan.

A second method which can be used in the case of glulam construction is to provide *additional* camber (see Chap. 5) so that the ¼-in./ft slope is built into supporting members. It should be emphasized that this slope camber is in addition to the camber provided to account for long-term (dead load) deflection.

2.8   Lateral Forces

The subject of lateral forces can easily fill several volumes. Wind and seismic are the two primary lateral forces considered in building design. Each has been the topic of countless research projects, and complete texts deal with the evaluation of these forces. Interest in the design for earthquake effects increased substantially in light of experience obtained in the San Fernando earthquake of 1971 and other recent well-documented earthquakes. In a similar manner, interest in designing structures to withstand extreme wind loads increased substantially following Hurricane Andrew in 1992 and other hurricanes in the mid-1990s.

The design criteria included in the IBC for wind and seismic forces will be summarized in the remainder of this chapter. The calculation of lateral forces for typical buildings using shearwalls and horizontal diaphragms is covered in Chap. 3.

In dealing with lateral forces, some consideration should be given as to what loads will act concurrently. For example, it is extremely unlikely that the maximum seismic force and the maximum wind force will act simultaneously. Consequently, the IBC simply requires that the seismic force *or* the wind force be used (in combination with other appropriate loads) in design. Of course, the loading which creates the more critical condition is the one which must be used. Regardless of whether wind or seismic forces create the greatest forces on the structure as a whole, the design needs to demonstrate that all elements and connections are adequate for each load type. There can be elements or connections controlled by seismic forces, even when the structure as a whole is governed by wind, and vice-versa. As an example, this could occur for anchorage of concrete or masonry walls to a wood diaphragm. This is an advanced concept that will be covered in Chap. 15.

Similarly, the IBC does not require that roof live loads (loads which act relatively infrequently) be considered simultaneously with snow loads. However, in areas subjected to snow loads, all or part (depending on local conditions) of the snow load must be considered simultaneously with lateral forces. For more information on Code-required load and force combinations, see Sec. 2.16.

Before moving on, factors used to modify loads and allowable stresses should be introduced. There are three modifications that will be discussed: a *load duration factor* (C_D), *an allowable stress increase* (ASI), and a *load combination factor* (LCF). The ASI is discussed primarily for historical reasons; it is not used for wood design in this text. This discussion is intended to provide an introduction to a complex subject. It is hoped that as use of these factors is demonstrated in later chapters, the concepts will become clear to the reader.

The *load duration factor* C_D reflects the unique ability of wood to support higher stresses for short periods of time, as well as lower stresses for extended periods of time. The C_D factor is not limited to wind or seismic loads, but is used as an allowable stress modification in all wood design calculations. In comparison, other materials such as structural steel and reinforced concrete exhibit very little variation in capacity for varying duration of load. The C_D factor is discussed at length in Chap. 4.

Traditionally the model building codes have permitted an *allowable stress increase* ASI of one-third (i.e., allowable stresses may be multiplied by 1.33) for all materials when design forces include wind or earthquake. The technical basis for the ASI is not completely clear. There are several theories regarding its origin. The first theory is that it accounts for the reduced probability that several transient (fluctuating) load types will act simultaneously at the full design load level (i.e. full floor live load acting simultaneously with full design wind load). The second theory is that slightly higher stresses and therefore lower factors of safety are acceptable when designing for wind and seismic forces due to their short duration. The exact justification for the ASI is not of great importance since this factor will not be used for wood design in this book. See further discussion below.

The *load combination factor* LCF has the same purpose as the first theory regarding the ASI. It is to account for the low probability of multiple transient (fluctuating) loads occurring simultaneously. The load equations (Sec. 2.16) recognize that dead load is permanent, not transient, by using a LCF of 1.0 for D. In contrast, combinations of multiple loads which will vary with time use a LCF of 0.75 since it is not likely that all loads will reach the full design value at the same time.

The IBC *basic load combinations* (Sec. 2.16) are considered the primary approach to load combinations and are used in examples in this book. With the *basic load combinations,* the IBC and NDS permit the use of a *load duration factor* (C_D). In addition, the *basic load combinations* include a *load combination factor* (LCF) of 0.75 with multiple transient loads (loads varying with time). The IBC *basic load combinations* are discussed in more detail in Sec. 2.16.

When using the IBC *alternative basic load combinations,* the permitted modification factors are slightly different. The allowable stresses are permitted to be modified using both the ASI and C_D simultaneously. There are no load combination factors other than reductions to individual loads used when wind and snow loads are combined. The *alternative basic load combinations* were the only allowable stress design load combinations included in the building codes prior to the mid-1990s.

The calculation of design loads in this book will be illustrated using the IBC *basic load combinations* which incorporate a *load combination factor* (LCF) for multiple transient loads. As specified in the NDS, a *load duration factor* (C_D) of 1.6 for wood will be used in design examples involving wind or seismic forces. In addition, for all materials, the IBC requires the use of special load combinations which have magnified seismic forces for a few specific elements and connections. This is an advanced topic that is covered in Chap. 16.

Another design method which is available for wood structures is *load and resistance factor design* (LRFD). LRFD is a strength level design method which compares an element demand (a load times a load factor greater than one) to an element capacity (the failure load of an element multiplied by a material reliability factor less than one). This design method is not covered in this text; the user may refer to ASCE 16-95 [Ref. 2.6]. A second version of this book, *Design of Wood Structures,* covering LRFD is being prepared for future release.

Load levels

The IBC incorporates another change from previous editions of the building codes that should be discussed in general terms before getting into detailed discussions of lateral forces. The wind forces calculated using the IBC equations are at an *allowable stress level,* as they have been in previous editions of the building codes. The seismic forces calculated using the IBC equations, however, have been modified to a higher *strength level*. The seismic forces calculated using the IBC will generally need to be multiplied by 0.7 or divided by a factor of 1.4 to return to an allowable stress level, but other factors are used in special circumstances.

Seismic design examples in this book will use *strength* (or ultimate) design forces in the calculations until the force in a particular element or fastener needs to be compared to an allowable stress. For comparison to an allowable stress, the force will be multiplied by 0.7 or divided by 1.4 (or other applicable factor) resulting in an allowable stress level force. Notations for strength or ultimate forces have a "*u*" subscript (to denote *ultimate*), while notations for allowable stress forces will not have a corresponding subscript. Therefore all forces not having a "*u*" subscript may be assumed to be allowable stress level. Example problems and figures that are conceptual in nature will not be identified as either allowable stress or strength. *Strength* versus *allowable stress* subscripts only need to be distinguished for *seismic forces*. All gravity and wind forces in this book use the allowable stress force level.

2.9   Wind Forces—Introduction

The wind force requirements in the IBC are based on procedures given in ASCE 7-02, *Minimum Design Loads for Buildings and Other Structures* (Ref. 2.5). The ASCE 7 provisions are based on the results of extensive research regarding wind loads on structures and components of various sizes and configurations in a wide variety of simulated exposure conditions. ASCE 7 presents two methods for analyzing wind loads on structures: a simplified method for determining wind forces on typical low-rise structures that are not located atop isolated hills and a more comprehensive analytical method for determining wind forces on structures of all sizes and configurations in any exposure conditions. IBC Sec. 1609.6 states that the simplified method may be used to determine wind forces on enclosed low-rise structures with a regular, approximately symmetrical shape; a flat, gable, or hip roof system; and mean roof height less than the least horizontal dimension of the building and not greater than 60 ft. The structure must not be located on the top half of an isolated hill where higher localized wind speeds are likely to occur. In addition, IBC Sec. 1609.6.1.1 states that the simplified method applies only for structures that have a natural frequency of vibration greater than 1 Hz and utilize floor and roof diaphragms to transmit lateral wind forces to vertical structural systems (such as shearwalls or building frames). Finally, IBC Sec. 1609.6.1.2 states that the simplified method for determining localized wind forces on individual structural elements is restricted to buildings with gable roofs sloped less than or equal to 45 degrees, hip roofs sloped less than or equal to 27 degrees, or flat roofs. Since many low-rise wood structures satisfy all of these criteria, the simplified method for determination of wind forces on buildings is the focus of discussion in this book. A number of tables and figures from the IBC and ASCE 7 are included in Appendix C and Appendix D of this book. The more comprehensive analytical procedures in ASCE 7 should be consulted for structures that do not satisfy the criteria stated in IBC Sec. 1609.6. Alternatively, IBC Sec. 1609.1.1 permits designers to use the provisions of the AF&PA *Wood Frame Construction Manual for One- and Two-Family Dwellings* (Ref. 2.2) for determination of wind forces on residential wood-frame structures.

The basic formulas for calculating design wind pressures p_s and p_{net} using the simplified method are

$$p_s = \lambda I_W p_{s30}$$ for primary structural systems such as diaphragms and shearwalls

$$p_{net} = \lambda I_W p_{net30}$$ for individual structural components such as rafters and studs

Each of the terms in these equations is defined as follows:

p_{s30} = simplified design wind pressure for main windforce-resisting systems. The simplified design wind pressure is defined as the wind pressure applied over the horizontal or vertical projection of the building surface at a height of 30 ft in Exposure B conditions (described in Sec. 2.5). Simplified design wind pressures for *main windforce-resisting systems* are provided in IBC Table 1609.6.2.1(1) as a function of the *basic wind speed,* the roof slope, and the region or zone on the building surface. Horizontal simplified design wind pressures combine the effects of windward and leeward pressures that occur on opposite sides of a structure exposed to wind loads. For roof slopes steeper than 25 degrees, two load cases must be considered for vertical loads applied to the horizontal projection of the roof surface.

The simplified design wind pressures for *main windforce-resisting systems* essentially apply when one is considering the structure as a whole in resisting wind forces. According to ASCE 7, main windforce-resisting systems typically support forces that are caused by loads on multiple surfaces of a structure. The lateral-force-resisting system used in typical wood-frame buildings is described in Chap. 3.

The basic wind speed V (in miles per hour) can be read from a map of the United States given in IBC Fig. 1609. Previous versions of wind speed maps were based on the "fastest mile" wind speed, which was defined as the highest recorded wind velocity averaged over the time it takes a mile of air to pass a given point. However, since short-term velocities due to gusts may be much higher, both the 2003 IBC and ASCE 7-02 provide maps based on 3-second gust wind speeds. In order to maintain continuity with historical wind speed maps, conversions from 3-second gust wind speeds to fastest mile wind speeds are provided in IBC Table 1609.3.1. The wind speed maps in the IBC and ASCE 7 show "special wind regions" which indicate that there may be the need to account for locally higher wind speeds in certain areas.

The basic wind speed is measured at a standard height of 33 ft above ground level with Exposure C (defined in Sec. 2.5) conditions and is associated with an annual probability of exceedence of 0.02 (mean recurrence interval of 50 years). The minimum velocity to be considered in designing for wind is 85 mph, and a linear interpolation between the wind speed contours in IBC Fig. 1609 may be used. Once the designer has determined the basic wind speed from IBC Fig. 1609 (or from the local building official in special wind regions), the simplified design wind pressure p_{s30} (in psf) can be read from IBC Table 1609.6.2.1(1).

$p_{net\,30}$ = net design wind pressure for components and cladding. The net design wind pressure is defined as the wind pressure applied normal to a building surface at a height of 30 ft in Exposure B conditions (described in Sec. 2.5). The net design wind pressure is used to determine wind forces on individual structural elements (*components and cladding*) that directly support a tributary area (*effective wind area*) of the building surface. Greater wind effects due to gusts tend to be concentrated on smaller tributary areas. Consequently, the magnitudes of p_{net30} for components and cladding are typically larger than the magnitudes of p_{s30} for main windforce-resisting systems. However, according to IBC Sec. 1609.2 the *effective wind area* for individual structural members (*components and cladding*) need not be taken smaller than the square of the span length divided by three.

Net design wind pressures for components and cladding are provided in IBC Tables 1609.6.2.1(2) and 1609.6.2.1(3) as a function of the basic wind speed V, the effective wind area supported by the structural element, the region or zone on the building surface (described in Example 2.10), and the roof slope. The net design wind pressures for components and cladding combine the effects of internal and external pressures that occur on individual structural elements. Two design wind pressures must be considered separately for each structural element: a positive wind pressure acting inward and a negative wind pressure acting outward. According to IBC Sec. 1609.6.2.2.1, the minimum positive net design wind pressure and the maximum negative net design wind pressure for components and cladding are $p_{net30} = \pm 10$ psf.

I_W = importance factor. As discussed in Sec. 2.5, the concept behind the importance factor is that certain structures should be designed for higher force levels than ordinary structures. Except for the default value of 1.0, note that the importance factors for snow, wind, and seismic forces are not equal. The importance factors for snow, wind, and seismic forces are provided in IBC Table 1604.5 for easy comparison and reference.

The I_W coefficient provides that essential facilities and hazardous facilities be designed to withstand higher wind forces than other structures. For both essential and hazardous facilities the importance factor is $I_w = 1.15$. This value of I_W was selected because it represents an approximate conversion of the 50-year wind-speed recurrence interval in IBC Fig. 1609 to a 100-year recurrence interval. The importance factor for "standard occupancy" for many wood structures is $I_W = 1.0$.

λ = height and exposure factor. As the name implies, the effects of building height and exposure to wind have been combined into one coefficient. Values of λ are obtained from IBC Table 1609.6.2.1(4) given the mean roof height above ground h_{mean} and the exposure condition of the site. The wind pressure increases with height above ground.

Turbulence caused by built-up or rough terrain can cause a substantial reduction in wind speed. The IBC references four types of exposure, which are intended to account for the effects of different types of terrain on wind and snow loads (see Sec. 2.5). However, only three of these (Exposures B, C, and D) have been incorporated into the IBC wind load criteria.

2.10   Wind Forces—Primary Systems

Two methods are given in the IBC and ASCE 7 for determining the design wind forces for the primary lateral-force-resisting system (LFRS). The two methods are a comprehensive *analytical method* and a *simplified method.*

The analytical method provides a more accurate description of the wind forces, but the simplified method produces satisfactory designs for many structures. A problem with the simplified method is that it gives incorrect joint moments in gable rigid frames. Consequently the simplified method is not applied to these types of structures (or to structures with mean roof height greater than 60 ft). Note that many wood-frame structures have a gable profile, but the primary LFRS is usually made up of a system of horizontal diaphragms and shearwalls. Therefore most wood-frame structures do not use gable *rigid frames,* and the simplified method can be applied. (A gable glulam arch is an example of a wood rigid frame structure which would require the analytical method to determine wind forces for main windforce-resisting systems.)

In the analytical method, inward pressures are applied to the windward wall, and outward pressures (suction forces) are applied to the leeward wall. The forces on a sloping roof are directed outward on the leeward side, and the force to the windward side will act either inward or outward, depending on the slope of the roof. In the simplified method, horizontal wind forces are applied to the vertical projected area of the building, and vertical forces are applied to the horizontal projected area of the building. See Example 2.8.

It should be noted that the simplified method for determination of design wind pressures applies only for fully enclosed structures. Partially enclosed and unenclosed (open) structures tend to have more complex wind pressure distributions and therefore must be designed to resist wind forces based on the comprehensive analytical method in ASCE 7. IBC Sec. 1609.2 provides extensive descriptions of partially enclosed buildings and open buildings.

Doors and windows in exterior walls are considered as openings unless they are protected by assemblies designed to resist the wind forces specified for elements and components (Sec. 2.11). Glazing for windows must either be certified as impact resistant or protected from impact (see IBC Sec. 1609.1.4).

Wind forces on main windforce-resisting systems are computed using the wind pressure formula introduced in Sec. 2.9 ($p_s = \lambda I_W p_{s30}$). The importance factor I_W specified in IBC Table 1604.5 applies for all wind forces on a given building. The height and exposure factor λ is provided in IBC Table 1609.6.2.1(4) based on the mean height of the roof h_{mean}.

The IBC simplified method of analysis for wind forces requires consideration of wind loads acting normal to the longitudinal walls of a building, as well as wind loads acting normal to the transverse walls (end walls) of a building. Since wind pressures vary with roof slope and may be larger near one end of a structure due to wind directionality and aerodynamic effects, IBC Fig. 1609.6.2.1 separates the vertical projected surface of a building into four zones for evaluating lateral wind forces on the main windforce-resisting system. Similarly, the horizontal projection of the roof surface is divided into four zones for evaluating vertical wind forces on the overall structure. Each zone may have a different magnitude wind pressure p_{s30} specified in IBC Table 1609.6.2.1(1), and the combined effects of wind pressures acting simultaneously in all eight zones must be considered in the design of the main windforce-resisting system for each direction of wind loading. IBC Table 1609.6.2.1(1) also specifies larger vertical wind pressures at overhanging eaves or rakes located on the windward side of structures.

Since simplified design wind pressures on the vertical projection of a sloped roof may sometimes be negative (outward pressure), a footnote to IBC Table 1609.6.2.1 indicates that the overall wind force due to lateral loads must also be checked with no horizontal pressure applied to the vertical projection of the roof. In addition, IBC Sec. 1609.6.2.1.1 states that the overall lateral wind force on main windforce-resisting systems must be no smaller than the force associated with a uniform horizontal design wind pressure of $p_s = 10$ psf applied over the entire vertical projected building surface.

The vertical projected area of a building is divided into Zones A, B, C, and D for the application of horizontal design wind pressures. Zones A and B are designated the wall end zone and the roof end zone, respectively, since they are located near one end of the structure. Zones C and D encompass the remainder of the vertical projected area of the structure and are designated the wall interior zone and the roof interior zone, respectively. Larger horizontal wind pressures are given in IBC Table 1609.6.2.1(1) for end Zones A and B, versus interior Zones C and D. End Zones A and B are assumed to include the portion of the vertical projected area located within a distance $2a$ from one end of the structure. The dimension a is defined as 0.1 times the least width of the structure or 0.4 times the mean roof height h_{mean}, whichever is smaller. However, the dimension a may not be taken less than 3 ft, or less than 0.04 times the least width of the structure.

The horizontal projected area of a building is divided into Zones E, F, G, and H for the application of vertical design wind pressures. Zones E and F are located near one end of the structure and are designated the roof end zones on the windward and leeward sides of the building, respectively. Zones G and H include the remainder of the horizontal projected area of the structure and are designated the roof interior zones on the windward and leeward sides of the building, respectively. Larger vertical wind pressures are given in IBC Table 1609.6.2.1(1) for end Zones E and F, versus interior Zones G and

H. As with wall end zones, roof end Zones E and F are assumed to include the portion of the horizontal projected area located within a distance $2a$ from one end of the structure. The dividing line between windward and leeward zones is located at the mid-length of the structure in the direction the wind is assumed to be blowing.

The comprehensive analytical method for evaluating wind forces in ASCE 7 explicitly addresses the fact that lateral wind pressure is lower for portions of the structure near ground level and increases with height above ground. However, in the IBC simplified method of analysis the design wind pressure p_s is assumed constant over the entire height of each zone on the projected building surface.

EXAMPLE 2.8   Wind Forces for Main Windforce-Resisting Systems

Determine the design wind pressures based on the IBC simplified method for the primary lateral-force-resisting system for the building in Fig. 2.9. This is a gable structure that uses a system of diaphragms and shearwalls for resisting lateral forces. The building is a standard occupancy enclosed structure located near Fort Worth, Texas. Exposure C is to be used.

Wind forces for designing main windforce-resisting systems are obtained based on p_{s30} from IBC Table 1609.6.2.1(1). End zone and interior zone locations to be considered for horizontal pressures on the vertical projection of the building surface include (Figs. 2.9a and b)

Zone A (wall end zone)

Zone B (roof end zone)

Zone C (wall interior zone)

Zone D (roof interior zone)

End zone and interior zone locations to be considered for vertical pressures on the horizontal projection of the building surface include (Figs. 2.9a and b)

Zone E (windward roof end zone)

Zone F (leeward roof end zone)

Zone G (windward roof interior zone)

Zone H (leeward roof interior zone)

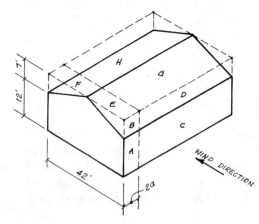

Figure 2.9a Wind pressure zones on vertical and horizontal projections of building surfaces for main windforce-resisting systems; wind direction parallel to transverse walls (end walls).

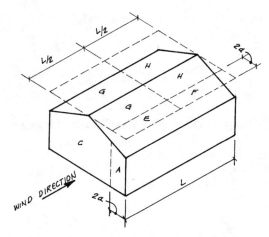

Figure 2.9b Wind pressure zones on vertical and horizontal projections of building surfaces for main windforce-resisting systems; wind direction perpendicular to transverse walls (end walls).

The building in this example does not have roof overhangs. Projected horizontal areas for overhangs at eaves or rakes on the windward side of a structure have higher vertical pressures, as specified in IBC Table 1609.6.2.1(1).

The design wind pressures in each zone for design of the main windforce-resisting system are determined as follows.

Wind speed:

$$V = 90 \text{ mph} \qquad \text{IBC Fig. 1609}$$

Importance factor:

$$I_W = 1.0 \qquad \text{IBC Table 1604.5}$$

Height and exposure factor:

The total height of the building is 19 ft.

The eave height is 12 ft.

Therefore the mean roof height is

$$h_{\text{mean}} = \frac{12 + 19}{2} = 15.5 \text{ ft} > 15 \text{ ft}$$

Using linear interpolation between $\lambda = 1.21$ and $\lambda = 1.29$ for mean roof heights of 15 ft and 20 ft [see IBC Table 1609.6.2.1(4)]:

$$\lambda = 1.22 \qquad \text{for } h_{\text{mean}} = 15.5 \text{ ft}$$

Roof slope is normally given as the rise that occurs in a 12-in. run. Convert a rise of 7 ft in a run of 21 ft to a standard roof slope:

$$\frac{\text{Rise}}{12 \text{ in.}} = \frac{7 \times 12}{21 \times 12}$$

$$\text{Rise} = 4 \text{ in.}$$

$$\therefore \text{Roof slope} = 4{:}12$$

Horizontal Wind Pressures on Vertical Projection of Building

Wall forces—End Zone A for roof slope of 4:12

$$p_{s30} = 17.8 \text{ psf}$$

Design wind pressure:

$$p_s = \lambda I_W p_{s30} = 1.0(1.22)(17.8) = 21.7 \text{ psf} \quad \text{(inward pressure)}$$

Wall forces—Interior Zone C for roof slope of 4:12

$$p_{s30} = 11.9 \text{ psf}$$

Design wind pressure:

$$p_s = \lambda I_W p_{s30} = 1.0(1.22)(11.9) = 14.5 \text{ psf} \quad \text{(inward pressure)}$$

Roof forces—End Zone B for roof slope of 4:12

$$p_{s30} = -4.7 \text{ psf}$$

Design wind pressure:

$$p_s = \lambda I_W p_{s30} = 1.0(1.22)(-4.7) = -5.7 \text{ psf} \quad \text{(outward pressure)}$$

Roof forces—Interior Zone D for roof slope of 4:12

$$p_{s30} = -2.6 \text{ psf}$$

Design wind pressure:

$$p_s = \lambda I_W p_{s30} = 1.0(1.22)(-2.6) = -3.2 \text{ psf} \quad \text{(outward pressure)}$$

Vertical Wind Pressures on Horizontal Projection of Building

Roof forces—Windward End Zone E for roof slope of 4:12

$$p_{s30} = -15.4 \text{ psf}$$

Design wind pressure:

$$p_s = \lambda I_W p_{s30} = 1.0(1.22)(-15.4) = -18.8 \text{ psf} \quad \text{(upward pressure)}$$

Roof forces—Leeward End Zone F for roof slope of 4:12

$$p_{s30} = -10.7 \text{ psf}$$

Design wind pressure:

$$p_s = \lambda I_W p_{s30} = 1.0(1.22)(-10.7) = -13.1 \text{ psf} \quad \text{(upward pressure)}$$

Roof forces—Windward Interior Zone G for roof slope of 4:12

$$p_{s30} = -10.7 \text{ psf}$$

Design wind pressure:

$$p_s = \lambda I_W p_{s30} = 1.0(1.22)(-10.7) = -13.1 \text{ psf} \quad \text{(upward pressure)}$$

Roof forces—Leeward Interior Zone H for roof slope of 4:12:

$$p_{s30} = -8.1 \text{ psf}$$

Design wind pressure:

$$p_s = \lambda I_W p_{s30} = 1.0(1.22)(-8.1) = -9.9 \text{ psf} \quad \text{(upward pressure)}$$

The distance from one end of the building for which higher end zone pressures are applicable is as follows:

$$0.4h_{\text{mean}} = 0.4(15.5) = 6.2 \text{ ft}$$

$$0.1 \text{ (least width of structure)} = 0.1b = 0.1(42) = 4.2 \text{ ft}$$

$$a = \text{lesser of } \{0.4h_{\text{mean}} \text{ or } 0.1b\} = 4.2 \text{ ft}$$

$$2a = 2(4.2) = 8.4 \text{ ft}$$

The wind pressures for design of main windforce-resisting systems in this structure are shown in Figs. 2.10a and 2.10b.

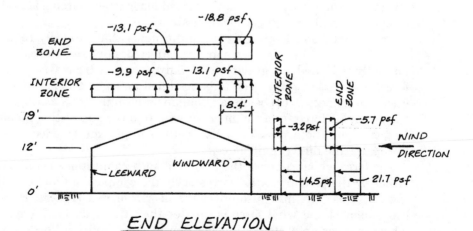

Figure 2.10a Wind pressures on vertical and horizontal projections of building surfaces; wind direction parallel to transverse walls (end walls).

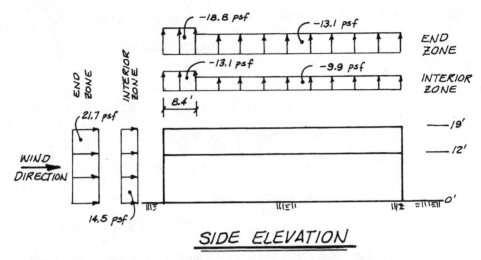

SIDE ELEVATION

Figure 2.10b Wind pressures on vertical and horizontal projections of building surfaces; wind direction perpendicular to transverse walls (end walls).

The upward forces shown in Fig. 2.10 are referred to as *uplift* forces. In addition to other considerations, both horizontal and uplift forces must be used in the moment stability analysis (known as a check on *overturning*) of the structure.

Wind uplift requires several considerations. The first could be classified as the direct transfer of the uplift forces from the roof down through the structure. Obviously, if the dead load of the roof structure exceeds the uplift force, little is required in the way of design for uplift. However, for partially enclosed structures and structures with light dead loads (these often go hand in hand), design for uplift may affect member sizes.

Connections and footing sizes are the items that typically require special attention even if member sizes are not affected. For example, connections are normally designed for gravity (vertically downward) loads. For large uplift forces, connections may need to be modified to act in tension. It may be necessary to connect a roof beam to a column, or a column to a footing, to transmit the net uplift force from the member on top to the supporting member below. In fact, it may be necessary to size footings to provide an adequate dead load to counter the direct uplift forces.

The force which controls the design of such connections could be governed either by the primary LFRS forces or by the components and cladding forces (Sec. 2.11). Uplift may also effect the design of roof trusses. If the vertical component of the wind force is greater than the dead load, truss members which are normally under tension due to dead and live loads may go into compression with wind uplift. The design of the truss would be controlled by the components and cladding forces discussed in Sec. 2.11.

A second uplift consideration relates to the moment stability of the structure. Depending on how a building is framed, the added requirement for the simultaneous application of the horizontal wind force and uplift wind force could substantially affect the design overturning requirements for a structure.

The design overturning moment (OM) is the difference between the gross OM and 60 percent of the resisting moment (RM). See Example 2.9. The IBC requires that $0.6 \times$ RM be greater than the OM. In other words, a factor of safety (FS) of $\frac{5}{3}$, or 1.67, is required for overturning stability. Notice that in this stability check, an overestimation of dead load tends to be unconservative (normally an overestimation of loading is considered conservative). To obtain the design OM, 60 percent of the RM is subtracted from the gross OM. Up to this point, the DL being used in the calculation of RM did not include the weight of the foundation.

Now, if the design OM is a positive value (i.e., the gross OM is more than 60 percent of the RM), the structure will have to be tied to the foundation. The design OM can be replaced by a couple (T and C). The tension force T must be developed by the connection to the foundation. This tension force is also known as the design *uplift* force. If the design OM is negative (i.e., the gross OM is less than or equal to $0.6 \times$ RM), there will be no uplift problem. Should an uplift problem occur, 60 percent of the DL of the foundation plus 60 percent of the DL of the building must be sufficient to counteract the gross OM.

EXAMPLE 2.9   Overall Moment Stability

Horizontal and vertical wind forces are shown acting on the shearwall in Fig. 2.11. In general, the vertical component may or may not occur, depending on how the roof is framed. Roof framing can transmit the uplift force to the wall or some other element in the structure.

In addition to the vertically upward wind pressure, the term *uplift* is sometimes used to refer to the anchorage tie-down force T.

$$\text{Gross overturning moment OM} = P(h) + U(l)$$

$$\text{Resisting moment RM} = W(l)$$

Required factor of safety for overall stability:

$$\text{Req'd FS} = \tfrac{5}{3} = 1.67$$

$\therefore$ For no uplift force T the following criterion must be satisfied:

$$\text{Gross OM} \leq 0.6 \text{ RM}$$

If this criterion is not satisfied, the design OM is obtained as follows:

$$\text{Design OM} = \text{gross OM} - 0.6 \text{ RM}$$

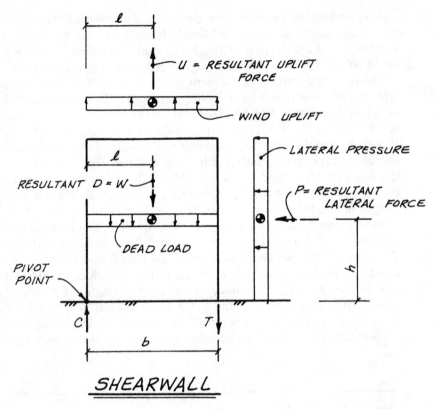

Figure 2.11

This moment can then be resolved into a couple (T and C):

$$\text{Uplift force } T = \frac{\text{design OM}}{b}$$

The design uplift force T is to be used for the design of the connection of the shearwall to the foundation. A subsequent stability check which includes the foundation weight in the resisting moment must satisfy the criterion

$$\text{Gross OM} \leq 0.6 \text{ RM}$$

The preceding discussion of overturning and the required factor of safety of 1.67 for stability applies to wind forces and seismic forces.

There are additional overturning provisions given in the IBC for seismic forces, but a comprehensive review of these details is beyond the scope of this introductory chapter. See Chap. 16 for a more detailed summary of the design requirements for overturning.

2.11 Wind Forces—Components and Cladding

The forces to be used in designing the *main windforce-resisting system* are described in Sec. 2.10. These are to be applied to the structure acting as a unit (i.e., to the horizontal diaphragms and shearwalls) in resisting lateral forces. The IBC and ASCE 7 wind force provisions require that special higher wind pressures be considered in the design of various *structural elements* (*components and cladding*) when considered *individually* (i.e., not part of the primary lateral-force-resisting system). In other words, when a roof beam or wall stud functions as part of the primary LFRS, the design forces will be determined in accordance with Sec. 2.10. However, when the design of these same members is considered independently, the higher wind pressures for components and cladding are to be used.

The forces on components and cladding are computed using the wind pressure formula introduced in Sec. 2.9 ($p_{net} = \lambda I_W p_{net30}$). The importance factor I_W will be the same for all wind forces on a given building. The height and exposure factor λ is again taken from IBC Table 1609.6.2.1(4) based on the mean height of the roof h_{mean}.

Proximity to *discontinuities* on the surface of a structure affects the magnitude of wind pressure for components and cladding. Surface discontinuities include changes in geometry of a structure such as wall corners, eaves, rakes (at the ends of gable roof systems), and ridges (for roofs sloped steeper than 7 degrees) that cause locally high wind pressures to develop. Wind tunnel tests and experience have shown that significantly larger wind pressures occur at these discontinuities versus at interior regions of wall and roof surfaces. The net design wind pressures p_{net30} in IBC Tables 1609.6.2.1(2) and 1609.6.2.1(3) range from lower pressures in interior zones (Zone 1 for roofs and Zone 4 for walls) to higher pressures in end zones (Zone 2 for roofs and Zone 5 for walls) and corner zones where end zones overlap (Zone 3 for roofs only). See Example 2.10. Net design wind pressures on roof elements also vary with roof slope. Particularly high wind pressures are specified in IBC Table 1609.6.2.1(3) for structural elements at overhanging eaves and rakes (at the ends of gable roof systems).

The areas to which the higher local wind pressures are applied for end zones and corner zones may or may not cover the entire tributary area of a member. The higher pressure is to be applied over a distance from the discontinuity of 0.1 times the least width of the structure or 0.4 times the mean roof height h_{mean}, whichever is smaller. However, the distance from a discontinuity may not be taken less than 3 ft, or less than 0.04 times the least width of the structure.

IBC and ASCE 7 wind load provisions recognize that wind pressures are larger on small surface areas due to localized gust effects. Net design wind pressures p_{net30} are provided in IBC Tables 1609.6.2.1(2) and 1609.6.2.1(3) for structural elements that support effective wind areas (tributary areas) of 10 ft^2, 20 ft^2, 50 ft^2, or 100 ft^2. Wind pressures are also listed for wall elements that support effective wind areas of 500 ft^2. Linear interpolation is permitted for intermediate areas. Structural roof elements supporting effective wind aeas in excess of 100 ft² should use the net design wind pressure for a 100 ft² area. Similarly, structural wall elements supporting effective wind areas in excess of 500 ft² should use the net design wind pressure for a 500 ft² area. The net design wind pressure for an effective wind area of 10 ft² applies to structural elements supporting smaller tributary areas. For more information regarding IBC and ASCE 7 wind forces, see Ref. 2.11.

EXAMPLE 2.10   Wind Forces—Components and Cladding

The basic wind pressure formula $(p_{net} = \lambda I_W p_{net30})$ is used to define forces for designing roof and wall elements and their connections. These pressures are larger than the pressures used to design the primary LFRS.

Wind forces for designing individual elements and components are based on net design wind pressures p_{net30} from IBC Tables 1609.6.2.1(2) and 1609.6.2.1(3). Typical locations to be considered are (Fig. 2.12a)

   $a.$ Roof area.

   $b.$ Wall area.

   $c.$ Wall area.

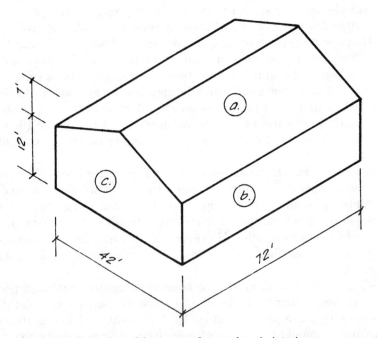

Figure 2.12a   General wind force areas for members in interior zones away from discontinuities.

Wind forces for designing individual elements near discontinuities are obtained based on p_{net30} from IBC Tables 1609.6.2.1(2) and 1609.6.2.1(3). End zone and corner zone locations to be considered include (Fig. 2.12*b*)

d. Wall corners (end zone 5)

e. Eaves without an overhang (end zone 2)

f. Rakes without an overhang (end zone 2)

g. Roof ridge when roof slope exceeds 7 degrees (end zone 2)

h. Overlap of roof ridge zone and roof rake zone when roof slope exceeds 7 degrees (corner zone 3)

i. Overlap of roof eave zone and roof rake zone (corner zone 3)

The building in this example does not have a roof overhang. Overhangs at the eaves or rakes represent additional areas that require larger design wind pressures in accordance with IBC Table 1609.6.2.1(3).

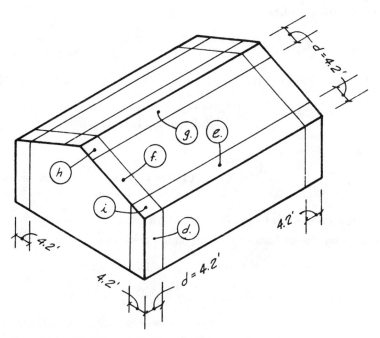

Figure 2.12b Wind force areas for members in end zones or corner zones at or near discontinuities.

Example 2.8 illustrated the computation of wind pressures for designing the *main windforce-resisting system* (see the summary in Fig. 2.10). Wind pressures required for the design of *components and cladding* in the same building are evaluated in the remaining portion of this example. The areas considered are those shown in Fig. 2.12*a* and *b*. Design conditions (location and exposure condition) are the same as in Example 2.8.

Information from previous example:

Exposure C

$I_W = 1.0$

$\lambda = 1.22$

Roof slope = 4:12 = 18.43 degrees

$h_{mean} = 15.5$ ft

The structure is an enclosed structure.

Components and Cladding—Away from Discontinuities

Roof forces—region a (interior zone 1) for effective wind area of 10 ft^2 and roof slope between 7 degrees and 27 degrees:

$$p_{net30} = \begin{cases} 8.4 \text{ psf} & \text{(inward pressure)} \\ -13.3 \text{ psf} & \text{(outward pressure)} \end{cases}$$

Design wind pressure:

$$p_{net} = \lambda I_W p_{net30} = \begin{cases} 1.0(1.22)(8.4) = 10.2 \text{ psf} & \text{(inward pressure)} \\ 1.0(1.22)(-13.3) = -16.2 \text{ psf} & \text{(outward pressure)} \end{cases}$$

Wall forces—regions b and c (interior zone 4) for effective wind area of 10 ft^2:

$$p_{net30} = \begin{cases} 14.6 \text{ psf} & \text{(inward pressure)} \\ -15.8 \text{ psf} & \text{(outward pressure)} \end{cases}$$

Design wind pressure:

$$p_{net} = \lambda I_W p_{net30} = \begin{cases} 1.0(1.22)(14.6) = 17.8 \text{ psf} & \text{(inward pressure)} \\ 1.0(1.22)(-15.8) = -19.3 \text{ psf} & \text{(outward pressure)} \end{cases}$$

Components and Cladding—Near Discontinuities

Distance from discontinuity for which higher pressures are applicable:

$$0.4h_{mean} = 0.4(15.5) = 6.2 \text{ ft}$$

$$0.1 \text{ (least width of structure)} = 0.1b = 0.1(42) = 4.2 \text{ ft}$$

$$d = \text{lesser of } \{0.4h_{mean} \text{ or } 0.1b\} = 4.2 \text{ ft}$$

Roof forces—regions e, f, and g (end zone 2) for effective wind area of 10 ft^2 and roof slope between 7 degrees and 27 degrees:

$$p_{net30} = \begin{cases} 8.4 \text{ psf} & \text{(inward pressure)} \\ -23.2 \text{ psf} & \text{(outward pressure)} \end{cases}$$

Design wind pressure:

$$p_{net} = \lambda I_W p_{net30} = \begin{cases} 1.0(1.22)(8.4) = 10.2 \text{ psf} & \text{(inward pressure)} \\ 1.0(1.22)(-23.2) = -28.3 \text{ psf} & \text{(outward pressure)} \end{cases}$$

Roof forces—regions h and i (corner zone 3) for effective wind area of 10 ft^2 and roof slope between 7 degrees and 27 degrees:

$$p_{net30} = \begin{cases} 8.4 \text{ psf} & \text{(inward pressure)} \\ -34.3 \text{ psf} & \text{(outward pressure)} \end{cases}$$

Design wind pressure:

$$p_{net} = \lambda I_W p_{net30} = \begin{cases} 1.0(1.22)(8.4) = 10.2 \text{ psf} & \text{(inward pressure)} \\ 1.0(1.22)(-34.3) = -41.8 \text{ psf} & \text{(outward pressure)} \end{cases}$$

Wall forces—region d (end zone 5) for effective wind area of 10 ft^2:

$$p_{net30} = \begin{cases} 14.6 \text{ psf} & \text{(inward pressure)} \\ -19.5 \text{ psf} & \text{(outward pressure)} \end{cases}$$

Design wind pressure:

$$p_{net} = \lambda I_W p_{net30} = \begin{cases} 1.0(1.22)(14.6) = 17.8 \text{ psf} & \text{(inward pressure)} \\ 1.0(1.22)(-19.5) = -23.8 \text{ psf} & \text{(outward pressure)} \end{cases}$$

All of the wind pressures determined in this example apply to structural elements supporting effective wind areas of 10 ft^2 or less. For larger effective wind areas, the pressures would be smaller in accordance with IBC Table 1609.6.2.1(2).

2.12 Seismic Forces—Introduction

Many designers have a good understanding of the types of loads and forces (gravity and wind) covered thus far. However, the forces that develop during an earthquake may not be as widely understood, and for this reason a fairly complete introduction to seismic forces is given.

The Structural Engineers Association of California (SEAOC) pioneered the work in the area of seismic design forces. Various editions of the SEAOC publication *Recommended Lateral Force Requirements and Commentary* [Ref. 2.13] (commonly referred to as the *Blue Book*), have served as the basis for earthquake design requirements for many editions of the *Uniform Building Code*. Starting in the 1980s, a second code resource document, the National Earthquake Hazard Reduction Program's (NEHRP) *Recommended Provisions for Seismic Regulations for New Buildings* [Ref. 2.8] was developed to address seismic hazards and design requirements on a nationwide basis. The 2000 and 2003 editions of the IBC merged the code provisions of the three national model building codes, the *Uniform Building Code* (UBC) [Ref. 2.10], the *Standard Building Code* (SBC) [Ref. 2.12], and the Building Officials and Code Administrator's (BOCA) *National Building Code* [Ref. 2.7]. The seismic design provisions of the 2000 and 2003 IBC substantially incorporate the 2000 NEHRP provisions. Although this may sound like a big change to designers who are accustomed to using the UBC provisions, there actually has been substantial alignment of the UBC provisions and NEHRP provisions that began with the 1997 editions of these documents. As a result, there are only limited technical changes between the 1997 UBC and the 2003 IBC. There is, however, a significant change in the presentation of seismic design provisions. Rather than including all of the seismic design provisions in the text of the IBC, large sections of IBC Chap. 16 provisions are referred to ASCE 7-02, *Minimum Design Loads for Buildings and Other Structures* [Ref. 2.5], which substantially incorporates the seismic force portions of the NEHRP provisions. ASCE 7 is viewed as the national consensus standard for seismic design forces, as well as other loads and forces. In order to design for forces in accordance with the 2003 IBC, it is necessary to use ASCE 7.

For designers using the provisions of the IBC, it is important to begin each design step by identifying the applicable sections of the IBC. Where the IBC adopts provisions from ASCE 7, it is important to identify what, if any, modifications the IBC makes to these ASCE 7 provisions. Modifications are clearly listed in the IBC following the reference to applicable ASCE 7 provisions. The designer may want to go so far as to note the IBC modifications within the ASCE 7 document.

Portions of the ASCE 7 seismic force provisions are reprinted in the IBC for use with the *simplified analysis procedure*. The reprinted portions may or may not be exactly the same as they appear in ASCE 7. Designers are encouraged to follow the intent of the IBC and use applicable ASCE 7 provisions.

The commentary to the SEAOC Blue Book remains a valuable resource for explanation of the many seismic design provisions that appear in both the 1997 UBC and the 2003 IBC, as well as material intending to set future directions for seismic design provisions. An updated edition of the SEAOC Blue Book should be available by 2004. The 2000 NEHRP Provisions Commentary [Ref. 2.8] is another excellent resource for the 2003 IBC seismic provisions.

The remaining portion of Chap. 2 deals with the basic concepts of earthquake engineering, and it is primarily limited to a review of the new seismic code as it applies to structurally *regular* wood-frame buildings. Many of the new requirements in the seismic code deal with added requirements for *irregular* structures. These more advanced topics are covered in later chapters (Chaps. 15 and 16) after the fundamentals of horizontal diaphragms and shearwalls are thoroughly understood.

Courses in structural dynamics and earthquake engineering deal at length with the subject of seismic forces. From structural dynamics it is known that a number of different forces act on a structure during an earthquake. These forces include inertia forces, damping forces, elastic forces, and an equivalent forcing function (mass times ground acceleration). The theoretical solution of the dynamic problem involves the addition of individual responses of a number of *modes of vibration*. Each mode is described by an equation of motion which includes a term reflecting each of the forces mentioned above.

In these types of theoretical studies, ground acceleration records from previous earthquakes are used as input, and the equations of motion are integrated numerically. This technique requires extensive computer time. A second theoretical method makes use of response spectra which eliminates the extensive numerical integration process.

Both techniques (integration of the equations of motion and response spectra studies) are forms of dynamic analysis. In the past, Code triggers for requiring the use of *dynamic analysis* have been as simple as building height limits for regular and irregular buildings. The specification of acceptable analysis methods is now a more complex function of the building use, the mapped seismic hazard at the building site, and the building period, as well as building irregularities. IBC Sec. 1616.6 addresses analysis and refers to Sec. 9.5.2.5.1 of ASCE 7, which in turn references Table 9.5.2.5.1. Table 9.5.2.5.1 of ASCE 7 provides a matrix of acceptable analysis procedures. In accordance with Table 9.5.2.5.1, dynamic analysis will generally not be required for buildings constructed entirely of light framing; however, in some cases it could be required in analysis of other buildings, including mixed structural systems.

For buildings which do not require a dynamic analysis, the Code provides a simplified method known as the *equivalent lateral force procedure.* In addition there is an even more simplified method called the *simplified design base shear.* The *simplified design base shear* may be used for light-frame structures, not exceeding three stories plus a basement, and other structures not exceeding two stories. In order to be simple, this method results in very conservative design forces in many cases. This text will illustrate the *equivalent lateral force procedure,* using the more detailed general method for calculating the base shear. The *simplified design base shear* is not covered in this book. The concept involved in the *equivalent lateral force procedure* is to design the structure for a set of Code-defined equivalent static forces.

Experience has proven that *regular structures* (i.e., symmetric structures and structures without discontinuities) perform much better in an earthquake than irregular structures. Therefore, even if the equivalent lateral force procedure is used in design, the Code penalizes *irregular structures* in areas of high seismic risk with *additional design requirements.* As previously noted, the definition of an irregular structure and a summary of some of the penalties that may be required in the design of an irregular wood building are covered in Chap. 16.

Rather than attempting to define all of the forces acting during an earthquake, the equivalent lateral force procedure given in the IBC takes a simplified approach. This empirical method is one that is particularly easy to visualize. The earthquake force is treated as an inertial problem only. Before the start of an earthquake, a building is in static equilibrium (i.e., it is at rest). Suddenly, the ground moves, and the structure attempts to remain stationary. The key to the problem is, of course, the length of time during which the movement takes place. If the ground displacement were to take place very slowly, the structure would simply ride along quite peacefully. However, because the ground movement occurs quickly, the structure lags behind and "seismic" forces are generated. See Example 2.11.

Seismic forces are generated by acceleration of the building mass. Typical practice is to consider that a lump of mass acts at each story level. This concept results in equivalent static forces being applied at each story level (i.e., at the roof and floors). Note that no such simplified forces are truly "equivalent" to the complicated combination of forces generated during an earthquake. For many buildings, however, it is felt that reasonable structural designs can be produced by designing to elastically resist the specified Code forces. Other seismic design methodologies are currently under development which take a more detailed look at the strength and failure mode of a structure. This book will, however, focus on the Code equivalent lateral force procedure, which at this time is by far the most commonly used and accepted approach to seismic design.

It should be realized that the forces given in the building Code are at a strength level and must be multiplied by 0.7 for use in allowable stress design. This adjustment occurs in the required load combinations. See Sec. 2.16 for the required load combinations.

EXAMPLE 2.11   Building Subjected to Earthquake

1. Original static position of the building before earthquake
2. Position of building if ground displacement occurs very slowly (i.e., in a static manner)
3. Deflected shape of building because of "dynamic" effects caused by rapid ground displacement

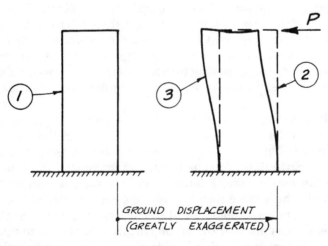

Figure 2.13

The force P in Fig. 2.13 is an "equivalent static" design force provided by the Code and can be used for certain structures in lieu of a more complicated dynamic analysis. This is common practice in wood buildings that make use of horizontal diaphragms and shearwalls.

The empirical forces given in the Code equivalent lateral force procedure are considerably lower than would be expected in a major earthquake. In a working stress approach, the structure is designed to remain *elastic* under the Code static forces. However, it is not expected that a structure will remain elastic in a major earthquake. The key in this philosophy is to *design* and *detail* the structure so that there is sufficient *system ductility* for the building to remain structurally safe when forced into the *inelastic* range in a major earthquake.

Therefore, in areas of high seismic risk, the seismic code has *detailing requirements* for all of the principal structural building materials (steel, concrete, masonry, and wood). The term "detailing" here refers to special connection design provisions and to a general tying together of the overall lateral-force-resisting system, so that there is a *continuous path* for the *transfer of lateral forces* from the top of the structure down into the foundation.

Anchorage is another term used to refer to the detailing of a structure so that it is adequately tied together for lateral forces. The basic seismic force requirements are covered in Chaps. 2 and 3, and the detailing and anchorage provisions as they apply to wood-frame structures are addressed in Chaps. 10, 15, and 16.

During an earthquake, vertical ground motion creates vertical forces in addition to the horizontal seismic forces discussed above. The vertical forces generated by earthquake ground motions are generally smaller than the horizontal forces. The 2003 IBC directly incorporates vertical ground motions into the seismic force equations. The result of including the vertical component of ground motion in the equations is a slight increase in net uplift and downward forces when considering overturning. Although the 1997 UBC was the first to directly incorporate vertical components of ground motion, an exception in the 1997 UBC permitted the vertical component to be taken as zero when using allowable stress design. In the 2003 IBC, the vertical component is required for allowable stress design.

The method used to calculate the Code horizontal story forces involves three parts. The first part is calculating the *base shear* (the horizontal force acting at the base of the building, V). The second part is assigning the appropriate percentages of this force to the various story levels throughout the height of the structure (story forces). The third part is to determine the forces on particular elements as a result of the story forces (element forces). As will be discussed later, there are several multiplying factors required to convert these *element forces* into design seismic forces for the elements at a story.

The *story forces* are given the symbol F_x (the force at level x). It should be clear that the sum of the F_x forces must equal the base shear V. See Fig. 2.14a. The formulas in the new seismic code for calculating F_x are examined in detail in Sec. 2.14.

Before the Code expressions for these forces are reviewed, it should be noted that the story forces are shown to increase with increasing height above the base of the building. The magnitude of the story forces depends on the mass (dead load) distribution throughout the height of the structure. In previous codes the vertical distribution has been such that if the dead load acting at each story level were equal, the distribution provided by the Code formula for F_x would result in a triangular distribution. The current vertical distribution provisions have deleted the concentrated story force at the top level F_t and, instead, have put an exponent of between one and two on the height term in the vertical distribution. Like the F_t term, this exponent accounts for the increased top-story forces that can occur in buildings with longer periods. The exponent is taken as one for structures with a calculated approximate period of 0.5 sec or less, which is applicable for virtually all wood-frame buildings. With the exponent taken as one, the resulting triangular distribution from earlier codes still holds true.

The reason for this distribution is that the Code bases its forces on the fundamental mode of vibration of the structure. The fundamental mode is also known as the *first mode* of vibration, and it is the significant mode for most structures.

To develop a feel for the above force distribution, the dynamic model used to theoretically analyze buildings should briefly be discussed. See Fig. 2.14b. In this model, the mass (weight) tributary to each story is assigned to that level. In other words, the weight of the floor and the tributary wall loads halfway between adjacent floors is assumed to be concentrated or "lumped" at the floor level. In analytical studies, this model greatly simplifies the solution of the dynamic problem.

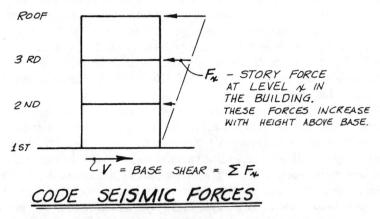

Figure 2.14a   Code seismic force distribution.

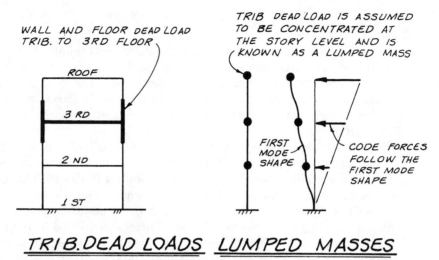

TRIB. DEAD LOADS   LUMPED MASSES

Figure 2.14b   Code seismic forces follow fundamental mode.

Now, with the term "lumped mass" defined, the concept of a mode shape can be explained. A *mode shape* is a simple displacement pattern that occurs as a structure moves when subjected to a dynamic force. The *first mode shape* is defined as the displacement pattern where all lumped masses are on one side of the reference axis. Higher mode shapes will show masses on both sides of the vertical reference axis. In a dynamic analysis, the complex motion of the complete structure is described by adding together the appropriate percentages of all of the modes of vibration. Again, the Code is essentially based on the first mode.

The point of this discussion is to explain why the F_x story forces increase with increasing height above the base. To summarize, the fundamental or first mode is the critical displacement pattern (deflected shape). The first mode shape shows all masses on one side of the vertical reference axis. Greater displacements and accelerations occur higher in the structure, and the F_x story forces follow this distribution.

2.13 Seismic Forces

IBC Sec. 1617.1 is the primary code section addressing seismic forces, and in turn references ASCE 7 Sec. 9.5.2.7. For the purposes of this text, Sec. 9.5.2.7 equations for E will be referred to as the Code seismic forces.

The Code seismic forces have gone through several major revisions in recent years. Starting with the 1997 UBC, the seismic force format was modified to recognize the consideration of both horizontal and vertical forces, and to incorporate a redundancy and reliability coefficient, ρ. In the 2003 IBC the variable E, the seismic force on an element of the structure, that appears in the basic load combinations (Sec. 2.16) is defined as:

$$E = \rho Q_E \pm 0.2 S_{DS} D$$

in which ρ is a factor representing redundancy and reliability, Q_E is the horizontal seismic force component, S_{DS} is the design spectral response acceleration at short periods, and D is the dead load. The first term represents horizontal forces, while the second term represents forces acting vertically, reducing dead load for overturning resistance, and increasing downward vertical reactions.

Use of the redundancy factor, ρ, is addressed in IBC Sec. 1617.2, which in turn references ASCE 7 Sec. 9.5.2.4. Section 9.5.2.4 defines the redundancy factor and specifies where the value can be set to 1.0. The 1997 UBC permitted ρ to be taken as 1.0 for Seismic Zones 0, 1, and 2. The IBC sets ρ equal to 1.0 for Seismic Design Categories A, B, and C. In addition, ρ is set to 1.0 for calculating story drift. Seismic Design Categories will be introduced shortly in conjunction with the IBC base shear formula variables. To provide consistency in calculations, ρ will always be included, whether it defaults to 1.0 or has a higher value. The seismic force on an element will always be multiplied by ρ. It also needs to be kept in mind that these seismic forces are at a *strength level*. When these forces are included in the ASD *basic load combinations* (Sec. 2.16), E will be multiplied by 0.7 to adjust to an allowable stress design level.

Redundancy/reliability factor

The redundancy/reliability factor, ρ, is used to encourage the designer to provide a reasonable number and distribution of lateral-force-resisting elements. In wood structures this usually means providing a sufficient number of shearwalls of reasonable length, well distributed throughout the building. The concept and calculation of ρ is very much the same as in the 1997 UBC, however, some terms have changed. In accordance with ASCE 7 Sec. 9.5.2.4, a redundancy factor ρ_x is calculated at each story, x, of the structure. The redundancy factor, ρ_x, is a function of the element-story shear ratio variable r_{max_x}, which is the maximum calculated value considering forces in each direction in a given story.

For those not familiar with the element-story shear ratio, it is suggested that each shearwall be systematically investigated to determine r_{max_x}. For shearwall buildings, the element-story shear ratio, r_{max_x} can be calculated as:

$$r_{max_x} = \frac{V_{wall}}{V_{story}} \left(\frac{10}{l_w} \right)$$

where r_{max_x} = the element-story shear ratio
 V_{wall} = the wall shear force in the wall with the highest unit shear (lb)
 V_{story} = the story shear force (lb)
 l_w = the length of the shearwall (ft)

The multiplier $10/l_w$ is intended to allow a wall longer than 10 ft to be counted as multiple walls, therefore incorporating the additional reliability provided by longer walls. This ratio is not intended to penalize shorter walls in light-frame construction, therefore the 2003 IBC and ASCE 7 specify that the ratio of $10/l_w$ need not be taken as greater than 1.0 for buildings of light-frame construction.

The value of r_{max_x} in each story, x, needs to be the largest for the two principal directions of the lateral-force-resisting system. From r_{max_x}, the variable ρ_x is determined for each story:

$$\rho_x = 2 - \frac{20}{r_{\text{max}_x} \sqrt{A_x}}$$

where A_x = the diaphragm area immediately above the story being considered (ft^2)

Finally, the redundancy factor, ρ, is taken as the largest of the calculated values of ρ_x. The Code does not permit a ρ of less than 1.0 or require a factor of greater than 1.5. It is intended that the designer try to provide adequate resisting elements to keep the value of ρ as close as possible to 1.0. Additional limitations on ρ apply for special moment-resisting frame systems.

Base shear calculation

As was the case in previous codes, the total horizontal base shear, V, is calculated from an expression which is essentially in the form:

$$F = Ma = \left(\frac{W}{g}\right) a = W \left(\frac{a}{g}\right)$$

where F = inertia force
  M = mass
  a = acceleration
  g = acceleration of gravity

The Code form of this expression is somewhat modified. The (a/g) term is replaced by a "seismic base shear coefficient."

For the *equivalent lateral force procedure,* Sec. 1617.4 of the IBC references Sec. 9.5.5 of ASCE 7, which specifies the base shear formula as:

$$V = C_s W$$

V = base shear. The *strength level* horizontal seismic force acting at the base of the structure (Fig. 2.14a).

W = **weight of structure.**   The total weight of the structure which is assumed to contribute to the development of seismic forces. For most structures, this weight is simply taken as the dead load. However, in structures where a large percentage of the live load is likely to be present at any given time, it is reasonable to include at least a portion of this live load in the value of *W*. ASCE 7 Sec. 9.5.3 lists four specific items that are to be included in the weight of the structure, *W*. For example, the Code specifies that in storage warehouses *W* is to include at least 25 percent of the floor live load. Other live loads are not covered specifically by the Code, and the designer must use judgment.

In offices and other buildings where the locations of *partitions* (nonbearing walls) are subject to relocation, IBC Chap. 16 requires that floors be designed for a *live load* of 20 psf. Use of this partition load was demonstrated in the summary of floor loads in Example 2.1 in Sec. 2.3. However, this 20-psf value is to account for localized partition loads, and it is intended to be used only for gravity load design. For seismic design it is recognized that the 20-psf loading does not occur at all locations at the same time. Consequently an *average floor load* of 10 psf may be used for the weight of partitions in determining *W* for seismic design.

Roof live loads need not be included in the calculation of *W*, but the Code does require that 20 percent of the snow load be included if it exceeds 30 psf.

C$_s$ = **the seismic response coefficient**

From ASCE 7 Sec. 9.5.5, the seismic response coefficient *C$_s$* is calculated as the greater of:

$$C_s = \frac{S_{DS}}{R/I}$$

or

$$C_s = \frac{S_{D1}}{T(R/I)}$$

C$_s$ cannot be taken as less than

$$C_s = 0.044 S_{DS} I$$

Nor, in Seismic Design Categories E and F (with near-fault ground motion amplification), can *C$_s$* be taken as less than

$$C_s = \frac{0.5 S_1}{R/I}$$

where S_{DS} = short-period design spectral response acceleration
S_{D1} = one-second design spectral response acceleration
R = response modification factor
I = importance factor
T = building period
S_1 = mapped one-second spectral acceleration

Design spectral response accelerations S_{DS} and S_{D1}

The design spectral response accelerations S_{DS} and S_{D1} are the primary variables defining an *IBC design response spectrum.* Although IBC refers to ASCE 7 for the *equivalent lateral load procedure,* the definitions of *site ground motions,* that define S_{DS} and S_{D1}, are given in IBC Sec. 1615. The first step in defining S_{DS} and S_{D1} is to read maximum considered earthquake (mapped) spectral response accelerations from the Code spectral response maps. Two spectral response acceleration maps now replace the single seismic zone map used in the 1997 UBC. The mapped short-period (0.2 sec) spectral acceleration, S_S, is used in determining the acceleration-controlled portion of the design spectra, while the mapped 1-sec spectral acceleration, S_1, determines the velocity-controlled portion. The variables S_S and S_1 are converted to maximum considered spectral response accelerations, S_{MS} and S_{M1}, by multiplying by site coefficients F_a and F_v, defined in Code tables. F_a and F_v are a function of Site (soil) Classes A through F. Site classes can be assigned as a function of three different soil parameters: shear wave velocity, penetration resistance, or undrained shear strength. Most building designers would need input from a geotechnical engineer in order to determine site class. As was true in the 1997 UBC, Site Class D is commonly assumed, provided that the building site is known to not have deep soft soils. Variables S_{MS} and S_{M1} are multiplied by ⅔ to convert from maximum considered to design spectral response accelerations for the acceleration and velocity-controlled regions, S_{DS} and S_{D1}, respectively.

The maximum considered earthquake ground motion maps incorporated into the IBC were developed through the National Seismic Hazard Mapping Project, conducted jointly by the United States Geological Survey (USGS), the Building Seismic Safety Council (BSSC), and the Federal Emergency Management Agency (FEMA). As part of this process, significant effort went into collection of available data and into workshops to receive input on a regional level. The maps contain acceleration values obtained from a combination of probabilistic and deterministic methods. The commentary to the NEHRP provisions contains a detailed discussion of the basis of the maps. Several changes are important for designers. First, where previous mapping reflected design-level earthquake data with a probability of exceedence of approximately 10 percent in 50 years (considered a design-level earthquake), the new mapping reflects the maximum considered earthquake, which is thought to represent for practical purposes the maximum earthquake that can be generated. These values are reduced to obtain design-level accelerations. Also of importance to the designer is that seismic hazard areas no longer follow state or county lines. This will mean that somewhat more effort will be required in reading maps to determine the seismic hazard for a particular site. Mapped short-period and one-second spectral response accelerations are also available on a compact disk, prepared by the National Earthquake Hazard Mapping Project, conducted jointly by the United States Geological Survey (USGS), the Federal Emergency Management Agency (FEMA), and the Building Seismic Safety Council (BSSC). (Disks are available with the IBC, and through the BSSC and FEMA web sites.) With the compact disk, the mapped spectral response accelerations can be accessed by either ZIP Code or location in degrees latitude and longitude. ZIP Codes were used for determining maximum considered spectral response accelerations in examples in this text.

The IBC design spectrum used for linear static design methods is a function of S_{DS} and S_{D1}/T (Fig. 2.18), where the UBC design spectrum used C_a and C_v/T. In addition, S_{DS} and S_{D1}/T define the response spectrum that can be used for linear dynamic analysis (response spectrum analysis) methods. The following discussion will introduce the dynamic properties of a structure and define the general concept of a response spectrum.

The first and most basic dynamic property of a structure is its fundamental period of vibration. To define the period, first assume that a one-story building has its mass tributary to the roof level assigned or "lumped" at that level. See Fig. 2.15. The dynamic model then becomes a flexible column with a single, concentrated mass at its top. If the mass is given some horizontal displacement (point 1) and then released, it will oscillate back and forth (i.e., from 1 to 2 to 3). This movement, with no externally applied load, is termed *free vibration*. The *period of vibration, T*, of this structure is defined as the length

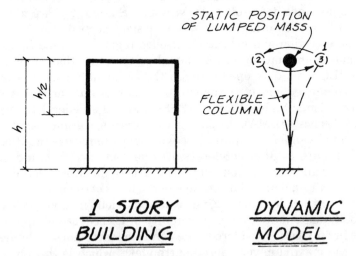

Figure 2.15 Period of vibration T is the time required for one cycle of free vibration. The shaded area represents the tributary wall and roof dead load, which is assumed to be concentrated at the roof level.

of time (in seconds) that it takes for one complete cycle of free vibration. The period is a characteristic of the structure (a function of mass and stiffness), and it is a value that can be calculated from dynamic theory.

When the multistory building of Fig. 2.14 was discussed (Sec. 2.12), the concept of fundamental mode of vibration was defined. Characteristic periods are associated with all of the modes of vibration. The fundamental period can be defined as the length of time (in seconds) that it takes for the first or fundamental mode (deflection shape) to undergo one cycle of free vibration. The fundamental period can be calculated from theory, or the Code's simple, normally conservative, method for the approximate period can be used.

In this latter approach, Sec. 9.5.5.3.2 of ASCE 7 (as referenced by IBC Sec. 1617.4) provides the following formula for the approximate period of vibration:

$$T_a = C_t h_n^x$$

where h_n = height of the highest (nth) level above the base, ft

    x = exponent dependent on structure type, from ASCE 7 Table 9.5.5.3.2

        = 0.80 for moment-resisting systems of steel

        = 0.90 for moment-resisting systems of concrete

        = 0.75 for eccentrically braced steel frames, and

        = 0.75 for all other structures

    C_t = coefficient dependent on structure type, from ASCE 7 Table 9.5.5.3.2

        = 0.028 for moment-resisting systems of steel

        = 0.016 for moment-resisting systems of concrete

        = 0.030 for eccentrically braced steel frames, and

        = 0.020 for all other structures

ASCE 7 provides optional alternative definitions for the approximate period T_a in structures with steel or concrete moment frames and in structures with concrete or masonry shearwalls. However, for simplicity, $T_a = (0.020)(h_n)^{0.75}$ is used for all buildings in this text. The approximate period calculated using this formula is conservative for most structures. A conservative period is one that falls within the level plateau of the design spectrum.

Damping is another dynamic property of the structure that affects earthquake performance. Damping can be defined as the resistance to motion provided by the building materials. Damping mechanisms can include friction, metal yielding, and wood crushing as the structure moves during an earthquake. Damping will slowly reduce the free-vibration displacement of the structure, eventually bringing it to a stop.

With the concepts of period of vibration and damping now defined, the idea of a response spectrum can be introduced. A *response spectrum* is defined as a plot of the maximum response (acceleration, velocity, displacement, or equivalent static force) versus the period of vibration. See Example 2.12 and Fig.

2.16. In a study of structural dynamics, it has been found that structures with the same period and the same amount of damping have essentially the same response to a given earthquake acceleration record.

EXAMPLE 2.12   Typical Theory Response Spectrum

The term *response spectrum* comes from the fact that *all building periods* are summarized on one graph (for a given earthquake record and a given percentage of critical damping). Figure 2.16 shows the complete *spectrum* of building periods. The curve shifts upward or downward for different amounts of damping.

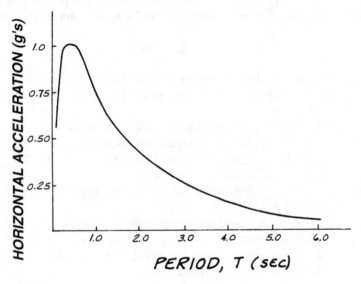

Figure 2.16

Earthquake records are obtained from strong-motion instruments known as accelerographs, which are triggered during an earthquake and record ground accelerations. Time histories of ground accelerations can then be used as the input for computer time-history analyses of a single degree of freedom model. Varying the period of the single degree of freedom model allows the development of a relationship between the period of vibration and the maximum response. A response spectrum can be determined from a single ground motion record or from a group of records. Once the response spectrum has been determined by analysis, it can be used to estimate the effect of the particular ground motion record, or group of records, on buildings. The information required to obtain values from a response spectrum is simply the fundamental or approximate period of the structure.

It should be pointed out that a number of earthquake ground acceleration records are available, and each record could be used to generate a unique response spectrum for a given damping level and soil condition. The IBC simplifies this process, for design, by providing coefficients S_{DS} and S_{D1}/T for construction of a smoothed design response spectrum which is based on an assumed damping coefficient and results from a large number of individual ground motion records. The spectrum is specific to the mapped spectral accelerations, S_S and S_1, and the particular site class. The development of the IBC design spectra includes a modest amount of damping that is applicable to all building types. The additional damping capacity of particular seismic bracing systems is further considered in development of the R-factors.

Now that the basic dynamic properties (*period* and *damping*) of a building and the concept of a *response spectrum* have been introduced, the formulation for the response spectrum values S_{DS} and S_{D1}/T can be reviewed. It should be clear that S_{D1}/T will depend on the period of vibration, T. Experience in several earthquakes has shown that local soil conditions can have a significant effect on earthquake response. The 1985 Mexico Earthquake shock is a prime example of earthquake ground motions being amplified by local soil conditions. See Example 2.13 and Fig. 2.17.

EXAMPLE 2.13   Effect of Local Soil Conditions

Soil-structure resonance is the term used to refer to the amplification of earthquake effects caused by local soil conditions (Fig. 2.17). The soil characteristics associated with a given building site (site-specific) are incorporated into the definition of site coefficients F_a and F_v.

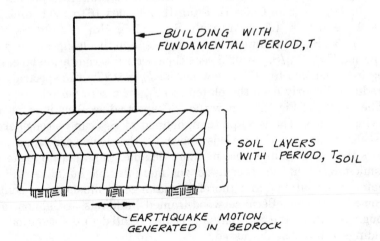

Figure 2.17   Geotechnical profile.

The Code establishes six *Site Classes* (Classes A through F, IBC Table 1615.1.1), and a different value of the site coefficients F_a and F_v are assigned to each class for each tabulated range of mapped spectral acceleration [IBC Table 1615.1.2(1) and 1615.1.2(2)]. If a structure is supported directly on hard rock (Site Class A), then F_a

and F_v are 0.8 for all mapped spectral accelerations. However, if the structure rests on softer soil, the earthquake ground motion originating in the bedrock may be amplified. In the absence of a geotechnical evaluation, Site Class D is the default site class normally permitted for use in determining the site coefficients F_a and F_v.

It is perhaps more difficult to visualize, but the soil layers beneath a structure have a period of vibration T_{soil} similar to the period of vibration of a building, T. Greater structural damage is likely to occur when the fundamental period of the structure is close to the period of the underlying soil. In these cases a quasi-resonance effect between the structure and the underlying soil develops. The conditions at a specific site are classified into one of six soil profile types, designated as Site Classes A through F. Site class and site coefficients are determined in accordance with IBC provisions, rather than referencing ASCE 7. IBC Sec. 1615.1.1 and IBC Table 1615.1.1 provide geotechnical definitions of site classes, which are then used to determine site coefficients F_a and F_v in accordance with IBC Tables 1615.1.2(1) and 1615.1.2(2). Additional near-source amplifiers, discretely considered in the 1997 UBC, are now directly included in the mapped accelerations, eliminating an additional step.

Finally, based on the discussion of the last several pages, the IBC design response spectrum curve needs to be created. A generic IBC design response spectrum curve is plotted in Fig. 2.18a. The response spectrum curve in Fig. 2.18b has been made specific to a location in California having $S_S = 1.5g$, $S_1 = 0.75g$, and Site Class D. From IBC Tables 1615.1.2(1) and 1615.1.2(2), $F_a = 1.0$ and $F_v = 1.5$. As a result, $S_{MS} = 1.5g$, $S_{M1} = 1.13g$, $S_{DS} = 1.0g$, and $S_{D1} = 0.75g$. S_{DS} determines the level plateau to the design response spectrum. At period $T_S = S_{D1}/S_{DS} = 0.75$ sec., the spectral acceleration level starts dropping in proportion to $1/T$. Below period $T_0 = 0.2T_S$, the spectral acceleration is reduced linearly from the plateau to S_{DS} at a zero period.

The graphs of Fig. 2.18 show the IBC spectral accelerations for buildings of varying periods. The level plateau, defined by S_{DS} can be considered to apply to stiffer buildings. For periods above T_S, the downward trend of the curve shows that as buildings become more flexible, they tend to experience lower seismic forces. On the other hand, the more flexible buildings will experience greater deformations, and therefore, damage to finishes and contents could become a problem. Because wood-framed buildings are almost always stiff enough to fall at the S_{DS} plateau, issues related to deformations of flexible buildings will not be discussed.

It was noted earlier that in a linear dynamic (response spectrum) analysis, the total response of a multidegree of freedom structure could be obtained by adding together appropriate percentages of spectral accelerations at a number of modes of vibration. However, the IBC design response spectrum, in addition to being a smoothed representation of multiple ground acceleration records, represents a multimode response spectrum envelope, modified to account for higher modes of vibration. These multimode effects are significant for relatively tall structures, which have correspondingly long periods. However, relatively low-rise structures are characterized by short periods of vibration. Consequently, it should be of little surprise that the flat plateau, defined by S_{DS} will apply to the buildings covered in this test. Numerical examples demonstrating this are given in Chap. 3.

Occupancy Importance Factor, *I*

An occupancy importance factor, *I*, was introduced into the seismic base shear formula as a result of failures which occurred in the 1971 San Fernando earthquake. IBC Sec. 1616.2 addresses the assignment of a Seismic Use Group and Importance Factor. Seismic Use Groups are related to Occupancy Categories defined in IBC Table 1604.5. In general, facilities that house large groups of occupants, or occupants that have reduced mobility, are assigned to Occupancy Category III and have a seismic importance factor of 1.25. This is intended to correspond to Seismic Use Group II. In general, facilities needed for emergency response and facilities that house significant quantities of hazardous materials are assigned to Occupancy Category IV and have a seismic importance factor of 1.5. This is intended to correspond to a Seismic Use Group III. Other occupancy types generally fall under Occupancy Categories I and II, with a seismic importance factor of 1.0. This is intended to correspond to Seismic Use Group I. Although the seismic importance factor uses the symbol

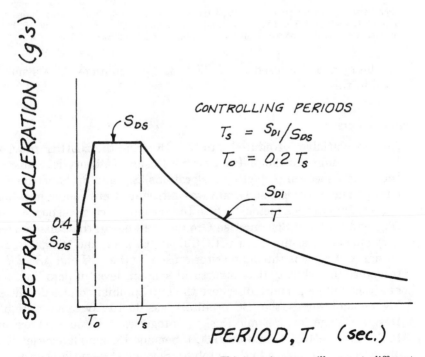

Figure 2.18a IBC design response spectra. This generic curve will generate different specific values depending on seismic zone and soil type.

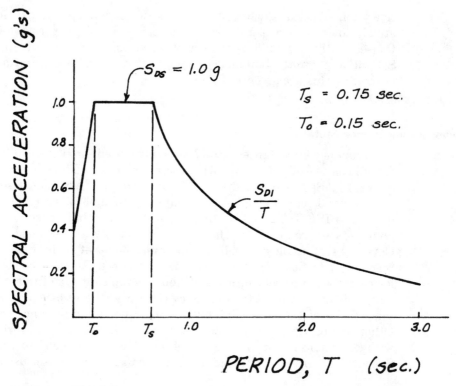

Figure 2.18b IBC design response spectrum using a California site with mapped spectral accelerations $S_S = 1.5$ and $S_1 = 0.75$, and site coefficients $F_a = 1.0$ and $F_v = 1.5$. This results in design spectral response accelerations of $S_{DS} = 1.0g$ and $S_{D1} = 0.75g$.

I_E, the symbol I is used in ASCE 7 design equations. The symbol I will be used in this text.

Seismic Design Category

Another variable introduced into the IBC from the NEHRP provisions is the *Seismic Design Category*. The Seismic Design Category is a function of both the design spectral response accelerations (S_{DS} and S_{D1}) and the Seismic Use Group. The Seismic Use Group was introduced previously in relation to the seismic importance factor, I. Once the design spectral response accelerations (S_{DS} and S_{D1}) and Seismic Use Groups are known, the Seismic Design Category may be determined from IBC Tables 1616.3(1) and 1616.3(2). The Seismic Design Category is the main criteria used by the IBC and ASCE 7 provisions to determine what type of structural system, level of detailing, and type of analysis will be permitted. Where the UBC prohibited less ductile systems in high seismic zones, the IBC prohibits less ductile systems in higher Seismic Design Categories. Seismic Design Categories A through D are included in IBC Tables 1616.3(1) and 1616.3(2). Seismic Design Categories E and F are discussed in footnote a to these tables. Seismic Design Categories E and F are applicable for sites where S_1, the mapped one-second spectral response, is equal to or greater than 0.75. This denotes a near-fault or near-source condition. In general, structures in Seismic Design Category A will have very few rules and restrictions, whereas structures in Category F will have significant restrictions.

Response Modification Factor, *R*

The response modification factor (*R*-factor) is found in ASCE 7 Table 9.5.2.2, as modified by IBC Sec. 1617.6.1.1. Of particular interest for users of this book, IBC increases by ½ the *R*-factors for light-framed walls sheathed with wood structural panels. The ASCE 7 *R*-factors, as modified by the IBC, will be designated as the *Code R-factors*. This factor reduces the design seismic forces as a function of the ductility and over-strength of the lateral force resisting system.

The premise of allowable stress design procedures, as has been discussed previously, is that stresses in elements resulting from expected loading are required to be less than the strength of the elements divided by a factor of safety. This results in element stresses remaining in the elastic range (stress proportional to strain) when subjected to design loads. If the premise of element stresses staying elastic were to be applied to seismic design, an *R* of approximately 1.0 would be used. This would mean that the full spectral acceleration plotted in the IBC design response spectrum would be used for design, resulting in many cases in design for seismic base shears in excess of 1.0g. Experience in past earthquakes, however, has demonstrated that buildings designed to a significantly lower base shear can adequately resist seismic forces without collapse. This experience provides the basis for use of the *R* factor. The reason for adequate performance at a lower base shear is thought to be the result of both extra or reserve strength in the structural system, and stable inelastic behavior of the structural elements. Based on reserve strength and inelastic behavior, the code allows use of *R* values significantly greater than 1.0, resulting in seismic base shears significantly lower than 1.0g.

The reserve strength in the structural system is called *over-strength* in the Code. The contribution of over-strength to the *response modification factor R* comes from several sources including *element over-strength* and *system over-strength*. The reader can visualize a wood structural panel shearwall. When the design seismic forces are applied, the wall stresses are well within the elastic range (stress nearly proportional to strain). More seismic force can be applied before the wall reaches what could be considered a yield stress (stresses no longer nearly proportional to strain). Yet more seismic force can usually be applied before the wall reaches its failure load and the strength starts decreasing. The difference between the initial design seismic force and the failure load is the *element over-strength*. The Ω_0 (W$_0$ in first ASCE printing) values tabulated in ASCE 7, Table 9.5.2.2 give an approximation of the expected *element over-strength* for each type of primary LFRS. The Ω_0 value and its use in ASCE 7 equations 9.5.2.7.1-1 and 9.5.2.7.1-2 will be discussed in Chap. 9, 10, and 16.

The system over-strength comes from the practice of designing a group of elements for the forces on the most highly loaded elements. This results in the less highly loaded elements in that group having extra capacity. Because the capacity of elements must generally be exceeded at more than one location in a system before a system failure occurs, the result is reserve capacity or over-strength in the system.

The ability of structural elements to withstand stresses in the inelastic range is call *ductility* in the Code.

In a major earthquake a structure will not remain *elastic,* but will be forced into the *inelastic* range. Inelastic action absorbs significantly more energy from the system. Therefore, if a structure is properly detailed and constructed so that it can perform in a ductile manner (i.e., deform in the inelastic range), it can be designed on a working-stress (elastic) basis for considerably smaller lateral forces (such as those given by the Code equivalent lateral force procedure).

Experience in previous earthquakes indicates that certain types of lateral-force-resisting systems (LFRSs) perform better than others. This better performance can be attributed to the *ductility* (the ability to deform in the inelastic range without fracture) of the system. The *damping* characteristics of the various types of structures also affect seismic performance.

The expected ductility and over-strength of each LFRS is taken into account in the Code R factors. The R term in the denominator of the seismic base shear formula is the empirical judgment factor that reduces the lateral seismic forces to an appropriate level for use in conventional working stress design procedures. Numerical values of R are assigned to the various LFRSs in ASCE Table 9.5.2.2, as modified by IBC Sec. 1617.6.1.1 (R-factors for nonbuilding structures are given in ASCE 7 Table 9.14.5.1.1).

The five basic structural systems recognized by the Code for conventional buildings are:

A. Bearing wall

B. Building frame

C. Moment-resisting frame

D. Dual (combined shearwall and moment-resisting frame)

E. Inverted pendulum and cantilevered column systems

For these systems, R-factors range from $1\frac{1}{4}$ to 8, an even greater range than had been included in recent editions of the UBC. Because R appears in the denominator of the base shear coefficient, more ductile performance is expected of systems with larger R-factors. One of the reasons for the wider range is the introduction of structural systems used regionally across the United States. A very important change in the R-factor table is the explicit listing of allowable Seismic Design Categories and height limits for each listed structural system. The lowest R-factors correspond to *ordinary plain concrete shearwalls,* to *ordinary plain masonry shearwalls,* and to *ordinary steel moment frames* when occurring as cantilevered columns. As less ductile systems, all of these are prohibited in Seismic Design Category D (SDC D), and some are also prohibited in Seismic Design Category C.

In previous codes the term "box system" was very descriptive of the LFRS used in typical wood-frame buildings with horizontal diaphragms and shearwalls. These structures are now classified as either a bearing wall system or a building frame system.

It is very common in a wood-frame building to have roof and floor beams resting on load-bearing stud walls. If a load-bearing stud wall is also a shearwall, the LFRS will be classified as a bearing wall system. For buildings with a bearing wall system the Code (ASCE 7 Table 9.5.2.2 modified by IBC Sec. 1617.6.1.1) assigns the following values of R:

Bearing wall system	R
Light-framed walls sheathed with wood structural panels rated for shear resistance or steel sheets	6½
Light-framed walls with shear panels of all other materials	2
Special reinforced concrete walls (permitted in SDC D)	5
Special reinforced masonry walls (permitted in SDC D)	5

Past editions of the UBC included a very modest difference between R-factors for structures with wood structural panel (plywood and OSB) bracing, and structures braced by other materials. Even this modest difference went away for buildings of four or more stories. The R-factors for wood structural panel sheathing and other bracing materials (gypsum wallboard, stucco, etc.) are now different by a factor of three. The very low R-factor is intended to reflect the perceived brittle nature of these materials and put their design on par with other brittle systems. Use of non-wood structural panel materials will significantly increase the design base shear, requiring not only additional bracing, but also additional fastening for shear transfer and overturning. The likely result is more extensive use of wood structural panels in order to qualify for a lower base shear. Special reinforced concrete and masonry shearwalls are included here because these systems are permitted in Seismic Design Category D. Additional system types are permitted in lower Seismic Design Categories. It should be noted that ASCE 7, like the NEHRP provisions, has linked structural systems and structural detailing requirements. Each separate title used in the ASCE 7 table denotes a system with specific detailing requirements.

A *building frame system* may also use horizontal diaphragms and shearwalls to carry lateral forces, but in this case gravity loads are carried by what the ASCE 7 definitions term "an essentially complete space frame." For example, vertical loads could be supported entirely by a wood or steel frame, and lateral forces could be carried by a system of non-load-bearing shearwalls. The term "non-load-bearing" indicates that these walls carry no gravity loads (other than their own dead load). The term "shearwall" indicates that the wall is a lateral-force-resisting element.

The distinction between a bearing wall system and a building frame system is essentially this: In a bearing wall system, the walls serve a dual function in that both gravity loads and lateral forces are carried by the same structural element. Here failure of an element in the LFRS during an earthquake could possibly compromise the ability of the system to support gravity loads. On the other hand, because of the separate vertical-load and lateral-force carrying elements in a building frame system, failure of a portion of the LFRS does not necessarily compromise the ability of the system to support gravity loads. Because of the expected better peformance, slightly larger *R*-factors are assigned to building frame systems than to bearing wall systems:

Building frame system	R
Light-framed walls sheathed with wood structural panels rated for shear resistance or steel sheets	7
Light-framed walls with shear panels of all other materials	2½
Special reinforced concrete walls (permitted in SDC D)	6
Special reinforced masonry walls (permitted in SDC D)	5½

Building frame systems, however, are not extremely common in wood light-frame construction.

Each of the coefficients in the base shear formula has been reviewed, so the designer should have a solid understanding of these terms.

2.14 Seismic Forces—Primary System

Seismic forces are calculated and distributed throughout the structure in the reverse order used for most other forces. In evaluating wind forces, e.g., the design pressures are calculated first. Later the shear at the base of the structure can be determined by summing forces in the horizontal direction. For earthquake forces the process is just the reverse. The shear at the base of the structure is calculated first, using the base shear formula for V (Sec. 2.13). Then total *story forces* F_x are assigned to the roof and floor levels by distributing the base shear vertically over the height of the structure. Finally, individual story forces are distributed horizontally at each level in accordance with the mass distribution of that level.

The reasoning behind the vertical distribution of seismic forces was given in Sec. 2.12. The general distribution was described, and it was seen that the shape of the first mode of vibration serves as the basis for obtaining the story forces acting on the *primary* lateral-force-resisting system (LFRS). When a part or portion of a building is considered, the seismic force F_p on the individual part may be larger than the seismic forces acting on the primary LFRS. Seismic forces on certain parts and elements of a structure are covered in Sec. 2.15. The methods used to calculate the distributed story forces on the primary lateral-force-resisting system is reviewed in the remaining portion of this section.

The primary LFRS is made up of both horizontal and vertical elements. In most wood-frame buildings, the horizontal elements are roof and floor systems that function as *horizontal diaphragms,* and the vertical elements are wall segments that function as *shearwalls.* A variety of other systems may be used (see Sec. 3.3 for a comparison of several types), but these alternative systems are more common in other kinds of structures (e.g., steel-frame buildings).

Another unique aspect of seismic force evaluation is that there are two different sets of story force distributions for the primary lateral-force-resisting system. One set of story forces is to be used in the design of the vertical elements in the LFRS, and the other set applies to the design of horizontal diaphragms. A different notation system is used to distinguish the two sets of story forces.

With the incorporation of the NEHRP provisions, the equations for calculation of horizontal story forces for design of diaphragms (F_{px}) now vary by Seismic Design Category. The variation is included with the intent of simplifying design procedures for lower Seismic Design Categories. A general discussion will be presented first, addressing why design story forces for horizontal elements (F_{px}) are different from design story forces for vertical element (F_x). A discussion of the simplified approach for lower Seismic Design Categories will follow.

The forces for designing the vertical elements (i.e., the shearwalls) are given the symbol F_x, and the forces applied to the design of horizontal diaphragms are given the symbol F_{px}. Both F_x and F_{px} are horizontal *story forces* applied to level x in the structure. Thus, the horizontal forces are assumed to be concentrated at the story levels in much the same manner as the masses tributary to a level are "lumped" or assigned to a particular story height.

Initially it may seem strange that the Code would provide two different distributions $(F_x$ and $F_{px})$ for designing the *primary LFRS,* but once the reasoning is understood, the concept makes sense. The rationale behind the F_x and F_{px} distributions has to do with the fact that the forces occurring during an earthquake change rapidly with time. Because of these rapidly changing forces and because of the different modes of vibration, it is likely that the maximum force on an individual horizontal diaphragm will not occur at the same instant in time as the maximum force on another horizontal diaphragm. Hence, the loading given by F_{px} is to account for the possible larger *instantaneous* forces that will occur on individual horizontal diaphragms. Therefore, the F_{px} story force is to be used in the design of individual horizontal diaphragms, diaphragm collectors (drag struts), and related connections. The design of horizontal diaphragms and the definition of terms (such as drag struts) are covered in detail in Chap. 9.

On the other hand, when all of the story forces are considered to be acting on the structure *concurrently,* it is reasonable to use the somewhat smaller distribution of earthquake forces given by F_x. The simultaneous application of all of the F_x story forces does not affect the design of individual horizontal diaphragms. Thus, F_x is used to design the vertical elements (shearwalls) in the primary lateral-force-resisting system. The connections anchoring the shearwall to the foundation, and the foundation system itself, are also to be designed for the accumulated effects of the F_x forces. The design of shearwalls is covered in Chap. 10, and a brief introduction to the foundation design problem for shearwalls is given in Chap. 16.

Now that the general concept of story forces F_x and F_{px} has been presented, the specific application in different Seismic Design Categories can be discussed. For design and detailing requirements and structural component load effects, IBC Sec. 1620.1 refers the user to Sec. 9.5.2.6 of ASCE 7. Section 9.5.2.6 organizes requirements by Seismic Design Category. As the Seismic Design Category increases, the detailing requirements and loading requirements also increase. This results in more attention to detailing being required in areas of highest seismic hazard, and lower requirements in areas of lower seismic hazard, where loading conditions other than seismic might be of greater concern. It is also important to note that requirements are cumulative. Each higher Seismic Design Category must meet the requirement of the next lowest, plus additional requirements.

Provisions for diaphragm design forces are given in ASCE 7 Sec. 9.5.2.6.2.7, which are applicable for Seismic Design Category B and higher. The diaphragm story force F_{px} is calculated as the greater of the vertical element story force F_x or $0.2S_{DS}Iw_p$. The formula given adds another term, V_{px}, which is intended to include any diaphragm forces created by the redistribution of forces between vertical elements (where rigid diaphragm assumptions are used in horizontal distribution of forces). While V_{px} is specifically noted in this one equation, redistributed forces should always be added in, no matter which equation is used. In Seismic Design Category B, the diaphragm will be designed for either the same story force as the vertical element, or the constant coefficient given by $0.2S_{DS}I$ times the story weight. Because diaphragm design is not modified by ASCE 7 Sec. 9.5.2.6.3 for Seismic Design Category C, the same equations apply to both Seismic Design Category B and C.

ASCE 7 Sec. 9.5.2.6.4.3 specifies different diaphragm story forces (F_{px}) for Seismic Design Category D, E, and F. These forces do recognize the higher instantaneous forces, and similar to prior editions of the UBC, require a second vertical distribution to account for these forces. The formulas for story force F_x for all Seismic Design Categories and for diaphragm story force F_{px} for Seismic Design Categories D, and higher, are given in Example 2.14. In practice, the F_x story forces must be determined first because they are then used to evaluate the F_{px} story forces. The F_x *story forces are to be applied simultaneously* to all levels in the primary LFRS for designing the vertical elements in the system. In contrast, the F_{px} *story forces are applied individually* to each level x in the primary LFRS for designing the horizontal diaphragms.

Although the purpose of the F_x forces is to provide the design forces for the shearwalls, the F_x forces are applied to the shearwalls through the horizontal diaphragms. Thus, both F_{px} and F_x are shown as uniform forces on the horizontal diaphragms in Fig. 2.19. To indicate that the diaphragm design forces are applied individually, only *one* of the F_{px} forces is shown with solid lines. In comparison the F_x forces act concurrently and are *all* shown with solid lines.

The formula for F_x will produce a triangular distribution of horizontal story forces if the masses (tributary weights) assigned to the various story levels are all equal (refer to Fig. 2.14a in Sec. 2.12). If the weights are not equal, some variation from the straight-line distribution will result, but the trend will follow the first-mode shape. Accelerations and, correspondingly, inertia forces ($F = Ma$) increase with increasing height above the base.

EXAMPLE 2.14 F_x and F_{px} Story Force Distributions

Two different distributions of seismic forces are used to define earthquake forces on the primary lateral-force-resisting system (Fig. 2.19). The *story forces* for the two major components of the primary LFRS are given by the following distributions.

F_x Distribution—Vertical Elements (Shearwalls)

All Seismic Design Categories

$$F_x = C_{vx} V$$

and
$$C_{vx} = \frac{w_x h_x^k}{\sum\limits_{i=1}^{n} w_i h_i^k}$$

where C_{vx} = vertical distribution factor
V = total base shear
w_i, w_x = tributary weights assigned to level i or x
h_i, h_x = height from the base of structure to level i or x, ft
k = an exponent related to the structural period
= 1 for structures having a period of 0.5 sec or less

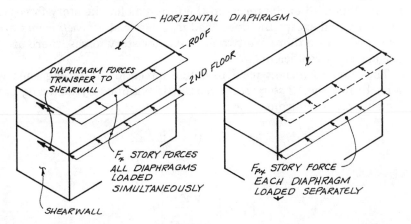

F_x FOR SHEARWALLS F_{px} FOR DIAPHRAGMS

Figure 2.19

F_{px} *Distribution—Horizontal Elements (Diaphragms)*

Seismic Design Categories D, E, and F

$$F_{px} = \frac{\displaystyle\sum_{i=x}^{n} F_i}{\displaystyle\sum_{i=x}^{n} w_i} w_{px}$$

and

$$0.2S_{DS}Iw_{px} \leq F_{px} \leq 0.4S_{DS}Iw_{px}$$

where F_{px} = horizontal force on primary LFRS at story level x for designing horizontal elements

F_i = lateral force applied to level i (this is story force determined in accordance with formula for F_x)

w_{px} = weight of diaphragm and elements tributary to diaphragm at level x

Other terms are as defined for F_x.

In the formulas for distributing the seismic force over the height of the structure, the superscript k is to account for whip action in tall, slender buildings and to allow for the effects of the higher modes (i.e., other than the first mode) of vibration. When the period of vibration is less than 0.5 sec, there is no whipping effect.

It may not be evident at first glance, but the formulas for F_x and F_{px} can be simplified to a form that is similar to the base shear expression. In other words, the earthquake force can be written as the mass (weight) of the structure multiplied by a seismic coefficient. For example,

$$V = (\text{seismic coefficient})W$$

The seismic coefficient in the formula for V is known as the *base shear coefficient*. When all of the terms in the formulas for the story forces (F_x and F_{px}) are evaluated except the dead load w, seismic story *coefficients* are obtained. Obviously, since there are two formulas for story forces, there are two sets of seismic story coefficients.

The story coefficient used to define forces for designing shearwalls is referred to as the F_x *story coefficient*. It is obtained by factoring out the story weight from the formula for F_x:

$$F_x = C_{vx}V = \frac{w_x h_x^k}{\displaystyle\sum_{i=1}^{n} w_i h_i^k} V$$

$$F_x = \left[\frac{V h_x^k}{\displaystyle\sum_{i=1}^{n} w_i h_i^k} \right] w_x$$

$$= (F_x \text{ story coefficient}) w_x$$

Likewise, the formula for F_{px} for use in diaphragm design can be viewed in terms of an F_{px} *story coefficient*. For Seismic Design Categories D, E, and F, the formula for F_{px} is initially expressed in this format:

$$F_{px} = \left[\frac{\sum\limits_{i=x}^{n} F_i}{\sum\limits_{i=x}^{n} w_i} \right] w_{px}$$

$$= (F_{px} \text{ story coefficient})w_{px}$$

It should be noted that a one-story building represents a special case for earthquake forces. In a one-story building, the diaphragm loads given by F_x and F_{px} are equal. In fact, the F_x and F_{px} story coefficients are the same as the base shear coefficient. In other words, for a *one-story building*:

Base shear coefficient = F_x story coefficient = F_{px} story coefficient

Having a single seismic coefficient for three forces greatly simplifies the calculation of seismic forces for one-story buildings. *Numerical examples* will greatly help to clarify the evaluation of lateral forces. Several one-story building examples are given in Chap. 3, and a comparison between the F_x and F_{px} force distributions for a two-story building is given in Example 3.10 in Sec. 3.6.

At this point one final concept needs to be introduced concerning the distribution of seismic forces. After the story force has been determined, it is distributed at a given level in proportion to the mass (dead load, D) distribution of that level. See Example 2.15.

The purpose behind this distribution goes back to the idea of an inertial force. If it is visualized that each square foot of dead load has a corresponding inertial force generated by an earthquake, then the loading shown in the sketches becomes clear. If each square foot of area has the same D, the distributed seismic force is in proportion to the length of the roof or floor that is parallel to the direction of the force. Hence the magnitude of the distributed force is large where the dimension of the floor or roof parallel to the force is large, and it is small where the dimension parallel to the force is small.

EXAMPLE 2.15   Distribution of Seismic Force at Story Level *x*

Transverse and Longitudinal Directions Defined

A lateral force applied to a building may be described as being in the transverse or longitudinal direction. These terms are interpreted as follows:

Transverse lateral force is parallel to the short dimension of the building.
Longitudinal lateral force is parallel to the long dimension of the building.

Buildings are designed for seismic forces applied independently in both the transverse and longitudinal directions.

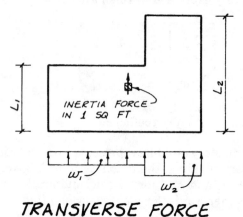

TRANSVERSE FORCE

Figure 2.20a Distribution of story force in transverse direction.

Each square foot of dead load, D, can be visualized as generating its own inertial force (Fig. 2.20a). If all of the inertial forces generated by these unit areas are summed in the transverse direction, the forces w_1 and w_2 are in proportion to lengths L_1 and L_2, respectively.

The sum of the distributed seismic forces w_1 and w_2 (i.e., the sum of their resultants) equals the transverse story force. For shearwall design the transverse story force is F_x, and for diaphragm design the transverse story force is F_{px}.

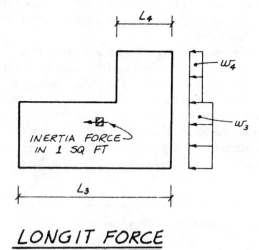

LONGIT FORCE

Figure 2.20b Distribution of story force in longitudinal direction.

In the longitudinal direction (Fig. 2.20b), L_3 and L_4 are measures of the distributed forces w_3 and w_4. The sum of these distributed seismic forces equals the story force in the longitudinal direction.

NOTE: The distribution of inertial forces generated by the dead load of the walls parallel to the direction of the earthquake is illustrated in Chap. 3.

The basic seismic forces acting on the primary lateral-force-resisting system of a *regular* structure have been described in this section. The Code requires that the designer consider the effects of structural *irregularities*. Section 9.5.2.3 of ASCE 7 identifies a number of these irregularities. In many cases, increased force levels and reduced stresses are required for the design of an irregular building.

It is important for the designer to be able to identify a structural irregularity and to understand the implications associated with the irregularity. However, a detailed study of these Code provisions is beyond the scope of Chap. 2. In fact, the majority of this book is written as an *introduction* to the basic principles of engineered wood structures. To accomplish this, most of the structures considered are rather simple in nature. Structural irregularities may be common occurrences in daily practice, but they can be viewed as advanced topics at this point in the study of earthquake design.

It is felt the reader should first develop a good understanding of the design requirements for regular structures. Therefore, the provisions for irregular structures are postponed to Chap. 16, after the principles of structural design for regular buildings have been thoroughly covered.

The seismic forces required for the design of elements and components that are not part of the primary LFRS are given in Sec. 2.15.

2.15 Seismic Forces—Wall Components

The seismic forces which have been discussed up to this point are those assumed to be developed in the primary lateral-force-resisting system of a building as it responds to an earthquake. However, when individual elements of the structure are analyzed separately, it may be necessary to consider different seismic effects. One reason for this is that certain elements which are attached to the structure respond dynamically to the motion of the structure rather than to the motion of the ground. Resonance between the structure and the attached element may occur.

The 2003 IBC and ASCE 7 have made a significant change in the calculation of seismic forces for out-of-plane loading on structural wall components. The IBC, like the UBC, has a section that addresses the design of components attached to the structure. For these items, a component seismic force F_p is used. In the IBC and ASCE 7, the F_p forces for exterior walls are very specifically addressing exterior nonstructural "skin" walls with discreet attachments to the main building structure, rather than structural walls that would be integral. An exception to this is that seismic F_p forces for cantilevered wall parapets are included. It could be interpreted that these forces apply to parapets on "structural" walls.

The basic equation for out-of-plane seismic forces on structural walls can be found in ASCE 7 Sec. 9.5.2.6.2.8, which is applicable to Seismic Design Categories B and up. The specified force is $F_c = 0.4S_{DS}IW_c$, but not less than $0.10W_c$. S_{DS} is the design spectral response acceleration, I is the importance factor used for the main structure (as opposed to a component importance factor I_p), and W_c is the weight of the wall being anchored. This is a *strength level* force that can be multiplied by 0.7 for allowable stress forces. Design seismic forces for wall anchorage vary by Seismic Design Category and will be addressed in a later chapter. Technically ASCE 7 only requires use of this equation for concrete and masonry walls. It is suggested, however, that this same equation be applied as a minimum to all structural walls, not specifically addressed by ASCE 7 Sec. 9.6. Seismic forces do not normally control design for exterior light-frame walls out-of-plane, but they may control design once veneers or other heavy finishes are used.

The seismic force for design of components will be introduced for calculation of parapet forces. The component forces also apply to a wide variety of architectural, mechanical, and electrical components. ASCE 7 Sec. 9.6.1 specifically exempts parapet walls supported on a bearing or shearwall from these design forces for Seismic Design Category B, in which case, the simpler equation of Sec. 9.5.2.6.2.8 should be applied to the parapet.

Requirements for design of components can be found in IBC Sec. 1621, which in turn adopts the provisions of ASCE 7, Sec. 9.6. IBC modifies several of the ASCE 7 provisions; however, the modifications do not apply to the current discussion. The titles of both the IBC and ASCE 7 sections refer to architectural, mechanical, and electrical components. It should be noted, however, that virtually everything on or attached to the structure requires design per this section.

The Code provides an "equivalent static" force, F_p, for various components of a structure. The F_p force will be calculated for an exterior structural wall parapet. By including the F_p forces, the Code takes into account the possible response of an element and the consequences involved if it collapses or fails. The force on a portion of the structure is given by the following formula:

$$F_p = \frac{0.4a_p S_{DS}W_p}{R_p/I_p}\left(1 + 2\frac{z}{h}\right)$$

with F_p limited to the range:

$$0.3S_{DS}I_p W_p \le F_p \le 1.6S_{DS}I_p W_p$$

where F_p = component seismic design force centered at the component's center of gravity and distributed relative to the component's mass distribution

S_{DS} = short-period design spectral acceleration, discussed in Sec. 2.13

a_p = component amplification factor per ASCE 7 Table 9.6.2.2 or 9.6.3.2

I_p = component importance factor per ASCE 7 Sec. 9.6.1.5

W_p = component operating weight

R_p = component response modification factor per ASCE 7 Table 9.6.2.2 or 9.6.3.2

z = height in structure of point of attachment of component with respect to the structure base. For items attached at or below the base, z/h need not exceed 1.0.

h = average roof height of structure with respect to the structure base

This equation for F_p is close in format to the equation used in the 1997 UBC, but does contain several changes. The equation was developed based on building acceleration data recorded during earthquakes. See the NEHRP commentary for further discussion. The term $1 + 2z/h$ allows F_p to vary from one value for components anchored at the ground level to three times that value for components anchored to the roof. This matches the general trend seen in recorded acceleration data. If the elevation at which the component will be anchored is not known, this term can default to three.

Prior to combining component seismic forces with dead and live loads, the seismic force will need to be calculated in accordance with ASCE 7 Sec. 9.5.2.7. This section combined horizontal and vertical seismic forces, and also includes the redundancy factor ρ. While it is not specifically stated, ρ is a function of the building primary structure and is not applied to components. For component forces F_p, ρ can therefore be taken as 1.0. As with the seismic base shear equations in Sec. 2.13, this component seismic force F_p is at a strength level. It will be multiplied by 0.7 in the load combination equations to convert it to an allowable stress design level.

EXAMPLE 2.16   Seismic Forces Normal to Wall

Determine the seismic design force normal to the wall for the building shown in Fig. 2.21. The wall spans vertically between the floor and the roof, which is 16 ft from ground level. The wall is constructed of reinforced brick masonry that weighs 90 psf. Known seismic information:

Mapped short- (0.2 sec) spectral acceleration, S_S = 150% g = 1.5g

Mapped 1.0-sec spectral acceleration, S_1 = 75% g = 0.75g

Site (soil) Class = D

Site coefficients F_a and F_v = 1.0 and 1.5, IBC Tables 1615.1.2(1) and 1615.1.2(2)

Seismic Design Category D

Compare the seismic force to the wind force on components and cladding. Assume an effective wind area of 10 ft^2. Known wind information:

Wind speed, V = 90 mph

Exposure B conditions

Importance factor, I_W = 1.0 for standard occupancy

Note that wind and seismic forces are not considered simultaneously.

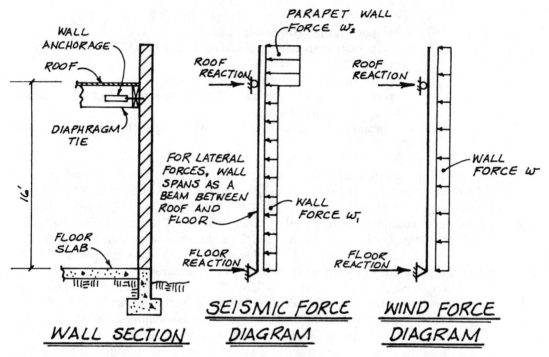

Figure 2.21

Seismic Forces for Design of Wall Element

The equations for calculation of component force, F_p, and wall out-of-plane force, F_c, are based on the design spectral response acceleration S_{DS}. S_{DS} is calculated as follows:

$$S_{MS} = F_a S_S = 1.0 \times 1.50g = 1.50g$$

$$S_{DS} = \tfrac{2}{3} \times S_{MS} = \tfrac{2}{3} \times 1.50g = 1.0g$$

Calculation of wall seismic forces

Forces w_{u1} and w_{u2} act normal to the wall in either direction (i.e., inward or outward). As was discussed in Sec. 2.8, these forces have the subscript u, denoting that the Code seismic forces are at a *strength* or *ultimate* level. Force w_{u1} will be calculated for wall out-of-plane design for portions of the wall other than the parapet. Force w_{u2} will be calculated for the parapet using the component force equations.

$$w_{u1} = 0.4 S_{DS} I w_c$$

Solving with $S_{DS} = 1.0$, $I = 1.0$, and $w_c = 90$ psf

$$w_{u1} = 0.4 w_c$$

$$w_{u1} = 0.4 \times 90 \text{ psf} = 36 \text{ psf}$$

For the parapet

$$w_{u2} = \frac{0.4a_p S_{DS} w_c}{R_p/I_p} \left(1 + 2\frac{z}{h}\right)$$

Solving with

$S_{DS} = 1.0,$
$a_p = 2.5,$ ASCE 7 Table 9.6.2.2
$R_p = 2.5,$ ASCE 7 Table 9.6.2.2
$I_p = 1.0,$ ASCE Sec. 9.6.1.5
$w_c = 90$ psf
$z = h = $ roof height

$$w_{u2} = \frac{0.4w_c}{1.0}(1 + 2)$$

$$w_{u2} = 1.2w_c = 108 \text{ psf}$$

with F_p limited to the range:

$$0.3S_{DS}I_p w_c \le F_p \le 1.6S_{DS}I_p w_c$$

$$0.3w_c \le F_p \le 1.6w_c$$

Before comparing these forces with allowable stress design wind forces, several adjustment factors need to be considered. The basic Code equation for E, earthquake forces is:

$$E = \rho Q_E \pm 0.2S_{DS}D$$

The variable ρ is permitted to be taken as 1.0 for design of components. The component $0.2S_{DS}D$ is acting vertically, and would be combined with the wall dead load, to determine a worst-case condition for axial plus flexural forces on the wall component. In addition the vertical component should be considered in checking anchorage of the component. For purposes of forces perpendicular to the wall surface, the vertical component has no effect.

In addition the strength-level seismic forces calculated need to be multiplied by 0.7, in accordance with the load combination equations, to obtain allowable stress design forces. This results in ASD level forces $w_1 = 25$ psf and $w_2 = 76$ psf.

These forces can now be compared to the wind forces.

Wind Forces

Height and exposure factor:

$$\lambda = 1.0 \quad \text{for } 0 \le h_{\text{mean}} \le 30 \text{ ft in Exposure B}$$

Wall forces—Interior Zone 4 for effective wind area of 10 ft^2:

$$p_{\text{net30}} = \begin{cases} 14.6 \text{ psf} & \text{(inward pressure)} \\ -15.8 \text{ psf} & \text{(outward pressure)} \end{cases}$$

Design wind pressure:

$$w = p_{\text{net}} = \lambda\, I_W\, p_{\text{net30}} = \begin{cases} 1.0(1.0)(14.6) = 14.6 \text{ psf} & \text{(inward pressure)} \\ 1.0(1.0)(-15.8) = -15.8 \text{ psf} & \text{(outward pressure)} \end{cases}$$

Wall forces—End Zone 5 for effective wind area of 10 ft^2:

$$p_{\text{net30}} = \begin{cases} 14.6 \text{ psf} & \text{(inward prerssure)} \\ -19.5 \text{ psf} & \text{(outward pressure)} \end{cases}$$

Design wind pressure:

$$w = p_{\text{net}} = \lambda I_W p_{\text{net30}} = \begin{cases} 1.0(1.0)(14.6) = 14.6 \text{ psf} & \text{(inward pressure)} \\ 1.0(1.0)(-19.5) = -19.5 \text{ psf} & \text{(outward pressure)} \end{cases}$$

Wind < seismic

∴ *seismic governs*

2.16   Load and Force Combinations

The IBC specifies a number of combinations that are to be considered in the design of a structure. These combinations define which loads and forces must be considered simultaneously. Obviously a given combination reflects the probability that various gravity loads and lateral forces will occur concurrently. Some of the probabilities of loading have been mentioned previously.

Load combinations are addressed in Sec. 1605 of the 2003 IBC. Section 1605.2 provides a set of strength design load combinations, while Sec. 1605.3 provides two sets of allowable stress design load combinations. The basic load combinations in Secs. 1605.2 and 1605.3.1 are based on the load combinations in Sec. 2.0 of ASCE 7, but with a slightly simplified presentation. The alternative basic load combinations in IBC Sec. 1605.3.2 are based on historically used load combinations.

This text will use the Sec. 1605.3.1 allowable stress design basic load combinations. There is a small change in these load combinations from the 1997 UBC. Instead of expressing allowable stress design seismic forces of $E/1.4$, the 2003 IBC uses $0.7E$. The IBC basic load combinations are:

D	(16-7)
D + L	(16-8)
D + L + (L$_r$ or S or R)	(16-9)
D + (W or 0.7E) + L + (L$_r$ or S or R)	(16-10)
0.6D + W	(16-11)
0.6D + 0.7E	(16-12)

IBC Sec. 1605.3.1.1 permits the effects of two or more transient loads to be multiplied by a load combination factor of 0.75. All loads except for dead load can be considered transient loads.

The equation numbers following each load combination are the equation numbers from the 2003 IBC which are reprinted in this book so that specific equations can be referred to in discussion. The basic load combinations are considered to be the preferred load combinations, and they are used throughout this book. A structure, and all elements and portions of a structure, must be designed to resist the most critical effects resulting from these load combinations. This means that the load on an element in a structure will need to be calculated using each of these equations unless the designer can tell by inspection that some of the load combinations will not control. The ability to eliminate load combinations by inspection will come with practice.

The IBC alternative *basic load combinations* are given below for information only. They represent a carry over from previous editions of the building codes, but are not used in examples in this book:

$$D + L + (L_r \text{ or } S \text{ or } R) \qquad\qquad (16\text{-}13)$$
$$D + L + (\omega W) \qquad\qquad (16\text{-}14)$$
$$D + L + \omega W + S/2 \qquad\qquad (16\text{-}15)$$
$$D + L + S + E/1.4 \qquad\qquad (16\text{-}16)$$
$$0.9D + E/1.4 \qquad\qquad (16\text{-}17)$$

where $\omega = 1.3$ for wind loads calculated in accordance with IBC Sec. 1609.6 or ASCE 7, and $\omega = 1.0$ for other wind loads.

Recall that three modification factors for loads and allowable stresses were introduced in Sec. 2.8: an *allowable stress increase* ASI a *load duration factor* C_D and a *load combination factor* LCF. Their use is reviewed here as part of this introduction to the Code required load combinations.

Per IBC Sec. 1605.3.1.1, the ASI does not apply to the IBC basic load combinations used in this text. The *load duration factor* C_D is a wood design adjustment factor and is covered in Chap. 4. While the 1997 UBC specifically modified the load duration factor to be used with wind and seismic forces, the IBC does not. As a result the load duration factors for IBC design correspond to those in the 2001 NDS (Ref. 2.1). Finally, the *load combination factor* (LCF) reflects the lower probability of obtaining the full design loads when multiple transient loads are considered simultaneously. Note that the multiplier of 0.7 applied to E in load combinations 16-10 and 16-12 and the divisor of 1.4 in equations 16-17 and 16-18 are not load combination factors. As discussed previously, these factors adjust forces from a strength level to an allowable stress design level.

Example 2.9 in Sec. 2.10 introduced the problem of overall moment stability under lateral forces. This is commonly referred to as a check on *overturning*. The IBC addresses overturning directly in load combinations 16-11 and 16-12, specifying a factor of 0.6 for dead loads that offset the overturning effects of wind loads or seismic loads.

A comprehensive summary of the IBC combinations for overturning and a comparison of wind and seismic provisions are given in Chap. 16.

2.17   References

[2.1]   American Forest and Paper Association (AF&PA). 2001. *National Design Specification for Wood Construction* and *Supplement,* 2001 ed., AF&PA, Washington, DC.

[2.2]   American Forest and Paper Association (AF&PA). 2001. *Wood Frame Construction Manual for One- and Two-Family Dwellings,* 2001 ed., AF&PA, Washington, DC.

[2.3]   American Institute of Steel Construction (AISC). 1989. *Manual of Steel Construction, Allowable Stress Design,* 9th ed., AISC, Chicago, IL.

[2.4]   American Institute of Timber Construction (AITC). 1994. *Timber Construction Manual,* 4th ed., John Wiley & Sons, New York, NY.

[2.5]   American Society of Civil Engineers (ASCE). 2003. *Minimum Design Loads for Buildngs and Other Structures* (ASCE 7-02), ASCE, Reston, VA.

[2.6]   American Society of Civil Engineers (ASCE). 1995. *Standard for Load and Resistance Factor Design (LRFD) for Engineered Wood Construction* (ASCE 16-95), ASCE, Reston, VA.

[2.7]   Building Officials and Code Administrators International (BOCA). 1999. *National Building Code,* 1999 ed., BOCA, Country Club Hills, IL.

[2.8]   Building Seismic Safety Council (BSSC). 2000. *National Earthquake Hazards Reduction Program (NEHRP) Recommended Provisions for Seismic Regulations for New Buildings* and *Commentary,* 2000 ed., BSSC, Washington, DC.

[2.9]   International Codes Council (ICC). 2003. *International Building Code,* 2003 ed., ICC, Falls Church, VA.

[2.10] International Conference of Building Officials (ICBO). 1997. *Uniform Building Code* (UBC), 1997 ed., ICBO, Whittier, CA.

[2.11] Mehta, K. C., and D. C. Perry. 2001. *Guide to the Use of the Wind Load Provisions of ASCE 7,* ASCE, Reston, VA.

[2.12] Southern Building Code Congress International (SBCCI). 1997. *Standard Building Code,* 1997 ed., SBCCI, Birmingham, AL.

[2.13] Structural Engineers Association of California (SEAOC). 1999. *Recommended Lateral Force Requirements and Commentary,* 7th ed., SEAOC, Sacramento, CA.

2.18   Problems

All problems are to be answered in accordance with the 2003 *International Building Code* (IBC). A number of Code tables are included in Appendix C.

2.1   *Given:*   The house framing section shown in Fig. 2.A

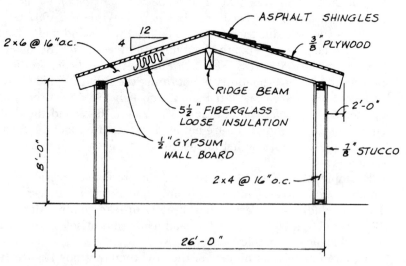

Figure 2.A

Find: a. Roof dead load D in psf on a horizontal plane
          b. Wall D in psf of wall surface area
          c. Wall D in lb/ft of wall
          d. Basic (i.e., consider roof slope but not trib. area) unit
              roof live load, L_r, in psf
          e. Basic unit roof L_r in psf if the slope is changed to $\frac{3}{12}$

2.2   *Given:*   The house framing section shown in Fig. 2.B. Note that a roofing square is equal to 100 ft².

       Find: a. Roof dead load D in psf on a horizontal plane
          b. Ceiling dead load D in psf
          c. Basic (i.e., consider roof slope but not trib. area) unit roof live load L_r in psf

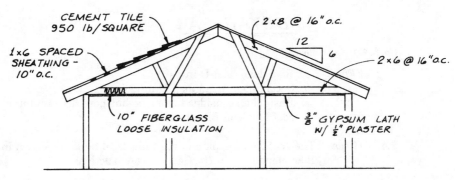

Figure 2.B

2.3   *Given:*   The building framing section shown in Fig. 2.C below and on next page.

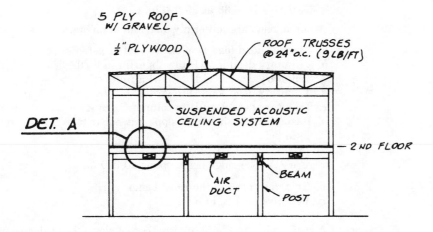

SECTION

Figure 2.C

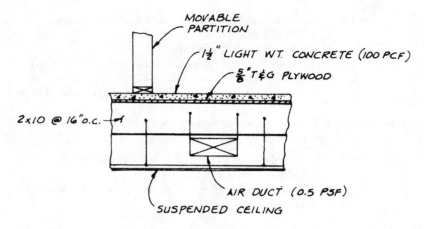

MOVABLE PARTITION

$1\frac{1}{2}''$ LIGHT WT. CONCRETE (100 PCF)

$\frac{5}{8}''$ T&G PLYWOOD

2×10 @ 16" O.C.

AIR DUCT (0.5 PSF)

SUSPENDED CEILING

DETAIL A

Figure 2.C *Continued.*

Find: a. Roof dead load D in psf
b. Second-floor dead load D in psf
c. Basic (i.e., consider roof slope but not trib. area) unit
roof live load L_r in psf

2.4 *Given:* The roof framing plan of the industrial building shown in Fig. 2.D. Roof slope is ¼ in./ft. General construction:

Roofing—5-ply felt
Sheathing—¹⁵⁄₃₂-in. plywood
Subpurlin—2 × 4 at 24 in. o.c.
Purlin—4 × 14 at 8 ft-0 in. o.c.
Girder—6¾ × 33 at 20 ft-0 in. o.c.

Assume loads are uniformly distributed on supporting members.

Find: a. Average dead load D of entire roof in psf
b. Tributary dead load D to subpurlin in lb/ft
c. Tributary dead load D to purlin in lb/ft
d. Tributary dead load D to girder in lb/ft
e. Tributary dead load D to column C1 in k
f. Basic (i.e., consider roof slope but not trib. area) unit roof live load L_r in psf

2.5 *Given:* Figure 2.A. The ridge beam spans 20 ft-0 in.

Find: a. Tributary area to the ridge beam
b. Roof live load L_r in lb/ft

2.6 *Given:* A roof similar to Fig. 2.A with ³⁄₁₂ roof slope. The ridge beam spans 22 ft-0 in.

Find: a. Tributary area to the ridge beam
b. Roof live load L_r in lb/ft

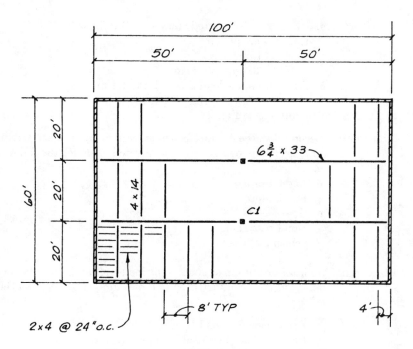

ROOF FRAMING PLAN

Figure 2.D

2.7 *Given:* Figure 2.B, standard residential occupancy, heated building, a ground snow load of 70 psf, and Exposure C terrain with a fully exposed roof

  Find: Design snow load S in psf on a horizontal plane

2.8 *Given:* A roof similar to Fig. 2.B with $^8/_{12}$ slope and a ground snow load of 90 psf, Exposure B terrain with a sheltered roof, standard residential occupancy, and a heated building

  Find: Design snow load S in psf on a horizontal plane

2.9 *Given:* The roof structure in Fig. 2.D

  Find: a. Unit roof live load L_r in psf for
      1. 2×4 subpurlin
      2. 4×14 purlin
      3. $6\frac{3}{4} \times 33$ glulam beam
    b. Uniformly distributed roof live loads in lb/ft for each of the members, using the unit L_r from (a)

2.10 *Given:* The roof structure in Fig. 2.D and a 25-psf design snow load specified by the building official

Find: a. Uniformly distributed snow load S in lb/ft for
1. 2 × 4 subpurlin
2. 4 × 14 purlin
3. 6¾ × 33 glulam beam
b. Tributary snow load S to column C1 in k

2.11   *Given:*   The building in Fig. 2.C

Find: Second-floor basic (i.e., consider occupancy but not tributary areas) unit floor live load L and concentrated loads for the following uses:
a. Offices
b. Light storage
c. Retail store
d. Apartments
e. Hotel restrooms
f. School classroom

2.12   *Given:*   An interior column supports only loads from the second floor of an office building. The tributary area to the column is 240 ft², and the dead load is 35 psf.

Find: a. Basic floor live load L_0 in psf
b. Reduced floor live load L in psf
c. Total load to column in k

2.13   *Given:*   An interior beam supports the floor of a classroom in a school building. The beam spans 26 ft. and the trib. width is 16 ft. Dead load = 20 psf.

Find: a. Basic floor live load L_0 in psf
b. Reduced floor live load L in psf
c. Uniformly distributed total load to the beam in lb/ft
d. Compare the loading in part *c* with the alternate concentrated load required by the Code. Which loading is more critical for bending, shear, and deflection?

2.14   *Given:*   IBC Table 1607.1

Find: a. Four occupancies where the unit floor live load L cannot be reduced. List the occupancy and the corresponding unit floor live load L.

2.15   *Given:*   IBC beam deflection criteria

Find: The allowable deflection limits for the following members. Beams are unseasoned wood members
a. Floor beam with 22-ft span
b. Roof rafter that supports a plaster ceiling below. Span = 12 ft

2.16   *Given:*   The *Timber Construction Manual* beam deflection recommendations in Fig. 2.8

Find: The allowable limits for the following beams. Beams are seasoned wood members that remain dry in service.

   a. Roof rafter in a commercial building that supports a gypsum board ceiling below. Span = 16 ft.

   b. Roof girder in an office building supporting an acoustic suspended ceiling. Span = 40 ft.

   c. Floor joist in the second floor of a residential building to be designed for "ordinary usage." Span = 20 ft. Tributary width = 4 ft. Floor dead load D = 16 psf. (Give beam loads in lb/ft for each deflection limit.)

   d. Girder in the second floor of a retail sales building. Increased floor stiffness is desired to avoid public concern about perceived excessive floor deflections. Span = 32 ft. Tributary width = 10 ft. Floor dead load D = 20 psf. (Give beam loads in lb/ft for each deflection limit.)

2.17 *Given:* The IBC wind force provisions

   Find: *a.* The expression for calculating the design wind pressure

   b. The section in the IBC where the terms of the expression are defined

   c. Distinguish between the following wind forces and the areas to which they are applied:
   1. Main windforce-resisting systems
   2. Components and cladding away from discontinuities
   3. Components and cladding at or near discontinuities

   d. Describe the three wind exposure conditions.

2.18 *Given:* The IBC wind and seismic force provisions

   Find: The required factor of safety against overturning for a structure subjected to wind forces

2.19 *Given:* The IBC wind force provisions

   Find: *a.* The mean recurrence interval for the wind speeds given in IBC Fig. 1609.

   b. The approximate mean recurrence interval associated with the wind pressure for essential and hazardous facilities

   c. The height used to determine the height and exposure factor for the design wind pressure

2.20 *Given:* An enclosed building in Tampa, Florida, that is an essential facility. The roof is flat and is 30 ft above grade. Exposure B applies.

   Find: *a.* Basic wind speed V

   b. Importance factor I_W

   c. Height and exposure factor λ

   d. The design wind pressures p_s in each zone for main wind force-resisting systems

   e. The design wind pressures p_{net} for components and cladding in zones near discontinuities and zones away from discontinuities

2.21 *Given:* The enclosed building in Fig. 2.E is a two-story essential facility located near Denver, Colorado. Wind Exposure C applies.

Find: The design wind pressures in both principal directions of the building for:
   a. Main windforce-resisting systems
   b. Design of individual structural elements having tributary areas of 20 ft² in the wall and roof systems away from discontinuities
   c. Design of individual structural elements in the roof near discontinuities having tributary areas of 50 ft². Sketch the wind pressures and show the areas over which they act.

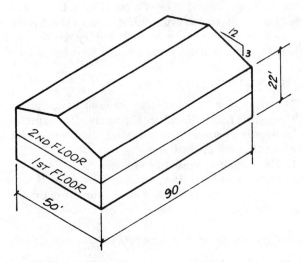

Figure 2.E

2.22   *Given:*   IBC seismic design force requirements

   Find: a. The formulas for the base shear. Give code reference.
   b. The maximum mapped maximum considered spectral response accelerations S_S and S_1 from seismic hazard maps. Cite Code reference. What is the physical significance of S_S and S_1?
   c. The maximum tabulated Site Class coefficients F_a and F_v. Cite Code reference. Explain the purpose of these coefficients.
   d. The maximum values of S_{MS}, S_{M1}, S_{DS}, and S_{D1}, based on previous values.
   e. Briefly describe the purpose of the R-factor. What value of R is used for a building with wood-frame bearing walls that are sheathed with wood structural panel sheathing?

2.23   *Given:*   IBC seismic design force requirements

   Find: a. The definition of period of vibration and the Code methods for estimating the fundamental period.
   b. How does period of vibration affect seismic forces?
   c. Describe the effects of the interaction of the soil and structure on seismic forces.
   d. What is damping, and how does it affect seismic forces? Do the Code criteria take damping into account?

2.24 *Given:*   IBC seismic design force requirements

        Find: *a.* Briefly describe the general distribution of seismic forces over the height of a multistory building.

            b. Describe differences in vertical distribution for vertical element and diaphragm forces between Seismic Design Categories B and D.

            c. Describe forces for out-of-plane design of wall components. Cite Code reference.

Behavior of Structures under Loads and Forces

3.1 Introduction

The loads and forces required by the IBC (Ref. 3.2) for designing a building were described in Chap. 2. Chapter 3 deals primarily with the transfer of these from one member to another throughout the structure. The distribution of *vertical loads* in a typical wood-frame building follows the traditional "post-and-beam" concept. This subject is briefly covered at the beginning of the chapter.

The distribution of *lateral forces* may not be as evident as the distribution of vertical loads. The majority of Chap. 3 deals with the transfer of lateral forces from the point of origin, through the building, and into the foundation. This subject is introduced by reviewing the three basic types of lateral-force-resisting systems (LFRSs) used in conventional rectangular-type buildings.

Shearwalls and horizontals diaphragms make up the LFRS used in most wood-frame buildings (or buildings with a combination of wood framing and concrete or masonry walls). The chapter concludes with two detailed examples of lateral force calculations for these types of buildings.

3.2 Structures Subject to Vertical Loads

The behavior of framing systems (post-and-beam type) under vertical loads is relatively straightforward. Sheathing (decking) spans between the most closely spaced beams; these short-span beams are given various names: stiffeners, rafters, joists, subpurlins. The reactions of these members in turn cause loads on the next set of beams in the framing system; these next beams may be referred to as beams, joists, or purlins. Finally, reactions of the second set of beams impose loads on the largest beams in the system. These large beams

are known as girders. The girders, in turn, are supported by columns. See
Example 3.1. and Fig. 3.1.

EXAMPLE 3.1   Typical Post-and-Beam Framing

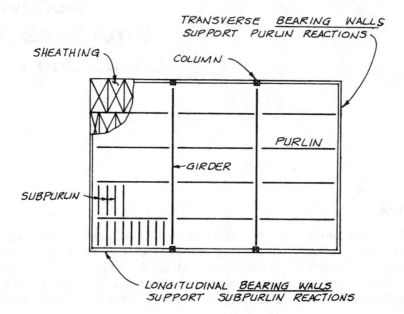

Figure 3.1

1. Sheathing spans between subpurlins.
2. Subpurlins span between purlins.
3. Purlins span between girders.
4. Girders span between columns.

Subpurlins and purlins are also supported by bearing walls. *Bearing walls* are defined
as walls that support vertical loads in addition to their own weight.

When this framing system is used for a roof, it is often constructed as a
panelized system. Panelized roofs typically use glulam girders spaced 18 to 40
ft on center, sawn lumber or glulam purlins at 8 ft on center, sawn lumber
subpurlins at 24 in. on center, and plywood sheathing. The name of the system
comes from the fact that 8-ft-wide roof *panels* are prefabricated and then lifted
onto preset girders using forklifts. See Fig. 3.2. The speed of construction and
erection makes panelized roof systems very economical. Panelized roofs are

Figure 3.2 Panelized roof system installed with forklift. (*Photo by Mike Hausmann.*)

widely used on large one-story commercial and industrial buildings. See Ref. 3.3 for more information on panelized roof structures.

Although the loads to successively larger beams are a result of reactions from lighter members, for structural design the loads on beams in this type of system are often assumed to be uniformly distributed. To obtain a feel for whether this approach produces conservative values for shear and moment, it is suggested that a comparison be made between the values of shear and moment obtained by assuming a *uniformly distributed load* and those obtained by assuming *concentrated loads* from lighter beams. The actual loading probably falls somewhere between the two conditions described. See Example 3.2.

Regardless of the type of load distribution used, it should be remembered that it is the *tributary area of the member being designed* (Sec. 2.3) which is used in establishing unit live loads, rather than the tributary area of the lighter members which impose the load. This concept is often confusing when it is first encountered.

EXAMPLE 3.2   Beam Loading Diagrams

Figure 3.3 shows the girder from the building in Fig. 3.1. The load to the girder can be considered as a number of concentrated reaction loads from the purlins. However, a more common design practice is to assume that the load is uniformly distributed. The uniformly distributed load is calculated as the unit load times the tributary width to the girder. As the number of concentrated loads increases, the closer the loading approaches the uniform load case.

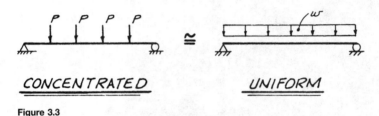

Figure 3.3

As an example, consider the design load for the girder in Fig. 3.1. Confusion may occur when the *unit* live load for the girder (based on a large tributary area) turns out to be less than the *unit* live load used in the design of the purlin. Obviously, the reaction of the purlin (using the higher live load) must be supported by the girder. Why is the lower live load used for design?

The reasoning is thus: The girder must be capable of supporting *individual* reactions from purlins calculated using the larger unit live load (obtained using the tributary area of the purlin). However, when the entire tributary area of the girder is considered loaded, the smaller unit live load may be used (this was discussed in detail in Chap. 2). Of course, each connection between the purlin and the girder must be designed for the higher unit live load, but not all purlins are subjected to this higher load simultaneously.

The spacing of members and the span lengths depend on the function and purpose of the building. Closer spacing and shorter spans require smaller member sizes, but short spans require closely spaced columns or bearing walls. The need for clear, unobstructed space must be considered when the framing system is first established. Once the layout of the building has been determined, dimensions for framing should be chosen which result in the best utilization of materials. For example, the standard size of a sheet of plywood is 4 ft by 8 ft, and a joist spacing should be chosen which fits this basic module. Spacings of 16, 24, and 48 in. o.c. (o.c. = on center and c.c. = center to center) all provide supports at the edge of a sheet of plywood.

Certainly an unlimited number of framing systems can be used, and the choice of the framing layout should be based on a consideration of the requirements of a particular structure. Several other examples of framing arrangements are shown in Fig. 3.4*a*, *b*, and *c*. These are given to suggest possible arrangements and are not intended to be a comprehensive summary of framing systems.

It should be noted that in the framing plans, a break in a member represents a simple end connection. For example, in Fig. 3.4*a* the lines representing joists are broken at the girder. If a continuous joist is to be shown, a solid line with no break at the girder would be used. This is illustrated in Fig. 3.4*c*

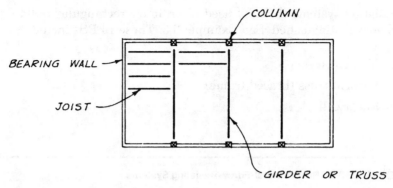

Figure 3.4a Alternate post-and-beam framing.

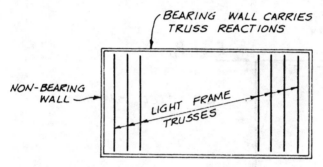

Figure 3.4b Light frame trusses.

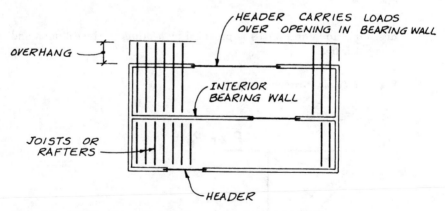

Figure 3.4c Roof framing with interior and exterior bearing walls.

where the joist is continuous at the rear wall overhang. Such points may seem obvious, but a good deal of confusion results if they are not recognized.

3.3 Structures Subject to Lateral Forces

The behavior of structures under lateral forces usually requires some degree of explanation. In covering this subject, the various types of lateral-force-resisting systems (LFRSs) used in ordinary rectangular buildings should be clearly distinguished. See Example 3.3. These LFRSs include

1. Moment frame
2. Vertical truss (braced frame)
3. Shearwall

EXAMPLE 3.3   Basic Lateral-Force-Resisting Systems

Moment Frame

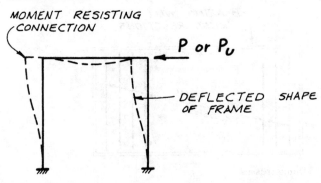

Figure 3.5a

Resistance to lateral forces is provided by *bending* in the columns and girders of the moment frame.

Vertical Truss (Braced Frame)

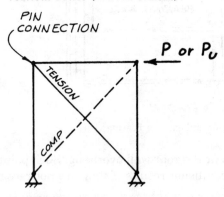

Figure 3.5b

Lateral forces develop *axial* forces in the vertical truss (braced frame).

Shearwall

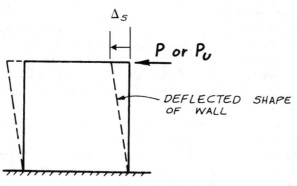

Figure 3.5c

Segments of walls can be designed to function as *shear*-resisting elements in carrying lateral forces. The deflected shape shows shear deformation rather than bending.

Note that two forces P and P_u are shown on Fig. 3.5a, b, and c. The designer needs to clearly understand the meaning of these two symbols (see Sec. 2.8). When the symbol for a force (or other property such as shear or moment) has the "u" subscript, the force (or other property) is at a *strength level*. On the other hand, when the symbol for a force (or other property) does not have a "u" subscript, it is at an *ASD level*. This notation is necessary because the wind forces W calculated using IBC equations are at an ASD level, noted by P, whereas the seismic forces E are at a strength level, noted by P_u. The wood design principles covered in this book are based on ASD procedures. Adjustment of seismic forces to an ASD level will be discussed further in this chapter as well as Chaps. 9, 10, 15, and 16. Because strength and ASD force levels differ substantially, it is important that the reader make a clear distinction between the two. Throughout Chap. 3, and all chapters dealing with lateral force design, the reader should pay close attention to whether a quantity is at a *strength level* or at an *ASD level*. The use of a "u" subscript is the notation convention used in this book to distinguish between the two.

Also shown in Fig. 3.5c is the shearwall horizontal deflection. In this book Δ_S is the symbol used to represent the horizontal deflection or drift of a shearwall. Shearwall drift is also closely related to the design story drift, Δ, the horizontal deflection of the full building, over the height of the story under consideration. The IBC places limits on the maximum acceptable story drift for seismic design, but does not limit story drift for wind design.

Moment frames, whether statically determinate or indeterminate, resist lateral forces by bending in the frame members. That is, the members have relatively small depths compared with their lengths, and the stresses induced as the structure deforms under lateral forces are essentially flexural. Some axial forces are also developed. In the United States, moment frames are often constructed of steel or reinforced concrete and are rarely constructed of wood. Earthquake experience with steel and reinforced concrete moment frames has shown that careful attention to detailing is required to achieve ductile behavior in these types of structures.

Vertical trusses or braced frames are analyzed in a manner similar to horizontal trusses: connections are assumed to be pinned, and forces are assumed to be applied at the joints. Vertical trusses often take the form of cross or X-bracing. For the braced frame in Fig. 3.5*b*, both diagonal members are generally designed to function simultaneously (one in tension and the other in compression). For a hand calculation, the forces in the diagonals can be assumed equal due to the symmetry of the brace. A computer analysis would show a very small difference in member forces.

Code provisions for X-bracing (Fig. 3.5*b*) with tension-only bracing members have varied significantly in recent years. In the past, the use of X-bracing tension-only systems was not uncommon. Because of their slenderness, the brace members would buckle in compression, resulting in the full load being carried by the opposing tie rod in tension. More recent editions of the *Uniform Building Code* had placed restrictions on the use of this type of system due to observed earthquake damage. The restrictions included limitation to one- and two-story buildings and increased design forces. The IBC now recognizes the use of a new flat-strap tension-only system, used with special design requirements.

Shearwall structures make use of specially designed wall sections to resist lateral forces. A shearwall is essentially a vertical cantilever with the span of the cantilever equal to the height of the wall. The depth of these members (i.e., the length of the wall element parallel to the applied lateral force) is large in comparison with the depth of the structural members in the moment frame LFRS. For a member with such a large depth compared with its height, *shear* deformation replaces bending as the significant action (hence, the name "shearwall").

It should be mentioned that the LFRSs in Example 3.3 are the *vertical resisting elements* of the system (i.e., the vertical components). Because buildings are three-dimensional structures, some horizontal system must also be provided to carry the lateral forces to the vertical elements. See Example 3.4. A variety of horizontal systems can be employed:

1. Horizontal wall framing

2. Vertical wall framing with horizontal trusses at the story levels

3. Vertical wall framing with horizontal diaphragms at the story levels

---

EXAMPLE 3.4   Horizontal Elements in the LFRS

Horizontal Wall Framing

Horizontal wall members are known as *girts* and distribute the lateral force to the vertical LFRS. Dashed lines represent the deflected shape of the girts. See Fig. 3.6*a*. The lateral force to the shearwalls is distributed over the height of the wall.

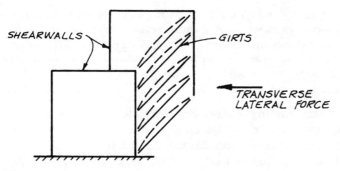

Figure 3.6a

Vertical Wall Framing

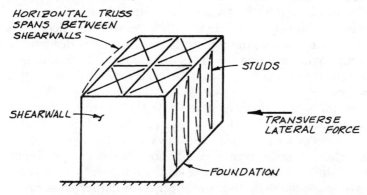

Figure 3.6b Vertical wall members are known as *studs*. The lateral force is carried by the studs to the roof level and the foundation. A horizontal truss in the plane of the roof distributes the lateral force to the transverse shearwalls. The diagonal members in the horizontal truss are sometimes steel rods which are designed to function in tension only.

Vertical Wall Framing and Horizontal Diaphragm

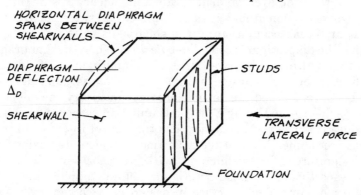

Figure 3.6c

In Fig. 3.6c the LFRS is similar to the system in Fig. 3.6b except that the horizontal truss in the plane of the roof is replaced with a *horizontal diaphragm*. The use of vertical wall framing and horizontal diaphragms is the most common system in wood-frame buildings because the roof sheathing can be designed economically to function as both a vertical-load- and a lateral-force-carrying element. The horizontal diaphragm is designed as a beam spanning between the shearwalls. The design requirements for horizontal diaphragms and shearwalls are given in Chaps. 9 and 10.

The first two framing systems in Example 3.4 are relatively easy to visualize. The third system is also easy to visualize once the concept of a diaphragm is understood. A diaphragm can be considered to be a large, thin structural element that is loaded in its plane. In Fig. 3.6c the vertical wall members develop horizontal reactions at the roof level and at the foundation level (studs are assumed to span as simple beams between these two levels). The reaction of the studs at the roof level provides a force in the plane of the roof. The diaphragm acts as a large horizontal beam.

In wood buildings, or buildings with wood roof and floor systems and concrete or masonry walls, the roof or floor sheathing is designed and connected to the supporting framing members to function as a diaphragm. In buildings with concrete roof and floor slabs, the concrete slabs are designed to function as diaphragms.

The stiffness of a diaphragm refers to the amount of horizontal deflection that occurs in the diaphragm as a result of the in-plane lateral force (Fig. 3.6c shows the deflected shape). In this book Δ_D is the symbol used to represent the deflection of the horizontal diaphragm. Because wood diaphragms are not nearly as stiff as concrete slabs, wood diaphragms have in the past been categorically considered *flexible* while concrete diaphragms have categorically been considered *rigid*. It is now recognized, however, that the stiffness of the diaphragm in comparison to the stiffness of the shearwalls (or other vertical elements) is of interest when evaluating the *diaphragm flexibility*.

The Code now has a criteria for determining when a diaphragm is flexible or rigid. This criteria involves a comparison of the diaphragm deflection Δ_D (Fig. 3.6c) with the average drift of adjoining shearwalls Δ_S (Fig. 3.5c). If the diaphragm deflection at mid-span Δ_D is more than twice the average shearwall drift Δ_S at the associated story (the shearwalls in the story below the diaphragm), the diaphragm is considered *flexible*. When the diaphragm deflection is less than twice the shearwall drift, the diaphragm is considered *rigid*. The identification of a diaphragm as *rigid* or *flexible* is applicable whether the forces being considered are seismic or wind. In theory the deflections Δ_S and Δ_D can be determined using a fictitious unit load rather than the Code design load, since it is the relative deflections that are of interest. In practice, however, it is recommended that the deflections be calculated using Code *strength level* design forces for seismic and *ASD forces* for wind. It should be noted that, irrespective of the Code criteria for defining diaphragms as rigid or flexible, the most prevalent practice in design is to categorically identify wood diaphragms as flexible. This is discussed further in Secs. 9.11 and 16.10.

For buildings having concrete or masonry shearwalls and a wood diaphragm, the diaphragm is almost always flexible when compared to the walls.

For buildings having wood shearwalls and wood diaphragms, the diaphragm can be rigid or flexible depending on the building configuration and the calculated values of Δ_S and Δ_D. A flexible diaphragm is modeled as a simple beam that spans between shearwalls, as depicted in Fig. 3.6c. The method of calculating diaphragm deflection Δ_D is discussed in Chap. 9 and the calculation of shearwall drift Δ_S is covered in Chap. 10.

In each of the sketches in Example 3.4, the transverse lateral force is distributed horizontally to end shearwalls. The same horizontal systems shown in these sketches can be used to distribute lateral forces to the other basic vertical LFRSs (i.e., moment frames or vertical trusses). Any combination of *horizontal* and *vertical* LFRSs can be incorporated into a given building to resist lateral forces.

The discussion of lateral forces in Example 3.4 was limited to forces in the transverse direction. In addition, a LFRS must be provided for forces in the longitudinal direction. This LFRS will consist of both horizontal and vertical components similar to those used to resist forces in the transverse direction.

Different types of *vertical* elements can be used to resist transverse lateral forces and longitudinal lateral forces in the same building. For example, rigid frames can be used to resist lateral forces in the transverse direction, and shearwalls can be used to resist lateral forces in the other direction. The choice of LFRS in one direction does not necessarily limit the choice for the other direction.

In the case of the *horizontal* LFRS, it is unlikely that the horizontal system used in one direction will be different from the horizontal system used in the other direction. If the sheathing is designed to function as a horizontal diaphragm for lateral forces in one direction, it probably can be designed to function as a diaphragm for forces applied in the other direction. On the other hand, if the roof or floor sheathing is incapable of functioning as a diaphragm, a system of horizontal trusses spanning in both the transverse and longitudinal directions appears to be the likely solution.

The common types of LFRSs for conventional buildings have been summarized as a general introduction and overview. It should be emphasized that the large majority of wood-frame buildings, or buildings with wood roof and floor framing and concrete or masonry walls, use a combination of

1. Horizontal diaphragms

2. Shearwalls

to resist lateral forces. Because of its widespread use, only the design of this type of system is covered in this book.

3.4   Lateral Forces in Buildings with Diaphragms and Shearwalls

The majority of wood structures use the sheathing, normally provided on floors, roofs, and walls, to form horizontal diaphragms and shearwalls that resist lateral wind and seismic forces. To function as a horizontal diaphragm or shearwall, the sheathing must be properly attached to the supporting members. The framing members must also be checked for additional stresses caused by the lateral forces. Furthermore, certain connections must be provided to transfer lateral forces down through the structure. The system must be tied together.

However, the economic advantage is clear. With the added attention to the framing and connection design, the usual sheathing material, which is required in any structure, can also be used to form the lateral-force-resisting system. In this way, one material serves two purposes (i.e., sheathing and lateral force resistance).

The following numerical examples illustrate how lateral forces are calculated and distributed in two different shearwall buildings. The first (Sec. 3.5) demonstrates the procedure for a one-story structure, and the second (Sec. 3.6) expands the system to cover a two-story building. The method used to calculate seismic forces in these two buildings is based on the same criteria, but the solution for the one-story structure is much more direct.

Before moving into design examples, an overview of seismic design forces is needed. There are six main types of seismic design forces which need to be considered for the primary lateral-force-resisting system (LFRS). All six are at a *strength level*. These are:

1. Seismic base shear force V.

    This is the total shear at the base of the structure. This is generally thought of as a seismic base shear coefficient times the building weight.

2. Seismic story (shearwall) force F_x.

    This is a set of forces, with one force for each story level above ground. These forces are used for *design of the vertical elements* of the lateral force resisting system (the shearwalls). The shearwalls at any story are designed to resist the sum of the story forces above that level. For one-story buildings, the story force F_x is equal to the base shear force V. For buildings with more than one story, a vertical distribution procedure is used to assign a seismic story force at each story level. The vertical distribution is given by the Code formula for F_x (Sec. 2.14). The subscript x in the notation is generally replaced with the specific story under consideration (i.e., F_r is the force at the roof level, F_3 is the force at the third floor level). The vertical distribution procedure is illustrated in Sec. 3.6. The seismic story (shearwall) force F_x is most often thought of as an F_x story (shearwall) coefficient times the weight tributary to the story w_x. The seismic story (shearwall) force F_x applies for Seismic Design Categories B and up.

3. Seismic story (diaphragm) force F_{px}.

    This is also a set of forces, one for each story level above ground. These forces are used for *design of the horizontal diaphragm*. The notation F_p is used by the IBC for Seismic Design Categories B and C, while the notation F_{px} is used for Seismic Design Categories D, E, and F. In order to clearly differentiate seismic story (diaphragm) forces from component forces, which also use the notation F_p, this book will use the notation F_{px} for story (diaphragm) forces, regardless of whether the formulas used are for Seismic Design Categories B and C, or D, E, and F. The x subscript is again replaced with the specific story under consideration.

For a one-story building in Seismic Design Categories D, E, or F, with wood diaphragms and wood shearwalls, the F_{px} force is equal to the base shear V and the seismic story (shearwall) force F_x. For a building with more than one story, a vertical distribution procedure is used to assign the seismic story force at each level. The Code formula for F_{px} was described in Sec. 2.14. In Seismic Design Categories B and C for buildings of one story or more, the F_{px} force is equal to the greater of $0.2S_{DS}Iw_p$, or F_x, where w_p is the weight of the diaphragm and attached structure. The seismic story (diaphragm) force F_{px} is most often thought of as an F_{px} story (diaphragm) coefficient times the weight tributary to the story w_x. The variables w_p in Seismic Design Categories B and C, and w_x in Seismic Design Categories D, E, and F, are representing the same weight. This book will use the notation w_x for all Seismic Design Categories.

Seismic force types one through three are used with ASCE 7 (Ref. 3.1) equations 9.5.2.7-1 and 9.5.2.7-2 as referenced in IBC Sec. 1617.1.1:

$$E = \rho Q_E + 0.2S_{DS}D$$

$$E = \rho Q_E - 0.2S_{DS}D$$

Q_E is the horizontal force from 1, 2, or 3 above, and ρ is the reliability/redundancy factor. The term ρQ_E is the seismic force acting horizontally. S_{DS} is the design spectral response acceleration, and D is the vertical dead load. The term $0.2S_{DS}D$ acts vertically. See Sec. 2.14. Both F_x and F_{px} forces are referred to as story forces, because there will be a force for each story in the building under design.

4. Wall component forces F_c.

These forces are used for concrete and masonry walls subject to out-of-plane seismic forces. Shearwalls acting in-plane are considered part of the primary lateral-force-resisting system; whereas, when loaded out of plane, shearwalls and other walls are treated in a manner similar to components. The force F_c is used for wall out-of-plane forces for concrete and masonry walls in Seismic Design Categories B and higher. This force is also recommended for exterior walls of other materials, although not specifically required. The force F_c used for design of concrete and masonry walls will be greater than the base shear force level. The force F_c is most often thought of as an F_c seismic coefficient times the unit weight of the wall. It should be noted that there is an additional set of force requirements for anchorage of concrete and masonry walls to diaphragms. The anchorage forces vary both by Seismic Design Category and diaphragm type.

5. Component forces F_p.

The force F_p, used for components anchored to the building (other than concrete and masonry walls), was introduced in Sec. 2.15. The force F_p used for design of components will be greater than the base shear force level. The force F_p is most often thought of as an F_p seismic coefficient times the unit weight of the component.

The primary difference between F_c and F_p forces is that F_p forces are magnified dramatically as the component is anchored higher on the structure, as discussed in Sec. 2.15; whereas, F_c forces do not get magnified with height. The variables used to define the F_c and F_p forces are distinctly different, with F_p using component variables I_p, a_p, and R_p.

6. Special seismic force.

This force type, from ASCE 7 equations 9.5.2.7.1-1 and 9.5.2.7.1-2 (as referenced in IBC Sec. 1617.1.1), defines a special magnified seismic force used for designing a limited number of structural elements, whose performance is viewed as critical to the performance of a building. The special seismic force is discussed in detail in Chaps. 9 and 16.

Designs using force types 1, 2, and 3 are illustrated in the remainder of this chapter. It is common for designers to think of these forces as a seismic co-efficient (g factor) times the weight that would be acting under seismic load-ing:

1. V = Seismic base shear coefficient $\times$ W

2. F_x = Seismic story (shearwall) coefficient $\times$ w_x

3. F_{px} = Seismic story (diaphragm) coefficient $\times$ w_x

Because there is significant repetition involved in the calculation of seismic coefficients, the examples in this book will use the approach of calculating the seismic coefficient and multiplying it by the applicable weight acting on the element or portion under consideration.

For a *single-story building* in Seismic Design Category D, E, or F, the seis-mic base shear force V, the story (shearwall) force F_x, and the story (dia-phragm) force F_{px} are all equal, as are the corresponding coefficients:

$$\text{Base shear force } V = \text{Story (shearwall) force } F_x$$

$$= \text{Story (diaphragm) force } F_{px}$$

and

$$\text{Base shear coefficient} = F_x \text{ story coefficient} = F_{px} \text{ story coefficient.}$$

In Seismic Design Category B or C it is possible that the story (diaphragm) coefficient will exceed the base shear and story (shearwall) coefficient.

Wood-frame structures have traditionally been limited to relatively low-rise (one- and two-story) structures, and these are the primary focus of this book. It is interesting to note that there have been an increasing number of three- and four-story (and even taller) wood-frame buildings constructed in recent years. See Fig. 3.7. The method of analysis given for the two-story example can be extended to handle the lateral force evaluation for taller multistory buildings.

Figure 3.7 Four-story wood-frame building with horizontal diaphragms and shearwalls of plywood. (*Photo courtesy of APA—The Engineered Wood Association.*)

Before proceeding with the one- and two-story building design examples, Example 3.5 will demonstrate the method used to compute some typical seismic base shear and story shear coefficients. Recall that the seismic base shear is the result of a fairly involved calculation. However, with a little practice it is fairly easy to determine the seismic coefficient for many common buildings which use diaphragms and shearwalls for the primary *lateral force resisting system* LFRS. These shearwall buildings have a low height-to-width ratio and are fairly rigid structures. As a result they tend to have low fundamental periods. For this reason it is common for low-rise shearwall buildings to use the maximum Code response spectrum value of $S_{DS}I/R$ for the seismic coefficient, as demonstrated in Example 3.5.

The coefficients (g factors) summarized in the table in Example 3.5 can be used to determine the seismic base shear V for both one-story and multistory buildings. In multistory buildings the base shear V is used to calculate the seismic story (shearwall) forces F_x and the F_x forces are then used to calculate the seismic story (diaphragm) forces F_{px}. Example 3.5 evaluates base shear coefficients for two common types of buildings ($R = 5$ and $R = 6.5$).

EXAMPLE 3.5   Seismic Coefficient Calculation

Develop a table of seismic base shear coefficients that will apply to commonly encountered buildings. Many buildings will meet the following three conditions:

1. The building has an importance factor I of 1.0.

   This is the case for the great majority of wood-framed buildings which have residential, office and commercial uses. Some uses where the importance factor might be greater than 1.0 include police and fire stations and schools. See Sec. 2.13.

2. The building has a short fundamental period T so the seismic forces are controlled by the S_{DS} plateau of the IBC design spectrum (Fig. 2.18 in Sec. 2.13).

   This is the case for practically all buildings braced by wood shearwalls. This condition could potentially not be met, however, if very tall slender shearwalls were to be used.

3. The Site Class is assumed to be D.

   Site Class D is often assumed in the absence of a geotechnical investigation.

For structures meeting all three of these conditions, the calculation of the *strength level* base shear coefficient can be simplified to S_{DS}/R. The two variables in this simplified expression for determining the seismic base shear V are the design short-period spectral response acceleration S_{DS} and the *response modification factor R*. In the past, only two R-factors were needed to address the great majority of wood-frame construction. In the IBC, due to the incorporation of systems used nationally, the number of systems and R-factors has increased. R-factors of interest for wood-frame construction include:

$R = 6.5$ is used for wood and metal light-frame shearwalls in bearing wall systems that are braced with wood structural panel sheathing. Recall that, although ASCE 7 Table 9.5.2.2 indicates the R-factor to be 6, IBC has modified it to be 6.5.

$R = 2$ is used for wood and metal light-frame shearwalls in bearing wall systems that are braced with shear panels other than wood structural panels.

$R = 5$ is used for special reinforced concrete and masonry shearwalls in bearing wall systems, which is permitted in all Seismic Design Categories.

$R = 4$ is used for ordinary reinforced concrete shearwalls in bearing wall systems, which is permitted for Seismic Design Categories B and C.

$R = 3.5$ is used for intermediate reinforced masonry shearwalls in bearing wall systems, which also is permitted for Seismic Design Categories B and C.

$R = 2$ is used for ordinary reinforced masonry shearwalls in bearing wall systems and is permitted in Seismic Design Categories B and C.

Other types of concrete and masonry shearwall systems are permitted in Seismic Design Category B. The terms *special, intermediate, ordinary,* and *plain,* when applied to concrete or masonry shearwalls, refer to the extent of seismic detailing required. Additional reinforcing and detailing is required for the higher Seismic Design Categories. Examples in this text will focus on $R = 6.5$ (for light-frame bearing wall buildings) and $R = 5$ (for concrete and masonry shearwall buildings), because they can be used across all Seismic Design Categories. Methods used to calculate seismic base shear forces do not change with Seismic Design Category, so the method could be applied equally to a system with any R-factor. The methods used to calculate diaphragm forces F_{px}, however, *do* vary by Seismic Design Category and require additional attention.

Sample Calculation

The following calculation is for a building in Hayward, California (ZIP Code 94541 for use with the "Earthquake Spectral Response Acceleration Maps" compact disk). From the maximum considered earthquake ground motion maps, at 37.7 degrees latitude and 122 degrees longitude, the mapped short-period spectral acceleration S_S, is $2.05g$. The mapped one-second spectral acceleration S_1, is $0.91g$.

The occupancy type is commercial, which is assigned Occupancy Category II in IBC Table 1604.5 and is assigned to Seismic Use Group I per IBC Sec. 1616.2.1.

The Site Class is assumed to be D. Based on the mapped short-period spectral acceleration and the Site Class, IBC Table 1615.1.2(1) assigns a site coefficient F_a of 1.0. From this, the design short-period spectral response acceleration S_{DS} can be calculated as $S_{DS} = S_{MS} \times \frac{2}{3} = F_a S_S \times \frac{2}{3} = 1.37g$.

IBC Tables 1616.3.1(1) and 1616.3.1(2) would identify the Seismic Design Category as E, based on footnote 1; however, an exception to Sec. 1616.3 allows the Seismic Design Category for this particular building to be reclassified as D. The ASCE 7 provisions do not include this exception, so if design were in accordance with ASCE 7 rather than IBC, Seismic Design Category E would apply. Because there are some differences, it is important to clearly identify which design provisions are being followed.

The structural system is light-frame bearing walls with wood structural panel shearwalls, using an R-factor of 6.5. This system is permitted in Seismic Design Category D.

The importance factor I is assumed to be 1.0, and is, therefore, dropped from the equation.

Seismic base shear coefficient

$= S_{DS}/R$

$= 1.37g/6.5$

$= 0.21g$

This *strength level* base shear coefficient is entered into the table. Additional values are computed and entered in a similar way, using the S_S ranges from IBC Table 1615.1.2(1). While the tabulated values provide an overview, the base shear coefficient will need to be calculated for virtually every building site, because mapped S_S and S_1 values vary for every site.

Example Seismic Coefficients

S_S (g)	Site Class D F_a	Site Class D S_{DS} (g)	Seismic base shear coefficient (g)	
			$R = 6.5$ (1)	$R = 5$ (2)
0.25	1.6	0.27	0.041	0.053
0.50	1.4	0.47	0.072	0.093
0.75	1.2	0.60	0.092	0.120
1.00	1.1	0.73	0.113	0.147
1.25	1.0	0.83	0.128	0.167
2.05	1.0	1.37	0.210	0.273

(1) The seismic coefficients in this table apply only to buildings that meet the three conditions listed above.

(2) To use this table, enter with the S_S value and read either S_{DS} or base shear coefficient. Base shear coefficients are at a *strength level*.

The coefficients in this table can be viewed as seismic base shear V coefficients for *both* one-story and multistory buildings.

For *one-story structures* the base shear V coefficient is also equal to the F_x story (shearwall) coefficient. In some buildings the base shear V coefficient also equals the F_{px} story (diaphragm) coefficient.

In *multistory structures* the base shear V must be determined first. However, in a multistory building the F_x story (shearwall) coefficients and the F_{px} story (diaphragm) coefficients vary from one-story level to another. These seismic coefficients must be evaluated using the appropriate Code equations (Sec. 2.13).

After the *strength level* forces are determined using the coefficients, two additional steps are required. 1. The seismic force must be multiplied by the redundancy/reliability factor ρ, and 2. The seismic force must be multiplied by 0.7 to reduce it to an allowable stress design ASD level. These adjustments are discussed in detail in Sec. 2.13.

3.5 Design Problem: Lateral Forces on One-Story Building

In this section a rectangular one-story building with a wood roof system and masonry walls is analyzed to determine both wind and seismic forces. The building chosen for this example has been purposely simplified so that the basic procedure is demonstrated without "complications." The structure is essentially defined in Fig. 3.8 with a plan view of the horizontal diaphragm and a typical transverse cross section.

The example is limited to the consideration of the lateral force in the transverse direction. This force is shown in both plan and section views. In the plan

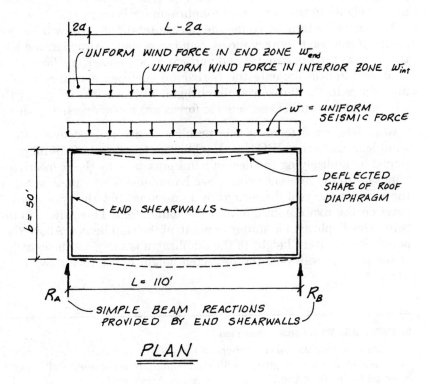

PLAN

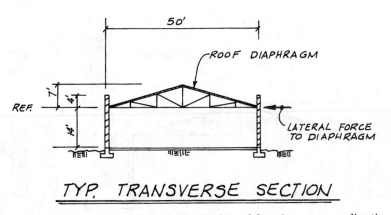

TYP. TRANSVERSE SECTION

Figure 3.8 One-story building subjected to lateral force in transverse direction.

view, it is seen as a uniformly distributed force to the horizontal diaphragm, and the diaphragm spans between the shearwalls. In the section view, the lateral force is shown at the point where the walls tie to the horizontal diaphragm. This height is taken as the *reference* location for evaluating the tributary heights to the horizontal diaphragm for both wind and seismic forces.

The critical lateral force for the horizontal diaphragm will be taken as the larger of the two tributary forces: wind or seismic. Although the IBC requires the effects of the horizontal and vertical wind and seismic forces to be considered simultaneously, only the horizontal component of the force affects the unit shear in the horizontal diaphragm. The possible effects of the vertical wind pressure (uplift) and seismic forces are not addressed in this example.

Wind. The wind force in this problem is determined using the simplified wind load method of IBC Sec. 1609.6. The force to the roof diaphragm is obtained by multiplying the design wind pressures by the respective wall areas tributary to the reference point. See Example 3.6. One method for calculating the force to the roof diaphragm is to assume that the wall spans vertically between the roof diaphragm and the foundation. Thus, the tributary height below the diaphragm is simply one-half of the wall height. Above the reference point, the tributary height to the diaphragm is taken as the cantilever height of the parapet wall plus the projected height of the roof above the top of the parapet.

EXAMPLE 3.6   Wind Force Calculation

Determine the horizontal component of the wind force tributary to the roof diaphragm. The basic wind speed is given as 85 mph. Standard occupancy and Exposure B apply. Selected tables from the IBC are included in Appendix C.

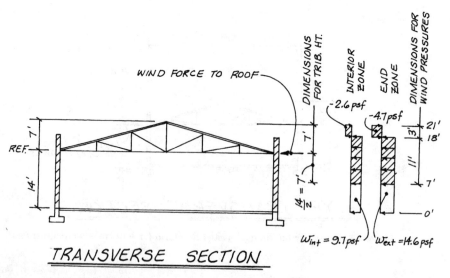

Figure 3.9 Wind pressures tributary to roof diaphragm.

Building Geometry

Roof slope = 7:25 = 15.64°

$$h_{mean} = \text{height to reference point} + \tfrac{1}{2}(\text{projected height of roof})$$

$$= 14 \text{ ft} + \tfrac{1}{2}(7 \text{ ft})$$

$$= 17.5 \text{ ft}$$

$$0.4h_{mean} = 0.4(17.5 \text{ ft}) = 7 \text{ ft}$$

Least horizontal building dimension = b = 50 ft

$$0.1b = 0.1(50 \text{ ft}) = 5 \text{ ft}$$

End zone dimension:

$$a = \text{lesser of } \{0.4h_{mean} \text{ or } 0.1b\}$$

$$= 5 \text{ ft}$$

Length of end zone = $2a$ = 10 ft

Wind Pressures

Wind pressure formula:

$$p_s = \lambda I_W p_{s30}$$

$$I_W = 1.0 \qquad \text{(IBC Table 1604.5)}$$

$$\lambda = 1.0 \qquad \text{for 0 to 30 ft (IBC Table 1609.6.2.1(4))}$$

Basic wind pressures p_{s30} from IBC Table 1609.6.2.1(1):

	Roof slope		Linear interpolation for 15.64° slope	$p_s = \lambda I_W p_{s30}$
	15°	20°		
Zone A	14.4 psf	15.9 psf	$p_{s30} = 14.6$ psf	14.6 psf
Zone B	−4.8 psf	−4.2 psf	$p_{s30} = -4.7$ psf	−4.7 psf
Zone C	9.6 psf	10.6 psf	$p_{s30} = 9.7$ psf	9.7 psf
Zone D	−2.7 psf	−2.3 psf	$p_{s30} = -2.6$ psf	−2.6 psf

These wind pressures are shown in the section view in Fig. 3.9.

Load to Diaphragm

$$\begin{Bmatrix} \text{Trib. height to} \\ \text{roof diaphragm} \end{Bmatrix} = \begin{Bmatrix} \text{Trib. wall height} \\ \text{below ref. point} \end{Bmatrix} + \begin{Bmatrix} \text{Height of} \\ \text{parapet wall} \end{Bmatrix} + \begin{Bmatrix} \text{Projected roof} \\ \text{height above} \\ \text{parapet wall} \end{Bmatrix}$$

Trib. wall height below ref. point = ½(14 ft) = 7 ft

Height of parapet wall = 4 ft

Projected roof height above parapet wall = 7 ft − 4 ft = 3 ft

∴ Trib. height to roof diaphragm = 7 ft + 4 ft + 3 ft = 14 ft

Since wind pressures are negative (outward) on the projected roof area above the parapet wall, the total wind force to the roof diaphragm will be greater if it is assumed that p_s = 0 psf in roof Zones B and D (per IBC Fig. 1609.6(1)). Thus,

$$w_{end} = 14.6 \text{ psf } (7 \text{ ft} + 4 \text{ ft}) + 0 \text{ psf } (3 \text{ ft}) = 160.6 \text{ lb/ft}$$

$$w_{int} = 9.7 \text{ psf } (7 \text{ ft} + 4 \text{ ft}) + 0 \text{ psf } (3 \text{ ft}) = 106.7 \text{ lb/ft}$$

$$W = w_{end}(2a) + w_{int}(L - 2a)$$

$$= 160.6(10) + 106.7(110 - 10)$$

$$= 12{,}276 \text{ lb}$$

Use static equilibrium equations to determine the larger reaction force (R_A or R_B) at either end of the roof diaphragm.

Summing moments about the reaction at B:

$$R_A = w_{end}(2a)(L - a)\left(\frac{1}{L}\right) + w_{int}(L - 2a)\left(\frac{L}{2} - a\right)\left(\frac{1}{L}\right)$$

$$= (160.6)(10)(110 - 5)\left(\frac{1}{110}\right) + (106.7)(110 - 10)\left(\frac{110}{2} - 5\right)\left(\frac{1}{110}\right)$$

$$= 6383 \text{ lb}$$

Summing forces:

$$R_B = W - R_A = 12{,}276 - 6383 = 5893 \text{ lb}$$

Check minimum load to diaphragm based on p_s = 10 psf throughout Zones A, B, C, and D.

$$w_{min} = p_s \text{ (Trib. height to roof diaphragm)} = 10 \text{ psf } (14 \text{ ft}) = 140 \text{ lb/ft}$$

$$W_{min} = w_{min}(L) = 140 \ (110) = 15{,}400 \text{ lb}$$

$$R_A = R_B = (½)(15{,}400) = 7700 \text{ lb}$$

∴ minimum wind pressure of 10 psf governs

Reaction forces due to wind load to diaphragm:

$$R = 7700 \text{ lb}$$

Seismic. Compared with the seismic analysis for multistory buildings, the calculation of earthquake forces on a one-story structure is greatly simplified. This is because no vertical distribution of the seismic forces occurs with a single-story building. For the masonry wall building in this example, the seismic coefficients for the base shear V and the story (shearwall) force F_x are the same. The seismic coefficient for the story (diaphragm) force varies by Seismic Design Category. For this example the story (diaphragm) force F_{px} coefficient will be the same as the base shear and story (shearwall) force coefficients. If the example building fell in Seismic Design Categories B or C, this would not be true.

The seismic force requirements in the IBC are somewhat more complex than in previous codes. It is the intent of the seismic example for this one story building to help the reader follow the somewhat involved path through this IBC criteria.

The direct evaluation of the uniform force on the diaphragm requires a clear understanding of the way inertial forces are distributed. To see how the earthquake forces work their way down through the structure, it is helpful to make use of the weight of a 1-ft-wide strip of dead load W_1 taken parallel to the direction of the earthquake being considered. For example, in the case of lateral forces in the transverse direction, the weight of a 1-ft-wide strip of dead load parallel to the short side of the building is used. See the 1-ft strip in Fig. 3.10. Only the dead loads tributary to the roof level are included in W_1.

The weight of this 1-ft-wide strip includes the *roof dead load* and the weight of the *walls* that are *perpendicular* to the direction of the earthquake force being considered. Thus, for seismic forces in the transverse direction, the tributary dead load D of the longitudinal walls is included in W_1. In the example being considered, there are only two walls perpendicular to the direction of the seismic force. The weight of interior partition walls (both parallel and perpendicular to the seismic force) as well as other non-structural items including mechanical equipment and ornamentation need to be considered in the weight of a one foot strip. These additional weights have not been included in this example for simplicity.

When shearwalls are wood framed, it is common to include the weight of the parallel as well as perpendicular shearwalls in the calculated roof unit weight. This makes the roof forces a bit conservative, but greatly streamlines the design calculations.

The forces determined in this manner satisfy the Code requirement that the seismic force be applied "in accordance with the mass distribution" of the level. The 1-ft strip of dead load can be viewed as the mass that causes inertial (seismic) forces to develop in the horizontal diaphragm. The weight of the transverse walls does not contribute to the seismic force in the horizontal diaphragm. The forces in the transverse shearwalls are handled in a later part of the example.

This example demonstrates the complete evaluation of the seismic coefficients for a particular structure. However, it will be recalled that Example 3.5 developed a table of example seismic coefficients that applies to buildings that meet the three criteria listed for a "typical" structure. Example 3.7 confirms the seismic coefficients for the table in Example 3.5.

The distribution of forces to the primary LFRS in Example 3.7 assumes that the longitudinal walls span between the roof diaphragm and the foundation. A similar loading for the seismic force on elements and components was shown in Fig. 2.21 (Sec. 2.15).

EXAMPLE 3.7   Seismic Force Calculation

Determine the story (diaphragm) force F_{px} and the unit shear (lb/ft) in the horizontal diaphragm. See Fig. 3.10. The "u" subscripts in Fig. 3.10 are a reminder that the IBC seismic forces are at a *strength level,* as was discussed in Sec. 3.3. Keep in mind that the seismic forces used for design of the diaphragm (F_{px}, Sec. 2.14) are different from those used to design the shearwalls (in-plane walls F_x, Sec. 2.14) and different from those used for wall out-of-plane design (F_c, Sec. 2.15).

As in Example 3.5, the building is located in Hayward, California, with a short-period spectral acceleration S_S of 2.05g, as determined from the maximum considered earthquake ground motion maps. Site Class D is assigned. The exception to IBC Sec. 1616.3 does not apply, because a flexible diaphragm is assumed. As a result, Seismic Design Category E is assigned. The structure is a bearing wall system with masonry shearwalls. Per ASCE 7 Table 9.5.2.2, these walls are required to be specially reinforced masonry shearwalls in Seismic Design Category E, resulting in an R-factor of 5.

The roof dead load has been determined by prior analysis. The roof dead load D of 10 psf has been converted to the load on a horizontal plane. The masonry walls are 8-in. medium-weight concrete block units with cells grouted at 16 in. o.c. For this construction the wall dead load D is 60 psf.

Building Period T_a and Design Spectral Response Accelerations S_{DS} and S_{D1}

To start, the building fundamental period will be estimated in accordance with the approximate method introduced in Sec. 2.13:

$$T_a = C_t h_n^x$$

$$T_a = 0.020(14)^{0.75} = 0.145 \text{ sec}$$

As in Example 3.5, for the building location, from the maximum considered earthquake ground motion maps, the mapped short-period spectral acceleration S_S is 2.05g, and the mapped one-second spectral acceleration S_1 is 0.91g.

For Site Class D, IBC Tables 1615.1.2(1) and 1615.1.2(2) assign values of 1.0 and 1.5 for site coefficients F_a and F_v, respectively. Using this information, the maximum considered spectral response accelerations S_{MS} and S_{M1}, and the design spectral response accelerations S_{DS} and S_{D1} can be calculated:

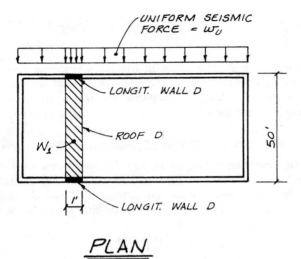

PLAN

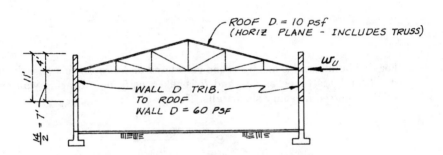

TRANSVERSE SECTION

Figure 3.10 Plan view shows a typical 1-ft-wide strip of dead load D in transverse direction. Weight of this strip W_1 generates a uniform seismic force on the horizontal diaphragm. Section view has mass of walls tributary to roof level indicated by cross-hatching. Both views show the force acting on the horizontal diaphragm.

$$S_{MS} = S_S \times F_a \qquad\qquad S_{M1} = S_1 \times F_v$$

$$= 2.05g \times 1.0 \qquad\qquad = 0.91g \times 1.5$$

$$= 2.05g \qquad\qquad\qquad = 1.37g$$

$$S_{DS} = \tfrac{2}{3} \times S_{MS} \qquad\qquad S_{D1} = \tfrac{2}{3} \times S_{M1}$$

$$= \tfrac{2}{3} \times 2.05g \qquad\qquad = \tfrac{2}{3} \times 1.37g$$

$$= 1.37g \qquad\qquad\qquad = 0.91g$$

The value of S_{DS} can be confirmed against the Example 3.5 table. In addition, from the S_{DS} and S_{D1} values, T_S can be calculated, and it can be verified that the approximate period T_a falls on the level plateau of the design response spectrum.

$$T_S = \frac{S_{D1}}{S_{DS}} = \frac{0.91g}{1.37g} = 0.66 \text{ sec}$$

Because T_a is less than T_S (the period at which the response spectrum plateau ends and the spectral acceleration starts dropping), the period is confirmed to fall on the plateau. As a result, the design short-period spectral response acceleration S_{DS} will define the seismic design forces.

Base Shear V, Story (Shearwall) Force F_x, and Story (Diaphragm) Force F_{px}

For this one-story building in Seismic Design Category E, the base shear, story (shearwall) force, and story (diaphragm) force will all use the same seismic force coefficient. This will not be true in Seismic Design Categories B and C, where the story (diaphragm) coefficient will be different. The seismic base shear can be calculated as:

$$V = C_S W$$

and, for short-period buildings

$$C_S = \frac{S_{DS}I}{R}$$

For this building, S_{DS} has just been calculated as $1.37g$. In this case the importance factor I (or I_E) is taken as 1.0, and R is taken as 5 for a bearing wall system with special reinforced masonry shearwalls. This results in

$$C_S = \frac{1.37g \times 1.0}{5}$$

$$= 0.274g$$

This is the *strength level* seismic coefficient for the base shear, the story (shearwall) forces, and the story (diaphragm) forces.

Seismic Force

For a one-story building, the uniform force to the diaphragm can be obtained by multiplying the seismic coefficient by the weight of a 1-ft-wide strip of dead load (W_1) tributary to the roof level.

$$
\begin{array}{llr}
W_1 = \text{roof dead load } D & = 10 \text{ psf} \times 50 \text{ ft} & = 500 \text{ lb/ft} \\
\quad + \text{ wall dead load } D & = 60 \text{ psf} \times 11 \text{ ft} \times 2 \text{ walls} & = 1320 \\
\hline
& W_1 & = 1820 \text{ lb/ft}
\end{array}
$$

$$W_u = 0.274W_1 = 0.274(1820)$$

$$W_u = 499 \text{ lb/ft} \quad (500 \text{ lb/ft will be used})$$

Redundancy/Reliability Factor ρ

The Code equations for seismic force are:

$$E = \rho Q_E + 0.2 S_{DS} D$$

and

$$E = \rho Q_E - 0.2 S_{DS} D$$

where ρQ_E is the force acting horizontally, and $0.2 S_{DS} D$ is a vertical force component. The redundancy/reliability factor ρ is a multiplier on the horizontal forces, including the base shear, the story (shearwall) force, and the story (diaphragm) force.

Assuming the building is 110 ft long and 50 ft wide and has 20 feet of door or window openings in each wall, the redundancy/reliability factor ρ is calculated as:

$$\rho_x = 2 - \frac{20}{r_{max_x} \sqrt{A_x}}$$

with

$$A_x = A_r = 110 \text{ ft}(50 \text{ ft}) = 5500 \text{ ft}^2$$

$$r_{max_x} = \frac{V_{wall}}{V_{story}} \left(\frac{10}{l_w} \right)$$

With this simple building configuration, it can reasonably be assumed that for loading in the transverse direction, each 50-ft transverse shearwall takes one-half of the base shear, giving $V_{wall}/V_{story} = 0.50$. A similar assumption may be made in the longitudinal direction. In each case a 20-ft opening length is subtracted from the overall wall length.

$$\text{Transverse:} \quad r_r = 0.50 \left(\frac{10}{50 - 20} \right) = 0.167$$

$$\text{Longitudinal:} \quad r_r = 0.50 \left(\frac{10}{110 - 20} \right) = 0.056$$

Therefore r_{max_r} is taken as the larger value, 0.167.

$$\rho_x = 2 - \frac{20}{r_{max_x} \sqrt{A_x}}$$

$$\rho_r = 2 - \frac{20}{0.167 \sqrt{5500}}$$

$$= 0.385$$

However, ρ may not be taken as less than 1.0, so 1.0 will be used for design.

Seismic Force Adjustments

The last two steps in determining the horizontal component of the seismic force on an element of the primary LFRS are multiplying by ρ and, as part of the basic load combinations, adjusting to an ASD level.

The corresponding modifications to the diaphragm design forces would be:

$$w_u = \text{E} = \rho Q_\text{E} = 1.0(500 \text{ lb/ft}) = 500 \text{ lb/ft}$$

and

$$w = 0.7\text{E} = 0.7(500) = 350 \text{ lb/ft}$$

In order to compare the story (diaphragm) seismic forces with wind loading, the reaction due to the seismic forces acting on the diaphragm needs to be summed.

The reaction due to seismic forces acting on the diaphragm can be calculated as:

$$R = 350 \text{ lb/ft}(110 \text{ ft}/2) = 19{,}250 \text{ lb}$$

The 19,250 lb reaction due to seismic forces acting on the diaphragm is significantly greater than the reaction of 7700 lb due to wind, so seismic forces control the diaphragm forces and in addition will control the base shear and shearwall forces.

One of the criteria used to design horizontal diaphragms and shearwalls is the unit shear. Although the design of diaphragms and shearwalls is covered in Chaps. 9 and 10, the calculation of unit shear is illustrated here. See Example 3.8 for the unit shear in the roof diaphragm.

For the building in this example subjected to lateral forces in the transverse direction, the only shearwalls are the exterior end walls. Because the wood diaphragm is flexible in comparison to the masonry walls, the diaphragm can be modeled as a simple beam spanning between exterior walls. The deflected shape of the roof diaphragm is again shown in Fig. 3.11. The reaction of the diaphragm on the transverse end walls is the reaction of a uniformly loaded simple beam with a span length equal to the distance between shearwalls. The shear diagram for a simple beam shows that the maximum internal shear is equal to the external reaction. The maximum total shear is converted to a unit shear by distributing it along the width of the diaphragm available for resisting the shear.

The reader is again cautioned to pay close attention to the subscripts for seismic loads. The IBC equations define *strength level* seismic forces which are noted in this book with a "u" subscript (e.g., $w_u = 500$ lb/ft). Seismic forces which have been reduced for use in ASD are shown in this book without the "u" subscript (e.g. $w = 350$ lb/ft). Other symbols with and without "u" subscripts have similar meanings in this book. For example, v_u indicates a unit shear calculated with a strength level seismic force, while v indicates an ASD unit shear.

EXAMPLE 3.8   Unit Shear in Roof Diaphragm

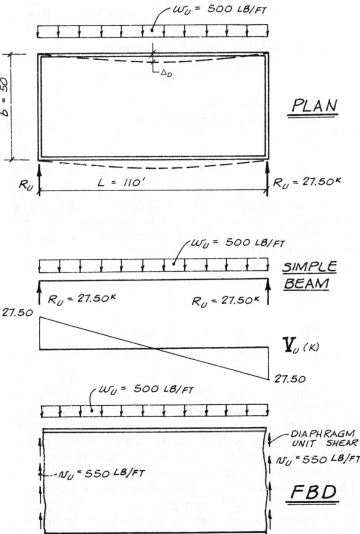

Figure 3.11   Diaphragm unit shear. For simplicity the calculations for unit shear are shown using the nominal length L and diaphragm width b (i.e., wall thickness is ignored).

The simple beam strength level loading diagram in Fig. 3.11 is a repeat of the loading on the horizontal diaphragm. The simple beam reactions R_u are shown along with the shear diagram.

The free-body diagram at the bottom of Fig. 3.11 is cut through the diaphragm a small distance away from the transverse shearwalls. The unit shear (lb/ft) is obtained by dividing the maximum total shear from the shear diagram by the width of the diaphragm b.

Diaphragm reaction:

$$R_u = \frac{w_u(L)}{2} = \frac{500(110)}{2} = 27{,}500 \text{ lb} = 27.50 \text{ k}$$

For a simple beam the shear equals the reaction:

$$V_u = R_u = 27.50 \text{ k}$$

The unit shear distributes the total shear over the width of the diaphragm. The conventional symbol for *total shear* is V_u, and the *unit shear* in the diaphragm is assigned the symbol v_u.

$$v_{u,\text{roof}} = \frac{V_u}{b} = \frac{27500}{50} = \boxed{550 \text{ lb/ft}}$$

The last two steps in determining a seismic force on an element of the primary LFRS are multiplying by ρ and, as part of the basic load combinations, adjusting to an ASD level. The value of $\rho = 1.0$ was determined in Example 3.7. The corresponding modifications to the roof diaphragm shear would be:

$$v_u = \text{E} = \rho Q_E = 1.0(550) = 550 \text{ lb/ft}$$

and

$$v = 0.7\text{E} = 0.7(550) = 385 \text{ lb/ft}$$

The next step in the lateral force design process is to consider similar quantities in the shearwalls. In determining the uniform force to the horizontal diaphragm, it will be recalled that only the dead load D of the roof and the longitudinal walls was included in the seismic force. The inertial force generated in the transverse walls was not included in the load to the roof diaphragm. The reason is that the shearwalls carry directly their own seismic force parallel to the wall. These forces do not, therefore, contribute to the force or shear in the horizontal diaphragm.

Several approaches are used by the designers to compute wall seismic forces. The Code requirement is that the wall be designed at the most critical shear location. In the more common method, the unit shear in the shearwall is evaluated at the *midheight* of the wall. See Example 3.9. This convention developed because the length of the shearwall b is often a minimum at this location. If the wall openings were different than shown, some location other than midheight might result in critical seismic forces. However, any openings in the wall (both doors and windows) are typically intersected by a horizontal line drawn at the midheight of the wall. In addition, using the midheight is consistent with the lumped-mass seismic model presented in Chap. 2.

The seismic force generated by the top half of the wall is given the symbol R_1. It can be computed as the dead load D of the top portion of the wall times the seismic coefficient. The total shear at the midheight of the wall is the sum of all forces above this level. For a one-story building, these forces include the reaction from the roof diaphragm R plus the wall seismic force R_1. The unit shear v may be computed once the total shear has been obtained.

The reader should note that, consistent with earlier examples, Example 3.9 computes the seismic forces to the wall at a strength level. At the end of the calculations, the strength level force is multiplied by ρ and also adjusted to an ASD force level.

EXAMPLE 3.9   Unit Shear in Shearwall

Determine the total shear and unit shear at the midheight of the shearwall in Fig. 3.12. For simplicity, ignore the reduction in wall dead load due to the opening (conservative).

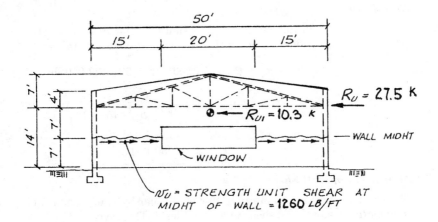

Figure 3.12   Shearwall unit shear. Maximum unit shear occurs at midheight of wall.

Roof Diaphragm Seismic Force

Compute the F_x story (shearwall) coefficient:

$$F_x \text{ story coefficient} = \frac{S_{DS}I}{R}$$

$$= \frac{1.37(1.0)}{5}$$

$$= 0.274g$$

Alternatively, this F_x coefficient could have been obtained from the seismic table in Example 3.5.

From Example 3.7, the weight of a 1-ft wide strip of dead load (W_1) tributary to the roof level is 1820 lb/ft. The strength level uniform seismic load to the diaphragm and resulting diaphragm reaction are:

$$w_u = 0.274W_1 = 0.274(1820) = 500 \text{ lb/ft}$$

$$R_u = w_u L/2 = 500(110 \text{ ft})/2 = 27,500 \text{ lb}$$

Wall Seismic Force

Seismic force generated by the top half of the wall (see Fig. 3.12):

$$\text{Wall area} = (11 \times 50) + \frac{1}{2}(3 \times 50) = 625 \text{ ft}^2$$

$$\text{Wall D} \quad = 625 \text{ ft}^2 \times 60 \text{ psf} = 37,500 \text{ lb} \qquad \text{(neglect window reduction)}$$

$$R_{u1} = 0.274W$$

$$= 0.274(37,500)$$

$$= 10,300 \text{ lb}$$

Wall Shear

$$\begin{Bmatrix} \text{Total shear at} \\ \text{midheight of wall} \end{Bmatrix} = \begin{Bmatrix} \text{sum of all forces on FBD of} \\ \text{shearwall above midheight} \end{Bmatrix}$$

$$V_u = R_u + R_{u1} = 27,500 + 10,300 = 37,800 \text{ lb}$$

$$\text{Unit wall shear} = v_u = \frac{V_u}{b} = \frac{37,800}{15 + 15}$$

$$\boxed{^\dagger v_u \text{ wall} = 1260 \text{ lb/ft}}$$

Adjustment of Wall Shear

The last two steps are to multiply this wall shear by ρ, and as part of the *basic load combinations,* adjust the shear to an ASD level. This gives:

$$v_u = E = \rho Q_E = 1.0(1260) = 1260 \text{ lb/ft}$$

$$v = 0.7E = 0.7(1260) = 882 \text{ lb/ft}$$

As mentioned earlier, the unit shear in the roof diaphragm and in the shear-wall constitute one of the main parameters in the design of these elements. There are additional design factors that must be considered, and these are covered in subsequent chapters.

These examples have dealt only with the transverse lateral forces, and a similar analysis is used for the longitudinal direction. Roof diaphragm shears are usually critical in the transverse direction, but both directions should be analyzed. Shearwalls may be critical in either the transverse or longitudinal directions depending on the size of the wall openings.

---

†For wall openings not symmetrically located, see Sec. 9.6.

3.6   Design Problem: Lateral Forces on Two-Story Building

A multistory building has a more involved analysis of seismic forces than a one-story structure. Once the seismic base shear V has been determined, the forces are distributed to the story levels in accordance with the Code formulas for F_x and F_{px}. These seismic forces were reviewed in Secs. 2.13 and 2.14. There it was noted that all three of the seismic forces on the primary LFRS (V, F_x, and F_{px}) could be viewed as a seismic coefficient times the appropriate mass or dead load of the structure:

$$V = \text{Base shear coefficient} \times W$$

$$F_x = F_x \text{ story coefficient} \times w_x$$

$$F_{px} = F_{px} \text{ story coeffcient} \times w_x$$

The purpose of the two-story building problem in Examples 3.10 and 3.11 is to compare the maximum wind and seismic forces and to evaluate the unit shears in the horizontal diaphragms and shearwalls. The base shear V, the F_x story (shearwall) forces, and the unit shears in shearwalls are covered in Example 3.10. The F_{px} story (diaphragm) forces and unit shears in diaphragms will be covered in Example 3.11. The wind pressures are slightly different from those in the previous one-story building of Example 3.5, due to the different wall height. Although the final objective of the earthquake analysis is to obtain numerical values of the design forces, it is important to see the overall process. To do this, the calculations emphasize the determination of the various seismic coefficients (g forces). Once the seismic coefficients have been determined, it is a simple matter to obtain the numerical values.

The one-story building example in Sec. 3.5 was divided into a number of separate problems. The two-story structure in Examples 3.10 and 3.11 is organized into two sets of design calculations which are more representative of what might be done in practice. However, sufficient explanation is provided to describe the process required for multistory structures.

EXAMPLE 3.10   Two-Story Lateral Force Calculation, Base Shear and Shearwalls

Determine the lateral wind and seismic forces in the transverse direction for the two-story office building in Fig. 3.13a. For the critical loading, determine the unit shear in the transverse shearwalls at the midheight of the first- and second-story walls. Assume that there are no openings in the masonry walls.

Wind forces are to be based on the simplified wind load provisions of IBC Sec. 1609.6. The basic wind speed is 85 mph. Standard occupancy and Exposure B apply.

The building is located in Pullman, Washington (ZIP Code 99163). Site Class D is assumed and $I = 1.0$. The following dead loads have been determined in a prior analysis: roof dead load D = 20 psf, floor dead load D = 12 psf, floor dead load D to account for the weight of interior wall partitions = 10 psf, and exterior wall dead load D = 60 psf.

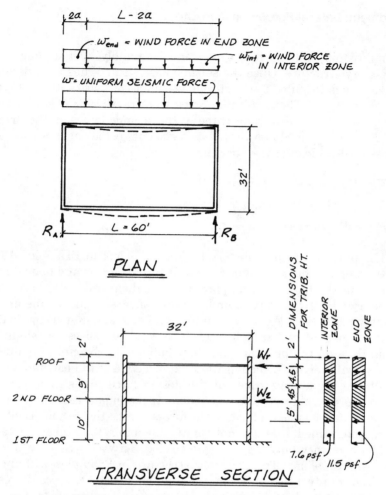

Figure 3.13a Wind pressures and tributary heights to roof and second-floor diaphragms.

NOTE: In buildings where the location of nonbearing walls and partitions is subject to change, the Code requires a partition dead load D of 20 psf for designing individual floor members for vertical loads. However, for evaluation of seismic design forces, an average floor dead load D of 10 psf is allowed (ASCE 7 Sec. 9.5.3 definition of W).

Wind Forces

Building Geometry

$$h_{mean} = 19 \text{ ft}$$

$$0.4h_{mean} = 0.4(19 \text{ ft}) = 7.6 \text{ ft}$$

Least horizontal building dimension = b = 32 ft

$$0.1b = 0.1(32 \text{ ft}) = 3.2 \text{ ft}$$

End zone dimension:

$$a = \text{lesser of } 0.4h_{\text{mean}} \text{ or } 0.1b$$

$$= 3.2 \text{ ft}$$

Length of end zone $= 2a = 6.4$ ft

Wind Pressures

Wind pressure formula:

$$p_s = \lambda I_W p_{s30}$$

$$I_W = 1.0 \qquad \text{(IBC Table 1604.5)}$$

$$\lambda = 1.0 \qquad \text{for 0 to 30 ft (IBC Table 1609.6.2.1(4))}$$

Basic wind pressures p_{s30} from IBC Table 1609.6.2.1(1) for a flat roof:

$$\text{Zone A} \qquad p_{s30} = p_s = 11.5 \text{ psf}$$
$$\text{Zone C} \qquad p_{s30} = p_s = 7.6 \text{ psf}$$

These wind pressures are shown in the section view in Fig. 3.13a.

Load to Diaphragms

Roof:

$$\left\{ \begin{array}{l} \text{Trib. height to} \\ \text{roof diaphragm} \end{array} \right\} = \left\{ \begin{array}{l} \text{Trib. wall height} \\ \text{below roof} \end{array} \right\} + \left\{ \begin{array}{l} \text{Height of} \\ \text{parapet wall} \end{array} \right\}$$

$$= \tfrac{1}{2}(9 \text{ ft}) + 2 \text{ ft} = 6.5 \text{ ft}$$

$$w_{\text{end}} = 11.5 \text{ psf}(6.5 \text{ ft}) = 69.0 \text{ lb/ft}$$

$$w_{\text{int}} = 7.6 \text{ psf}(6.5 \text{ ft}) = 49.4 \text{ lb/ft}$$

$$W_r = w_{\text{end}}(2a) + w_{\text{int}}(L - 2a)$$

$$= 69.0(6.4) + 49.4(60 - 6.4)$$

$$= 3089 \text{ lb}$$

Use static equilibrium equations to determine the larger reaction force (R_{rA} or R_{rB}) at either end of the roof diaphragm.

Summing moments about the reaction at B:

$$R_{rA} = w_{\text{end}}(2a)(L - a)\left(\frac{1}{L}\right) + w_{\text{int}}(L - 2a)\left(\frac{L}{2} - a\right)\left(\frac{1}{L}\right)$$

$$= (69.0)(6.4)(60 - 3.2)\left(\frac{1}{60}\right) + (49.4)(60 - 6.4)\left(\frac{60}{2} - 3.2\right)\left(\frac{1}{60}\right)$$

$$= 1601 \text{ lb}$$

Summing forces:

$$R_{rB} = W_r - R_{rA} = 3089 - 1601 = 1488 \text{ lb}$$

Check minimum load to roof diaphragm based on $p_s = 10$ psf throughout Zones A and C.

$$w_{min} = p_s(\text{Trib. height to roof diaphragm}) = 10 \text{ psf}(6.5 \text{ ft}) = 65 \text{ lb/ft}$$

$$W_{min} = w_{min}(L) = 65(60) = 3900 \text{ lb}$$

$$R_{rA} = R_{rB} = (½)(3900) = 1950 \text{ lb}$$

$$\therefore \text{ minimum wind pressure of 10 psf governs}$$

Reaction forces due to wind load to roof diaphragm:

$$\boxed{R_r = 1950 \text{ lb}}$$

Second floor:

$$\begin{Bmatrix} \text{Trib. height to 2nd} \\ \text{floor diaphragm} \end{Bmatrix} = \begin{Bmatrix} \text{Trib. wall height} \\ \text{above 2nd floor} \end{Bmatrix} + \begin{Bmatrix} \text{Trib. wall height} \\ \text{below 2nd floor} \end{Bmatrix}$$

$$= ½(9 \text{ ft}) + ½(10 \text{ ft}) = 9.5 \text{ ft}$$

$$w_{end} = 11.5 \text{ psf}(9.5 \text{ ft}) = 109.25 \text{ lb/ft}$$

$$w_{int} = 7.6 \text{ psf}(9.5 \text{ ft}) = 72.2 \text{ lb/ft}$$

$$W_2 = w_{end}(2a) + w_{int}(L - 2a)$$

$$= 109.25(6.4) + 72.2(60 - 6.4)$$

$$= 4569 \text{ lb}$$

Summing moments about the reaction at B:

$$R_{2A} = w_{end}(2a)(L - a)\left(\frac{1}{L}\right) + w_{int}(L - 2a)\left(\frac{L}{2} - a\right)\left(\frac{1}{L}\right)$$

$$= (109.25)(6.4)(60 - 3.2)\left(\frac{1}{60}\right) + (72.2)(60 - 6.4)\left(\frac{60}{2} - 3.2\right)\left(\frac{1}{60}\right)$$

$$= 2390 \text{ lb}$$

Summing forces:

$$R_{2B} = W_2 - R_{2A} = 4569 - 2390 = 2179 \text{ lb}$$

Check minimum load to second-floor diaphragm based on $p_s = 10$ psf throughout Zones A and C.

$$w_{\min} = p_s(\text{Trib. height to roof diaphragm}) = 10 \text{ psf}(9.5 \text{ ft}) = 95 \text{ lb/ft}$$

$$W_{\min} = w_{\min}(L) = 95 (60) = 5700 \text{ lb}$$

$$R_{2A} = R_{2B} = (\tfrac{1}{2})(5700) = 2850 \text{ lb}$$

$$\therefore \text{ minimum wind pressure of 10 psf governs}$$

Reaction forces due to wind load to second-floor diaphragm:

$$\boxed{R_2 = 2850 \text{ lb}}$$

Seismic Forces

Building Period T_a and Design Spectral Response Accelerations S_{DS} and S_{D1}

To start, the building fundamental period will be estimated in accordance with the approximate method introduced in Sec. 2.13:

$$T_a = C_t h_n^x$$

$$T_a = 0.020(19)^{0.75} = 0.182 \text{ sec}$$

For the building location in Pullman, Washington, from the maximum considered earthquake ground motion maps, the mapped short-period spectral acceleration S_S is $0.286g$, and the mapped one-second spectral acceleration S_1 is $0.089g$.

For Site Class D, IBC Tables 1615.1.2(1) and 1615.1.2(2) assign values of 1.57 and 2.4 for site coefficients F_a and F_v, respectively. The F_a value of 1.57 is found by interpolating between the tabulated values of 1.6 and 1.4. Using this information, the maximum considered spectral response accelerations S_{MS} and S_{M1}, and the design spectral response accelerations S_{DS} and S_{D1} can be calculated:

$$
\begin{aligned}
S_{MS} &= S_S \times F_a & S_{M1} &= S_1 \times F_v \\
&= 0.286g \times 1.57 & &= 0.089g \times 2.4 \\
&= 0.449g & &= 0.214g
\end{aligned}
$$

$$
\begin{aligned}
S_{DS} &= \tfrac{2}{3} \times S_{MS} & S_{D1} &= \tfrac{2}{3} \times S_{M1} \\
&= \tfrac{2}{3} \times 0.449g & &= \tfrac{2}{3} \times 0.214g \\
&= 0.299g & &= 0.143g
\end{aligned}
$$

From the S_{DS} and S_{D1} values, T_S can be calculated, and it can be verified that the approximate period T_a falls on the level plateau of the design response spectrum.

$$T_S = \frac{S_{D1}}{S_{DS}} = \frac{0.143g}{0.299g} = 0.48 \text{ sec}$$

Because T_a is less than T_S (the period at which the response spectrum plateau ends and spectral acceleration starts dropping), the period is confirmed to fall on the plateau. As a result, the design short-period spectral response acceleration S_{DS} will define the seismic design forces.

Redundancy/Reliability Factor ρ

For this example, there will be a number of element forces that will need to be adjusted using the ρ factor. Assuming that the building is 60 ft long and 32 ft wide, has no door or window openings in the transverse walls, and has ten feet of opening in each longitudinal wall, the redundancy factor can be calculated as:

$$\rho_x = 2 - \frac{20}{r_{\max_x} \sqrt{A_x}}$$

with

$$A_x = 60 \text{ ft}(32 \text{ ft}) = 1920 \text{ ft}^2$$

$$r_{\max_x} = \frac{V_{\text{wall}}}{V_{\text{story}}} \left(\frac{10}{l_w}\right)$$

With this simple building configuration, it can reasonably be assumed that for loading in the transverse direction, each 32-ft transverse shear wall takes one-half of the base shear, giving $V_{\text{wall}}/V_{\text{story}} = 0.50$. A similar assumption may be made in the longitudinal direction. In the longitudinal walls, a 10-ft opening length is subtracted from the overall wall length. The calculation of r needs to be repeated for both the roof and floor diaphragms and supporting walls. In this case, the roof and floor calculations would be the same:

Transverse: $\qquad r_2 = r_r = 0.50\left(\frac{10}{32}\right) = 0.156$

Longitudinal: $\qquad r_2 = r_r = 0.50\left(\frac{10}{60 - 10}\right) = 0.100$

Therefore $r_{\max_r}$ is taken as the larger value, 0.156.

$$\rho_x = 2 - \frac{20}{r_{\max_x} \sqrt{A_x}}$$

$$\rho_2 = \rho_r = 2 - \frac{20}{0.156\sqrt{1920}}$$

$$= -0.926$$

However, ρ may not be taken as less than 1.0, so 1.0 will be used for design.

Seismic Base Shear Coefficient

The seismic base shear V will be used to calculate the story (shearwall) forces F_x, which are used to design the vertical elements of the lateral-force-resisting system. The seismic base shear can be calculated as:

$$V = C_S W$$

where

$$C_S = \frac{S_{D1}I}{RT} \leq \frac{S_{DS}I}{R}$$

For this building, S_{DS} has just been calculated as 0.299g, and S_{D1} as 0.143g. A previous calculation identified the period T_S up to which the plateau defined by S_{DS} controls. Because the approximate period for this building T_a is less than T_S, calculation of C_S can be simplified to:

$$C_S = \frac{S_{DS}I}{R}$$

In this case I (or I_E) is taken as 1.0, and R is taken as 5 for a bearing wall system with special reinforced masonry shearwalls. This reuslts in:

$$C_S = \frac{0.299g \times 1.0}{5}$$

$$C_S = 0.060g$$

This is the *strength level* seismic coefficient for the base shear. This example again demonstrates the complete calculation of the seismic coefficient.

Tributary Roof Dead Loads

The total dead load for the structure is all that is required to complete computation of the base shear V. However, in the process of developing the total dead load, it is beneficial to summarize the weight tributary to the roof diaphragm and second-floor diaphragm using the idea of a 1-ft-wide strip. Recall from Example 3.7 that W_1 represents the mass or weight that will cause a uniform seismic force to be developed in a horizontal diaphragm.

The values of W_1, tributary to the roof and second floor, will eventually be used to determine the distributed story forces. It is recommended that the reader sketch the 1-ft-wide strip on the plan view in Fig. 3.13a. The tributary wall heights are shown on the section view.

Weight of 1-ft-wide strip tributary to roof:

PARAPET

Roof dead load D $= (20 \text{ psf})(32 \text{ ft}) = 640 \text{ lb/ft}$

$+ \text{ Wall dead load D (2 longit. walls)} = 2(60 \text{ psf})\left(\frac{9}{2} + 2\right) = 780$

_____

Dead load D of 1-ft strip at roof $= W_1 = 1420 \text{ lb/ft}$

The mass that generates the entire seismic force in the roof diaphragm is given the symbol W_r'. It is the sum of all the W_1 values at the roof level.

$$W_r' = \Sigma W_1 = 1420 \text{ lb/ft}(60 \text{ ft}) = 85.2 \text{ k}$$

To obtain the total mass tributary to the roof level, the weight of the top half of the transverse shearwalls is added to W_r'. The total dead load D tributary to the roof level is given the symbol W_r.

$$\text{Dead load D of 2 end walls} = 2(60 \text{ psf})(32)\left(\frac{9}{2} + 2\right) = 25.0 \text{ k}$$

$$\text{Total dead load D trib. to roof} = W_r = 85.2 + 25.0 = 110.2 \text{ k}$$

Similar quantities are now computed for the second floor.

Tributary Second-Floor Dead Loads

Weight of 1-ft-wide strip tributary to second floor:

Second-floor dead load D	$= (12 \text{ psf})(32 \text{ ft})$	$= 384 \text{ lb/ft}$
+ Partition dead load D	$= (10 \text{ psf})(32 \text{ ft})$	$= 320$
+ Wall dead load D (2 longit. walls)	$= 2(60 \text{ psf})\left(\dfrac{9}{2} + \dfrac{10}{2}\right)$	$= 1140$
Dead load D of 1-ft strip at second floor $= W_1$		$= 1844 \text{ lb/ft}$

The mass that generates the entire seismic force in the second floor diaphragm is W_2'.

$$W_2' = \Sigma W_1 = 1844 \text{ lb/ft}(60 \text{ ft}) = 110.6 \text{ k}$$

The total mass tributary to the second-floor level is the sum of W_2' and the tributary weight of the transverse shearwalls. The total dead load D tributary to the second-floor level is given the symbol W_2.

$$\text{Dead load D of 2 end walls} = 2(60 \text{ psf})(32)\left(\frac{9}{2} + \frac{10}{2}\right) = 36.5 \text{ k}$$

$$\text{Total dead load D trib. to second floor} = W_2 = 110.6 + 36.5 = 147.1 \text{ k}$$

Seismic Tables

Calculations of seismic forces for multistory buildings are conveniently carried out in tables. Tables are not only convenient for bookkeeping, but also provide a comparison of the F_x and F_{px} story coefficients.

Tables and the necessary formulas can easily be stored in equation solving computer software. Once stored on a computer, tables serve as a template for future problems. In this way, the computer can be used to handle repetitive calculations and problem formatting (e.g., setting up the table), and the designer can concentrate on the best way to solve the problem at hand. Tables can be expanded to take into account taller buildings and to include items such as overturning moments.

For a building in Seismic Design Category D, E, or F, the calculation of F_x story (shearwall) forces and F_{px} story (diaphragm) forces can be combined in a single table because both types of forces are based on the same base shear V. For the building in Example 3.10 in Seismic Design Category C a different approach needs to be used to calculate the F_{px} forces. This is most conveniently done in a separate table. The balance of Example 3.10 will look at the F_x story (shearwall) forces and the resulting unit shears in the shearwalls. The F_{px} story (diaphragm) forces and resulting unit shears in the diaphragms are calculated in Example 3.11.

The F_x table below is shown completely filled out. However, at this point in the solution of the problem, only the first four columns can be completed. Columns 1, 2, and 3 simply list the story levels, heights, and masses (dead load Ds). The values in column 4 are the products of the respective values in columns 2 and 3. The sum of the story masses at the bottom of column 3, Σw_x, is the weight of the structure W to be used in the calculation of base shear.

The steps necessary to complete the remaining columns in the F_x table are given in the two sections immediately following the table.

F_x Story (Shearwall) Force Table—$R = 5$

1	2	3	4	5	6	7
Story	Height h_x	Weight w_x	$w_x h_x$	Story force $F_x = .0043 h_x w_x$	F_x Coef.	Story shear V_x
R	19	110.2	2094	$F_r = 9.00$ k	0.0817	9.00 k
2	10	147.1	1471	$F_2 = 6.33$ k	0.0430	15.33 k
1	0					
Sum		257.3 k	3565 k-ft	$V = .060W = 15.4$ k		

Base Shear

The *strength level* seismic base shear coefficient for the F_x forces was determined previously to be 0.060g. The *strength level* base shear for F_{px} forces will be calculated in Example 3.11. The total base shear for the building is 0.060 times the total weight from column 3.

$$V = 0.060(257.3) = 15.4 \text{ k}$$

The story coefficients for distributing the seismic force over the height of the structure can now be determined. The distribution of forces to the vertical elements in the primary LFRS is given by the Code formula for F_x.

F_x Story (Shearwall) Coefficients

In Chap. 2 it was noted that the formula for F_x can be written as an F_x story coefficient times the mass tributary to level x, w_x:

$$F_x = C_{vx} V = \left[\frac{w_x h_x^k}{\sum\limits_{i=1}^{n} w_i h_i^k} \right] V$$

$$F_x = (F_x \text{ story coefficient}) w_x = \left[\frac{V h_x^k}{\sum\limits_{i=1}^{n} w_i h_i^k} \right] w_x$$

The exponent k is related to the building period, and can be taken as 1.0 for buildings with an estimated period of less than 0.5 sec. The period has been estimated to be 0.182 sec for this building, so the story force coefficient can be simplified to:

$$F_x = (F_x \text{ story coefficient})w_x = \left[\frac{Vh_x}{\sum\limits_{i=1}^{n} w_i h_i}\right] w_x$$

The *strength level* F_x story coefficients will now be evaluated. The base shear V is known. The summation term in the denominatory is obtained as the last item in column 4 of the seismic table.

$$F_x = \left[\frac{Vh_x}{\sum\limits_{i=1}^{n} w_i h_i}\right] w_x = \left[\frac{(15.4)h_x}{3565}\right] w_x$$

This general formula for F_x is entered at the top of column 5 as

$$F_x = (0.0043h_x)w_x$$

Individual F_x story coefficients follow this entry. At the roof level F_x is given the symbol F_r, and at the second-floor level the symbol is F_2.

Roof:

$$F_r = (0.0043h_r)w_r = (0.0043)(19)w_r = 0.0817w_r$$

The numerical value for the *strength level* seismic force at the roof level is added to column 5 next to the F_x story coefficient:

$$F_r = 0.0817w_r = 0.0817(110.2 \text{ k}) = 9.00 \text{ k}$$

Second floor:

$$F_2 = (0.0043h_2)w_2 = (0.0043)(10)w_2 = 0.043w_2$$

The numerical value for the seismic force at the second-floor level is also added to column 5.

$$F_2 = 0.043w_2 = 0.043(147.1 \text{ k}) = 6.33 \text{ k}$$

The summation at the bottom of column 5 serves as a check on the numerical values. The *sum* of all the F_x story forces must equal the total base shear.

$$V = \Sigma F_x = F_r + F_2 = 9.00 + 6.33 = 15.33 \text{ k} \qquad OK$$

The values in column 7 of the seismic table represent the total *strength level* story shears between the various levels in the structure. The story shear can be obtained as the sum of all the F_x story forces above a given section.

In a simple structure of this nature, the story shears from column 7 may be used directly in the design of the vertical elements (i.e., the shearwalls). However, as the structure becomes more complicated, a more progressive distribution of seismic forces from the diaphragms to the vertical elements may be necessary. Both approaches are illustrated in this example.

With all of the F_x story coefficients determined, the individual distributed forces for designing the shearwalls can be evaluated.

Uniform Forces to Diaphragms Using F_x Story Coefficients

For shearwall design, the forces to the diaphragms are based on the F_x story coefficients. See Fig. 3.13b. These uniformly distributed forces will be used to compute the forces in the shearwalls following the progressive distribution in Method 1 (described later in this example).

Load to roof diaphragm:

The load to the roof diaphragm that is used for design of the shearwalls needs to be based on the F_x story forces from the F_x Seismic Story (Shearwall) Force table. The strength level roof diaphragm reaction can be calculated as follows:

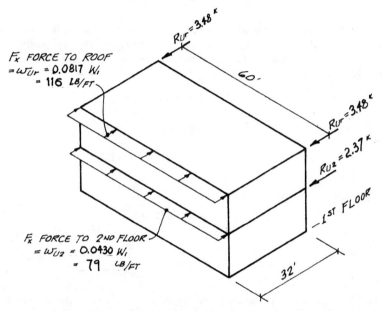

Figure 3.13b Seismic forces to roof diaphragm w_{ur} and second-floor diaphragm w_{u2} are for designing the vertical elements in the LFRS. Concentrated forces on shearwalls are diaphragm reactions R_{ur} and R_{u2}.

$$w_{ur} = 0.0817(1420) = 116 \text{ lb/ft}$$

$$R_{ur} = \frac{w_{ur}L}{2} = \frac{116(60)}{2} = 3.48 \text{ k}$$

Load from second-floor diaphragm:

The uniform force on the second-floor diaphragm is also determined using the F_x story coefficient from column 6 of the F_x Seismic Story Force Table.

$$w_{u2} = 0.0430W_1 = 0.0430(1844)$$

$$= 79 \text{ lb/ft}$$

The reaction of the second-floor diaphragm on the shearwall is

$$R_{u2} = \frac{w_{u2}L}{2} = \frac{79(60)}{2} = 2.37 \text{ k}$$

Comparison to Wind

In order to determine whether wind or seismic forces govern shearwall design, the reaction forces due to seismic loads are compared to the reaction forces due to wind loads, as calculated at the beginning of this example. The reaction forces at the roof and second-floor diaphragms due to wind loads are:

$$R_r = 1950 \text{ lb}$$

$$R_2 = 2850 \text{ lb}$$

In order to compare wind and seismic forces to the shearwalls, the seismic force reactions are multiplied by the redundancy factor ρ, and multiplied by 0.7 to convert to an ASD level. The factor ρ has been determined to be 1.0 at the start of this example. The ASD seismic forces are:

$$R_r = w_{ur}(L/2)\rho(0.7)$$

$$= 116(60/2)1.0(0.7)$$

$$= 2440 \text{ lb} > 1950 \text{ lb}$$

$$\text{seismic governs}$$

$$R_2 = w_{u2}(L/2)\rho(0.7)$$

$$= 79(60/2)1.0(0.7)$$

$$= 1660 \text{ lb} < 2850 \text{ lb}$$

Here it is important to note that although the wind force to the second-floor diaphragm is greater than the seismic, this is not a definitive check on whether wind or seismic forces will govern design of the first-story shearwalls. For wind forces, the first-story shearwalls must resist the sum of the wind reaction at the roof and second floor:

$$R_r + R_2 = 1950 + 2850 = 4830 \text{ lb wind}$$

For seismic forces, the first-story shearwalls must resist the sum of the diaphragm forces at the roof and second floor plus the weight of the transverse walls above midheight of the first story. This total is the same as one-half of the calculated seismic base shear V, which is 15,400 lb/2 = 7700 lb at a *strength level* or 5390 lb at an *ASD level*. With this closer look, it is determined that seismic forces control design of first-story shearwalls as well as second-story shearwalls.

Shear at Midheight of Second-Story Walls (Using F_x Story Shearwall Forces)

Two methods for evaluating the shear in the shearwalls are illustrated. The first method demonstrates the progressive distribution of the forces from the horizontal diaphragms to the shearwalls. Understanding Method 1 is essential to the proper use of the F_x story forces for more complicated shearwall arrangements. Method 2 can be applied to simple structures where the distribution of seismic forces to the shearwalls can readily be seen.

METHOD 1

For the shear between the second floor and the roof, the free-body diagram (FBD) of the wall includes two seismic forces. See Fig. 3.13c. One force is the reaction from the roof diaphragm (from Fig. 3.13b), and the other is the inertial force developed by the mass of the top half of the shearwall.

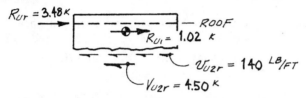

Figure 3.13c FBD of shearwall cut midway between second-floor and roof levels.

Force from top half of shearwall:

The seismic force generated by the top half of the second-story shearwall is given the symbol R_{u1}. This force is obtained by multiplying the dead load of the wall by the F_x story coefficient for the *roof* level.

$$R_{u1} = 0.0817 w_u = 0.0817\left[(60 \text{ psf})\left(\frac{9}{2} + 2\right)(32 \text{ ft})\right]$$

$$= 1.02 \text{ k}$$

The shear in the wall between the second floor and the roof is given the symbol V_{u2r}, and it is obtained by summing forces in the x direction.

$$\Sigma F_x = 0$$

$$V_{u2r} = R_{ur} + R_{u1} = 3.48 + 1.02 = 4.50 \text{ k}$$

Strength level unit shear in wall:

$$v_{u2r} = \frac{V_{u2r}}{b} = \frac{4.50}{32} = \boxed{140 \text{ lb/ft}}$$

METHOD 2

For this simple rectangular building with two equal-length transverse shearwalls, the shear in one wall V_{u2r} can be obtained as one-half of the total story shear from column 7 of the F_x Story (Shearwall) Force Table.

$$\text{Wall shear } V_{u2r} = \tfrac{1}{2}(\text{story shear } V_{u2r})$$

$$= \tfrac{1}{2}(9.00) = 4.50 \text{ k} \qquad (\text{same as Method 1})$$

For other shearwall arrangements, including interior shearwalls, the progressive distribution of forces using Method 1 is required.

Seismic Force Adjustments

The last two steps in determining the seismic force on an element of the primary LFRS are multiplying by ρ and, as part of the basic load combinations, adjusting to an ASD level.

The corresponding modifications to the second story shearwall shear would be:

$$v_u = \text{E} = \rho Q_\text{E} = 1.00(140) = 140 \text{ lb/ft}$$

and

$$v = 0.7\text{E} = 0.7(140) = 98 \text{ lb/ft}$$

Shear at Midheight of First-Story Walls (Using F_x Story (Shearwall) Forces)

METHOD 1

The shear in the walls between the first and second floors is obtained from the FBD in Fig. 3.13d. The two forces on the top are the forces from Fig. 3.13c. The load R_{u2} is the reaction from the second-floor diaphragm (from Fig. 3.13b). The final seismic force is the second force labeled R_{u1}. This represents the inertial force generated by the mass of the shearwall tributary to the second floor.

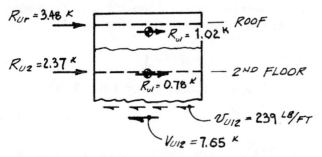

Figure 3.13d FBD of shearwall cut midway between first-floor and second-floor levels.

Force from wall mass tributary to second-floor level:

The R_{u1} force for the middle portion of the shearwall uses the F_x story coefficient for the second-floor level:

$$R_{u1} = 0.0430 w_u = 0.0430 \left[(60 \text{ psf}) \left(\frac{9}{2} + \frac{10}{2} \right) (32 \text{ ft}) \right]$$

$$= 0.78 \text{ k}$$

The shear between the first- and second-floor levels is given the symbol V_{12}. It is obtained by summing forces in the x direction (Fig. 3.13d):

$$\Sigma F_x = 0$$

$$V_{u12} = R_{ur} + R_{u1} + R_{u2} + R_{u1}$$

$$= 3.48 + 1.02 + 2.37 + 0.78$$

$$= 7.65 \text{ k}$$

Unit shear in wall between first and second floor:

$$v_{u12} = \frac{V_{u12}}{b} = \frac{7650}{32} = \boxed{239 \text{ lb/ft}}$$

METHOD 2

Again for a simple rectangular building with two exterior equal-length shearwalls, the total shear in a wall can be determined as one-half of the story shear from column 7 of the F_x Story (Shearwall) Force Table.

$$\text{Wall shear } V_{u12} = \tfrac{1}{2}(\text{story shear } V_{u12})$$

$$= \tfrac{1}{2}(15.33) = 7.67 \text{ k} \qquad (\text{same as Method 1})$$

Seismic Force Adjustments

The last two steps in determining the seismic force on an element are multiplying by ρ and, as part of the basic load combinations, adjusting to an ASD level.

The corresponding modifications to the first-story shearwall shear is:

$$v_u = E = \rho Q_E = 1.0(239) = 239 \text{ lb/ft}$$

and

$$v = 0.7E = 0.7(239) = 167 \text{ lb/ft}$$

The above analysis is for lateral forces in the transverse direction. A similar analysis is required in the longitudinal direction.

Example 3.11 continues design calculations for the two-story building from Example 3.10. In Example 3.10 the applied seismic and wind forces used for design of *shearwalls* were calculated and compared. The shearwall unit shears were then calculated based on the more critical seismic forces, determined using F_x story (shearwall) coefficients.

Example 3.11 shifts the focus from *shearwall* design forces to *diaphragm* design forces. This is done by calculating the F_{px} story (diaphragm) forces and diaphragm unit shears due to seismic forces and comparing them to the diaphragm unit shears from wind forces.

EXAMPLE 3.11   Two-Story Lateral Force Calculation, Diaphragm Forces

Determine the wind and seismic diaphragm forces for the two-story office building from Example 3.10. Evaluate the unit shears in the roof and second-floor horizontal diaphragms. The wind and seismic criteria remain unchanged from Example 3.10.

F_{px} Story (Diaphragm) Coefficient

The F_{px} coefficients will be used to calculate the forces for design of the diaphragms. The seismic story (diaphragm) forces F_{px} were introduced in Sec. 2.14, where it was noted that different equations apply for Seismic Design Categories B and C, than for D, E, and F. For Seismic Design Categories B and C, the F_{px} forces can be calculated as:

$$F_{px} = 0.2S_{DS}Iw_p$$

but not less than F_x. It is important to recognize that this P_{px} equation does not include a vertical redistribution, and only varies as a function of the weight of the diaphragm and the attached structure. It is convenient to put the seismic force coefficients for F_x and F_{px} in a table to compare and identify which will control design of the diaphragms.

F_{px} Story (Diaphragm) Force Table—Seismic Design Categories B and C

1	2	3	4	5
Diaphragm level	S_{DS}	F_{px} coefficient $= 0.2S_{DS}I$	F_x coefficient	Controlling F_{px} coefficient
Roof, F_{pr}	$0.299g$	0.0598	0.0817	0.0817
2nd floor, F_{p2}	$0.299g$	0.0598	0.0430	0.0598

Column 5 identifies the controlling story (diaphragm) force coefficient. Note that the F_x coefficient controls for design of the second-floor diaphragm, while the F_x coefficient controls for design of the roof diaphragm. This approach works when shearwall locations are the same at the first and second stories. Where shearwall locations change, additional diaphragm forces will occur due to transfer of force into and out of shearwalls. This is specifically discussed for Seismic Design Categories B and C in ASCE 7 Sec. 9.5.2.6.2.7. It is implied that these forces should be included at an F_x coefficient level. These forces should also be included in diaphragm forces in Seismic Design Categories D, E, and F.

With calculation of both the F_x and F_{px} forces, some observations can be made:

1. The maximum story force coefficients at the roof level exceed the magnitude of the base shear coefficient (0.060g). This will always be true for a building with more than one story.

2. The minimum value for the story (shearwall) force coefficient F_x at the second-floor level is less than the base shear coefficient. This will always be true of the lowermost stories in multistory buildings.

The fact that the F_{px} story (diaphragm) force coefficient at the second floor is essentially the same as the base shear coefficient is incidental in this case. The F_x forces include an R-factor, while the F_{px} force equation does not.

Repeat of F_{px} Story (Diaphragm) Coefficient for Seismic Design Category D

Because the method of calculating F_{px} forces is different in Seismic Design Design Categories D, E, and F, this F_{px} calculation method will be illustrated. From Sec. 2.14, the F_{px} story (diaphragm) force is calculated as:

$$F_{px} = \frac{\sum_{i=c}^{n} F_i}{\sum_{i=x}^{n} w_i} w_{px}$$

and

$$0.2 S_{DS} I w_{px} \le F_{px} \le 0.4 S_{DS} I w_{px}$$

This calculation makes use of information in the story (shearwall) force coefficient table, so it makes sense to repeat this information and add the F_{px} forces.

F_{px} **Story (Shearwall) Table—Seismic Design Categories D, E, and F**

1	2	3	4	5	6	7	8
Story	Height h_x	Weight w_x	Story (shearwall) force F_x	$\sum_{i=x}^{n} F_i$	$\sum_{i=x}^{n} w_i$	Story (diaphragm) force F_{px}	Story (diaphragm) force F_{px} coefficient
R	19	110.2	9.00	9.00	110.20	9.00	0.0817
2	10	147.1	6.33	15.33	257.30	8.76	0.0596
1	0						
Sum		257.3 k	15.33 k				

The story (diaphragm) force in column 6 is calculated as:

$$F_{px} = \frac{\sum_{i=x}^{n} F_i}{\sum_{i=x}^{n}} w_{px}$$

At the roof diaphragm, the sum of F_x = 9.00, and the sum of w_x = 110.2 = w_{px}, giving:

$$F_{pr} = \frac{9.00}{110.2} \, 110.2 = 9.00 \text{ k}$$

Note that the sum of F_x and w_x is calculated from the topmost level down to the level being considered. The upper and lower limits for F_{pr} also need to be checked:

$$0.2S_{DS}Iw_{px} \leq F_{px} \leq 0.4S_{DS}Iw_{px}$$

$$0.2(0.299)1.00(110.2) \leq F_{pr} \leq 0.4(0.299)1.00(110.2)$$

$$6.58 \text{ k} \leq F_{pr} \leq 13.18 \text{ k}$$

F_{pr} of 9.00 falls between these limits, so 9.00 is the controlling value. In column 7, this value is converted to a coefficient by dividing by $w_r = 110.2$ k.

At the second-floor diaphragm, summing from the top level down, the sum of $F_x = 9.00 + 6.33 = 15.33$, and the sum of $w_x = 110.2 + 147.1 = 257.3$, giving:

$$F_{p2} = \frac{15.33}{257.3} \, 147.1 = 8.76 \text{ k}$$

The upper and lower limits for F_{p2} need to be checked:

$$0.2S_{DS}Iw_{px} \leq F_{px} \leq 0.4S_{DS}Iw_{px}$$

$$0.2(0.299)1.00(147.1) \leq F_{p2} \leq 0.4(0.299)1.00(147.1)$$

$$8.79 \text{ k} \leq F_{p2} \leq 17.59 \text{ k}$$

F_{p2} of 8.76 k falls just below the lower limit of 8.79 k, so 8.79 k is the controlling value. In column 7, this value is converted to a coefficient by dividing by $w_2 = 147.1$ k.

Again, it is appropriate to make some observations regarding the base shear coefficient, the story (shearwall) force coefficients F_x, and the story (diaphragm) force coefficients F_{px}. The relationships observed in Seismic Design Categories D, E, and F are slightly different than in B and C, due to the different F_{px} formula:

1. The coefficients F_x and F_{px} are the same at the roof level.
2. The maximum story coefficients (at the roof level) exceed the magnitude of the base shear coefficient.
3. The minimum value for the F_x story (shearwall) coefficient (at the second-floor level) is less than the base shear coefficient.
4. The minimum value for the F_{px} story (diaphragm) coefficient (at the second-floor level) is equal to the magnitude of the seismic base shear coefficient.

These rules are not limited to two-story structures, and they hold true for multistory buildings in general in Seismic Design Categories D, E, and F.

With all of the *strength level* F_{px} story coefficients determined, the individual distributed forces for designing the horizontal diaphragms can be evaluated. The forces are considered in the following order: roof diaphragm (using F_{px}) and second-floor diaphragm (using F_{px}).

Shear in Roof Diaphragm Using F_{px} Forces

Compare the F_{px} seismic force at the roof level with the wind force to determine which is critical. The uniformly distributed *strength level* seismic force is determined by multiplying the F_{px} story coefficient at the roof level by the weight of a 1-ft-wide strip of roof dead load D. The weight W_1 at the roof level was determined in Example 3.10.

$$w_{upr} = 0.0817W_1 = 0.0817(1420) = 116 \text{ lb/ft}$$

The roof diaphragm is treated as a simple beam spanning between transverse end shearwalls. See Fig. 3.14a. For a simple span the shear is equal to the beam reaction. The unit shear in the roof diaphragm is the total shear in the diaphragm divided by the width of the diaphragm.

$$V_{ur} = R_{ur} = \frac{w_{upr}L}{2} = \frac{116(60)}{2} = 3480 \text{ lb}$$

$$v_{ur} = \frac{V_{ur}}{b} = \frac{3480}{32} = \boxed{109 \text{ lb/ft}}$$

This unit shear may be used with the information in Chap. 9 to design the roof diaphragm.

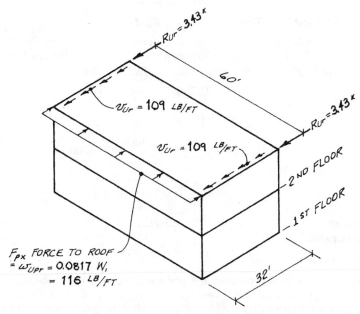

Figure 3.14a Roof diaphragm strength level design force w_{upr} and the corresponding unit shear in the roof diaphragm v_{ur}.

Seismic Force Adjustments

The last two steps in determining the seismic force on an element of the primary LFRS are multiplying by ρ and, as part of the *basic load combinations,* adjusting to an ASD level. The redundancy factor ρ was calculated in Example 3.10.

The corresponding modifications to the roof diaphragm unit shear are:

$$v_{ur} = E = \rho Q_E = 1.0(109) = 109 \text{ lb/ft}$$

and

$$v_r = 0.7(E) = 0.7(109) = 76 \text{ lb/ft}$$

It is important that the designer pay particular attention to whether or not the element force has been adjusted. For this reason, the adjustment is best done at the very end of a problem.

Comparison to Wind Load

The adjusted seismic unit shear in the roof diaphragm can be compared to the corresponding unit shear from the wind load calculated in Example 3.10:

Wind

$$V_r = \frac{W_r}{2} = \frac{3900}{2} = 1950 \text{ lb}$$

$$v_r = \frac{V_r}{b} = \frac{1950}{32} = 61 \text{ lb/ft}$$

$$61 \text{ lb/ft} < 76 \text{ lb/ft}$$

seismic governs

Shear in Second-Floor Diaphragm Using F_{px} Forces

The second-floor diaphragm is analyzed in a similar manner. See Fig. 3.14b. The seismic force is again obtained by multiplying the F_{px} story coefficient from column 7 by the dead load D of a 1-ft-wide strip. The weight W_1 comes from Example 3.10.

$$w_{up2} = 0.0596W_1 = 0.0596(1844)$$

$$= 110 \text{ lb/ft}$$

$$V_{u2} = R_{u2} = \frac{w_{up2}L}{2} = \frac{110(60)}{2} = 3300 \text{ lb}$$

$$v_{u2} = \frac{V_{u2}}{b} = \frac{3300}{32} = \boxed{103 \text{ lb/ft}}$$

This unit shear may be used with the information in Chap. 9 to design the second-floor diaphragm.

Seismic Force Adjustments

The last two steps in determining the seismic force on an element are multiplying by ρ and, as part of the *basic load combinations*, adjusting to an ASD level.

The corresponding modifications to the second-floor diaphragm unit shear are:

$$v_{u2} = E = \rho Q_E = 1.0(103) = 103 \text{ lb/ft}$$

and

$$v_2 = 0.7E = 0.7(103) = 72 \text{ lb/ft}$$

Comparison to Wind Load

The adjusted seismic unit shear in the second-floor diaphragm can be compared to the corresponding unit shear from the wind load, as calculated in Example 3.10.

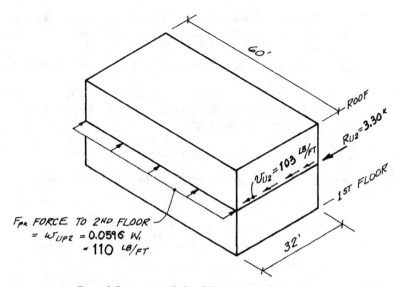

Figure 3.14b Second-floor strength level diaphragm design force w_{up2} and the corresponding unit shear in the second-floor diaphragm v_{u2}.

Wind

$$V_2 = W_2/2 = 5700/2 = 2850 \text{ lb}$$

$$v_2 = V_2/b = 2850/32 = 89 \text{ lb/ft}$$

$$89 \text{ lb/ft} > 72 \text{ lb/ft seismic}$$

> wind governs

In this instance the wind and seismic ASD diaphragm shear stresses v have been compared. In Example 3.10 the wind and seismic loads were compared using the unit applied forces w. Both of these comparisons are valid, and either could have been used in each of these examples. The reader is reminded that it is critical that the proper seismic design force is used in the comparison: F_x for shearwall design and F_{px} for diaphragm design.

3.7 References

[3.1] American Society of Civil Engineers (ASCE). 2002. *Minimum Design Loads for Buildings and Other Structures (ASCE 7-02)*, ASCE, New York, NY.
[3.2] International Codes Council (ICC). 2003. *International Bulding Code*, 2003 ed., ICC, Falls Church, VA.
[3.3] Rood, Roy. 1991. "Panelized Roof Structures," *Wood Design Focus*, Vol. 2, No. 3, Forest Products Society, Madison WI.

3.8 Problems

All problems are to be answered in accordance with the 2003 International Building Code (IBC). A number of Code tables are included in Appendix C for reference.

3.1 The purpose of this problem is to compare the design values of shear and moment for a girder with different assumed load configurations (see Fig. 3.3 in Example 3.2).

Given: The roof framing plan in Fig. 3.A with girders G1, G2, and G3 supporting loads from purlin P1. Roof dead load D = 13 psf. Roof live load L_r is to be obtained from IBC Sec. 1607.11.

Find: *a.* Draw the shear and moment diagrams for girder G1 (D + L_r), assuming
1. A series of concentrated reaction loads from the purlin P1.
2. A uniformly distributed load over the entire span (unit load times the tributary width).
b. Rework part *a* for girder G2.
c. Rework part *a* for girder G3.

3.2   This problem is the same as Prob. 3.1 except that the roof dead load D = 23 psf.

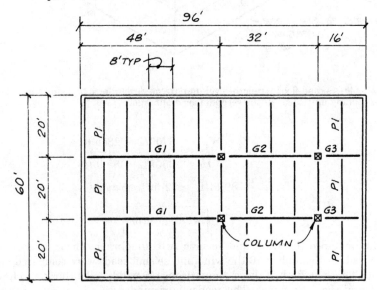

ROOF FRAMING PLAN

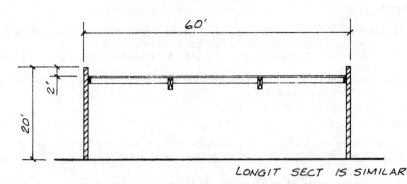

LONGIT SECT IS SIMILAR

TRANSVERSE SECTION

Figure 3.A

3.3 *Given:* IBC Chap. 16 lateral force requirements

   Find: The definition of
   - *a.* Building frame system
   - *b.* LFRS
   - *c.* Shearwall
   - *d.* Braced frame
   - *e.* Bearing wall system

3.4 *Given:* The plan and section of the building in Fig. 3.B. The basic wind speed is 100 mph, and Exposure B applies. The building is enclosed and has a standard occupancy classification. Roof dead load D = 15 psf on a horizontal plane. Wind forces to the primary LFRS are to be in accordance with IBC Sec. 1609.6.

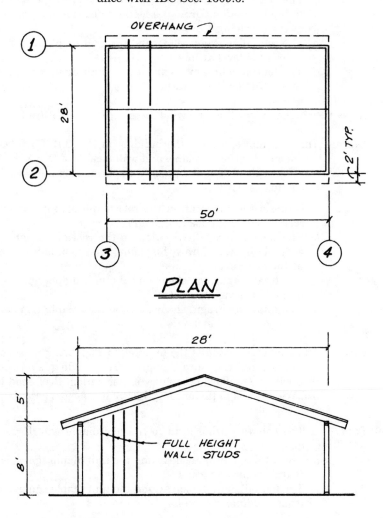

PLAN

SECTION

Figure 3.B

Find: *a.* The wind force on the roof diaphragm in the transverse direction. Draw the loading diagram.
   b. The wind force distribution on the roof diaphragm in the longitudinal direction. Draw the loading diagram.
   c. The total diaphragm shear and the unit diaphragm shear at line 1
   d. The total diaphragm shear and the unit diaphragm shear at line 4

3.5   *Given:*   The plan and section of the building in Fig. 3.B. Roof dead load D = 15 psf on a horizontal plane, and wall dead load D = 12 psf. The seismic diaphragm force F_{px} coefficient has been calculated as 0.200.

Find: *a.* Uniform seismic force on the roof diaphragm in the transverse direction. Draw the loading diagram.
   b. The seismic force distribution on the roof diaphragm in the longitudinal direction. Draw the loading diagram noting the lower force at the overhang.
   c. The total diaphragm shear and the unit diaphragm shear adjusted to an ASD level at line 1
   d. The total diaphragm shear and the unit diaphragm shear adjusted to an ASD level at line 4

3.6   Repeat Prob. 3.4 except that the wind forces are for Exposure C.

3.7   *Given:*   The plan and section of the building in Fig. 3.B. Roof dead load D = 10 psf on a horizontal plane, and wall dead load D = 8 psf. The seismic base shear coefficient and seismic diaphragm force coefficient have been calculated as 0.200.

Find: *a.* Uniform seismic force on the roof diaphragm in the transverse direction. Draw the loading diagram.
   b. The seismic force distribution on the roof diaphragm in the longitudinal direction. Draw the loading diagram, noting the lower force at the overhang.
   c. The total diaphragm shear and the unit diaphragm shear adjusted to an ASD level at line 1
   d. The total diaphragm shear and the unit diaphragm shear adjusted to an ASD level at line 4

3.8   *Given:*   The plan and section of the building in Fig. 3.A. The basic wind speed is 85 mph, and Exposure C applies. The building is an enclosed structure with a standard occupancy classification. Roof dead load D = 13 psf. Wind forces to the primary LFRS are to be in accordance with IBC Sec. 1609.6.

Find: *a.* The tributary wind force to the roof diaphragm. Draw the loading diagram
   b. The total diaphragm shear and the unit diaphragm shear at the 60-ft transverse end walls
   c. The total diaphragm shear and the unit diaphragm shear at the 96-ft longitudinal side walls

3.9   Repeat Prob. 3.8 except that the wind forces are for 120 mph.

3.10 *Given:* The plan and section of the building in Fig. 3.A. The basic wind speed is 85 mph, and Exposure C applies. The building is an enclosed structure with an essential occupancy classification. Roof dead load D = 13 psf.

  Find: *a.* The wind pressure (psf) for designing components and cladding in the roof system away from discontinuities
  b. The tributary wind force to a typical purlin using the load from part *a*. Draw the loading diagram
  c. The wind pressure (psf) for designing an element in the roof system near an eave

3.11 *Given:* The plan and section of the building in Fig. 3.A. Roof dead load D = 10 psf, and the walls are 7½-in.-thick concrete. The building has a bearing wall system, braced with special reinforced concrete shearwalls. The building location is Charleston, South Carolina (ZIP Code 29405), where the mapped short-period spectral acceleration S_S is 1.50g, and the mapped one-second spectral acceleration S_1 is 0.42g. Site Class D, Occupancy Category II, and Seismic Use Group I should be assumed.

  Find: For the transverse direction:
  a. The seismic base shear coefficient and the seismic diaphragm force coefficient.
  b. The uniform force to the roof diaphragm in lb/ft. Draw the loading diagram
  c. The total diaphragm shear and the unit diaphragm shear adjusted to an ASD level adjacent to the transverse walls
  d. The total shear and the unit shear adjusted to an ASD level at the midheight of the transverse shearwalls

3.12 Repeat Prob. 3.11 except that the longitudinal direction is to be considered.

3.13 *Given:* The plan and section of the building in Fig. 3.A. Roof dead load D = 12 psf, and the walls are 6-in.-thick concrete. The building has a bearing wall system, braced with special reinforced concrete shearwalls. The building location is Memphis, Tennessee, where the mapped short-period spectral acceleration S_S is 1.26g, and the mapped one-second spectral acceleration S_1 is 0.38g. Site Class D, Occupancy Category IV, and Seismic Use Group III should be assumed.

  Find: *a.* The seismic base shear and diaphragm force coefficients
  b. The uniform force to the roof diaphragm in lb/ft. Draw the loading diagram
  c. The total diaphragm shear and the unit diaphragm shear adjusted to an ASD level adjacent to the transverse walls
  d. The total shear and the unit shear adjusted to an ASD level at the midheight of the transverse shearwalls

3.14 Repeat Prob. 3.13 except that the longitudinal direction is to be considered.

3.15 *Given:* The elevation of the end shearwall of a building as shown in Fig. 3.C. The force from the roof diaphragm to the shearwall is 10 k. The wall dead load D = 20 psf, and the seismic coefficient is 0.200.

Find: The total shear and the unit shear at the midheight of the wall adjusted to an ASD level

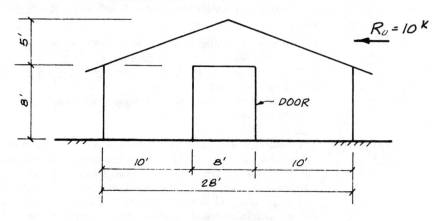

$$R_U = 10^{\,k}$$

DOOR

10'   8'   10'

28'

ELEVATION

Figure 3.C

3.16 *Given:*   The elevation of the side shearwall of a building as shown in Fig. 3.D. The force from the roof diaphragm to the shearwall is 50 k. The wall dead load D = 65 psf, and the seismic base shear coefficient is 0.244.

Find: The total shear and the unit shear adjusted to an ASD level at the midheight of the wall

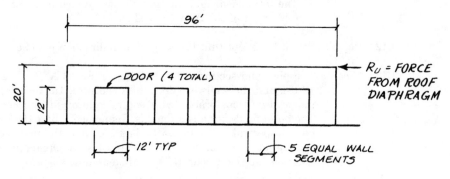

96'

DOOR (4 TOTAL)

$$R_U = \text{FORCE}$$
FROM ROOF
DIAPHRAGM

20'   12'

12' TYP   5 EQUAL WALL
               SEGMENTS

ELEVATION

Figure 3.D

3.17 *Given:*   The elevation of the side shearwall of a building as shown in Fig. 3.D. The force from the roof diaphragm to the shearwall is 43 k. The wall dead load D = 75 psf, and the seismic base shear coefficient is 0.244.

Find: The total shear and the unit shear at the midheight of the wall

3.18 *Given:*   The plan and section of the building in Fig. 3.E. Roof dead load D = 15 psf, floor dead load D = 20 psf (includes an allowance for interior walls), exterior wall dead load D = 53 psf. Basic wind speed = 100

mph. Exposure C and IBC Sec. 1609.6 are specified. The building has a bearing wall system, braced with special reinforced masonry shearwalls. The building is located in Sacramento, California (ZIP Code 95814), with a mapped short-period spectral acceleration of 0.56g and a mapped one-second spectral acceleration of 0.22g. Site Class B, Occupancy Category IV, and Seismic Use Group III should be assumed.

Find: For the transverse direction adjusted to an ASD level:
   a. The unit shear in the roof diaphragm
   b. The unit shear in the floor diaphragm
   c. The unit shear in the second-floor shearwall
   d. The unit shear in the first-floor shearwall

3.19   Repeat Prob. 3.18 except that the longitudinal direction is to be considered.

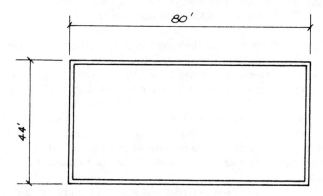

PLAN

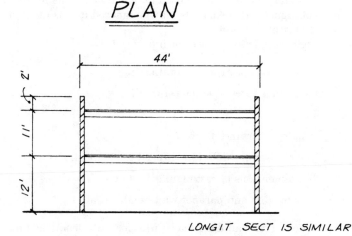

LONGIT SECT IS SIMILAR

TRANSVERSE SECTION

Figure 3.E

3.20 *Given:* The plan and section of the building in Fig. 3.E. Roof dead load D = 10 psf, floor dead load D = 18 psf plus 10 psf for interior partitions, exterior wall dead load D = 16 psf. Basic wind speed = 85 mph. Exposure C and IBC Sec. 1609.6 are specified. Enclosed bearing wall structure has standard occupancy classification. The building has a bearing wall system, braced with wood structural panel shearwalls. The building is located in Sacramento, California, with a mapped short-period spectral acceleration of 0.56g and a mapped one-second spectral acceleration of 0.22g. Site Class B, Occupancy Category IV, and Seismic Use Group III should be assumed. Neglect any wall openings

Find: For the transverse direction adjusted to an ASD level:
   a. The unit shear in the roof diaphragm
   b. The unit shear in the floor diaphragm
   c. The unit shear in the second-floor shearwall
   d. The unit shear in the first-floor shearwall

3.21 Repeat Prob. 3.20 except that the longitudinal direction is to be considered.

3.22 Use a microcomputer spreadsheet to set up the solution of seismic forces for primary lateral-force-resisting system for a multistory building up to four stories.

The LFRS to be considered consists of horizontal diaphragms and shearwalls. The structural systems may be limited to bearing wall systems with wood-frame roof, floor, and wall construction *or* wood-frame roof and floor construction and masonry or concrete walls. Thus, an R of special reinforced 5 or 6.5 (ASCE 7 Table 9.5.2.2) apply.

The spreadsheet is to handle structures without "complications." For example, buildings will be limited to structures that are seismically *regular*. In addition, only the exterior walls will be used for shearwalls, and openings in the horizontal diaphragms and shearwalls may be ignored in this assignment. Wind forces are not part of this problem.

The following is to be used for input:

   Short-period spectral acceleration S_S

   One-second spectral acceleration S_1

   Site Class

   Seismic importance factor I

   Type of bearing wall system (used to establish R)

   Plan dimensions of rectangular building

   Story heights and parapet wall height (if any)

Roof, floor, and wall dead loads; interior wall dead loads may be handled by increasing the floor dead loads.

The spreadsheet is to do the following:
   a. Evaluate the seismic base shear coefficient and numerical value of base shear.
   b. Evaluate the seismic diaphragm force coefficient, if different from the base shear coefficient.

c. Generate the *seismic tables* summarizing the F_x and F_{px} story coefficients.

d. Compute the w_{ux} and w_{upx} uniformly distributed seismic forces to the horizontal diaphragms.

e. Determine the design unit shears in the horizontal diaphragms adjusted to an ASD level.

f. Determine the total shear and unit shear adjusted to an ASD level in the exterior shearwalls between each story level.

Properties of Wood and Lumber Grades

4.1 Introduction

The designer should have a basic understanding of the characteristics of wood, especially as they relate to the functioning of structural members. The terms *sawn lumber* and *solid sawn lumber* are often used to refer to wood members that have been manufactured by cutting a member directly from a log. Other structural members may start as lumber and then undergo additional fabrication processes. For example, small pieces of lumber can be graded into laminating stock then glued and laid up to form larger wood members, known as *glued laminated timbers,* or *glulams.*

Many other wood-based products are available for use in structural applications. Some examples include solid members such as wood poles and timber piles; fabricated components such as trusses, wood I joists, and box beams; and other manufactured products such as wood structural panels (e.g., plywood and oriented strand board), and structural composite lumber (e.g., laminated veneer lumber and parallel strand lumber). A number of these products are recent developments in the wood industry. They are the result of new technology and the economic need to make use of different species and smaller trees that cannot be used to produce solid sawn lumber.

This chapter introduces many of the important physical and mechanical properties of wood. In addition, the sizes and grades of sawn lumber are covered. A number of other wood products are addressed later in this book. For example, glulam is covered in Chap. 5, and the properties and grades of plywood and other wood structural panels are reviewed in Chap. 8. Additionally, structural composite lumber and several types of manufactured components are described, in part, in Chap. 6.

4.2   Design Specification

The *2001 National Design Specification for Wood Construction* (Ref. 4.4) is the basic specification in the United States for the design of wood structures. All or part of the NDS is usually incorporated into the IBC.

Traditionally the NDS has been updated on a 3- to 5-year cycle. Although there have been significant changes from time to time, the usual revisions often involved minor changes and the clarification of certain design principles. The reader should have a copy of the 2001 NDS to follow the discussion in this book. Having a copy of the NDS in order to learn timber design is analogous to having a copy of the steel manual in order to learn structural-steel design. One can read about the subject, but it is difficult to truly develop a feel for the material without both an appropriate text *and* the basic industry publication. In the case of wood design, the NDS is the basic industry document.

The NDS is actually the formal *design section* of what is a series of interrelated design documents. Together, this series of documents comprises a package referred to as the *Allowable Stress Design Manual for Engineered Wood Construction* (Ref. 4.1). The ASD Manual includes what was traditionally considered the two core components of the NDS:

1. The *National Design Specification for Engineered Wood Construction* and
2. The *NDS Supplement* (design values for wood construction)

In addition to this, the ASD Manual includes a collection of additional *supplements* covering:

1. Structural lumber
2. Structural glued laminated timber
3. Timber poles and piles
4. Wood structural panels
5. Wood structural panel shearwalls and diaphragms

A collection of *guidelines* are included in the ASD Manual. *Guidelines* differ from *supplements* in that guidelines address the design and use of proprietary products (fabricated components). The following guidelines are included with the ASD manual:

1. Wood I-joists
2. Structural composite lumber
3. Metal plate connected wood trusses
4. Pre-engineered metal connectors

The ASD package includes what is specifically referred to as the *Manual*. The Manual provides additional guidance for the design of some of the most commonly used components of wood-frame buildings. The last component of the ASD package is a separate supplement dealing with *special design provisions for wind and seismic design*.

The *NDS design section* covers the basic principles of wood engineering that are applied to *all* products and species groups. The design section is written and published by the American Forest and Paper Association (AF&PA) with input from the wood industry, government agencies, universities, and the structural engineering profession.

Many of the chapters in *this* book deal with the provisions in the design section of the NDS, including procedures for beams, columns, members with combined stress, and connections. The study of each of these topics should be accompanied by a review of the corresponding section in the NDS. To facilitate this, a number of sections and tables in the 2001 NDS are referenced throughout the discussion in this book.

The designer should be aware of changes and additions for any new editions of a design specification or code. For the 2001 edition of the NDS, some of the revisions include the following: new chapters covering prefabricated wood I-joists, structural composite lumber, wood structural panels, shearwalls and diaphragms, and fire design. Prior to the 2001 NDS, design using prefabricated I-joists, structural composite lumber, and wood structural panels was not explicitly included in the NDS. Furthermore, the design of shearwalls and diaphragms, while comprising the primary lateral-force-resisting system (LFRS) in most wood structures, was also not explicitly part of the NDS. These additions allow the designer one primary industry document to cover the design of wood structures. In addition to these new chapters, the 2001 NDS has significantly revised chapters on connection design. While the basis for design with mechanical fasteners remains largely unchanged, the equations and procedures have been unified. Previously, the designer was presented with a set of design equations for each connector type (e.g., nails, screws, bolts, etc.). With the latest revision of the NDS, all dowel-type connections are designed using the same set of equations (see Chaps. 11 through 13).

The second part of the NDS is traditionally referred to simply as the *NDS Supplement*, even though there are now several supplements to the NDS. The NDS Supplement contains the numerical values of design stresses for the various species groupings of structural lumber and glued-laminated timber. Although the Supplement is also published by AF&PA, the mechanical properties for sawn lumber are obtained from the agencies that write the grading rules for structural lumber. See Fig. 4.1. There are currently seven rules-writing agencies for visually graded lumber that are certified by the American Lumber Standards Committee. The design values in the NDS Supplement are reviewed and approved by the American Lumber Standards Committee.

The NDS Supplement provides *tabulated design values* for the following mechanical properties:

Bending stress F_b

Tension stress parallel to grain F_t

1. Northeastern Lumber Manufacturers Association (NELMA)
2. Northern Softwood Lumber Bureau (NSLB)
3. Redwood Inspection Service (RIS)
4. Southern Pine Inspection Bureau (SPIB)
5. West Coast Lumber Inspection Bureau (WCLIB)
6. Western Wood Products Association (WWPA)
7. National Lumber Grades Authority (NLGA), a Canadian agency

Figure 4.1 List of rules-writing agencies for visually graded structural lumber. The addresses of the American Lumber Standards Committee (ALSC) and the seven rules-writing agencies are listed in the NDS Supplement.

Shear stress F_v

Compression stress parallel to grain F_c

Compression stress perpendicular to grain $F_{c\perp}$

Modulus of elasticity E (some publications use the notation MOE)

An important part of timber design is being able to locate the proper design value in the tables. The reader is encouraged to verify the numerical values in the examples given throughout this book. As part of this process, it is suggested that tabulated values be checked against the NDS Supplement. Some of the stress adjustments are found in the NDS Supplement, and others are in the NDS design section. An active review of the numerical examples will require use of both.

The material in the 2001 NDS design section represents the latest in wood design principles for allowable stress design. Likewise, the design values in the 2001 NDS Supplement are the most recent structural properties. It should be understood that these two sections of the NDS are an integrated package. In other words, both parts of the 2001 NDS should be used together, and the user should not mix the design section from one edition with the supplement from another.

Traditionally, a third part of the NDS is known as the Commentary (Ref. 4.2). First introduced in 1993 for the 1991 NDS, the Commentary provides background information regarding the provisions of the NDS. Included are discussions of the historical development of NDS provisions, example problems, and tables comparing current design provisions with provisions of earlier editions of the NDS. Currently, the NDS Commentary is applicable to the 1997 NDS and is, therefore, not included as part of the ASD Manual package.

In addition to the traditional core components of the NDS (the design section and the supplement with design values), the 2001 ASD Manual includes five product design supplements, four product guidelines, a manual providing additional guidance for the design of some of the most commonly used components of wood-frame buildings, and a separate supplement dealing specifically with special provisions for wind and seismic design. With the exception of the wind and seismic supplement, these supplements and guidelines are organized according to specific product lines. Each document contains design information and details specific to each product type. The supplements are documents that include complete design information, including design values, which allow the designer to fully design the specific product or system in accordance with the provisions of the NDS. The guidelines cover the design of proprietary product lines. As such, the guidelines do not contain specific design values, which must be obtained from the product supplier, but otherwise the guidelines include information that is required to design with the various products.

The ASD Manual was first introduced in 1999 for the 1997 NDS and for the first time brought together all necessary information required for the design of wood structures. Previous to this, the designer referred to the NDS for the design of solid sawn lumber and glulam members, as well as the design of many connection details. For the design of other wood components and systems, the designer was required to look elsewhere. For example, the design of shearwalls and diaphragms, which comprise the primary lateral-force-resisting system (LFRS) in wood structures, was not covered in the NDS, but was covered through various other sources including publications by APA— The Engineered Wood Association. The 2002 ASD Manual represents a major advancement for the design of engineered wood structures by bringing together all of the necessary design information into a single package.

The reader is cautioned that the ASD Manual and the NDS represent *recommended* design practice by the wood industry, and it does not have *legal authority* unless it becomes part of a local building code. The code change process can be lengthy, and some codes may not accept all of the industry recommendations. Consequently, it is recommended that the designer verify local code acceptance before using the 2001 NDS.

4.3 Methods of Grading Structural Lumber

The majority of sawn lumber is graded by visual inspection, and material graded in this way is known as *visually graded* structural lumber. As the lumber comes out of the mill, a person familiar with the lumber grading rules examines each piece and assigns a grade by stamping the member. The *grade stamp* includes the grade, the species or species group, and other pertinent information. See Fig. 4.2a. If the lumber grade has recognized mechanical properties for use in structural design, it is referred to as a *stress grade*.

The lumber grading rules establish limits on the size and number of growth (strength-reducing) characteristics that are permitted in the various stress grades. A number of the growth characteristics found in full-size pieces of lumber and their effect on strength are discussed later in this chapter.

The term *resawn lumber* is applied to smaller pieces of wood that are cut from a larger member. The resawing of previously graded structural lumber invalidates the initial grade stamp. The reason for this is that the acceptable size of a defect (e.g., a knot) in the original large member may not be permitted in the same grade for the smaller resawn size. The primary example is the changing of a centerline knot into an edge knot by resawing. Restrictions on

(a)

Figure 4.2a Typical grade stamps for visually graded structural lumber. Elements in the grade stamp include (*a*) lumber grading agency (e.g., WWPA), (*b*) mill number (e.g., 12), (*c*) lumber grade (e.g., Select Structural and No. 3), (*d*) commercial lumber species (e.g., Douglas Fir-Larch and Western Woods), (*e*) moisture content at time of surfacing (e.g., S-GRN, and S-DRY). (*Courtesy of WWPA*)

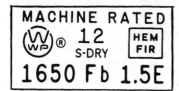

(b)

Figure 4.2b Typical grade stamp for machine stress-rated (MSR) lumber. Elements in the grade stamp include (*a*) MSR marking (e.g., MACHINE RATED), (*b*) lumber grading agency (e.g., WWPA), (*c*) mill number (e.g., 12), (*d*) nominal bending stress (e.g., 1650 psi) and modulus of elasticity (e.g., 1.5×10^6 psi), (*e*) commercial lumber species (e.g., Hem-Fir), (*f*) moisture content at time of surfacing (e.g., S-DRY). (*Courtesy of WWPA*)

edge knots are more severe than those on centerline knots. Thus, if used in a structural application, resawn lumber must be *regraded.*

The designer should be aware that more than one set of grading rules can be used to grade some commercial species groups. For example, Douglas Fir-Larch can be graded under Western Wood Products Association (WWPA) rules or under West Coast Lumber Inspection Bureau (WCLIB) rules. There are some differences in allowable stresses between the two sets of rules. The tables of design properties in the NDS Supplement have the grading rules clearly identified (e.g., WWPA and/or WCLIB). The differences in allowable stresses occur only in large-size members known as *Timbers,* and allowable stresses are the same under both sets of grading rules for *Dimension lumber.* The sizes of Timbers and Dimension lumber are covered later in this chapter.

Because the designer usually does not have control over which set of grading rules will be used, the lower allowable stress should be used in design when conflicting values are listed in NDS tables. The higher allowable stress is justified only if a grade stamp associated with the higher design value actually appears on a member. This situation could arise in reviewing the capacity of an existing member.

Although most lumber is visually graded, a small percentage of lumber is *machine stress rated* by subjecting each piece of wood to a nondestructive test. The nondestructive test is typically highly automated, and the process takes very little time. As lumber comes out of the mill, it passes through a series of rollers. In this process, a bending load is applied about the minor axis of the cross section, and the modulus of elasticity of each piece is measured. In addition to the nondestructive test, machine stress rated lumber is subjected to a visual check. Because of the testing procedure, machine stress rating (MSR) is limited to thin material (2 in. or less in thickness). Lumber graded in this manner is typically known as *MSR lumber*.

Each piece of MSR lumber is stamped with a grade stamp that allows it to be fully identified, and the grade stamp for MSR lumber differs from the stamp for visually graded lumber. The grade stamp for MSR lumber includes a numerical value of nominal bending stress F_b and modulus of elasticity E. A typical grade stamp for MSR lumber is shown in Fig. 4.2b. Tabulated stresses for MSR lumber are given in NDS Supplement Table 4C.

MSR lumber has less variability in some mechanical properties than visually graded lumber. Consequently, MSR lumber is often used to fabricate engineered wood products. For example, MSR lumber is used for laminating stock for some glulam beams. Another application is in the production of wood components such as light frame trusses and wood I joists.

A more recent development in the sorting of lumber by nondestructive measurement of its properties is known as *machine evaluated lumber* (MEL). This process typically employs radiographic (x-ray) inspection to measure density, and allows a greater mix of F_b/E combinations than is permitted in MSR lumber. As with MSR lumber, MEL is subjected to a supplementary visual check. Design values for machine evaluated lumber are also listed in NDS Supplement Table 4C.

4.4   In-Grade Versus Clear-Wood Design Values

Prior to the 1991 edition of the NDS, tabulated design values were determined using clear-wood test procedures. However, design values tabulated in the 1991 NDS were based in part on *clear-wood test procedures* and in part on the full-size lumber *in-grade test methods*. This practice continues for the 2001 NDS.

There are two broad size classifications of sawn lumber:

1. Dimension lumber
2. Timbers

Dimension lumber is the smaller (thinner) sizes of structural lumber. Dimension lumber usually range in size from 2 × 2 through 4 × 16. In other words, dimension lumber constitutes any material that has a nominal *thickness* of 2 to 4 in. Note that in lumber grading terminology *thickness* refers to the smaller cross-sectional dimension of a piece of wood and *width* refers to the larger dimension. The availability of lumber in the wider widths varies with species, and not all sizes are available in all species.

Timbers are the larger sizes and have a 5-in. minimum nominal dimension. Thus, practically speaking, the smallest size timber is a 6 × 6, and any member larger than a 6 × 6 is classified as a timber. There are additional size categories within both Dimension lumber and Timbers, and the further subdivision of sizes is covered later in this chapter.

In the 2001 NDS Supplement, the design properties for visually graded sawn lumber are based on two different sets of ASTM standards:

1. *In-grade* procedures (ASTM D 1990), applied to Dimension lumber (Ref. 4.15)

2. *Clear-wood* procedures (ASTM D 2555 and D 245), applied to Timbers (Ref. 4.13 and 4.14)

The method of establishing design values for visually graded sawn lumber for *Timbers* is based on the *clear-wood strength* of the various species and species combinations. The clear-wood strength is determined by testing small, clear, straight-grained specimens of a given species. For example, the clear-wood bending strength test is conducted on a specimen that measures $2 \times 2 \times 30$ in. The testing methods to be used on small, clear-wood specimens are given in ASTM D 143 (Ref. 4.16). The unit strength (stress) of a small, clear, straight-grained piece of wood is much greater than the unit strength of a full-size member.

After the clear-wood strength properties for the species have been determined, the effects of the natural growth characteristics that are permitted in the different grades of full-size members are taken into account. This is accomplished by multiplying the clear-wood values by a reduction factor known as a *strength ratio*. In other words, the strength ratio takes into account the various strength-reducing defects (e.g., knots) that may be present.

As noted, the procedure for establishing tabulated design values using the clear-wood method is set forth in ASTM Standards D 245 and D 2555 (Refs. 4.14 and 4.13). Briefly, the process involves the following:

1. A statistical analysis is made of a large number of clear-wood strength values for the various commercial species. With the exception of $F_{c\perp}$ and E, the 5 *percent exclusion value* serves as the starting point for the development of allowable stresses. The 5 percent exclusion value represents a strength property (e.g., bending strength). Out of 100 clearwood specimens, 95 could be expected to fail at or above the 5 percent exclusion value, and 5 could be expected to fail below this value.

2. For Dimension lumber, the 5 percent exclusion value for an unseasoned specimen is then increased by an appropriate seasoning adjustment factor to a moisture content of 19 percent or less. This step is not applicable to Timbers since design values for Timbers are for unseasoned conditions.

3. Strength ratios are used to adjust the clear-wood values to account for the strength-reducing defects permitted in a given stress grade.

4. The stresses are further reduced by a general adjustment factor which accounts for the duration of the test used to establish the initial clearwood values, a manufacture and use adjustment, and several other factors.

The combined effect of these adjustments is to provide an average factor of safety on the order of 2.5. Because of the large number of variables in a wood member, the factor of safety for a given member may be considerably larger or smaller than the average. However, for 99 out of 100 pieces, the factor of safety will be greater than 1.25, and for 1 out of 100, the factor of safety will exceed 5. References 4.20, 4.21, and 4.25 give more details on the development of mechanical properties using the clear-wood strength method.

In 1978 a large research project, named the *In-Grade Testing Program,* was undertaken jointly between the lumber industry and the U.S. Forest Products Laboratory (FPL). The purpose of the In-Grade Program was to test full-size Dimension lumber that had been graded in the usual way. The grading rules for the various species did not change, and as the name "In-Grade" implies, the members tested were representative of lumber available in the marketplace. Approximately 73,000 pieces of full-size *Dimension lumber* were tested in bending, tension, and compression parallel to grain in accordance with ASTM D 4761 (Ref. 4.17). Relationships were also developed between mechanical properties and moisture content, grade, and size.

The objective of the In-Grade Program was to verify the published design values that had been determined using the clear-wood strength method. Although some of the values from the In-Grade Testing Program were close, there were enough differences between the In-Grade results and the clear-wood strength values that a new method of determining allowable stress properties was developed. These procedures are given in ASTM Standard D 1990 (Ref. 4.15).*

This brief summary explains why the design values for *Dimension lumber* and *Timbers* are published in separate tables in the NDS Supplement. As a practical matter, the designer does not need the ASTM standards to design a wood structure. The ASTM standards simply document the methods used by the rules-writing agencies to develop the tabulated stress properties listed in the NDS Supplement.

4.5   Species and Species Groups

A large number of *species* of trees can be used to produce structural lumber. As a general rule, a number of species are grown, harvested, manufactured, and marketed together. From a practical standpoint, the structural designer uses lumber from a commercial *species group* rather than a specific individual species. The same grading rules, tabulated stresses, and grade stamps are applied to all species in the species group. Tabulated stresses for a species group were derived using statistical procedures that ensure conservative values for all species in the group.

In some cases, the mark of one or more *individual* species may be included in the grade stamp. When one or more species from a species *group* are identified in the grade stamp, the allowable stresses for the species group are the appropriate stresses for use in structural design. In other cases, the grade

---

*ASTM D 1990 does not cover shear and compression perpendicular to grain. Therefore, values of F_v and $F_{c\perp}$ for Dimension lumber are obtained from ASTM D 2555 and D 245.

stamp on a piece of lumber may reflect only the name of the *species group,* and the actual species of a given piece will not be known. Special knowledge in wood identification would be required to determine the individual species.

The 2001 NDS Supplement contains a complete list of the species groups along with a summary of the various individual species of trees that may be included in each group. Examples of several commonly used species groups are shown in Fig. 4.3. Individual species as well as the species groups are shown.

It should be noted that there are a number of species groups that have similar names [e.g., Douglas Fir-Larch *and* Douglas Fir-Larch (N), Hem-Fir *and* Hem-Fir (N), and Spruce-Pine-Fir *and* Spruce-Pine-Fir (S)]. It is important to understand that each is a separate and distinct species group, and there are different sets of tabulated stresses for each group. Different properties may be the result of the trees being grown in different geographical locations. However, there may also be different individual species included in combinations with similar names.

The choice of species for use in design is typically a matter of economics. For a given location, only a few species groups will be available, and a check with local lumber distributors or a wood products agency will narrow the selection considerably. Although the table in Fig. 4.3 identifies only a small number of the commercial lumber species, those listed account for much of the total volume of structural lumber in North America.

The species of trees used for structural lumber are classified as hardwoods and softwoods. These terms are not necessarily a description of the physical properties of the wood, but are rather classifications of trees. *Hardwoods* are broadleafed deciduous trees. *Softwoods,* on the other hand, have narrow, needlelike leaves, are generally evergreen, and are known as conifers. By far the large majority of structural lumber comes from the softwood category.

For example, Douglas Fir-Larch and Southern Pine are two species groups that are widely used in structural applications. Although these contain species that are all classified as softwoods, they are relatively dense and have structural properties that exceed those of many hardwoods.

It has been noted that the lumber grading rules establish the limits on the strength-reducing characteristics permitted in the various lumber grades. Before discussing the various stress grades, a number of the natural growth characteristics found in lumber will be described.

4.6   Cellular Makeup

As a biological material, wood represents a unique structural material because its supply can be renewed by growing new trees in forests which have been harvested. Proper forest management is necessary to ensure a continuing supply of lumber.

Wood is composed of elongated, round, or rectangular tubelike cells. These cells are much longer than they are wide, and the length of the cells is essentially parallel with the length of the tree. The cell walls are made up of *cellulose,* and the cells are bound together by material known as *lignin.*

Species Group Name and Group Name Mark that may appear in grade stamp	Individual Species that may be included in the Species Group and Individual Species Mark that may appear in grade stamp	Notes
Douglas Fir-Larch **DOUG. FIR-L**	Douglas Fir Western Larch	Individual species mark for Douglas Fir may also appear as "DOUGLAS FIR" or "D. FIR"
Douglas Fir-Larch (N) **D. FIR(N)**	Douglas Fir[1] Western Larch[1]	(N) - indicates a Canadian species group
Douglas Fir South	Douglas Fir South	South indicates Douglas Fir grown in Arizona, Colorado, Nevada, New Mexico, and Utah
Hem-Fir **HEM FIR**	California Red Fir[1] Grand-Fir[1] Noble Fir[1] Pacific Silver Fir[1] Western Hemlock HEM White Fir[1]	
Hem-Fir (N) **HEM-FIR-N**	Amabilis Fir Western Hemlock	(N) - indicates a Canadian species group
Southern Pine **SYP**	Loblolly Pine[1] Longleaf Pine[1] Shortleaf Pine[1] Slash Pine[1]	Group mark is not used when graded under Southern Pine Inspection Bureau - grade stamp will show: **SPIB**
Spruce-Pine-Fir **S-P-F**	Alpine Fir[1] Balsam Fir[1] Black Spruce[1] Englemann Spruce[1] Jack Pine[1] Lodgepole Pine[1] Red Spruce[1] White Spruce[1]	Canadian species group
Spruce-Pine-Fir (S) **SPF[S]**	Balsam Fir[2] Eastern Spruce[2] Englemann Spruce ES Jack Pine[1] Lodgepole Pine LP Red Pine[2] Sitka Spruce	(S) - indicates USA species group (established 1991). Eastern Spruce is any combination of Black Spruce, Red Spruce, and White Spruce.

Figure 4.3 Typical species groups of structural lumber. These and a number of additional species groups are given in the NDS Supplement. The species groups listed here account for a large percentage of the structural lumber sold in the United States. Also shown are the individual species that may be included in a given species group.

If the cross section of a log is examined, concentric rings are seen. One ring represents the amount of wood material which is deposited on the outside of the tree during one growing season. One ring then is termed an *annual ring*. See Fig. 4.4.

The annual rings develop because of differences in the wood cells that are formed in the early portion of the growing season compared with those formed toward the end of the growing season. Large, thin-walled cells are formed at the beginning of the growing season. These are known as *early-wood* or *spring-wood* cells. The cells deposited on the outside of the annual ring toward the end of the growing season are smaller, have thicker walls, and are known as *latewood* or *summerwood* cells. It should be noted that annual rings occur only in trees that are located in climate zones which have distinct growing seasons. In tropical zones, trees produce wood cells which are essentially uniform throughout the entire year.

Because summerwood is denser than springwood, it is stronger (the more solid material per unit volume, the greater the strength of the wood). The annual rings, therefore, provide one of the *visual* means by which the strength of a piece of wood may be evaluated. The more summerwood in relation to the amount of springwood (other factors being equal), the stronger the piece of lumber. This comparison is normally made by counting the number of growth rings per unit width of cross section.

In addition to annual rings, two different colors of wood may be noticed in the cross section of the log. The darker center portion of the log is known as *heartwood*. The lighter portion of the wood near the exterior of the log is known as *sapwood*. The relative amount of heartwood compared with sapwood varies with the species of tree. Heartwood, because it occurs at the center of the tree, is obviously much older than sapwood, and, in fact, heart-wood represents wood cells which are inactive. These cells, however, provide strength and support to the tree. Sapwood, on the other hand, represents both living and inactive wood cells. Sapwood is used to store food and transport water. The strength of heartwood and sapwood is essentially the same. Heartwood

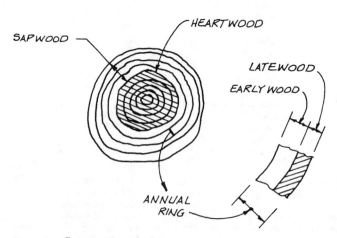

Figure 4.4 Cross section of a log.

is somewhat more decay-resistant than sapwood, but sapwood more readily accepts penetration by wood-preserving chemicals.

4.7 Moisture Content and Shrinkage

The *solid* portion of wood is made of a complex cellulose-lignin compound. The cellulose comprises the framework of the cell walls, and the lignin cements and binds the cells together.

In addition to the solid material, wood contains moisture. The moisture content (MC) is measured as the percentage of water to the oven dry weight of the wood:

$$MC = \frac{\text{moist weight} - \text{oven dry weight}}{\text{oven dry weight}} \times 100 \text{ percent}$$

The moisture content in a living tree can be as high as 200 percent (i.e., in some species the weight of water contained in the tree can be 2 times the weight of the solid material in the tree). However, the moisture content of structural lumber in service is much less. The average moisture content that lumber assumes in service is known as the *equilibrium moisture content* (EMC). Depending on atmospheric conditions, the EMC of structural framing lumber in a covered structure (dry conditions) will range somewhere between 7 and 14 percent. In most cases, the MC at the time of construction will be higher than the EMC of a building (perhaps 2 times higher). See Example 4.1.

EXAMPLE 4.1   Bar Chart Showing Different MC Conditions

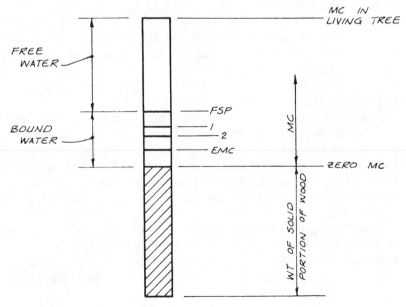

Figure 4.5

Figure 4.5 shows the moisture content in lumber in comparison with its solid weight. The values indicate that the lumber was manufactured (point 1) at an MC below the fiber saturation point. Some additional drying occurred before the lumber was used in construction (point 2). The EMC is shown to be less than the MC at the time of construction. This is typical for most buildings.

Moisture is held within wood in two ways. Water contained in the cell cavity is known as *free water*. Water contained within the cell walls is known as *bound water*. As wood dries, the first water to be driven off is the free water. The moisture content that corresponds to a complete loss of free water (with 100 percent of the bound water remaining) is known as the *fiber saturation point* (FSP). No loss of bound water occurs as lumber dries above the fiber saturation point. In addition, no volume changes or changes in other structural properties are associated with changes in moisture content above the fiber saturation point.

However, with moisture content changes below the fiber saturation point, bound water is lost and volume changes occur. If moisture is lost, wood shrinks; if moisture is gained, wood swells. Decreases in moisture content below the fiber saturation point are accompanied by increases in strength properties. Prior to the In-Grade Testing Program, it was generally believed that the more lumber dried, the greater would be the increase in strength. However, results from the In-Grade Program show that strength properties peak at around 10 to 15 percent MC. For a moisture content below this, member strength capacities remain about constant.

The fiber saturation point varies with species, but the *Wood Handbook* (Ref. 4.22) indicates that 30 percent is average. Individual species may differ from the average. The drying of lumber in order to increase its structural properties is known as *seasoning*. As noted, the MC of lumber in a building typically decreases after construction until the EMC is reached. Although this drying in service can be called seasoning, the term *seasoning* often refers to a controlled drying process. Controlled drying can be performed by air or kiln drying (KD), and both increase the cost of lumber.

From this discussion it can be seen that there are opposing forces occurring as wood dries below the fiber saturation point. On one hand, shrinkage decreases the size of the cross section with a corresponding reduction in section properties. On the other hand, a reduction in MC down to approximately 15 percent increases most structural properties. The net effect of a decrease in moisture content in the 10 to 30 percent range is an overall increase in structural capacity.

Shrinkage can also cause cracks to form in lumber. As lumber dries, the material near the surface of the member loses moisture and shrinks before the wood at the inner core. Longitudinal cracks, known as *seasoning checks,* may occur near the neutral axis (middle of wide dimension) of the member as a result of this nonuniform drying process. See Fig. 4.6*a* in Example 4.2. Cracking of this nature causes a reduction in shear strength which is taken into account in the lumber grading rules and tabulated design values. This type of behavior is more common in thicker members.

EXAMPLE 4.2   Shrinkage of Lumber

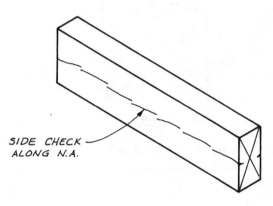

Figure 4.6*a* Seasoning checks may occur in the wide side of a member at or near the neutral axis. These cracks form because wood near the surface dries and shrinks first. In larger pieces of lumber, the inner core of the member loses moisture and shrinks much slower. Checking relieves the stresses caused by nonuniform drying.

SIDE CHECK
ALONG N.A.

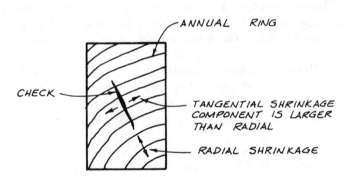

ANNUAL RING

CHECK

TANGENTIAL SHRINKAGE
COMPONENT IS LARGER
THAN RADIAL

RADIAL SHRINKAGE

LUMBER SECTION

Figure 4.6*b* Tangential shrinkage is greater than radial shrinkage. This promotes the formation of radial cracks known as *end checks*.

The *Wood Handbook* (Ref. 4.22) lists average clear-wood shrinkage percentages for many individual species of wood. Tangential shrinkage is greatest. Radial shrinkage is on the order of one-half of the tangential value, but is still significant. Longitudinal shrinkage is very small and is usually disregarded.

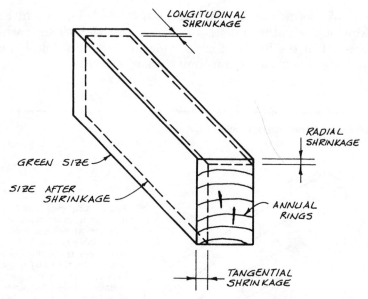

Figure 4.6c Tangential, radial, and longitudinal shrinkage.

Another point should be noted about the volume changes associated with shrinkage. The dimensional changes as the result of drying are not uniform. Greater shrinkage occurs parallel (tangent) to the annual ring than normal (radial) to it. See Fig. 4.6b. These nonuniform dimensional changes may cause radial checks.

In practice, the orientation of growth rings in a member will be arbitrary. In other words, the case shown in Fig. 4.6c is a rather unique situation with the annual rings essentially parallel and perpendicular to the sides of the member. Annual rings can be at any angle with respect to the sides of the member.

It is occasionally necessary for the designer to estimate the amount of shrinkage or swelling that may occur in a structure. The more common case involves shrinkage of lumber as it dries in service. Several approaches may be used to estimate shrinkage.

One method comes from the *Wood Handbook* (Ref. 4.22). Values of tangential, radial, and volumetric shrinkage from clear-wood samples are listed for many individual species. The shrinkage percentages are assumed to take place from no shrinkage at a nominal FSP of 30 percent to full shrinkage at zero MC. A linear interpolation is used for shrinkage at intermediate MC values. See Fig. 4.6c. The maximum shrinkage can be estimated using the tangential shrinkage, and the minimum can be evaluated with the radial value. Thus, the method from the *Wood Handbook* can be used to bracket the probable shrinkage.

A second approach to shrinkage calculations is given in Ref. 4.23. It provides formulas for calculating the percentage of shrinkage for the *width* and *thickness* of a piece of lumber. This method was used for shrinkage adjustments for the In-Grade Test data and is included in the appendix to ASTM D 1990 (Ref. 4.15).

In structural design, there are several reasons why it may be more appropriate to apply a simpler method for estimating the shrinkage than either of the two methods just described:

1. Shrinkage is a variable property. The shrinkage that occurs in a given member may be considerably different from those values obtained using the published *average* radial and tangential values.

2. Orientation of the annual rings in a real piece of lumber is unknown. The sides of a member are probably not parallel or perpendicular to the growth rings.

3. The designer will probably know only the *species group,* and the *individual species* of a member will probably not be known.

For these and perhaps other reasons, a very simple method of estimating shrinkage in structural lumber is recommended in Ref. 4.28. In this third approach, a constant shrinkage value of 6 percent is used for both the width and the thickness of a member. The shrinkage is taken as zero at a FSP of 30 percent, and the full 6 percent shrinkage is assumed to occur at a MC of zero. A linear relationship is used for MC values between 30 and 0. See Example 4.3.

Although Ref. 4.28 deals specifically with western species lumber, the recommended general shrinkage coefficient should give reasonable estimates of shrinkage in most species.

To carry out the type of shrinkage estimate illustrated in Example 4.3, the designer must be able to establish reasonable values for the initial and final moisture content for the lumber. The *initial moisture content* is defined to some extent by the specification for the lumber for a particular job. The general MC range at the time of manufacture is shown in the grade stamp, and this value needs to be reflected in the lumber specification for a job.

The *grade stamp* on a piece of lumber will contain one of three MC designations, which indicates the condition of the lumber at the time of manufacture. *Dry lumber* is defined as lumber having a moisture content of 19 percent or less. Material with a moisture content of over 19 percent is defined as *unseasoned* or *green lumber.*

EXAMPLE 4.3   Simplified Method of Estimating Shrinkage

Estimate the shrinkage that will occur in a four-story wood-frame wall that uses Hem-Fir framing lumber. Consider a decrease in moisture content from 15 to 8 percent.

Framing is typical platform construction with 2×12 floor joists resting on bearing walls. Wall framing is conventional $2 \times$ studs with a typical single $2 \times$ bottom plate and double $2 \times$ top plates. See Fig. 4.7.

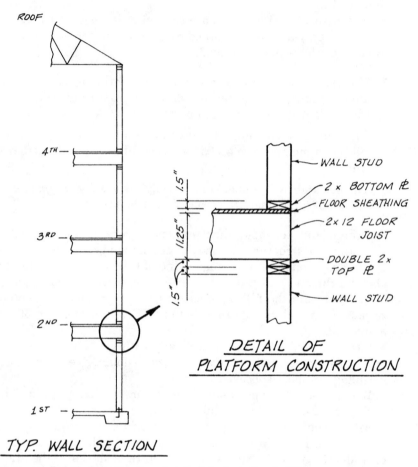

Figure 4.7 Details for estimating shrinkage in four-story building.

The species group of Hem-Fir is given. The list in Fig. 4.3 indicates that any one of six species may be grade-marked with the group name of Hem-Fir. If the individual species is known, the shrinkage coefficients from *Wood Handbook* (Ref. 4.22) could be used to bracket the total shrinkage, using tangential and radial values. However, for practical design purposes, the simplified approach from Ref. 4.28 is used to develop a design estimate of the shrinkage.

A shrinkage of 6 percent of the member dimension is assumed to occur between MC = 30 percent and MC = 0 percent. Linear interpolation allows the shrinkage value (SV) per unit (percent) change in moisture content to be calculated as

Shrinkage value SV = $\frac{6}{30}$ = 0.2 percent per 1 percent change in MC

$$= 0.002 \text{ in./in. per 1 percent change in MC}$$

The shrinkage S that occurs in the dimension d of a piece is calculated as the shrinkage value times the dimension times the change in moisture content:

$$\text{Shrinkage } S = \text{ SV} \times d \times \Delta_{\text{MC}}$$

$$= 0.002 \times d \times \Delta_{\text{MC}}$$

Shrinkage in the depth of one 2×12 floor joist:

$$S_{\text{floor}} = 0.002 \times d \times \Delta_{\text{MC}}$$

$$= 0.002 \times 11.25 \times (15 - 8) = 0.158 \text{ in.}$$

Shrinkage in the thickness* of one $2 \times$ wall plate.

$$S_{\text{plate}} = 0.002 \times d \times \Delta_{\text{MC}}$$

$$= 0.002 \times 1.5 \times (15 - 8) = 0.021 \text{ in.}$$

Shrinkage in the length of a stud: The longitudinal shrinkage of a piece of lumber is small.

$$S_{\text{stud}} \approx 0$$

The first floor is a concrete slab. The second, third, and fourth floors each use 2×12 floor joists (three total). There is a $2 \times$ bottom plate on the first, second, third, and fourth floors (four total). There is a double $2 \times$ plate on top of the first-, second-, third-, and fourth-floor wall studs (a total of eight $2 \times$ top plates).

$$\text{Total } S = \sum S = 3(S_{\text{floor}}) + 12(S_{\text{plate}})$$

$$= 3(0.158) + (4 + 8)(0.021)$$

$$\boxed{\text{Total } S = 0.725 \text{ in.} \approx \tfrac{3}{4} \text{ in.}}$$

When unseasoned lumber is grade-stamped, the term "S-GRN" (surfaced green) will appear. "S-DRY" (surfaced dry) or "KD" (kiln dried) indicates that the lumber was manufactured with a moisture content of 19 percent or less. Refer to the sample grade stamps in Fig. 4.2a for examples of these markings. Some smaller lumber sizes may be seasoned to 15 percent or less in moisture

---

*In lumber terminology, the larger cross-sectional dimension of a piece of wood is known as the width, and the smaller is the thickness.

content and marked "MC 15," or "KD 15." It should be understood that larger-size wood members (i.e., Timbers) are not produced in a dry condition. The large cross-sectional dimensions of these members would require an excessive amount of time for seasoning.

In addition to the MC range reflected in the grade stamp, the initial moisture content of lumber in place in a structure is affected by a number of variables including the size of the members, time in transit to the job site, construction delays, and time for construction. Reference 4.28 recommends that in practical situations the following assumptions can reasonably be made:

Moisture designation in grade stamp	Initial moisture content assumed in service
S-GRN (MC greater than 19 percent at time of manufacture)	19 percent
S-DRY or KD (MC of 19 percent or less at time of manufacture)	15 percent

It should be noted that these recommendations are appropriate for relatively thin material (e.g., the 2 × floor joists and wall plates in Example 4.3). However, larger-size members will dry slower. The designer should take this and other possible factors into consideration when estimating the initial moisture content for shrinkage calculations.

The *final moisture content* can be taken as the equilibrium moisture content (EMC) of the wood. Various surveys of the moisture content in existing buildings have been conducted, and it was previously noted that the EMC in most buildings ranges between 7 and 14 percent.

Reference 4.9 gives typical EMC values of several broad atmospheric zones. The average EMC for framing lumber in the "dry southwestern states" (eastern California, Nevada, southern Oregon, southwest Idaho, Utah, and western Arizona) is given as 9 percent. The MC in most covered structures in this area is expected to range between 7 and 12 percent. For the remainder of the United States, the average EMC is given as 12 percent with an expected range of 9 to 14 percent. These values basically agree with the 8 to 12 percent MC suggested in Ref. 4.28.

The average values and the MC ranges can be used to estimate the EMC for typical buildings. Special conditions must be analyzed individually. As an alternative, the moisture content of wood in an existing structure can be measured with a portable, hand-held moisture meter.

The discussion of moisture content and shrinkage again leads to an important conclusion that was mentioned earlier: Wood is a unique structural material, and its behavior must be understood if it is to be used successfully. Wood is not a static material, and significant changes in dimensions can result because of atmospheric conditions.

Even if shrinkage calculations are not performed, the designer should allow for the movement (shrinkage or swelling) that may occur. This may be necessary in a number of cases. A primary concern in structural design is the potential splitting of wood members.

Wood is very weak in tension perpendicular to grain. The majority of shrinkage occurs across the grain, and connection details must accommodate this movement. If a connection does not allow lumber to shrink freely, tension stresses perpendicular to grain may develop. Splitting of the member will be the likely result. The proper detailing of connections to avoid built-in stresses due to changes in moisture content is covered in Chaps. 13 and 14.

In addition to the structural failures that may result from cross-grain tension, there are a number of other practical shrinkage considerations. Although these may not affect structural safety, they may be crucial to the proper functioning of a building. Consider the shrinkage evaluated in the multistory structure in Example 4.3. Several types of problems could occur.

For example, consider the effect of ceiling joists, trusses, or roof beams supported by a four-story wood-frame wall on one end *and* a concrete or masonry wall on the other end. Shrinkage will occur in the wood wall but not in the concrete or masonry. Thus, one end of the member in the top level will eventually be ¾ in. lower than the other end. This problem occurs as the result of *differential* movement.

Even greater differential movement problems can occur. Consider an all-wood-frame building again, of the type in Example 4.3. In all-wood construction, the shrinkage will be uniform throughout. However, consider the effect of adding a short-length concrete block wall (say, a stair tower enclosure) in the middle of one of the wood-frame walls. The differential movement between the wood-frame wall and the masonry wall now takes place in a very short distance. Distress of ceiling, floor, and wall sheathing will likely develop.

Other potential problems include the possible buckling of finish wall siding. Even if the shrinkage is uniform throughout a wall, there must be sufficient clearance in wall covering details to accommodate the movement. This may require slip-type architectural details (for example, Z flashing). In addition, plumbing, piping, and electrical and mechanical systems must allow for the movement due to shrinkage. This can be accomplished by providing adequate clearance or by making the utilities flexible enough to accommodate the movement without distress. See Ref. 4.28 for additional information.

4.8 Effect of Moisture Content on Lumber Sizes

The moisture content of a piece of lumber obviously affects the cross-sectional dimensions. The width and depth of a member are used to calculate the section properties used in structural design. These include area A, moment of inertia I, and section modulus S.

Fortunately for the designer, it is not necessary to compute section properties based on a consideration of the initial MC and EMC and the resulting shrinkage (or swelling) that occurs in the member. Grading practices for *Dimension lumber* have established the *dry* size (MC $\leq$ 19 percent) of a member as the basis for structural calculations. This means that only one set of cross-sectional properties needs to be considered in design.

This is made possible by manufacturing lumber to different cross-sectional dimensions based on the moisture content of the wood at the time of manufacture. Therefore, lumber which is produced from green wood will be somewhat larger at the time of manufacture. However, when this wood reaches a dry moisture content condition, the cross-sectional dimensions will closely coincide with those for lumber produced in the dry condition. Again, this discussion has been based on the manufacturing practices for dimension lumber.

Because of their large cross-sectional dimensions, *Timbers* are not produced in a dry condition since an excessive amount of time would be required to season these members. For this reason, cross-sectional dimensions that correspond to a *green* (MC > 19 percent) condition have been established as the basis for design calculations for these members. In addition, tabulated stresses have been adjusted to account for the higher moisture content of timbers.

4.9   Durability of Wood and the Need for Pressure Treatment

The discussion of the moisture content of lumber often leads to concerns about the durability of wood structures and the potential for decay. However, the record is clear. If wood is used properly, it can be a permanent building material. If wood is used incorrectly, major problems can develop, sometimes very rapidly. Again, understanding the material is the key to its proper use. The performance of many classic wood structures (Ref. 4.11) is testimony to the durability of wood in properly designed structures.

Generally, if it is protected (i.e., not exposed to the weather or not in contact with the ground) and is used at a relatively low moisture content (as in most covered structures), wood performs satisfactorily without chemical treatment. Wood is also durable when *continuously* submerged in *fresh* water. However, if the moisture content is high and varies with time, or if wood is in contact with the ground, the use of an appropriate preservative treatment should be considered.

High MC values can occur in wood roof systems over swimming pools and in processing plants with high-humidity conditions. *High* moisture content is generally defined as exceeding 19 percent in sawn lumber and as being 16 percent or greater in glulam. Problems involving high moisture content can also occur in geographic locations with high humidity.

In some cases, moisture can become entrapped in roof systems that have *below-roof insulation*. This type of insulation can create dead-air spaces, and moisture from condensation or other sources may lead to decay. Moisture-related problems have occurred in some flat or nearly flat roofs. To create air movement, a minimum roof slope of $\frac{1}{2}$ in./ft is now recommended for panelized roofs that use below-roof insulation. A number of other recommendations have been developed by the industry to minimize these types of problems. See Refs. 4.8, 4.18, and 4.27.

The issue of mold in buildings, particularly wood-frame buildings, has received considerable attention in recent years. Mold and mildew are often present in buildings where there is excessive moisture. Moist, dark, or low-light environments with stagnant airflow contribute to active mold and mildew growth. Such conditions are common in foundations and basements, but also susceptible are exterior walls and roof or attic areas. Industry associations such as APA—The Engineered Wood Association have produced considerable technical and nontechnical literature for designers seeking to mitigate mold- and mildew-related problems (see Ref. 4.19).

When required for new construction, chemicals can be impregnated into lumber and other wood products by a pressure-treating process. The chemical preservatives prevent or effectively retard the destruction of wood. Pressure treating usually takes place in a large steel cylinder. The wood to be treated is transported into the cylinder on a tram, and the cylinder is closed and filled with a preservative. The cylinder is then subjected to pressure which forces the chemical into the wood.

The chemical does not saturate the complete cross section of the member. Therefore, field cutting and drilling of holes for connections after treating should be minimized. It is desirable to carry out as much fabrication of structural members as possible before the members are treated. The depth of penetration is known as the *treated zone*. The retention of the chemical treatment is measured in lb/ft^3 in the treated zone. The required retention amounts vary with the end use and type of treatment.

Many species, most notably the southern pines, readily accept preservative treatments. Other species, however, do not accept pressure treatments as well and require *incising* to make the treatment effective. In effect, incised lumber has small cuts, or incisions, made into all four sides along its length. The incisions create more surface area for the chemicals to penetrate the wood member, thereby increasing the effectiveness of the pressure treating. Incising, while increasing the effectiveness of preservative treatment, adversely affects many mechanical properties. When incised lumber is used, modification of modulus of elasticity and allowable bending, tension and compression parallel to grain design values must be made. See Section 4.20.

Rather than focusing only on moisture content, a more complete overview of the question of long-term performance and durability recognizes that several instruments can destroy wood. The major ones are

1. Decay

2. Termites

3. Marine borers

4. Fire

Each of these is addressed briefly in this section, but a comprehensive review of these subjects is beyond the scope of this book. Detailed information is available in Refs. 4.20, 4.22, and 4.26. In the case of an existing wood structure that has been exposed to some form of destruction, guidelines are available for its evaluation, maintenance, and upgrading (Ref. 4.10).

Decay is caused by fungi which feed on the cellulose or lignin of the wood. These fungi must have food, moisture (MC greater than approximately 20 percent), air, and favorable temperatures. All of these items are required for decay to occur (even so-called dry rot requires moisture).

If any of the requirements is not present, decay will not occur. Thus, untreated wood that is continuously dry (MC < 20 percent, as in most covered structures), or continuously wet (submerged in fresh water—no air), will not decay. Exposure to the weather (alternate wetting and drying) can set up the conditions necessary for decay to develop. Pressure treatment introduces chemicals that poison the food supply of the fungi.

Termites can be found in most areas of the United States, but they are more of a problem in the warmer-climate areas. Subterranean termites are the most common, but drywood and dampwood species also exist. Subterranean termites nest in the ground and enter wood which is near or in contact with damp ground. The cellulose forms the food supply for termites.

The IBC (Sec. 2304.11.2.1) requires a minimum clearance of 18 in. between the bottom of unprotected floor joists (12 in. for girders) and grade. Good ventilation of crawl spaces and proper drainage also aid in preventing termite attack. Lumber which is near or in contact with the ground, and wall plates on concrete ground-floor slabs and footings, must be pressure-treated to prevent termite attack. (Foundation-grade redwood has a natural resistance and can be used for wall plates.) The same pressure treatments provide protection against decay and termites.

Marine borers are found in salt waters, and they present a problem in the design of marine piles. Pressure-treated piles have an extensive record in resisting attack by marine borers.

A brief introduction to the fire-resistive requirements for buildings was given in Chap. 1. Where necessary to meet building code requirements, or where the designer decides that an extra measure of fire protection is desirable, *fire-retardant-treated wood* may be used. This type of treatment involves the use of chemicals in formulations that have fire-retardant properties. Some of the types of chemicals used are preservatives and thus also provide decay and termite protection. Fire-retardant treatment, however, requires higher concentrations of chemicals in the treated zone than normal preservative treatments.

The tabulated design values in the NDS Supplement apply to both untreated and pressure-preservative-treated lumber. In other words, there is no required stress modification for preservative treatments. The exception to this is if the lumber is incised to increase the penetration of the preservatives and thereby increasing the effectiveness of the pressure treatment. Incising effectively decreases the strength and stiffness and must be accounted for in design when incised lumber is used. *Preservative treatments* are those that guard against decay, termites, and marine borers. The high concentrations of chemicals used in fire-retardant treated lumber will probably require that allowable design values be reduced. However, the reduction coefficients vary with the treating process, and the NDS refers the designer to the company providing the fire-retardant treatment and redrying service for the appropriate factors. The three basic types of pressure preservatives are

1. Creosote and creosote solutions

2. Oilborne treatments (pentachlorophenol and others dissolved in one of four hydrocarbon solvents)

3. Waterborne oxides

There are a number of variations in each of these categories. The choice of the preservative treatment and the required retentions depend on the application. Detailed information on pressure treatments and their uses can be obtained from the American Wood Preservers Institute (AWPI). For the address of AWPI, see the list of organizations in the Nomenclature section.

An introduction to pressure treatments is given in Ref. 4.9. This reference covers fire-resistive requirements and fire hazards as well as preservative and fire-retardant treatments. Reference 4.7 provides a concise summary of preservative treatments. See Ref. 4.26 for additional information on the use of wood in adverse environments.

4.10   Growth Characteristics of Wood

Some of the more important growth characteristics that affect the structural properties of wood are density, moisture content, knots, checks, shakes, splits, slope of grain, reaction wood, and decay. The effects of density, and how it can be measured visually by the annual rings, were described previously. Likewise, moisture content and its effects have been discussed at some length. The remaining natural growth characteristics also affect the strength of lumber, and limits are placed on the size and number of these structural defects permitted in a given stress grade. These items are briefly discussed here.

Knots constitute that portion of a branch or limb that has been incorporated into the main body of the tree. See Fig. 4.8. In lumber, knots are classified by form, size, quality, and occurrence. Knots decrease the mechanical properties of the wood because the knot displaces *clear wood* and because the slope of the grain is forced to deviate around the knot. In addition, stress concentrations occur because the knot interrupts wood fibers. Checking also may occur around the knot in the drying process. Knots have an effect on both tension and compression capacity, but the effect in the tension zone is greater. Lumber grading rules for different species of wood describe the size, type, and distribution (i.e., location and number) of knots allowed in each stress grade.

Checks, shakes, and *splits* all constitute separations of wood fibers. See Fig. 4.9. Checks have been discussed earlier and are radial cracks caused by nonuniform volume changes as the moisture content of wood decreases (Sec. 4.7). Recall that the outer portion of a member shrinks first, which may cause longitudinal cracks. In addition, more shrinkage occurs tangentially to the annual ring than radially. Checks therefore are seasoning defects. Shakes, on

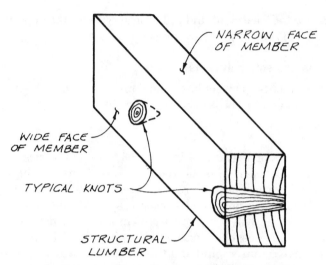

Figure 4.8 Examples of knots. Lumber grading rules for the commercial species have different limits for knots occurring in the wide and narrow faces of the member.

the other hand, are cracks which are usually parallel to the annual ring and develop in the standing tree. Splits represent complete separations of the wood fibers through the thickness of a member. A split may result from a shake or seasoning or both. Splits are measured as the penetration of the split from the end of the member parallel to its length. Again, lumber grading rules provide limits on these types of defects.

The term *slope of grain* is used to describe the deviation of the wood fibers from a line that is parallel to the edge of a piece of lumber. Slope of grain is expressed as a ratio (for example, 1 in 8, 1 in 15, etc.). See Fig. 4.10. In structural lumber, the slope of grain is measured over a sufficient length and area to be representative of the general slope of wood fibers. Local deviations, such as around knots, are disregarded in the general slope measurement. Slope of grain has a marked effect on the structural capacity of a wood member. Lumber grading rules provide limits on the slope of grain that can be tolerated in the various stress grades.

Reaction wood (known as compression wood in softwood species) is abnormal wood that forms on the underside of leaning and crooked trees. It is hard and brittle, and its presence denotes an unbalanced structure in the wood. Compression wood is not permitted in readily identifiable and damaging form in stress grades of lumber.

Decay is a degradation of the wood caused by the action of fungi. Grading rules establish limits on the decay allowed in stress-grade lumber. Section 4.9 describes the methods of preserving lumber against decay attack.

4.11 Sizes of Structural Lumber

Structural calculations are based on the *standard net size* of a piece of lumber. The effects of moisture content on the size of lumber are discussed in Sec. 4.8.

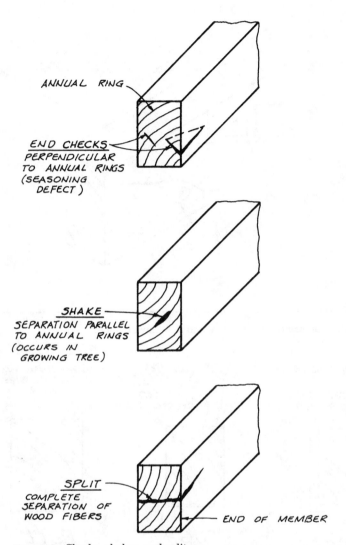

ANNUAL RING

END CHECKS
PERPENDICULAR
TO ANNUAL RINGS
(SEASONING
DEFECT)

SHAKE
SEPARATION PARALLEL
TO ANNUAL RINGS
(OCCURS IN
GROWING TREE)

SPLIT
COMPLETE
SEPARATION OF
WOOD FIBERS

END OF MEMBER

Figure 4.9 Checks, shakes, and splits.

The designer may have to allow for shrinkage when detailing connections, but standard dimensions are accepted for stress calculations.

Most structural lumber is *dressed lumber*. In other words, the lumber is *surfaced* to the standard net size, which is less than the *nominal* (stated) size. See Example 4.4. Lumber is dressed on a planing machine for the purpose of obtaining smooth surfaces and uniform sizes. Typically lumber will be S4S (surfaced four sides), but other finishes can be obtained (for example, S2S1E indicates surfaced two sides and one edge).

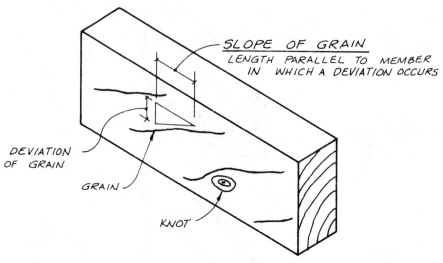

Figure 4.10 Slope of grain.

EXAMPLE 4.4   Dressed, Rough-Sawn, and Full-Sawn Lumber

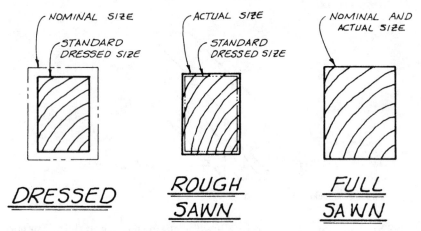

Figure 4.11

Consider an 8 × 12 member (nominal size = 8 in. × 12 in.).

1. *Dressed lumber.* Standard net size = 7½ in. × 11½ in. Refer to NDS Supplement Tables 1A and 1B for dressed lumber sizes.

2. *Rough-sawn lumber.* Approximate size = 7⅝ in. × 11⅝ in. Rough size is approximately ⅛ in. larger than the dressed size.

3. *Full-sawn lumber.* Minimum size 8 in. × 12 in. Full-sawn lumber is not generally available.

Dressed lumber is used in many structural applications, but large timbers are commonly *rough sawn* to dimensions that are close to the standard net sizes. The textured surface of rough-sawn lumber may be desired for architectural purposes and may be specially ordered in smaller sizes. The cross-sectional dimensions of rough-sawn lumber are approximately ⅛ in. larger than the standard dressed size. A less common method of obtaining a rough surface is to specify *full-sawn* lumber. In this case, the actual size of the lumber should be the same as the specified size. Cross-sectional properties for rough-sawn and full-sawn lumber are not included in the NDS because of their relatively infrequent use.

The terminology in the wood industry that is applied to the dimensions of a piece of lumber differs from the terminology normally used in structural calculations. The grading rules refer to the thickness and width of a piece of lumber. It was previously stated that the *thickness* is the smaller cross-sectional dimension, and the *width* is the larger.

However, in the familiar case of a beam, design calculations usually refer to the *width* and *depth* of a member. The width is parallel to the neutral axis of the cross section, and the depth is perpendicular. In most beam problems, the member is loaded about the strong or x axis of the cross section. Therefore, the width of a beam is usually the smaller cross-sectional dimension, and the depth is the larger. Naturally, the strong axis has larger values of section modulus and moment of inertia. Loading a beam about the strong axis is also described as having the *load applied to the narrow face of the beam.*

Another type of beam loading is less common. If the bending stress is about the weak axis or y axis, the section modulus and moment of inertia are much smaller. *Decking* is an obvious application where a beam will have the *load applied to the wide face of the member.* In this case the width is the larger cross-sectional dimension, and the depth is the smaller. As with all structural materials, the objective is to make the most efficient use of materials. Thus, a wood beam is used in bending about the strong axis whenever possible.

The dimensions of sawn lumber are given in the 2001 NDS Supplement Table lA, *Nominal and Minimum Dressed Sizes of Sawn Lumber.* However, a more useful table for design is the list of cross-sectional properties in the NDS Supplement Table lB, *Section Properties of Standard Dressed (S4S) Sawn Lumber.* The properties include nominal and dressed dimensions, area, and section modulus and moment of inertia for both the x and y axes. The section properties for a typical sawn lumber member are verified in Example 4.5. The weight per linear foot for various densities of wood is also given in Table 1B.

EXAMPLE 4.5   Section Properties for Dressed Lumber

Show calculations for the section properties of a 2×8 sawn lumber member. Use standard net sizes for dressed (S4S) lumber, and verify the section properties in NDS Table 1B.

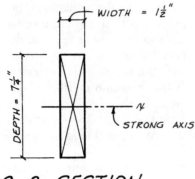

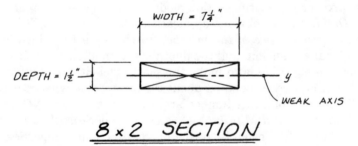

Figure 4.12a Dimensions for section properties about strong x axis of 2×8.

$$2 \times 8 \quad SECTION$$

Section Properties for x Axis

$$A = bd = 1\frac{1}{2} \times 7\frac{1}{4} = 10.875 \text{ in.}^2$$

$$S_x = \frac{bd^2}{6} = \frac{1.5(7.25)^2}{6} = 13.14 \text{ in.}^3$$

$$I_x = \frac{bd^3}{12} = \frac{1.5(7.25)^3}{12} = 47.63 \text{ in.}^4$$

The section properties for the x axis agree with those listed in the NDS Supplement.

$$8 \times 2 \quad SECTION$$

Figure 4.12b Dimensions for section properties about weak y axis of 2×8.

Section Properties for y Axis

$$S_y = \frac{bd^2}{6} = \frac{7.25(1.5)^2}{6} = 2.719 \text{ in.}^3$$

$$I_y = \frac{bd^3}{12} = \frac{7.25(1.5)^3}{12} = 2.039 \text{ in.}^4$$

The section properties for the y axis agree with those listed in the NDS Supplement.

4.12   Size Categories and Stress Grades

The lumber grading rules which establish allowable stresses for use in structural design have been developed over many years. In this development process, the relative size of a piece of wood was used as a guide in anticipating the application or "use" that a member would receive in the field. For example, pieces of lumber with rectangular cross sections make more efficient beams than members with square (or approximately square) cross sections. Thus, if the final application of a piece of wood were known, the stress-grading rules would take into account the primary function (e.g., axial strength or bending strength) of the member. See Example 4.6.

EXAMPLE 4.6   Size and Use categories

There are three main *size categories* of lumber. The categories and nominal size ranges are:

Boards	1 to 1½ in. thick
	2 in. and wider
Dimension lumber	2 to 4 in. thick
	2 in. and wider
Timbers	5 in. and thicker
	5 in. and wider

A number of additional subdivisions are available within the main size categories. Each represents a size and use category in the lumber grading rules. The primary *size and use categories* for stress-graded (structural) lumber are as follows:

Boards
   Stress-Rated Board (SRB)
Dimension lumber
   Structural Light Framing (SLF)
   Light Framing (LF)
   Studs
   Structural Joists and Planks (SJ&P)
   Decking
Timbers
   Beams and Stringers (B&S)
   Posts and Timbers (P&T)

Stress-Rated Boards may be used in structural applications. However, because they are relatively thin pieces of lumber, Stress-Rated Boards are not commonly used for structural framing. Therefore, the remaining discussion is limited to a consideration of Dimension lumber and Timbers. Sizes in the seven basic subcategories of structural lumber are summarized in the following table.

Symbol	Name	Nominal dimensions		Examples of sizes
		Thickness	Width	
LF	Light Framing and			
SLF	Structural Light Framing	2 to 4 in.	2 to 4 in.	2 × 2, 2 × 4, 4 × 4
SJ&P	Structural Joist and Plank	2 to 4 in.	5 in. and wider	2 × 6, 2 × 14, 4 × 10
	Stud	2 to 4 in.	2 in. and wider	2 × 4, 2 × 6, 4 × 6 (lengths limited to 10 ft and shorter)
	Decking*	2 to 4 in.	4 in. and wider	2 × 4, 2 × 8, 4 × 6
B&S	Beams and Stringers	5 in. and thicker	More than 2 in. greater than thickness	6 × 10, 6 × 14, 12 × 16
P&T	Posts and Timbers	5 in. and thicker	Not more than 2 in. greater than thickness	6 × 6, 6 × 8, 12 × 14

*Decking is normally stressed about its minor axis. In this book, all other bending members are assumed to be stressed about the major axis of the cross section, unless otherwise noted.

It has been noted that size and use are related. However, in the process of determining the allowable stresses for a member, the structural designer needs to place emphasis on understanding the *size* classifications. The reason is that *different allowable stresses* apply to the *same stress grade name* in the *different size categories*. For example, Select Structural (a stress grade) is available in SLF, SJ&P, B&S, and P&T size categories. Allowable stresses for a given commercial species of lumber are generally different for Select Structural in all of these size categories. See Example 4.7.

EXAMPLE 4.7   Stress Grades

Typical stress grades vary within the various size and use categories. The stress grades shown are for Douglas Fir-Larch.

1. *Structural Light Framing* (SLF)
   Select Structural
   No. 1 and Better
   No. 1
   No. 2
   No. 3

2. *Light Framing* (LF)
   Construction
   Standard
   Utility

3. *Structural Joist and Plank* (SJ&P)
   Select Structural
   No. 1 and Better
   No. 1
   No. 2
   No. 3

4. *Stud*
   Stud

5. *Decking*
   Select Decking
   Commercial Decking

6. *Beams and Stringers* (B&S)
   Dense Select Structural
   Select Structural
   Dense No. 1
   No. 1
   Dense No. 2
   No. 2

7. *Posts and Timbers* (P&T)
   Dense Select Structural
   Select Structural
   Dense No. 1
   No. 1
   Dense No. 2
   No. 2

NOTE: The stress grades listed are intended to be representative, and they are not available in all species groups. For example, *No. 1 and Better* is available only in DF-L and Hem-Fir. Southern Pine has a number of additional *dense* and *nondense* stress grades.

Several important points should be made about the *size and use categories* given in Example 4.6 *and* the *stress grades* listed in Example 4.7:

1. *Decking* is normally stressed in bending about the minor axis of the cross section, and allowable stresses for Decking are listed in a separate table. See NDS Supplement Table 4E, *Design Values for Visually Graded Decking.*

2. Allowable stresses for Dimension lumber (except Decking) are given in a number of separate tables. In these tables the stress grades are grouped together regardless of the size and use subcategory.

Allowable stresses for *Dimension lumber* are listed in the following tables in the 2001 NDS Supplement:

Table 4A. *Base Design Values for Visually Graded Dimension Lumber* (2″–4″ thick) (*All Species except Southern Pine*)

Table 4B. *Design Values for Visually Graded Southern Pine Dimension Lumber* (2″–4″ thick)

Table 4C. *Design Values for Mechanically Graded Dimension Lumber*

Table 4F. *Basic Design Values for Non-North American Visually Graded Dimension Lumber* (2″–4″ thick)

The simplification of the allowable stresses in Tables 4A and 4B requires the use of several adjustment factors (Sec. 4.13).

3. Allowable stresses for Beams and Stringers (B&S) and Posts and Timbers (P&T) are given in NDS Supplement Table 4D, *Design Values for Visually Graded Timbers* (5″ × 5″ *and larger*). Table 4D covers all species groups including Southern Pine.

Allowable stresses for B&S are generally different from the allowable stresses for P&T. This requires a complete listing of values for all of the stress grades for both of these size categories. Furthermore, it should be noted that there are two sets of design values for both B&S and P&T in three species groups: Douglas Fir-Larch, Hem-Fir, and Western Cedars. This is the result of differences in grading rules from two agencies.

As noted, the lumber grading rules reflect the *anticipated use* of a wood member based on its size, but no such restriction exists for the *actual use* of the member by the designer. In other words, lumber that falls into the B&S size category was originally anticipated to be used as a bending member. As a rectangular member, a B&S bending about its strong axis is a more efficient beam (because of its larger section modulus) than a square (or essentially square) member such as a P&T. However, allowable stresses are tabulated for tension, compression, and bending for *all size categories*. The designer may, therefore, use a B&S in any of these applications.

Although size and use are related, it must be emphasized again that the *allowable stresses depend on the size* of a member rather than its *use*. Thus, a member in the P&T size category is always graded as a P&T even though it could possibly be used as a beam. Therefore, if a 6 × 8 is used as a beam, the allowable bending stress for a P&T applies. Similarly, if a 6 × 10 is used as a column, the compression value for a B&S must be used.

The general notation used in the design of wood structures is introduced in the next section. This is followed by a review of a number of the adjustment factors required in wood design.

4.13   Notation for ASD

The design of wood structures under the 2001 NDS follows the principles known as *allowable stress design* (ASD). The *load and resistance factor design* (LRFD) method for engineered wood construction (Ref. 4.3) is beyond the scope of this text. It is expected that ASD will continue to be the popular method in the near future. However, the wood industry is in a transition period when both methods may be applied in design practice. It is expected that eventually the LRFD method will become the primary design technique.

The notation system for stress calculations in ASD for wood structures is very similar to that used in the design of steel structures according to the ASD steel manual (Ref. 4.5). However, wood is a unique structural material, and its proper use may require a number of adjustment factors. Although the basic concepts of timber design are very straightforward, the many possible adjustment factors can make wood design cumbersome in the beginning. The conversion of the NDS to an *equation format* has provided much better organization of this material.

In allowable stress design, actual stresses in a member are computed as the structure is subjected to a set of Code-required loads. Generally speaking, the forces and stresses in wood structures are computed according to the principles of engineering mechanics and strength of materials. The same basic linear elastic theory is applied in the design of wood beams as is applied to the design of steel members in ASD. The unique properties of wood members and the differences in behavior are usually taken into account with adjustment factors.

For consistency, it is highly recommended that the adjustment factors for wood design be kept as multiplying factors for allowable stresses. An alternative approach of using the stress adjustments to modify design loads can lead to confusion. The modification of design loads with wood design adjustment factors is not recommended. The general notation system for use in ASD for wood structures is summarized in Example 4.8.

EXAMPLE 4.8   Symbols for Stresses and Adjustment Factors

Symbols for use in wood design are standardized in the 2001 NDS.

Actual Stresses

Actual stresses are calculated from known loads and member sizes. These stresses are given the symbol of lowercase f, and a subscript is added to indicate the type of stress. For example, the axial tension stress in a member is calculated as the force divided by the cross-sectional area. The notation is

$$f_t = \frac{P}{A}$$

Tabulated Stresses

The stresses listed in the tables in the NDS Supplement are referred to as *tabulated stresses* or *tabulated design values*. All the tabulated stresses (except modulus of elasticity) include reductions for safety. The values of modulus of elasticity listed in the tables are average values and do not include reductions for safety. Tabulated stresses are given the symbol of an uppercase F, and a subscript is added to indicate the type of stress. For example, F_t represents the tabulated tension stress parallel to grain. The modulus of elasticity is assigned the traditional symbol E.

Allowable Stresses

Tabulated stresses for wood simply represent a starting point in the determination of the allowable stress for a particular design. *Allowable stresses* are determined by multiplying the tabulated stresses by the appropriate adjustment factors. The term "allowable design *value*" is perhaps more general than "allowable *stress*" in that it can properly be applied to quantities that are not actually stresses such as modulus of elasticity and connection capacity.

It is highly desirable to have a notation system that permits the designer to readily determine whether a design value in a set of calculations is a *tabulated* or an *allowable* property. A prime is simply added to the symbol for the tabulated stress to indicate that the necessary adjustments have been applied to obtain the allowable stress. For example, the allowable tension stress is obtained by multiplying the tabulated value for tension by the appropriate adjustment factors:

$$F'_t = F_t \times \text{(product of adjustment factors)}$$

For a design to be acceptable, the actual stress must be less than or equal to the allowable stress:

$$f_t \leq F'_t$$

On the other hand, if the actual stress exceeds the allowable stress, the design needs to be revised.

The following design values are included in the NDS Supplement:

Design value	Symbol for tabulated design value	Symbol for allowable (adjusted) design value
Bending stress	F_b	F_b'
Tension stress parallel to grain	F_t	F_t'
Shear stress parallel to grain	F_v	F_v'
Compression stress perpendicular to grain	$F_{c\perp}$	$F_{c\perp}'$
Compression stress parallel to grain	F_c	F_c'
Modulus of elasticity	E	E'

Adjustment Factors

The adjustment factors in wood design are usually given the symbol of an uppercase C, and one or more subscripts are added to indicate the purpose of the adjustment. Some of the subscripts are uppercase letters, and others are lowercase. Therefore, it is important to pay close attention to the form of the subscript, because simply changing from an uppercase to a lowercase subscript can change the meaning of the adjustment factor. Some of the possible adjustment factors for use in determining allowable design values are

$$C_D = \text{load duration factor}$$
$$C_M = \text{wet service factor}$$
$$C_F = \text{size factor}$$
$$C_{fu} = \text{flat use factor}$$
$$C_f = \text{form factor}$$
$$C_i = \text{incising factor}$$
$$C_t = \text{temperature factor}$$
$$C_r = \text{repetitive member factor}$$

These adjustment factors do not apply to all tabulated design values. In addition, other adjustments may be necessary in certain types of problems. For example, the column stability factor C_P is required in the design of wood columns. The factors listed here are simply representative, and the additional adjustment factors are covered in the chapters where they are needed.

The large number of factors is an attempt to remind the designer to not overlook something that can affect the performance of a structure. However, in many practical design situations, a number of adjustment factors may have a value of 1.0. In such a case, the adjustment is said to *default to unity*. Thus, in many common designs, the problem will not be as complex as the long list of adjustment factors would make it appear.

Tables summarizing the adjustment factors for various products are given in specific tables in the NDS. For example, NDS Table 4.3.1 provides a summary of the *Applicability of Adjustment Factors for Sawn Lumber*. Other NDS tables provide similar information for glued laminated timber (Table 5.3.1), round timber poles and piles (Table 6.3.1), wood I-joists (Table 7.3.1), structural composite lumber (Table 8.3.1), and wood structural panels (Table 9.3.1). A summary of the factors for use in the design of mechanical fasteners is given in NDS Table 10.3.1, *Applicability of Adjustment Factors for Connections.*

Some of the adjustment factors will cause the tabulated stress to decrease, and others will cause the stress to increase. When factors that reduce strength are considered, a larger member size will be required to support a given load. On the other hand, when circumstances exist which produce increased strength, smaller, more economical members can result if these factors are taken into consideration. The point here is that a number of items can affect the strength of wood. These items *must* be considered in design when they result in a reduction of member capacity. Factors which increase the calculated strength of a member *may* be considered in the design.

This discussion emphasizes that a *conservative* approach (i.e., in the direction of greater safety) in structural design is the general rule. Factors which cause member sizes to increase *must* be considered. Factors which cause them to decrease *may* be considered or ignored. The question of whether the latter should be ignored has to do with economics. It may not be practical to ignore reductions in member sizes that result from a beneficial set of conditions.

Most adjustments for wood design are handled as a string of multiplying factors that are used to convert tabulated stresses to allowable stresses for a given set of design circumstances. However, to avoid an excessive number of coefficients, often only those coefficients which have an effect on the final design are shown in calculations. In other words, if an adjustment has no effect on a stress value (i.e., it defaults to $C = 1.0$), the factor is often omitted from design calculations.

It should be noted that a number of adjustment factors have been in the NDS for many years. One adjustment factor, however, that has been discontinued with the 2001 NDS is the *shear stress factor* C_H. Tabulated values for shear stress formerly reflected the assumption that the member may be split along its full length. ASTM procedures for establishing allowable design values required two separate adjustments for the possible presence of splits, checks, and/or shakes. However, in 2000, ASTM Standard D245 (Ref. 4.14) was revised, and one of the two adjustments for splits, checks, and/or shakes was eliminated. This resulted in an increase of nearly two for allowable shear design values. These new design values assume that members include representative splitting, rather than conservatively assuming, as in previous editions of the NDS, that all members were split along their full length. For less severe splitting, designers were allowed to increase the allowable shear design value by up to a factor of two. With the changes in ASTM D245, the reason for a shear stress factor that allowed an increase in the shear capacity for less severe splitting is nullified, and thus the factor has been dropped from the NDS.

In this book the adjustment factors will generally be shown, including those with values of unity. A general summary of adjustment factors is usually part of a computer evaluation of allowable stresses. The adjustment factors mentioned in Example 4.8 are described in the remainder of this chapter. Others are covered in the chapters that deal with specific problems.

4.14   Wet Service Factor C_M

The moisture content of wood and its relationship to strength were described in Sec. 4.7. Tabulated design stresses in the NDS Supplement generally apply to wood that is used in a *dry condition,* as in most covered structures.

For sawn lumber, the tabulated values apply to members with an equilibrium moisture content (EMC) of 19 percent or less. Values apply whether the lumber is manufactured S-DRY, KD, or S-GRN. If the moisture content in service will exceed 19 percent for an extended period of time, the tabulated values are to be multiplied by an appropriate *wet service factor C_M.* Note that the subscript M refers to moisture.

For *member stresses in sawn lumber,* the appropriate values of C_M are obtained from the summary of *Adjustment Factors* at the beginning of each table in the NDS Supplement (i.e., at the beginning of Tables 4A to 4F). In most cases, C_M is less than 1.0 when the moisture content exceeds 19 percent. The exceptions are noted in the tables for C_M. For lumber used at a moisture content of 19 percent or less, the default value of $C_M = 1.0$ applies.

Prior to the 1991 NDS, lumber grade marked MC15 was permitted use of a C_M greater than 1.0, but as a result of the In-Grade Program, this has been deleted. For some grades of Southern Pine the wet service factor has been incorporated into the tabulated values, and for these cases the use of an additional C_M is not appropriate.

For *connection design,* the moisture content at the time of fabrication of the connection *and* the moisture content in service are both used to evaluate C_M. Values of C_M for connection design are summarized in NDS Table 10.3.3.

For *glulam members* (Chap. 5), tabulated stresses apply to MC values of less than 16 percent (that is, $C_M = 1.0$). For a MC of 16 percent or greater, use of C_M less than 1.0 is required. Values of C_M for softwood glulam members are given in the summary of *Adjustment Factors* preceding the NDS Supplement Tables 5A and 5B, and in Tables 5C and 5D for hardwood glulam members.

4.15   Load Duration Factor C_D

Wood has a unique structural property. It can support higher stresses if the loads are applied for a short period of time. This is particularly significant when one realizes that if an overload occurs, it is probably the result of a temporary load.

All tabulated design stresses and nominal fastener design values for connections apply to *normal* duration loading. In fact, the tables in the NDS generally remind the designer that the published values apply to "normal load duration and dry service conditions." This, together with the equation format of the NDS, should highlight the need for the designer to account for other conditions. The load duration factor C_D is the adjustment factor used to convert *tabulated* stresses and nominal fastener values to *allowable* values based on the expected duration of full design load.

In other words, C_D converts values for normal duration to design values for other durations of loading. Normal duration is taken as 10 years, and floor live loads are conservatively associated with this time of loading. Because tabulated stresses apply directly to floor live loads, $C_D = 1.0$ for this type of loading. For other loads, the duration factor lies in the range $0.9 \leq C_D \leq 2.0$. It should be noted that C_D applies to all tabulated design values except compression perpendicular to grain $F_{c\perp}$ and modulus of elasticity E. In the case of pressure-preservative-treated and fire-retardant-treated wood, the NDS limits the load duration factor to a maximum of 1.6 ($C_D \leq 1.6$). This is due to a tendency for treated material to become less resistant to impact loading.

The historical basis for the load duration factor is the curve shown in Fig. 4.13. See Example 4.9. The load duration factor is plotted on the vertical axis versus the accumulated duration of load on the horizontal axis. This graph appears in the *Wood Handbook* (Ref. 4.22) and in the NDS Appendix B. Over the years this plot has become known as the *Madison Curve* (the FPL is located in Madison, Wisconsin), and its use has been integrated into design practice since the 1940s. The durations associated with the various design loads are shown on the graph and in the summary below the graph.

EXAMPLE 4.9   Load Duration Factor

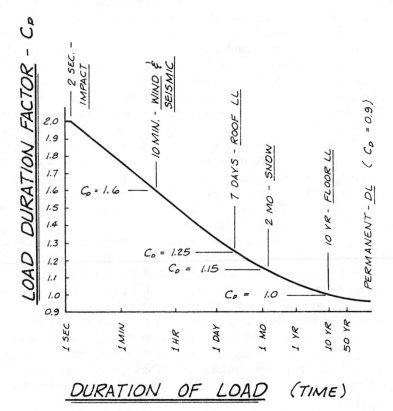

Figure 4.13 Madison curve.

Shortest-duration load in combination	C_D
Dead load	0.9
Floor live load	1.0
Snow load	1.15
Roof live load	1.25
Wind or seismic force	1.6
Impact	2.0

NOTES: 1. Check all Code-required load and force combinations. 2. The C_D associated with the shortest-duration load or force in a given combination is used to adjust the tabulated stress. 3. The critical combination of loads and forces is the one that requires the largest-size structural member.

The term "duration of load" refers to the total *accumulated* length of time that the full design load is applied during the life of a structure. Furthermore, in considering duration, it is indeed the *full design load* that is of concern, and not the length of time over which a portion of the load may be applied. For example, it is obvious that some wind or air movement is almost always present. However, in assigning C_D for wind, the duration is taken as the total length of time over which the design *maximum* wind force will occur.

A major change was introduced in the 1991 edition of the NDS (see Ref. 4.2) regarding the load duration factor assigned to *wind and seismic* forces (NDS Sec. 2.3.2). In the past a 1-day duration was conservatively assumed for wind and seismic forces, and a corresponding duration factor of $C_D = 1.33$ was the traditional value. Wind forces in the IBC and in the other model building codes are now based on the wind force provisions in load standard, ASCE 7-02 (Ref. 4.12). Research indicates that the peak wind forces in ASCE 7 have a cumulative duration of a few seconds. In addition, strong-motion earthquake effects are typically less than a minute in duration. Because of these duration studies, the NDS adopted an accumulated duration of 10 minutes for wind and seismic forces. This shifts the load duration factor for wind and seismic forces to $C_D = 1.6$ on the Madison Curve.

It was previously recommended that adjustment factors, including C_D, be applied as multiplying factors for adjusting tabulated design values. Modifications of actual stresses or modifications of applied loads should not be used to account for duration of load. A consistent approach in the application of C_D to the tabulated design value will avoid confusion.

The stresses that occur in a structure are usually not the result of a single applied load (see Sec. 2.16 for a discussion of Code-required load combinations). Quite to the contrary, they are normally caused by a combination of loads and forces that act simultaneously. The question then arises about which load duration factor should be applied when checking a stress caused by a given combination. It should be noted that the load duration factor applies to the *entire combination of loads* and not just to that portion of the stress caused by a load of a particular duration. The C_D to be used is the one associated with the *shortest-duration* load or force in a given combination.

For example, consider the possible load combinations on a floor beam that also carries a column load from the roof. What are the appropriate load duration factors for the various load combinations? If stresses under the dead load alone are checked, $C_D = 0.9$. If stresses under (D + L) are checked, the shortest-duration load in the combination is floor live load, and $C_D = 1.0$. For (D + L + S), $C_D = 1.15$. If the structure is located in an area where snow loads do not occur, the last combination becomes (D + L + L_r), and $C_D = 1.25$.

In this manner, it is possible for a smaller load of longer duration (with a small C_D) to be more critical than a larger load of shorter duration (with a large C_D). Whichever combination of loads, together with the appropriate load duration factor, produces the largest required member size is the one that must be used in the design of the structure. It may be necessary, therefore, to check several different combinations of loads to determine which combination governs the design. With some practice, the designer can often tell by inspection which combinations need to be checked. In many cases, only one or possibly two combinations need be checked. See Example 4.10.

EXAMPLE 4.10   Comparison of Load Combinations

Determine the design loads and the critical load combination for the roof beam in Fig. 4.14. The tributary width to the beam and the design unit loads are given.

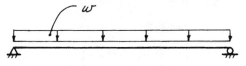

Figure 4.14

$$\text{Tributary width} = 10 \text{ ft}$$

$$D = 20 \text{ psf}$$

$$L_r = 16 \text{ psf}$$

Part a

Load combination 1 (D alone):

$$w_D = 20 \times 10 = 200 \text{ lb/ft}$$

$$C_D = 0.90$$

Load combination 2 (D + L_r):

$$w_{\text{TL}} = (20 + 16)10 = 360 \text{ lb/ft}$$

$$C_D = 1.25$$

The tabulated stress for the beam is to be multiplied by 0.90 for load combination 1 and 1.25 for load combination 2. Theoretically both load combinations must be considered. However, with some practice, the designer will be able to tell from the relative magnitude of the loads which combination is critical. For example, 360 lb/ft is so large in comparison with 200 lb/ft that load combination 2 will be critical. Therefore, calculations for load combination 1 are not required. If it cannot be determined by inspection which loading is critical, calculations for both load cases should be performed.

In some cases calculations for two or more cases *must* be performed. Often this occurs in members with combined axial and bending loads. These types of problems are considered in Chap. 7.

Part b

Show calculations which verify the critical load case for the beam in part *a* without complete stress calculations. Remove "duration" by dividing the design loads by the appropriate C_D factors.*

Load combination 1:

$$\frac{w_D}{C_D} = \frac{200}{0.9} = 222 \text{ lb/ft\dag}$$

Load combination 2:

$$\frac{w_{\text{TL}}}{C_D} = \frac{360}{1.25} = 288 \text{ lb/ft\dag}$$

$$222 < 288$$

$$\therefore \text{ load combination 2 governs}$$

When designers first encounter the adjustment for duration of load, they like to have a *system* for determining the critical loading combination. See Example 4.10, part *b*. Essentially the system involves removing the question of load duration from the problem. If the sum of the loads in a given combination is divided by the C_D for the combination, duration is removed from the

---

*This method is appropriate for short columns and beams with full lateral support. The effect of C_D decreases as the unbraced length of these members increases. When Euler-type buckling governs, the loads should be compared without dividing by C_D.

†These modified loads are used to determine the critical load combination only. Actual design loads (for example, $w = 360$ lb/ft) should be used in calculations, and $C_D = 1.25$ should be applied to tabulated stresses.

load. If this is done for each required load combination, the resulting loads can be compared. The largest modified load represents the critical combination.

This method is not foolproof. The C_D has full effect for short columns and no effect for very long columns. Thus, the method is accurate for short columns, and it becomes less appropriate as the length increases. A similar caution applies to laterally unsupported beams.

There is a second objection to the system just described. It runs counter to the recommendation that adjustment factors be applied to the tabulated stress and not to the design loads. Thus, if this analysis is used, the calculations should be done separately (perhaps on scrap paper). Once the critical combination is known, the actual design loads (not modified) can be used in formal calculations, and C_D can be applied to the tabulated stresses in the usual manner.

4.16 Size Factor C_F

It has been known for some time that the size of a wood member has an effect on its unit strength (stress). This behavior is taken into account by the *size factor* C_F (NDS Sec. 4.3.6). The size-effect factors are based on the size classification.

Visually graded Dimension lumber. The size factors C_F for most species of visually graded Dimension lumber are summarized in the *Adjustment Factor* section that precedes NDS Supplement Tables 4A and 4F. The size factors for F_b, F_t, and F_c are given in a table which depends on the stress *grade* and the *width* (depth) of the piece of lumber. For bending stress, the *thickness* of the member also affects the size factor.

Tabulated stresses for use with these expanded size factors are termed "*base design values*" in the title of NDS Supplement Tables 4A and 4F. In other words, the allowable stresses for a given piece of Dimension lumber are obtained by multiplying *base values* by the appropriate size factors. The concept of base design values lends itself to the evaluation of allowable stresses in a computer program or microcomputer spreadsheet template.

The size factors for Southern Pine Dimension lumber are handled somewhat differently. For Southern Pine, a number of the size factors have been incorporated into the tabulated values given in NDS Supplement Table 4B. Thus, the tabulated values for Southern Pine are said to be "size-specific," and the concept of base design values is not included in the table for Southern Pine.

Unfortunately, the size-specific tables for Southern Pine do not completely avoid the use of a C_F multiplier. Bending values in Table 4B apply to lumber that has a nominal thickness of 2 in. A size factor of $C_F = 1.1$ is provided for F_b if the lumber being considered has a nominal thickness of 4 in. instead of 2 in. In addition, a size factor of $C_F = 0.9$ is provided for F_b, F_t, and F_c for Dimension lumber that has a width greater than 12 in.

Refer to NDS Supplement Tables 4A and 4B for values of C_F for Dimension lumber and for a comparison of base design values with size-specific design values.

Timbers. For Timbers the size factor C_F applies only to F_b. Essentially the size factor reflects the fact that as the *depth* of a beam increases, the *unit strength* (and correspondingly the allowable stress) decreases. When the depth d of a timber exceeds 12 in., the size factor is defined by the expression

$$C_F = \left(\frac{12}{d}\right)^{1/9}$$

For members that are less than 12 in. deep, the size factor defaults to unity: $C_F = 1.0$.

At one time this size factor was also used for glulam beams. However, the size factor for glulams has been replaced with the volume factor C_V (see Chap. 5).

4.17 Repetitive Member Factor C_r

Many wood structures have a series of closely spaced parallel members. The members are often connected together by sheathing or decking. In this arrangement, the performance of the *system* does not depend solely on the capacity of an individual member. This can be contrasted to an engineered wood structure with relatively large structural members spaced a greater distance apart. The failure of one large member would essentially be a failure of the system.

The system performance of a series of small, closely spaced wood members is recognized in the NDS by providing a 15 percent increase in the tabulated bending stress F_b. This increase is provided by the *repetitive-member factor C_r* (NDS Sec. 4.3.9). It applies only to F_b and only to Dimension lumber used in a repetitive system. A *repetitive-member system* is defined as one that has

1. Three or more parallel members of Dimension lumber

2. Members spaced not more than 24 in. on center

3. Members connected together by a load-distributing element such as roof, floor, or wall sheathing

For a repetitive-member system, the tabulated F_b may be multiplied by $C_r = 1.15$. For all other framing systems and stresses $C_r = 1.0$.

The repetitive-member factor recognizes system performance. If one member should become overloaded, parallel members come into play. The load is distributed by sheathing to adjacent members, and the load is shared by a number of beams. The repetitive-member factor is not applied to the larger sizes of wood members (i.e., Timbers and glulams) because these large members are not normally spaced closely enough together to qualify as a repetitive member.

When a concentrated load is supported by a deck which distributes the load to parallel beams, the entire concentrated load need not be assumed to be supported by one member. NDS Sec. 15.1 provides a method for the *Lateral Distribution of a Concentrated Load* to adjacent parallel beams. According to Ref. 4.24, the single-member bending stress (that is, $C_r = 1.0$) applies if the load distribution in NDS Sec. 15.1 is used.

4.18 Flat Use Factor C_{fu}

Except for decking, tabulated bending stresses for Dimension lumber apply to wood members that are stressed in flexure about the strong axis of the cross section. The NDS refers to this conventional type of beam loading as *edgewise* use or *load applied to narrow face* of the member.

In a limited number of situations, Dimension lumber may be loaded in bending about the minor axis of the cross section. The terms *flatwise use* and *load applied to wide face* describe this application. When members are loaded in bending about the weak axis, the tabulated bending stresses F_b may be increased by multiplying by the *flat-use factor* C_{fu} (NDS Sec. 4.3.7). Numerical values for C_{fu} are given in the *Adjustment Factor* sections of NDS Supplement Tables 4A, 4B, 4C, and 4F.

Tabulated bending stresses for Beams and Stringers also apply to the usual case of bending about the x axis of the cross section. The NDS does not provide a flat-use factor for bending about the y axis. It is recommended that the designer contact the appropriate rules-writing agency for assistance if this situation is encountered. For example, WWPA and WCLIB provide flat use factors for Beams and Stringers for western species for both bending stress and modulus of elasticity.

The tabulated bending stress for a glulam beam that is stressed in bending about the weak axis is given the symbol F_{by}. Values of F_{by} apply to glulams that have a cross-sectional dimension *parallel to the wide face of the laminations* of at least 12 in. For beams that are less than 12 in. wide, the value of F_{by} may be increased by a flat-use factor (NDS Sec. 5.3.7). For values of C_{fu} for glulams, see NDS Supplement Tables 5A, 5B, 5C, and 5D.

4.19 Temperature Factor C_t

The strength of a member is affected by the temperature of the wood in service. Strength is increased as the temperature cools below the normal temperature range found in most buildings. On the other hand, the strength decreases as temperatures are increased. The *temperature factor* C_t is the multiplier that is used to reduce tabulated stresses if prolonged exposure to higher than normal temperatures are encountered in a design situation.

Tabulated design values apply to wood used in the ordinary temperature range and perhaps *occasionally* heated up to 150°F. Prolonged exposure to temperatures above 150°F may result in a *permanent* loss of strength. Reductions in strength caused by heating below 150°F are generally reversible. In other words, strength is recovered as the temperature is reduced to the normal range.

Values of C_t are given NDS Sec. 2.3.3. The first temperature range that requires a reduction in design values is $100°F < T < 125°F$. At first this seems to be a rather low temperature range. After all, temperatures in many of the warmer areas of the country often exceed 100°F. In these locations, is it necessary to reduce the tabulated design stresses for members in a wood roof system? The answer is that it is generally *not* considered necessary. Members in roof structures subjected to *temporarily* elevated temperatures are not usually subjected to the *full* design load under these conditions. For example, snow loads will not be present at these elevated temperatures, and roof live loads occur infrequently. Furthermore, any loss in strength should be regained when the temperature returns to normal.

For these and other reasons, $C_t = 1.0$ is normally used in the design of ordinary wood-frame buildings. However, in an industrial plant there may be operations that cause temperatures to be consistently elevated. Structural members in these types of situations may require use of a temperature factor less than 1. Additional information is given in NDS Appendix C.

4.20   Incising Factor C_i

The incising adjustment factor was first introduced in the 1997 NDS. Many species, most notably the Southern pines, readily accept preservative treatments, while other species do not accept pressure treatments as well. For species that are not easily treated, incising is often used to make the treatment effective. See Sec. 4.9. If incising is used to increase the penetration of the preservatives, some design values in the NDS Supplement must be adjusted (NDS Sec. 4.3.8). For the modulus of elasticity, $C_i = 0.95$, and for bending stress, tension stress, and compression parallel to grain, $C_i = 0.80$. For shear and compression perpendicular to grain, as well as for non-incised treated lumber, the incising factor is taken as 1.0.

4.21   Form Factor C_f

The form factor C_f has been in the specification for wood design for many years, but its use is very limited. The purpose of the form factor is to adjust the tabulated bending stress F_b for certain nonrectangular cross sections. The NDS provides two form factors (NDS Sec. 3.3.4): one for circular cross sections, and one for a square beam loaded in the plane of the diagonal (i.e., a beam with a diamond cross section).

Circular cross sections are common in wood design in the case of round timber piles and poles. Timber piles are used for foundation structures, and the most common application of poles is for utility structures. Poles are also used as the supporting frame for both vertical loads and lateral forces in *pole buildings*.

Pole buildings originated in farm applications such as sheds. Other uses of pole buildings include elevated housing in coastal areas that are subject to flooding. The framing for the lower floor level in this type of housing is chosen to be above high-water level during storm conditions. The floor and roof framing is attached to vertical poles that are embedded in earth.

Another use of pole framing for housing is on property with a fairly steep slope. Again, floor and roof framing is attached to vertical poles. In this case, the advantage of pole construction is that it does not require extensive grading of the property and the construction of retaining walls.

Timber piles and poles usually involve wood that is in contact with soil or concrete. Consequently these members are usually treated with a pressure preservative (Sec. 4.9). Properties for treated round timber piles are given in NDS Table 6A *Design Values for Treated Round Timber Piles.* It should be noted that the bending stresses in NDS Table 6A already have the form factor C_f included (NDS Sec. 6.3.8). Therefore, the designer should not apply $C_f = 1.18$ for circular cross sections to the values of F_b given in NDS Table 6A. For untreated piles see NDS Sec. 6.3.5. The design of round timber poles and piles is beyond the scope of this book.

4.22   Design Problem: Allowable Stresses

If one examines NDS Tables 4.3.1, 5.3.1, 6.3.1, 7.3.1, 8.3.1, and 9.3.1, *Applicability of Adjustment Factors,* it will be noticed that the temperature factor C_t applies to all design values. However, it should be realized that this table shows what adjustments *may* be required under certain conditions. It simply serves as a reminder to the designer to not overlook a necessary adjustment. The table says nothing about the *frequency of use* of an adjustment factor.

It was observed in Sec. 4.19 that the temperatures in most wood buildings do not require a reduction in design values. Thus, C_t is a factor that is rarely used in the design of typical wood buildings, and the default value of $C_t = 1.0$ often applies. On the other hand, the load duration factor C_D is a common adjustment factor that is used in the design of practically all wood structures. In this case, $C_D = 1.0$ only when the floor live load controls the design.

Several examples are given to illustrate the use of the NDS tables and the required adjustment factors. Complete design problems are given later in this book, and the current examples simply emphasize obtaining the correct *allowable* stress. The first requirement is to obtain the correct *tabulated* value from the NDS Supplement for the given size, grade, and species group. The second step is to apply the appropriate *adjustment* factors.

Example 4.11 deals with four different sizes of the same stress grade (No. 1) in a single species group (Douglas Fir-Larch). Dimensions are obtained from NDS Supplement Tables 1A and 1B. The example clearly shows the effect of a number of variables. Several different loading conditions and stress adjustment factors are illustrated. The reader is encouraged to verify the tabulated design values and the adjustment factors in the NDS.

Some stress adjustment factors in NDS Table 4.3.1 are not shown in this example. These factors do not apply to the given problem. Other factors may have a default value of unity and are shown for information purposes. Note that C_D does not apply to $F_{c\perp}$ or E.

EXAMPLE 4.11   Determination of Allowable Stresses

Determine the *allowable* design stresses for the four members given below. All members are No. 1 DF-L. Bending loads will be about the strong axis of the cross section (load applied to narrow face). Bracing conditions are such that buckling is not a concern. Consider dry-service conditions (EMC ≤ 19 percent) unless otherwise indicated. Normal temperature conditions apply.

For each member a *single* load duration factor C_D will be used to adjust the design values for the given load combination. In practice, a number of loading conditions must be checked, and each load case will have an appropriate C_D. Limiting each member to a single-load case is done for simplicity in this example.

Part a

Roof rafters are 2 × 8 at 24 in. o.c., and they directly support the roof sheathing. Loads are (D + L_r).

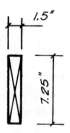

1.5″

7.25″

Figure 4.15a A 2 × 8 is a Dimension lumber size.

Tabulated design values of visually graded DF-L *Dimension lumber* are obtained from NDS Supplement Table 4A. The framing arrangement qualifies for the 15 percent increase in bending stress for repetitive members. The load duration factor is 1.25 for the combination of (D + L_r). Dimension lumber requires a size-effect factor for F_b, F_t, and F_c.

$$F_b' = F_b(C_D \times C_M \times C_t \times C_F \times C_r \times C_i)$$

$$= 1000(1.25 \times 1.0 \times 1.0 \times 1.2 \times 1.15 \times 1.0)$$

$$= 1725 \text{ psi}$$

$$F_t' = F_t(C_D \times C_M \times C_t \times C_F \times C_i)$$

$$= 675(1.25 \times 1.0 \times 1.0 \times 1.2 \times 1.0) = 1012 \text{ psi}$$

$$F_v' = F_v(C_D \times C_M \times C_t \times C_i)$$

$$= 180(1.25 \times 1.0 \times 1.0 \times 1.0) = 225 \text{ psi}$$

$$F'_{c\perp} = F_{c\perp}(C_M \times C_t \times C_i)$$

$$= 625(1.0 \times 1.0 \times 1.0) = 625 \text{ psi}$$

$$F'_c = F_c(C_D \times C_M \times C_t \times C_F \times C_i)$$

$$= 1500(1.25 \times 1.0 \times 1.0 \times 1.05 \times 1.0) = 1968 \text{ psi}$$

$$E' = E(C_M \times C_t \times C_i)$$

$$= 1,700,000(1.0 \times 1.0 \times 1.0) = 1,700,000 \text{ psi}$$

Part b

Roof beams are 4 × 10 at 4 ft-0 in. o.c. Loads are (D + S).

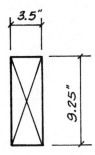

Figure 4.15b A 4 × 10 is a Dimension lumber size.

Design values for visually graded DF-L *Dimension lumber* are again obtained from NDS Supplement Table 4A. A 4-ft framing module exceeds the 24-in. spacing limit for repetitive members, and $C_r = 1.0$. The load duration factor is 1.15 for the combination of (D + S).

$$F'_b = F_b(C_D \times C_M \times C_t \times C_F \times C_r \times C_i)$$

$$= 1000(1.15 \times 1.0 \times 1.0 \times 1.2 \times 1.0 \times 1.0) = 1380 \text{ psi}$$

$$F'_t = F_t(C_D \times C_M \times C_t \times C_F \times C_i)$$

$$= 675(1.15 \times 1.0 \times 1.0 \times 1.1 \times 1.0) = 854 \text{ psi}$$

$$F'_v = F_v(C_D \times C_M \times C_t \times C_i)$$

$$= 180(1.15 \times 1.0 \times 1.0 \times 1.0) = 207 \text{ psi}$$

$$F'_{c\perp} = F_{c\perp}(C_M \times C_t \times C_i)$$

$$= 625(1.0 \times 1.0 \times 1.0) = 625 \text{ psi}$$

$$F'_c = F_c(C_D \times C_M \times C_t \times C_F \times C_i)$$

$$= 1500(1.15 \times 1.0 \times 1.0 \times 1.05 \times 1.0) = 1725 \text{ psi}$$

$$E' = E(C_M \times C_t \times C_i)$$

$$= 1,700,000(1.0 \times 1.0 \times 1.0) = 1,700,000 \text{ psi}$$

Part c

A 6 × 16 floor beam supports loads from both the floor and the roof. Several load combinations have been studied, and the critical loading is (D + L + L_r).

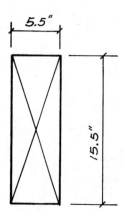

Figure 4.15c A 6 × 16 is a Beams and Stringers size.

A B&S has a minimum cross-sectional dimension of 5 in., and the width is *more* than 2 in. larger than the thickness. *Beams and Stringers* sizes are described in Example 4.6 (Sec. 4.12). Tabulated stresses are obtained from NDS Supplement Table 4D. To be conservative, take the smaller tabulated stresses listed for the two sets of grading rules (WCLIB and WWPA). In this problem the values are the same for both.

The load duration factor for the combination of loads is based on the shortest-duration load in the combination. Therefore, $C_D = 1.25$. Unlike Dimension lumber, large members have one size factor, and it applies to bending stress only. When the depth of a Timber exceeds 12 in., the size factor is given by the following expression

$$C_F = \left(\frac{12}{d}\right)^{1/9} = \left(\frac{12}{15.5}\right)^{1/9} = 0.972$$

$$F'_b = F_b(C_D \times C_M \times C_t \times C_F \times C_i)$$

$$= 1350(1.25 \times 1.0 \times 1.0 \times 0.972 \times 1.0) = 1640 \text{ psi}$$

$$F'_t = F_t(C_D \times C_M \times C_t \times C_i)$$

$$= 675(1.25 \times 1.0 \times 1.0 \times 1.0) = 844 \text{ psi}$$

$$F'_v = F_v(C_D \times C_M \times C_t \times C_i)$$

$$= 170(1.25 \times 1.0 \times 1.0 \times 1.0) = 212 \text{ psi}$$

$$F'_{c\perp} = F_{c\perp}(C_M \times C_t \times C_i)$$

$$= 625(1.0 \times 1.0 \times 1.0) = 625 \text{ psi}$$

$$F'_c = F_c(C_D \times C_M \times C_t \times C_i)$$

$$= 925(1.25 \times 1.0 \times 1.0 \times 1.0) = 1156 \text{ psi}$$

$$E' = E(C_M \times C_t \times C_i)$$

$$= 1{,}600{,}000(1.0 \times 1.0 \times 1.0) = 1{,}600{,}000 \text{ psi}$$

Part d

A 6 × 8 is used as a column to support a roof. It also supports tributary wind forces, and the critical loading condition has been determined to be D + 0.75(L + W). High-humidity conditions exist, and the moisture content of this member may exceed 19 percent. The member is incised.

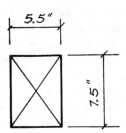

Figure 4.15d A 6 × 8 is a Posts and Timbers size.

A P&T has a minimum cross-sectional dimension of 5 in., and the width is *not more* than 2 in. larger than the thickness. *Posts and Timbers* sizes are described in Example 4.6 (Sec. 4.12). Tabulated stresses are obtained from NDS Supplement Table 4D. To be conservative, take the smaller tabulated stresses listed for the two sets of grading rules (WCLIB and WWPA). In this problem the values are the same for both.

The load duration factor for the combination of loads is based on the shortest-duration load in the combination. Therefore, $C_D = 1.6$. The depth of this member is less than 12 in., and C_F defaults to unity. Recall that for Timbers the size factor applies only to F_b.

$$F'_b = F_b(C_D \times C_M \times C_t \times C_F)$$

$$= 1200(1.6 \times 1.0 \times 1.0 \times 1.0) = 1920 \text{ psi}$$

$$F'_t = F_t(C_D \times C_M \times C_t)$$

$$= 825(1.6 \times 1.0 \times 1.0) = 1320 \text{ psi}$$

$$F'_v = F_v(C_D \times C_M \times C_t)$$

$$= 170(1.6 \times 1.0 \times 1.0) = 272 \text{ psi}$$

$$F'_{c\perp} = F_{c\perp}(C_M \times C_t)$$

$$= 625(0.67 \times 1.0) = 419 \text{ psi}$$

$$F'_c = F_c(C_D \times C_M \times C_t)$$

$$= 1000(1.6 \times 0.91 \times 1.0) = 1456 \text{ psi}$$

$$E' = E(C_M \times C_t)$$

$$= 1,600,000(1.0 \times 1.0) = 1,600,000 \text{ psi}$$

4.23   Future Directions in Wood Design

The wood industry is not a static business. If there is a better way of doing something, a better way of doing it will be found. In this book the question of "better" generally refers to developing more accurate methods of structural design, but there is an underlying economic force that drives the system.

It was noted at the beginning of Chap. 4 that many wood-based products that are in widespread use today were unavailable only a few years ago. These include a number of structural-use panels, wood I joists, resawn glulam beams, laminated veneer lumber (LVL), and more recently parallel strand lumber (PSL). A number of these developments, especially in the area of reconstituted wood products, are the result of new technology, and they represent an economic response to environmental concerns and resource constraints. With these products the move is plainly in the direction of *engineered* wood construction.

The design profession is caught in the middle of this development spiral. Anyone who is at all familiar with previous editions of the NDS will testify to the broad changes to the wood design criteria. Recent changes included new lumber values originating from the In-Grade Test Program, new column and laterally unbraced beam formulas, new interaction equation for members with combined stresses, an engineering mechanics approach to the design of wood connections, and now the inclusion of various wood product lines and systems formerly not covered by the NDS, including shearwalls, diaphragms, structural wood panels, structural composite lumber, I-joists, and metal plate connected wood trusses.

The current NDS is based on a deterministic method known as *allowable stress design* (ASD). Some argue that the method should be referred to more appropriately as *working stress design* (WSD) because the stresses that are computed are based on working or service loads. Both names have been used in the past, but ASD seems to be the term most widely used today.

Another approach to design is based on reliability theory. As with ASD, various terms are used to refer to this alternative system. These include *reliability-based design, probability-based design, limit states design,* and *load and resistance factor design* (LRFD). Of these, *load and resistance factor design* is the approach that is generally agreed upon by the profession as the appropriate technique for use in structural design. Reinforced concrete has operated under this general design philosophy in the "strength" method for quite some time. The structural-steel industry has recently transitioned from ASD to the LRFD format (Refs. 4.5 and 4.6).

The wood industry is in the process of moving to an LRFD format. In mid-1991, the wood industry completed a 3-year project to develop a *draft LRFD Specification for Engineered Wood Construction.* This document was developed by a team from the wood industry, university faculty, and the design profession, and was subjected to trial use by professionals knowledgeable in the area of timber design. The final LRFD specification was published by the American Society of Civil Engineers (ASCE), and a joint ASCE/AF&PA standards committee is responsible for the specification and will oversee its ongoing revision process. Already, all model building codes, including the IBC, recognize the LRFD specification as an alternate design procedure. The wood industry completed development of a comprehensive *LRFD Manual for Engineered Wood Construction* (Ref. 4.3) based on the provisions of ASCE Standard.

Currently, the AF&PA committee responsible for maintaining the NDS is developing a new version of the NDS that presents both design methods, ASD and LRFD, side by side in a single design specification. This is possible, since both the ASD and LRFD design specifications are based on the same behavior equations, and because considerable effort was made to keep both documents fully compatible. While many are concerned that having both design methods presented in the same specification may be confusing, others feel that this will ensure continued compatibility between the methods. Regardless, it appears that the next NDS may be a dual format design specification, presenting both ASD and LRFD methodologies.

A brief description about the differences between ASD and LRFD is given now. In ASD, actual stresses are checked to be less than or equal to allowable stresses:

$$\text{Actual stress} \leq \text{allowable stress}$$

It has been noted that a 5 percent exclusion value is the basis for most allowable stresses. Although this approach generally produces safe designs, the reliability differs with each structure. Some wood-based products are highly variable, and others are much less variable. Two examples from this chapter are visually graded sawn lumber and MSR lumber. Some design values for visually graded lumber are more variable than for MSR lumber, and many other examples can be cited. In addition, current design practice assumes that dead loads, live loads, and lateral forces are known with equal reliability. Obviously this is not the case, because it is possible to predict some loads (e.g., dead loads) much more accurately than others. These differences are not reflected directly in ASD.

The title "load and resistance factor design" is a good description of the reliability-based design procedure. As the name implies, service loads (loads expected under normal service conditions) are multiplied by appropriate *load factors,* and the nominal resistance of the structure (such as the calculated moment or shear capacity of a member) is multiplied by appropriate strength-reduction factors (*resistance factors*).

For a structure to be useful, the factored resistance offered by the structure must equal or exceed the factored load effects:[†]

$$\phi R \geq \sum \gamma Q$$

where ϕ = resistance factor
R = nominal calculated resistance of the structure (such as shear or moment capacity)
γ = load factor
Q = effects of service loads (such as the applied shear or moment in a member)

Resistance factors ϕ are numerically less than unity and are designed to reduce the calculated nominal resistance (capacity) of a structure. The purpose of this reduction in calculated capacity is to account for resistance uncertainties such as material properties and variability. On the other side of the equation, load factors γ are intended to account for the uncertainties in magnitudes and combinations of loading. Thus the load factor for dead load is smaller than the load factor for live loads. Designers who are familiar with the strength design of reinforced concrete should have a feel for load and resistance factors.

4.24 References

[4.1] American Forest and Paper Association (AF&PA). 2001. *Allowable Stress Design Manual for Engineered Wood Construction and Supplements and Guidelines,* 2001 ed., AF&PA, Washington, DC.

[4.2] American Forest and Paper Association (AF&PA). 1999. *Commentary on the National Design Specification for Wood Construction,* 1997 ed., AF&PA, Washington DC.

[4.3] American Forest and Paper Association (AF&PA). 1996. *Load and Resistance Factor Design Manual for Engineered Wood Construction and Supplements.* 1996 ed., AF&PA, Washington DC.

[4.4] American Forest and Paper Association (AF&PA). 2001. *National Design Specification for Wood Construction and Supplement.* 2001 ed., AF&PA, Washington DC.

[†]In practice, service loads are multiplied by the appropriate load factors, and the factored loads, in turn, increase the *effect* of the applied loads.

[4.5]   American Institute of Steel Construction (AISC). 1989. *Manual of Steel Construction,* 9th ed., AISC, Chicago, Il.
[4.6]   American Institute of Steel Construction (AISC). 2001. *Manual of Steel Construction—Load and Resistance Factor Design,* 3rd ed., AISC, Chicago, IL.
[4.7]   American Institute of Timber Construction (AITC). 1990. *Standard for Preservative Treatment of Structural Glued Laminated Timber, AITC 109-90,* AITC, Englewood, CO.
[4.8]   American Institute of Timber Construction (AITC). 1992. *Guidelines to Minimize Moisture Entrapment in Panelized Wood Roof Systems, AITC Technical Note 20,* AITC, Englewood, CO.
[4.9]   American Society of Civil Engineers (ASCE). 1975. *Wood Structures: A Design Guide and Commentary,* ASCE, New York, NY.
[4.10] American Society of Civil Engineers (ASCE). 1982. *Evaluation, Maintenance and Upgrading of Wood Structures,* ASCE, New York, NY.
[4.11] American Society of Civil Engineers (ASCE). 1989. *Classic Wood Structures,* ASCE, New York, NY.
[4.12] American Society of Civil Engineers (ASCE). 2003. *Minimum Design Loads for Buildings and Other Structures, ASCE 7-02,* ASCE, New York, NY.
[4.13] American Society for Testing and Materials (ASTM). 1998. "Standard Test Methods for Establishing Clear-Wood Strength Values," ASTM D2555-98, *Annual Book of Standards, Vol. 04.10 Wood,* ASTM, Philadelphia, PA.
[4.14] American Society for Testing and Materials (ASTM). 2000. "Standard Practice for Establishing Structural Grades and Related Allowable Properties for Visually Graded Lumber," ASTM D245-00e1, *Annual Book of Standards, Vol. 04.10 Wood,* ASTM, Philadelphia, PA.
[4.15] American Society for Testing and Materials (ASTM). 2000. "Standard Practice for Establishing Allowable Properties for Visually-Graded Dimension Lumber from In-Grade Tests of Full-Size Specimens," ASTM D1990-00e1, *Annual Book of Standards, Vol. 04.10 Wood,* ASTM, Philadelphia, PA.
[4.16] American Society for Testing and Materials (ASTM). 1997. "Standard Methods of Testing Small Clear Specimens of Timber," ASTM D143-94, *Annual Book of Standards, Vol. 04.10 Wood,* ASTM, PHiladelphia, PA.
[4.17] American Society for Testing and Materials (ASTM). 1997. "Standard Test Methods for Mechanical Properties of Lumber and Wood-Base Materials," ASTM D4761-93, *Annual Book of Standards, Vol. 04.10 Wood,* ASTM, Philadelphia, PA.
[4.18] APA—The Engineered Wood Association. 1992. *Moisture Control in Load Slope Roofs, Technical Note EWS R525,* APA—The Engineered Wood Association, Engineered Wood Systems, Tacoma, WA.
[4.19] APA—The Engineered Wood Association. 2001. *Controlling Mold and Mildew. Form A525.* APA—The Engineered Wood Association, Tacoma, WA.
[4.20] Dietz, A.G., Schaffer, E.L., and Gromala, D.S. (eds.). 1982. *Wood as a Structural Material,* Clark C. Heritage Memorial Series, vol. 2, Pennsylvania State University, University Park, PA.
[4.21] Faherty, K.F., and Williamson, T.G. (eds.). 1995. *Wood Engineering and Construction Handbook,* 2nd ed., McGraw-Hill, New York, NY.
[4.22] Forest Products Laboratory (FPL). 1999. *Wood Handbook: Wood as an Engineering Material, Technical Report 113,* FPL, Forest Service, U.S.D.A., Madison, WI.
[4.23] Green, D.W. 1989. "Moisture Content and the Shrinkage of Lumber," Research Paper FPL-RP-489, Forest Products Laboratory, Forest Service, U.S.D.A., Madison, WI.
[4.24] Gurfinkel, G. 1981. *Wood Engineering,* 2nd ed., Kendall/Hunt Publishing (available through Southern Forest Products Association, Kenner, LA).
[4.25] Hoyle, R.J., and Woeste, F.E. 1989. *Wood Technology in the Design of Structures,* 5th ed., Iowa State University Press, Ames, IA.
[4.26] Meyer, R.W., and Kellogg, R.M. (eds.). 1982. *Structural Use of Wood in Adverse Environments,* Van Nostrand Reinhold, New York, NY.
[4.27] Pneuman, F.C. 1991. "Inspection of a Wood-Framed-Warehouse-Type Structure," *Wood Design Focus,* vol. 2, no. 2.
[4.28] Rummelhart, R., and Fantozzi, J.A. 1992. "Multistory Wood-Frame Structures: Shrinkage Considerations and Calculations," *Proceedings of the 1992 ASCE Structures Congress,* American Society of Civil Engineers, New York, NY.

4.25   Problems

Design values and adjustment factors in the following problems are to be taken from the 2001 NDS and ASD Manual. Assume wood will be used in dry-service conditions and at normal temperatures unless otherwise noted.

4.1   *a.* Describe softwoods.
b. Describe hardwoods.
c. What types of trees are used for most structural lumber?

4.2   Sketch the cross section of a log. Label and define the following items:
a. Annual ring
b. The two types of wood cells
c. Heartwood and sapwood

4.3   Define the following terms:
a. Moisture content
b. Fiber saturation point
c. Equilibrium moisture content

4.4   Give the moisture content ranges for:
a. Dry lumber
b. Green lumber

4.5   What is the average EMC for an enclosed building in southern California? Cite reference.

4.6   List the components of the *Allowable Stress Design Manual for Engineered Wood Construction.*

4.7   *a.* List the *supplements* of the ASD Manual.
b. List the *guidelines* of the ASD Manual.
c. What is the difference between a *supplement* and a *guideline* in the ASD Manual?
d. Why is the supplement for *special design provisions for wind and seismic* separated from the other supplements?

4.8   Determine the dressed size, area, moment of inertia, and section modulus for the following members. Give values for both axes. Tables may be used (cite reference).
a. 2 × 4
b. 8 × 8
c. 4 × 10
d. 6 × 16

4.9   *a.* Give the range of sizes of lumber in Dimension lumber.
b. Give the range of sizes of lumber in Timbers.
c. Briefly summarize why the design values in the NDS Supplement for members in these broad categories are given in separate tables. What tables apply to Dimension lumber and what tables apply to Timbers?

4.10 Give the range of sizes for the following size and use subcategories. In addition, indicate whether these categories are under the general classification of *Dimension lumber* or *Timbers.*
a. Beams and Stringers
b. Structural Light Framing
c. Decking
d. Structural Joists and Planks
e. Posts and Timbers
f. Light Framing
g. Stud

4.11 Briefly describe what is meant by the terms "visually graded sawn lumber" and "machine stress-rated (MSR) lumber." What tables in the NDS Supplement give design values for each? Are there any size distinctions? Explain.

4.12 Assume that the following members are visually graded lumber from a species group other than Southern Pine. Indicate whether the members are a size of Dimension lumber, Beams and Stringers (B&S), or Posts and Timbers (P&T). Also give the appropriate table in the NDS Supplement for obtaining design values. The list does not include material that is graded as Decking.
a. 10 × 12   e. 2 × 12
b. 14 × 14   f. 6 × 12
c. 4 × 8   g. 8 × 12
d. 4 × 4   h. 8 × 10

4.13 Repeat Prob. 4.12 except the material is Southern Pine.

4.14 What stress grades are listed in the NDS Supplement for visually graded Hem-Fir in the following size categories? Give table reference.
a. Dimension lumber
b. Beams and Stringers (B&S)
c. Posts and Timbers (P&T)

4.15 What stress grades are listed in the NDS Supplement for visually graded Southern Pine in the following size categories? Give table reference.
a. Dimension lumber
b. Beams and Stringers (B&S)
c. Posts and Timbers (P&T)

4.16 Give the tabulated design values for No.1 DF-L for the following sizes. List values for F_b, F_t, F_v, $F_{c\perp}$, F_c, and E. Give table reference.
a. 10 × 10   e. 2 × 10
b. 12 × 14   f. 6 × 12
c. 4 × 16   g. 6 × 8
d. 4 × 4   h. 10 × 14

4.17 Give the notation for the following stress adjustment factors. In addition, list the design properties (that is, F_b, F_t, F_v, $F_{c\perp}$, F_c, or E) that may require adjustment (NDS Table 4.3.1) by the respective factors.

        a. Size factor   *e.* Temperature factor
        b. Form factor   *f.* Wet service factor
        c. Load duration factor   *g.* Flat use factor
        d. Repetitive member factor

4.18 Briefly describe the following adjustments. To what design values do they apply? Give NDS reference for numerical values of adjustment factors.
        a. Load duration factor
        b. Wet service factor
        c. Size factor
        d. Repetitive member factor

4.19 Briefly describe why the shear stress adjustment factor has been eliminated from the 2001 NDS.

4.20 Regarding wind and seismic forces, distinguish between the terms *load duration factor* and load *combination* factor. Refer to Sec. 2.8 to help answer this question.

4.21 Give the load duration factor C_D associated with the following loads:
        a. Snow
        b. Wind
        c. Floor live load
        d. Roof live load
        e. Dead load

4.22 What tabulated design values for wood, if any, are not subject to adjustment for duration of loading?

4.23 Under what conditions is a *reduction* in tabulated values for wood design required based on duration of loading?

4.24 Above what moisture content is it necessary to reduce the allowable stresses for most species of (*a*) sawn lumber and (*b*) glulam?

4.25 Under what conditions is it necessary to adjust the allowable stresses in wood design for temperature effects? Cite NDS reference for the temperature modification factors.

4.26 Distinguish between pressure-preservative-treated wood and fire-retardant-treated wood. Under what conditions is it necessary to adjust allowable stresses in wood design for the effects of pressure-impregnated chemicals? Where are adjustment factors obtained?

4.27 Should lumber be pressure-treated if it is to be used in an application where it will be continuously submerged in fresh water? Salt water? Explain.

4.28 *Given:*   A column in a building is subjected to several different loads, including roof loads of D = 3 k and L_r = 5 k; floor loads of D = 6 k and L = 10 k; and W = 10 k (resulting from overturning forces on the lateral-force-resisting system). Assume that the column is a short column with full lateral support, and, therefore, the load duration factor C_D applies. Consider the load combinations described in Sec. 2.16.

Find: The critical combination of loads.

4.29   *Given:*   A column in a building is subjected to several different loads, including roof loads of D = 10 k and L_r = 2 k; floor loads of D = 8 k and L = 10 k; and W = 6 k (resulting from overturning forces on the lateral-force-resisting system). Assume that the column is a short column with full lateral support, and, therefore, the load duration factor C_D applies. Consider the load combinations described in Sec. 2.16.

Find: The critical combination of loads.

4.30   *Given:*   A column in a building is subjected to several different loads, including roof loads of D = 5 k and L_r = 7 k; floor loads of D = 6 k and L = 15 k; S = 18 k; and W = 10 k and E = 12 k (resulting from overturning forces on the lateral-force-resisting system). Assume that the column is a short column with full lateral support, and, therefore, the load duration factor C_D applies. Consider the load combinations described in Sec. 2.16.

Find: The critical combination of loads.

4.31   A column in a structure supports a water tank. The axial load from the tank plus water is P_w, and the axial load resulting from the lateral overturning force is P_l. Because the contents of the tank are present much of the time, the load P_w is considered a permanent load.

Find: The critical load combination for each of the following loadings. Assume that the column is a short column with full lateral support, and, therefore, the load duration factor C_D applies.
    a. P_w = 60 k; P_l = 10.5 k
    b. P_w = 60 k; P_l = 40.3 k
    c. P_w = 60 k; P_l = 55.1 k

4.32   Determine the tabulated and allowable design values for the following members and loading conditions. All members are No. 2 Hem-Fir. Bending occurs about the strong axis.
    a. Roof joists are 2 × 10 at 16 in. o.c. which directly support the roof sheathing. Loads are (D + S).
    b. A 6 × 14 carries an equipment load that can be considered a permanent load.
    c. Purlins in a roof are 4 × 14 at 8 ft o.c. Loads are (D + L_r).
    d. Floor beams are 4 × 6 at 4 ft o.c. Loads are (D + L). High-humidity conditions exist, and the moisture content may exceed 19 percent.

4.33   Determine the tabulated and allowable design values for the following members and loading conditions. All members are Select Structural Southern Pine. Bending occurs about the strong axis.
    a. Roof joists are 2 × 6 at 24 in. o.c. which directly support the roof sheathing. Loads are (D + S).
    b. A 4 × 12 supports (D + L + L_r).
    c. Purlins in a roof are 2 × 10 at 4 ft o.c. Loads are (D + L_r).
    d. Floor beams are 4 × 10 at 4 ft o.c. Loads are (D + L + W).

4.34   Estimate the amount of shrinkage that will occur in the depth of the beam in Fig. 4.A. Use the simplified shrinkage approach recommended in Ref. 4.28. Assume an initial moisture content of 19 percent and a final moisture content of 10 percent.

NOTE: The top of the wood beams should be set higher than the top of the girder by an amount equal to the estimated shrinkage. After shrinkage, the roof sheathing will be supported by beams and girders that are all at the same elevation. Without this allowance for shrinkage, a wave or bump may be created in the sheathing where it passes over the girder.

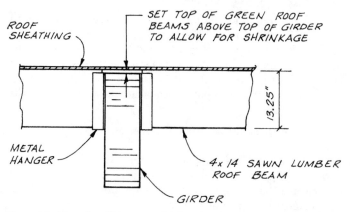

Figure 4.A   Top of roof beams set higher for shrinkage.

4.35   Estimate the total shrinkage that will occur in a four-story building similar to the one in Example 4.3. Floor joists are 2 × 10's instead of 2 × 12's. The initial moisture content can be taken as 19 percent, and the final moisture content is assumed to be 9 percent. All other information is the same as in Example 4.3.

4.36   Use a personal computer spreadsheet or a database to input the tabulated design values for one or more species (as assigned) of sawn lumber. Include values for both Dimension lumber, Beams-and-Stringers, and Posts-and-Timbers sizes.

　　　The purpose of the spreadsheet is to list *tabulated* design values (F_b, F_t, F_v, $F_{c\perp}$, F_c, or E) as output for a specific problem with the following input being provided by the user:

a. Species (if values for more than one species are in spreadsheet or database)
b. Stress grade of lumber (e.g., Select Structural, No. 1, etc.)
c. Nominal size of member (for example, 2 × 4, 6 × 12, 6 × 6, etc.)

4.37   Expand or modify the spreadsheet from Prob. 4.36 to develop *allowable* design values. The spreadsheet should be capable of applying all the adjustment factors introduced in Chap. 4 except C_{fu}, C_H, and C_t. The input should be expanded to provide sufficient information to the spreadsheet template so that the appropriate adjustment factors can be computed or drawn from a database or table.

　　　Output should include a summary of the adjustment factors and the final design values F_b', F_t', F_v', $F_{c\perp}'$, F_c', or E'. Default values of unity may be listed for any adjustment factor that does not apply.

5

Structural
Glued Laminated
Timber

5.1 Introduction

Sawn lumber is manufactured in a large number of sizes and grades (Chap. 4) and is used for a wide variety of structural members. However, the cross-sectional dimensions and lengths of these members are limited by the size of the trees available to produce this type of lumber.

When the span becomes long or when the loads become large, the use of sawn lumber may become impractical. In these circumstances (and possibly for architectural reasons) structural glued laminated timber (*glulam*) can be used.

Glulam members are fabricated from relatively thin laminations (nominal 1 and 2 in.) of wood. These laminations can be end-jointed and glued together in such a way to produce wood members of practically any size and length. Lengths of glulam members are limited by handling systems and length restrictions imposed by highway transportation systems rather than by the size of the tree.

This chapter provides an introduction to glulam timber and its design characteristics. The similarities and differences between glulam and sawn wood members are also noted.

5.2 Sizes of Glulam Members

The specifications for glulam permit the fabrication of a member of any width and any depth. However, standard practice has resulted in commonly accepted widths and thicknesses of laminations (see Ref. 5.6). The generally accepted dimensions for glulams fabricated from the Western Species are slightly different from those for Southern Pine glulams as given in NDS Table 5.1.3 (Ref.

5.1). See Fig. 5.1. Because of surfacing requirements, Southern Pine lamina-tions are usually thinner and narrower, although they can be manufactured to the same net sizes as Western Species if necessary. The dimensions given in Fig. 5.1 are net sizes, and the total depth of a member will be a multiple of the lamination thickness.

Straight or slightly curved glulams will be fabricated with 1½-in. (or 1⅜-in.) laminations. If a member is sharply curved, thinner (¾-in. or less) lami-nations should be used in the fabrication because smaller built-in, or residual, stresses will result. These thinner laminations are not used for straight or slightly curved glulams because cost is heavily influenced by the number of

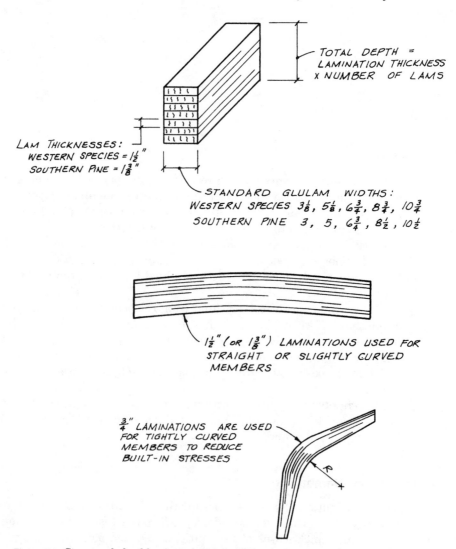

Figure 5.1 Structural glued laminated timber (glulam).

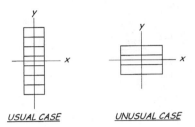

USUAL CASE   UNUSUAL CASE

Figure 5.2 The orientation of the x and y axes for glulams is actually related to the orientation of the laminations, not the strong and weak directions. The usual case is for a rectangular beam with its depth significantly greater than its width, and thus the x axis is parallel to the laminations and is the strong axis of the section. The unusual case is where the depth is less than the width. In this unusual situation, the x axis for the glulam, being parallel to the laminations, is not the strong axis of the section.

glue lines in a member. Only the design of straight and slightly curved rectangular members is included in this text. The design of tapered members and curved members (including arches) is covered in the *Timber Construction Manual* (TCM, Ref. 5.7).

The sizes of glulam members are called out on plans by giving their net dimensions (unlike sawn lumber which uses "nominal" sizes). Cross-sectional properties for glulams are listed in the 2001 NDS Supplement Table 1C, Section Properties of *Western Species* Glued Laminated Timber, and Table 1D, Section Properties of *Southern Pine* Glued Laminated Timber. Section properties include

1. Cross-sectional area A (in.^{2})

2. Section modulus about the strong axis S_x (in.^{3})

3. Moment of inertia about the strong axis I_x (in.^{4})

4. Radius of gyration about the strong axis r_x (in.)

5. Section modulus about the weak axis S_y (in.^{3})

6. Moment of inertia about the weak axis I_y (in.^{4})

7. Radius of gyration about the weak axis r_y (in.)

Strictly speaking, the x and y axes are not always the strong and weak axes as indicated in the above listing. For glulams, the x and y orientation is actually related to the orientation of the laminations, not the strong and weak axes of the section. See Fig. 5.2. For the vast majority of glulams produced and used in structural applications, the *usual* case orientation (Fig. 5.2), of the x axis being parallel to the laminations *and* being the strong axis, holds true. Therefore, the above definitions for the section properties will be used throughout this book.

Section properties for glulam members are determined using the same basic principles illustrated in Example 4.5 (Sec. 4.11). The approximate weight per linear foot for a given size glulam can be obtained by converting the cross-sectional area in NDS Supplement Table 1C or 1D from in.^{2} to ft^2 and multiplying by the following unit weights:

Type of glulam	Unit weight
Southern Pine	36 pcf
Western Species	
Douglas Fir-Larch	35 pcf
Alaska Cedar	35 pcf
Hem-Fir and California Redwood	27 pcf

5.3   Resawn Glulam

In addition to the standard sizes of glulams shown in Fig. 5.1, NDS Supplement Tables 1C and 1D give section properties for a narrower width glulam. Glulams that are 2½ in. wide are obtained by ripping a glulam manufactured from nominal 2 × 6 laminations into two pieces. The relatively narrow beams that are produced in this way are known as *resawn glulams*. See Fig. 5.3. Although section properties are listed only for 2½ in.-wide beams, wider resawn members can be produced from glulams manufactured from wider laminations.

The resawing of a glulam to produce two narrower members introduces some additional manufacturing controls that are not required on the production of a normal-width member which is not going to be resawn. For example, certain strength-reducing characteristics (such as knot size or location) may be permitted in a 5⅛-in. glulam that is not going to be resawn. If the member is going to be resawn, a more restrictive set of grading limitations apply.

Resawn glulams are a fairly recent development in the glulam industry. These members can have large depths. With a narrow width and a large depth, resawn glulams produce beams with very efficient cross sections. In other words, the section modulus and moment of inertia for the strong axis (that is, S_x and I_x) are large for the amount of material used in the production

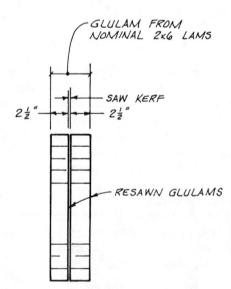

Figure 5.3 Resawn glulams are obtained by longitudinally cutting standard-width glulams to form narrower members. For example, a 2½-in.-wide member is obtained by resawing a glulam manufactured from nominal 2 × 6 laminations. A resawn glulam is essentially an industrial-use (i.e., not architectural) beam because three sides of the members are finished and one side is sawn.

of the member. On the negative side, very narrow members are weak about the minor axis (that is, S_y and I_y are small). The relatively thin nature of these members requires that they be handled properly in the field to ensure that they are not damaged during construction. In addition, it is especially important that the compression edge of a deep, narrow beam be properly braced so that the member does not buckle when a load is applied. Bracing of the tension edge, other than at the supports, is not necessary except in cases where moment reversal is anticipated.

Resawn glulams are used as an alternative to certain sizes of sawn lumber. They also provide an alternative to wood I joists in some applications. Resawn beams are normally used where appearance is not a major concern.

5.4 Fabrication of Glulams

Specifications and guidelines covering the design and fabrication of glulam members (Refs. 5.4, 5.5, 5.8, 5.10, 5.11), are published by the American Institute of Timber Construction (AITC) and Engineered Wood Systems (EWS), a related corporation of APA—The Engineered Wood Association. AITC and EWS are technical trade associations representing the structural glued laminated timber industry. AITC also publishes the *Timber Construction Manual* (TCM, Ref. 5.7), which was introduced in Chap. 1 and is referenced throughout this book. The TCM is a wood engineering handbook that can be considered the basic reference on glulam (for convenience, it also includes information on other structural wood products such as sawn lumber).

Most structural glulam members are produced using Douglas Fir or Southern Pine. Hem-Fir, Spruce-Pine-Fir, Alaska Cedar, and various other species including hardwood species can also be used. Quality control standards ensure the production of a reliable product. In fact, the structural properties of glulam members in most cases exceed the structural properties of sawn lumber.

The reason that the structural properties for glulam are so high is that the material included in the member can be selected from relatively high-quality laminating stock. The growth characteristics that limit the structural capacity of a large solid sawn wood member can simply be excluded in the fabrication of a glulam member.

In addition, laminating optimizes material use by dispersing the strength-reducing defects in the laminating material throughout the member. For example, consider the laminations that are produced from a sawn member with a knot that completely penetrates the member at one section. See Fig. 5.4. If this member is used to produce laminating stock which is later reassembled in a glulam member, it is unlikely that the knot defect will be reassembled in all the laminations at exactly the same location in the glulam member. Therefore, the reduction in cross-sectional properties at any section consists only of a portion of the original knot. The remainder of the knot is distributed to other locations in the member.

Besides dispersing strength-reducing characteristics, the fabrication of glulam members makes efficient use of available structural materials in another

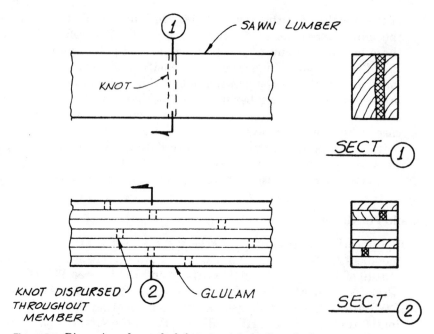

Figure 5.4 Dispersion of growth defects in glulam. Growth characteristics found in sawn lumber can be eliminated or (as shown in this sketch) dispersed throughout the member to reduce the effect at a given cross section.

way. High-quality laminations are located in the portions of the cross section which are more highly stressed. For example, in a typical glulam beam, wood of superior quality is located in the outer tension and compression zones. This coincides with the location of maximum bending stresses under typical loading. See Example 5.1. Although the maximum bending compressive and tensile stresses are equal, research has demonstrated that the outer laminations in the tension zone are the most critical laminations in a beam. For this reason, additional grading requirements are used for the outer tension laminations.

The different grades of laminations over the depth of the cross section really make a glulam a composite beam. Recall from strength of materials that a composite member is one that is made up of more than one material with different values of modulus of elasticity. Composite members are analyzed using the transformed section method. The most obvious example of a composite member in building construction is a reinforced-concrete beam, but a glulam is also a composite member because the different grades of laminations have different E's.

However, from a designer's point of view, a glulam beam can be treated as a homogeneous material with a rectangular cross section. Allowable stresses have been determined in accordance with ASTM D 3737 (Ref. 5.9) using transformed sections. All glulam design values have been mathematically transformed to allow the use of apparent rectangular section properties. Thus, ex-

EXAMPLE 5.1   Distribution of Laminations in Glulam Beams

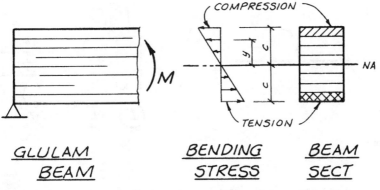

Figure 5.5

Bending stress calculation:

$$\text{Arbitrary point} \qquad f_b = \frac{My}{I}$$

$$\text{Maximum stress} \qquad f_b = \frac{Mc}{I}$$

In glulam beams, high-quality laminations are located in areas of high stress (i.e., near the top and bottom of the beam). Lower-quality wood is placed near the neutral axis where the stresses are lower. The outer tension laminations are critical and require the highest-grade stock.

cept for differences in design values and section properties, a glulam design is carried out in much the same manner as the design of a solid sawn beam.

Glulam beams are usually loaded in bending about the strong axis of the cross section. Large section properties and the distribution of laminations over the depth of the cross section make this an efficient use of materials. This is the loading condition assumed in Example 5.1, and bending about the strong axis should be assumed unless otherwise noted. In the tables for glulam design values, bending about the *strong axis* is described as the transverse load being applied *perpendicular to the wide face of the laminations*. See Example 5.2. Loading about the minor axis is also possible, but it is much less common. One common example of glulam beams loaded about the minor axis is bridge decks. Different tabulated stresses apply to members loaded about the x and y axes.

EXAMPLE 5.2   Bending of Glulams

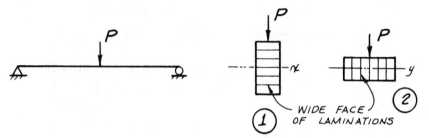

Figure 5.6

Bending can occur about either the x or y axis of a glulam, or both. In section 1 the load is perpendicular to the wide faces of the laminations, and bending occurs about the major axis of the member. This is the more common situation. In section 2 the load is parallel to the wide faces of the laminations, and bending occurs about the weak or minor axis of the member.

Laminations are selected and dried to a moisture content of 16 percent or less before gluing. Differences in moisture content for the laminations in a member are not permitted to exceed 5 percent in order to minimize internal stresses and checking. Because of the relatively low moisture content (MC) of glulam members at the time of fabrication, the change in moisture content in service (i.e., the initial MC minus the EMC) is generally much smaller for glulam than it is for sawn lumber. Thus glulams are viewed as being more dimensionally stable. Even though the percent change in MC is normally less, the depth of a glulam is usually much larger than that of a sawn lumber member. Thus the possible effects of shrinkage need to be considered in glulam design. See Sec. 4.7 for more discussion of shrinkage and Chap. 14 for recommendations about how to avoid shrinkage-related problems in connections.

Traditionally, two types of glue have been permitted in the fabrication of glulam members: (1) dry-use adhesives (casein glue) and (2) wet-use adhesives (usually phenolresorcinol-base, resorcinol-base, or melamine-base adhesives). While both types of glue are capable of producing joints which have horizontal shear capabilities in excess of the capacity of the wood itself, today only wet-use adhesives are permitted in glulam manufacturing, according to ANSI/ AITC 190.1-2002 (Ref. 5.3). Wet-use adhesives have been used almost exclusively for a number of years and only recently have their use become required. The increased, and now required, use of wet-use adhesives was made possible with the development of room-temperature-setting glues for exterior use. Wet-use adhesives, as the name implies, can withstand severe conditions of exposure.

The laminations run parallel to the length of a glulam member. The efficient use of materials and the long length of many glulam members require that effective end splices be developed in a given lamination. While several different configurations of lamination *end-joint* splices are possible including *finger* and *scarf* joints (see Fig. 5.7), virtually all glued laminated timber produced in North America uses some form of finger joint.

Finger joints produce high-strength joints when the fingers have relatively flat slopes. The fingers have very small blunt tips to ensure adequate bearing pressure. Finger joints also make efficient use of laminating stock because the lengths of the fingers are usually short in comparison with the lengths of scarf joints. With scarf joints, the flatter the slope of the joint, the greater the strength of the connection. Scarf slopes of 1 in 5 or flatter for compression and 1 in 10 or flatter for tension are recommended (Ref. 5.12).

If the width of the laminating stock is insufficient to produce the required net width of glulam, more than one piece of stock can be used for a lamination. The *edge joints* in a lamination can be glued. However, the vast majority of glulam producers do not edge-glue laminates. Rather, the edge joints are staggered in adjacent laminations.

Although one should be aware of the basic fabrication procedures and concepts outlined in this section, the building designer does not have to be con-

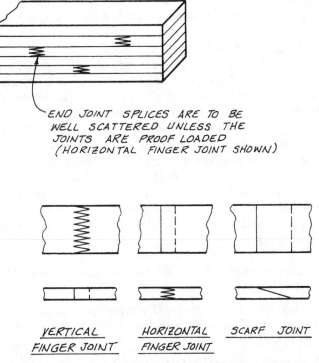

Figure 5.7 End-joint splices in laminating stock. Most glulam fabricators use either the vertical or horizontal finger joints for end-joint splices. In addition, proof loading of joints is common, and in this case the location of end joints is not restricted.

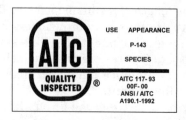 

Figure 5.8 Typical grade stamps for glulam (*courtesy of AITC and EWS*).

cerned about designing the individual laminations, splices, and so on. In fact AITC has separated its laminating specification into two parts. One gives structural engineering properties and is titled *Design* (Ref. 5.4), and the other deals with the fabrication requirements for glulam and is titled *Manufacturing* (Ref. 5.5). The design values from Ref. 5.4 are reproduced in the NDS Supplement for convenience, and problems in this book refer to the NDS tables. In practice, designers involved with glulam on a regular basis should obtain copies of Refs. 5.4 and 5.5 as well as a number of other AITC and EWS publications.

The manufacturing standards for glulam are based on ANSI/AITC A190.1, *Structural Glued Laminated Timber* (Ref. 5.3), and implementation is ensured through a quality control system. Quality assurance involves the inspection and testing of glulam production by a qualified agency. The majority of glulam produced in the United States is inspected by two agencies: AITC Inspection Bureau and Engineered Wood Systems (EWS), a related corporation of the APA. Each glulam is grade-stamped for identification purposes. See Fig. 5.8. In addition, because of the importance of the tension laminations, the top of a glulam bending member using an unbalanced layup is also marked with a stamp. This identification allows construction personnel in the field to orient the member properly in the structure (i.e., get it right side up). If a glulam were inadvertently turned upside down, the compression laminations would be stressed in tension and the strength of the member could be greatly reduced. For applications such as continuous or cantilevered beams, the designer should specify a balanced layup that has high quality tension laminations on both the top and bottom of the member and therefore has equivalent positive and negative moment capacities. Some of the items in the grade stamp include

1. Quality control agency (e.g., American Institute of Timber Construction or Engineered Wood Systems)

2. Structural use (possible symbols: *B*, simple span bending member; *C*, compression member; *T*, tension member; and CB, continuous or cantilever bending member)

3. Appearance grade (FRAMING, framing; IND, industrial; ARCH, architectural; PREM, premium).

4. Plant or mill number (for example, 143 and 0000 shown)

5. Standard for structural glued laminated timber (i.e., ANSI/AITC A190.1-1992)

6. Laminating specification and combination symbol (for example, 117-1993 24F-1.8E)

7. Species/species group of the glulam

8. Proof-loaded end joints, if used during the manufacturing process

A complete list of all required markings is provided in ANSI/AITC A190.1 (Ref. 5.3).

5.5 Grades of Glulam Members

For strength, grades of glulam members traditionally have been given as combinations of laminations. The two main types are *bending combinations* and *axial combinations*. New to the 2001 NDS is the *Stress Class System* for softwood glulam bending combinations. The new Stress Class System is recommended by the glulam industry for specifying beams, as it will greatly benefit designers and manufacturers alike. Softwood glulam bending combinations with similar properties have been grouped into stress classes, markedly reducing the complexity in selecting an appropriate grade combination. With the new Stress Class System, the number of tabulated values has also been reduced, to simplify things for the designer.

In addition to grading for strength, glulam members are graded for appearance. One of the four appearance grades (Framing, Industrial, Architectural, and Premium) should be specified along with the strength requirements to ensure that the member furnished is appropriate for the intended use. It is important to understand that the selection of an appearance grade does not affect the strength of a glulam (see Ref. 5.2 for additional information), but can increase costs due to the required additional finishing to achieve the higher quality appearance.

Members that are stressed principally in bending and loaded in the normal manner (i.e., with the applied load perpendicular to the wide faces of the laminations) are produced from the bending combinations. Bending combinations are defined by a *combination symbol* and the *species* of the laminating stock. The combination symbol is made up of two parts. The first is the allowable bending stress for the grade in hundreds of psi followed by the letter F. For example, 24F indicates a bending combination with a tabulated bending stress of 2400 psi for normal duration of loading and dry-service conditions. Bending combinations that are available include 16F, 20F, 22F, 24F, and 26F flexural stress levels for most species. Additionally, 28F and 30F combinations are available for Southern Pine glulams.

It should be noted that a number of combinations of laminations can be used to produce a given bending stress level. Therefore, there is an abbreviation that follows the bending stress level which gives the distribution of laminating stock to be used in the fabrication of a member. Two basic abbreviations are used in defining the combinations: one is for visually graded laminating stock (for example, 24F-V3), and the other is for laminating stock that is mechanically graded, or E-rated, for stiffness (for example, 22F-E5).

In addition to the combination symbol, the species of wood is required to define the grade. The symbols for the species are DF for Douglas Fir-Larch, DFS for Douglas Fir South, HF for Hem-Fir, SW for Softwoods,* and SP for Southern Pine. Section 5.4 indicated that higher-quality laminating stock is located at the outer faces of a glulam bending combination, and lower-quality stock is used for the less highly stressed inner zone. In a similar manner, the laminating specifications allow the mixing of more than one species of wood in certain combinations. The idea is again to make efficient use of raw materials by allowing the use of a strong species for the outer laminations and a weaker species for the center core.

If more than one species of wood is used in a member, both species are specified (for example, DF/HF indicates DF outer laminations and HF inner core laminations). If only one species is used throughout the member, the species symbol is repeated (for example, DF/DF). Although the laminating specification allows the mixing of more than one species, most glulam production currently uses laminating stock from only one species of wood for a given member.

As mentioned, the new *Stress Class System* for softwood glulam bending combinations has been introduced in the 2001 NDS. The basic premise of the Stress Class System is to simplify the designer's choices and give the manufacturer more flexibility to meet the needs of the designer. The glulam industry has had a longstanding position that designers should specify by required design stresses, rather than by combination symbol. This was intended to give manufacturers flexibility in choosing combinations to fit their resource. Unfortunately, the design community has never broadly followed this practice. Designers are trained to choose a material and then size the member appropriately, so engineers continue to specify by combination symbol, leaving the manufacturer with no flexibility. This old system of specifying combinations and species was also far from simple for a new designer. In looking at the glulam design tables in the NDS, it was easy to become overwhelmed with the choices. The new Stress Class System should improve the design process for the designer and the ability of the manufacturer to meet the design requirements.

Softwood bending combinations with similar properties are now grouped into stress classes. All of the design properties in a higher stress class equal or exceed those from the lower stress class. This allows designers and manufacturers the ability to substitute higher grades based on availability. The number of tabulated values has also been reduced to simplify things for the designer. Since glulam combinations are derived from broad ranges of species

---

*Formerly, the symbol WW was used for Western Species. This was changed in AITC 117-2001 to SW for softwoods, because the species allowed for this glulam combination do not match the lumber grading agencies' definitions of "Western Woods."

groups, some anomalies are bound to be present. Accordingly, unusual cases have been footnoted in the design tables instead of being separately tabulated. This is intended to further simplify the design process.

As with the bending combinations, stress classes are defined by a two part *stress class symbol*. The first part of the symbol is the allowable bending stress for the class in hundreds of psi followed by the letter F. For example, 24F indicates a stress class with a tabulated bending stress of 2400 psi for normal duration of loading and dry-service conditions. Stress classes that are readily available are 16F, 20F, and 24F. Higher stress classes of 26F, 28F, and 30F may be available from some manufacturers.

The second part of the stress class symbol also provides information about a design value of the glulam. Recall that for a traditional bending combination, the second part of the combination symbol indicates the distribution of laminating stock used in the fabrication of the member (for example, V3 for visually-graded laminating stock or E5 for E-rated laminating stock). With the Stress Class System, the second part of the symbol is the bending modulus of elasticity in millions of psi. For example, 24F-1.8E indicates a stress class with a tabulated bending stress of 2400 psi and modulus of elasticity of 1.8×10^6 psi.

The 2001 NDS has the stress classes in Table 5A and the individual combinations in each stress class are shown in Table 5A Expanded. It is the intent of the glulam industry to transition from NDS Table 5A Expanded to NDS Table 5A. However, AITC and APA-EWS will likely continue to list the design values for the individual combinations after this transition, because there may be special cases where use of a particular combination is desirable.

Members which are principally axial-load-carrying members are identified with a numbered combination symbol such as 1, 2, 3, and so on. The new Stress Class System applies only to softwood glulam members, used primarily in bending. Members stressed primarily in axial tension or compression are designed using numbered combinations. See NDS Supplement Table 5B. Because axial load members are assumed to be uniformly stressed throughout the cross section, the distribution of lamination grades is uniform across the member section, compared with the distribution of lamination quality used for beams.

Glulam combinations are, in one respect, similar to the "use" categories of sawn lumber. The *bending combinations* anticipate that the member will be used as a beam, and the *axial combinations* assume that the member will be loaded axially.

Bending combinations are often fabricated with higher-quality laminating stock at the outer fibers, and consequently they make efficient beams. This fact, however, does not mean that a bending combination cannot be loaded axially. Likewise, an axial combination can be designed for a bending moment. The combinations, then, have to do with efficiency, but they do not limit the use of a member. The ultimate use is determined by stress calculations.

Design values for glulams are listed in the following tables in the 2001 NDS Supplement:

Table 5A *Design Values for Structural Glued Laminated Softwood Timber (Members stressed primarily in bending)*. These are the bending stress classes.

Table 5A Expanded *Design Values for Structural Glued Laminated Softwood Timber (Members stressed primarily in bending)*. These are the bending combinations that meet the requirements of each stress class in Table 5A.

Table 5B *Design Values for Structural Glued Laminated Softwood Timber (Members stressed primarily in axial tension and compression)*. These are the axial load combinations.

Table 5C *Design Values for Structural Glued Laminated Hardwood Timber (Members stressed primarily in bending)*. These are the bending combinations.

Table 5D *Design Values for Structural Glued Laminated Hardwood Timber (Members stressed primarily in axial tension and compression)*. These are the axial load combinations.

Tabulated design values for glulam members are also available in a number of other publications including Refs. 5.4, 5.7, 5.8, 5.10, and 5.11. The tables include the following properties:

Bending stress F_{bx} and F_{by}

Tension stress parallel to grain F_t

Shear stress parallel to grain F_{vx} and F_{vy}

Compression stress parallel to grain F_c

Compression stress perpendicular to grain $F_{c\perp x}$ and $F_{c\perp y}$

Modulus of elasticity E_x, E_y, and E_{axial}

Tabulated design values for glulam are the same basic stresses that are listed for solid sawn lumber, but the glulam tables are more complex. The reason for this is the way glulams are manufactured with different grades of laminations. As a result, different design properties apply for bending loads about both axes, and a third set is provided for axial loading. It is suggested that the reader accompany this summary with a review of NDS Supplement Tables 5A, 5B, 5C, and 5D.

For softwood glulam members the stresses for the more common application of a glulam member are listed first in the tables. For example, a *member stressed primarily in bending* will normally be used as a beam loaded about the strong axis, and design values for loading about the x axis are the first values given in NDS Table 5A and Table 5A Expanded. These are followed by values for loading about the y axis and for axial loading.

For loading about the x axis, two values of F_{bx} are listed. The first value represents the more efficient use of a glulam, and consequently it is the more frequently used stress in design. F_{bx}^+ indicates that the high-quality tension laminations are stressed in tension (i.e., tension zone stressed in tension). If, for example, a glulam were installed upside down, the second value of F_{bx} would apply. In other words, F_{bx}^- indicates that the lower-quality compression laminations are stressed in tension (i.e., compression zone stressed in tension).

The real purpose for listing F_{bx}^- is not to analyze beams that are installed improperly (although that is one possible use). An application of this stress in a beam that is properly installed is given in Chap. 6. The reason for mentioning the case of a beam being installed upside down is to simply illustrate why the two values for F_{bx} can be so different.

Three values of modulus of elasticity are given in Tables 5A and 5A Expanded: E_x, E_y, and E_{axial}. Values of E_x and E_y are for use in beam deflection calculations about the x and y axes, respectively. They are also used in stability calculations for columns and laterally unbraced beams. On the other hand, E_{axial} is to be used for deformation calculations in members subjected to axial loads, such as the shortening of a column or the elongation of a tension member.

Two design values for compression perpendicular to grain $F_{c\perp x}$ are listed in NDS Table 5A Expanded. One value applies to bearing on the face of the outer tension lamination, and the other applies to bearing on the compression face. The allowable bearing stress may be larger for the tension face because of the higher-quality laminations in the tension zone. Under the new Stress Class System, a single design value for compression perpendicular to grain is listed in NDS Table 5A. The value for $F_{c\perp x}$ listed in Table 5A is taken as the minimum design value for the group of combinations comprising the stress class. Because compression perpendicular to grain rarely governs a design, this conservative approach simplifies the design value table with minimal impact.

Design values listed in NDS Table 5B are for *axial combinations* of glulams, and therefore the properties for axial loading are given first in the table. The distribution of laminations for the axial combinations does not follow the distribution for beams given in Example 5.1. Consequently, values for F_{bx}^+ and F_{bx}^- do not apply to axial combinations. The design stresses for a glulam from an axial combination depend on the number of laminations in a particular member.

Design values for hardwood glulam members are provided in Tables 5C and 5D of the NDS Supplement. These tables are identical in format to Tables 5A Expanded and 5B, which provide design values for softwood glulam members. Prior to the 2001 NDS, the design of hardwood glulam members was significantly different from that of softwood glulam members. The reason for this past difference was that softwood glulam members were more popular and were used to such an extent that establishing various combinations (and now stress classes) was warranted. However, for hardwood glulam members, the use was not as extensive, and the design of the lamination layup was less standardized. Now the use of hardwood glulam members has increased, and the design approach has been unified with that of softwood glulam members. Regardless, the remainder of this chapter focuses on the design of softwood glulam members.

All of the tables for glulam have an extensive set of footnotes which should be consulted for possible modification of design values. Additional information on ordering and specifying glulam members can be obtained from AITC or APA-EWS.

5.6   Stress Adjustments for Glulam

The notation for tabulated stresses, adjustment factors, and allowable stresses is essentially the same for glulam and for sawn lumber. Refer to Sec. 4.13 for a review of the notation used in wood design. The basic system involves the determination of an allowable stress by multiplying the tabulated stress by a series of adjustment factors

$$F' = F \times (\text{product of } C \text{ factors})$$

The tabulated design values for glulams are generally larger than similar properties for sawn lumber. This is essentially a result of the selective placement of laminations and the dispersion of imperfections. However, glulams are a wood product, and they are subject to many of the stress adjustments described in Chap. 4 for sawn lumber. Some of the adjustment factors are numerically the same for glulam and sawn lumber, and others are different. In addition, some adjustments apply only to sawn lumber, and several other factors are unique to glulam design.

The general summary of adjustment factors for use in glulam design is given in NDS Table 5.3.1, *Applicability of Adjustment Factors for Glued Laminated Timber.* Several adjustment factors for glulam were previously described for sawn lumber. A brief description of the similarities and differences for glulam and sawn lumber is given here. Where appropriate the reader is referred to Chap. 4 for further information. The TCM (Ref. 5.7) also includes adjustment factors and design procedures for glulam not covered in the NDS, such as tapered beams which are quite common with glulam.

Wet service factor (C_M)

Tabulated design values for glulam are for dry conditions of service. For glulam, dry is defined as MC < 16 percent. For moisture contents of 16 percent or greater, tabulated stresses are multiplied by C_M. Values of C_M for glulam are given in the Adjustment Factors section preceding NDS Supplement Tables 5A through 5D. When a glulam member is used in high moisture conditions, the need for pressure treatment (Sec. 4.9) should be considered.

Load duration factor (C_D)

Tabulated design values for glulam are for normal duration of load. Normal duration is defined as 10 years and is associated with floor live loads. Loads and load combinations of other durations are taken into account by multiplying by C_D. The same load duration factors are used for both glulam and sawn lumber. See Sec. 4.15 for a complete discussion.

Temperature factor (C_t)

Tabulated design values for glulam are for use at normal temperatures. Section 4.19 discussed design values for other temperature ranges.

Flat use factor (C_{fu})

The flat use factor is somewhat different for sawn lumber and for glulam. For sawn lumber, tabulated values for F_b apply to bending about the x axis. When bending occurs about the y axis, tabulated values of F_b are multiplied by C_{fu} (Sec. 4.18) to convert the value to a property for the y axis.

On the other hand, glulam members have tabulated bending values for both the x and y axes (that is, F_{bx} and F_{by} are both listed). When the depth of the member for bending about the y axis (i.e., the cross-sectional dimension parallel to the wide faces of the laminations) is less than 12 in., the tabulated value of F_{by} may be increased by multiplying by C_{fu}. Values of C_{fu} for glulam are found in the Adjustment Factors section preceding NDS Supplement Tables 5A through 5D.

Because most beams are stressed about the strong axis and not about the y axis, the flat use factor is not a commonly applied adjustment factor. In addition, C_{fu} exceeds unity, and it can conservatively be ignored.

Volume factor (C_v)

It has been noted that the allowable stress in a wood member is affected by the relative size of the member. This general behavior is termed *size effect*. In sawn lumber, the size effect is taken into account by the size factor C_F.

In the past, the same size factor was applied to F_b for glulam that is currently applied to the F_b for sawn lumber in the Timber sizes. Full-scale test data indicate that the size effect in glulam is related to the *volume* of the member rather than to only its depth.

Therefore, the volume factor C_V replaces the size factor C_F for use in glulam design. Note that C_V applies only to bending stress. Tabulated values of F_b apply to a standard-size glulam beam with the following base dimensions: width = 5⅛ in., depth = 12 in., length = 21 ft. The volume factor C_V is used to obtain the allowable bending stress for other sizes of glulams. See Example 5.3. It has been shown that the volume effect is less significant for Southern pine than for other species, and the volume factor is thus species-dependent.

EXAMPLE 5.3   Volume Factor C_V for Glulam

Tabulated values of F_b apply to a glulam with the dimensions shown in Fig. 5.9.

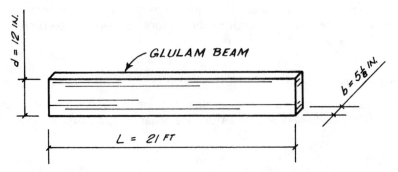

Figure 5.9 *Base dimensions* for tabulated bending design value in glulam.

The allowable bending stress for a glulam of another size is obtained by multiplying the tabulated stress (and other adjustments) by the volume factor.

$$F'_b = F_b \times C_V \times \ldots$$

For Western Species of glulam

$$C_V = \left(\frac{21 \text{ ft}}{L}\right)^{1/10} \left(\frac{12 \text{ in.}}{d}\right)^{1/10} \left(\frac{5.125 \text{ in.}}{b}\right)^{1/10} \le 1.0$$

or

$$C_V = \left(\frac{15,498 \text{ in.}^3}{V}\right)^{1/10} \le 1.0$$

For Southern Pine glulam

$$C_V = \left(\frac{21 \text{ ft}}{L}\right)^{1/20} \left(\frac{12 \text{ in.}}{d}\right)^{1/20} \left(\frac{5.125 \text{ in.}}{b}\right)^{1/20} \le 1.0$$

or

$$C_V = \left(\frac{15,498 \text{ in.}^3}{V}\right)^{1/20} \le 1.0$$

where L = length of beam between points of zero moment, ft
d = depth of beam, in.
b = width of beam, in. (*Note:* For laminations that consist of more than one piece, b is the width of widest piece in layup.)
V = volume of beam between points of zero moment, in.³
$= (L \times 12 \text{ in./ft}) \times d \times b$

The application of the volume-effect factor is shown in Sec. 5.7. Other modification factors for glulam design are introduced as they are needed.

5.7 Design Problem: Allowable Stresses

The allowable stresses for a glulam member are evaluated in Example 5.4. As with sawn lumber, the first step is to obtain the correct tabulated design values from the NDS Supplement. The second step is to apply the appropriate adjustment factors. A primary difference between a glulam problem and a sawn lumber problem is the use of the volume factor instead of the size factor.

In this example, a single load combination is given and one load duration factor C_D is used to adjust allowable stresses. It is recognized that in practice a number of different loading combinations must be considered (Sec. 2.16), and the same load duration factor may not apply to all load cases. Appropriate loading combinations are considered in more complete problems later in this book. A single C_D is used in Example 5.4 for simplicity.

EXAMPLE 5.4   Determination of Allowable Design Values for a Glulam

A glulam beam is shown in Fig. 5.10. The member is Douglas Fir from the 24F-1.7E Stress Class. From the sketch the bending load is about the x axis of the cross section. Loads are [D + 0.75(S + W)]. Use a single C_D based on the shortest duration in the combination. Bracing conditions are such that buckling is not a concern. Consider dry-service application (EMC < 16 percent). Normal temperature conditions apply.

Determine the following allowable stresses:

Positive bending stress about the strong axis $F_{bx}^{+\prime}$

Negative bending stress about the strong axis $F_{bx}^{-\prime}$

Tension stress parallel to grain F_t'

Compression stress parallel to grain F_c'

Compression stress perpendicular to grain under concentrated load $F_{c\perp}'$ on compression face

Compression stress perpendicular to grain at support reactions $F_{c\perp}'$ on tension face

Shear stress parallel to grain F_{vx}' (with bending about the strong axis)

Modulus of elasticity for deflection calculations (beam loaded about strong axis) E_x'

Douglas Fir is a Western Species glulam. The Stress Class 24F-1.7E is recognized as a *bending combination*. Tabulated properties are taken from NDS Supplement Table 5A. Bending is about the strong axis of the member. The member is properly installed (top side up), and the tension laminations are on the bottom of the beam. The moment diagram is positive throughout, and bending tension stresses are on the bottom of the member. It is thus confirmed that the normally used bending stress is appropriate (that is, the "tension zone stressed in tension" design value $F_{bx}^{+\prime}$ applies to the problem at hand; but for illustrative purposes, the "compression zone stressed in tension" design value $F_{bx}^{-\prime}$ will also be determined in this example).

The shortest duration load in the combination is wind, and $C_D = 1.6$. Any stress adjustment factors in NDS Table 5.3.1 that are not shown in this example do not apply to the given problem or have a default value of unity. Recall that C_D does not apply to $F_{c\perp}$ or to E.

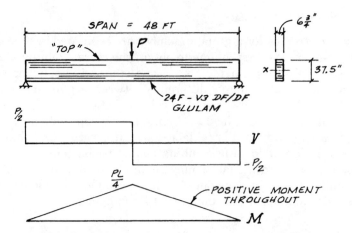

Figure 5.10 Load, shear, and moment diagrams for glulam beam.

Volume Factor C_V

The dimensions of the given member do not agree with the base dimensions for the standard-size glulam in Example 5.3. Therefore the bending stress will be multiplied by C_V. The length L in the formula is the distance between points of zero moment, which in this case is the span length of 48 ft.

$$C_V = \left(\frac{21}{L}\right)^{1/10} \left(\frac{12}{d}\right)^{1/10} \left(\frac{5.125}{b}\right)^{1/10}$$

$$= \left(\frac{21}{48}\right)^{0.1} \left(\frac{12}{37.5}\right)^{0.1} \left(\frac{5.125}{6.75}\right)^{0.1} = 0.799$$

Allowable Design Values

$$F_{bx}^{+\prime} = F_{bx}(C_D \times C_M \times C_t \times C_V) = 2400(1.6 \times 1.0 \times 1.0 \times 0.799) = 3068 \text{ psi}$$

$$F_{bx}^{-} = F_{bx}^{-\prime}(C_D \times C_M \times C_t \times C_V) = 1450(1.6 \times 1.0 \times 1.0 \times 0.799) = 1854 \text{ psi}$$

(NOTE: If a balanced section were required (that is, $F_{bx}^{-} = F_{bx}^{+}$), the designer must specify that a balanced layup is required. See footnote 1 of NDS Supplement Table 5A.)

$$F_t' = F_t(C_D \times C_M \times C_t) = 775(1.6 \times 1.0 \times 1.0) = 1240 \text{ psi}$$

$$F_c' = F_c(C_D \times C_M \times C_t) = 1000(1.6 \times 1.0 \times 1.0) = 1600 \text{ psi}$$

$$F_{c\perp}' = F_{c\perp}(C_M \times C_t) = 500(1.0 \times 1.0) = 500 \text{ psi}$$

$$F_{vx}' = F_{vx}(C_D \times C_M \times C_t) = 190(1.6 \times 1.0 \times 1.0) = 304 \text{ psi}$$

$$E_x' = E(C_M \times C_t) = 1,700,000(1.0 \times 1.0) = 1,700,000 \text{ psi}$$

5.8 References

[5.1]   American Forest and Paper Association (AF&PA). 2001. *National Design Specification for Wood Construction and Supplement.* 2001 ed., AF&PA, Association, Washington, DC.

[5.2]   American Institute of Timber Construction (AITC). 2001. *Standard Appearance Grades for Structural Glued Laminated Timber, AITC 110-2001,* AITC, Englewood, CO.

[5.3]   American Institute of Timber Construction (AITC). 2002. *Structural Glued Laminated Timber, ANSI / AITC Standard 190.1-2002,* AITC, Englewood, CO.

[5.4]   American Institute of Timber Construction (AITC). 2001. *DESIGN Standard Specifications for Structural Glued Laminated Timber of Softwood Species, AITC 117-2001,* AITC, Englewood, CO.

[5.5]   American Institute of Timber Construction (AITC). 2001. *MANUFACTURING Standard Specifications for Structural Glued Laminated Timber of Softwood Species, AITC 117-2001,* AITC, Englewood, CO.

[5.6]   American Institute of Timber Construction (AITC). 2001. *Standard Dimensions for Structural Glued Laminated Timber, AITC 113-2001,* AITC, Englewood, CO.

[5.7]   American Institute of Timber Construction (AITC). 1994. *Timber Construction Manual,* 4th ed., AITC, Englewood, CO.

[5.8]   American Institute of Timber Construction (AITC). 1996. *Standard Specifications for Structural Glued Laminated Timber of Hardwood Species, AITC 119-96,* AITC, Englewood, CO.

[5.9]   American Society for Testing and Materials (ASTM). 2001. "Standard Practice for Establishing Stresses for Structural Glued Laminated Timber (Glulam)," ASTM D3737-01a, *Annual Book of Standards, Vol. 04.10 Wood,* ASTM, Philadelphia, PA.

[5.10] APA—The Engineered Wood Association. 1997. *Data File: Glued Laminated Beam Design Tables, EWS S475,* APA—The Engineered Wood Association, Engineered Wood Systems, Tacoma, WA.

[5.11] APA—The Engineered Wood Association. 1997. *Glulam Product and Application Guide, EWS Q455,* APA—The Engineered Wood Association, Engineered Wood Systems, Tacoma, WA.

[5.12] Forest Products Laboratory (FPL). 1999. *Wood Handbook: Wood as an Engineering Material, Technical Report 113,* FPL, Forest Service, U.S.D.A., Madison, WI.

5.9 Problems

Design values and adjustment factors in the following problems are to be taken from the 2001 NDS. Assume that glulams will be used in dry-service conditions and at normal temperatures unless otherwise noted.

5.1   What is the usual thickness of laminations used to fabricate glulam members from
  a. Western Species?
  b. Southern Pine?
  c. Under what conditions would thinner laminations be used?

5.2   What are the usual widths of glulam members fabricated from:
  a. Western Species
  b. Southern Pine

5.3   How are the *strength grades* denoted for a glulam that is
  a. Primarily a bending member fabricated with visually graded laminations?
  b. Primarily a bending member fabricated with *E*-rated laminations?
  c. Primarily an axial-load-carrying member?
  d. What are the *appearance grades* of glulam members, and how do they affect the grading for strength?

5.4 Briefly describe what is meant by resawn glulam. What range of sizes is listed in NDS Supplement Tables 1C and 1D for resawn glulam?

5.5 What is the most common type of lamination end-joint splice used in glulam members? Sketch the splice.

5.6 If the width of a lamination in a glulam beam is made up of more than one piece of wood, must the edge joint between the pieces be glued?

5.7 Describe the distribution of laminations used in the fabrication of a glulam to be used principally as an axial load member.

5.8 Describe the distribution of laminations used in the fabrication of a glulam member that is used principally as a bending member.

5.9 Briefly describe the meaning of the following glulam designations:
  a. Stress Class 20F-1.5E
  b. Stress Class 24F-1.8E
  c. Stress Class 30F-2.1E SP
  d. Combination 20F-V8 DF/DF
  e. Combination 24F-V5 DF/HF
  f. Combination 24F-E11 HF/HF
  g. Combination 22F-V2 SP/SP
  h. Combination 5 DF
  i. Combination 32 DF
  j. Combination 48 SP

5.10 Tabulated values of F_{bx} for a member stressed primarily in bending apply to a glulam of a "standard" size. What are the dimensions of this hypothetical beam? Describe the adjustment that is required if a member of another size is used.

5.11 *Given:* A $5\frac{1}{8} \times 28.5$ 24F-1.8E Douglas Fir glulam is used to span 32 ft, carrying a load of (D + S). The load is a uniform load over a simple span, and the beam is supported so that buckling is prevented.

  Find: *a.* Sketch the beam and the cross section. Show calculations to verify the section properties S_x and I_x for the member, and compare with values in NDS Supplement Table 1C.
  b. Determine the allowable stresses associated with the section properties in part *a*. These include $F_{bx}^{+\prime}$, $F_{c\perp x}^{\prime}$, $F_{vx}^{\prime}$, and $E_x^{\prime}$.
  c. Repeat part *b* except the moisture content of the member may exceed 16 percent.

5.12 Repeat Prob. 5.11 except the member is a 24F-V4 Douglas Fir-Larch glulam

5.13 *Given:* Assume that the member in Prob. 5.11 may also be loaded about the minor axis.

  Find: *a.* Show calculations to verify the section properties S_y and I_y for the member. Compare with values in NDS Supplement Table 1C.

      b. Determine the allowable stresses associated with the section properties in part *a*. These include F'_{by}, $F'_{c\perp y}$, F'_{vy}, and E'_y.

      c. Repeat part *b*, except the moisture content of the member may exceed 16 percent.

5.14 *Given:* Assume that the member in Prob. 5.11 may also be subjected to an axial tension or compression load.

     Find: *a.* Show calculations to verify the cross-sectional area *A* for the member. Compare with the value in NDS Supplement Table 1C.

          b. Determine the allowable stresses associated with the section properties in part *a*. These include F'_t, F'_c, and E'_{axial}.

          c. Repeat part *b*, except the moisture content of the member may exceed 16 percent.

5.15 Repeat Prob. 5.11 except the member is a 5×33 26F-1.9E Southern Pine glulam.

5.16 Explain why the allowable stress tables for glulam bending combinations (NDS Supplement Table 5A Expanded) list two values of compression perpendicular to grain for loads normal to the *x* axis ($F_{c\perp x}$).

5.17 List the load duration factors C_D associated with the design of glulam members for the following loads:

    a. Dead load
    b. Snow
    c. Wind
    d. Floor live load
    e. Seismic
    f. Roof live load

5.18 Over what moisture content are the tabulated stresses in glulam to be reduced by a wet-service factor C_M?

5.19 List the wet-service factors C_M to be used for designing glulam beams with high moisture contents.

6.1 Introduction

The design of rectangular sawn wood beams and straight or slightly curved rectangular glulam beams is covered in this chapter. Glulam members may be somewhat more complicated than sawn lumber beams, and the special design procedures that apply only to glulam design are noted. Where no distinction is made, it may be assumed that essentially the same procedures apply to both sawn lumber and glulam design.

Glulam beams are sometimes tapered and/or curved for architectural considerations, to improve roof drainage, or to lower wall heights. The design of these types of members requires additional considerations beyond the information presented in this book. For the additional design considerations for these advanced subjects, see the *Timber Construction Manual* (TCM) (Ref. 6.5).

The design of wood beams follows the same basic overall procedure used in the design of beams of other structural materials. The factors that need to be considered are

1. Bending (including lateral stability)
2. Shear
3. Deflection
4. Bearing

The first three items can govern the size of a wood member. The fourth item must be considered in the design of the supports. In many beams the bending stress is the critical design item. For this reason, a trial size is often obtained from bending stress calculations. The remaining items are then simply checked using the trial size. If the trial size proves inadequate in any of the checks, the design is revised.

Computer solutions to these problems can greatly speed up the design process, and with the use of the computer, much more thorough beam deflection studies are possible. However, the basic design process needs to be fully understood first.

Designers are cautioned about using canned programs in a *blackbox* approach. Any program used should be adequately documented and sufficient output should be available so that results can be verified by hand solutions. The emphasis throughout this book is on understanding the design criteria. Modern spreadsheet or equation-solving software can be effective tools in design. With such an application program, the user can tailor the solution to meet a variety of goals. With very little computer training, the designer can develop a *template* to solve a basic problem. A basic template can serve as the starting point for more sophisticated solutions.

6.2 Bending

In discussing the strength of a wood beam, it is important to understand that the bending stresses are parallel to the length of the member and are thus parallel to the grain of the wood. This is the common beam design problem (Fig. 6.1a), and it is the general subject of this section. See Example 6.1.

Occasionally, however, bending stresses across the grain (Fig. 6.1b) are developed, and the designer needs to recognize this situation. It has been noted previously that wood is relatively weak in *tension perpendicular to grain*. This is true whether the cross-grain tension stress is caused by a direct tension force perpendicular to grain or by loading that caused cross-grain bending. Cross-grain tension should generally be avoided.

EXAMPLE 6.1   Bending in Wood Members

Longitudinal Bending Stresses—(Parallel to Grain)

Ordinarily, the bending stress in a wood beam is parallel to the grain. The free-body diagram (FBD) in Fig. 6.1a shows a *typical beam* cut at an arbitrary point. The internal forces V and M are required for equilibrium. The bending stress diagram indicates that the stresses developed by the moment are longitudinal stresses, and they are, therefore, *parallel to grain*. Bending is shown about the strong or x axis of the member.

Cross-Grain Bending—Not Allowed

Section 1 in Fig. 6.1b shows a concrete wall connected to a wood horizontal diaphragm. The lateral force is shown to be transferred from the wall through the *wood ledger* by means of anchor bolts and nailing.

Section 2 indicates that the ledger cantilevers from the anchor bolt to the diaphragm level. Section 3 is an FBD showing the internal forces at the anchor bolt and the bending stresses that are developed in the ledger. The bending stresses in the ledger are *across the grain* (as opposed to being parallel to the grain). Wood is very weak in cross-grain bending and tension. This connection is introduced at this point to define the cross-grain bending problem. Tabulated bending stresses for wood design apply to longitudinal bending stresses only.

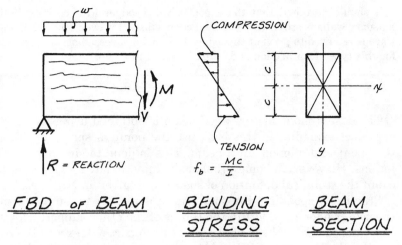

Figure 6.1a Bending stress is parallel to grain in the usual beam design problem.

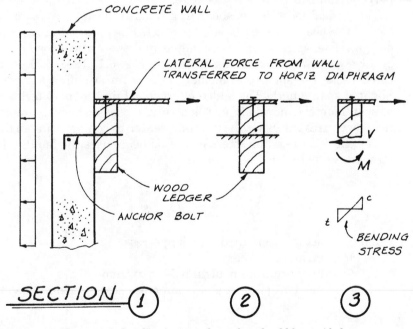

Figure 6.1b Cross-grain bending in a wood member should be avoided.

Because of failures in some ledger connections of this type, cross-grain bending and cross-grain tension are not permitted by the IBC for the anchorage of seismic forces. Even for other loading conditions, designs should generally avoid stressing wood in bending or tension across the grain.

It should be noted that the use of a wood ledger in a building with concrete or masonry walls is still a very common connection. However, additional anchorage hardware is required to prevent the ledger from being stressed across the grain. Anchorage for this type of connection is covered in detail in Chap. 15.

The design moment in a wood beam is obtained using ordinary elastic theory. Most examples in this book use the nominal span length for evaluating the shear and moment in a beam. This is done to simplify the design calculations. However, in some problems it may be advantageous to take into account the technical definition of *span length* given in NDS Sec. 3.2.1 (Ref. 6.2).

Practically speaking, the span length is usually taken as the distance from the center of one support to the center of the other support. However, in most cases the *furnished* bearing length at a support will exceed the *required* bearing length. Thus, the NDS permits the designer to consider the span to be the *clear* distance between supports *plus* one-half of the *required* bearing length at each end. The required bearing length is a function of compression stress perpendicular to grain $F_{c\perp}$ (Sec. 6.8).

The critical location for shear in a wood beam is at a distance d from the face of the beam support (a similar practice is followed in reinforced-concrete design). The span length for bending and the critical loading condition for shear are shown later in this chapter in Fig. 6.13 (Sec. 6.5). Again, for hand calculations the shear and moment in a beam are often determined using a nominal span length. The added effort to obtain the more technical definition of span length is normally justified only in cases where the member appears to be overstressed using the nominal center-to-center span length.

The check for bending stress in a wood beam uses the familiar formula from strength of materials

$$f_b = \frac{Mc}{I} = \frac{M}{S} \leq F'_b$$

where f_b = actual (computed) bending stress
M = moment in beam
c = distance from neutral axis to extreme fibers
I = moment of inertia of beam cross section about axis of bending
$S = \dfrac{I}{c}$
    = section modulus of beam cross section about axis of bending
F'_b = allowable bending stress

According to allowable stress design (ASD) principles, this formula says that the *actual* (computed) bending stress must be less than or equal to the *allowable* bending stress. The allowable stress takes into account the necessary adjustment factors to tabulated stresses that may be required for a wood member.

Most wood beams are used in an efficient manner. In other words, the *moment is applied about the strong axis (x axis) of the cross section.* From an engineering point of view, this seems to be the most appropriate description of the common loading situation. However, other terms are also used in the wood industry to refer to bending about the strong axis. For solid *sawn* lumber of rectangular cross section, the terms *loaded edgewise, edgewise bending,* and *load applied to the narrow face* of the member all refer to bending about the *x* axis. For *glulam* beams, the term *load applied perpendicular to the wide face* of the laminations is commonly used.

As wood structures become more highly engineered, there is a need to generalize the design expressions to handle a greater variety of situations. In a general approach to beam design, the moment can occur about either the *x* or *y* axis of the beam cross section. See Example 6.2. For sawn lumber, the case of bending about the weak axis (*y* axis) is described as *loaded flatwise, flatwise bending,* and *load applied to the wide face of the member.* For glulam, it is referred to as *load applied parallel to the wide face of the laminations.* In engineering terms, *weak-axis bending* and *bending about the y axis* are probably better descriptions.

Throughout this book the common case of bending about the strong axis is assumed, unless otherwise noted. Therefore, the symbols f_b and F'_b imply bending about the *x* axis and thus represent the values f_{bx} and F'_{bx}. Where needed, the more complete notation of f_{bx} and F'_{bx} is used for clarity. (An exception to the general rule of bending about the strong axis is Decking, which is normally stressed about the *y* axis.)

EXAMPLE 6.2   Strong- and Weak-Axis Bending

The large majority of wood beams are rectangular in cross section and are loaded as efficient bending members. See Fig. 6.2*a*. This common condition is assumed, unless otherwise noted.

The bending stress in a beam about the strong axis (Fig. 6.2*a*) is

$$f_{bx} = \frac{M_x}{S_x} = \frac{M_x}{bd^2/6} \leq F'_{bx}$$

A less efficient (and therefore less common) type of loading is to stress the member in bending about the minor axis. See Fig. 6.2*b*. Although it is not common, a structural member may occasionally be loaded in this manner.

The bending stress in a beam loaded about the weak axis (Fig. 6.2*b*) is

$$f_{by} = \frac{M_y}{S_y} = \frac{M_y}{bd^2/6} \leq F'_{by}$$

The designer must be able to recognize and handle either bending application.

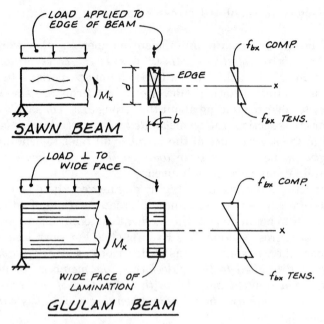

Figure 6.2a Most wood beams have bending about the strong axis. For sawn lumber, *loaded edgewise*. For glulam, *load perpendicular to wide face of laminations*.

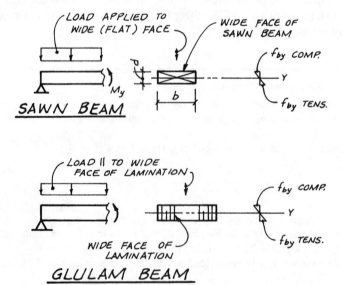

Figure 6.2b Occasionally beams have bending about the weak axis. For sawn lumber, *loaded flatwise*. For glulam, *load parallel to wide face of laminations*.

The formula from engineering mechanics for bending stress f_b was developed for an ideal material. Such a material is defined as a solid, homogeneous, isotropic (having the same properties in all directions) material. In addition, plane sections before bending are assumed to remain plane during bending, and stress is assumed to be linearly proportional to strain.

From the discussion of some of the properties of wood in Chap. 4, it should be clear that wood does not fully satisfy these assumptions. Wood is made up of hollow cells which generally run parallel to the length of a member. In addition, there are a number of growth characteristics and service conditions such as annual rings, knots, slope of grain, and moisture content. However, adequate beam designs are obtained by applying the ordinary bending formula and adjusting the allowable stress to account for the unique characteristics of wood beams.

The starting point is to obtain the correct *tabulated* bending stress for the appropriate species and grade of member. Values of F_b are listed in NDS Supplement Tables 4A to 4F for sawn lumber and NDS Supplement Tables 5A to 5D for glulam. NDS Table 4.3.1, *Applicability of Adjustment Factors for Sawn Lumber* and Table 5.3.1, *Applicability of Adjustment Factors for Glued Laminated Timber,* then provide a string of multiplying factors to obtain the allowable bending stress once the tabulated stress is known. The *allowable bending stress* is defined as

$$F'_b = F_b(C_D)(C_M)(C_t)(C_L)(C_F)(C_V)(C_{fu})(C_r)(C_c)(C_f)(C_i)$$

where F'_b = allowable bending stress
  F_b = tabulated bending stress
  C_D = load duration factor (Sec. 4.15)
  C_M = wet service factor (Sec. 4.14—note that subscript M stands for moisture)
  C_t = temperature factor (Sec. 4.19)
  C_L = beam stability factor (consider when lateral support to compression side of beam may permit beam to buckle laterally—Sec. 6.3)
  C_F = size factor (Sec. 4.16)
  C_V = volume factor (Sec. 5.6)
  C_{fu} = flat use factor (Sec. 4.18)
  C_r = repetitive member factor (Sec. 4.17)
  C_c = curvature factor [Apply only to curved glulam beams; C_c = 1.0 for straight and cambered (slightly curved) glulams. The design of curved beams is beyond the scope of this book.]
  C_f = form factor (Sec. 4.21)
  C_i = incising factor for sawn lumber (Sec. 4.20)

The reader is referred to the appropriate sections in Chaps. 4 and 5 for background on the adjustment factors discussed previously. Lateral stability is an important consideration in the design of a beam. The beam stability factor C_L is an adjustment factor that takes into account a reduced moment capacity if lateral torsional buckling can occur. Initially it is assumed that buckling is prevented, and C_L defaults to unity. See Example 6.3.

It should be realized that the long list of adjustment factors for determining F'_b is basically provided as a reminder that a number of special conditions may require an adjustment of the tabulated value. However, in many practical design situations, a number of the possible adjustment factors will default to 1.0. In addition, not all of the possible adjustments apply to all types of wood beams. Section 6.4 shows how the string of adjustment factors can be greatly reduced for practical beam design.

EXAMPLE 6.3   Full Lateral Support a Beam

The analysis of bending stresses is usually introduced by assuming that lateral torsional buckling of the beam is prevented. Continuous support of the compression side of a beam essentially prevents the member from buckling (Fig. 6.3a).

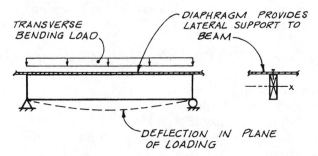

Figure 6.3a Direct attachment of roof or floor diaphragm provides full lateral support to top side of a beam. When subjected to transverse loads, a beam with full lateral support is stable, and it will deflect only in its plane of loading.

A beam with positive moment everywhere has *compressive* bending stresses on the top side of the member throughout its length. An effective connection (proper nailing) of a roof or floor diaphragm (sheathing) to the top side of such a beam reduces the unbraced length to zero ($l_u = 0$). Technically the unbraced length is the *spacing of the nails* through the sheathing and into the compression side of the beam. For most practical diaphragm construction and most practical beam sizes, the unbraced length can be taken as zero.

Many practical wood structures have *full* or *continuous* lateral support as part of their normal construction. See Fig. 6.3b. Closely spaced beams in a repetitive framing arrangement are shown. However, a roof or floor diaphragm can also be used to provide lateral support to larger beams and girders, and the concept is not limited to closely spaced members.

Figure 6.3b When plywood sheathing is properly attached to framing, a diaphragm is formed that provides stability to beams. (*Photo courtesy of APA.*)

With an unbraced length of zero, lateral buckling is eliminated, and the beam stability factor C_L defaults to unity. For other conditions of lateral support, C_L may be less than 1.0. The stability of laterally unbraced beams is covered in detail in Sec. 6.3.

Several points should be mentioned concerning the tabulated bending stress for different kinds of wood beams. Unlike glulam, the tables for sawn lumber do not list separate design properties for bending about the x and y axes. Therefore, it is important to understand which axis is associated with the tabulated values.

Tabulated bending stresses F_b for *visually graded sawn lumber* apply to both the x and y axes *except* for Beams and Stringers and Decking. Because Decking is graded with the intent that the member will be used flatwise (i.e., weak-axis bending), the tabulated value in the NDS Supplement is F_{by}. A flat-use factor C_{fu} has already been incorporated into the tabulated value, and the designer should not apply C_{fu} to Decking. The use of Decking is mentioned only briefly, and it is not a major subject in this book.

For members in the B&S size category, the tabulated bending stress applies to the x axis only (i.e., $F_b = F_{bx}$). However, for other members including Dimension lumber and Posts and Timbers sizes, the tabulated bending stress applies to both axes (i.e., $F_b = F_{bx} = F_{by}$). To obtain the allowable bending stress for the y axis, F_b must be multiplied by the appropriate flat-use factor C_{fu}.

In the infrequent case that a member in the B&S size category is loaded in bending about the minor axis, the designer should use the size adjustment factors provided with Table 4D in the NDS Supplement to determine the tabulated bending stress for the y axis.

Another point needs to be understood about the allowable bending stresses in the B&S size category. ASTM D 245 allows the application of a less restrictive set of grading criteria to the outer thirds of the member length. This practice anticipates that the member will be used in a *simple beam* application. It further assumes that the length of the member will not be reduced substantially by sawing the member into shorter lengths.

Therefore, if a B&S is used in some other application where the maximum bending stress does not occur in the middle third of the original member length (e.g., a cantilever beam or a continuous beam), the designer should specify that the grading provisions applicable to the middle third of the length shall be applied to the entire length. See Example 6.4.

EXAMPLE 6.4   Allowable Bending Stresses for Beams and Stringers

Tabulated bending stresses for B&S sizes are for bending about the x axis of the cross section. Lumber grading agencies may apply less restrictive grading rules to the outer thirds of the member length. This assumes that the maximum moment will be located in the middle third of the member length. The common uniformly loaded simply supported beam is the type of loading anticipated by this grading practice.

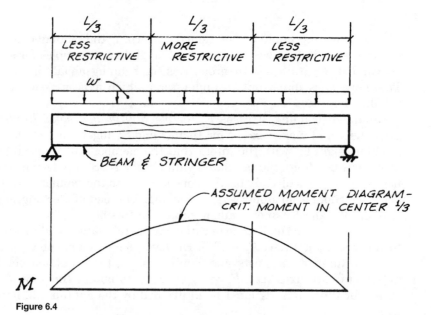

Figure 6.4

If the loading or support conditions result in a moment diagram which does not agree with the assumed distribution, the designer should specify that the grading rules normally applied to the middle third shall be applied to the entire length.

A similar problem develops if a long B&S member is ordered and then cut into shorter lengths (see NDS Sec. 4.1.7). A note on the plans should prohibit cutting beams of this type, *or* full-length grading should be specified.

NOTE: A way to reduce the length of a B&S without affecting its stress grade is to cut approximately equal lengths from *both* ends.

A brief introduction to tabulated bending stresses for glulams was given in Chap. 5. Recall that *two* values of F_{bx} are listed for the softwood glulam bending combinations in NDS Supplement Table 5A along with a value of F_{by}. It should be clear that F_{by} is for the case of bending about the weak axis of the member, but the two values for F_{bx} require further explanation. Although the computed bending stresses in a rectangular beam are equal at the extreme fibers, tests have shown the outer *tension laminations* are critical. Therefore, high-grade tension laminations are placed in the outer tension zone of the beam. The *top* of a glulam beam is marked in the laminating plant so that the member can be identified at the job site and oriented properly in the structure.

If the beam is loaded so that the tension laminations are stressed in tension, the appropriate bending stress is F_{bx} *tension zone stressed in tension*. In this book the following notation is used to indicate this value: F_{bx}^+. In most cases a glulam beam is used in an efficient manner, and F_{bx} is normally F_{bx}^+. In other words, F_{bx} is assumed to be F_{bx}^+ unless otherwise indicated. On the other hand, if the member is loaded in such a way that the compression laminations are stressed in bending tension, the tabulated value known as F_{bx} *compression zone stressed in tension* (F_{bx}^-) is the corresponding tabulated bending stress.

A review of the NDS Supplement for glulam shows that the two tabulated stresses for F_{bx} just described can vary by a factor of 2. Accordingly, depending on the combination, the calculated bending strength of a member could be 50 percent less than expected if the beam were inadvertently installed upside down. Thus, it is important that the member be installed properly in the field.

Simply supported beams under gravity loads have positive moment throughout, and bending tensile stresses are everywhere on the bottom side of the member. Here the designer is just concerned with F_{bx} *tension zone stressed in tension*. See Example 6.5.

In the design of beams with both positive and negative moments, both values of F_{bx} need to be considered. In areas of negative moment (tension on the top side of the beam), the value of F_{bx} *compression zone stressed in tension* applies. When the negative moment is small, the reduced allowable bending stress for the compression zone stressed in tension may be satisfactory. Small negative moments may occur, for example, in beams with relatively small cantilever spans.

EXAMPLE 6.5 F_{bx} in Glulam Bending Combinations

Some glulam beams are fabricated so that the allowable bending tensile stress is the same for both faces of the member. Others are laid up in such a manner that the allowable bending tensile stresses are not the same for both faces of the beam. Two different allowable bending stresses are listed in the glulam tables:

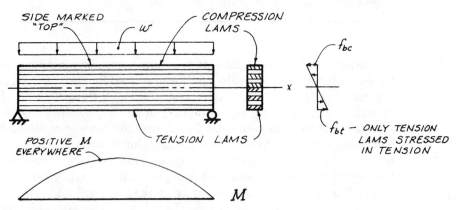

Figure 6.5a   Glulam with positive moment everywhere.

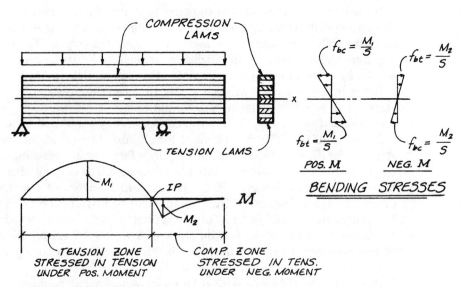

Figure 6.5b   Glulam with positive and negative moments.

1. F_{bx} tension zone stressed in tension F_{bx}^+
2. F_{bx} compression zone stressed in tension F_{bx}^- (this value never exceeds F_{bx}^+, and it may be much less)

In Fig. 6.5a the designer needs to consider only F_{bx}^+ because there is tension everywhere on the bottom side of the beam.

However, when positive and negative moments occur (Fig. 6.5*b*), both values of F_{bx} need to be considererd.

In Figure 6.5*b* F_{bx}^{+} applies to M_1, and F_{bx}^{-} is used for M_2. If M_1 and M_2 are equal, a bending combination can be used which has equal values for the two F_{bx} stresses.

Figure 6.5c Large glulam beam in manufacturing plant undergoing finishing operation. A stamp is applied to the "top" of a glulam so that field crews will install the member right side up. (*Photo courtesy of FPL.*)

On the other hand, when the negative moment is large, the designer is not limited to a small value of F_{bx}^{-}. The designer can specify that tension zone grade requirements, including end-joint spacing, must be applied to both sides of the member. In this case the higher allowable bending stress F_{bx}^{+} may be used to design for both positive and negative moments. Large negative moments often occur in cantilever beam systems (Sec. 6.16). For additional information regarding F_{bx} tension zone stressed in tension and F_{bx} compression zone stressed in tension, see Refs. 6.4 and 6.5.

A final general point should be made about the strength of a wood beam. The *notching of structural members* to accommodate piping or mechanical systems is the subject of considerable concern in the wood industry. The notching and cutting of members in residential construction is fairly common practice. Although this may not be a major concern for members in repetitive systems which are lightly loaded, it can cause serious problems in other situations. Therefore, a note on the building plans should prohibit the cutting or notching of any structural member unless it is specifically detailed on the structural plans.

The effects of notching in areas of *bending stresses* are often addressed separately from the effects of notching on the *shear capacity* at the end of a beam.

The discussion in the remaining portion of this section deals primarily with the effects of a notch where a bending moment exists. For shear considerations see Sec. 6.5.

The effect of a notch on the bending strength of a beam is not fully understood, and convenient methods of analyzing the bending stress at a notch are not currently available. However, it is known that the critical location of a notch is in the bending tension zone of a beam. Besides reducing the depth available for resisting the moment, stress concentrations are developed. Stress concentrations are especially large for the typical square-cut notch.

To limit the effect in sawn lumber, NDS Sec. 4.4.3 limits the maximum depth of a notch to one-sixth the depth of the member and states that the notch shall not be located in the middle third of the span. The NDS further limits notches at ends of the member for bearing over a support to no more than one-fourth the depth. Although not stated, it is apparent that this latter criterion applies to simply supported beams because of the high bending stresses in this region of the span. Except for notches at the ends of a member, the NDS prohibits the notching of the tension side of beams when the nominal width of the member is 4 in. or greater.

Notches are especially critical in glulams because of the high-quality laminations at the outer fibers. Again, the tension laminations are the most critical and are located on the bottom of a beam that is subjected to a positive moment. For a glulam beam, NDS Sec. 5.4.4 prohibits notching in the tension face, except at the ends of the beam for bearing over a support. Even where allowed at the ends, the NDS limits the maximum depth of a notch in the tension face to one-tenth the depth of the glulam. The NDS further prohibits notching in the compression face in the middle third of the span and limits the depth of a notch in the compression face at the end of a beam to two-fifths the depth of the member.

These negative statements about the use of notches in wood beams should serve as a warning about the potential hazard that can be created by stress concentrations due to reentrant corners. Failures have occurred in beams with notches located some distance from the point of maximum bending and at a load considerably less than the design load.

The problem is best handled by avoiding notches. In the case of an existing notch, some strengthening of the member at the notch may be advisable.

6.3 Lateral Stability

When a member functions as a beam, a portion of the cross section is stressed in compression and the remaining portion is stressed in tension. If the compression zone of the beam is not braced to prevent lateral movement, the member may buckle at a bending stress that is less than the allowable stress defined in Sec. 6.2. The allowable bending stress described in Sec. 6.2 assumed that lateral torsional buckling was prevented by the presence of adequate bracing.

The bending compressive stress can be thought of as creating an equivalent column buckling problem in the compressive half of the cross section. Buckling in the plane of loading is prevented by the presence of the stable tension portion of the cross section. Therefore, if buckling of the compression side occurs, movement will take place *laterally* between points of lateral support. See Example 6.6.

EXAMPLE 6.6   Lateral Buckling of Bending Member

Unlike the beam in Example 6.3, the girder in Fig. 6.6 does not have full lateral support.

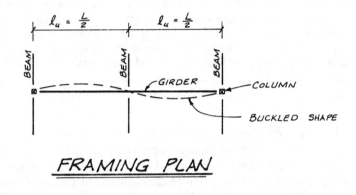

FRAMING PLAN

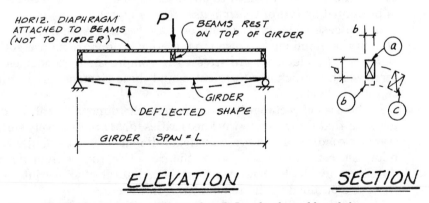

ELEVATION   SECTION

Figure 6.6   Bending member with span length L and unbraced length l_u.

1. The distance between points of lateral support to the compression side of a bending member is known as the *unbraced length l_u* of the beam. The beams that frame into the girder in Fig. 6.6 provide lateral support of the compression (top) side of the girder at a spacing of $l_u = L/2$.
2. It is important to realize that the *span* of a beam and the *unbraced length* of a beam are two different items. They may be equal, but they may also be quite dif-

ferent. The span is used to calculate stresses and deflections. The unbraced length, together with the cross-sectional dimensions, is used to analyze the stability of a bending member.

In other words, the span L gives the actual bending stress f_b, and the unbraced length l_u defines the allowable stress F'_{bx}.

3. The section view in Fig. 6.6 shows several possible conditions:

   a. The unloaded position of the girder.

   b. The deflected position of the girder under a vertical load with no instability. Vertical deflection occurs if the girder remains stable.

   c. The buckled position. If the unbraced length is excessive, the compression side of the member may buckle laterally in a manner similar to a slender column. Buckling takes place between points of lateral support. This buckled position is also shown in the plan view.

4. When the top of a beam is always in compression (positive moment everywhere) and when roof or floor sheathing is effectively connected directly to the beam, the unbraced length approaches zero. Such a member is said to have full lateral support. When lateral buckling is prevented, the strength of the beam depends on the bending strength of the material and not on stability considerations.

In many practical situations, the equation of lateral instability is simply elimated by providing lateral support to the compression side of the beam at close intervals. It has been noted that an effective connection (proper nailing) of a roof or floor *diaphragm* (or sheathing) to the compression side of a beam causes the unbraced length to approach zero ($l_u = 0$), and lateral instability is prevented by *full* or *continuous* lateral support.

In the case of laterally unbraced *steel beams* (W shapes), the problem of stability is amplified because cross-sectional dimensions are such that relatively *slender* elements are stressed in compression. Slender elements have large width-to-thickness (b/t) ratios, and these elements are particularly susceptible to buckling.

In the case of rectangular *wood beams,* the dimensions of the cross section are such that the depth-to-thickness ratios (d/b) are relatively small. Common framing conditions and cross-sectional dimensions cause large reductions in allowable bending stresses to be the exception rather than the rule. Procedures are available, however, for taking lateral stability into account, and these are outlined in the remainder of this section.

Two methods of handling the lateral stability of beams are currently in use. One method is based on *rules of thumb* that have developed over time. These rules are applied to the design of sawn lumber beams. In this approach the required type of lateral support is specified on the basis of the depth-to-thickness ratio d/b of the member.

These rules are outlined in NDS Sec. 4.4.1, *Stability of Bending Members* for sawn lumber. As an example, the rules state that if $d/b = 6$, bridging, full-depth solid blocking, or diagonal cross-bracing is required at intervals of 8 ft-0 in. maximum, full lateral support must be provided for the compression edge, and the beam must be supported at bearing points such that rotation is prevented. The requirement for bridging, blocking, or cross-bracing can be omitted if both edges are held in line for their entire length. See Example 6.7.

EXAMPLE 6.7   Lateral Support of Beams—Approximate Method

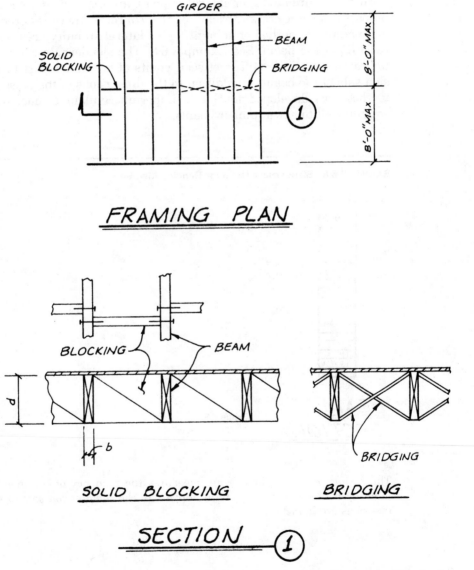

Figure 6.7 Solid (full depth) blocking or bridging for lateral stability based on traditional rules involving (d/b) ratio of beam.

When $d/b = 6$, lateral support can be provided by full-depth solid blocking, diagonal cross bracing or by bridging spaced at 8 ft-0 in. maximum. *Solid blocking* must be the same depth as the beams. Adjacent blocks may be staggered to facilitate construction (i.e., end nailing through beam). *Bridging* is cross-bracing made from wood (typically 1 × 3 or 1 × 4) or light-gauge steel (available prefabricated from manufacturers of hardware for wood construction).

These requirements for lateral support are *approximate* because only the proportions of the cross section (i.e., the d/b ratio) are considered. The second, more accurate method of accounting for lateral stability uses the *slenderness ratio* R_B of the beam. See Example 6.8. The slenderness ratio considers the unbraced length (distance between points of lateral support to the compression side of the beam) in addition to the dimensions of the cross section. This method was developed for large, important glulam beams, but it applies equally well to sawn lumber beams.

EXAMPLE 6.8   Slenderness Ratio for Bending Members

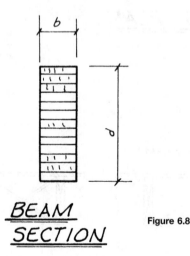

BEAM
SECTION

Figure 6.8

The *slenderness ratio* for a beam measures the tendency of the member to buckle laterally between points of lateral support to the *compression side* of the beam. Dimensions are in inches.

$$R_B = \sqrt{\frac{l_e d}{b^2}}$$

where R_B = slenderness ratio for a bending member
b = beam width
d = beam depth
l_u = unbraced length of beam (distance between points of lateral support as in Fig. 6.6)
l_e = effective unbraced length

The effective unbraced length is a function of the type of span, loading condition, and l_u/d ratio of the member. Several definitions of l_u are given here for common beam configurations. NDS Table 3.3.3., *Effective Length, l_e, for Bending Members,* summarizes these and a number of additional loading conditions involving multiple concentrated loads.

	Cantilever Beam	
Type of load	When $l_u/d < 7$	When $l_u/d \geq 7$
Uniformly distributed load	$l_e = 1.33l_u$	$l_e = 0.90l_u + 3d$
Concentrated load at free end	$l_e = 1.87l_u$	$l_e = 1.44l_u + 3d$
	Single-Span Beam	
Type of load	When $l_u/d < 7$	When $l_u/d \geq 7$
Uniformly distributed load	$l_e = 2.06l_u$	$l_e = 1.63l_u + 3d$
Concentrated load at midspan with no lateral support at center	$l_e = 1.80l_u$	$l_e = 1.37l_u + 3d$
Concentrated load at center with lateral support at center	$l_e = 1.11l_u$	
Two equal concentrated loads at one-third points and lateral support at one-third points	$l_e = 1.68l_u$	

NOTE: For a cantilever, single-span, or multiple-span beam with any loading, the following values of l_e may conservatively be used:

$$l_e = \begin{cases} 2.06l_u & \text{when } l_u/d < 7 \\ 1.63l_u + 3d & \text{when } 7 \leq l_u/d \leq 14.3 \\ 1.84l_u & \text{when } l_u/d > 14.3 \end{cases}$$

In calculating the beam slenderness ratio R_B, the *effective unbraced length* is defined in a manner similar to the effective length of a column (Chap. 7). For a beam, the effective length l_e depends on the end conditions (span type) and type of loading. In addition the ratio of the unbraced length to the beam depth l_u/d may affect the definition of effective length.

Once the slenderness ratio of a beam is known, the effect of lateral stability on the allowable bending stress may be determined. For large slenderness ratios, the allowable bending stress is reduced greatly, and for small slenderness ratios, lateral stability has little effect. At a slenderness ratio of zero, the beam can be considered to have full lateral support, and the allowable bending stress is as defined in Sec. 6.2 with $C_L = 1.0$.

The maximum beam slenderness is $0 \leq R_B \leq 50$. The effect of lateral stability on the bending strength of a beam is best described on a graph of the allowable bending stress F'_{bx} plotted against the beam slenderness ratio R_B. See Example 6.9. The NDS formula for evaluating the effect of lateral stability on beam capacity gives a continuous curve for F'_{bx} over the entire range of beam slenderness ratios.

EXAMPLE 6.9   Allowable Bending Stress Considering Lateral Stability

The NDS has a continuous curve for evaluating the effects of lateral torsional buckling on the bending strength of a beam. See Fig. 6.9a. Lateral torsional buckling may occur betweeen points of lateral support to the compression side of a beam as the member is stressed in bending about the x axis of the cross section.

The tendency for a beam to buckle is eliminated if the moment occurs about the weak axis of the member. Therefore, the allowable stress reduction given by the curve in Fig. 6.9a is limited to bending about the x axis, and the bending stress is labeled F'_{bx}. However, the x subscript is often omitted, and it is understood that the reduced allowable bending stress is about the x axis (that is, $F'_b = F'_{bx}$).

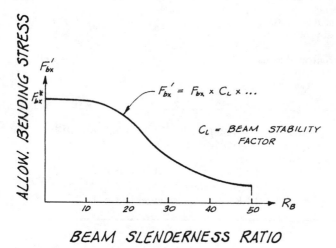

Figure 6.9a   Typical plot of allowable bending stress about the x axis F'_{bx} versus beam slenderness ratio R_B.

Allowable Bending Stress

The allowable bending stress curve in Fig. 6.9a is obtained by multiplying the tabulated bending stress by the beam stability factor C_L *and* all other appropriate adjustment factors.

$$F'_{bx} = F_{bx}\,(C_L) \times \cdots$$

where F'_{bx} = allowable bending stress for x axis
$\quad\quad F_{bx}$ = tabulated bending stress for x axis
$\quad\quad C_L$ = beam stability factor (defined below)
$\quad\quad \times \cdots$ = product of other appropriate adjustment factors

Beam Stability Factor C_L

$$C_L = \frac{1 + F_{bE}/F^*_{bx}}{1.9} - \sqrt{\left(\frac{1 + F_{bE}/F^*_{bx}}{1.9}\right)^2 - \frac{F_{bE}/F^*_{bx}}{0.95}}$$

where F_{bE} = Euler-based critical buckling stress for bending members
$\quad\quad = \dfrac{K_{bE}E'_y}{R_B^2}$
$\quad F^*_{bx}$ = tabulated bending stress for x axis multiplied by certain adjustment factors
$\quad\quad = F_{bx} \times$ (product of all adjustment factors except C_{fu}, C_V, and C_L)
$\quad K_{bE}$ = 0.439 for visually graded lumber
$\quad\quad = 0.561$ for MEL
$\quad\quad = 0.610$ for products with less variability such as MSR lumber and glulam.
$\quad\quad$ See NDS Appendices D and F.2 additional information.
$\quad E'_y$ = modulus of elasticity associated with lateral torsional buckling
$\quad\quad =$ modulus of elasticity about y axis multiplied by all appropriate adjustment factors. Recall that C_D does not apply to E. For sawn lumber, $E_y = E_x$. For glulam, E_x and E_y may be different.
$\quad\quad = E_y(C_M)(C_t)$
$\quad R_B$ = slenderness ratio for bending member (Example 6.8)

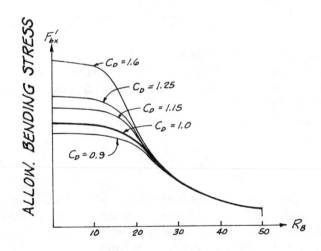

Figure 6.9b Effect of load duration factor on F'_{bx} governed by lateral stability.

In lateral torsional buckling, the bending stress is about the x axis. With this mode of buckling, instability is related to the y axis, and E_y is used to evaluate F_{bE}.

The load durration factor C_D has full effect on allowable bending stress in a beam that has full lateral support. On the other hand, C_D had no influence on the allowable bending stress when instability predominates. A transition between C_D having full effect at a slenderness ratio of 0 and C_D having no effect at a slenderness of 50 is automatically provided in the definition of C_L. This relationship is demonstrated in Fig. 6.9b.

___

The form of the expression for the *beam stability factor* C_L is the same as the form of the *column stability factor* C_P. The column stability factor is presented in Sec. 7.4 on column design. Both expressions serve to reduce the allowable stress based on the tendency of the member to buckle. For a beam, C_L measures the effects of lateral torsional buckling, and for a member subjected to axial compression, C_P evaluates column buckling.

The general form of the beam and column buckling expressions is the result of column studies by Ylinen. They were confirmed by work done at the Forest Products Laboratory (FPL) as part of a unified treatment of combined axial and bending loads for wood members (Ref. 6.13). The beam stability factor and the column stability factor provide a continuous curve for allowable stresses.

The expressions for C_L and C_P both make use of an elastic buckling stress divided by a factor of safety (FS). The Euler critical buckling stress is the basis of the elastic buckling stress

$$F_{\text{Euler}} = F_E = \frac{\pi^2 E}{(\text{slenderness ratio})^2 \times \text{FS}}$$

Recall that values of modulus of elasticity listed in the NDS Supplement are average values. The factor of safety in the F_E formula includes an adjustment which converts the average modulus of elasticity to a 5 percent exclusion value on pure bending modulus of elasticity.

For beam design, the elastic buckling stress is stated as

$$F_{bE} = \frac{K_{bE} E'_y}{R_B^2}$$

When a value of $K_{bE} = 0.439$ is used in this expression, the allowable bending stress F'_{bx} for *visually graded sawn lumber* includes a factor of safety of 1.66. Visually graded sawn lumber is generally more variable than other wood products that are used as beams (e.g., MEL, MSR lumber and glulam). For these less variable materials, a factor of safety of 1.66 is maintained when $K_{bE} = 0.561$ for MEL and $K_{bE} = 0.610$ for MSR lumber and glulam are used to compute F_{bE}. Use of $K_{bE} = 0.439$ for glulam and MSR lumber, for example, would represent less than a 0.01 percent lower exclusion value with a factor of safety of 1.66. (See NDS Appendices D and F.)

In the lateral torsional analysis of beams, the bending stress about the x axis is the concern. However, instability with this mode of buckling is associated with the y axis (see the section view in Fig. 6.6), and E_y is used to compute F_E. For glulams, the values of E_x and E_y may not be equal, and the designer should use E_y from the glulam tables to evaluate the Euler stress for beam buckling.

For the beam and column stability factors, the elastic buckling value F_E is divided by a materials strength property to form a ratio that is used repeatedly in the formulas for C_L and C_P. For beams the material strength property is given the notation F_b^*. In this book subscript x is sometimes added to this notation. This is a reminder that F_b^* is the tabulated bending stress for the x axis F_{bx} multiplied by certain adjustment factors. Again, the ratio F_{bE}/F_b^* is used a number of times in the Ylinen formula.

From strength of materials it is known that the Euler formula defines the critical buckling stress in long slender members. The effect of beam and column stability factors (C_L and C_P) is to define an allowable stress curve that converges on the Euler curve for large slenderness ratios.

Several numerical examples are given later in this chapter that demonstrate the application of C_L for laterally unbraced beams. Section 6.4 summarizes the adjustments for allowable bending stress for different types of beam problems.

6.4 Allowable Bending Stress Summary

The comprehensive listing of adjustment factors given in Sec. 6.2 for determining F_b' is a general summary, and not all of the factors apply to all beams. The purpose of this section is to identify the adjustment factors required for specific applications. In addition, a number of the adjustment factors that frequently default to unity are noted.

Some repetition of material naturally occurs in a summary of this nature. The basic goal, however, is to simplify the long list of possible adjustment factors and to provide a concise summary of the factors relevant to a particular type of beam problem. The objective is to have a complete outline of the design criteria, without making the problem appear overly complicated. Knowing what adjustment factors default to unity for frequently encountered design problems should help in the process.

The allowable bending stresses for sawn lumber are given in Example 6.10. The example covers *visually graded sawn lumber,* and bending stresses apply to all size categories except Decking. The grading rules for Decking presume that loading will be about the minor axis, and published values are F_{by}. The flat use factor C_{fu} has already been applied to the tabulated F_b for Decking.

EXAMPLE 6.10   Allowable Bending Stress—Visually Graded Sawn Lumber

The allowable bending stresses for sawn lumber beams of rectangular cross section are summarized in this example. The common case of bending about the strong axis is covered first. See Fig. 6.10a. The appropriate adjustment factors are listed, and a brief comment is given as reminder about each factor. Certain common default values are suggested (e.g., dry-service conditions and normal temperatures, as found in most covered structures).

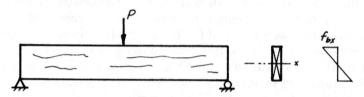

Figure 6.10a Sawn lumber beam with moment about strong axis.

Allowable Bending Stress for Strong Axis

$$F'_{bx} = F_{bx}(C_D)(C_M)(C_t)(C_L)(C_F)(C_r)(C_i)$$

where F'_{bx} = allowable bending stress about x axis

$F_{bx} = F_b$ = tabulated bending stress. Recall that for sawn lumber, tabulated values of bending stress apply to x axis (except Decking). Values are listed in NDS Supplement Tables 4A, 4B, 4C, and 4F for Dimension lumber and in Table 4D for Timbers.

C_D = load duration factor (Sec. 4.15)

C_M = wet-service factor (Sec. 4.14)

= 1.0 for MC ≤ 19 percent (as in most covered structures)

C_t = temperature factor (Sec. 4.19)

= 1.0 for normal temperature conditions

C_L = beam stability factor

= 1.0 for continuous lateral support of compression face of beam. For other conditions compute C_L in accordance with Sec. 6.3.

C_F = size factor (Sec. 4.16). Obtain values from Adjustment Factors section of NDS Supplement Tables 4A, 4B and 4F for Dimension lumber and in Table 4D for Timbers.

C_r = repetitive member factor (Sec. 4.17)

= 1.15 for Dimension lumber applications that meet the definition of a repetitive member

= 1.0 for all other conditions

C_i = incising factor (Sec. 4.20)

= 0.8 for incised dimension lumber

= 1.0 for dimension lumber not incised (whether the member is treated or untreated)

Although bending about the strong axis is the common bending application, the designer should also be able to handle problems when the loading is about the weak axis. See Fig. 6.10b.

Figure 6.10b Sawn lumber beam with moment about weak axis.

Allowable Bending Stress for Weak Axis

$$F'_{by} = F_{by}(C_D)(C_M)(C_t)(C_F)(C_{fu})(C_i)$$

where F'_{by} = allowable bending stress about y axis

$F_{by} = F_b$ = tabulated bending stress. Recall that tabulated values of bending stress apply to y axis for all sizes of sawn lumber except Beams and Stringers. Values of F_b are listed in NDS Supplement Tables 4A, 4B, 4C, and 4F for Dimension lumber and in Table 4D for Timbers. For F_{by} in a B&S size, a size factor is provided for bending about the y-axis.

C_D = load duration factor (Sec. 4.15)

C_M = wet service factor (Sec. 4.14)

= 1.0 for MC ≤ 19 percent (as in most covered structures)

C_t = temperature factor (Sec. 4.19)

= 1.0 for normal temperature conditions

C_F = size factor (Sec. 4.16). Obtain values from Adjustment Factors section of NDS Supplement Tables 4A, 4B, and 4F for Dimension lumber and in Table 4D for Timbers.

C_{fu} = flat use factor (Sec. 4.18). Obtain values from Adjustment Factors section of NDS Supplement Tables 4A, 4B, 4C, and 4F for Dimension lumber.

C_i = incising factor (Sec. 4.20)

= 0.8 for incised dimension lumber

= 1.0 for dimension lumber not incised (whether the member is treated or untreated)

A summary of the appropriate adjustment factors for a glulam beam is given in Example 6.11. Note that the size factor C_F that is used for sawn lumber beams is replaced by the volume factor C_V in glulams. However, in glulams the volume factor C_V is not applied simultaneously with the beam stability factor C_L. The industry position is that volume factor C_V is a bending stress coefficient that adjusts for strength in the tension zone of a beam. Therefore, it is not applied concurrently with the beam stability factor C_L, which is an adjustment related to the bending strength in the compression zone of the beam.

EXAMPLE 6.11   Allowable Bending Stress—Glulam

The allowable bending stresses for straight or slightly curved glulam beams of rectangular cross section are summarized in this example. The common case of bending about the strong axis with the tension laminations stressed in tension is covered first. See

Fig. 6.11a. This summary is then revised to cover the case of the compression laminations stressed in tension. As with the sawn lumber example, the appropriate adjustment factors are listed along with a brief comment.

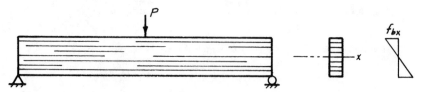

Figure 6.11a Glulam beam with moment about strong axis.

Allowable Bending Stress for Strong Axis

A glulam beam *bending combination* is normally stressed about the x axis. The usual case is with the tension laminations stressed in tension. The notation F_{bx} typically refers to this loading situation. The allowable bending stress is taken as the smaller of the following two values:

$$F'_{bx} = F^{+\prime}_{bx} = F_{bx}(C_D)(C_M)(C_t)(C_L)$$

and

$$F'_{bx} = F^+_{bx} = F^+_{bx}(C_D)(C_M)(C_t)(C_V)$$

where $F'_{bx} = F^{+\prime}_{bx}$ = allowable bending stress about x axis with high-quality tension laminations stressed in tension

$F_{bx} = F^+_{bx}$ = tabulated bending stress about x axis tension zone stressed in tension. Values are listed in NDS Supplement Table 5A for softwood glulam.

C_D = load duration factor (Sec. 4.15)

C_M = wet service factor (Sec. 4.14)

= 1.0 for MC < 16 percent (as in most covered structures)

C_t = temperature factor (Sec. 4.19)

= 1.0 for normal temperature conditions (as in most covered structures)

C_L = beam stability factor

= 1.0 for continuous lateral support of compression face of beam. For other conditions of lateral support C_L is evaluated in accordance with Sec. 6.3.

C_V = volume factor (Sec. 5.6)

Glulam beams are sometimes loaded in bending about the x axis with the compression laminations stressed in tension. The typical application for this case is in a beam with a relatively short cantilever (Fig. 6.5b). The allowable bending stress is taken as the smaller of the following two values:

$$F^{-\prime}_{bx} = F^-_{bx}(C_D)(C_M)(C_t)(C_L)$$

and

$$F^{-\prime}_{bx} = F^-_{bx}(C_D)(C_M)(C_t)(C_V)$$

where $F_{bx}^{-\,\prime}$ = allowable bending stress about x axis with compression laminations stressed in tension

F_{bx}^{-} = tabulated bending stress about x axis with compression zone stressed in tension. Values are listed in NDS Supplement Table 5A for softwood glulam.

Other terms are as defined above.

Although loading about the strong axis is the common application for a bending combination, the designer may occasionally be required to handle problems with bending about the weak axis See Fig. 6.11b.

Figure 6.11b Glulam beam with moment about weak axis.

Allowable Bending Stress For Weak Axis

$$F_{by}' = F_{by}(C_D)(C_M)(C_t)(C_{fu})$$

where F_{by}' = allowable bending stress about y axis

F_{by} = tabulated bending stress about y axis. Values are listed in NDS Supplement Table 5A for softwood glulam.

C_D = load duration factor (Sec. 4.15)

C_M = wet service factor (Sec. 4.14)

  = 1.0 for MC < 16 percent (as in most covered structures)

C_t = temperature factor (Sec. 4.19)

  = 1.0 for normal temperature conditions

C_{fu} = flat use factor (Sec. 4.18). Obtain values from Adjustment Factors section of NDS Supplement Table 5A for softwood glulam. Flat use factor may conservatively be taken equal to 1.0.

Example 6.11 deals with the most common type of glulam which is a softwood bending combination. A glulam constructed from an axial load combination does not have the distribution of laminations that is used in a bending combination. Therefore, only one value of F_{bx} is tabulated for axial combination glulams, and the distinction between F_{bx}^{+} and F_{bx}^{-} is not required. Other considerations for the allowable bending stress in an axial combination glulam are similar to those in Example 6.11. Tabulated values and adjustment factors for axial combination softwood glulams are given in NDS Supplement Table 5B. Design values for all hardwood glulam combinations are given in NDS Supplement Tables 5C and 5D for bending and axial combinations, respectively.

The designer should not be overwhelmed by the fairly extensive summary of allowable bending stresses. Most sawn lumber and glulam beam applications involve bending about the strong axis, and most glulams have the tension zone stressed in tension. The other definitions of allowable bending stress are simply provided to complete the summary and to serve as a reference in the cases when they may be needed.

Numerical examples later in this chapter will demonstrate the evaluation of allowable bending stresses for both visually graded sawn lumber and glulams.

6.5   Shear

The shear stress in a beam is often referred to as *horizontal* shear. From strength of materials it will be recalled that the shear stress at any point in the cross section of a beam can be computed by the formula

$$f_v = \frac{VQ}{Ib}$$

Recall also that the horizontal and vertical shear stresses at a given point are equal. The shear strength of wood parallel to the grain is much less than the shear strength across the grain, and in a wood beam the grain is parallel with the longitudinal axis. In the typical horizontal beam, then, the horizontal shear is critical.

It may be helpful to compare the shear stress distribution given by VQ/Ib for a typical steel beam and a typical wood beam. See Example 6.12. Theoretically the formula applies to the calculation of shear stresses in both types of members. However, in design practice the shear stress in a steel W shape is approximated by a nominal (average web) shear calculation.

The average shear stress calculation gives reasonable results in typical steel beams, but it does not apply to rectangular wood beams. The maximum shear in a rectangular beam is 1.5 times the average shear stress. This difference is significant and cannot be disregarded.

EXAMPLE 6.12   Horizontal Shear Stress Distribution

Steel Beam

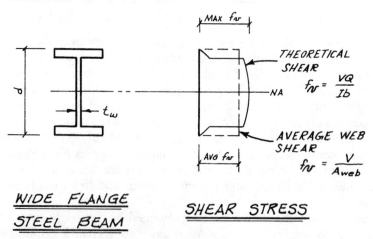

Figure 6.12a   Theoretical shear and average web shear in a steel beam.

For a steel W shape, a nominal check on shear is made by dividing the total shear by the cross-sectional area of the web:

$$\text{Avg. } f_v = \frac{V}{A_{\text{web}}} = \frac{A}{dt_w} \approx \text{max. } f_v$$

Wood Beam

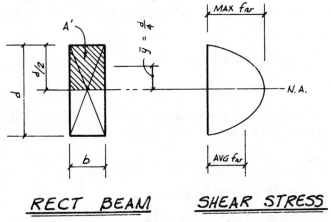

Figure 6.12b Shear stress distribution in a typical wood beam (rectangular section).

For rectangular beams the theoretical *maximum* "horizontal" shear *must* be used. The following development shows that the maximum shear is 1.5 times the average:

$$\text{Avg. } f_v = \frac{V}{A}$$

$$\text{Max. } f_v = \frac{VQ}{Ib} = \frac{VA'\bar{y}}{Ib} = \frac{V(bd/2)(d/4)}{(bd^3/12) \times b}$$

$$= \frac{3v}{2bd} = 1.5\,\frac{V}{A} = 1.5\,(\text{avg. } f_v)$$

A convenient formula for horizontal shear stresses in a rectangular beam is developed in Example 6.12. For wood cross sections of other configurations, the distribution of shear stresses will be different, and it will be necessary to use the basic shear stress formula or some other appropriate check, depending on the type of member involved. The check on shear for a rectangular wood beam is

$$f_v = \frac{1.5V}{A} \le F'_v$$

where f_v = actual (computed) shear stress in beam
V = maximum design shear in beam
A = cross-sectional area of beam
F'_v = allowable shear stress
  = $F_v(C_D)(C_M)(C_t)(C_i)$
F_v = tabulated shear stress

The terms used to evaluate the allowable shear stress were introduced in Chap. 4. The adjustment factors and typical values for frequently encountered conditions are

C_D = load duration factor (Sec. 4.15)

C_M = wet service factor (Sec. 4.14)

  = 1.0 for dry-service conditions, as in most covered structures. Dry-service conditions are defined as

  MC $\leq$ 19 percent for sawn lumber

  MC $<$ 16 percent for glulam

C_t = temperature factor (Sec. 4.19)

  = 1.0 for normal temperature conditions

C_i = incising factor (Sec. 4.20)

  = 1.0 for sawn lumber (*Note:* The incising factor is only applicable to dimension lumber and is *not* applicable to glulam.)

As discussed in Sec. 4.13, one adjustment factor that has been discontinued with the 2001 NDS is the *shear stress factor* C_H. This factor reflected the fact that the presence of splits, checks, and shakes in a wood member will, of course, reduce the horizontal shear capacity. Tabulated values for shear stress formerly reflected the assumption that the member will have extensive splitting. Published values of F_v were conservative since the ASTM procedures for establishing allowable design values required two separate adjustments for the possible presence of splits, checks, and/or shakes. When the length of a split or check at the end of a sawn lumber member was known, and it was judged that the length would not increase, values of the shear stress factor C_H greater than 1.0, and up to a maximum value of 2.0, could be used.

However, in 2000, ASTM Standard D245 (Ref. 6.7) was revised and one of the two adjustments for splits, checks, and/or shakes was eliminated. This resulted in an increase of nearly two for tabulated allowable shear design values. These new design values assume members include representative splitting rather than conservatively assuming, as in previous editions of the NDS, that all members were split along their full length. With the changes in ASTM D245, the reason for a shear stress factor (allowing increase in the shear capacity for less severe splitting) is nullified, and thus the factor has been dropped from the NDS.

In beams that are not likely to be critical in shear, the value of V used in the shear stress formula is often taken as the maximum shear from the shear diagram. However, the NDS Sec. 3.4.3.1 permits the maximum design shear to be reduced in stress calculations. To take this reduction into account, the load must be applied to one face of the beam, and the support reactions are on the opposite face. This is the usual type of loading. The reduction does not apply, for example, to the case where the loads are hung or suspended from the bottom face of the beam.

The reduction in shear is accomplished by neglecting or removing all uniformly distributed loads within a distance d (equal to the depth of the beam) from the face of the beam supports. Concentrated loads within a distance d from the face of the supports *cannot* be simply neglected. Rather, concentrated loads can be reduced by a factor x/d, where x is the distance from the face of the beam support to the load. Therefore, a concentrated load *can* be ignored if located at the face of the support, but must be fully considered if located a distance d from the face of the support, and be considered with a linear reduction in magnitude if located anywhere else within a distance d from the support. See Example 6.13. In the case of a single moving concentrated load, a reduced shear may be obtained by locating the moving load at a distance d from the support (rather than placing it directly at the support). These reductions in computed shear stress can be applied in the design of both glulam and sawn lumber beams.

Prior to the 2001 NDS, concentrated loads were permitted to be ignored if they were located within a distance d from the face of the supports; identical to uniformly distributed loads. The change to a linear reduction for concentrated loads was in response to the increases in tabulated shear design values and elimination of the shear stress factor.

EXAMPLE 6.13   Reduction in Loads for Horizontal Shear Calculations

1. The maximum design shear may be reduced by:
   a. Omitting uniformly distributed load within a distance d (the depth of the beam) from the face of the support, and
   b. Reducing the magnitude of a concentrated load by a factor x/d, where x is the distance from the face of the beam support to the load. Concentrated loads can be ignored ($x/d = 0$) if located at the face of the support ($x = 0$); must be fully considered ($x/d = 1$) if located a distance $x = d$ from the face of the support; and be considered with a linear reduction in magnitude, if located anywhere else within a distance d from the support.

2. The modified loads are *only for horizontal shear stress* calculations in wood (sawn and glulam) beams. The full design loads must be used for other design criteria.

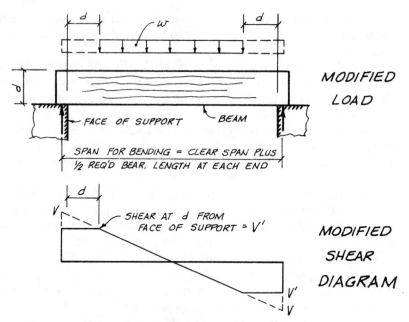

Figure 6.13 Permitted reduction in shear for calculating f_v.

3. The concept of omitting or reducing loads within d from the support is based on an assumption that the loads are applied to one side of the beam (usually the top) and the member is supported by bearing on the opposite side (usually the bottom). In this way the loads within d from the support are transmitted to the supports by diagonal compression. A similar type of adjustment for shear is used in reinforced-concrete design.

The *span length for bending* is defined in Sec. 6.2 and is shown in Fig. 6.13 for information. It is taken as the clear span plus one-half of the required bearing length at each end (NDS Sec. 3.2.1). Although this definition is permitted, it is probably more common (and conservative) in practice to use the distance between the centers of bearing. The span for bending is normally the length used to construct the shear and moment diagrams.

When the details of the beam support conditions are fully known, the designer may choose to calculate the shear stress at a distance d from the *face* of the support. However, in this book many of the examples do not have the support details completely defined. Consequently, if the reduction for shear is used in an example, the loads are conservatively considered within a distance d from the reaction point in the "span for bending." The designer should realize that the point of reference is technically the face of the support and a somewhat greater reduction in calculated shear may be obtained.

Since a higher actual shear stress will be calculated without this modification, it is conservative not to apply it. It is convenient in calculations to adopt a notation which indicates whether the reduced shear from Example 6.13 is being used. Here V represents a shear which is not modified, and V'

is used in this book to indicate a shear which has been reduced. Similarly, f_v is the shear stress calculated using V, and f'_v is the shear stress based on V'.

$$f'_v = \frac{1.5V'}{A} \le F'_v$$

Other terms are as previously defined. The modified load diagram is to be used for horizontal shear stress calculations only. Reactions and moments are to be calculated using the full design loads.

It was noted earlier that bending stresses often govern the size of a beam, but secondary items, such as shear, *can* control the size under certain circumstances. It will be helpful if the designer can learn to recognize the type of beam in which shear is critical. As a general guide, shear is critical on relatively *short, heavily loaded spans*. With some experience, the designer will be able to identify by inspection what probably constitutes a "short, heavily loaded" beam. In such a case the design would start by obtaining a trial beam size which satisfies the horizontal shear formula. Other items, such as bending and deflection, would then be checked.

If a beam is notched at a support, the shear at the notch must be checked (NDS Sec. 3.4.3.2). To do this, the theoretical formula for horizontal shear is applied with the actual depth at the notch d_n used in place of the total beam depth d. See Example 6.14. For square-cut notches in the tension face, the calculated stress must be increased by a stress concentration factor which is taken as the ratio of the total beam depth to the net depth at the notch $(d/d_n)^2$. Notches of other configurations which tend to relieve stress concentrations will have lower stress concentration factors. *The notching of a beam in areas of bending tensile stresses is not recommended* (see Sec. 6.2 for additional comments).

EXAMPLE 6.14   Shear in Notched Beams

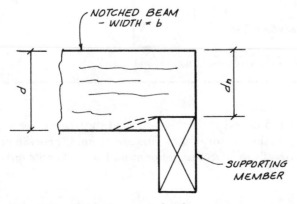

Figure 6.14a   Notch at supported end.

For square-cut notches at the *end* of a beam on the tension side, the NDS provides an expression for the design shear, V'_r, which must be greater than or equal to the calculated shear force on the member, V or V' (see Example 6.13):

$$V'_r = \frac{2}{3} F'_v b d_n \left(\frac{d_n}{d}\right)^2 \geq V$$

The shear force must then be less than the design shear, as determined from the above equation. The approach outlined here, where the design shear is compared to the applied shear force, is not typical in ASD. It is, however, quite common when the load and resistance factor design (LRFD) method is used. For allowable stress design, it is much more common to compare the allowable stress to the calculated or working stress. In this case, the above equation can be rewritten in terms of stresses as follows:

$$f_v = \frac{1.5V}{b d_n} \left(\frac{d}{d_n}\right)^2 \leq F'_v$$

Notches in the tension face of a beam induce tension stresses perpendicular to the grain. These interact with horizontal shear to cause a splitting tendency at the notch. Tapered notches can be used to relieve stress concentrations (dashed lines in Fig. 6.14a).

Mechanical reinforcement such as the fully threaded lag bolt in Fig. 6.14b can be used to resist splitting.

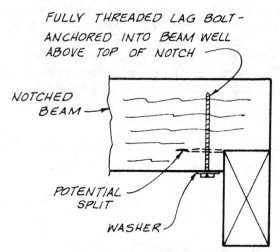

Figure 6.14b Mechanical reinforcement at notched end.

When designing a notched glulam beam, the tabulated shear value must be multiplied by a reduction factor of 0.8. This reduction is specified as a footnote in Tables 5A and 5B of the NDS Supplement and is applicable only to softwood glulam members.

Notches at the end of a beam in the compression face are less critical than notches in the tension side. NDS Section 3.4.3.2(e) provides a method for analyzing the effects of reduced stress concentrations for notches in the compression side. Additional provisions for horizontal shear at bolted connections in beams are covered in Sec. 13.9.

6.6 Deflection

The deflection limits for wood beams required by the IBC and the additional deflection limits recommended by AITC are discussed in Sec. 2.7. Actual deflections for a trial beam size are calculated for a known span length, support conditions, and applied loads. Deflections may be determined from a traditional deflection analysis, from standard beam formulas, or from a computer analysis. The actual (calculated) deflections should be less than or equal to the allowable deflections given in Chap. 2. See Example 6.15.

EXAMPLE 6.15   Beam Deflection Criteria and Camber

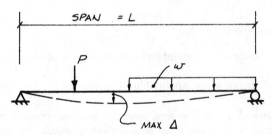

Figure 6.15a   Deflected shape of beam.

Actual Deflection

The maximum deflection is a function of the loads, type of span, moment of inertia, and modulus of elasticity:

$$\text{Max. } \Delta = f\left(\frac{P, w, L}{I, E'}\right)$$

where E' = allowable (i.e., adjusted) modulus of elasticity

$$= E(C_M)(C_t)(C_T)(C_i)$$

Other terms for beam deflection analysis are as normally otherwise defined.

The adjustment factors for evaluating the allowable (i.e., adjusted) modulus of elasticity are introduced in Chap. 4. The factors and typical values for frequently encountered conditions are

E = tabulated modulus of elasticity

  = E_x for usual case of bending about strong axis

C_M = wet service factor (Sec. 4.14)

  = 1.0 for dry-service conditions as in most covered structures.

  Dry-service conditions are defined as

  MC ≤ 19 percent for sawn lumber

  MC < 16 percent for glulam

C_t = temperature factor (Sec. 4.19)

  = 1.0 for normal temperature conditions

C_T = buckling stiffness factor

  = 1.0 for beam deflection calculations. (*Note:* A buckling stiffness factor other than unity may be applied to E for column stability calculations in certain light wood truss applications. See NDS Sec. 4.4.2.)

C_i = incising factor (Sec. 4.20)

  = 0.95 for incised dimension lumber

  = 1.0 for dimension lumber not incised (whether the member is treated or untreated).
  (*Note:* The incising factor is *not* applicable to glulam.)

Deflections are often checked under live load alone Δ_L and under total load Δ_TL (dead load plus live load). Recall that in the total load deflection check, the dead load may be reduced by a factor of 0.5, if the wood member has a moisture content of less than 16 percent at the time of installation and is used under dry conditions (IBC Table 1604.3 footnote d, Ref. 6.10). See Sec. 2.7 for additional information.

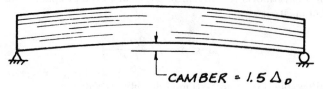

Figure 6.15b Typical camber built into glulam beam is 1.5 times dead load deflection.

Deflection Criteria

$$\text{Max. } \Delta_\text{L} \leq \text{allow. } \Delta_\text{L}$$

$$\text{Max. } \Delta_\text{TL} \leq \text{allow. } \Delta_\text{TL}$$

If these criteria are not satisfied, a new trial beam size is selected using the moment of inertia and the allowable deflection as a guide.

Camber

Camber is initial curvature built into a member which is opposite to the deflection under gravity loads.

The material property that is used to evaluate beam deflection is the adjusted modulus of elasticity E'. The NDS refers to this as the allowable modulus of elasticity. The modulus of elasticity has relatively few adjustment factors. The few adjustments that technically apply to E default to unity for many common beam applications. Note that the load duration factor C_D does not apply to modulus of elasticity (Sec. 4.15).

It will be recalled that the tabulated modulus of elasticity is an average value. It is common design practice to evaluate deflections using the average E. However, in certain cases deflection may be a critical consideration, and NDS Appendix F may be used to convert the average E to a lower-percentile modulus of elasticity. Depending on the required need, the average modulus of elasticity can be converted to a value that will be exceeded by either 84 percent or 95 percent of the individual pieces. These values are given the symbols $E_{0.16}$ and $E_{0.05}$ and are known as the 16 percent and 5 percent lower exclusion values, respectively. See NDS Appendix F for additional information.

In the design of glulam beams and wood trusses, it is common practice to call for a certain amount of camber to be built into the member. Camber is defined as an initial curvature or reverse deflection which is built into the member when it is fabricated. In glulam design, the typical camber is 1.5 Δ_D. This amount of camber should produce a nearly level member under long-term deflection, including creep. Additional camber may be required to improve appearance or to obtain adequate roof slope to prevent ponding (see Ref. 6.5).

See Chap. 2 for more information on deflection, specifically Fig. 2.8.

6.7 Design Summary

One of the three design criteria discussed in the previous sections (bending, shear, and deflection) will determine the required size of a wood beam. In addition, consideration must be given to the type of lateral support that will be provided to prevent lateral instability. If necessary, the bending stress analysis will be expanded to take the question of lateral stability into account. With some practice, the structural designer may be able to tell which of the criteria will be critical by inspection.

The sequence of the calculations used to design a beam has been described in the above sections. It is repeated here in summary.

For many beams the bending stress is the critical design item. Therefore, a trial beam size is often developed from the bending stress formula

$$\text{Req'd } S = \frac{M}{F_b'}$$

A trial member is chosen which provides a furnished section modulus S that is greater than the required value. Because the magnitude of the size factor C_F or the volume factor C_V is not definitely known until the size of the beam has been chosen, it may be helpful to summarize the actual versus allowable bending stresses after a size has been established:

$$f_b = \frac{M}{S} \leq F_b'$$

After a trial size has been established, the remaining items (shear and deflection) should be checked. For a rectangular beam, the shear is checked by the expression

$$f_v = \frac{1.5V}{A} \leq F_v'$$

In this calculation a reduced shear V' can be substituted for V, and f_v' becomes the computed shear in place of f_v. If this check proves unsatisfactory, the size of the trial beam is revised to provide a sufficient area A so that the shear is adequate.

The deflection is checked by calculating the actual deflection using the moment of inertia for the trial beam. The actual deflection is then compared with the allowable deflection:

$$\Delta \leq \text{allow. } \Delta$$

If this check proves unsatisfactory, the size of the trial beam is revised to provide a sufficient moment of inertia I so that the deflection criteria are satisfied.

It is possible to develop a trial member size by starting with something other than bending stress. For example, for a beam with a short, heavily loaded span, it is reasonable to establish a trial size using the shear calculation

$$\text{Req'd } A = \frac{1.5V}{F_v'}$$

The trial member should provide an area A which is greater than the required area.

If the structural properties of wood are compared with the properties of other materials, it is noted that the modulus of elasticity for wood is relatively low. For this reason, in fairly long span members, deflection can control the design. Obviously, if this case is recognized, or if more restrictive deflection criteria are being used in design, the trial member size should be based on satisfying deflection limits. Then the remaining criteria of bending and shear can be checked.

The section properties such as section modulus and moment of inertia increase rapidly with an increase in depth. Consequently, narrow and deep cross sections are more efficient beams.

In lieu of other criteria, the most economical beam for a given grade of lumber is the one that satisfies all stress and deflection criteria with the minimum cross-sectional area. Sawn lumber is purchased by the board foot (a board foot is a volume of wood based on nominal dimensions that corresponds to a 1 × 12 piece of wood 1 ft long). The number of board feet for a given member is obviously directly proportional to the cross-sectional area of a member.

A number of factors besides minimum cross-sectional area can affect the final choice of a member size. First, there are detailing considerations in which a member size must be chosen which fits in the structure and accommodates other members and their connections. Second, a member size may be selected that is uniform with the size of members used elsewhere in the structure. This may be convenient from a structural detailing point of view, and it also can simplify material ordering and construction. Third, the availability of lumber sizes and grades must also be considered. However, these other factors can be considered only with knowledge about a specific job, and the general practice in this book is to select the beam with the least cross-sectional area.

The design summary given above is essentially an outline of the process that may be used in a hand solution. Computer solutions can be used to automate the process. Generally computer designs will be more direct in that the required section properties for bending, shear, and deflection (that is, S, A, and I) will be computed directly with less work done by trial and error. However, even with computer solutions, wood design often involves iteration to some extent in order to obtain a final design.

The designer is encouraged to start using the computer by developing simple spreadsheet or equation-solving software templates for beam design calculations. If a dedicated computer program is used, the designer should ensure that sufficient output and documentation are available for verifying the results by hand.

6.8   Bearing Stresses

Bearing stresses perpendicular to the grain of wood occur at beam supports or where loads from other members frame into the beam. See Example 6.16. The actual bearing stress is calculated by dividing the load or reaction by the contact area between the members or between the member and the connection bearing plate. The actual stress must be less than the allowable bearing stress

$$f_{c\perp} = \frac{P}{A} \leq F'_{c\perp}$$

The allowable compressive stress perpendicular to grain is obtained by multiplying the tabulated value by a series of adjustment factors

$$F'_{c\perp} = F_{c\perp}(C_M)(C_t)(C_i)(C_b)$$

where $F'_{c\perp}$ = allowable compressive (bearing) stress perpendicular to grain
$F_{c\perp}$ = tabulated compressive (bearing) stress perpendicular to grain
C_M = wet service factor (Sec. 4.14)
   = 1.0 for dry-service conditions, as in most covered structures. Dry-service conditions are defined as
   MC ≤ 19 percent for sawn lumber
   MC < 16 percent for glulam
C_t = temperature factor (Sec. 4.19)
   = 1.0 for normal temperature conditions
C_i = incising factor (Sec. 4.20)
   = 1.0 for sawn lumber. (*Note:* The incising factor is not applicable to glulam.)
C_b = bearing area factor (defined below)
   = 1.0 is conservative for all cases

For *sawn lumber* a single value of $F_{c\perp}$ is listed for individual stress grades in the NDS Supplement. For *glulams* a number of different tabulated values of $F_{c\perp}$ are listed. For a glulam bending combination stressed about the x axis, the value of $F_{c\perp x}$ to be used depends on whether the bearing occurs on the *compression* laminations or on the higher-quality *tension* laminations. For the common case of a beam with a positive moment, the compression laminations are on the top side of the member, and the tension laminations are on the bottom.

The bearing area factor C_b is used to account for an effective increase in bearing length. The bearing length l_b (in.) is defined as the dimension of the contact area measured *parallel* to the grain. The bearing area factor C_b may be used to account for additional wood fibers beyond the actual bearing length l_b that develop normal resisting force components. Under the conditions shown in Fig. 6.16*b*, a value of C_b greater than 1.0 is obtained by adding ⅜ in. to the actual bearing length.

Note that C_b is always greater than or equal to 1.0. It is, therefore, conservative to disregard the bearing area factor (i.e., use a default value of unity). Values of C_b may be read from NDS Table 3.10.4, or they may be calculated as illustrated in Example 6.16.

Compression perpendicular to grain is generally not considered to be a matter of life safety. Instead, it relates to the amount of deformation that is acceptable in a structure. Currently published values of bearing perpendicular to grain $F_{c\perp}$ are *average* values which are based on a *deformation limit* of 0.04 in. when tested in accordance with ASTM D 143 (Ref. 6.6). This deformation limit has been found to provide adequate service in typical wood-frame construction.

One of the most frequently used adjustment factors in wood design is the load duration factor C_D (Sec. 4.15), and it should be noted that C_D is not applied to compression perpendicular to grain design values. In addition, tabulated values of $F_{c\perp}$ are generally lower for glulam than for sawn lumber of the same deformation limit (for a discussion of these differences see Ref. 6.5).

EXAMPLE 6.16   Bearing Perpendicular to Grain

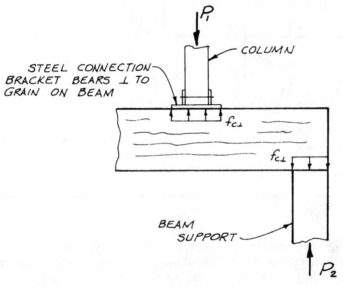

Figure 6.16a Compression perpendicular to grain.

Bearing stress calculation:

$$f_{c\perp} = \frac{P}{A} \leq F'_{c\perp}$$

where $f_{c\perp}$ = actual (computed) bearing stress perpendicular to grain
 P = applied load or reaction (force P_1 or P_2 in Fig. 6.16a)
 A = contact area
 $F'_{c\perp}$ = allowable bearing stress perpendicular to grain

Adjustment Based on Bearing Length

When the bearing length l_b (Fig. 6.16b) is less than 6 in. *and* when the distance from the end of the beam to the contact area is more than 3 in., the allowable bearing stress may be increased (multiplied) by the bearing area factor C_b.

Essentially, C_b increases the effective bearing length by ⅜ in. This accounts for the additional wood fibers that resist the applied load after the beam becomes slightly indented.

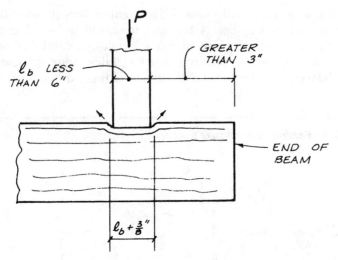

Figure 6.16*b* Required conditions to use C_b greater than 1.0.

Bearing area factor:

$$C_b = \frac{l_b + 0.375}{l_b}$$

In design applications where deformation may be critical, a reduced value of $F_{c\perp}$ may be appropriate. The following expressions are recommended when a deformation limit of 0.02 in. (one-half of the limit associated with the tabulated value) is desired.

$$F_{c\perp 0.02} = 0.73 F_{c\perp}$$

where $F_{c\perp 0.02}$ = reduced compressive stress perpendicular to grain value at deformation limit of 0.02 in.

$F_{c\perp}$ = tabulated compressive stress perpendicular to grain (deformation limit of 0.04 in.)

The other adjustments described previously for $F_{c\perp}$ also apply to $F_{c\perp 0.02}$.

The bearing stress discussed thus far has been perpendicular to the grain in the wood member. A second type of bearing stress is known as the bearing stress parallel to grain (NDS Sec. 3.10.1). It applies to the bearing that occurs on the end of a member, and it is not to be confused with the compressive stress parallel to the grain that occurs away from the end (e.g., column stress in Sec. 7.4). The bearing stress parallel to the grain assumes that the member is adequately braced and that buckling does not occur. The actual bearing stress parallel to the grain is not to exceed the allowable stress

$$f_c = \frac{P}{A} \leq F_c^*$$

where f_c = actual (computed) bearing stress parallel to grain

P = load parallel to grain on end of wood member

A = net bearing area

F_c^* = allowable compressive (bearing) stress parallel to grain on end of wood member including all adjustments *except* column stability

$= F_c(C_D)(C_M)(C_t)(C_F)(C_i)$

F_c = tabulated compressive (bearing) stress parallel to grain

C_D = load duration factor (Sec. 4.15)

C_M = wet service factor (Sec. 4.14)

$= 1.0$ for dry-service conditions, as in most covered structures. Dry-service conditions are defined as

MC $\leq$ 19 percent for sawn lumber

MC $<$ 16 percent for glulam

C_t = temperature factor (Sec. 4.19)

$= 1.0$ for normal temperature conditions

C_F = size factor (Sec. 4.16). Obtain values from Adjustment Factors section of NDS Supplement Tables 4A, 4B, and 4F for Dimension lumber and in Table 4D for Timbers.

C_i = incising factor (Sec. 4.20)

$= 0.8$ for incised dimension lumber

$= 1.0$ for dimension lumber not incised (whether the member is treated or untreated)

Bearing parallel to grain applies to two wood members bearing end to end as well as end bearing on other surfaces. Member ends are assumed to be accurately cut square. When f_c exceeds $0.75F_c^*$, bearing is to be on a steel plate or other appropriate rigid bearing surface. When required for end-to-end bearing of two wood members, the rigid insert shall be at least a 20-gage metal plate with a snug fit between abutting ends.

A comparison of the tabulated bearing stresses parallel to grain F_c and perpendicular to grain $F_{c\perp}$ shows that the values differ substantially. To make this comparison, refer to NDS Supplement Tables 4A to 4D and 4F.

It is also possible for bearing stresses in wood members to occur at some angle other than 0 to 90 degrees with respect to the direction of the grain. In this case, an allowable bearing stress somewhere between F_c' and $F_{c\perp}'$ is determined from the Hankinson formula (NDS Sec. 3.10.3). See Example 6.17.

EXAMPLE 6.17   Bearing at an Angle to Grain

Bearing at some angle to grain (Fig. 6.17) other than 0 to 90 degrees:

$$f_\theta = \frac{P}{A} \leq F_\theta'$$

where f_θ = actual bearing stress at angle to grain θ
$\quad P$ = applied load or reaction
$\quad A$ = contact area
$\quad F'_\theta$ = allowable bearing stress at angle to grain θ

Hankinson Formula

The allowable stress at angle to grain θ is given by the Hankinson formula (NDS equation 3.10-1)

$$F'_\theta = \frac{F_c^* F'_{c\perp}}{F'_c \sin^2 \theta + F'_{c\perp} \cos^2 \theta}$$

where F_c^* = allowable bearing stress parallel to grain
$\quad F'_{c\perp}$ = allowable bearing stress perpendicular to grain

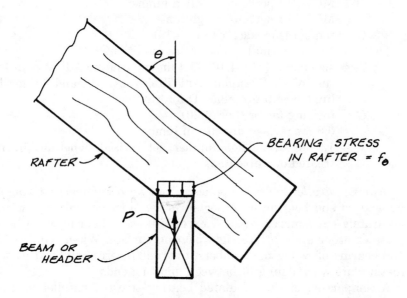

RAFTER CONNECTION

Figure 6.17 Bearing stress in two wood members. Bearing in rafter is at an angle to grain θ. Bearing in the supporting beam or header is perpendicular to grain.

This formula can probably best be solved mathematically, but the graphical solution in NDS Appendix J, *Solution of Hankinson Formula*, may be useful in visualizing the effects of angle of load to grain.

NOTE: The connection in Fig. 6.17 is given to illustrate bearing at an angle θ. For the condition shown, bearing stresses may be governed by compression perpendicular to the grain $f_{c\perp}$ in the beam supporting the rafter, rather than by f_θ in the rafter. If $f_{c\perp}$ in the beam is excessive, a bearing plate between the rafter and the beam can be used to reduce the bearing stress in the beam. The bearing stress in the rafter would not be relieved by use of a bearing plate.

As indicated in Example 6.17, the allowable stress adjustments are applied individually to F_c and $F_{c\perp}$ *before* F'_θ is computed using the Hankinson formula.

A number of examples are now given to illustrate the design procedures for beams. A variety of sawn lumber and glulam beams are considered with different support conditions and types of loading.

6.9   Design Problem: Sawn Beam

In this beam example and those that follow, the span lengths for bending and shear are, for simplicity, taken to be the same length. However, the designer may choose to determine the *design moment* based on the clear span plus one-half the required bearing length at each end (Sec. 6.2) and the *design shear* at a distance d from the support (Sec. 6.5). These different span length considerations are described in Example 6.13 (Sec. 6.5) for a simply supported beam.

In Example 6.18 a typical sawn lumber beam is designed for a roof that is essentially flat. Minimum slope is provided to prevent ponding. An initial trial beam size is determined from bending stress calculations. The extensive list of possible adjustment factors for bending stress is reduced to seven for the case for a visually graded sawn lumber beam with bending about the strong axis (see Example 6.10 in Sec. 6.4).

The beam in this problem is used in dry-service conditions and at normal temperatures, and C_M and C_t both default to unity. In addition, the roof sheathing provides continuous lateral support to the compression side of the beam. Consequently, there is no reduction in moment capacity due to lateral stability, and C_L is unity. The beam is not incised for pressure treatment since it is protected from exposure to moisture in service. Accordingly, C_i also defaults to unity. Therefore, the potential number of adjustment factors for allowable bending stress is reduced to three in this typical problem. The allowable bending stress is affected by the load duration factor C_D, the size factor for Dimension lumber C_F, and the repetitive-member factor C_r.

After selection of a trial size, shear and deflection are checked. The shear stress is not critical, but the second deflection check indicates that deflection under $(D + L_r)$ is slightly over the recommended allowable deflection. The decision of whether to accept this deflection is a matter of judgment. In this case it was decided to accept the deflection, and the trial size was retained for the final design.

EXAMPLE 6.18   Sawn Beam Design

Design the roof beam in Fig. 6.18 to support the given loads. Beams are spaced 16 in. o.c. (1.33 ft), and sufficient roof slope is provided to prevent ponding. The ceiling is gypsum wallboard. Plywood roof sheathing prevents lateral buckling. Material is No. 1 Douglas Fir-Larch (DF-L).

D = 14 psf, and L_r = 20 psf. The MC ≤ 19 percent, and normal temperature conditions apply. Tabulated stresses and section properties are to be taken from NDS Supplement.

Loads

Uniform loads are obtained by multiplying the given design loads by the tributary width.

$$w_D = 14 \text{ psf} \times 1.33 \text{ ft} \qquad = 18.67 \text{ lb/ft}$$

$$\underline{w_L = 20 \times 1.33 \qquad\qquad = 26.67}$$

$$\text{Total load } w_{TL} \qquad = 45.33 \text{ lb/ft}$$

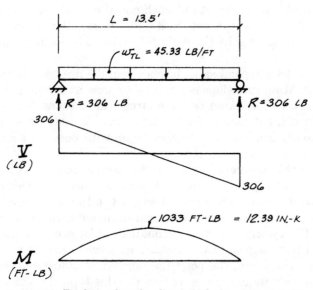

Figure 6.18 Trial size from bending calculations is 2 × 6 (Dimension lumber size).

The required load combinations are D alone with C_D = 0.9, and (D + L_r) with C_D = 1.25. By a comparison of the loads and the load duration factors (Example 4.10 in Sec. 4.15), it is determined that the critical load combination is D + L_r (i.e., total load governs).

Determine a trial size based on bending, and then check other criteria.

Bending

The span length and load for this beam are fairly small. It is assumed that the required beam size is from the range of sizes known as Dimension lumber. Tabulated stresses are found in NDS Supplement Table 4A. The beam qualifies for the repetitive-member stress increase of 15 percent. A size factor of $C_F = 1.2$ is initially assumed, and the true size factor is confirmed after a trial beam is developed. C_M, C_t, C_L and C_i default to 1.0.

$$F'_b = F'_{bx} = F_{bx}(C_D)(C_M)(C_t)(C_L)(C_F)(C_r)(C_i)$$

$$= 1000(1.25)(1.0)(1.0)(1.0)(1.2)(1.15)(1.0)$$

$$= 1725 \text{ psi}$$

$$\text{Req'd } S = \frac{M}{F'_b} = \frac{12,390}{1725} = 7.18 \text{ in.}^3$$

A trial beam size is obtained by reviewing the available sizes in NDS Supplement Table 1B. The general objective is to choose the member with the least area that furnishes a section modulus greater than that required. However, certain realities must also be considered. For example, a 1-in. nominal board would not be used for this type of beam application.

$$Try \quad 2 \times 6 \quad S = 7.56 \text{ in.}^3 \; > \; 7.18 \quad OK$$

The trial size of a 2×6 was determined using an assumed value for C_F. The size factor can now be verified in the Adjustment Factors section of NDS Supplement Table 4A:

$$C_F = 1.3 > 1.2 \quad OK$$

At this point the member has been shown to be adequate for bending stresses. However, it is often convenient to compare the actual stress and the allowable stress in a summary.

$$f_b = \frac{M}{S} = \frac{12,390}{7.56} = 1640 \text{ psi}$$

$$F'_b = F_b(C_D)(C_M)(C_t)(C_L)(C_F)(C_r)(C_i)$$

$$= 1000(1.25)(1.0)(1.0)(1.0)(1.3)(1.15)(1.0)$$

$$= 1870 \text{ psi} > 1640$$

$$\therefore \text{Bending} \quad OK$$

NOTE: A 2×5 can be checked with $C_F = 1.4$, but the reduced section modulus causes f_b to exceed F'_b. Furthermore, 2×5's are not commonly available.

Shear

Because it is judged that the shear stress for this beam is likely not to be critical, the maximum shear from the shear diagram is used without modification. C_M, C_t, and C_i are all set equal to 1.0.

$$f_v = \frac{1.5V}{A} = \frac{1.5(306)}{8.25} = 55.6 \text{ psi}$$

$$F_v' = F_v(C_D)(C_M)(C_t)(C_i)$$

$$= 180(1.25)(1.0)(1.0)(1.0)$$

$$= 225 \text{ psi} > 55.6$$

$$\therefore \text{ Shear} \quad OK$$

Deflection

The IBC specifies deflection criteria for roof beams (IBC Table 1604.3, Ref. 6.10). The calculations below use these recommended deflection criteria.

Recall that the modulus of elasticity for a wood member is not subject to adjustment for load duration, and the buckling stiffness factor C_T does not apply to deflection calculations. The adjustment factors for E in this problem all default to unity.

$$E' = E(C_M)(C_t)(C_i)$$

$$= 1,700,000(1.0)(1.0)(1.0) = 1,700,000 \text{ psi}$$

$$\Delta_L = \frac{5w_L L^4}{384E'I} = \frac{5(26.7)(13.5)^4(1728 \text{ in.}^3/\text{ft}^3)}{384(1,700,000)(20.8)} = 0.56 \text{ in.}$$

$$\text{Allow. } \Delta_L = \frac{L}{240} = \frac{13.5 \times 12}{240} = 0.67 \text{ in.} > 0.56 \quad OK$$

Deflection under total load can be calculated using the same beam deflection formula, or it can be figured by proportion.

$$\Delta_{TL} = \Delta_L\left(\frac{w_{TL}}{w_L}\right) = 0.56\left(\frac{45.3}{26.7}\right) = 0.95 \text{ in.}$$

$$\text{Allow. } \Delta = \frac{L}{180} = \frac{13.5 \times 12}{180} = 0.90 \text{ in.} < 0.95$$

In the second deflection check, the actual deflection is slightly over the allowable.

NOTE: In accordance with the IBC, the full live load was included since the moisture content was specified as MC ≤ 19 percent. If the member was specified as less than 16 percent moisture content at the time of installation and maintained in a dry condition, then only half of the live load would be required when checking against the $L/180$ limit. Using D + 0.5L, the Δ_{TL} = 0.67 in. < 0.90. As discussed in Sec. 4.7, the moisture content of S-DRY or KD lumber at the time of manufacture is 19 percent or less. However, the initial moisture content assumed in service is 15 percent, which is below the limit set by the IBC for the reduced load.

After consideration of the facts concerning this particular building design, assume that it is decided to accept the trial size.

> *Use*   2 × 6 No. 1 DF-L
> MC ≤ 19 percent

Bearing

Evaluation of bearing stresses requires knowledge of the support conditions. Without such information, the minimum bearing length will simply be determined. Recall that C_D does not apply to $F_{c\perp}$.

$$F'_{c\perp} = F_{c\perp}(C_M)(C_t)(C_b) = 625(1.0)(1.0)(1.0) = 625 \text{ psi}$$

$$\text{Req'd } A = \frac{R}{F'_{c\perp}} = \frac{306}{625} = 0.49 \text{ in.}^2$$

$$\text{Req'd } l_b = \frac{A}{b} = \frac{0.49}{1.5} = 0.33 \text{ in.}$$

All practical support conditions provide bearing lengths in excess of this minimum value.

6.10   Design Problem: Rough-Sawn Beam

In this example, a large rough-sawn beam with a fairly short span is analyzed. The cross-sectional properties for dressed lumber (S4S) are smaller than those for rough-sawn lumber, and it would be conservative to use S4S section properties for this problem. However, the larger section properties obtained using the rough-sawn dimensions are used in this example. Refer to Sec. 4.11 for information on lumber sizes.

In this problem the basic shear adjustment of neglecting any loads within a distance d from the support is used. See Example 6.19.

The importance of understanding the size categories for sawn lumber is again emphasized. The member in this problem is a Beams and Stringers size, and tabulated design values are taken from NDS Supplement Table 4D.

EXAMPLE 6.19   Rough-Sawn Beam

Determine if the 6 × 14 rough-sawn beam in Fig. 6.19*a* is adequate to support the given loads. The member is Select Structural DF-L. The load is a combination of (D + L). Lateral buckling is prevented. The beam is used in dry-service conditions (MC

≤ 19 percent) and at normal temperatures. The beam is not incised. Allowable stresses are to be taken from the NDS Supplement Table 4D.

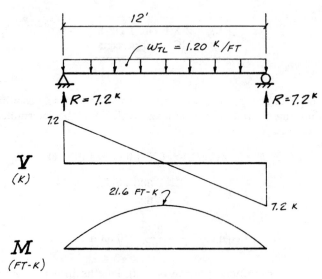

Figure 6.19a Simply supported floor beam.

Section Properties

The dimensions of rough-sawn members are approximately ⅛ in. larger than standard dressed sizes.

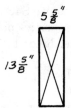

Figure 6.19b Rough-sawn 6 × 14. For a member in the *Beams and Stringers* size category, the smaller cross-sectional dimension (i.e., the thickness) is 5 in. or larger, and the width is more than 2 in. greater than the thickness.

$$A = bd = (5\tfrac{5}{8})(13\tfrac{5}{8}) = 76.64 \text{ in.}^2$$

$$S = \frac{bd^2}{6} = \frac{(5\tfrac{5}{8})(13\tfrac{5}{8})^2}{6} = 174.0 \text{ in.}^3$$

$$I = \frac{bd^3}{12} = \frac{(5\tfrac{5}{8})(13\tfrac{5}{8})^3}{12} = 1185 \text{ in.}^4$$

Bending

$$M = 21.6 \times 12 = 259 \text{ in.-k}$$

$$f_b = \frac{M}{S} = \frac{259{,}000}{174.0} = 1488 \text{ psi}$$

The size factor for a sawn member in the Beams and Stringers category is given by the formula

$$C_F = \left(\frac{12}{d}\right)^{1/9} = \left(\frac{12}{13.625}\right)^{1/9} = 0.986$$

The load duration factor for the combination of (D + L) is 1.0. All of the adjustment factors for allowable bending stress default to unity except C_F.

$$F'_b = F_b(C_D)(C_M)(C_t)(C_L)(C_F)(C_r)(C_i)$$

$$= 1600(1.0)(1.0)(1.0)(1.0)(0.986)(1.0)(1.0)$$

$$= 1577 \text{ psi} > 1488 \quad OK$$

Shear

Consider the shear diagram given in Fig. 6.19a.

$$f'_v = \frac{1.5V}{A} = \frac{1.5(7200)}{76.64} = 140.9 \text{ psi}$$

$$F'_v = F_v(C_D)(C_M)(C_t)$$

$$= 170(1.0)(1.0)(1.0)$$

$$= 170 \text{ psi} > 140.9 \quad OK$$

If the calculated actual shear stress had exceeded the allowable shear, the beam may not have been inadequate. Recall that the NDS permits the uniform load within a distance d from the face of the support to be ignored for shear calculations (see Sec. 6.5). Therefore, a modified shear V' could be determined and used to compute a reduced shear stress f'_v.

Deflection

Because the percentages of D and L were not given, only the total load deflection is calculated.

$$E' = E(C_M)(C_t)(C_i)$$

$$= 1{,}600{,}000(1.0)(1.0)(1.0) = 1{,}600{,}000 \text{ psi}$$

$$\Delta_{TL} = \frac{5w_{TL}L^4}{384E'I} = \frac{5(1200)(12)^4(12 \text{ in./ft})^3}{384(1{,}600{,}000)(1185)} = 0.30 \text{ in.}$$

$$\text{Allow. } \Delta_{\text{TL}} = \frac{L}{240} = \frac{12(12)}{240} = 0.60 > 0.30 \quad OK$$

Bearing

$$F'_{c\perp} = F_{c\perp}(C_M)(C_t)(C_b) = 625(1.0)(1.0)(1.0) = 625 \text{ psi}$$

$$\text{Req'd } A_b = \frac{R}{F'_{c\perp}} = \frac{7200}{625} = 11.52 \text{ in.}^2$$

$$\text{Req'd } l_b = \frac{A_b}{b} = \frac{11.52}{5.625} = 2.05 \text{ in. minimum}$$

> 6 × 14 rough-sawn
> Sel. Str. DF-L beam is   *OK*

The importance of understanding the size categories of sawn lumber can be seen by comparing the allowable stresses shown above for No. 1 DF-L Beams and Stringers (B&S) with those in Example 6.18 for No. 1 DF-L Dimension lumber. For a given grade, the allowable stresses depend on the size category.

6.11   Design Problem: Notched Beam

Example 6.19 will be reevaluated assuming that the ends of a deeper section are notched at the supports to provide the same member depth over the support. In Example 6.19, the adequacy of a 6 × 14 rough-sawn beam was determined for the given loads. In this example, a notched 6 × 18 rough-sawn member is used as the beam in Fig. 6.19a. Since the notches are at the supports only, they will only impact the shear calculations.

EXAMPLE 6.20   Notched Beam

Determine if a 6 × 18 rough-sawn beam with notched supports is adequate, considering shear only, to support the given loads. The notches are 4-in. deep and sufficiently long to provide adequate bearing length at the support. The notch does not extend past the face of the support. See Fig. 6.20. The depth of the notch allows the remaining depth of the member to be compatible with the 6 × 14 beams supported on the same level.

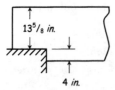

Figure 6.20   Notch detail at support for rough-sawn beam.

Notch Size

NDS Sec. 4.4.3 limits the size of notches at the ends of members for bearing over a support. The depth of such notches cannot exceed one-fourth of the total depth d of the beam. In this example, $d = 17\frac{5}{8}$ in. The maximum notch depth is $4\frac{13}{32}$ in., which is greater than the 4-in. notch specified in Fig. 6.20. The notch is permissible by the NDS.

Shear Check

$$f'_v = \frac{1.5V'}{bd_n}\left(\frac{d}{d_n}\right)^2 = \frac{1.5(7200)}{(5.625)(13.625)}\left(\frac{17.625}{13.625}\right)^2 = 235.8 \text{ psi}$$

$$F'_v = F_v(C_D)(C_M)(C_t)$$

$$= 170(1.0)(1.0)(1.0)$$

$$= 170 \text{ psi} < 235.8 \text{ psi} \qquad \textit{No Good}$$

From Example 6.19, the shear stress for the unnotched 6×14 section was 140.9 psi. This illustrates the effect of notching and the stress concentration resulting from it.

NOTE: For notched sections, the NDS *requires* that all loads, including a uniform load within a distance d from the face of the support, are to be considered for shear calculations (see NDS Sec. 3.4.3.2). Therefore, a modified shear V' *cannot* be used to compute a reduced shear stress f'_v for notched sections.

Sometimes notching is required to maintain uniform depth above a support for a series of members. As seen by this example, the effect of notching is significant on the shear strength of a beam and can cause shear to control the design. Consider the general case of a maximum depth notch with $d_n = 0.75d$. The stress increase caused by this maximum notch depth at a support is $(d/d_n)^2 = 1.78$.

6.12 Design Problem: Sawn-Beam Analysis

The two previous examples have involved beams in the Dimension lumber and Beams and Stringers size categories. Example 6.21 is provided to give additional practice in determining allowable stresses. The member is again Dimension lumber, but the load duration factor, the wet service factor, and the size factor are all different from those in previous problems. Wet service factors and the size factor are obtained from the Adjustment Factors section in the NDS Supplement.

EXAMPLE 6.21   Sawn-Beam Analysis

Determine if the 4×16 beam given in Fig. 6.21 is adequate for a dead load of 70 lb/ft and a snow load of 180 lb/ft. Lumber is stress grade No. 1 and Better, and the species group is Hem-Fir. The member is not incised. Adequate bracing is provided, so that lateral stability is not a concern.

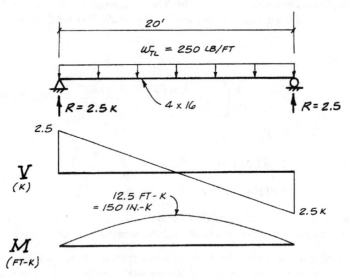

Figure 6.21   A 4 × 16 beam is in the *Dimension lumber* size category.

This beam is used in a factory where the EMC will exceed 19 percent,* but temperatures are in the normal range. Beams are 4 ft-0 in. o.c. The minimum roof slope for drainage is provided so that ponding need not be considered. Allowable deflection limits for this design are assumed to be $L/360$ for snow load and $L/240$ for total load.
    Allowable stresses and section properties are to be in accordance with the NDS.

Bending

Section properties for a 4 × 16 are listed in NDS Supplement Table 1B.

$$f_b = \frac{M}{S} = \frac{150,000}{135.7} = 1105 \text{ psi}$$

The load duration factor is $C_D = 1.15$ for the load combination of (D + S). Beam spacing does not qualify for the repetitive-member stress increase, and $C_r = 1.0$. Lateral stability is not a consideration, and $C_L = 1.0$.
    For a 4 × 16 the size factor is read from Table 4A:

$$C_F = 1.0$$

---

*The need for pressure-treated lumber to prevent decay should be considered (Sec. 4.9).

Also from Table 4A the wet-service factor for bending is given as

$$C_M = 0.85$$

Except that, when $F_b(C_F) \leq 1150$ psi, $C_M = 1.0$.
In the case of 4×16 No. 1 & Btr Hem-Fir:

$$F_b(C_F) = 1100(1.0) < 1150 \text{ psi}$$

$$\therefore C_M = 1.0$$

The coefficients for determining F'_{bx} for a sawn lumber beam are obtained from the summary in Example 6.10 in Sec. 6.4 (see also NDS Table 4.3.1). In the bending stress summary given below, most of the adjustment factors default to unity. However, it is important for the designer to follow the steps leading to this conclusion.

$$F'_b = F_b(C_D)(C_M)(C_t)(C_L)(C_F)(C_r)(C_i)$$

$$= 1100(1.15)(1.0)(1.0)(1.0)(1.0)(1.0)(1.0)$$

$$= 1265 \text{ psi} > 1105 \text{ psi} \quad OK$$

Shear

$$f_v = \frac{1.5V}{A} = \frac{1.5(2500)}{53.375} = 70.3 \text{ psi}$$

$$F'_v = F_v(C_D)(C_M)(C_t)$$

$$= 150(1.15)(0.97)(1.0)$$

$$= 167.3 \text{ psi} > 70.3 \text{ psi†} \quad OK$$

Deflection

$$E' = E(C_M)(C_t)(C_i) = 1{,}500{,}000(0.9)(1.0)(1.0) = 1{,}350{,}000 \text{ psi}$$

$$\Delta_S = \frac{5wL^4}{384E'I}$$

$$= \frac{5(180)(20)^4(1728)}{384(1{,}350{,}000)(1034)} = 0.46 \text{ in.}$$

$$\text{Allow. } \Delta_S = \frac{L}{360} = \frac{20 \times 12}{360} = 0.67 > 0.46 \quad OK$$

---

†If f_v had exceeded F'_v, the design shear could have been reduced in accordance with Sec. 6.5.

By proportion,

$$\Delta_{TL} = \left(\frac{250}{180}\right) 0.46 = 0.64 \text{ in.}$$

$$\text{Allow. } \Delta_{TL} = \frac{L}{240} = \frac{20 \times 12}{240} = 1.00 > 0.64 \qquad OK$$

> 4 × 16 No. 1 & Btr
> Hem-Fir beam   *OK*

6.13   Design Problem: Glulam Beam with Full Lateral Support

The examples in Secs. 6.13, 6.14, and 6.15 all deal with the design of the same glulam beam, but different conditions of lateral support for the beam are considered in each problem. the first example deals with the design of a beam that has *full lateral support* to the compression side of the member, and lateral stability is simply not a concern. See Example 6.22.

EXAMPLE 6.22   Glulam Beam—Full Lateral Support

Determine the required size of a Western Species 24F-1.8E stress class glulam for the simple-span roof beam shown in Fig. 6.22. Assume dry-service conditions and normal temperature range. D = 200 lb/ft, and S = 800 lb/ft. Use the IBC-required deflection limits for a roof beam in a commercial building supporting a nonplaster ceiling. By inspection the critical load combination is

$$D + S = 200 + 800 = 1000 \text{ lb/ft}$$

A number of adjustment factors for determining allowable stresses can be determined directly from the problem statement. For example, the load duration factor is $C_D = 1.15$ for the combination of (D + S). In addition, the wet service factor is $C_M = 1.0$ for a glulam with MC < 16 percent, and the temperature factor is $C_t = 1.0$ for members used at normal temperatures.

Bending

The glulam beam will be loaded such that the tension laminations will be stressed in tension, and the tabulated stress F_{bx}^+ applies. The summary for glulam beams in Example 6.11 in Sec. 6.4 (and NDS Table 5.3.1) indicates that there are two possible definitions of allowable bending stress. One considers the effects of lateral stability as measured by the beam stability factor C_L. The other evaluates the effect of beam width, depth, and length as given by the volume factor C_V.

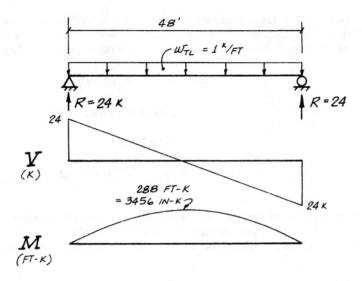

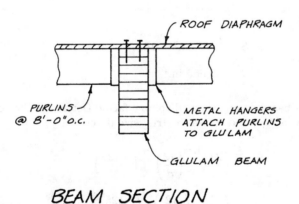

BEAM SECTION

Figure 6.22 Glulam beam with span of 48 ft and full lateral support to the compression side of the member provided by roof diaphragm.

Lateral stability: $\qquad F'_{bx} = F_{bx}(C_D)(C_M)(C_t)(C_L)$

Volume effect: $\qquad F'_{bx} = F_{bx}(C_D)(C_M)(C_t)(C_V)$

The sketch of the beam cross section shows that the compression side of the beam (positive moment places the top side in compression) is restrained from lateral movement by an effective connection to the roof diaphragm. The unbraced length is zero, and the beam slenderness ratio is zero. Lateral buckling is thus prevented, and the beam stability factor $C_L = 1.0$. In this case, only the allowable bending stress using the volume factor needs to be considered.

Before the volume factor can be evaluated, a trial beam size must be established. This is done by assuming a value for C_V which will be later verified. Assume $C_V = 0.82$. Tabulated stresses are obtained from NDS Supplement Table 5A.

$$F'_{bx} = F_{bx}(C_D)(C_M)(C_t)(C_V) ,$$

$$= 2400(1.15)(1.0)(1.0)(0.82)$$

$$= 2263 \text{ psi}$$

$$\text{Req'd } S = \frac{M}{F'_b} = \frac{3,456,000}{2263} = 1527 \text{ in.}^3$$

As in most beam designs, the objective is to select the member with the least cross-sectional area that provides a section modulus greater than that required. This can be done using the S_x column for Western Species glulams in NDS Supplement Table 1C.

$$Try \quad 6\frac{3}{4} \times 37\frac{1}{2} \text{ (twenty-five } 1\frac{1}{2}\text{-in. lams)}$$

$$A = 253.1 \text{ in.}^2$$

$$S = 1582 \text{ in.}^3 > 1527 \quad OK$$

$$I = 29,660 \text{ in.}^4$$

The trial size was based on an assumed volume factor. Determine the actual C_V (see Sec. 5.6 for a review of C_V), and revise the trial size if necessary.

$$C_V = \left(\frac{21}{L}\right)^{1/10}\left(\frac{12}{d}\right)^{1/10}\left(\frac{5.125}{b}\right)^{1/10}$$

$$= \left(\frac{21}{48}\right)^{0.1}\left(\frac{12}{37.5}\right)^{0.1}\left(\frac{5.125}{6.75}\right)^{0.1}$$

$$= 0.799 < 0.82$$

Because the assumed value of C_V was not conservative, the actual and allowable stresses will be compared in order to determine if the trial size is adequate.

$$f_b = \frac{M}{S} = \frac{3,456,000}{1582} = 2185 \text{ psi}$$

$$F'_b = F_b(C_D)(C_M)(C_t)(C_V)$$

$$= 2400(1.15)(1.0)(1.0)0.799)$$

$$= 2205 \text{ psi} > 2185 \quad OK$$

Shear

Ignore the reduction of shear given by V' (conservative).

$$f_v = \frac{1.5V}{A} = \frac{1.5(24{,}000)}{253.1} = 142 \text{ psi}$$

$$F'_{vx} = F_v(C_D)(C_M)(C_t) = 240(1.15)(1.0)(1.0)$$

$$= 276 \text{ psi} > 142 \quad OK$$

Deflection

$$E'_x = E_x(C_M)(C_t)$$

$$= 1{,}800{,}000(1.0)(1.0)$$

$$= 1{,}800{,}000 \text{ psi}$$

$$\Delta_{\text{TL}} = \frac{5w_{\text{TL}}L^4}{384E'I} = \frac{5(1000)(48)^4(12 \text{ in./ft})^3}{384(1{,}800{,}000)(29{,}660)} = 2.24 \text{ in.}$$

$$\frac{\Delta_{\text{TL}}}{L} = \frac{2.24}{48 \times 12} = \frac{1}{257} < \frac{1}{180} \quad OK$$

NOTE: The total load was conservatively used for this deflection check. According to footnote d in the IBC Table 1604.3, the dead load component may be reduced by a factor of 0.5. If this deflection check had not been satisfied, then the deflection caused by 0.5D + S could be checked against $L/240$.

By proportion,

$$\Delta_{\text{S}} = \left(\frac{800}{1000}\right)\Delta_{\text{TL}} = 0.8(2.24) = 1.79 \text{ in.}$$

$$\frac{\Delta_{\text{S}}}{L} = \frac{1.79}{48 \times 12} = \frac{1}{321} < \frac{1}{240} \quad OK$$

$$\text{Camber} = 1.5\Delta_{\text{D}} = 1.5\left(\frac{200}{1000}\right)(2.24) = 0.67 \text{ in.}$$

Bearing

The support conditions are unknown, so the required bearing length will simply be determined. Use $F'_{c\perp}$ for bearing on the tension face of a glulam bending about the x axis. Recall that C_D does not apply to $F_{c\perp}$.

$$F'_{c\perp} = F_{c\perp}(C_M)(C_t)(C_b)$$

$$= 650(1.0)(1.0)(1.0) = 650 \text{ psi}$$

$$\text{Req'd } A = \frac{R}{F'_{c\perp}} = \frac{24{,}000}{650} = 36.9 \text{ in.}^2$$

$$\text{Req'd } l_b = \frac{36.9}{6.75} = 5.47 \qquad \text{Say } l_b = 5\tfrac{1}{2} \text{ in. min.}$$

> *Use* $6\tfrac{3}{4} \times 37\tfrac{1}{2}$ (twenty-five $1\tfrac{1}{2}$-in. lams)
> 24F-1.8E glulam—camber 0.67 in.

In Sec. 6.14 this example is reworked with lateral supports at 8 ft-0 in. o.c. This spacing is obtained when the purlins rest on top of the glulam. With this arrangement the sheathing is separated from the beam, and the distance between points of lateral support becomes the spacing of the purlins.

In Sec. 6.15, the beam is analyzed for an unbraced length of 48 ft-0 in. In other words, only the ends of the beam are braced against translation and rotation. This condition would exist if no diaphragm action developed in the sheathing (i.e., the sheathing for some reason was not capable of functioning as a diaphragm), or if no sheathing or effective bracing is present along the beam. Fortunately, this situation is not common in ordinary building design.

6.14   Design Problem: Glulam Beam with Lateral Support at 8 ft-0 in.

In order to design a beam with an unbraced compression zone, it is necessary to check both *lateral stability* and *volume effect*. To check lateral stability, a trial beam size is required so that the beam slenderness ratio R_B can be computed. This is similar to column design, where a trial size is required before the column slenderness ratio and the strength of the column can be evaluated.

All criteria except unbraced length are the same for this example and the previous problem. Therefore, initial trial beam size is taken from Example 6.22. The $6\tfrac{3}{4} \times 37\tfrac{1}{2}$ trial represents the minimum beam size based on the volume-effect criterion. Because all other factors are the same, only the lateral stability criteria are considered in this example. See Example 6.23. The calculations for C_L indicate that lateral stability is less critical than the volume effect for this problem. The trial size, then, is adequate.

EXAMPLE 6.23   Glulam Beam—Lateral Support at 8 ft-0 in.

Rework Example 6.22, using the modified lateral support condition shown in the beam section view in Fig. 6.23. All other criteria are the same. See Fig. 6.22 for the load, shear, and moment diagrams.

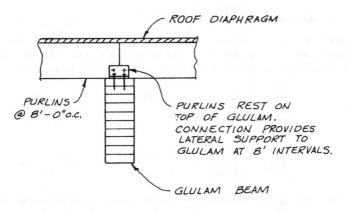

BEAM SECTION

Figure 6.23 Beam from Example 6.22 with revised lateral support conditions.

Bending

The allowable stresses for a glulam beam bending about the x axis are summarized in Example 6.11. Separate allowable stresses are provided for lateral stability and volume effect:

Lateral stability: $F'_{bx} = F_{bx}(C_D)(C_M)(C_t)(C_L)$

Volume effect: $F'_{bx} = F_{bx}(C_D)(C_M)(C_t)(C_V)$

See Example 6.22 for the development of a trial size based on the volume effect. This trial size will now be analyzed for the effects of lateral stability using an unbraced length of 8 ft-0 in.

Try $6\frac{3}{4} \times 37\frac{1}{2}$ 24F-1.8E Stress Class glulam

Slenderness ratio for bending member R_B

Unbraced length $l_u = 8$ ft $= 96$ in.

Effective unbraced lengths are given in Example 6.8 and in NDS Table 3.3.3. For a single-span beam with a uniformly distributed load, the definition of l_e depends on the l_u/d ratio

$$\frac{l_u}{d} = \frac{96}{37.5} = 2.56 < 7$$

$$\therefore l_e = 2.06 l_u = 2.06(96) = 198 \text{ in.}$$

$$R_B = \sqrt{\frac{l_e d}{b^2}} = \sqrt{\frac{198(37.5)}{(6.75)^2}} = 12.76$$

Coefficients for computing beam stability factor C_L

A beam subject to lateral torsional buckling is governed by stability about the y axis, and the modulus of elasticity for use in determining the beam stability factor is E_y'.

The Euler critical buckling stress for a glulam beam uses the coefficient $K_{bE} = 0.610$.

$$E_y' = E_y(C_M)(C_t) = 1,600,000(1.0)(1.0)$$

$$= 1,600,000 \text{ psi}$$

$$F_{bE} = \frac{K_{bE}E_y'}{R_B^2} = \frac{0.610(1,600,000)}{12.76^2} = 5994 \text{ psi}$$

The tabulated bending stress about the x axis modified by all factors except C_V and C_L is given the notation F_b^*

$$F_{bx}^* = F_{bx}(C_D)(C_M)(C_t)$$

$$= 2400(1.15)(1.0)(1.0) = 2760 \text{ psi}$$

$$\frac{F_{bE}}{F_{bx}^*} = \frac{5994}{2760} = 2.172$$

$$\frac{1 + F_{bE}/F_{bx}^*}{1.9} = \frac{1 + 2.172}{1.9} = 1.669$$

Beam stability factor

$$C_L = \frac{1 + F_{bE}/F_{bx}^*}{1.9} - \sqrt{\left(\frac{1 + F_{bE}/F_{bx}^*}{1.9}\right)^2 - \frac{F_{bE}/F_{bx}^*}{0.95}}$$

$$= 1.669 - \sqrt{1.669^2 - 2.172/0.95}$$

$$= 0.962$$

From Example 6.22, the volume factor for this beam is

$$C_V = 0.799 < C_L$$

∴ Volume effect governs over lateral stability.

The allowable bending stress for the beam with lateral support to the compression side at 8 ft-0 in. is the same as that for the beam in Example 6.22:

$$F_b' = 2205 \text{ psi} > 2185 \qquad OK$$

Use   6¾ × 37½
24F-1.8E glulam

The beam in Example 6.23 is seen to be unaffected by an unbraced length of 8 ft. The beam slenderness ratio R_B is the principal measure of lateral stability, and R_B is a function of the unbraced length, beam depth, and beam width. The slenderness ratio is especially sensitive to beam *width* because of the square in the denominator.

A large slenderness ratio is obtained in Example 6.24 by increasing the unbraced length from 8 to 48 ft.

6.15 Design Problem: Glulam Beam with Lateral Support at 48 ft-0 in.

The purpose of this brief example is to illustrate the impact of a very long unbraced length and a correspondingly large beam slenderness ratio. See Example 6.24. As with the previous example, the initial trial size is taken from Example 6.22 because a trial size is required in order to calculate the beam slenderness ratio.

This example illustrates why it is desirable to have at least some intermediate lateral bracing. The very long unbraced length causes the trial size to be considerably overstressed, and a new trial beam size is required.

The problem is not carried beyond the point of checking the initial trial beam because the purpose of the example is simply to demonstrate the effect of lateral buckling. A larger trial size would be evaluated in a similar manner.

EXAMPLE 6.24   Glulam Beam—Lateral Support at 48 ft-0 in.

Rework the beam design problem in Examples 6.22 and 6.23 with lateral supports at the ends of the span only. See Fig. 6.22 for the load, shear, and moment diagrams.

Bending

The allowable stresses for a glulam beam are

Lateral stability: $$F'_{bx} = F_{bx}(C_D)(C_M)(C_t)(C_L)$$

Volume effect: $$F'_{bx} = F_{bx}(C_D)(C_M)(C_t)(C_V)$$

The size in Example 6.22 was based on the volume factor C_V. This member will now be checked for the effects of lateral stability with an unbraced length of 48 ft-0 in.

$$Try \quad 6\frac{3}{4} \times 37\frac{1}{2} \; 24F\text{-}1.8E \; \text{Stress Class glulam}$$

Slenderness ratio for bending member R_B

$$\text{Unbraced length } l_u = 48 \text{ ft} = 576 \text{ in.}$$

Effective unbraced lengths are given in Example 6.8 and in NDS Table 3.3.3. For a single-span beam with a uniformly distributed load, the definition of l_e depends on the l_u/d ratio

$$\frac{l_u}{d} = \frac{576}{37.5} = 15.36 > 7$$

$$\therefore l_e = 1.63 l_u + 3d = 1.63(576) + 3(37.5) = 1051 \text{ in.}$$

$$R_B = \sqrt{\frac{l_e d}{b^2}} = \sqrt{\frac{1051(37.5)}{(6.75)^2}} = 29.42$$

Coefficients for computing beam stability factor C_L

$$F_{bE} = \frac{K_{bE} E_y'}{R_B^2} = \frac{0.610(1,600,000)}{(29.42)^2} = 1127 \text{ psi}$$

$$F_{bx}^* = F_{bx}(C_D)(C_M)(C_t)$$

$$= 2400(1.15)(1.0)(1.0) = 2760 \text{ psi}$$

$$\frac{F_{bE}}{F_{bx}^*} = \frac{1127}{2760} = 0.408$$

$$\frac{1 + F_{bE}/F_{bx}^*}{1.9} = \frac{1 + 0.408}{1.9} = 0.741$$

Beam stability factor

$$C_L = \frac{1 + F_{bE}/F_{bx}^*}{1.9} - \sqrt{\left(\frac{1 + F_{bE}/F_{bx}^*}{1.9}\right)^2 - \frac{F_{bE}/F_{bx}^*}{0.95}}$$

$$= 0.741 - \sqrt{0.741^2 - 0.408/0.95}$$

$$= 0.395$$

From Example 6.22, the volume factor for this beam is

$$C_V = 0.799 > C_L$$

$$\therefore \text{ Lateral stability governs over the volume factor.}$$

The allowable bending stress for the beam with lateral support to the compression side at 48 ft-0 in. is

$$F_b' = F_b(C_D)(C_M)(C_t)(C_L)$$

$$= 2400(1.15)(1.0)(1.0)(0.395)$$

$$= 1090 \text{ psi}$$

$$f_b = 2185 \text{ psi} > 1090 \qquad \text{NG}$$

The trial size of a $6\frac{3}{4} \times 37\frac{1}{2}$ is considerably overstressed in bending and is no good (NG). A revised trial size is thus required and is left as an exercise for the reader.

6.16 Design Problem: Glulam with Compression Zone Stressed in Tension

Some glulam beams have *balanced* combinations of laminations. These have the same allowable bending stress on the top and bottom faces of the member. Other combinations have tension lamination requirements only on one side of the beam. For this latter case there are two different values of allowable bending stress:

1. F_{bx} tension zone stressed in tension (F_{bx}^+)

2. F_{bx} compression zone stressed in tension (F_{bx}^-)

The beam in Example 6.25 has a large positive moment and a small negative moment. The member in this example involves a combination that is not balanced. See Example 6.25. The beam is first designed for the large positive moment using F_{bx}^+. The bending stress that results from the negative moment is then checked against the smaller allowable bending stress F_{bx}^-. The cantilever beam system in Example 6.29 uses a balanced bending combination.

EXAMPLE 6.25   Compression Zone Stressed in Tension

The roof beam in Fig. 6.24 is a Western Species 24F-1.8E glulam. The design load includes a concentrated load and a uniformly distributed load. Loads are a combination of (D + L_r). Lateral support is provided to the top face of the beam by the roof sheathing. However, the bottom face is laterally unsupported in the area of negative moment except at the reaction point. The beam is used in dry-service conditions and at normal temperatures. The minimum roof slope is provided so that ponding need not be considered. For this problem consider bending stresses only.

Positive Moment (Tension Zone Stressed in Tension)

In the area of positive bending moment, the allowable bending stress is F_{bx}^+. Allowable stresses for a glulam beam are

Lateral stability: $F_b' = F_{bx}^{+\prime} = F_{bx}^+(C_D)(C_M)(C_t)(C_L)$

Volume effect: $F_b' = F_{bx}^+ = F_{bx}^+(C_D)(C_M)(C_t)(C_V)$

Also in the area of positive bending moment, the unbraced length is $l_u = 0$ because continuous lateral support is provided to the top side of the beam by the roof sheathing. Therefore C_L defaults to unity, and lateral stability *does not govern* (DNG).

Develop a trial beam size, using an assumed value for the volume factor, and check the actual C_V later. The load duration factor is $C_D = 1.25$ for the combination of (D + L_r). Both C_M and C_t default to unity. Tabulated stress are given in NDS Supplement Table 5A.

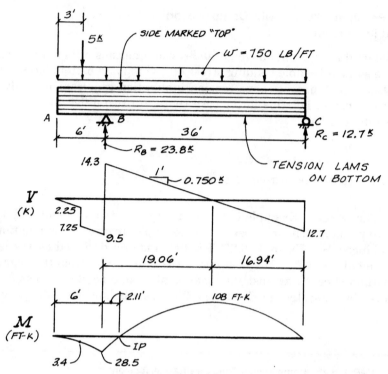

Figure 6.24 Glulam beam with small cantilever.

Assume $C_V = 0.90$:

$$F'_b = F^{+'}_{bx} = F^+_{bx}(C_D)(C_M)(C_t)(C_V)$$

$$= 2400(1.25)(1.0)(1.0)(0.90) = 2700 \text{ psi}$$

Max. $M = 108$ ft-k $= 1295$ in.-k (from Fig. 6.23)

$$\text{Req'd } S = \frac{M}{F'_b} = \frac{1,295,000}{2700} = 480 \text{ in.}^3$$

Refer to NDS Supplement Table 1C, and choose the smallest Western Species glulam size that furnishes a section modulus greater than the required.

$$Try \quad 5\tfrac{1}{8} \times 24 \quad 24F\text{-}1.8E \quad \text{glulam}$$

$$S = 492 \text{ in.}^3 > 480 \quad OK$$

Verify C_V.

The volume factor is a function of the length, depth, and width of a beam. The length is to be taken as the distance between points of zero moment in Fig. 6.24 ($L = 36 - 2.11 = 33.89$ ft). However, it is simple and conservative to use the full span length of 36 ft.

$$C_V = K_L \left(\frac{21}{L}\right)^{1/10} \left(\frac{12}{d}\right)^{1/10} \left(\frac{5.125}{b}\right)^{1/10} \leq 1.0$$

$$= 1.0 \left(\frac{21}{36}\right)^{0.1} \left(\frac{12}{24}\right)^{0.1} \left(\frac{5.125}{5.125}\right)^{0.1}$$

$$= 0.884 < 1.0$$

The assumed value of $C_V = 0.90$ was not conservative. Therefore, compare the actual and allowable bending stresses in a summary:

$$f_b = \frac{M}{S} = \frac{1,295,000}{492} = 2630 \text{ psi}$$

$$F_b' = F_{bx}^{+\prime} = F_{bx}(C_D)(C_M)(C_t)(C_V)$$

$$= 2400(1.25)(1.0)(1.0)(0.884)$$

$$= 2650 \text{ psi} > 2630$$

$$\therefore \text{ Positive moment} \quad OK$$

Negative Moment (Compression Zone Stressed in Tension)

The trial beam size remains the same, and the computed bending stress is

$$\text{Neg. } M = 28.5 \text{ ft-k} = 342 \text{ in.-k}$$

$$f_b = \frac{M}{S} = \frac{342,000}{492} = 695 \text{ psi}$$

In the area of negative bending moment, the tabulated bending stress is $F_{bx}^- = 1450$ psi. The allowable stress is the smaller value determined from the two criteria

Lateral stability: $F_b' = F_{bx}^{-\prime} = F_{bx}^-(C_D)(C_M)(C_t)(C_L)$

Volume effect: $F_b' = F_{bx}^{-\prime} = F_{bx}^-(C_D)(C_M)(C_t)(C_V)$

Lateral stability

The possibility of lateral buckling needs to be considered because the bottom side of the beam does not have continuous lateral support.

Slenderness ratio for beam R_b:

To the left of the support:

$$l_u = 6 \text{ ft} = 72 \text{ in.}$$

To the right of the support to the inflection point (IP):

$$l_u = 2.11 \text{ ft} = 25.3 \text{ in.} \quad \text{(not critical)}$$

Effective unbraced lengths are given in Example 6.8 (Sec. 6.3) and in NDS Table

3.3.3. For a cantilever beam with any loading, the definition of l_e depends on the l_u/d ratio

$$\frac{l_u}{d} = \frac{72}{24} = 3.0 < 7$$

$$\therefore l_e = 2.06 l_u = 2.06(72) = 148 \text{ in.}$$

$$R_B = \sqrt{\frac{l_e d}{b^2}} = \sqrt{\frac{148(24)}{(5.125)^2}} = 11.64$$

Coefficients for computing beam stability factor C_L:

$$E_y' = E_y(C_M)(C_t) = 1{,}600{,}000(1.0)(1.0)$$

$$= 1{,}600{,}000 \text{ psi}$$

$$F_{bE} = \frac{K_{bE}E_y'}{R_B^2} = \frac{0.610(1{,}600{,}000)}{11.64^2} = 7204 \text{ psi}$$

$$F_{bx}^* = F_{bx}(C_D)(C_M)(C_t)$$

$$= 1450(1.25)(1.0)(1.0) = 1812 \text{ psi}$$

$$\frac{F_{bE}}{F_{bx}^*} = \frac{7204}{1812} = 3.976$$

$$\frac{1 + F_{bE}/F_{bx}^*}{1.9} = \frac{1 + 3.976}{1.9} = 2.619$$

Beam stability factor

$$C_L = \frac{1 + F_{bE}/F_{bx}^*}{1.9} - \sqrt{\left(\frac{1 + F_{bE}/F_{bx}^*}{1.9}\right)^2 - \frac{F_{bE}/F_{bx}^*}{0.95}}$$

$$= 2.619 - \sqrt{2.619^2 - 3.976/0.95}$$

$$= 0.984$$

Volume effect

The length to compute the volume factor is defined as the distance between points of zero moment ($L = 6 + 2.11 = 8.11$ ft).

$$C_V = K_L \left(\frac{21}{L}\right)^{1/10} \left(\frac{12}{d}\right)^{1/10} \left(\frac{5.125}{b}\right)^{1/10} \le 1.0$$

$$= 1.0 \left(\frac{21}{8.11}\right)^{0.1} \left(\frac{12}{24}\right)^{0.1} \left(\frac{5.125}{5.125}\right)^{0.1}$$

$$= 1.026 > 1.0$$

$$\therefore C_V = 1.0$$

The lateral stability factor governs over the volume factor.

$$F'_b = F_{bx}^{-'} = F_{bx}^{-}(C_D)(C_M)(C_t)(C_L)$$

$$= 1450(1.25)(1.0)(1.0)(0.984)$$

$$= 1784 \text{ psi}$$

$$f_b = 695 \text{ psi} < 1784 \qquad OK$$

$5\frac{1}{8} \times 24$ 24F-1.8E   Stress Class glulam   *OK* for bending

6.17 Cantilever Beam Systems

Cantilever beam systems that have an internal hinge connection are often used in glulam construction. The reason for this is that a smaller-size beam can generally be used with a cantilever system compared with a series of simply supported beams. The cantilever length L_c in the cantilever beam system is an important variable. See Example 6.26. A cantilever length can be established for which an optimum beam size can be obtained.

EXAMPLE 6.26   Cantilever Beam Systems

The *bending strength* of a cantilever beam system can be optimized by choosing the cantilever length L_c so that the local maximum moments M_1, M_2, and M_3 will all be equal. For the two-equal-span cantilever system shown in Fig. 6.25, with a constant uniform load on both spans, the cantilever length

$$L_c = 0.172L$$

gives equal local maximum moments

$$M_1 = M_2 = M_3 = 0.086wL^2$$

This maximum moment is considerably less than the maximum moment for a uniformly loaded simple beam:

$$M = \frac{wL^2}{8} = 0.125wL^2$$

Recommended cantilever lengths for a number of cantilever beam systems are given in the TCM (Ref. 6.5).

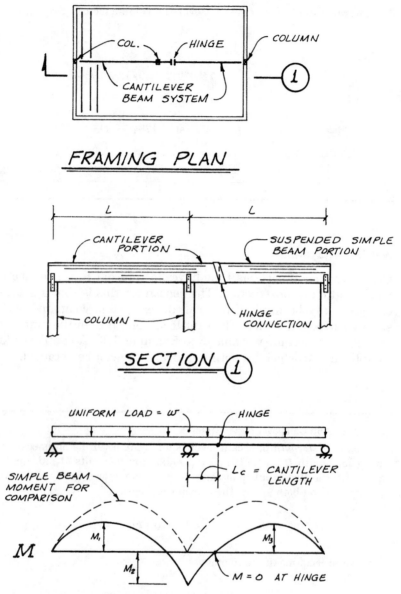

Figure 6.25 Two span cantilever beam system.

Cantilever beam systems are not recommended for floors. Proper cambering is difficult, and cantilever beam systems in floors may transmit vibrations from one span to another. AITC recommends the use of simply supported beams for floors.

For the design of both roofs and floors, IBC Chap. 16 requires that the case of dead load on all spans plus roof live load on alternate spans (unbalanced L_r) must be considered in addition to full $(D + L_r)$ on all spans. See Example 6.27.

EXAMPLE 6.27   Load Cases for a Two-Span Cantilever Beam System

Load Case 1: (D + L on All Spans)

This load constitutes the maximum total load and can produce the critical design moment, shear, and deflection. See Fig. 6.26.

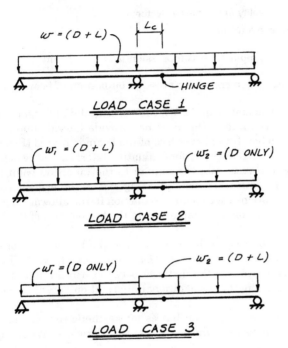

Figure 6.26   When $L_r < 20$ psf, *full* and *unbalanced* live load analyses are required.

Load Case 2: (D + Unbalanced L on Left Span)

When unbalanced live load is required, this load will produce the critical positive moment in the left span.

Load Case 3: (D + Unbalanced L on Right Span)

This load case will produce the same maximum negative moment as load case 1. It will also produce the maximum length from the interior support to the inflection point on the moment diagram for the left span. Depending on bracing conditions this length could be critical for lateral stability. In addition, this load case will produce the minimum reaction at the left support. For a large live load and a long cantilever length, it is possible to develop an uplift reaction at this support.

The case of unbalanced live loads can complicate the design of cantilever beam systems. This is particularly true if deflections are considered. When unbalanced live loads are required, the optimum cantilever span length L_c will be different from those established for the same uniform load on all spans.

In a cantilever system the compression side of the member is not always on the top of the beam. This will require a *lateral stability* analysis of bending stresses even though the top of the girder may be connected to the horizontal diaphragm. See Example 6.28.

EXAMPLE 6.28   Lateral Stability of Cantilever Systems

Moment diagram sign convention:

Positive moment = compression on top of beam

Negative moment = compression on bottom of beam

In areas of negative moment (Fig. 6.27a), the horizontal diaphragm is connected to the tension side of the beam, and this does not provide lateral support to the compression side of the member. If the lower face of the beam is braced (Fig. 6.27b) at the interior column, the unbraced length l_u for evaluating lateral stability is the cantilever length L_c, or it is the distance from the column to the inflection point (IP). For the given beam these unbraced lengths are equal (Fig. 6.27a) under balanced loading. If lateral stability considerations cause a large reduction in the allowable bending stress, additional diagonal braces from the diaphragm to the bottom face of the beam may be required.

Several types of knee braces can be used to brace the bottom side of the beam. A *prefabricated metal knee brace* and a *lumber knee brace* are shown in Fig. 6.27b. The distance between knee braces, or the distance between a brace and a point of zero moment, is the unbraced length. For additional information on unbraced lengths, see Ref. 6.5.

In order to avoid the use of diagonal braces for aesthetic reasons, some designers use a beam-to-column connection which is designed to provide lateral support to the

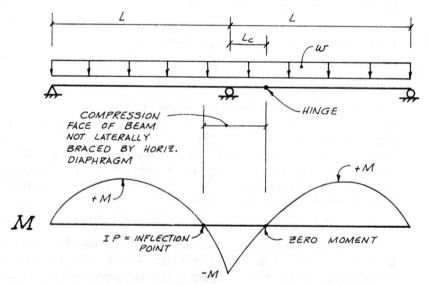

Figure 6.27a Unbraced length considerations for negative moment.

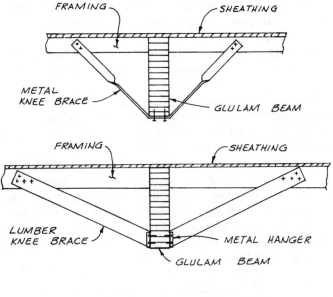

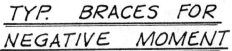

Figure 6.27b   Methods of bracing bottom side of beam.

bottom face of the beam. Considerable care and engineering judgment must be exercised in the design of this type of connection to ensure effective lateral restraint.

6.18   Design Problem: Cantilever Beam System

In this example a cantilever beam system with two equal 50-ft spans is designed. See Example 6.29. The initial step is to determine the cantilever length L_c.

The girder is designed for a reduced roof live load, and this requires that both full and unbalanced loading be considered. For this loading, L_c is taken as $0.2L$. Two different beam sizes are developed because the three local maximum moments are not equal for the required load cases. The larger beam is required for the cantilever beam member AD, and the smaller size is for the suspended beam member DF. For this example, a specific glulam combination (NDS Supplement Table 5A Expanded) will be specified rather than using the Stress Class System (NDS Supplement Table 5A). The combination used in the example, a 22F-V8 Douglas Fir glulam, is a "balanced" section—meaning the layup provides equal positive and negative moment capacity. Other combinations are "unbalanced"—meaning the positive moment capacity is greater than the negative moment capacity. It was noted in Chap. 5 that, under the new Stress Class system, the standard layup is unbalanced as listed in Table

5A, but all Stress Class layups can be specified by the designer as balanced with the negative moment capacity equaling the published positive moment capacity, or $F_{bx}^- = F_{bx}^+$.

For the cantilever member AD, it is necessary to check lateral stability for the portion of the member where negative moment occurs (compression on the bottom of the beam).

The final part of the example considers the camber provisions for the girder. Hand calculations are shown for the deflection analysis under dead loads. However, this is done for illustration purposes only, and it is recognized that deflection calculations will normally be done by computer.

In cambering members, most glulam manufacturers are able to set jigs at 4-ft intervals. However, the designer in most cases does not have to specify the camber settings at these close intervals. Typically the required camber would be specified at the midspans, at the internal hinge points, and perhaps at the point of inflection. The manufacturer, then, would establish the camber at various points along the span, using a parabolic or circular curve. Camber tolerance is roughly $\pm \frac{1}{4}$ in.

EXAMPLE 6.29   Cantilever Beam System

Design a cantilever roof beam system, using IBC roof design loads (see Sec. 2.4). Determine the optimum location of the hinge. Use 22F-V8 Douglas Fir glulam. Tributary width to the girder is 20 ft. Roof dead load = 14 psf, including an estimated 2 psf (40 lb/ft) for the weight of the girder. There is no snow load, and the beam does not support a plastered ceiling. The member is used in a dry-service condition ($C_M = 1.0$) and at normal temperatures ($C_t = 1.0$). Allowable stresses and section properties are obtained from the NDS Supplement.

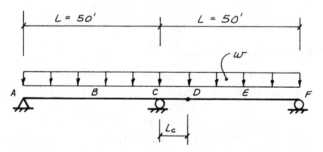

Figure 6.28a

Loads

As noted in Sec. 6.17, the IBC requires that both balanced and unbalanced roof live load be considered, whichever produces the greatest effect. That is, the design is made using a reduced roof live load, considering (D + L_r) on all spans or (D + unbalanced L_r), whichever is critical.

The tributary area is

$$A = (20 \text{ ft})(50 \text{ ft}) = 1000 \text{ ft}^2 > 600 \text{ ft}^2$$

$$L_r = 20 R_1 R_2$$

$$R_1 = 0.6; \quad R_2 = 1 \text{ (see Sec. 2.4)}$$

$$L_r = 20(0.6)(1) = 12 \text{ psf}$$

$$w_D = 14 \times 20 = 280 \text{ lb/ft}$$

$$\underline{w_L = 12 \times 20 = 240}$$

$$w_{TL} = 520 \text{ lb/ft}$$

In Example 6.26 a cantilever length of

$$L_c = 0.172L$$

was recommended for a two-span cantilever system with a constant uniform load on both spans. When unbalanced live load is also considered, a different cantilever length will give approximately equal positive and negative moments for the cantilever segment. This length is

$$L_c = 0.2L = 0.2(50) = 10 \text{ ft}$$

With the cantilever length known, the shear and moment diagrams for the three loading conditions can be drawn (see Fig. 6.28b, c, and d).

Load Case 1: (D + L$_r$ _on All Spans_)

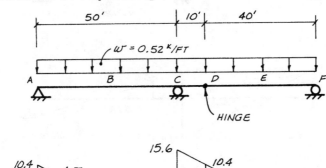

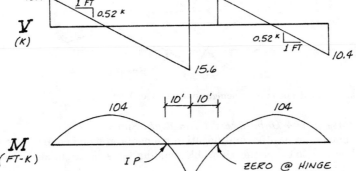

Figure 6.28b Load, shear, and moment diagrams for load case 1.

Member AD

BENDING:

The glulam combination 22F-V8 DF is "balanced" to provide equal positive and negative moment capacity. In other words, F_{bx} *tension zone stressed in tension* and F_{bx} *compression zone stressed in tension* are equal for this combination.

Maximum moments from load cases 1, 2, and 3:

$$\text{Max.} + M \approx \text{max.} - M = 130 \text{ ft-k} = 1560 \text{ in.-k}$$

NOTE: For comparison, the moment for a simple beam is

$$M = \frac{wL^2}{8} = \frac{0.52(50)^2}{8} = 162 \text{ ft-k} > 130$$

Load Case 2: (D + Unbalanced L_r on Left Span)

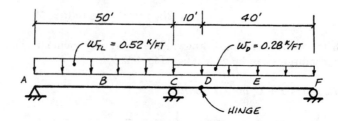

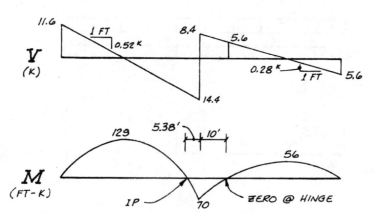

Figure 6.28c Load, shear, and moment diagrams for load case 2.

Load Case 3: (D + Unbalanced L_r on Right Span)

The maximum positive and negative moments in member *AD* are seen to be equal. It will be recalled that the allowable bending stress in a glulam is the smaller value given by two criteria

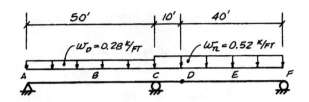

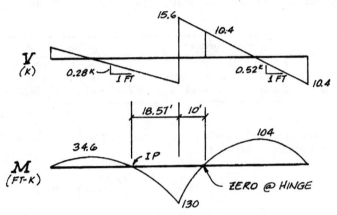

Figure 6.28d Load, shear, and moment diagrams for load case 3.

Volume effect: $F'_b = F_{bx}(C_D)(C_M)(C_t)(C_V)$

Lateral stability: $F'_b = F_{bx}(C_D)(C_M)(C_t)(C_L)$

A trial beam size will be developed using the volume factor. This size will then serve as the basis for the check on lateral stability.

Volume effect

Start by assuming a value for C_V, and verify it later. The load duration factor is $C_D = 1.25$ for the combination of (D + L$_r$). Both C_M and C_t default to unity. Tabulated stresses are given in NDS Supplement Table 5A Expanded.

Assume $C_V = 0.90$:

$$F'_b = F_{bx}(C_D)(C_M)(C_t)(C_V)$$

$$= 2200(1.25)(1.0)(1.0)(0.90) = 2475 \text{ psi}$$

$$\text{Req'd } S = \frac{M}{F'_b} = \frac{1,560,000}{2475} = 630 \text{ in.}^3$$

Refer to NDS Supplement Table 1C, and choose the glulam (Western Species) with the smallest area that furnishes a section modulus greater than the required.

$$Try \quad 5\frac{1}{8} \times 28\frac{1}{2} \quad 22F\text{-}V8 \quad DF \text{ glulam}$$

$$S = 693.8 \text{ in.}^3 > 630 \quad OK$$

Verify C_V.

The volume factor is a function of the length, depth, and width. The length is to be taken as the distance between points of zero moment. The distance between points of zero moment for member AD is summarized for the three load cases:

Load case	Positive moment	Negative moment
1 (Fig. 6.28b)	$L = 50 - 10 = 40$ ft	$L = 10 + 10 = 20$ ft
2 (Fig. 6.28c)	$L = 50 - 5.38 = 44.62$ ft	$L = 5.38 + 10 = 15.38$ ft
3 (Fig. 6.28d)	$L = 50 - 18.57 = 31.43$ ft	$L = 18.57 + 10 = 28.57$ ft

The maximum distance between points of zero moment is 44.62 ft. (Note that $L = 50$ ft could conservatively be used.)

$$C_V = \left(\frac{21}{L}\right)^{1/10}\left(\frac{12}{d}\right)^{1/10}\left(\frac{5.125}{b}\right)^{1/10} \leq 1.0$$

$$= \left(\frac{21}{44.62}\right)^{0.1}\left(\frac{12}{28.5}\right)^{0.1}\left(\frac{5.125}{5.125}\right)^{0.1}$$

$$= 0.851 < 1.0$$

The assumed value of $C_V = 0.90$ was not conservative. Therefore, verify trial size by comparing the actual and allowable bending stresses:

$$f_b = \frac{M}{S} = \frac{1,560,000}{693.8} = 2250 \text{ psi}$$

$$F'_b = F_{bx}(C_D)(C_M)(C_t)(C_V)$$

$$= 2200(1.25)(1.0)(1.0)(0.851)$$

$$= 2340 \text{ psi} > 2250 \text{ psi}$$

$\therefore$ Bending stress for trial beam size as defined by volume factor is *OK*.

Lateral stability

In the area of positive bending moment, the roof diaphragm will be continuously attached to the top side of the girder. Thus, there is full lateral support where there is positive moment, and lateral stability is not a consideration.

However, in the area of negative bending moment, the compression (bottom) side of the member is laterally unsupported between the hinge and the column and between the column and the inflection point. The distance between points of lateral support for member AD is summarized for the three load cases:

Load case	Negative moment
1 (Fig. 6.28b)	$l_{u\,max} = 10$ ft
2 (Fig. 6.28c)	$l_{u\,max} = 10$ ft
3 (Fig. 6.28d)	$l_{u\,max} = 18.57$ ft

The maximum unbraced length is 18.57 ft. An evaluation of the lateral stability factor C_L for an unbraced length of 18.57 ft was done separately and is not shown. The lateral stability factor for $l_u = 18.57$ ft causes the allowable bending stress F_b' to be reduced substantially below the actual bending stress f_b. To solve this problem, an additional diagonal brace (Fig. 6.27b) will be provided between the column and the inflection point. Locate this intermediate brace so that l_u to the left of the column is 10 ft or less. Therefore, the maximum unbraced length to the left and right of the column is 10 ft.

Show calculations to determine the effect of an unbraced length of 10 ft on allowable bending stress.

$$l_u = 10 \text{ ft} = 120 \text{ in.}$$

Slenderness ratio for beam R_B:

Effective unbraced lengths are given in Example 6.8 (Sec. 6.3) and in NDS Table 3.3.3. For a cantilever on single-span beam with any loading, the definition of l_e depends on the l_u/d ratio.

$$\frac{l_u}{d} = \frac{120}{28.5} = 4.21 < 7$$

$$\therefore l_e = 2.06 l_u = 2.06(120) = 247 \text{ in.}$$

$$R_B = \sqrt{\frac{l_e d}{b^2}} = \sqrt{\frac{247(28.5)}{(5.125)^2}} = 16.38$$

Coefficients for computing beam stability factor C_L

$$E_y' = E_y(C_M)(C_t) = 1{,}600{,}000(1.0)(1.0)$$

$$= 1{,}600{,}000 \text{ psi}$$

$$F_{bE} = \frac{K_{bE}E_y'}{R_B^2} = \frac{0.610(1{,}600{,}000)}{16.38^2} = 3638 \text{ psi}$$

$$F_b^* = F_{bx}(C_D)(C_M)(C_t)$$

$$= 2200(1.25)(1.0)(1.0) = 2750 \text{ psi}$$

$$\frac{F_{bE}}{F_b^*} = \frac{3638}{2750} = 1.323$$

$$\frac{1 + F_{bE}/F_b^*}{1.9} = \frac{1 + 1.323}{1.9} = 1.223$$

Beam stability factor

$$C_L = \frac{1 + F_{bE}/F_b^*}{1.9} - \sqrt{\left(\frac{1 + F_{bE}/F_b^*}{1.9}\right)^2 - \frac{F_{bE}/F_b^*}{0.95}}$$

$$= 1.223 - \sqrt{1.223^2 - 1.323/0.95}$$

$$= 0.903$$

Allowable bending stress

$$F_b' = F_{bx}(C_D)(C_M)(C_t)(C_L)$$

$$= 2200(1.25)(1.0)(1.0)(0.903)$$

$$= 2480 \text{ psi} > f_b = 2250 \qquad OK$$

∴ The allowable bending stress given by volume factor *and* lateral stability factor are both satisfactory.

SHEAR:

Max. $V = 15.6$ Neglect reduced shear V' (conservative).

$$f_v = \frac{1.5V}{A} = \frac{1.5(15,600)}{146.1} = 160 \text{ psi}$$

$$F_v' = F_V(C_D)(C_M)(C_t)$$

$$= 240(1.25)(1.0)(1.0)$$

$$= 300 \text{ psi} > 160 \qquad OK$$

Member *AD* trial size 5⅛ × 28½ is adequate for bending and shear.

Member DF

BENDING:

Member *DF* has positive moment everywhere, and the compression side of the member has continuous lateral support. Therefore, $l_u = 0$, and lateral stability need not be considered. Determine a trial size, using the volume factor.

$$\text{Max. } M = 104 \text{ ft-k} = 1248 \text{ in.-k}$$

Assume $C_V = 0.87$:

$$F_b' = F_{bx}(C_D)(C_M)(C_t)(C_V)$$

$$= 2200(1.25)(1.0)(1.0)(0.87) = 2390 \text{ psi}$$

$$\text{Req'd } S = \frac{M}{F_b'} = \frac{1,248,000}{2390} = 522 \text{ in.}^3$$

Select a 5⅛ in. wide trial size glulam from NDS Supplement Table 1C.

Try $5\frac{1}{8} \times 25\frac{1}{2}$ 22F-V8   DF glulam

$$S = 555.4 \text{ in.}^3 > 522 \quad OK$$

Verify C_V.

$$C_V = K_L \left(\frac{21}{L}\right)^{1/10} \left(\frac{12}{d}\right)^{1/10} \left(\frac{5.125}{b}\right)^{1/10} \leq 1.0$$

$$= 1.0 \left(\frac{21}{40}\right)^{0.1} \left(\frac{12}{25.5}\right)^{0.1} \left(\frac{5.125}{5.125}\right)^{0.1}$$

$$= 0.870 < 1.0$$

The actual value and the assumed value of C_V are equal, and the trial size for bending is adequate. Show a comparison of actual and allowable bending stresses anyway:

$$f_b = \frac{M}{S} = \frac{1{,}248{,}000}{555.4} = 2245 \text{ psi}$$

$$F_b' = F_{bx}(C_D)(C_M)(C_t)(C_V)$$

$$= 2200(1.25)(1.0)(1.0)(0.870)$$

$$= 2390 \text{ psi} > 2245 \quad OK$$

SHEAR:

$$\text{Max. } V = 10.4 \text{ k} \qquad \text{Neglect reduction.}$$

$$f_v = \frac{1.5V}{A} = \frac{1.5(10{,}400)}{130.7} = 119 \text{ psi}$$

$$F_v' = F_v(C_D)(C_M)(C_t)$$

$$= 2.40(1.25)(1.0)(1.0)$$

$$= 300 \text{ psi} > 119 \quad OK$$

Member *DF* trial size $5\frac{1}{8} \times 25\frac{1}{2}$ is adequate for bending and shear.

Deflection and Camber

Trial sizes for members *AD* and *DF* have been determined considering bending and shear stresses. Attention is now turned to deflections.

If done by hand, a comprehensive deflection analysis for a cantilever beam system can be a cumbersome calculation. This is especially true with unbalanced loads being involved. Computer solutions can greatly reduce the design effort in analyzing deflections.

To simplify this example, only the dead load deflection calculation is illustrated. This is required in order to determine the camber for the beam (camber = $1.5\Delta_D$).

Various methods of calculating deflection can be used. Here the dead load deflection is calculated by the superposition of handbook deflection formulas (Refs. 6.5 and 6.3). The deflection is evaluated at three points:

At the center of span AC (point B)

At the hinge (point D)

At the midspan of the suspended beam (point E)

Bending is about the x axis, and the modulus of elasticity for deflection calculations is

$$E'_x = E_x(C_M)(C_t)$$

$$= 1,700,000(1.0)(1.0) = 1,700,000 \text{ psi}$$

Section properties:

Member AD: $I_x = 9887 \text{ in.}^4$

Member DF: $I_x = 7082 \text{ in.}^4$

NOTE: The camber provisions included in this example are for long-term deflection. A minimum roof slope of ¼ in./ft (in addition to long-term dead load deflection considerations) is required to provide drainage and avoid ponding.

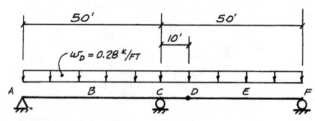

Figure 6.28e Loading for camber.

CAMBER AT B:

Deflection at B due to uniform load on member AD:

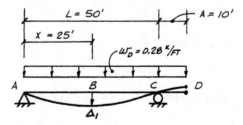

Figure 6.28f

$$\Delta_1 = \frac{wx}{24E'IL}(L^4 - 2L^2x^2 + Lx^3 - 2A^2L^2 + 2A^2x^2)$$

$$= \frac{(0.28)(25)(12 \text{ in./ft})^3}{24(1700)(9887)(50)}[(50)^4 - 2(50)^2(25)^2 + 50(25)^3 - 2(10)^2(50)^2 + 2(10)^2(25)^2]$$

$$= 2.12 \text{ in.} \text{down}$$

Deflection at B due to load on DF:

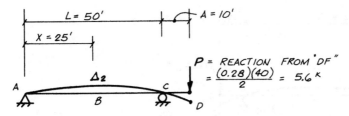

Figure 6.28g

$$\Delta_2 = \frac{PAx}{6E'IL}(L^2 - x^2)$$

$$= \frac{(5.6)(10)(25)(12)^3}{6(1700)(9887)(50)}[(50)^2 - (25)^2]$$

$$= 0.90 \text{ in. up}$$

$$\Delta_D = \Delta_1 + \Delta_2 = 2.12 - 0.90 = 1.22 \text{ in. down}$$

$$\text{Camber at } B = 1.5\Delta_D = 1.5(1.22) = 1.83 \approx 1\%\text{ in. up}$$

CAMBER AT HINGE D:

Deflection at D due to uniform load on AD:

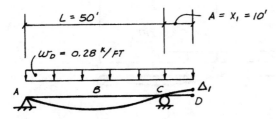

Figure 6.28h

$$\Delta_1 = \frac{wx_1}{24E'I}(4A^2L - L^3 + 6A^2x_1 - 4Ax_1^2 + x_1^3)$$

$$= \frac{(0.28)(10)(12 \text{ in./ft})^3}{24(1700)(9887)}[(4)(10)^2(50) - (50)^3 + 6(10)^2(10) - 4(10)^2 + (10)^3]$$

$$= 1.22 \text{ in. up}$$

Deflection at D due to load on member DF:

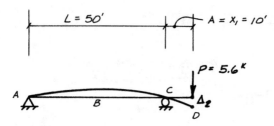

Figure 6.28*i*

$$\Delta_2 = \frac{Px_1}{6E'I}(2AL + 3Ax_1 - x_1^2)$$

$$= \frac{(5.6)(10)(12)^3}{6(1700)(9887)}[(2)(10)(50) + 3(10)(10) - (10)^2]$$

$$= 1.15 \text{ in.} \quad \text{down}$$

$$\Delta_D = \Delta_1 + \Delta_2 = -1.22 + 1.15 = -0.07 \text{ in.} \quad \text{very small}$$

Specify *no camber* at hinge *D*.

CAMBER AT *E*:

The left support of member *DF* (i.e., the hinge) has been found to have a very small deflection. The dead load deflection calculation for point *E* is a simple beam deflection calculation.

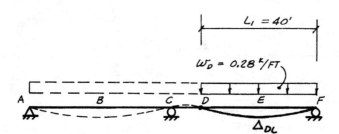

Figure 6.28*j*

$$\Delta_D = \frac{5wL_1^4}{384E'I} = \frac{5(0.28)(40)^4(12)^3}{384(1700)(7082)} = 1.34 \text{ in. down}$$

$$\text{Camber at } E = 1.5\Delta_D = 1.5 \times 1.34 \approx 2 \text{ in.} \quad \text{up}$$

> *Use* $5\frac{1}{8} \times 28\frac{1}{2}$ for member *AD*
> $5\frac{1}{8} \times 25\frac{1}{2}$ for member *DF*
> 22F-V8   DF glulam
>
> Camber: $1\frac{7}{8}$ in. up at point *B*
>   Zero camber at hinge *D*
>   2 in. up at point *E*

NOTE: Again, a complete deflection analysis, including the unbalanced loading, would be required and is best done using computer solutions. This exercise is left for the reader.

In the previous example, a glulam combination was specified from NDS Supplement Table 5A Expanded. Prior to the new Stress Class System, designers selected combinations such as was done in this example. The basic premise with the Stress Class System is to simplify design choices and give the manufacturers more flexibility. Combinations with similar properties have been grouped into stress classes as noted in the NDS Supplement Table 54 Expanded. For a cantilevered beam system, a "balanced" layup, providing equal positive and negative moment capacity, is most appropriate. The new Stress Class System allows both *balanced* and *unbalanced* sections for each stress class. Footnote (1) of NDS Supplement Table 5A states that the designer simply must specify when balanced layups are required and use F_{bx}^- equal to F_{bx}^+. For Example 6.29, either a slightly larger 20F-1.5E or slightly smaller 24F-1.7E layup may be considered.

6.19 Lumber Roof and Floor Decking

Lumber sheating (1-in. nominal thickness) can be used to span between closely spaced roof or floor beams. However, plywood and other wood structural panels are often used for this application. Plywood and other panel products are covered in Chap. 8.

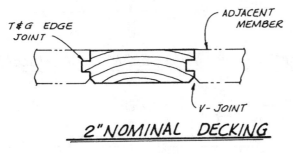

2" NOMINAL DECKING

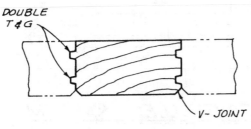

3" & 4" NOMINAL DECKING

Figure 6.29 Solid lumber decking. Decking can be obtained with various surface patterns if the bottom side of the decking is architecturally exposed. These sketches show a V-joint pattern.

Timber decking is used for longer spans. It is available as *solid decking* or *laminated decking*. Solid decking is made from dry lumber and is available in several grades in a number of commercial wood species. Common sizes are 2 × 6, 3 × 6, and 4 × 6 (nominal sizes). Various types of edge configurations are available, but tongue-and-groove (T&G) edges are probably the most common. See Fig. 6.29. A single T&G is used on 2-in.-nominal decking, and a double T&G is used on the larger thicknesses.

Glued laminated decking is fabricated from three or more individual laminations. Laminated decking also has T&G edge patterns.

Decking essentially functions as a series of parallel beams that span between the roof or floor framing. Bending stresses or deflection criteria usually govern the allowable loads on decking. Spans range from 3 to 20 ft and more depending on the load, span type, grade, and thickness of decking. The *layup* of decking affects the load capacity. See Example 6.30. It has been noted elsewhere that decking is graded for bending about the minor axis of the cross section.

EXAMPLE 6.30   Layup of Decking

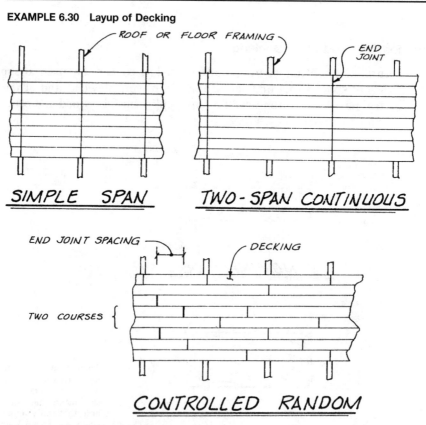

Figure 6.30   Three forms of layup for decking.

Layup refers to the arrangement of end joints in decking. Five different layups are defined in Ref. 6.5, and three of these are shown in Fig. 6.30. Controlled random layup is economical and simply requires that end joints in adjacent courses be well staggered. Minimum end-joint spacing is 2 ft for 2-in. nominal decking and 4 ft for 3- and 4-in. nominal decking. In addition, end joints that occur on the same transverse line must be separated by at least two courses. End joints may be mechanically interlocked by matched T&G ends or by wood or metal splines to aid in load transfer. For other requirements see Ref. 6.5.

The TCM gives bending and deflection coefficients for the various types of layups. These can be used to calculate the required thickness of decking. However, the designer can often refer to allowable span and load tables for decking requirements. IBC Table 2308.10.9 gives the allowable span for 2-in. T&G decking. Reference 6.5 includes allowable load tables for simple span and controlled random layups for a variety of thicknesses.

6.20   Fabricated Wood Components

Several fabricated wood products are covered in considerable detail in this book. These include glulam (Chap. 5) and plywood and other wood structural panel products (Chap. 8). In addition to these, a number of other fabricated wood elements can be used as beams in a roof or floor system. Many of these components are produced as proprietary products, and consequently design criteria and material properties may vary from manufacturer to manufacturer.

The purpose of this section is simply to describe some of the wood components that may be used in typical wood-frame buildings. The structural design of some of these products may be performed by the manufacturer. For example, the design engineer for a building may decide to use a certain system in a roof application. After the spacing of the members has been established and the loading has been determined, the engineering staff of the supplier may design the component to perform in the specified manner.

For other components the project engineer may use certain information supplied by the manufacturer to determine the size of the required structural member. The information provided by the manufacturer can take the form of load/span tables or allowable stresses and effective section properties. Cooperation between the designer and the supplier is recommended in the early planning stages. The designer should also verify local building code recognition of the product and the corresponding design criteria.

The fabricated wood components covered in this section are

1. Structural composite lumber (SCL)
   a. Laminated veneer lumber (LVL)
   b. Parallel strand lumber (PSL)

2. Prefabricated wood I-joists

3. Light-frame wood trusses

4. Fiber-reinforced glulam

Structural composite lumber (SCL) refers to engineered lumber that is produced in a manufacturing plant. Although glulam was described in Chap. 5
as a composite material, the term *structural composite lumber* generally refers
to a reconstituted wood product made from much smaller pieces of wood. It is
fabricated by gluing together thin pieces of wood that are dried to a low moisture content. The glue is a waterproof adhesive. As a result of the manufacturing process, SCL is dimensionally stable and has less variability than sawn
lumber.

The allowable stresses for glulams are generally higher than those for solid
sawn lumber, and allowable stresses for structural composite lumber are
higher than those for glulam. Tabulated bending stresses F_b for SCL range
from 2300 to 3200 psi, and tabulated shear stresses F_v range from 150 to 290
psi. Current practice involves production of two general types of SCL which
are known as laminated veneer lumber and parallel or oriented strand lumber.
The basic design process for SCL is covered in the ASD Manual Guideline
"Structural Composite Lumber" (Ref. 6.1).

Laminated veneer lumber (LVL) is similar in certain respects to glulam and
plywood. It is fabricated from veneer similar to that used in plywood. The
veneer typically ranges in thickness between $\frac{1}{10}$ and $\frac{1}{6}$ in. and is obtained
from the rotary cutting process illustrated in Fig. 8.3. Laminated veneer lumber generally makes use of the same species of wood used in the production
of structural plywood (i.e., Douglas Fir-Larch and Southern Pine).

Unlike plywood which is cross-laminated, the veneers in LVL are laid up
with the wood fibers all running in one direction (i.e., parallel to the length
of the member). The parallel orientation of the wood fiber is one reason for
the high allowable stresses in LVL. Selective grading of veneer and the dispersion of defects as part of the manufacturing process (similar to the dispersion of defects in glulam, see Fig. 5.3) are additional reasons for the higher
stress values. The layup of veneer for LVL can also follow a specific pattern
similar to glulam to meet strength requirements.

LVL is produced in either a continuous-length manufacturing operation or
in fixed lengths. Fixed lengths are a function of the press size in a manufacturing plant. However, any desired length can be obtained by end jointing
members of fixed lengths. Laminated veneer lumber is produced in boards or
billets that can range from $\frac{3}{4}$ to $3\frac{1}{2}$ in. thick and may be 4 ft wide and 80 ft
long. A billet is then sawn into sizes as required for specific applications. See
Example 6.31.

EXAMPLE 6.31   Laminated Veneer Lumber

Laminated veneer lumber is fabricated from sheets of veneer that are glued into panels
called billets. Unlike the cross-lamination of veneers in plywood (Sec. 8.3), LVL has
the direction of the wood grain in all veneers running parallel with the length of the
billet. Pieces of LVL are trimmed from the billet for use in a variety of applications.
See Fig. 6.31*b*.

DIRECTION OF WOOD GRAIN
IN ALL VENEERS

TYPICAL LVL BILLET

Figure 6.31a Billet of laminated veneer lumber.

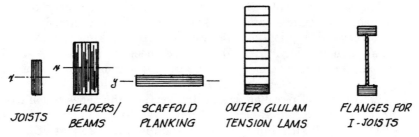

JOISTS   HEADERS/ BEAMS   SCAFFOLD PLANKING   OUTER GLULAM TENSION LAMS   FLANGES FOR I-JOISTS

Figure 6.31b Typical uses of LVL.

Uses of laminated veneer lumber include beams, joists, headers, and scaffold planking. Beams and headers may require multiple thicknesses of LVL to obtain the necessary member width. LVL can also be used for the higher-quality tension laminations in glulams. Additional applications include flanges of prefabricated wood I joists and chords of trusses. Two LVL beams are shown in Fig. 6.31c.

Figure 6.31c Laminated veneer lumber beams. (*Photo courtesy of Trus Joist—A Weyerhaeuser Company.*)

The use of LVL is economical where the added expense of the material is offset by its increased strength and greater reliability.

There are two types of *parallel strand lumber* (PSL) currently in production. One type is made from the same species of wood used for plywood (i.e., Douglas Fir-Larch and Southern Pine). It starts with a sheet of veneer, which is clipped into narrow *strands* that are approximately ½ in. wide and up to 8 ft long. The strands are dried, coated with a waterproof adhesive, and bonded together under pressure and heat. The strands are aligned so that the wood grain is parallel to the length of the member (hence the name).

The second type of PSL is also known as *oriented strand lumber* (OSL) and is made from small-diameter trees of Aspen that previously could not be used in structural applications because of size. Flaking machines are used for small-diameter logs (instead of veneer peelers) to produce wood flakes that are approximately ½ in. wide, 0.03 in. thick, and 1 ft long. The flakes are also glued and bonded together under pressure and heat.

Both forms of parallel strand lumber (i.e., the types made from *strands* or *flakes*) result in a final piece called a billet. Billets of PSL are similar to those of LVL (Fig. 6.31*a*), but the sizes are different. Billets of PSL can be as large as 12 in. wide, 17 in. deep, and 60 ft long. Again, final sizes for field applications are obtained by sawing the billet. Parallel strand lumber may be used alone as high-grade structural lumber for beams and columns. See Example 6.32. It may also be used in the fabrication of other structural components similar to the LVL applications in Example 6.31.

EXAMPLE 6.32   Parallel Strand Lumber

Figure 6.32*a* Parallel strand lumber. (*Photo courtesy of Trus Joist—A Weyerhaeuser Company.*)

Parallel strand lumber is manufactured from strands or flakes of wood with the grain parallel to the length of the member. High-quality wood members in large sizes are possible with this form of SCL. See Fig. 6.32a. Applications include beams and columns which can be left architecturally exposed. See Fig. 6.32b.

Figure 6.32b Beams and columns of PSL. (*Photo courtesy of Trus Joist—A Weyerhaeuser Company.*)

The use of prefabricated wood components has increased substantially in recent years. The most widely used form of these composite members is the wood *I-joist*. See Example 6.33. Wood I-joists are efficient bending members for two reasons. First, the cross section is an efficient shape. The most popular steel beams (W shapes and S shapes) have a similar configuration. The large flange areas are located away from the neutral axis of the cross section, thus increasing the moment of inertia and section modulus. In other words, the shape is efficient because the flanges are placed at the point in the cross section where the material does the most good: At the point of maximum flexural stress. The relatively thin web is satisfactory as long as it has adequate shear strength.

Second, wood I-joists are efficient from a material usage standpoint. The flanges are stressed primarily in tension and compression as the result of the flexural stresses in the member. The material used for the flanges in wood I-joists has high tensile and compressive strengths. Some manufacturers use sawn lumber flanges, but laminated veneer lumber flanges are common.

Although the bending moment is primarily carried by the flanges, it will be recalled that the shear in the I-beam is essentially carried by the web. Wood I-joists also gain efficiency by using web materials that are strong in shear. Plywood and oriented strand board panels are used in other high shear applications such as horizontal diaphragms (Chap. 9) and shearwalls (Chap. 10), in addition to being used as the web material in fabricated wood beams.

Additional information on the design of wood composite I-joists is provided in the ASD Manual Guideline "Wood I-Joists" (Ref. 6.1). Supplemental design considerations (specific to wood I-joists) and recommended installation details are also provided in this Guideline.

EXAMPLE 6.33   Prefabricated Wood I-Joists

Figure 6.33a Typical prefabricated wood I-joists. (*Photo courtesy of Louisiana-Pacific Corporation.*)

Initially, prefabricated wood I-joists were constructed with solid sawn lumber flanges and plywood webs. However, more recently I-joists are produced from some of the newer wood products. For example, laminated veneer lumber (LVL) is used for flanges and oriented strand board (OSB) for web material (Fig. 6.33a).

Figure 6.33b Wood I-joists supported on LVL header. (*Photo courtesy of Trus Joist—A Weyerhaeuser Company.*)

Prefabricated wood I-joists have gained wide acceptance in certain areas of the country for repetitive framing applications (Fig. 6.33b and c). Web stiffeners for wood I-joists may be required to transfer concentrated loads or reactions in bearing through the flange and into the web. Prefabricated metal hardware is available for a variety of connection applications. Because of the slender cross section of I-joists, particular attention must be paid to stabilizing the members against translation and rotation. The manufacturer's recommendations for bracing and blocking should be followed in providing stability for these members.

Figure 6.33c Wood I-joists as part of a wood roof system in a building with masonry walls. (*Photo courtesy of Trus Joist—A Weyerhaeuser Company.*)

Wood I-joists make efficient use of materials, and because of this they are relatively lightweight and easy to handle by crews in the field. In addition to strength, the depth of the cross section provides members that are relatively stiff for the amount of material used. Wood I-joists can be used to span up to 40 or 50 ft, but many applications are for shorter spans. Wood I-joists can be deep and slender, and care should be taken in the installation of these members to ensure adequate stability. Information on the design of lumber and plywood beams is available in Refs. 6.8 and 6.9. Additional information on the design and installation of wood I-joists is available from individual manufacturers.

Wood trusses represent another common type of fabricated wood component. Heavy wood trusses have a long history of performance, but light wood trusses are more popular today. The majority of residential wood structures, and many commercial and industrial buildings, use some form of closely spaced light wood trusses in roof and floor systems. Common spans for these trusses range up to 75 ft, but larger spans are possible. The spacing of trusses is on the order of 16 to 24 in. o.c. for floors and up to 8 ft o.c. in roof systems.

Some manufacturers produce trusses that have wood top and bottom chords and steel web members. However, the majority of truss manufacturers use light-gage toothed metal plates to connect wood chords and wood web members. See Example 6.34. The metal plates have teeth which are produced by stamping the metal plates. The metal plates are placed over the members to be connected together and the teeth are pressed into the wood.

EXAMPLE 6.34   Light-Frame Wood Trusses

Figure 6.34a Wood trusses with tubular steel webs. Trusses in the foreground are supported on a wood member attached to a steel W shape beam. In the background, trusses rest on glulam. (*Photo courtesy of Trus Joist—A Weyerhaueser Company*).

Figure 6.34b Metal plate connected trusses being placed in roof system supported on masonry walls. (*Photo courtesy of Alpine Engineered Products, Inc.*)

Trusses can be manufactured with sawn lumber or LVL chords and steel web members (Fig. 6.34*a*). Trusses can be supported in a variety of ways. The top or bottom chord of a truss can bear on wood walls or beams, on steel beams, or on top of concrete or masonry walls. Another method is to suspend the truss from a ledger attached to a concrete or masonry wall.

Metal-plate-connected trusses (Fig. 6.34*b*) use toothed or barbed plates to connect the truss members. See Fig. 11.3 in Sec. 11.2 for a photograph of metal plate connectors. Typically the metal plates are assigned a unit load capacity (lb/in.^{2} of contact area). Thus the required plate size is determined by dividing the forces to be transferred through the connection by the unit load capacity of the metal plate.

Light wood trusses are rather limber elements perpendicular to their intended plane of loading. Because of this flexibility, proper handling procedures are required in the field to avoid damage to the truss during erection. The use of strongbacks with a sufficient number of pickup points for lifting the trusses into place will avoid buckling of the truss about its weak axis during installation. Once a truss is properly positioned, it must be braced temporarily until the sheathing and permanent bracing are in place (Ref. 6.11). Trusses which are not adequately braced can easily buckle or rotate. However, once the bracing is in place, the trusses provide a strong, stiff, and economical wood framing system.

The Truss Plate Institute (TPI) is the technical trade association of the metal plate truss industry. TPI, along with the Wood Truss Council of America (WTCA), prepared the ASD Manual Guideline "Metal Plate Connected Wood Trusses" (Ref. 6.1), which provides fundamental concepts of metal-plate-connected wood truss design. TPI also publishes the more comprehensive "National Design Specification for Metal Plate Connected Wood Trusses" (Ref. 6.12), which provides detailed design requirements for these trusses. Among other considerations, this specification requires that the continuity of the chords at the joints in the truss be taken into account. In addition to its design specification, TPI publishes other pertinent literature such as the "Commentary and Recommendations for Handling, Installing, and Bracing Metal Plate Connected Wood Trusses" (Ref. 6.11). Additional information on wood trusses and bracing can be obtained from individual truss manufacturers.

Fiber-reinforced glulam is one of the newer products to be developed. While, in general, many different materials can be used to reinforce glulam, the use of *fiber-reinforced polymers* (FRPs) has proven to be the most effective reinforcement. FRPs consist of synthetic fibers (including glass, carbon, and graphite) and a thermoplastic polymer that serves as a binder, holding the fibers together.

Fiber-reinforced glulam is actually a modification to traditional glulam where FRP sheets are placed between laminations, particularly in the tension region, to improve performance. See Example 6.35. Specifically, the advantages of integrating FRP sheets into a glulam layup include increased strength and stiffness, increased ductility, reduced creep, and reduced overall variability.

EXAMPLE 6.35   Fiber-Reinforced Glulam

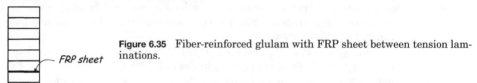

Figure 6.35   Fiber-reinforced glulam with FRP sheet between tension laminations.

As discussed in Chap. 5, glulam is an engineered wood composite that allows more efficient use of materials and provides larger size members. With the addition of FRP sheets, an increase in strength of the section is possible, but the primary advantages are increased stiffness, increased ductility, reduced creep, and reduced overall variability. Since deflections can oftentimes control the size of a member, the increase in stiffness and reduced creep allow for smaller sections to be required. The reduction in variability can result in higher design values as well. In addition, the increase in ductility provides for a potentially safer failure mechanism. These same attributes can also allow the utilization of low-quality wood in glulam without reduction in overall design values. That is, even though the use of low-quality wood in a glulam may translate to a reduced set of design values, this reduction may be offset by the improved performance resulting from the addition of FRP.

Fiber-reinforced glulam is a proprietary product that is not covered in the NDS or ASD Manual. Additional information on the use and design of fiber-reinforced glulam can be obtained from APA—The Engineered Wood Association, from AITC, as well as from the manufacturer.

The use of FRPs to improve the performance of wood materials is not limited to glulam. The combination of FRPs with SCL, I-joists, and wood panels have all been used with varying degrees of success.

6.21   References

[6.1]   American Forest and Paper Association (AF&PA). 2001. *Allowable Stress Design Manual for Engineered Wood Construction and Supplements and Guidelines,* 2001 ed., AF&PA, Washington DC.

[6.2]   American Forest and Paper Association (AF&PA). 2001. *National Design Specification for Wood Construction and Supplement.* 2001 ed., AF&PA, Washington DC.

[6.3]   American Institute of Steel Construction (AISC). 2001. *Manual of Steel Construction—Load and Resistance Factor Design,* 3rd ed., AISC, Chicago, IL.

[6.4]   American Institute of Timber Construction (AITC). 2001. *Design Standard Specifications for Structural Glued Laminated Timber of Softwood Species, AITC 117-2001,* AITC, Englewood, CO.

[6.5]   American Institute of Timber Construction (AITC). 1994. *Timber Construction Manual,* 4th ed., AITC, Englewood, CO.

[6.6]   American Society for Testing and Materials (ASTM). 1997. "Standard Methods of Testing Small Clear Specimens of Timber," ASTM D143-94, *Annual Book of Standards, Vol. 04.09 Wood,* ASTM, Philadelphia, PA.

[6.7]   American Society for Testing and Materials (ASTM). 2000. "Standard Practice for Establishing Structural Grades and Related Allowable Properties for Visually Graded Lumber," ASTM D245-00e1, *Annual Book of Standards, Vol. 04.09 Wood,* ASTM, Philadelphia, PA.

[6.8]   APA—The Engineered Wood Association. 1995. *Plywood Design Specification, APA Form Y510,* APA—The Engineered Wood Association, Engineered Wood Systems, Tacoma, WA.

[6.9]   APA—The Engineered Wood Association. 1995. *Plywood Design Specification Supplements 1-5, APA Forms S811, S812, U812, U814, H815,* APA—The Engineered Wood Association, Engineered Wood Systems, Tacoma, WA.

[6.10] International Codes Council (ICC). 2003. *International Building Code,* 2003 ed., ICC, Falls Church, VA.

[6.11] Truss Plate Institute (TPI). 1991. *Commentary and Recommendations for Handling, Installing and Bracing Metal Plate Connected Wood Trusses, HIB-91,* TPI, Madison, WI.

[6.12] Truss Plate Institute (TPI). 2002. *National Design Standard for Metal Plate Connected Wood Truss Construction, ANSI/TPI 1-2002,* TPI, Madison, WI.

[6.13] Zahn, J.J. 1991. "Biaxial Beam-Column Equation for Wood Members," *Proceedings of Structures Congress '91,* American Society of Civil Engineers, pp. 56–59.

6.22   Problems

Allowable stresses and section properties for the following problems are to be in accordance with the 2001 NDS. Dry-service conditions, normal temperatures, and bending about the strong axis apply unless otherwise indicated.

Some problems require the use of spreadsheet or equation-solving software. Problems that are solved on a spreadsheet or equation-solving software can be saved and used as a *template* for other similar problems. Templates can have many degrees of sophistication. Initially, a template may only be a hand (i.e., calculator) solution worked on a computer. In a simple template of this nature, the user will be required to provide many of the "lookup" functions for such items as

Tabulated stresses

Lumber dimensions

Load duration factor

Wet-service factor

Size factor

Volume factor

As the user gains experience with the chosen software, the template can be expanded to perform lookup and decision-making functions that were previously done manually.

Advanced computer programming skills are not required to create effective spreadsheet or equation-solving templates. Valuable templates can be created by designers who normally do only hand solutions. However, some programming techniques are helpful in automating *lookup* and *decision-making* steps.

The first requirement is that a template operate correctly (i.e., calculate correct values). Another major consideration is that the input and output be structured in an orderly manner. A sufficient number of intermediate answers should be displayed and labeled so that the solution can be verified by hand.

6.1   *Given:*   The roof beam in Fig. 6.A with the following information:

Load: $P = 2$ k

Load combination: $D + L_r$

Span:	$L = 8$ ft
Member size:	4×8
Stress grade and species:	No. 1 DF-L
Unbraced length:	$l_u = 0$
Moisture content:	$MC \leq 19$ percent
Deflection limit:	Allow. $\Delta \leq L/360$

Find: a. Size category (Dimension lumber, B&S, or P&T)
  b. Tabulated stresses: F_b, F_v, and E
  c. Allowable stresses: F'_b, F'_v, and E'
  d. Actual stresses and deflection: f_b, f_v, and Δ
  e. Compare the actual and allowable design values, and determine if the member is adequate.

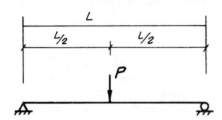

Figure 6.A

6.2   Repeat Prob. 6.1 except the moisture content exceeds 19 percent.

6.3   Prob. 6.1 except the unbraced length is $l_u = L/2 = 4$ ft. $C_M = 1.0$.

6.4   Use the hand solution to Probs. 6.1 and 6.3 as a guide to develop a personal computer template to solve similar problems.
  a. Consider only the specific criteria given in Probs. 6.1 and 6.3.
  b. Expand the template to handle any

  Span length L

  Magnitude load P

  Unbraced length l_u

  Sawn lumber trial member size

  The template is to include a list (i.e., database) of tabulated stresses for all size categories (Dimension lumber B&S, P&T) of No. 1 DF-L.
  c. Expand the database in part b to include all stress grades of DF-L from No. 2 through Select Structural.

6.5   Repeat Prob. 6.1 except the unbraced length is $l_u = L = 8$ ft.

6.6   *Given:* The roof beam in Fig. 6.A with the following information:

Load:	$P = 1.5$ k
Load combination:	D + S
Span:	$L = 24$ ft
Member size:	$3\frac{1}{8} \times 21$
Bending Stress Class glulam:	24F-1.8E
Unbraced length:	$l_u = 0$
Moisture content:	$MC < 16$ percent
Deflection limit:	Allow. $\Delta \leq L/240$

Find: *a.* Tabulated stresses: F_b, F_v, E_x, and E_y
       b. Allowable stresses: F'_b, F'_v, E'_x, and E'_y
       c. Actual stresses and deflection: f_b, f_v, and Δ
       d. Compare the actual and allowable design values and determine if the member is adequate. How much camber should be provided if the dead load on the beam is 35 percent of the given total load?

6.7   Repeat Prob. 6.6 except the moisture content exceeds 16 percent.

6.8   Repeat Prob. 6.6 except the unbraced length is $l_u = L/2 = 12$ ft. $C_M = 1.0$.

6.9   Use the hand solution to Probs. 6.6 and 6.8 as a guide to develop a personal computer template to solve similar problems.
a. Consider only the specific criteria given in Probs. 6.6 and 6.8.
b. Expand the template to handle any

      Span length L

      Magnitude load P

      Unbraced length l_u

      Size Western Species bending combination glulam

    The template is to include a list (i.e., database) of tabulated stresses for glulam bending Stress Classes 16F-1.3E, 20F-1.5E, and 24F-1.8E.

6.10 *Given:* The roof beam in Fig. 6.B with the following information:

Load:	$w_D = 200$ lb/ft
	$w_L = 250$ lb/ft
	$w_{TL} = 450$ lb/ft
Load combination:	D + L$_r$
Span:	$L = 10$ ft
Member size:	4 × 10
Stress grade and species:	Sel. Str. Hem-Fir
Unbraced length:	$l_u = 0$
Moisture content:	MC ≤ 19 percent
Deflection limit:	Allow. $\Delta_L \leq L/360$
	Allow. $\Delta_{(D+L)} \leq L/240$

Find: *a.* Size category (Dimension lumber, B&S, or P&T)
       b. Tabulated stresses: F_b, F_v, and E
       c. Allowable stresses: F'_b, F'_v, and E'
       d. Actual stresses and deflection: f_b, f_v, and Δ
       e. Compare the actual and allowable design values, and determine if the member is adequate.

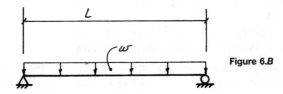

Figure 6.B

6.11   Repeat Prob. 6.10 except the moisture content exceeds 19 percent.

6.12 Repeat Prob. 6.10 except the unbraced length is $l_u = L/2 = 5$ ft.

6.13 Use the hand solution to Probs. 6.10 and 6.12 as a guide to develop a personal computer template to solve similar problems.
   a. Consider only the specific criteria given in Probs. 6.10 and 6.12.
   b. Expand the template to handle any

   Span length L
   Magnitude load w
   Unbraced length l_u
   Sawn lumber trial member size

   The template is to include a list (i.e., database) of tabulated stresses for all size categories (Dimension lumber, B&S, P&T) of Sel. Str. Hem-Fir.
   c. Expand the database in part *b* to include all stress grades of Hem-Fir from No. 2 through Select Structural.

6.14 *Given:* The roof beam in Fig. 6.B with the following information:

Load:	$w_D = 200$ lb/ft
	$w_S = 300$ lb/ft
	$w_{TL} = 500$ lb/ft
Load combination:	D + S
Span:	$L = 20$ ft
Member size:	$5 \times 19\frac{1}{4}$
Glulam bending Stress Class:	24F-1.7E SP
Unbraced length:	$l_u = 0$
Moisture content:	MC < 16 percent
Deflection limit:	Allow. $\Delta_S \leq L/360$
	Allow. $\Delta_{(D+S)} \leq L/240$

   Find: *a.* Tabulated stresses: F_b, F_v, E_x, and E_y.
   b. Allowable stresses: F'_b, F'_v, E'_x, and E'_y
   c. Actual stresses and deflection: f_b, f_v, and Δ
   d. Compare the actual and allowable design values, and determine if the member is adequate. How much camber should be provided?

6.15 Repeat Prob. 6.14 except the moisture content exceeds 16 percent.

6.16 Repeat Prob. 6.14 except the unbraced length is $l_u = L/2 = 10$ ft. $C_M = 1.0$.

6.17 Use the hand solution to Probs. 6.14 and 6.16 as a guide to develop a personal computer template to solve similar problems.
   a. Consider only the specific criteria given in Probs. 6.14 and 6.16.
   b. Expand the template to handle any

   Span length L
   Magnitude load w
   Unbraced length l_u
   Size Southern Pine bending combination glulam

   The template is to include a list (i.e., database) of tabulated stresses for glulam bending Stress Class 16F-1.3E, 20F-1.5E, and 24F-1.8E SP.

6.18 *Given:*   The floor beam in Fig. 6.C with the following information:

Load:	$P = 2$ k
Load combination:	D + L
Span:	$L_1 = 8$ ft
	$L_2 = 4$ ft
Member size:	4 × 12
Stress grade and species:	Sel. Str. SP
Unbraced length:	$l_u = 0$
Moisture content:	MC ≤ 19 percent
Deflection limit:	Allow. $\Delta_{\text{free end}} \leq 2(L_2/360)$
	Allow. $\Delta_{\text{between supports}} \leq L_1/360$

Find: *a.* Size category (Dimension lumber, B&S, or P&T)
     b. Tabulated stresses: F_b, F_v, and E
     c. Allowable stresses: F_b', F_v', and E'
     d. Actual stresses and deflection: f_b, f_v, and Δ
     e. Compare the actual and allowable design values, and determine if
         the member is adequate.

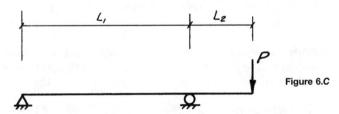

Figure 6.C

6.19 Repeat Prob. 6.18 except the moisture content exceeds 19 percent.

6.20 Repeat Prob. 6.18 except lateral support is provided at the vertical supports and at the free end.

6.21 Use the hand solution to Probs. 6.18 and 6.20 as a guide to develop a personal computer template to solve similar problems.
    a. Consider only the specific criteria given in Probs. 6.18 and 6.20.
    b. Expand the template to handle any

       Span lengths L_1 and L_2
       Magnitude load P
       Unbraced length l_u
       Sawn lumber trial member size

       The template is to include a list (i.e., database) of tabulated stresses for all size categories (Dimension lumber, B&S, P&T) of Sel. Str. SP.
    c. Expand the database in part *b* to include the stress grades of No. 2, No. 1, and Select Structural Southern Pine.

6.22 A series of closely spaced floor beams is to be designed. Loading is similar to Fig. 6.B. The following information is known:

Load: $w_D = 18$ psf
                                             $w_L = 50$ psf
Load combination: D + L
Span: $L = 14$ ft
Member spacing: Trib. width $= b = 16$ in. o.c.
Stress grade and species: No. 1   Hem-Fir
Unbraced length: $l_u = 0$
Moisture content: MC ≤ 19 percent
Deflection limit: Allow. $\Delta_L \leq L/360$
                                             Allow. $\Delta_{(D+L)} \leq L/240$

Find: Minimum beam size. As part of the solution also give
   a.   Size category (Dimension lumber, B&S, or P&T)
   b.   Tabulated stresses: F_v, F_b, and E
   c.   Allowable stresses: F_b', F_v', and E'
   d.   Actual stresses and deflection: f_b, f_v, and Δ

6.23   For the beam designed in Prob. 6.22, determine the size of notches allowed by the NDS on both the tension and compression side at (a) the supports or ends of the member and (b) in the interior of the span. Determine the shear capacity at the support, if a 1-in. deep notch were assumed at the support.

6.24   If the member designed in Prob. 6.22 was a glulam, determine the size of notches allowed by the NDS on both the tension and compression side at (a) the supports or ends of the member and (b) in the interior of the span.

6.25   *Given:*   The roof rafters in Fig. 6.D are 24 in o.c. Roof dead load is 12 psf along the roof, and roof live load is in accordance with the IBC. See Example 2.4. Calculate design shear and moment, using the horizontal plane method of Example 2.6 (see Fig. 2.6b). Lumber is No. 2 DF-L. Lateral stability is not a problem. Disregard deflection. $C_M = 1.0$ and $C_t = 1.0$.

   Find: Minimum rafter size. As part of the solution also give
   a.   Size category (Dimension lumber, B&S, or P&T)
   b.   Tabulated stresses: F_b and F_v
   c.   Allowable stresses: F_b' and F_t'
   d.   Actual stresses: f_b and f_v

6.26   Repeat Prob. 6.25 except that the rafters are spaced 6 ft-0 in. o.c.

6.27   *Given:*   The roof rafters in Fig. 6.D are 24 in. o.c. The roof dead load is 15 psf along the roof, and the design snow load is 50 psf. Calculate the design shear and moment, using the horizontal plane method of Example 2.6 (see Fig. 2.6b). Disregard deflection. Lateral stability is not a problem. Lumber is No. 1 DF-L. $C_M = 1.0$ and $C_t = 1.0$.

   Find: The minimum rafter size. As part of the solution also give
   a.   Size category (Dimension lumber, B&S, or P&T)
   b.   Tabulated stresses: F_b and F_v
   c.   Allowable stresses: F_b' and F_v'
   d.   Actual stresses: f_b and f_v

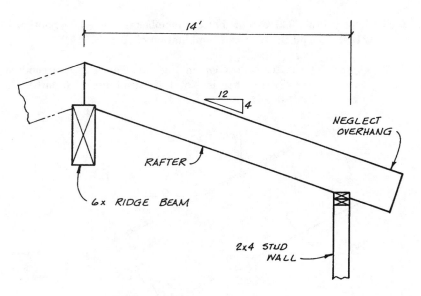

SECTION

Figure 6.D

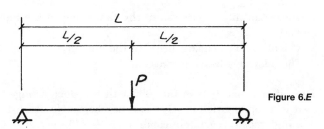

P

Figure 6.E

6.28 *Given:* The beam in Fig. 6.E is supported laterally at the ends only. The span length is $L = 25$ ft. The member is a 6 × 14 DF-L Select Structural. The load is a combination of (D + L). $C_M = 1.0$ and $C_t = 1.0$.

  Find: The allowable bending moment in ft-k and the corresponding allowable load P in k.

6.29 *Given:* The beam in Fig. 6.E has the compression side of the member supported laterally at the ends and midspan only. The span length is $L = 25$ ft. The member is a $3\frac{1}{8}$ × 18 DF glulam Stress Class 24F-1.8E. The load is a combination of (D + L). $C_M = 1.0$ and $C_t = 1.0$.

  Find: The allowable bending moment in ft-k and the corresponding allowable load P in k.

6.30 *Given:* The beam in Fig. 6.E has the compression side of the member supported laterally at the ends and the quarter points. The span length is $L = 24$ ft. The member is a resawn glulam $2\frac{1}{2}$ × $19\frac{1}{2}$ DF 24F-1.8E. The load is a combination of (D + L_r). $C_M = 1.0$ and $C_t = 1.0$.

  Find: The allowable bending moment in ft-k and the corresponding allowable load P in k.

6.31   Repeat Prob. 6.29 except that the member is a $3 \times 17\frac{7}{8}$ Southern Pine glulam 24F-1.8E, and the load is a combination of (D + S).

6.32   *Given:*   The rafter connection in Fig. 6.F. The load is a combination of roof (D + S). Lumber is No. 1 Spruce-Pine-Fir (South). $C_M = 1.0$ and $C_t = 1.0$.

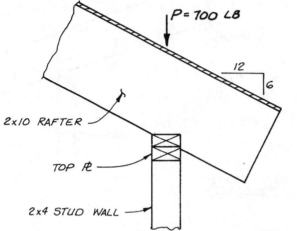

P = 700 LB

12
6

2x10 RAFTER

TOP PL

2x4 STUD WALL

Figure 6.F

Find: a. The actual bearing stress in the rafter and in the top plate of the wall.
b. The allowable bearing stress in the top plate.
c. The allowable bearing stress in the rafter.

6.33   *Given:*   The rafter connection in Fig. 6.F with the slope changed to $^{12}\!/_{12}$. The load is a combination of (D + L_r). Lumber is No. 2 DF-L that is used in a high-moisture-content condition (MC > 19 percent). $C_t = 1.0$.
Find: a. The actual bearing stress in the rafter and in the top plate of the wall.
b. The allowable bearing stress in the top plate.
c. The allowable bearing stress in the rafter.

6.34   *Given:*   The beam-to-column connection in Fig. 6.G. The gravity reaction from the simply supported beam is transferred to the column by bearing (not by the bolts). Assume the column and the metal bracket have adequate strength to carry the load. $C_t = 1.0$.
Find: The maximum allowable beam reaction governed by bearing stresses for the following conditions:
a. The beam is a 4×12 No. 1 DF-L. MC $\leq$ 19 percent, and the dimensions are $A = 12$ in. and $B = 5$ in. Loads are (D + S).
b. The beam is a $5\frac{1}{8} \times 33$ DF glulam Combination 24F-V4. MC = 18 percent, and the dimensions are $A = 0$ and $B = 12$ in. Loads are (D + S).
c. The beam is a 6×16 No. 1 DF-L. MC = 20 percent, and the dimensions are $A = 8$ in. and $B = 10$ in. Loads are (D + L_r).

   d. What deformation limit is associated with the bearing stresses used in parts *a* to *c*?

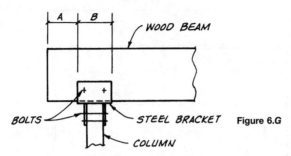

BOLTS — WOOD BEAM — STEEL BRACKET — COLUMN — **Figure 6.G**

6.35 Repeat Prob. 6.34 except a deformation limit of 0.02 in. is to be used.

6.36 *Given:*   The beam in Fig. 6.H is a 5⅛ × 19.5 DF glulam 16F-1.3E. The load is (D + S). MC < 16 percent. Lateral support is provided to the top side of the member by roof sheathing. $C_t = 1.0$.

   Find: Check the given member for bending and shear stresses.

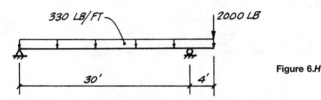

330 LB/FT   2000 LB   30'   4'   **Figure 6.H**

6.37 *Given:*   The roof framing plan of the commercial building in Fig. 6.I. There is no ceiling. The *total* dead loads to the members are

   Subpurlin (2 × 4 at 24 in. o.c.) = 7.0 psf

   Purlins (4 × 14 at 8 ft-0 in. o.c.) = 8.5

   Girder = 10.0

   The roof is flat except for a minimum slope of ¼ in./ft to prevent ponding. Roof live loads are to be in accordance with the IBC. See Example 2.4. The roof diaphragm provides continuous lateral support to the top side of all beams. $C_M = 1.0$ and $C_t = 1.0$.

   Find: a. Check the subpurlins, using No. 1 & Btr DF-L. Are the AITC-recommended deflection criteria satisfied?
   b. Check the purlins, using No. 1 & Btr DF-L. Are the AITC deflection limits met?
   c. Design the girder, using 24F-1.8E DF glulam. Determine the minimum size, considering both strength and stiffness.

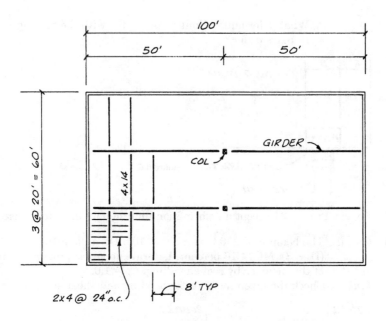

ROOF FRAMING PLAN

Figure 6.I

6.38 *Given:*   The girder in the roof framing plan in Fig. 6.J is to be designed using the optimum cantilever length L_c. The girder is 20F-1.5E DF glulam. D = 16 psf. The top of the girder is laterally supported by the roof sheathing. Deflection need not be checked, but camber requirements are to be determined. The roof is flat except for a minimum slope of ¼ in./ft to prevent ponding. C_M = 1.0 and C_t = 1.0.

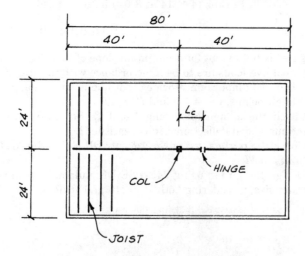

ROOF FRAMING PLAN

Figure 6.J

Find: a. The minimum required beam size if the girder is designed for a 20-psf roof live load with no reduction for tributary area.

b. The minimum required beam size if the girder is designed for a basic 20-psf roof live load that is to be adjusted for tributary area. Roof live load is to be determined in accordance with the IBC. See Example 2.4. For the roof live load reduction, consider the tributary area of the suspended portion of the cantilever system.

Axial Forces and Combined Bending and Axial Forces

7.1 Introduction

An axial force member has the load applied parallel to the longitudinal axis through the centroid of the cross section. The axial force may be either tension or compression. Because of the need to carry vertical gravity loads down through the structure into the foundation, columns are more often encountered than tension members. Both types of members, however, see widespread use in structural design in such items as trusses and diaphragms.

In addition to the design of axial force members, this chapter covers the design of members with a more complicated loading condition. These include members with bending (beam action) occurring simultaneously with axial forces (tension or compression). This type of member is often referred to as a *combined stress* member. A combination of loadings is definitely more critical than the case of the same forces being applied individually. The case of compression combined with bending is probably encountered more often than tension plus bending, but both types of members are found in typical wood-frame buildings.

To summarize, the design of the following types of members is covered in this chapter:

1. Axial tension
2. Axial compression
3. Combined bending and tension
4. Combined bending and compression

The design of axial force tension members is relatively straightforward, and the required size of a member can be solved for directly. For the other three types of members, however, a trial-and-error solution is the typical design approach. Trial-and-error solutions may seem awkward in the beginning. However, with a little practice the designer will be able to pick an initial trial size which will be relatively close to the required size. The final selection can often be made with very few trials. Several examples will illustrate the procedure used in design.

As noted, the most common axial load member is probably the column, and the most common combined stress member is the beam-column (combined bending and compression). See Fig. 7.1. In this example an axial load is assumed to be applied to the interior column by the girder. For the exterior column there is both a lateral force that causes bending and a vertical load that causes axial compression.

The magnitude of the lateral force (wind or seismic) to the column depends on the unit design force and how the wall is framed. The wall may be framed horizontally to span between columns, or it may be framed vertically to span between story levels (Example 3.4 in Sec. 3.3).

Numerous other examples of axial force members and combined load members can be cited. However, the examples given here are representative, and they adequately define the type of members and loadings that are considered in this chapter.

7.2   Axial Tension Members

Wood members are stressed in tension in a number of structural applications. For example, trusses have numerous axial force members, and roughly half of these are in tension. It should be noted that unless the loads frame directly into the joints in the truss and unless the joints are pinned, bending stresses will be developed in addition to the axial stresses obtained in the standard truss analysis.

Axial tension members also occur in the chords of horizontal and vertical diaphragms. In addition, tension members are used in diaphragm design when the length of the horizontal diaphragm is greater than the length of the shearwall to which it is attached. This type of member is known as a collector or drag strut, and it is considered in Chap. 9.

The check for the axial tension stress in a member of known size uses the formula

$$f_t = \frac{P}{A_n} \le F_t'$$

where f_t = actual (computed) tension stress parallel to grain
$\quad\quad P$ = axial tension force in member
$\quad\, A_n$ = net cross-sectional area
$\quad\quad\quad$ = $A_g - \Sigma A_h$
$\quad\, A_g$ = gross cross-sectional area

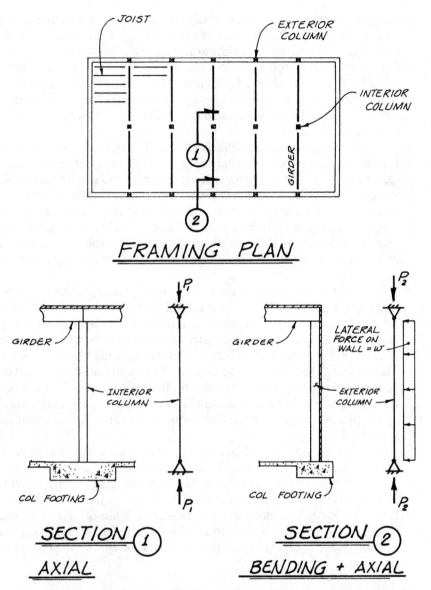

Figure 7.1 Examples of columns and beam-columns. Both of the members in this example are vertical, but horizontal or inclined members with this type of loading are also common.

ΣA_h = sum of projected area of holes at critical section
F'_t = allowable tension stress parallel to grain (defined below)

The formula for comparing the actual stress in a member with the allowable stress is usually referred to as an analysis expression. In other words, an *analysis* problem involves *checking* the adequacy of a member of known size. In a *design* situation, the size of the member is unknown, and the usual objective is to establish the minimum required member size.

In a tension member design, the axial stress formula can be rewritten to solve for the required area by dividing the load by the allowable tension stress. Although certain assumptions may be involved, this can be described as a *direct solution.* The member size determined in this way is usually something close to the final solution.

It was previously noted that the design of axial tension members is the only type of problem covered in Chap. 7 that can be accomplished by direct solution. Columns, for example, involve design by trial and error because the allowable stress depends on the column slenderness ratio. The slenderness ratio, in turn, depends on the size of the column, and it is necessary to first establish a trial size. The adequacy of the trial size is evaluated by performing an analysis. Depending on the results of the analysis, the trial size is accepted or adjusted up or down. Members with combined stresses are handled in a similar way.

It should be emphasized that the tension stress problems addressed in this chapter are for *parallel-to-grain loading.* The weak nature of wood in tension perpendicular to grain is noted throughout this book, and the general recommendation is again to avoid stressing wood in tension across the grain.

There are a variety of fasteners that can be used to connect wood members. The projected area of holes or grooves for the installation of fasteners is to be deducted from the *gross area* to obtain the *net area.* Some frequently used fasteners in wood connections include nails, bolts, lag bolts, split rings, and shear plates, and the design procedures for these are covered in Chaps. 11 through 13.

In determining the net area of a tension member, the projected area for nails is usually disregarded. The projected area of a bolt hole is a rectangle. A split ring or shear plate connector involves a dap or groove cut in the face of the wood member plus the projected area of the hole for the bolt (or lag bolt) that holds the assembly together. See Example 7.1. The projected area removed from the cross section for the installation of a lag bolt is determined by the shank diameter and the diameter of the lead or pilot hole for the threads.

EXAMPLE 7.1   Net Areas at Connections

The gross area of a wood member is the width of the member times its depth:

$$A_g = b \times d$$

The standard net dimensions and the gross cross-sectional areas for sawn lumber are given in NDS Supplement Table 1B. Similar properties for glulam members are listed in NDS Supplement Tables 1C and 1D.

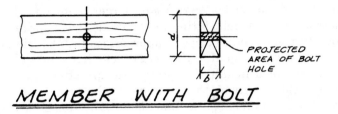

MEMBER WITH BOLT

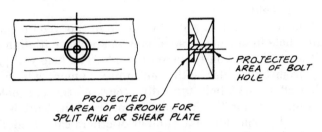

MEMBER WITH CONNECTOR

IN ONE FACE

Figure 7.2 *Net-section* through two wood members. One member is shown cut at a bolt hole. The other is at a joint with a split ring or shear plate connector in one face plus the projected area of a bolt. The bolt is required to hold the entire assembly (wood members and connectors) together. Photographs of split ring and shear plate connectors are included in Chap. 13 (Fig. 13.26*a* and *b*).

The projected areas for fasteners to be deducted from the gross area are as follows:

Nail holes—disregarded.

Bolt holes—computed as the hole diameter times the width of the wood member. The hole diameter is between $\frac{1}{32}$ and $\frac{1}{16}$ in. larger than the bolt diameter (NDS Sec. 11.1.2.2). In this book the bolt hole, for strength calculation purposes, is taken as the bolt diameter plus $\frac{1}{16}$ in.

Lag bolt holes—a function of the connection details. See NDS Appendix L for lag bolt dimensions. Drill diameters for lead holes and shank holes are given in NDS Sec. 11.1.3.

Split ring and shear plate connectors—a function of the connection details. See NDS Appendix K for the projected areas of split rings and shear plates.

If more than one fastener is used, the sum of the projected areas of all the fasteners *at the critical section* is subtracted from the gross area. For staggered fastener pattern, see NDS Sec. 3.1.2.

Perhaps the most common situation that requires a reduction of area for tension member design is a bolted connection. The NDS (Ref. 7.2) requires that the hole diameter be $\frac{1}{32}$ to $\frac{1}{16}$ in. larger than the bolt diameter. It also recommends against tight-fitting installations that require forcible driving of the bolt. In *ideal* conditions it is appropriate to take the hole diameter for calculation purposes equal to the actual hole diameter.

In practice, ideal installation procedures are often viewed as goals. There are many field conditions that may cause the actual installation to be less than perfect. For example, a common bolt connection is through two steel side plates with the wood member between the two metal plates. Holes in the steel plates are usually punched in the shop, and holes in the wood member are drilled in the field.

It is difficult to accurately drill the hole in the wood member from one side (through a hole in one of the steel plates) and have it align perfectly with the hole in the steel plate on the opposite side. The hole will probably be drilled partially from both sides with some misalignment where they meet. Some oversizing of the bolt hole typically occurs as the two holes are reamed to correct alignment for the installation of the bolt. This is one example of a practical field problem, and a number of others can be cited.

In this book the hole diameter for net-area calculations will be taken as the bolt diameter plus $\frac{1}{16}$ in. as specified in the NDS. See Chap. 13 for more information on bolted connections.

The allowable tension stress in a wood member is determined by multiplying the tabulated tension stress by the appropriate adjustment factors:

$$F'_t = F_t(C_D)(C_M)(C_t)(C_F)(C_i)$$

where F'_t = allowable tension stress parallel to grain
F_t = tabulated tension stress parallel to grain
C_D = load duration factor (Sec. 4.15)
C_M = wet service factor (Sec. 4.14)
 = 1.0 for dry service conditions (as in most covered structures). Dry service is defined as
 MC $\leq$ 19 percent for sawn lumber
 MC $<$ 16 percent for glulam
C_t = temperature factor (Sec. 4.19)
 = 1.0 for normal temperature conditions
C_F = size factor (Sec. 4.16) for sawn lumber in tension. Obtain values for visually-graded Dimension lumber from the Adjustment Factors section of NDS Supplement Tables 4A, 4B, and 4F.
 = 1.0 for sawn lumber in B&S and P&T sizes and MSR lumber (*Note:* The size factor is not applicable for glulam.)
C_i = incising factor for sawn lumber (Sec. 4.20)
 = 0.80 for incised sawn lumber
 = 1.0 for sawn lumber not incised (whether the member is treated or untreated)
 (*Note:* The incising factor is not applicable for glulam.)

It can be seen that the usual adjustment factors for load duration, moisture content, temperature, size effect, and incising apply to tension stresses parallel to grain. Numerical values for the load duration factor depend on the shortest-duration load in a given combination of loads. Values of C_M, C_t, and C_i frequently default to unity, but the designer should be aware of conditions that may require an adjustment. The size factor for tension applies to visually-graded Dimension lumber only, and values are obtained from NDS Supplement Tables 4A, 4B, and 4F.

7.3 Design Problem: Tension Member

In this example the required size for the lower chord of a truss is determined. The loads are assumed to be applied to the top chord of the truss only. See Example 7.2.

To determine the axial forces in the members using simple truss analysis techniques, it is useful to assume the loads to be applied to the joints. Loads for the truss analysis are obtained by taking the tributary width to one joint times the uniform load. Because the actual loads are applied uniformly to the top chord, these members will have combined stresses. Other members in the truss will have only axial forces if the joints are pinned.

The tension force in the bottom chord is obtained through a standard truss analysis (method of joints). The member size is determined by calculating the required net area and adding to it the area removed by the bolt hole.

EXAMPLE 7.2   Tension Chord

Determine the required size of the lower (tension) chord in the truss in Fig. 7.3a. The loads are (D + S), and the effects of roof slope on the magnitude of the snow load have already been taken into account. Joints are assumed to be pinned.

Connections will be made with a single row of ¾-in.-diameter bolts. Trusses are 4 ft-0 in. o.c. Lumber is No. 1 Spruce-Pine-Fir (South) [abbreviated S-P-F(S)]. MC ≤ 19 percent, and normal temperatures apply. Allowable stresses and cross-sectional properties are to be taken from the NDS Supplement.

Loads

$$D = 14 \text{ psf} \quad \text{horizontal plane}$$

$$\text{Reduced } S = 30 \text{ psf}$$

$$TL = 44 \text{ psf}$$

$$w_{TL} = 44 \times 4 = 176 \text{ lb/ft}$$

[handwritten annotation: spacing of trusses / tributary area]

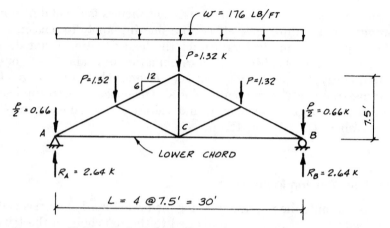

Figure 7.3a   Uniform load on top chord converted to concentrated joint loads.

For truss analysis (load to joint),

$$P = 176 \times 7.5 = 1320 \text{ lb}$$

Force in Lower Chord
Use method of joints.

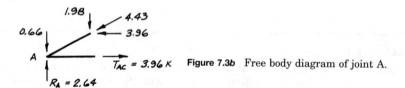

$T_{AC} = 3.96$ K   **Figure 7.3b**   Free body diagram of joint A.

Determine Required Size of Tension Member
The relatively small tension force will require a Dimension lumber member size. The allowable tension stress is obtained from NDS Supplement Table 4A for S-P-F(S). A value for the size factor for tension stress parallel to grain will be assumed and checked later.

Assume $C_F = 1.3$.

$$F'_t = F_t(C_D)(C_M)(C_t)(C_F)(C_i)$$

$$= 400(1.15)(1.0)(1.0)(1.3)(1.0) = 598 \text{ psi}$$

$$\text{Req'd } A_n = \frac{P}{F'_t} = \frac{3960}{598} = 6.62 \text{ in.}^2$$

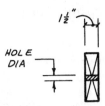

Figure 7.3c Net section of tension member.

The actual hole diameter is to be $\frac{1}{32}$ to $\frac{1}{16}$ in. larger than the bolt size. For net-area calculations, arbitrarily assume that the bolt hole is $\frac{1}{16}$ in. larger than the bolt (for stress calculations only). Select a trial size from NDS Supplement Table 1B.

$$\text{Req'd } A_g = A_n + A_h$$

$$= 6.62 + 1.5(\tfrac{3}{4} + \tfrac{1}{16}) = 7.84 \text{ in.}^2$$

Try 2 × 6:

$$A = 8.25 \text{ in.}^2 > 7.84 \text{ in.}^2 \quad OK$$

Verify the size factor for tension in NDS Supplement Table 4A for a 6-in. nominal width:

$$C_F = 1.3 \quad (\text{same as assumed}) \quad OK$$

$$\boxed{Use \quad 2 \times 6 \quad \text{No. 1} \quad \text{S-P-F(S)}}$$

NOTE: The simplified truss analysis used in this example applies only to trusses with pinned joints. If some form of toothed metal plate connector (see Section 11.2) is used for the connections, the design should conform to Ref. 7.7 or applicable building code standard.

7.4 Columns

In addition to being a compression member, a *column* generally is sufficiently long that the possibility of buckling needs to be considered. On the other hand, the term *short column* usually implies that a compression member will not buckle, and its strength is related to the crushing capacity of the material.

In order to evaluate the tendency of a column to buckle, it is necessary to know the size of the member. Thus in a design situation, a trial size is first established. With a known size or a trial size, it is possible to compare the actual stress with the allowable stress. Based on this comparison, the member size will be accepted or rejected.

The check on the capacity of an axially loaded wood column of known size uses the formula

$$f_c = \frac{P}{A} \leq F'_c$$

where f_c = actual (computed) compressive stress parallel to grain
    P = axial compressive force in member
    A = cross-sectional area
    F'_c = allowable compressive stress parallel to grain as defined later in this section

In the calculation of actual stress f_c, the cross-sectional area to be used will be either the gross area A_g of the column or the net area A_n at some hole in the member. The area to be used depends on the location of the hole along the length of the member and the tendency of the member at that point to buckle laterally. If the hole is located at a point which is braced, the gross area of the member may be used in the check for column stability between brace locations. Another check of f_c at the reduced cross section (using the net area) should be compared with the allowable compressive stress for a short column with no reduction for stability at the braced location. See Example 7.3. The other possibility is that some reduction of column area occurs in the laterally unbraced portion of the column. In the latter case, the net area is used directly in the stability check.

EXAMPLE 7.3   Actual Stresses in a Column

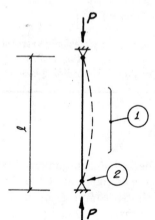

Figure 7.4 Pinned end column.

Actual Stresses

In Fig. 7.4 it is assumed that there are no holes in the column except at the supports (connections). Check the following column stresses:

1. Away from the supports

$$f_c = \frac{P}{A_g} \leq F'_c \qquad \text{as determined by column stability (using } C_P)$$

2. At the connection where buckling is not a factor

$$f_c = \frac{P}{A_n} \le F'_c \quad \text{as determined for a \textit{short} column (without } C_P)$$

The allowable stress in a column reflects many of the familiar adjustment factors in addition to column stability:

$$F'_c = F_c(C_D)(C_M)(C_t)(C_F)(C_P)(C_i)$$

where F'_c = allowable compressive stress parallel to grain
F_c = tabulated compressive stress parallel to grain
C_D = load duration factor (Sec. 4.15)
C_M = wet service factor (Sec. 4.14)
= 1.0 for dry service conditions as in most covered structures. Dry service conditions are defined as
MC ≤ 19 percent for sawn lumber
MC < 16 percent for glulam
C_t = temperature factor (Sec. 4.19)
= 1.0 for normal temperature conditions
C_F = size factor (Sec. 4.16). Obtain values for visually graded Dimension lumber from Adjustment Factors section of NDS Supplement Tables 4A, 4B, and 4F.
= 1.0 for Timbers
= 1.0 for MSR and MEL lumber
(*Note:* The size factor is not applicable for glulam.)
C_P = column stability factor
= 1.0 for fully supported column
C_i = incising factor for sawn lumber (Sec. 4.20)
= 0.80 for incised sawn lumber
= 1.0 for sawn lumber not incised (whether the member is treated or untreated)
(*Note:* The incising factor is not applicable for glulam.)

The size factor for compression applies only to Dimension lumber sizes, and C_F defaults to 1.0 for other members. The column stability factor takes buckling into account, and the slenderness ratio is the primary measure of buckling. The column slenderness ratio and C_P are the subjects of the remainder of this section.

In its traditional form, the slenderness ratio is expressed as the effective unbraced length of a column divided by the least radius of gyration, l_e/r. For the design of rectangular wood columns, however, the slenderness ratio is modified to a form that is somewhat easier to apply. Here the slenderness ratio is the effective unbraced length of the column divided by the least dimension of the cross section, l_e/d.

Use of this modified slenderness ratio is possible because the radius of gyration r can be expressed as a direct function of the width of a rectangular column. See Example 7.4. The constant in the conversion of the modified slenderness ratio is simply incorporated into the allowable stress column design formulas.

Most wood columns have rectangular cross sections, and the allowable stress formula given in this chapter is for this common type of column. However, a column of nonrectangular cross section may be analyzed by substituting $r\sqrt{12}$ in place of d in the formula for rectangular columns. For round columns see NDS Sec. 3.7.3. A much more detailed analysis of the slenderness ratio for columns is given in the next section.

EXAMPLE 7.4   Column Slenderness Ratio—Introduction

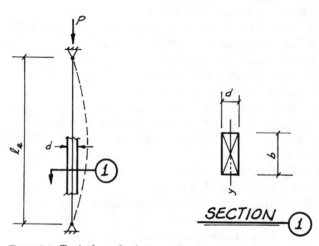

Figure 7.5 Typical wood column with rectangular cross section.

Column stability is measured by the slenderness ratio.

$$\text{General slenderness ratio} = \frac{l_e}{r}$$

where l_e = effective unbraced length of column
  $r = r_y$ = least radius of gyration of column cross section

$$\text{Slenderness ratio for } rectangular\ columns = \frac{l_e}{d}$$

where l_e = effective unbraced length of column
  d = least cross-sectional dimension of column*

For a rectangular cross section, the dimension d is directly proportional to the radius of gyration.

---

*In beam design, d is normally associated with the strong axis.

$$r_y = \sqrt{\frac{I_y}{A}} = \sqrt{\frac{bd^3/12}{bd}} = \sqrt{\frac{d^2}{12}} = d\sqrt{\frac{1}{12}}$$

$$\therefore \quad r \propto d$$

A more complete review of column slenderness ratio is given in Sec. 7.5.

The column stability factor C_P was shown previously as an adjustment factor for obtaining the allowable compressive stress in a column. The treatment of C_P as another multiplying factor is convenient from an organizational or bookkeeping point of view. However, the expression for C_P times F_c essentially defines the *column curve* or *column equation* for wood design. The other coefficients in the expression for F_c' are more in keeping with the general concept of adjustment factors, and the column equation as given by $F_c \times C_P$ is probably more basic to column behavior than the term of *adjustment factor* implies.

The column equation in the NDS provides a continuous curve over the full range of slenderness ratios. See Example 7.5. The column expression in the NDS was originally developed by Ylinen and was verified by studies at the Forest Products Laboratory. The Ylinen formula also serves as the basis for the expression used for laterally unsupported beams in Sec. 6.3.

Zahn (Ref. 7.9) explained that the *behavior of wood columns* as given by the Ylinen formula is the result of the *interaction* of two modes of failure: *buckling and crushing*. Pure *buckling* is defined by the Euler critical buckling stress

$$F_{cr} = \frac{\pi^2 E}{(l_e/d)^2}$$

For use in allowable stress design (ASD) the Euler stress is divided by an appropriate factor of safety and is expressed in the NDS as

$$F_{cE} = \frac{K_{cE}E'}{(l_e/d)^2}$$

The K_{cE} term incorporates π^2 divided by the factor of safety. The Euler column stress F_{cE} is graphed in Fig. 7.6a. It will be noted that the Ylinen formula converges to the Euler-based formula for columns with large slenderness ratios.

The second mode of failure is crushing of the wood fibers. When a compression member fails by pure crushing, there is no column buckling. Therefore in ASD, *crushing* is measured by the tabulated compressive stress parallel to grain multiplied by all applicable adjustment factors except C_P. This value is given the symbol F_c^* and is defined mathematically as

$$F_c^* = F_c(C_D)(C_M)(C_t)(C_F)(C_i)$$

The value of F_c^* is the limiting value of allowable column stress for a slenderness ratio of zero.

Again, column behavior is defined by the *interaction* of the buckling and crushing modes of failure, and the ratio F_{cE}/F_c^* appears several times in the Ylinen formula. The coefficient c in the Ylinen formula is viewed by Zahn as a generalized interaction parameter. The value of c lies in the range

$$0 \leq c \leq 1.0$$

A value of $c = 1.0$ is an upper bound of column behavior and can only be met by an ideal material, loaded under ideal conditions. Because practical columns do not satisfy these idealizations, the value of c for wood compression members is less than one. The more a wood column deviates from the ideal situation, the smaller c becomes. Glulam members are generally thought to be straighter and more homogeneous than sawn lumber, and consequently glulam is assigned a larger value for c. The effects of different values of c are shown in Fig. 7.6b.

As the slenderness ratio increases, the column expression makes a transition from an allowable stress based on the crushing strength of wood (at a zero slenderness ratio) to an allowable stress based on the Euler curve (for large slenderness ratios). The Ylinen column curve more closely fits the results of column tests. Compared with previously used column formulas, the Ylinen equation gives slightly more conservative values of allowable compressive stress for members with intermediate slenderness ratios.

EXAMPLE 7.5   Ylinen Column Equation

The NDS uses a continuous curve for evaluating the effects of column buckling. The allowable column stress given by the Ylinen equation is plotted versus column slenderness ratio in Fig. 7.6a.

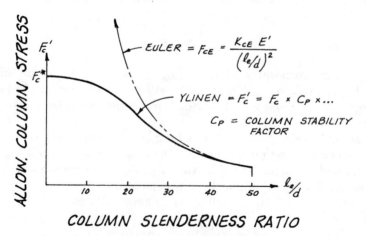

Figure 7.6a Ylinen column curve: plot of F_c' versus l_e/d.

Allowable Column Stress

The allowable column stress curve in Fig. 7.6a is obtained by multiplying the tabulated compressive stress parallel to grain F_c by the column stability factor C_P and all other appropriate factors.

$$F'_c = F_c(C_P) \times \cdots$$

where F'_c = allowable compressive stress in a column
F_c = tabulated compressive stress parallel to grain
C_P = column stability factor (defined below)
$\times \cdots$ = product of other appropriate adjustment factors

Column Stability Factor

$$C_P = \frac{1 + F_{cE}/F_c^*}{2c} - \sqrt{\left(\frac{1 + F_{cE}/F_c^*}{2c}\right)^2 - \frac{F_{cE}/F_c^*}{c}}$$

where F_{cE} = Euler critical buckling stress for columns

$= \dfrac{K_{cE}E'}{(l_e/d)^2}$

F_c^* = limiting compressive stress in column at zero slenderness ratio
= tabulated compressive stress parallel to grain multiplied by all adjustment factors except C_P
$= F_c(C_D)(C_M)(C_t)(C_F)(C_i)$

K_{cE} = 0.3 for visually graded lumber
= 0.384 for MEL
= 0.418 for products with less variability such as MSR lumber and glulam
See NDS Appendix F.2 for additional information

E' = modulus of elasticity associated with the axis of column buckling (see Sec. 7.4). Recall that C_D does not apply to E. For sawn lumber, $E_x = E_y$. For glulam, E_x and E_y may be different.
$= E(C_M)(C_t)(C_T)(C_i)$

c = buckling and crushing interaction factor for columns
= 0.8 for sawn lumber columns
= 0.85 for round timber poles and piles
= 0.9 for glulam or structural composite lumber columns

C_T = buckling stiffness factor for 2 × 4 *and smaller compression chords in trusses* with ⅜-in. or thicker plywood nailed to narrow face of member
= 1.0 for all other members

C_F = size factor (Sec. 4.16) for compression. Obtain values for visually graded Dimension lumber from the Adjustment Factors section of NDS Supplement Tables 4A, 4B, and 4F.
= 1.0 for sawn lumber in B&S and P&T sizes and MSR and MEL lumber
(*Note:* The size factor is not applicable for glulam.)

Other factors are as previously defined.

The interaction between column buckling and crushing of wood fibers in a compression member is measured by parameter c. The effect of several different values of c is illustrated in Fig. 7.6b.

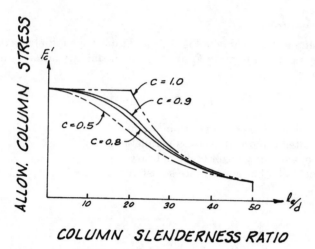

Figure 7.6b Plot of F'_c versus l_e/d showing the effect of different values of c. The parameter c measures the interaction between crushing and buckling in wood columns.

A value of $c = 1.0$ applies to idealized column conditions. Practical wood columns have $c < 1.0$:

For sawn lumber: $\qquad\qquad\qquad\qquad\qquad\qquad\qquad c = 0.8$

For glulam and structural composite lumber: $\quad c = 0.9$

The effect of the load duration factor varies depending on the mode of column failure that predominates. See Fig. 7.6c.

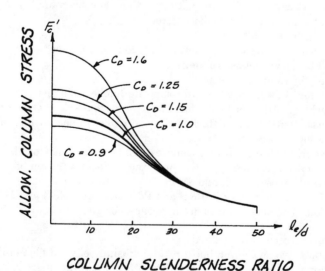

Figure 7.6c Plot of F'_c versus l_e/d showing the effect of load duration on allowable column stress.

The load duration factor C_D has full effect on allowable column stress when crushing controls (i.e., at a slenderness ratio of zero). On the other hand, C_D has no influence on the allowable column stress when instability predominates. A transition between C_D having full effect at $l_e/d = 0$, and C_D having no effect at the maximum slenderness ratio of 50, is automatically provided in the definition of C_P.

The designer should have some understanding about the factor of safety provided by the column formula. It was noted in an earlier chapter that the values of modulus of elasticity listed in the NDS Supplement are average values. Furthermore, the tabulated values have been modified to account for shear deformation.

The pure bending modulus of elasticity is obtained by multiplying the value of E listed in NDS Supplement Tables 4A through 4F by a factor of 1.03 and Tables 5A through 5D by 1.05. In addition, the formula for F_{cE} includes an adjustment which converts the average modulus of elasticity to a 5 percent exclusion value. When a value of $K_{cE} = 0.3$ is used for *sawn lumber,* the allowable column stress F'_c includes a *factor of safety* of 1.66 at an approximate 5 percent lower exclusion value.

MEL, MSR lumber, and glulam are less variable than sawn lumber. For these less variable materials, a factor of safety of 1.66 is maintained at a 5 percent lower exclusion value when $K_{cE} = 0.384$ for MEL or $K_{cE} = 0.418$ for MSR or glulam is used to compute F_{cE}. If a value of $K_{cE} = 0.3$ is retained for glulam and MSR lumber, the corresponding allowable stress represents less than a one percent lower exclusion value with a factor of safety of 1.66 (see NDS Appendices F and H).

The design of a glulam column follows essentially the same procedure as that used for a sawn lumber column. In addition to the values of c and K_{cE} in the expressions for C_P, the basic difference for glulam is in the tabulated design values. Recall that glulam is available in either bending or axial combinations. Although pure columns are axial force members, bending combinations may also be loaded in compression. For these members the designer will have to select the appropriate value(s) of modulus of elasticity (E_x and/or E_y) for column analysis from the glulam tables. This is demonstrated in Example 7.7 in Sec. 7.7.

7.5 Detailed Analysis of Slenderness Ratio

The concept of the slenderness ratio was briefly introduced in Sec. 7.4. There it was stated that the *least* radius of gyration is used in the l_e/r ratio and that the *least* dimension of the column cross section is used in the l_e/d ratio.

These statements assume that the unbraced length of the column is the same for both the x and the y axes. In this case the column, if loaded to failure, would buckle about the weak y axis. See Fig. 7.7a. Note that if buckling occurs about the y axis, the column moves in the x direction. Figure 7.7a illustrates

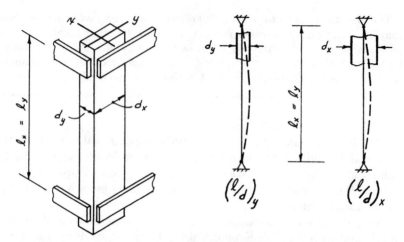

Figure 7.7a By inspection the slenderness ratio about the y axis is larger and is therefore critical.

this straightforward case of column buckling. Only the slenderness ratio about the weak axis needs to be calculated.

Although this concept of column buckling applies in many situations, the designer should have a deeper insight into the concept of the slenderness ratio. Conditions can exist under which the column may actually buckle about the *strong* axis of the cross section rather than the weak axis.

In this more general sense, the column can be viewed as having two slenderness ratios. One slenderness ratio would evaluate the tendency of the column to buckle about the *strong* axis of the cross section. The other would measure the tendency of buckling about the *weak* axis. For a rectangular column these slenderness ratios would be written

$$\left(\frac{l_e}{d}\right)_x \quad \text{for buckling about strong } x \text{ axis (column movement in } y \text{ direction)}$$

$$\left(\frac{l_e}{d}\right)_y \quad \text{for buckling about weak } y \text{ axis (column movement in } x \text{ direction)}$$

If the column is loaded to failure, buckling will occur about the axis that has the larger slenderness ratio. In design, the larger slenderness ratio is used to calculate the allowable compressive stress. (Note that it is conceivable in a glulam column with different values for E_x and E_y that a slightly smaller l_e/d could produce the critical F'_c.)

The reason that the strong axis can be critical can be understood from a consideration of the *bracing* and *end conditions* of the column. The *effective unbraced length* is the length to be used in the calculation of the slenderness ratio. It is possible to have a column with different unbraced lengths for the x and y axes. See Fig. 7.7b. In this example the unbraced length for the x axis is twice as long as the unbraced length for the y axis. In practice, bracing can occur at any interval. The effect of column end conditions is explained later.

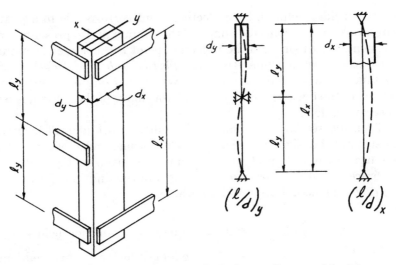

Figure 7.7b Different unbraced lengths for both axes. Because of the intermediate bracing for the y axis, the critical slenderness ratio cannot be determined by inspection. Both slenderness ratios must be calculated, and the larger value is used to determine F'_c.

Another case where the unbraced lengths for the x and y axes are different occurs when sheathing is attached to a column. If the sheathing is attached to the column with an effective connection, buckling about an axis that is perpendicular to the sheathing is prevented. See Fig. 7.8. The most common example of this type of column is a stud in a bearing wall. The wall sheathing can prevent column buckling about the weak axis of the stud, and only the slenderness ratio about the strong axis of the member needs to be evaluated.

The final item regarding the slenderness ratio is the effect of column *end conditions*. The length l used in the slenderness ratio is theoretically the unbraced length of a *pinned-end* column. For columns with other end conditions, the length is taken as the distance between inflection points (IPs) on a sketch

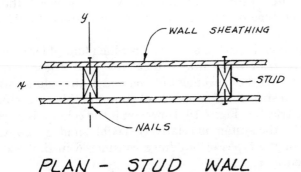

Figure 7.8 Column braced by sheathing. Sheathing attached to stud prevents column buckling about the weak (y) axis of the stud. Therefore, consider buckling about the x axis only.

of the buckled column. An inflection point corresponds to a point of reverse curvature on the deflected shape of the column and represents a point of zero moment. For this reason the inflection point is considered as a pinned end for purposes of column analysis. The *effective unbraced length* is taken as the distance between inflection points. When only one inflection point is on the sketch of the buckled column, the mirror image of the column is drawn to give a second inflection point.

Typically six "ideal" column end conditions are identified in various fields of structural design. See Fig. 7.9*a*. The recommended effective length factors for use in the design of wood columns are given in Fig. 7.9*b*. The effective unbraced length can be determined by multiplying the effective length factor K_e times the actual unbraced length.

$$\text{Effective length} = \text{distance between inflection points}$$
$$= \text{effective length factor} \times \text{unbraced length}$$
$$l_e = K_e \times l$$

The effective lengths shown on the column sketches in Fig. 7.9*a* are *theoretical* effective lengths, and practical field column end conditions can only approximate the ideal pinned and fixed column end conditions. The recommended *design* effective length factors from NDS Appendix G are to be used for practical field end conditions.

In practice, the designer must determine which "ideal" column most closely approximates the actual end conditions for a given column. Some degree of judgment is required for this evaluation, but several key items should be considered in making this comparison.

First, note that three of the ideal columns undergo sidesway, and the other three do not. *Sidesway* means that the top of the column is relatively free to displace laterally with respect to the bottom of the column. The designer must be able to identify which columns will undergo sidesway and, on the other hand, what constitutes restraint against sidesway. In general, the answer to this depends on the type of lateral-force-resisting system (LFRS) used (Sec. 3.3).

Usually, if the column is part of a system in which lateral forces are resisted by *bracing* or by *shearwalls,* sidesway will be prevented. These types of LFRSs are relatively rigid, and the movement of one end of the column with respect to the other end is restricted. If an overload occurs in this case, column buckling will be symmetric. See Fig. 7.10. However, if the column is part of a *rigid frame* type of LFRS, the system is relatively flexible, and sidesway can occur. Typical columns for the types of buildings considered in this text will have sidesway prevented.

It should also be noted that columns with *sidesway prevented* have an effective length which is *less than or equal to* the actual unbraced length

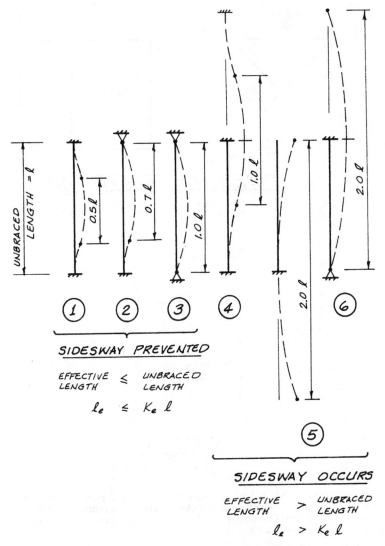

Figure 7.9a Six typical idealized columns showing buckled shapes and theoretical effective lengths.

Column No.	1	2	3	4	5	6
Theoretical effective length factors	0.5	0.7	1.0	1.0	2.0	2.0
Recommended design K_e	0.65	0.8	1.0	1.2	2.1	2.4

Figure 7.9b Table of theoretical and recommended effective length factors K_e. Values of recommended K_e are from NDS Appendix G.

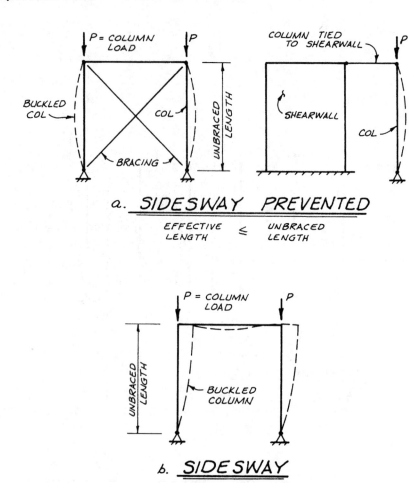

Figure 7.10 Columns with and without sidesway. (*a*) Braced frames or buildings with shearwalls limit the displacement of the top end of the column so that sidesway does not occur. (*b*) Columns in rigid frames (without bracing) will undergo sidesway if the columns buckle.

($K_e \leq 1.0$). A common and conservative practice is to consider the effective length equal to the unbraced length for these columns ($K_e = 1.0$).

For columns where *sidesway* can occur, the effective length is *greater than* the actual unbraced length ($K_e > 1.0$). For these types of columns, the larger slenderness ratio causes the allowable axial load to be considerably less than the allowable load on a column with both ends pinned and braced against sidesway.

In addition to answering the question of sidesway, the comparison of an actual column to an ideal column should evaluate the effectiveness of the column connections. Practically all wood columns have *square-cut ends*. For structural design purposes, this type of column end condition is normally assumed to be pinned. Square-cut column ends do offer some restraint against column end rotation. However, most practical column ends are not exactly square, and some *accidental eccentricity* may be present due to nonuniform bearing. These effects are often assumed to be compensating. Therefore, columns in a typical wood-frame building with shearwalls are usually assumed to be type 3 in Fig. 7.9a, and the effective length factor is taken to be unity.

It is possible to design moment-resisting connections in wood members, but they are the exception rather than the rule. As noted, the majority of connections in ordinary wood buildings are "simple" connections, and it is generally conservative to take the effective length equal to the unbraced length. However, the designer should examine the actual *bracing conditions* and *end conditions* for a given column and determine whether or not a larger effective length should be used.

7.6 Design Problem: Axially Loaded Column

The design of a column is a trial-and-error process because, in order to determine the allowable column stress F'_c, it is first necessary to know the slenderness ratio l_e/d. In the following example, only two trials are required to determine the size of the column. See Example 7.6.

Several items in the solution should be emphasized. First, the importance of the size category should be noted. The initial trial is a Dimension lumber size, and the second trial is a Posts and Timbers size. Tabulated stresses are different for these two size categories.

The second item concerns the load duration factor. This example involves $(D + L_r)$. It will be remembered that roof live load is an arbitrary minimum load required by the Code, and C_D for this combination is 1.25. For many areas of the country the design load for a roof will be $(D + S)$, and the corresponding C_D for snow would be 1.15. The designer is cautioned that Example 7.6 is illustrative in that the critical load combination $(D + L_r)$ was predetermined. To be complete, the designer should also check D only with the corresponding C_D for permanent load of 0.9.

EXAMPLE 7.6   Sawn Lumber Column

Design the column in Fig. 7.11a, using No. 1 Douglas Fir-Larch. Bracing conditions are the same for buckling about the x and y axes. The load is combined dead load and roof live load. MC ≤ 19 percent (C_M = 1.0), the member is not treated or incised (C_i = 1.0) and normal temperatures apply (C_t = 1.0). Allowable stresses are to be in accordance with the NDS.

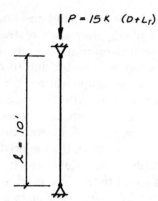

Figure 7.11a Elevation view of column.

Trial 1

Try 4 × 6 (Dimension lumber size category). From NDS Supplement Table 4A

$$F_c = 1500 \text{ psi}$$

$$C_F = 1.1 \quad \text{size factor for compression}$$

$$E = 1,700,000 \text{ psi}$$

$$A = 19.25 \text{ in.}^2$$

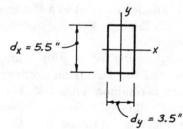

Figure 7.11b Cross section of 4 × 6 trial column.

Determine capacity using Ylinen column equation (Example 7.5):

$$\left(\frac{l_e}{d}\right)_{\text{max}} = \left(\frac{K_e l}{d}\right)_y = \frac{1 \times 10 \text{ ft} \times 12 \text{ in./ft}}{3.5 \text{ in.}} = 34.3$$

$$E' = E(C_M)(C_t)(C_T)(C_i) = 1,700,000(1.0)(1.0)(1.0)(1.0)$$

$$= 1,700,000 \text{ psi}$$

For visually graded sawn lumber,

$$K_{cE} = 0.3$$

$$c = 0.8$$

$$F_{cE} = \frac{K_{cE}E'}{(l_e/d)^2} = \frac{0.3(1,700,000)}{(34.3)^2} = 434 \text{ psi}$$

$$F_c^* = F_c(C_D)(C_M)(C_t)(C_F)(C_i)$$

$$= 1500(1.25)(1.0)(1.0)(1.1)(1.0) = 2062 \text{ psi}$$

$$\frac{F_{cE}}{F_c^*} = \frac{434}{2062} = 0.210$$

$$\frac{1 + F_{cE}/F_c^*}{2c} = \frac{1 + 0.210}{2(0.8)} = 0.756$$

$$C_P = \frac{1 + F_{cE}/F_c^*}{2c} - \sqrt{\left(\frac{1 + F_{cE}/F_c^*}{2c}\right)^2 - \frac{F_{cE}/F_c^*}{c}}$$

$$= 0.756 - \sqrt{(0.756)^2 - 0.210/0.8} = 0.200$$

$$F_c' = F_c(C_D)(C_M)(C_t)(C_F)(C_P)(C_i)$$

$$= 1500(1.25)(1.0)(1.0)(1.1)(0.200)(1.0)$$

$$= 412 \text{ psi}$$

Allow. $P = F_c'A = 0.412(19.25) = 7.94 \text{ k} < 15$ *NG*

Trial 2

Try 6×6 (P&T size category). Values from NDS Supplement Table 4D:

$$F_c = 1000 \text{ psi}$$

$$E = 1,600,000 \text{ psi}$$

Size factor for compression defaults to unity for all sizes except Dimension lumber ($C_F = 1.0$).

$$d_x = d_y = 5.5 \text{ in.}$$

$$A = 30.25 \text{ in.}^2$$

Determine column capacity and compare with the given design load.

$$\left(\frac{l_e}{d}\right)_{max} = \frac{1.0(10 \text{ ft} \times 12 \text{ in./ft})}{5.5 \text{ in.}} = 21.8$$

$$E' = E(C_M)(C_t)(C_i) = 1,600,000(1.0)(1.0)(1.0)$$

$$= 1,600,000 \text{ psi}$$

$$K_{cE} = 0.3$$

$$c = 0.8$$

$$F_{cE} = \frac{K_{cE}E'}{(l_e/d)^2} = \frac{0.3(1.600,000)}{(21.8)^2} = 1008 \text{ psi}$$

$$F_c^* = F_c(C_D)(C_M)(C_t)(C_F)(C_i)$$

$$= 1000(1.25)(1.0)(1.0)(1.0)(1.0) = 1250 \text{ psi}$$

$$\frac{F_{cE}}{F_c^*} = \frac{1008}{1250} = 0.807$$

$$\frac{1 + F_{cE}/F_c^*}{2c} = \frac{1 + 0.807}{2(0.8)} = 1.129$$

$$C_P = \frac{1 + F_{cE}/F_c^*}{2c} - \sqrt{\left(\frac{1 + F_{cE}/F_c^*}{2c}\right)^2 - \frac{F_{cE}/F_c^*}{c}}$$

$$= 1.129 - \sqrt{(1.129)^2 - 0.807/0.8} = 0.613$$

$$F_c' = F_c(C_D)(C_M)(C_t)(C_F)(C_P)(C_i)$$

$$= 1000(1.25)(1.0)(1.0)(1.0)(0.613)(1.0)$$

$$= 766 \text{ psi}$$

Allow. $P = F_c'A = 0.766(30.25) = 23.2 \text{ k} > 15$ *OK*

> *Use* 6 × 6 column No. 1 DF-L

In this example, the load combination of (D + L$_r$) was predetermined to be critical. For comparative purposes, the D-only allowable load (C_D = 0.9) for the 6 × 6 column is 19.8 k.

7.7 Design Problem: Capacity of a Glulam Column

This example determines the axial load capacity of a glulam column that is fabricated from a bending combination. See Example 7.7. Usually if a glulam member has an axial force only, an axial combination (rather than a bending combination) will be used. However, this example demonstrates the proper selection of E_x and E_y in the evaluation of a column. (Note that for an axial combination $E_x = E_y = E_{axial}$ and the proper selection of modulus of elasticity is automatic.)

To make the use of a glulam bending combination a practical problem, one might consider the given loading to be one possible load case. Another load case could involve the *axial* force *plus* a transverse *bending* load. This second load case would require a combined stress analysis (Sec. 7.12). Two different unbraced lengths are involved in this problem, and the designer must determine the slenderness ratio for the x and y axes.

EXAMPLE 7.7 Capacity of a Glulam Column

Determine the axial compression load capacity of the glulam column in Fig. 7.12. The column is a $6\frac{3}{4} \times 11$ 24F-1.7E Southern Pine glulam. It is used in an industrial plant where the MC will exceed 16 percent. Normal temperatures apply. Loads are (D + S). The designer is cautioned that the critical load combination (D + S) was predetermined. To be complete, the designer should also check D alone with the corresponding C_D for permanent load of 0.9.

Glulam properties are from the NDS Supplement.

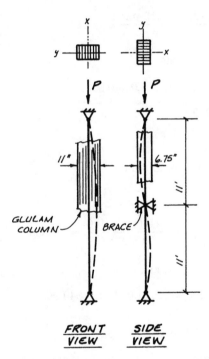

Figure 7.12 Front and side elevation views of glulam column showing different bracing conditions for column buckling about x and y axes. Also shown are section views above the respective elevations.

Tabulated stresses from NDS Supplement Table 5A:

When the moisture content of glulam is 16 percent or greater, a wet use factor C_M less than one is required and the need for pressure treatment should be considered.

$$F_c = 1000 \text{ psi} \qquad C_M = 0.73$$

$$E_x = 1{,}700{,}000 \text{ psi} \qquad C_M = 0.833$$

$$E_y = 1{,}300{,}000 \text{ psi} \qquad C_M = 0.833$$

$$E_{\text{axial}} = 1{,}400{,}000 \text{ psi} \qquad C_M = 0.833$$

NOTE: E_x and E_y are used in beam deflection calculations and for stability analysis, and E_{axial} is used for axial deformation computations.

In this problem there are different unbraced lengths about the x and y axes. Therefore, the effects of column buckling about both axes of the cross section are evaluated. In a member with E_x equal to E_y, this analysis would simply require the comparison of the slenderness ratios for the x and y axes [that is, $(l_e/d)_x$ and $(l_e/d)_y$]. The load capacity of the column would then be evaluated using the larger slenderness ratio $(l_e/d)_{max}$. However, in the current example there are different material properties for the x and y axes, and a full evaluation of the column stability factor is given for both axes.

Analyze Column Buckling About x Axis

$$\left(\frac{l_e}{d}\right)_x = \frac{1.0(22 \text{ ft} \times 12 \text{ in./ft})}{11 \text{ in.}} = 24.0$$

Use E_x to analyze buckling about the x axis.

$$E'_x = E_x(C_M)(C_t) = 1{,}700{,}000(0.833)(1.0)$$
$$= 1{,}416{,}000 \text{ psi}$$

Column stability factor x axis:

$$K_{cE} = 0.418 \qquad \text{for glulam}$$

$$c = 0.9$$

$$F_{cE} = \frac{K_{cE}E'_x}{[(l_e/d)_x]^2} = \frac{0.418(1{,}416{,}100)}{(24.0)^2} = 1027 \text{ psi}$$

$$F^*_c = F_c(C_D)(C_M)(C_t)(C_F)$$
$$= 1000(1.15)(0.73)(1.0)(1.0) = 840 \text{ psi}$$

$$\frac{F_{cE}}{F^*_c} = \frac{1027}{840} = 1.22$$

$$\frac{1 + F_{cE}/F^*_c}{2c} = \frac{1 + 1.22}{2(0.9)} = 1.235$$

For x axis $\quad C_P = \frac{1 + F_{cE}/F^*_c}{2c} - \sqrt{\left(\frac{1 + F_{cE}/F^*_c}{2c}\right)^2 - \frac{F_{cE}/F^*_c}{c}}$

$$= 1.235 - \sqrt{(1.235)^2 - 1.22/0.9} = \boxed{0.823}$$

Analyze Column Buckling About y Axis

$$\left(\frac{l_e}{d}\right) = \frac{1.0(11 \text{ ft} \times 12 \text{ in./ft})}{6.75 \text{ in.}} = 19.6$$

Use E_y to analyze buckling about the y axis.

$$E_y' = E_y(C_M)(C_t) = 1,300,000(0.833)(1.0)$$

$$= 1,082,900 \text{ psi}$$

$$F_{cE} = \frac{K_{cE}E_y'}{[(l_e/d)_y]^2} = \frac{0.418(1,082,900)}{(19.6)^2} = 1178 \text{ psi}$$

$$F_c^* = F_c(C_D)(C_M)(C_t)(C_F)$$

$$= 1000(1.15)(0.73)(1.0)(1.0) = 840 \text{ psi}$$

$$\frac{F_{cE}}{F_c^*} = \frac{1178}{840} = 1.402$$

$$\frac{1 + F_{cE}/F_c^*}{2c} = \frac{1 + 1.402}{2(0.9)} = 1.334$$

$$\text{For } y \text{ axis} \quad C_p = \frac{1 + F_{cE}/F_c^*}{2c} - \sqrt{\left(\frac{1 + F_{cE}/F_c^*}{2c}\right)^2 - \frac{F_{cE}/F_c^*}{c}}$$

$$= 1.334 - \sqrt{(1.334)^2 - 1.402/0.9} = \boxed{0.863}$$

The x axis produces the smaller value of the column stability factor, and the x axis is critical for column buckling.

$$F_c' = F_c(C_D)(C_M)(C_t)(C_F)(C_P)$$

$$= 1000(1.15)(0.73)(1.0)(1.0)(0.823)$$

$$= 691 \text{ psi}$$

$$\text{Allow. } P = F_c'A = 691(74.25) = 51.3 \text{ k}$$

$$\boxed{\text{Allow. } P = 51.3 \text{ k}}$$

7.8 Design Problem: Capacity of a Bearing Wall

An axial compressive load may be applied to a wood-frame wall. For example, an interior bearing wall may support the reactions of floor or roof joists (Fig. 3.4c). Exterior bearing walls also carry reactions from joists and rafters, but, in addition, exterior walls usually must be designed to carry lateral wind forces.

In Example 7.8 the vertical load capacity of a wood-frame wall is determined. Two main factors should be noted about this problem. The first relates to the column capacity of a stud in a wood-frame wall. Because sheathing is attached to the stud throughout its height, continuous lateral support is provided in the x direction. Therefore, the possibility of buckling about the weak y axis is prevented. The column capacity is evaluated by the slenderness ratio about the strong axis of the stud, $(l_e/d)_x$.

The second factor to consider is the bearing capacity of the top and bottom wall plates. It is possible that the vertical load capacity of a bearing wall may be governed by compression perpendicular to the grain on the wall plates rather than by the column capacity of the studs. This is typically not a problem with major columns in a building because steel bearing plates can be used to distribute the load perpendicular to the grain on supporting members. However, in a standard wood-frame wall, the stud bears *directly* on the horizontal wall plates.

EXAMPLE 7.8   Capacity of a Stud Wall

Determine the vertical load capacity of the stud shown in Fig. 7.13a. There is no bending. Express the allowable load in pounds per lineal foot of wall. Lumber is Standard-grade Hem-Fir. Load is (D + S). $C_M = 1.0$, $C_t = 1.0$, and $C_i = 1.0$.

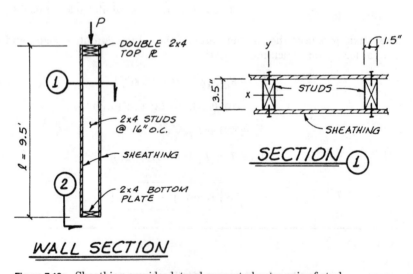

Figure 7.13a Sheathing provides lateral support about y axis of stud.

Wall studs are 2 × 4 (Dimension lumber size category). Values from NDS Supplement Table 4A:

$$F_c = 1300 \text{ psi}$$

$$C_F = 1.0 \quad \text{for compression}$$

$$E = 1,200,000 \text{ psi}$$

$$F_{c\perp} = 405 \text{ psi} \quad \text{bearing capacity of wall plate}$$

Column Capacity of Stud

Buckling about the weak axis of the stud is prevented by the sheathing, and the only slenderness ratio required is for the x axis.

$$\left(\frac{l_e}{d}\right)_x = \frac{1.0(9.5 \text{ ft} \times 12 \text{ in./ft})}{3.5 \text{ in.}} = 32.6$$

$$E' = E(C_M)(C_t)(C_T)(C_i) = 1,200,000(1.0)(1.0)(1.0)(1.0)$$

$$= 1,200,000 \text{ psi}$$

For visually graded sawn lumber

$$K_{cE} = 0.3$$

$$c = 0.8$$

$$F_{cE} = \frac{K_{cE}E'}{[(l_e/d)_x]^2} = \frac{0.3(1,200,000)}{(32.6)^2} = 339 \text{ psi}$$

$$F_c^* = F_c(C_D)(C_M)(C_t)(C_F)(C_i)$$

$$= 1300(1.15)(1.0)(1.0)(1.0)(1.0) = 1495 \text{ psi}$$

$$\frac{F_{cE}}{F_c^*} = \frac{339}{1495} = 0.227$$

$$\frac{1 + F_{cE}/F_c^*}{2c} = \frac{1 + 0.227}{2(0.8)} = 0.767$$

$$C_p = \frac{1 + F_{cE}/F_c^*}{2c} - \sqrt{\left(\frac{1 + F_{cE}/F_c^*}{2c}\right)^2 - \frac{F_{cE}/F_c^*}{c}}$$

$$= 0.767 - \sqrt{(0.767)^2 - 0.227/0.8} = 0.215$$

$$F_c' = F_c(C_D)(C_M)(C_t)(C_F)(C_p)(C_i)$$

$$= 1300(1.15)(1.0)(1.0)(1.0)(0.215)(1.0)$$

$$= 322 \text{ psi}$$

$$\text{Allow. } P = F_c'A = 322(5.25) = 1690 \text{ lb}$$

$$\text{Allow. } w = \frac{1690 \text{ lb}}{1.33 \text{ ft}} = \boxed{1270 \text{ lb/ft}}$$

Bearing Capacity of Wall Plates

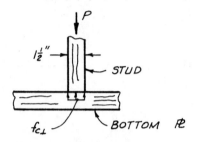

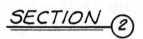

Figure 7.13*b* Bearing on bottom wall plate.

The conditions necessary to apply the bearing area factor are summarized in Fig. 6.16*b* (Sec. 6.8). Since the bottom plate will typically be composed of multiple pieces of sawn lumber placed end-to-end, it is possible that some studs will be located within 3 in. of the cut end of the wall plate. Therefore, the bearing area factor conservatively defaults to 1.0. Recall that C_D does not apply to $F_{c\perp}$.

$$F'_{c\perp} = F_{c\perp}(C_M)(C_t)(C_b)$$

$$= 405(1.0)(1.0)(1.0)$$

$$= 405 \text{ psi} > 322$$

$$F'_{c\perp} > F'_c$$

$\therefore$ column capacity governs over bearing perpendicular to the grain.

7.9 Built-up Columns

A *built-up* column is constructed from several parallel wood members which are nailed or bolted together to function as a composite column. These are to be distinguished from *spaced columns,* which have specially designed timber connectors to transfer shear between the separate parallel members.

The NDS includes criteria for designing spaced columns (NDS Sec. 15.2). Spaced columns can be used to increase the allowable load in compression members in heavy wood trusses. They are, however, relatively expensive to fabricate and are not often used in ordinary wood buildings. For this reason spaced columns are not covered in this text.

Built-up columns see wider use because they are fairly easy to fabricate. Their design is briefly covered here. The combination of several members in a built-up column results in a member with a larger cross-sectional dimension d and, correspondingly, a smaller slenderness ratio l_e/d.

With a smaller slenderness ratio, a larger allowable column stress can be used. Therefore, the allowable load on a built-up column is larger than the allowable load for the same members used individually. However, the fasteners connecting the members do not fully transfer the shear between the various pieces, and the capacity of a *built-up* column is less than the capacity of a *solid* sawn or glulam column of the same size and grade.

The capacity of a built-up column is determined by first calculating the column capacity of an equivalent solid column. This value is then reduced by an adjustment factor K_f that depends on whether the built-up column is fabricated with nails or bolts. This procedure is demonstrated in Example 7.9.

Recall that, in general, a column has two slenderness ratios: one for possible buckling about the x axis, and another for buckling about the y axis. In the typical problem, the allowable column stress is simply evaluated using the larger of the two slenderness ratios.

For a built-up column this may or may not be the controlling condition. Because the reduction factor K_f measures the effectiveness of the shear transfer between the individual laminations, the K_f factor applies only to the column slenderness ratio for the axis *parallel to the weak axis of individual laminations.* In other words, the column slenderness ratio parallel to the strong axis of the individual laminations does not require the K_f reduction. Depending on the relative magnitude of $(l_e/d)_x$ and $(l_e/d)_y$, the evaluation of *one* or possibly *two* allowable column stresses may be required.

The design of built-up columns is covered in NDS Sec. 15.3. The procedure applies to columns that are fabricated from two to five full-length parallel members that are nailed or bolted together. Details for the nailing or bolting of the members in order to qualify as a built-up column are given in the NDS.

Work has been done at the Forest Products Laboratory (Ref. 7.4) on the effect of built-up columns fabricated with members that are not continuous over the full length of the column. Information regarding design recommendations for nail-laminated posts with butt joints may be obtained from the FPL.

EXAMPLE 7.9   Strength of Built-up Column

Determine the allowable axial load on the built-up column in Fig. 7.14. Lumber is No. 1 DF-L. Column length is 13 ft-0 in. $C_D = 1.0$, $C_M = 1.0$, $C_t = 1.0$, and $C_i = 1.0$.

The 2 × 6s comprising the built-up column are in the Dimension lumber size category. Design values are obtained from NDS Supplement Table 4A:

$$F_c = 1500 \text{ psi}$$

$$C_F = 1.1 \qquad \text{for compression}$$

$$E = 1,700,000 \text{ psi}$$

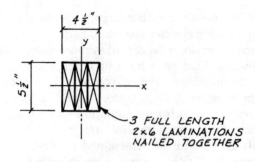

3 FULL LENGTH
2×6 LAMINATIONS
NAILED TOGETHER

Figure 7.14 Cross section of nailed built-up column. For nailing requirements see NDS Sec. 15.3.3.

BUILT-UP COLUMN

Column Capacity

$$\left(\frac{l_e}{d}\right)_x = \frac{1.0(13.0 \text{ ft} \times 12 \text{ in./ft})}{5.5} = 28.4$$

$$\left(\frac{l_e}{d}\right)_y = \frac{1.0(13.0 \text{ ft} \times 12 \text{ in./ft})}{4.5 \text{ in.}} = 34.7$$

$$\left(\frac{l_e}{d}\right)_{max} = \left(\frac{l_e}{d}\right)_y$$

The adjustment factor K_f applies to the allowable stress for the column axis that is parallel to the weak axis of the individual laminations. Therefore, K_f applies to the column stress based on $(l_e/d)_y$. In this problem the y axis is critical for both column buckling as well as the reduction for built-up columns, and only one allowable column stress needs to be evaluated.

$$E' = E(C_M)(C_t)(C_T)(C_i) = 1,700,000(1.0)(1.0)(1.0)(1.0)$$

$$= 1,700,000 \text{ psi}$$

For visually graded sawn lumber

$$K_{cE} = 0.3$$

$$c = 0.8$$

$$F_{cE} = \frac{K_{cE}E'}{(l_e/d)^2} = \frac{0.3(1,700,000)}{(34.7)^2} = 424 \text{ psi}$$

$$F_c^* = F_c(C_D)(C_M)(C_t)(C_F)(C_i)$$

$$= 1500(1.0)(1.0)(1.0)(1.1)(1.0) = 1650 \text{ psi}$$

$$\frac{F_{cE}}{F_c^*} = \frac{424}{1650} = 0.257$$

$$\frac{1 + F_{cE}/F_c^*}{2c} = \frac{1 + 0.257}{2(0.8)} = 0.756$$

For nailed built-up columns

$$K_f = 0.6$$

$$C_P = K_f \left[\frac{1 + F_{cE}/F_c^*}{2c} - \sqrt{\left(\frac{1 + F_{cE}/F_c^*}{2c} \right)^2 - \frac{F_{cE}/F_c^*}{c}} \right]$$

$$= 0.6[0.756 - \sqrt{(0.756)^2 - 0.257/0.8}]$$

$$= 0.6(0.256) = 0.153$$

$$F_c' = F_c(C_D)(C_M)(C_t)(C_F)(C_P)(C_i)$$

$$= 1500(1.0)(1.0)(1.0)(1.1)(0.153)(1.0)$$

$$= 252 \text{ psi}$$

Capacity of a nailed built-up (three 2×6's) column:

Allow. $P = F_c'A = (0.252 \text{ ksi})(3 \times 8.25 \text{ in.}^2)$

$$\boxed{\text{Allow. } P = 6.24 \text{ k}}$$

For the nailing requirements of a mechanically laminated built-up column, see NDS Sec. 15.3.3.

In Example 7.9, the y axis gave the maximum slenderness ratio, and the y axis also required the use of the reduction factor K_f. In such a problem, only one allowable column stress needs to be evaluated. However, another situation could require the evaluation of a second allowable column stress. For example, if an additional lamination is added to the column in Fig. 7.14, the maximum slenderness ratio would become $(l_e/d)_x$, and an allowable column stress would be determined using $(l_e/d)_x$ *without* K_f. Another allowable stress would be evaluated using $(l_e/d)_y$ *with* K_f. The smaller of the two allowable stresses would then govern the capacity of the built-up member.

7.10 Combined Bending and Tension

When a bending moment occurs simultaneously with an axial tension force, the effects of combined stresses must be taken into account. The distribution of axial tensile stresses and bending stresses can be plotted over the depth of the cross section of a member. See Example 7.10.

From the plots of combined stress (Fig. 7.15a) it can be seen that on one side of the member the axial tensile stresses and the bending tensile stresses add. On the opposite face, the axial tensile stresses and bending compressive stresses cancel. Depending on the magnitude of the stresses involved, the resultant stress on this face can be *either* tension or compression.

Therefore, the capacity of a wood member with this type of combined loading can be governed by either a combined *tension* criterion or a net *compression* criterion. These criteria are given in NDS Sec. 3.9.1, *Bending and Axial Tension*.

Combined axial tension and bending tension

First, the combined *tensile* stresses are analyzed in an *interaction equation*. In this case the interaction equation is a straight-line expression (Fig. 7.15b) which is made up of two terms known as *stress ratios*. The first term measures the effects of axial tension, and the second term evaluates the effects of bending. In each case the actual stress is divided by the corresponding allowable stress. The allowable stresses in the denominators are determined in the usual way with one exception. Because the actual bending stress is the *bending tensile* stress, the allowable bending stress does not include the lateral stability factor C_L. In other words, F_b^* is F_b' determined with C_L set equal to unity.

It may be convenient to think of the stress ratios in the interaction equation as percentages or fractions of total member capacity. For example, the ratio of actual tension stress to allowable tension stress, f_t/F_t', can be viewed as the fraction of total member capacity that is used to resist axial tension. The ratio of actual bending stress to allowable bending stress f_b/F_b^* then represents the fraction of total member capacity used to resist bending. The sum of these fractions must be less than the total member capacity, 1.0.

Net compressive stress

Second, the combined stresses on the opposite face of the member are analyzed. If the sense of the combined stress on this face is tension, no additional work is required. However, if the combined stress at this point is *compressive*, a bending analysis is required. The allowable bending stress in the denominator of the second combined stress check must reflect the lateral stability of the compression side of the member. This is done by including the lateral stability factor C_L in evaluating F_b'. Refer to Sec. 6.3 for information on C_L.

EXAMPLE 7.10   Criteria for Combined Bending and Tension

The axial tensile stress and bending stress distributions are shown in Fig. 7.15a. If the individual stresses are added algebraically, one of two possible combined stress distributions will result. If the axial tensile stress is larger than the bending stress, a *trapezoidal* combined stress diagram results in tension everywhere throughout the depth of the member (combined stress diagram 1). If the axial tensile stress is smaller than the bending stress, the resultant combined stress diagram is *triangular* (combined stress diagram 2).

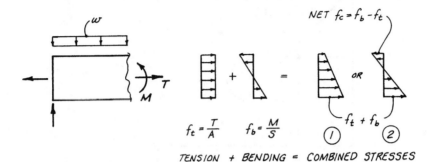

Figure 7.15a Combined bending and axial tension stresses.

Theoretically the following stresses in Fig. 7.15a are to be analyzed:

1. *Combined axial tension and bending tension stress.* This is done by using the straight-line interaction equation described below. These combined stresses are shown at the bottom face of the member in stress diagrams 1 and 2.

2. *Net compressive stress.* This stress is shown on the top surface of the member in stress diagram 2.

Combined Axial Tension and Bending Tension

The basic straight-line interaction equation is used for combined axial tensile and bending tensile stresses (NDS equation 3.9-1). The two stress ratios define a point on the graph in Fig. 7.15b. If the point lies on or below the line representing 100 percent of member strength, the interaction equation is satisfied.

INTERACTION EQUATION

$$\frac{f_t}{F_t'} + \frac{f_b}{F_b^*} \leq 1.0$$

where f_t = actual (computed) tensile stress parallel to grain
$\quad\quad = T/A$
$\quad F_t'$ = allowable tensile stress (Sec. 7.2)
$\quad\quad = F_t(C_D)(C_M)(C_t)(C_F)(C_i)$ for sawn lumber
$\quad\quad = F_t(C_D)(C_M)(C_t)$ for glulam
$\quad f_b$ = actual (computed) bending tensile stress. For usual case of bending about x axis, this is f_{bx}.
$\quad\quad = M/S$
$\quad F_b^*$ = F_b' allowable bending *tensile* stress without the adjustment for lateral stability. For the usual case of bending about the x axis of a rectangular cross section, the allowable bending stress is
$\quad\quad = F_b(C_D)(C_M)(C_t)(C_F)(C_r)(C_i)$ for sawn lumber
$\quad\quad = F_b(C_D)(C_M)(C_t)(C_V)$ for glulam

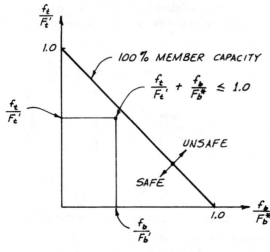

BASIC STRAIGHT LINE
INTERACTION FORMULA

Figure 7.15b Interaction curve for axial tension plus bending tension.

Net Compressive Stress

When the bending compressive stress exceeds the axial tensile stress, the following stability check is given by NDS equation 3.9-2.

$$\frac{\text{Net } f_c}{F'_b} = \frac{f_b - f_t}{F_b^{**}} \leq 1.0$$

where net f_c = net compressive stress

f_b = actual (computed) bending *compressive* stress. For usual case of bending about x axis, this is f_{bx}.

= M / S

f_t = actual (computed) axial tensile stress parallel to grain

= T / A

F_b^{**} = F'_b allowable bending *compressive* stress (Sec. 6.3). The beam stability factor C_L applies, but the volume factor C_V does not. For usual case of bending about x axis of rectangular cross section, allowable bending stress is

= $F_b(C_D)(C_M)(C_t)(C_L)(C_F)(C_r)(C_i)$ for sawn lumber

= $F_b(C_D)(C_M)(C_t)(C_L)$ for glulam

The designer should use a certain degree of caution in applying the criterion for the *net compressive stress*. Often, combined stresses of this nature are the result of different loadings, and the maximum bending compressive stress may occur *with* or *without* the axial tensile stress.

For example, the bottom chord of the truss in Fig. 7.16a will always have the dead load moment present regardless of the loads applied to the top chord. Thus f_b is a constant in this example. However, the tension force in the member varies depending on the load applied to the top chord. The tensile stress f_t will be small if dead load alone is considered, and it will be much larger under dead load plus snow load. To properly check the net compressive stress, the designer must determine the *minimum* f_t that will occur simultaneously with the bending stress f_b.

On the other hand, a simple and conservative approach for evaluating bending compressive stresses is to ignore the reduction in bending stress provided by the axial tensile stress. In this case the compressive stress ratio becomes

$$\frac{\text{Gross } f_c}{F_b'} = \frac{f_b}{F_b^{**}} \le 1.0$$

This check on the *gross* bending compressive stress can be rewritten as

$$f_b \le F_b' = F_b^{**}$$

where $F_b' = F_b^{**}$ is the allowable bending stress considering the effects of lateral stability and other applicable adjustments. This more conservative approach is used for the examples in this book. Under certain circumstances, however, the designer may wish to consider the *net* compressive stress. For example, the designer may be overly conservative to use the gross bending compressive stress rather than the net compressive stress when tension and bending are caused by the same load, such as in the design of wall studs for wind-induced bending and uplift in high wind zones.

It should be noted that NDS Sec. 3.9.1 introduces special notation for F_b' for use in the combined bending plus axial tension interaction equations. The symbols F_b^* and F_b^{**} are allowable bending stress values obtained with certain adjustment factors deleted. These are noted in Example 7.10.

The design expressions given in Example 7.10 are adequate for most combined bending and axial tension problems. However, as wood structures become more highly engineered, there may be the need to handle problems involving axial tension plus bending about both the x and y axes. In this case the expanded criteria in Example 7.11 may be used.

EXAMPLE 7.11   Generalized Criteria for Combined Bending and Tension

The problem of biaxial bending plus axial tension has not been studied extensively. However, the following interaction equations are extensions of the NDS criteria for the general axial tension plus bending problem.

Combined Axial Tension and Bending Tension

$$\frac{f_t}{F'_t} + \frac{f_{bx}}{F'_{bx}} + \frac{f_{by}}{F'_{by}} \leq 1.0$$

The three terms all have the same sign. The allowable stresses include applicable adjustments (the adjustment for lateral stability C_L does not apply to bending tension).

Net Compressive Stress

For the check on *net* bending compressive stress the tension term is negative:

$$-\frac{f_t}{F'_t} + \frac{f_{bx}}{F'_{bx}} + \frac{f_{by}}{F'_{by}} \leq 1.0$$

Depending on the magnitude of the stresses involved, or reasons of simplicity, the designer may prefer to omit the negative term in the expression.

An even more conservative approach is to apply the general interaction formula for the net compressive stress in Example 7.14 (Sec. 7.12) with the axial component of the expression set equal to zero. This provides a more conservative biaxial bending (i.e., bending about the x and y axes) interaction formula. The allowable bending stress terms in the biaxial bending interaction equation would include all appropriate adjustment factors. Because the focus is on compression, the effect of lateral torsional buckling is to be taken into account with the beam stability factor C_L, but the volume factor C_V does not apply.

***

7.11   Design Problem: Combines Bending and Tension

The truss in Example 7.12 is similar to the truss in Example 7.2. The difference is that in the current example, an additional load is applied to the bottom chord. This load is uniformly distributed and represents the weight of a ceiling supported by the bottom chord of the truss.

The first part of the example deals with the calculation of the axial force in member *AC*. In order to analyze a truss using the method of joints it is necessary for the loads to be resolved into joint loads. The tributary width to the three joints along the top chord is 7.5 ft, and the tributary width to the joint at the midspan of the truss on the bottom chord is 15 ft. The remaining loads (both top and bottom chord loads) are tributary to the joints at each support.

The design of a combined stress member is a trial-and-error procedure. In this example, a 2 × 8 bottom chord is the initial trial, and it proves satisfactory. *Independent* checks on the tension and bending stresses are first completed. The independent check on the bending stress automatically satisfies the check on the *gross* bending compression discussed in Sec. 7.10. Finally, the combined effects of tension and bending are evaluated.

It should be noted that the load duration factor used for the independent check for axial tension is $C_D = 1.15$ for combined (D + S). Dead load plus snow causes the axial force of 4.44 k. The independent check of bending uses

$C_D = 0.9$ for dead load, because only the dead load of the ceiling causes the bending moment of 10.8 in.-k.

In the combined stress check, however, $C_D = 1.15$ applies to *both* the axial and the bending portions of the interaction formula. Recall that the C_D to be used in checking stresses caused by a *combination* of loads is the one associated with the shortest-duration load in the combination. For combined stresses, then, the same C_D applies to both terms.

EXAMPLE 7.12   Combined Bending and Tension

Design the lower chord of the truss in Fig. 7.16a. Use No. 1 and Better Hem-Fir. MC $\leq$ 19 percent, and normal temperature conditions apply. Connections will be made with a single row of ¾-in.-diameter bolts. Connections are assumed to be pinned. Trusses are 4 ft-0 in. o.c. Loads are applied to both the top and bottom chords. Assume that lateral buckling is prevented by the ceiling. Allowable stresses are from the NDS Supplement.

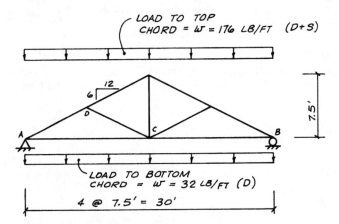

Figure 7.16a Loading diagram for truss. The uniformly distributed loads between the truss joints cause bending stresses in top and bottom chords, in addition to axial truss forces. Bottom chord has combined bending and axial tension.

Loads

TOP CHORD:

$$D = 14 \text{ psf} \quad \text{(horizontal plane)}$$

$$S = 30 \text{ psf} \quad \text{(reduced snow load based on roof slope)}$$

$$TL = 44 \text{ psf}$$

$$w_{TL} = 44 \times 4 = 176 \text{ lb/ft to truss}$$

Load to joint for truss analysis:

$$P_T = 176 \times 7.5 = 1320 \text{ lb/joint}$$

BOTTOM CHORD:

$$\text{Ceiling } D = 8 \text{ psf}$$

$$w_D = 8 \times 4 = 32 \text{ lb/ft}$$

Load to joint for truss analysis:

$$P_B = 32 \times 15 = 480 \text{ lb/joint}$$

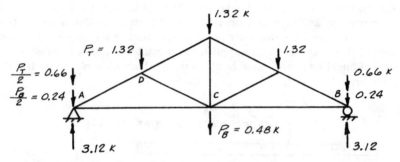

Figure 7.16b Loading diagram for truss. The distributed loads to the top and bottom chords are converted to concentrated joint forces for conventional truss analysis.

Force in lower chord (method of joints):

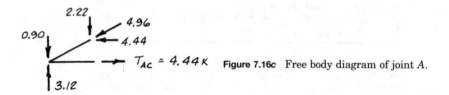

Figure 7.16c Free body diagram of joint A.

Load diagram for tension chord AC:

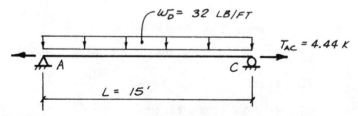

Figure 7.16d Loading diagram for member AC. The tension force is obtained from the truss analysis, and the bending moment is the result of the transverse load applied between joints A and C.

Member Design

Try 2 × 8 No. 1 & Btr Hem-Fir (Dimension lumber size category).
Values from NDS Supplement Table 4A:

$$F_b = 1100 \text{ psi}$$

$$F_t = 725 \text{ psi}$$

Size factors:

$$C_F = 1.2 \text{ for bending}$$

$$C_F = 1.2 \text{ for tension}$$

Section properties:

$$A_g = 10.875 \text{ in.}^2$$

$$S = 13.14 \text{ in.}^3$$

AXIAL TENSION:

1. Check axial tension at the net section. Because this truss is assumed to have pinned connections, the bending moment is theoretically zero at this point. Assume the hole diameter is $\frac{1}{16}$ in. larger than the bolt diameter (for stress calculations only).

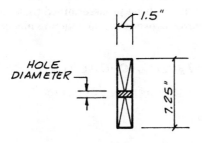

Figure 7.16e Net section for tension member.

$$A_n = 1.5[7.25 - (0.75 + 0.0625)] = 9.66 \text{ in.}^2$$

$$f_t = \frac{T}{A_n} = \frac{4440}{9.66} = 460 \text{ psi}$$

$$F_t' = F_t(C_D)(C_M)(C_t)(C_F)(C_i)$$

$$= 725(1.15)(1.0)(1.0)(1.2)(1.0) = 1000 \text{ psi}$$

$$1000 \text{ psi} > 460 \text{ psi} \quad OK$$

2. Determine tension stress at the point of maximum bending stress (midspan) for use in the interaction formula.

$$f_t = \frac{T}{A_g} = \frac{4440}{10.875} = 408 \text{ psi} < 1000 \quad OK$$

BENDING:

For a simple beam with a uniform load,

$$M = \frac{wL^2}{8} = \frac{32(15)^2}{8} = 900 \text{ ft-lb} = 10,800 \text{ in.-lb}$$

$$f_b = \frac{M}{S} = \frac{10,800}{13.14} = 822 \text{ psi}$$

The problem statement indicates that lateral buckling is prevented. In addition, the truss spacing exceeds the limit for repetitive members. Therefore, the beam stability factor C_L and the repetitive-member factor C_r are both 1.0.

The bending stress of 822 psi is caused by dead load alone. Therefore, use $C_D = 0.9$ for an independent check on bending. Later use $C_D = 1.15$ for the combined stress check in the interaction formula.

$$F_b' = F_b(C_D)(C_M)(C_t)(C_L)(C_F)(C_r)(C_i)$$

$$= 1100(0.9)(1.0)(1.0)(1.0)(1.2)(1.0)(1.0)$$

$$= 1188 \text{ psi} > 822 \quad OK$$

In terms of NDS notation, this value of F_b' is F_b^{**}.

COMBINED STRESSES:

1. Axial tension plus bending. Two load cases should be considered for axial tension plus bending: D-only and D + S. It has been predetermined that the D + S load combination with a $C_D = 1.15$ governs over dead load acting alone with a $C_D = 0.9$.

$$F_b^* = F_b' = F_b(C_D)(C_M)(C_t)(C_F)(C_r)(C_i)$$

$$= 1100(1.15)(1.0)(1.0)(1.2)(1.0)(1.0)$$

$$= 1518 \text{ psi}$$

$$\frac{f_t}{F_t'} + \frac{f_{bx}}{F_{bx}'} = \frac{408}{1000} + \frac{822}{1518} = 0.95$$

$$0.95 < 1.0 \quad OK$$

NOTE: It can be determined that $f_t = 160$ psi results from D-only and $F_t' = 783$ psi with $C_D = 0.9$. The f_b and F_b' values were determined previously for D-only bending. The D-only axial tension plus bending interaction results in a value of 0.90 vs. 0.95 for D + S. If the specified snow load had been slightly smaller, then D acting alone would have governed the design of the lower chord.

2. Net bending compressive stress. The *gross* bending compressive stress was shown to be not critical in the independent check on bending ($f_b = 822$ psi $< F_b' = 1188$). Therefore, the *net* bending compressive stress is automatically *OK*. The simpler, more conservative check on compression is recommended in this book.

The 2 × 8 No. 1 & Btr Hem-Fir member is seen to pass the combined stress check. The combined stress ratio of 0.95 indicates that the member is roughly overdesigned by 5 percent (1.0 corresponds to full member capacity or 100 percent of member strength).

$$(1.0 - 0.95)100 = 5\% \text{ overdesign}$$

Because of the inaccuracy involved in estimating design loads and because of variations in material properties, a calculated overstress of 1 or 2 percent (i.e., combined stress ratio of 1.01 or 1.02) is considered by many designers to still be within the "spirit" of the design specifications. Judgments of this nature must be made individually with knowledge of the factors relating to a particular problem. However, in this problem the combined stress ratio is less than 1.0, and the trial member size is acceptable.

> *Use* 2 × 8 No. 1 & Btr Hem-Fir

NOTE: The simplified analysis used in this example applies only to trusses with pinned joints. For a truss connected with toothed metal plate connectors, a design approach should be used which takes the continuity of the joints into account (Ref. 7.7).

7.12 Combined Bending and Compression

Structural members that are stressed simultaneously in bending and compression are known as *beam-columns*. These members occur frequently in wood buildings, and the designer should have the ability to handle these types of problems. In order to do this, it is first necessary to have a working knowledge of laterally unsupported beams (Sec. 6.3) and axially loaded columns (Secs. 7.4 and 7.5). The interaction formulas presented in this section can then be used to handle the combination of these stresses.

The straight-line interaction equation was introduced in Fig. 7.15*b* (Sec. 7.10) for combined bending and axial tension. At one time a similar straight-line equation was also used for the analysis of beam-columns. More recent editions of the NDS used a modified version of the basic equation.

There are many variables that affect the strength of a beam-column. The NDS interaction equation for the analysis of beam-columns was developed by Zahn (Ref. 7.8). It represents a unified treatment of

1. Column buckling

2. Lateral torsional buckling of beams

3. Beam-column interaction

The Ylinen buckling formula was introduced in Sec. 7.4 for column buckling and in Sec. 6.3 for the lateral buckling of beams. The allowable stresses F'_c and F'_{bx} determined in accordance with these previous sections are used in the Zahn interaction formula to account for the first two items. The added considerations for the *simultaneous* application of beam and column loading can be described as beam-column interaction. These factors are addressed in this section.

When a bending moment occurs simultaneously with an axial compressive force, a more critical combined stress problem exists in comparison with combined bending and tension. In a beam-column, an additional bending stress is created which is known, as the $P\text{-}\Delta$ *effect*. The $P\text{-}\Delta$ effect can be described in this way. First consider a member without an axial load. The bending moment developed by the *transverse* loading causes a deflection Δ. When the axial force P is then applied to the member, an *additional* bending moment of $P \times \Delta$ is generated. See Example 7.13. The $P\text{-}\Delta$ moment is known as a *second-order effect* because the added bending stress is not calculated directly. Instead, it is taken into account by *amplifying* the computed bending stress in the interaction equation.

The most common beam-column problem involves axial compression combined with a bending moment about the *strong* axis of the cross section. In this case, the actual bending stress f_{bx} is multiplied by an amplification factor that reflects the magnitude of the load P and the deflection Δ. This concept should be familiar to designers who also do structural steel design. The amplification factor in the NDS is similar to the one used for beam-columns in the AISC steel specification (Ref. 7.3).

The amplification factor is a number greater than 1.0 given by the following expression:

$$\text{Amplification factor for } f_{bx} = \left(\frac{1}{1 - f_c/F_{cEx}} \right)$$

This amplification factor is made up of two terms that measure the $P\text{-}\Delta$ effect for a bending moment about the strong axis (x-axis). The intent is to have the amplification factor increase as

1. Axial force P increases.

2. Deflection Δ due to bending about the x axis increases.

Obviously the compressive stress $f_c = P/A$ increases as the load P increases. As f_c becomes larger, the amplification factor will increase.

The increase in the amplification factor due to an increase in Δ may not be quite as clear. The increase for Δ is accomplished by the term F_{cE}. F_{cE} is defined as the value obtained from the Euler buckling stress formula evaluated using the column slenderness ratio for the axis about which the bending moment is applied. Thus, if the transverse loads cause a moment about the x axis, the *slenderness ratio about the x axis* is used to determine F_{cE}. The notation used in this book for this quantity is F_{cEx}.

Figure 7.6a (Sec. 7.4) shows both the Ylinen column equation and the Euler equation. For purposes of beam-column analysis, it should be understood that the allowable column stress F'_c is defined by the Ylinen formula, but the amplification factor for P-Δ makes use of the Euler formula. For use in the amplification factor, the value given by the expression for F_{cEx} is applied over the entire range of slenderness ratios. In other words, F_{cEx} goes to ∞ as the slenderness ratio becomes small, and F_{cEx} approaches 0 as $(l_e/d)_x$ becomes large.

The logic in using F_{cEx} in the amplification factor for P-Δ is that the deflection will be large for members with a large slenderness ratio. Likewise, Δ will be small as the slenderness ratio decreases. Thus, F_{cEx} produces the desired effect on the amplification factor.

It is necessary for the designer to clearly understand the reasoning behind the amplification factor. In a general problem there are two slenderness ratios: one for the x axis $(l_e/d)_x$ and one for the y axis $(l_e/d)_y$. In order to analyze combined stresses, the following convention should be applied:

1. Column buckling is governed by the larger slenderness ratio, $(l_e/d)_x$ or $(l_e/d)_y$, and the allowable column stress F'_c is given by the Ylinen formula.

2. When the bending moment is about the x axis, the value of F_{cE} for use in the amplification factor is to be based on $(l_e/d)_x$.

EXAMPLE 7.13   Interaction Equation for Beam-Column with Moment about x Axis

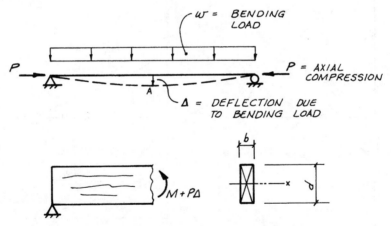

Figure 7.17a Deflected shape of beam showing P-Δ moment. The computed bending stress f_b is based on the moment M from the moment diagram. The moment diagram considers the effects of the transverse load w, but does not include the secondary moment $P \times \Delta$. The P-Δ effect is taken into account by amplifying the computed bending stress f_b.

By cutting the beam-column in Fig. 7.17a and summing moments at point A, it can be seen that a moment of $P \times \Delta$ is created which *adds* to the moment M caused by the transverse load. In a beam with an axial tension force, this moment *subtracts* from the normal bending moment, and may be conservatively ignored.

The general interaction formula (Eq. 3.9-3 in the NDS) reduces to the following form for the common case of an axial compressive force combined with a bending moment about the x axis:

$$\left(\frac{f_c}{F_c'}\right)^2 + \left(\frac{1}{1 - f_c/F_{cEx}}\right)\frac{f_{bx}}{F_{bx}'} \leq 1.0$$

where f_c = actual (computed) compressive stress

= P/A

F_c' = allowable column stress as given by Ylinen formula (Secs. 7.4 and 7.5). Consider critical slenderness ratio $(l_e/d)_x$ or $(l_e/d)_y$. The critical slenderness ratio produces the smaller value of F_c'.

= $F_c(C_D)(C_M)(C_t)(C_F)(C_P)(C_i)$

f_{bx} = actual (computed) blending stress about x axis

= M_x/S_x

F_{bx}' = allowable bending stress about x axis considering effects of lateral torsional buckling (Sec. 6.3)

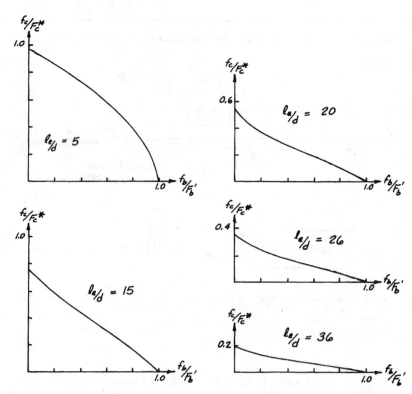

Figure 7.17b Five interaction curves for beam-columns with different slenderness ratios.

For sawn lumber:

$$F'_{bx} = F_b(C_D)(C_M)(C_t)(C_F)(C_L)(C_i)$$

For glulam the smaller of the following bending stress values should be used:

$$F'_{bx} = F_b(C_D)(C_M)(C_t)(C_L)$$
$$F'_{bx} = F_b(C_D)(C_M)(C_t)(C_V)$$

F_{cEx} = Euler-based elastic buckling stress. Because transverse loads cause a bending moment about the x axis, F_{cE} is based on the slenderness ratio for the x axis, i.e., $(l_e/d)_x$.

$$= \frac{K_{cE}E'_x}{[(l_e/d)_x]^2}$$

The interaction formula for a beam-column takes into account a number of factors, including column buckling, lateral torsional buckling, and the P-Δ effect. It is difficult to show on a graph, or even a series of graphs, all of the different variables in a beam-column problem. However, the interaction plots in Fig. 7.17b are helpful in visualizing some of the patterns. The graphs are representative only, and results vary with specific problems. The ordinate on the vertical axis shows the effect of different slenderness ratios in that the ratio of f_c/F'_c decreases as the slenderness ratio increases.

The interaction formula in Example 7.13 covers the common problem of axial compression with a bending moment about the strong axis. This can be viewed as a special case of the Zahn general interaction equation. The general formula has a third term which provides for consideration of a bending moment about the y axis. See Example 7.14.

The concept of a general interaction equation can be carried one step further. The problems considered up to this point have involved axial compression plus bending caused by transverse loads. Although this is a very comprehensive design expression, Zahn's expanded equation permits the compressive force in the column to be applied with an eccentricity. Thus in the general case, the bending moments about the x and y axes can be the result of transverse bending loads *and* an eccentrically applied column force.

The general loading condition is summarized as follows:

1. Compressive force in member = P

2. Bending moment about x axis
   a. Moment due to transverse loads = M_x
   b. Moment due to eccentricity about x axis = $P \times e_x$

3. Bending moment about y axis
   a. Moment due to transverse loads = M_y
   b. Moment due to eccentricity about y axis = $P \times e_y$

A distinction is made between the moments caused by transverse loads and the moments from an eccentrically applied column force. This distinction is necessary because the bending stresses that develop as a result of the eccentric load are subject to an *additional* P-Δ amplification factor. The general interaction formula may appear overly complicated at first glance, but taken term by term, it is straightforward and logical. The reader should keep in mind that greatly simplified versions of the interaction formula apply to most practical loading conditions (e.g., the version in Example 7.13). The simplified expression is obtained by setting the appropriate stress terms in the general interaction equation equal to zero.

EXAMPLE 7.14   General Interaction Formula for Combined Compression and Bending

The member in Fig. 7.18a has an *axial* compression load, a transverse load causing a moment about the x axis, and a transverse load causing a moment about the y axis. The following interaction formula from the NDS (NDS Sec. 3.9.2) is used to check this member:

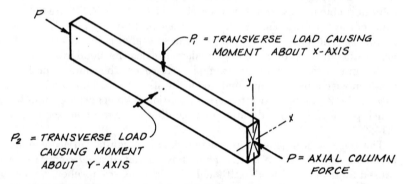

Figure 7.18a Axial compression plus bending about x and y axes.

$$\left(\frac{f_c}{F'_c}\right)^2 + \frac{f_{bx}}{F'_{bx}(1 - f_c/F_{cEx})} + \frac{f_{by}}{F'_{by}[1 - f_c/F_{cEy} - (f_{bx}/F_{bE})^2]} \leq 1.0$$

where f_{by} = actual (computed) bending stress about y axis
   = M_y/S_y
 F'_{by} = allowable bending stress about y axis (Sec. 6.4)
 F_{cEy} = Euler elastic buckling stress based on the slenderness ratio for the y axis $(l_e/d)_y$
   = $\dfrac{K_{cE}E'_y}{[(l_e/d)_y]^2}$
 F_{bE} = elastic buckling stress considering lateral torsional buckling of beam; F_{bE} based on the beam slenderness factor R_B (Sec. 6.3)
   = $\dfrac{K_{bE}E'_y}{R_B^2}$

Other terms are as previously defined.

If one or more of the loads in Fig. 7.18a do not exist, the corresponding terms in the interaction formula are set equal to zero. For example, if there is no load causing a moment about the y axis, f_{by} is zero and the third term in the interaction formula drops out. With $f_{by} = 0$, the interaction formula reduces to the form given in Example 7.13.

On the other hand, if the axial column force does not exist, f_c becomes zero, and the general formula becomes an interaction formula for biaxial bending (i.e., simultaneous bending about the x and y axes).

In the interaction problems considered thus far, the bending stresses have been caused only by transverse applied loads. In some cases, bending stresses may be the result of an eccentric column force. See Fig. 7.18b. The development of a generalized interaction formula for beam-columns with transverse *and* eccentric bending stresses is shown below. In the general formula, the amplification for P-Δ effect introduced in Example 7.13 is applied the same way. However, the bending stresses caused by the *eccentric* column force are subject to an *additional* P-Δ amplification.

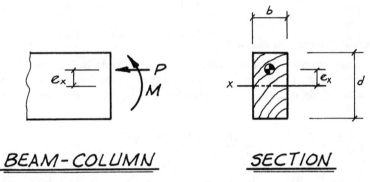

BEAM-COLUMN   SECTION

Figure 7.18b Bending moment M due to transverse loads plus eccentric column force $P \times e$. Note that the eccentric moment $P \times e$ is a computed (first-order) bending moment, and it should not be confused with the second-order P-Δ moment. The eccentric moment in Fig. 7.18b causes a bending stress about the x axis.

Cross-Sectional Properties

$$A = bd$$

$$S = \frac{bd^2}{6}$$

Bending Stresses

$$\text{Total moment} = \text{transverse load } M + \text{eccentric load } M$$
$$= M + Pe$$

Bending stress due to transverse bending loads:

$$f_b = \frac{M}{S}$$

Bending stress due to eccentrically applied column force:

Computed stress:

$$\frac{Pe}{S} = \frac{Pe}{bd^2/6} = f_c\left(\frac{6e}{d}\right)$$

Amplified eccentric bending stress:

$$f_c\left(\frac{6e}{d}\right) \times \text{(amplification factor)} = f_c\left(\frac{6e}{d}\right)\left[1 + 0.234\left(\frac{f_c}{F_{cE}}\right)\right]$$

Bending stress due to transverse loads + amplified eccentric bending stress

$$= f_b + f_c\left(\frac{6e}{d}\right)\left[1 + 0.234\left(\frac{f_c}{F_{cE}}\right)\right]$$

General Interaction Formula

The interaction formula is expanded here to include the effects of eccentric bending stresses. Subscripts are added to the eccentric terms to indicate the axis about which the eccentricity occurs. This general form of the interaction formula is given in NDS Sec. 15.4.

$$\left(\frac{f_c}{F'_c}\right)^2 + \frac{f_{bx} + f_c(6e_x/d_x)[1 + 0.234(f_c/F_{cEx})]}{F'_{bx}(1 - f_c/F_{cEx})}$$

$$+ \frac{f_{by} + f_c(6e_y/d_y)\left[1 + 0.234(f_c/F_{cEy}) + 0.234\left[\dfrac{f_{bx} + f_c(6e_x/d_x)}{F_{bE}}\right]^2\right]}{F'_{by}\left[1 - f_c/F_{cEy} - \left[\dfrac{f_{bx} + f_c(6e_x/d_x)}{F_{bE}}\right]^2\right]} \leq 1.0$$

In the general interaction formula in Example 7.14, it is assumed that the eccentric load is applied at the *end* of the column. In some cases an eccentric compression force may be applied through a *side bracket* at some point between the ends of the column. The reader is referred to NDS Sec. 15.4.2 for an approximate method of handling beam-columns with side brackets.

Several examples of beam-column problems are given in the remaining portion of this chapter.

7.13   Design Problem: Beam-Column

In the first example, the top chord of the truss analyzed in Example 7.12 is considered. The top chord is subjected to bending loads caused by (D + S) being applied along the member. The top chord is also subjected to axial compression, which is obtained from a truss analysis using tributary loads to the truss joints.

A 2 × 8 is selected as the trial size. In a beam-column problem, it is often convenient to divide the stress calculations into three subproblems. In this approach, somewhat independent checks on axial, bending, and combined stresses are performed. See Example 7.15.

The first check is on *axial* stresses. Because the top chord of the truss is attached directly to the roof sheathing, lateral buckling about the weak axis of the cross section is prevented. Bracing for the strong axis is provided at the truss joints by the members that frame into the top chord.

The compressive stress is calculated at two different locations along the length of the member. First, the allowable column stress F'_c adjusted for column stability is checked away from the joints, using the gross area in the calculation of f_c. The second calculation involves the stress at the net section at a joint compared with the allowable stress F'_c without the reduction for stability.

In the *bending* stress calculation, the moment is determined by use of the horizontal span of the top chord and the load on a horizontal plane. It was shown in Example 2.5 (Sec. 2.5) that the moment obtained using the *horizontal plane method* is the same as the moment obtained using the inclined span length and the normal component of the load.

The final step is the analysis of *combined* stresses. The top chord has axial compression plus bending about the strong axis, and the simple interaction formula from Example 7.13 applies. The trial member is found to be acceptable.

EXAMPLE 7.15   Beam-Column Design

Design the top chord of the truss shown in Fig. 7.16a in Example 7.12. The axial force and bending loads are reproduced in Fig. 7.19a. Use No. 1 Southern Pine. MC ≤ 19 percent, and normal temperatures apply.

Connections will be made with a single row of ¾-in. diameter bolts. The top chord is stayed laterally throughout its length by the roof sheathing. Trusses are 4 ft.-0 in. o.c. Allowable stresses and section properties are to be obtained from the NDS Supplement.

NOTE: Two load combinations must be considered in this design: D-only and D + S. It has been predetermined that the D + S combination controls the design and only those calculations associated with this combination are included in the example.

Try 2 × 8. Tabulated stresses in NDS Supplement Table 4B for Southern Pine in the Dimension lumber size category are size-specific:

$$F_c = 1650 \text{ psi}$$

$$F_b = 1500 \text{ psi}$$

$$E = 1,700,000 \text{ psi}$$

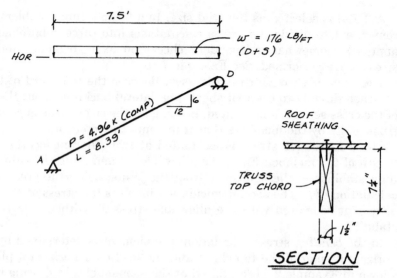

Figure 7.19a Loading diagram for top chord of truss. Section view shows lateral support by roof sheathing.

Because the tabulated values are size-specific, most stress grades of Southern Pine have the appropriate size factors already incorporated into the published values. For these grades, the size factors can be viewed as defaulting to unity:

$$C_F = 1.0 \text{ for compression parallel to grain}$$

$$C_F = 1.0 \text{ for bending}$$

Some grades of Southern Pine, however, require size factors other than unity. Section properties:

$$A = 10.875 \text{ in.}^2$$

$$S = 13.14 \text{ in.}^3$$

Axial

1. *Stability check.* Column buckling occurs away from truss joints. Use gross area.

$$f_c = \frac{P}{A} = \frac{4960}{10.875} = 456 \text{ psi}$$

$(l_e/d)_y = 0$ because of lateral support provided by roof diaphragm

$$\left(\frac{l_e}{d}\right)_x = \frac{8.39 \text{ ft} \times 12 \text{ in./ft}}{7.25 \text{ in.}} = 13.9$$

$$E' = E(C_M)(C_t) = 1,700,000(1.0)(1.0)$$

$$= 1,700,000 \text{ psi}$$

For visually graded sawn lumber:

$$K_{cE} = 0.3$$

$$c = 0.8$$

$$F_{cE} = \frac{K_{cE}E'}{[(l_e/d)_{max}]^2} = \frac{0.3(1,700,000)}{(13.9)^2} = 2645 \text{ psi}$$

$$F_c^* = F_c(C_D)(C_M)(C_t)(C_F)(C_i)$$

$$= 1650(1.15)(1.0)(1.0)(1.0)(1.0) = 1898 \text{ psi}$$

$$\frac{F_{cE}}{F_c'} = \frac{2645}{1898} = 1.394$$

$$\frac{1 + F_{cE}/F_c^*}{2c} = \frac{1 + 1.394}{2(0.8)} = 1.496$$

$$C_P = \frac{1 + F_{cE}/F_c^*}{2c}$$

$$- \sqrt{\left(\frac{1 + F_{cE}/F_c^*}{2c}\right)^2 - \frac{f_{cE}/F_c^*}{c}}$$

$$= 1.496 - \sqrt{(1.496)^2 - 1.394/0.8} = 0.792$$

$$F_c' = F_c(C_D)(C_M)(C_t)(C_F)(C_P)(C_i)$$

$$= 1650(1.15)(1.0)(1.0)(1.0)(0.792)(1.0)$$

$$= 1502 \text{ psi} > 456 \quad OK$$

2. *Net section check.* Assume the hole diameter is $\frac{1}{16}$ in. larger than the bolt (for stress calculations only).

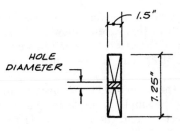

HOLE DIAMETER

1.5"

7.25"

Figure 7.19b Net section of top chord at connection.

$$A_n = 1.5(7.25 - 0.8125) = 9.66 \text{ in.}^2$$

$$f_c = \frac{P}{A_n} = \frac{4960}{9.66} = 514 \text{ psi}$$

At braced location there is no reduction for stability.

$$F'_c = F^*_c = F_c(C_D)(C_M)(C_t)(C_F)(C_i)$$

$$= 1650(1.15)(1.0)(1.0)(1.0)(1.0)$$

$$= 1898 \text{ psi} > 514 \text{ psi} \quad OK$$

Bending

Assume simple span (no end restraint). Take span and load on horizontal plane (refer to Example 2.6 in Sec. 2.5).

$$M = \frac{wL^2}{8} = \frac{0.176(7.5)^2}{8} = 1.24 \text{ ft-k} = 14.85 \text{ in.-k}$$

$$f_b = \frac{M}{S} = \frac{14,850}{13.14} = 1130 \text{ psi}$$

The beam has full lateral support. Therefore l_u and R_B are zero, and the lateral stability factor is $C_L = 1.0$. In addition, the spacing of the trusses is 4 ft o.c., and the allowable bending stress does not qualify for the repetitive-member increase, and $C_r = 1.0$.

$$F'_b = F_b(C_D)(C_M)(C_t)(C_L)(C_F)(C_r)(C_i)$$

$$= 1500(1.15)(1.0)(1.0)(1.0)(1.0)(1.0)(1.0)$$

$$= 1725 \text{ psi} > 1130 \quad OK$$

Combined Stresses

There is no bending stress about the y axis, and $f_{by} = 0$. Furthermore, the column force is concentric, and the general interaction formula reduces to

$$\left(\frac{f_c}{F'_c}\right)^2 + \frac{f_{bx}}{F'_{bx}(1 - f_c/F_{cEx})} \leq 1.0$$

The load duration factor C_D for use in the interaction formula is based on the shortest-duration load in the combination, which in this case is snow load. A C_D of 1.15 for snow was used in the individual checks on the axial stress and bending stress. Therefore, the previously determined values of F'_c and F'_b are appropriate for use in the interaction formula.

In addition to the allowable stresses, the combined stress check requires the elastic buckling stress F_{cE} for use in evaluating the amplification factor. The bending moment is about the strong axis of the cross section, and the P-Δ effect is measured by the slenderness ratio about the x axis [that is, $(l_e/d)_x = 13.9$].

The value of F_{cE} determined earlier in the example was for the column buckling portion of the problem. Column buckling is based on $(l_e/d)_{max}$. The fact that $(l_e/d)_{max}$ and $(l_e/d)_x$ are equal, is a coincidence in this problem. In other words, F_{cE} for the column portion of the problem is based on $(l_e/d)_{max}$, and F_{cE} for the P-Δ analysis is based on the axis about which the bending moment occurs, $(l_e/d)_x$. In this example, the two values of F_{cE} are equal, but in general they could be different.

$$F_{cEx} = F_{cE} = 2645 \text{ psi}$$

$$\left(\frac{f_c}{F_c'}\right)^2 + \left(\frac{1}{1 - f_c/F_{cEx}}\right)\frac{f_{bx}}{F_{bx}'} = \left(\frac{456}{1502}\right)^2 + \left(\frac{1}{1 - 456/2645}\right)\frac{1130}{1725}$$

$$(0.304)^2 + 1.21(0.655) = 0.884 < 1.0 \qquad OK$$

> Use 2×8 No. 1   SP

NOTE 1:   The load combination of D + S was predetermined to control the design and for brevity only those calculations associated with this combination were provided. It can be determined that $f_c = 179$ psi and $f_b = 360$ psi results from D-*only*, and $F_c' = 1257$ psi and $F_b' = 1350$ psi with $C_D = 0.9$. The D-*only* axial tension plus bending interaction results in a value of 0.306 vs. 0.884 for D + S.

NOTE 2:   The simplified analysis used in this example applies only to trusses with pinned joints. For a truss connected with metal plate connectors, a design approach should be used which takes the continuity and partial rigidity of the joints into account (Ref. 7.7).

It should be noted that all stress calculations in this example are for loads caused by the design load of (D + S). For this reason the load duration factor for snow (1.15) is applied to each *individual* stress calculation as well as to the *combined* stress check. The designer is cautioned that the critical load combination (D + S) was predetermined in this example. To be complete, the designer should also check D-*only* with the corresponding C_D for permanent load of 0.9.

In some cases of combined loading, the *individual* stresses (axial and bending) may not both be caused by the same load. In this situation, the respective C_D values are used in evaluating the individual stresses. However, in the *combined* stress calculation, the same C_D is used for all components. Recall that the C_D for the *shortest*-duration load applies to the entire combination. The appropriate rules for applying C_D should be followed in checking both individual and combined stresses. The application of different C_D's in the individual stress calculations is illustrated in some of the following examples.

7.14 Design Problem: Beam-Column Action in a Stud Wall

A common occurrence of beam-column action is found in an exterior bearing wall. Axial column stresses are developed in the wall studs by vertical gravity loads. Bending stresses are caused by lateral wind or seismic forces.

For typical wood-frame walls, the wall dead load is so small that the design wind force usually exceeds the seismic force F_p. This applies to the normal wall force only, and seismic may be critical for parallel-to-wall (i.e., shearwall) forces. In buildings with large wall dead loads, the F_p seismic force can exceed the wind force. Large wall dead loads usually occur in concrete and masonry (brick and concrete block) buildings.

In the two-story building of Example 7.16, the studs carry a number of axial compressive loads including dead, roof live, and floor live loads. In *load case 1*, the various possible combinations of gravity loads are considered.

In Example 4.10 part *b* (Sec. 4.15), a "system" was introduced to determine the critical load combination. This system breaks down in the analysis of certain members because of the effect of the load duration factor. Figure 7.6*c*, earlier in this chapter, shows that C_D has a varied effect on F'_c depending on the slenderness ratio of the column. Therefore, the idea of dividing out the load duration factor is inappropriate except for *short* columns.

As a result, each vertical load combination should theoretically be checked using the appropriate C_D. In this example, the load case of (D-*only*) can be eliminated by inspection. The results of the other two vertical loadings are close, and the critical combination can be determined by evaluating the ratio f_c/F'_c for each combination. The loading with the larger stress ratio is critical. Only the calculations for the critical combination of (D + L + L$_r$) are shown in the example for load case 1.

Load case 2 involves both vertical loads and lateral forces. According to the IBC basic load combinations, roof live load and floor live load are considered simultaneously with lateral forces (Ref. 7.6 and Sec. 2.16).

Notice that C_D for wind (the shortest-duration load in the combination) applies to both components of stress in the interaction formula.

EXAMPLE 7.16   Combined Bending and Compression in a Stud Wall

Check the 2 × 6 stud in the first-floor bearing wall in the building shown in Fig. 7.20*a*. Consider the given vertical loads and lateral forces. Lumber is No. 2 DF-L. MC ≤ 19 percent and normal temperatures apply. Allowable stresses are to be in accordance with the NDS.

The following gravity loads are given:

$$\text{Roof: D} = 10 \text{ psf}$$

$$\text{L}_r = 20 \text{ psf}$$

$$\text{Wall: D} = 7 \text{ psf}$$

$$\text{Floor: D} = 8 \text{ psf}$$

$$\text{L} = 40 \text{ psf}$$

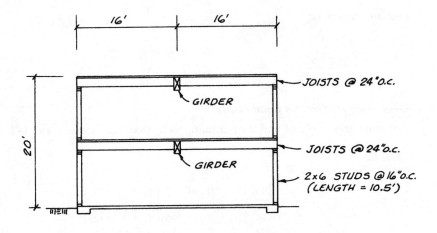

TRANSVERSE SECTION

Figure 7.20a   Transverse section showing exterior bearing walls.

The following lateral forces are also given:

$$W = 27.8 \text{ psf horizontal}$$

$$E = 12.3 \text{ psf}$$

The IBC basic load combinations consider (W or 0.7E),

$$0.7E = 0.7(12.3 \text{ psf}) = 8.6 \text{ psf} < 27.8 \text{ psf}$$

$$\therefore \text{ Wind governs.}$$

Try 2×6 No. 2 DF-L (Dimension lumber size):

Values from NDS Supplement Table 4A:

$$F_b = 900 \text{ psi}$$

$$F_c = 1350 \text{ psi}$$

$$F_{c\perp} = 625 \text{ psi}$$

$$E = 1,600,000 \text{ psi}$$

Size factors:

$$C_F = 1.3 \qquad \text{for bending}$$

$$C_F = 1.1 \qquad \text{for compression parallel to grain}$$

Section properties:

$$A = 8.25 \text{ in.}^2$$

$$S = 7.56 \text{ in.}^3$$

Load Case 1:   Gravity Loads Only

Tributary width of roof and floor framing to the exterior bearing wall is 8 ft.

Dead loads:

$$\text{Roof D} = 10 \text{ psf} \times 8 \text{ ft} = 80 \text{ lb/ft}$$

$$\text{Wall D} = 7 \text{ psf} \times 20 \text{ ft} = 140 \text{ lb/lft}$$

$$\underline{\text{Floor D} = 8 \text{ psf} \times 8 \text{ ft} = 64 \text{ lb/ft}}$$

$$w_D = 284 \text{ lb/ft}$$

Live loads:

$$\text{Roof L}_r = 20 \text{ psf} \times 8 \text{ ft} = 160 \text{ lb/ft}$$

$$\text{Floor L} = 40 \text{ psf} \times 8 \text{ ft} = 320 \text{ lb/ft}$$

LOAD COMBINATIONS:

Calculate the axial load on a typical stud:

$$D\text{-}only = (284 \text{ lb/ft})(1.33 \text{ ft}) = 378 \qquad C_D = 0.9$$

$$D + L = (284 + 320)1.33 = 803 \qquad C_D = 1.0$$

$$D + L + L_r = (284 + 320 + 160)1.33 = 1016 \text{ lb} \qquad C_D = 1.25$$

The combination of D-*only* can be eliminated by inspection. When the C_D's are considered for the second two combinations, the net effects are roughly the same. However, $(D + L + L_r)$ was determined to be the critical vertical loading, and stress calculations for this combination only are shown. The designer is responsible for determining the critical load combinations, including the effects of the load duration factor C_D. The axial stress in the stud and the bearing stress on the wall plate are equal.

$$f_c = f_{c\perp} = \frac{P}{A} = \frac{1016}{8.25} = 123 \text{ psi}$$

COLUMN CAPACITY:

Sheathing provides lateral support about the weak axis of the stud. Therefore, check column buckling about the *x* axis only ($L = 10.5$ ft and $d_x = 5.5$ in.):

$$\left(\frac{l_e}{d}\right)_y = 0 \quad \text{because of sheathing}$$

$$\left(\frac{l_e}{d}\right)_{max} = \left(\frac{l_e}{d}\right)_x = \frac{10.5 \text{ ft} \times 12 \text{ in./ft}}{5.5 \text{ in.}} = 22.9$$

$$E' = E(C_M)(C_t)(C_i) = 1{,}600{,}000(1.0)(1.0)(1.0)$$

$$= 1{,}600{,}000 \text{ psi}$$

For visually graded sawn lumber:

$$K_{cE} = 0.3$$

$$c = 0.8$$

$$F_{cE} = \frac{K_{cE}E'}{(l_e/d)^2} = \frac{0.3(1{,}600{,}000)}{(22.9)^2} = 915 \text{ psi}$$

$$F_c^* = F_c(C_D)(C_M)(C_t)(C_F)(C_i)$$

$$= 1350(1.25)(1.0)(1.0)(1.1)(1.0) = 1856 \text{ psi}$$

$$\frac{F_{cE}}{F_c^*} = \frac{915}{1856} = 0.493$$

$$\frac{1 + F_{cE}/F_c^*}{2c} = \frac{1 + 0.493}{2(0.8)} = 0.933$$

$$C_P = \frac{1 + F_{cE}/F_c^*}{2c} - \sqrt{\left(\frac{1 + F_{cE}/F_c^*}{2c}\right)^2 - \frac{F_{cE}/F_c^*}{c}}$$

$$= 0.933 - \sqrt{(0.933)^2 - 0.493/0.8} = 0.429$$

$$F_c' = F_c(C_D)(C_M)(C_t)(C_F)(C_P)(C_i)$$

$$= 1350(1.25)(1.0)(1.0)(1.1)(0.429)(1.0)$$

$$= 796 \text{ psi} > 123 \text{ psi} \quad OK$$

BEARING OF STUD ON WALL PLATES:

For a bearing length of 1½ in. on a stud more than 3 in. from the end of the wall plate:

$$C_b = \frac{l_b + 0.375}{l_b} = \frac{1.5 + 0.375}{1.5} = 1.25$$

$$F_{c\perp}' = F_{c\perp}(C_M)(C_t)(C_b) = 625(1.0)(1.0)(1.25)$$

$$= 781 \text{ psi} > 123 \quad OK$$

$$f_c < F_c' \quad \text{and} \quad f_c < F_{c\perp}'$$

$$\therefore \text{Vertical loads} \quad OK$$

Load Case 2:   Gravity Loads + Lateral Forces

BENDING:

Wind governs over seismic. Force to one stud:

$$\text{Wind} = 27.8 \text{ psf}$$

$$w = 27.8 \text{ psf} \times 1.33 \text{ ft} = 37.0 \text{ lb/ft}$$

$$M = \frac{wL^2}{8} = \frac{37.0(10.5)^2}{8} = 510 \text{ ft-lb} = 6115 \text{ in.-lb}$$

$$f_b = \frac{M}{S} = \frac{6115}{7.56} = 809 \text{ psi}$$

The stud has full lateral support provided by sheathing. Therefore, l_u and R_B are zero, and the lateral stability factor is $C_L = 1.0$. The load duration factor for wind is $C_D = 1.6$, and the repetitive-member factor is 1.15.

$$F'_b = F_b(C_D)(C_M)(C_t)(C_L)(C_F)(C_r)(C_i)$$

$$= 900(1.6)(1.0)(1.0)(1.0)(1.3)(1.15)(1.0)$$

$$= 2152 \text{ psi} > 809 \quad OK$$

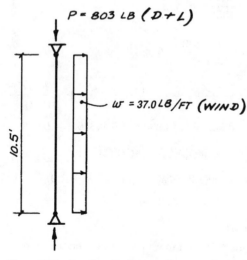

Figure 7.20b Loading for beam-column analysis.

AXIAL:

Two load combinations must be considered: D + W and D + 0.75(L + L$_r$ + W). Upon first inspection, the D + 0.75(L + L$_r$ + W) may appear to govern the design. However, as will be shown when the interaction of axial compression and bending is evaluated, the D + W actually is the critical load combination.

$$D + W: \quad f_c = \frac{P}{A} = \frac{378}{8.25} = 46 \text{ psi}$$

$$D + 0.75(L + L_r + W): \quad f_c = \frac{P}{A} = \frac{378 + 0.75(427 + 213 + 0)}{8.25} = 104 \text{ psi}$$

Again, in the combined stress calculation, a single C_D is used throughout. Hence, F_c' is determined here using $C_D = 1.6$.

The slenderness ratio about the y axis is zero because of the continuous support provided by the sheathing. The column slenderness ratio and the elastic buckling stress that were determined previously apply to the problem at hand:

$$\left(\frac{l_e}{d}\right)_{max} = \left(\frac{l_e}{d}\right)_x = 22.9$$

$$F_{cE} = 915 \text{ psi}$$

$$F_c^* = F_c(C_D)(C_M)(C_t)(C_F)(C_i)$$

$$= 1350(1.6)(1.0)(1.0)(1.1)(1.0) = 2376 \text{ psi}$$

$$\frac{F_{cE}}{F_c^*} = \frac{915}{2376} = 0.385$$

$$\frac{1 + F_{cE}/F_c^*}{2c} = \frac{1 + 0.385}{2(0.8)} = 0.866$$

$$C_p = \frac{1 + F_{cE}/F_c^*}{2c} - \sqrt{\left(\frac{1 + F_{cE}/F_c^*}{2c}\right)^2 - \frac{F_{cE}/F_c^*}{c}}$$

$$= 0.866 - \sqrt{(0.866)^2 - 0.385/0.8} = 0.348$$

$$F_c' = F_c(C_D)(C_M)(C_t)(C_F)(C_p)(C_i)$$

$$= 1350(1.6)(1.0)(1.0)(1.1)(0.348)(1.0)$$

$$F_c' = 826 \text{ psi} > 104 \text{ psi} \quad OK$$

COMBINED STRESS:

The simplified interaction formula from Example 7.13 (Sec. 7.12) applies:

$$\left(\frac{f_c}{F_c'}\right)^2 + \frac{f_{bx}}{F_{bx}'(1 - f_c/F_{cEx})} \le 1.0$$

Recall that the allowable column stress F_c' is determined using the maximum slenderness ratio for the column $(l_e/d)_{max}$, and the Euler buckling stress F_{cEx} for use in evaluating the P-Δ effect is based on the slenderness ratio for the axis with the bending moment. In this problem the bending moment is about the strong axis of the cross section, and $(l_e/d)_x$, coincidentally, controls both F_c' and F_{cEx}. In general, one slenderness ratio does not necessarily define these two quantities.

The value of F_{cE} determined earlier in the example using $(l_e/d)_x$ is also F_{cEx}:

$$F_{cEx} = F_{cE} = 915 \text{ psi}$$

D + W:

In this load combination, D produces the axial stress f_c and W results in the bending stress f_{bx}.

$$\left(\frac{f_c}{F'_c}\right)^2 + \left(\frac{1}{1 - f_c/F_{cEx}}\right)\frac{f_{bx}}{F'_{bx}} = \left(\frac{46}{826}\right)^2 + \left(\frac{1}{1 - 46/915}\right)\frac{809}{2152} = 0.399 < 1.0 \qquad OK$$

D + 0.75(L + L$_r$ + W):

In this load combination, the axial stress f_c results from D + 0.75(L + L$_r$) and the bending stress f_{bx} is caused by 0.75W.

$$\left(\frac{f_c}{F'_c}\right)^2 + \left(\frac{1}{1 - f_c/F_{cEx}}\right)\frac{f_{bx}}{F'_{bx}} = \left(\frac{104}{826}\right)^2 + \left(\frac{1}{1 - 104/915}\right)\frac{(0.75)(809)}{2152}$$

$$= 0.334 < 1.0 \qquad OK$$

2 × 6   No. 2   DF-L   exterior bearing wall   OK

Although several load cases were considered, the primary purpose of Example 7.15 is to illustrate the application of the interaction formula for beam-columns applied to a stud wall. The reader should understand that other load cases, including uplift due to wind, may be required in the analysis of a bearing wall subject to lateral forces.

7.15   Design Problem: Glulam Beam-Column

In this example a somewhat more complicated bracing condition is considered. The column is a glulam that supports both roof dead and live loads as well as lateral wind forces. See Example 7.17.

In load case 1 the vertical loads are considered, and (D + L$_r$) is the critical loading. The interesting aspect of this problem is that there are different unbraced lengths for the x and y axes. Lateral support for the strong axis is provided at the ends only. However, for the weak axis the unbraced length is the height of the window.

In load case 2 the vertical dead load and lateral wind force are considered. Bending takes place about the strong axis of the member. The bending analysis includes a check of lateral stability using the window height as the unbraced length. In checking combined stresses, a C_D of 1.6 for wind is applied to all components of the interaction formula. Note that C_D appears several times in the development of the allowable column stress, and F'_c must be reevaluated for use in the interaction formula.

The example makes use of a member that is an axial-load glulam combination. Calculations show that the bending stress is more significant than the axial stress, and it would probably be a more efficient design to choose a member from the glulam bending combinations instead of an axial combination. However, with a combined stress ratio of 0.682, the given member is considerably understressed.

EXAMPLE 7.17   Glulam Beam-Column

Check the column in the building shown in Fig. 7.21a for the given loads. The column is an axial combination 2 DF glulam (combination symbol 2) with tension laminations (F_{bx} = 2000 psi). The member supports the tributary dead load, roof live load, and lateral wind force. The wind force is transferred to the column by the window framing in the wall.

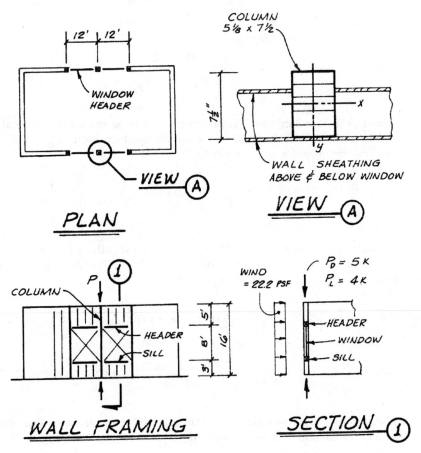

Figure 7.21a Glulam column between windows subject to bending plus axial compression.

The lateral force is the inward or outward wind pressure:

$$W = 22.2 \text{ psf}$$

The seismic force is not critical. Tabulated glulam design values are to be taken from the NDS Supplement Table 5B. $C_M = 1.0$, and $C_t = 1.0$.

Glulam Column

$5\frac{1}{8} \times 7\frac{1}{2}$	*Axial combination 2 DF glulam:*
$A = 38.4 \text{ in.}^2$	$F_c = 1950 \text{ psi}$
$S_x = 48 \text{ in.}^3$	$F_{bx} = 2000 \text{ psi}$ (requires tension laminations)
	$E_x = E_y = 1{,}600{,}000 \text{ psi}$

Load Case 1: Gravity Loads

$$D = 5 \text{ k}$$

$$D + L_r = 5 + 4 = 9 \text{ k}$$

The $(D + L_r)$ combination was predetermined to be the critical vertical loading condition, and the load duration factor for the combination is $C_D = 1.25$. (The D-only combination should also be checked.)

$$f_c = \frac{P}{A} = \frac{9000}{38.4} = 234 \text{ psi}$$

Neglect the column end restraint offered by wall sheathing for column buckling about the y axis. Assume an effective length factor (Fig. 7.9) of $K_e = 1.0$ for both the x and y axes.

$$\left(\frac{l_e}{d}\right)_x = \frac{1(16 \text{ ft} \times 12 \text{ in./ft})}{7.5 \text{ in.}} = 25.6$$

$$\left(\frac{l_e}{d}\right)_y = \frac{1(8 \text{ ft} \times 12 \text{ in./ft})}{5.125 \text{ in.}} = 18.7 < 25.6$$

The larger slenderness ratio governs the allowable column stress. Therefore, the strong axis of the column is critical, and $(l_e/d)_x$ is used to determine F_c'.

$$E_x = E_y = 1,600,000 \text{ psi}$$

$$E' = E(C_M)(C_t) = 1,600,000(1.0)(1.0)$$

$$= 1,600,000 \text{ psi}$$

For glulam:

$$K_{cE} = 0.418$$

$$c = 0.9$$

$$F_{cE} = \frac{K_{cE}E'}{[(l_e/d)_{max}]^2} = \frac{0.418(1,600,000)}{(25.6)^2} = 1021 \text{ psi}$$

$$F'_c = F_c(C_D)(C_M)(C_t)$$

$$= 1950(1.25)(1.0)(1.0) = 2438 \text{ psi}$$

$$\frac{F_{cE}}{F'_c} = \frac{1021}{2438} = 0.419$$

$$\frac{1 + F_{cE}/F^*_c}{2c} = \frac{1 + 0.419}{2(0.9)} = 0.788$$

$$C_P = \frac{1 + F_{cE}/F^*_c}{2c} - \sqrt{\left(\frac{1 + F_{cE}/F^*_c}{2c}\right)^2 - \frac{F_{cE}/F^*_c}{c}}$$

$$= 0.788 - \sqrt{(0.788)^2 - 0.419/0.9} = 0.394$$

$$F'_c = F_c(C_D)(C_M)(C_t)(C_P)$$

$$= 1950(1.25)(1.0)(1.0)(0.394)$$

$$= 960 \text{ psi} > 234 \text{ psi} \quad OK$$

The member is adequate for vertical loads.

Load Case 2: D + W

The two applicable load combinations for this design are (D + W) and D + 0.75(L + L_r + W). It has been predetermined that D + W governs the interaction, and only those calculations associated with this combination are provided. The load duration factor for (D + W) is taken as 1.6 throughout the check for combined stresses.

AXIAL (DEAD LOAD):

$$f_c = \frac{P}{A} = \frac{5000}{38.4} = 130 \text{ psi}$$

From load case 1:

$$\left(\frac{l_e}{d}\right)_{max} = \left(\frac{l_e}{d}\right)_x = 25.6$$

$$F_{cE} = \frac{K_{cE}E'}{[(l_e/d)_{max}]^2} = 1021 \text{ psi}$$

$$F_c^* = F_c(C_D)(C_M)(C_t)$$

$$= 1950(1.6)(1.0)(1.0) = 3120 \text{ psi}$$

$$\frac{F_{cE}}{F_c^*} = \frac{1021}{3120} = 0.327$$

$$\frac{1 + F_{cE}/F_c^*}{2c} = \frac{1 + 0.327}{2(0.9)} = 0.737$$

$$C_P = \frac{1 + F_{cE}/F_c^*}{2c} - \sqrt{\left(\frac{1 + F_{cE}/F_c^*}{2c}\right)^2 - \frac{F_{cE}/F_c^*}{c}}$$

$$= 0.737 - \sqrt{(0.737)^2 - 0.327/0.9} = 0.313$$

$$F_c' = F_c(C_D)(C_M)(C_t)(C_P)$$

$$= 1950(1.6)(1.0)(1.0)(0.313)$$

$$= 977 \text{ psi}$$

$$\text{Axial stress ratio} = \frac{f_c}{F_c'} = \frac{130}{977} = 0.133$$

BENDING (WIND):

The window headers and sills span horizontally between columns. Uniformly distributed wind forces to a typical header and sill are calculated using a 1-ft section of wall and the tributary heights shown in Fig. 7.21b.

$$\text{Wind on header} = w_1 = (22.2 \text{ psf})(6.5 \text{ ft}) = 144 \text{ lb/ft}$$

Horizontal reaction of two headers on center column (Fig. 7.21c):

$$P_1 = (144 \text{ lb/ft})(12 \text{ ft}) = 1728 \text{ lb}$$

$$\text{Wind on sill} = w_2 = (22.2 \text{ psf})(5.5 \text{ ft}) = 122 \text{ lb/ft}$$

Horizontal reaction of two sills on center column:

$$P_2 = (122 \text{ lb/ft})(12 \text{ ft}) = 1464 \text{ lb}$$

From the moment diagram in Fig. 7.21c,

$$M_x = 7318 \text{ ft-lb} = 87.8 \text{ in.-k}$$

$$f_b = \frac{M}{S} = \frac{87,816}{48} = 1830 \text{ psi}$$

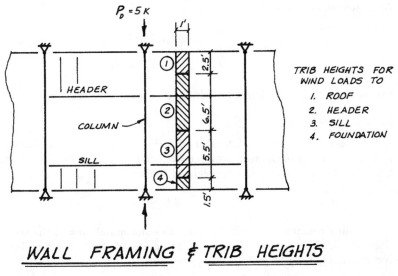

WALL FRAMING & TRIB HEIGHTS

Figure 7.21b Wall framing showing tributary wind pressure heights of 6.5 ft and 5.5 ft to window header and sill, respectively.

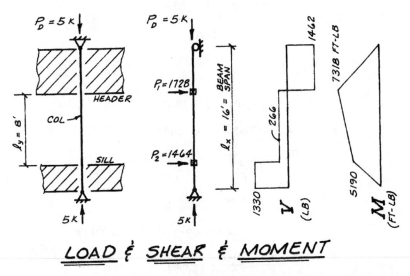

LOAD & SHEAR & MOMENT

Figure 7.21c Load, shear, and moment diagrams for center column subject to lateral wind forces. Concentrated forces are header and sill reactions from window framing.

Bending is about the strong axis of the cross section. The allowable bending stress for a glulam is governed by the smaller of two criteria: volume effect or lateral stability (see Example 6.11 in Sec. 6.4). The wind pressure can act either inward or outward, and tension laminations are required on both faces of the glulam.

Allowable stress criteria:

$$F'_b = F_b(C_D)(C_M)(C_t)(C_L)$$

$$F'_b = F_b(C_D)(C_M)(C_t)(C_V)$$

Compare C_L and C_V to determine the critical design criteria.

Beam stability factor C_L

If the beam-column fails in lateral torsional buckling as a beam, the cross section will move in the plane of the wall between the window sill and header. Thus, the unbraced length for beam stability is the height of the window.

$$l_u = 8 \text{ ft} = 96 \text{ in.}$$

The loading condition for this member does not match any of the conditions in NDS Table 3.3.3. However, the effective length given in the footnote to this table may be conservatively used for any loading.

$$\frac{l_u}{d} = \frac{96}{7.5} = 12.8$$

$$7 \le 12.8 \le 14.3$$

$$\therefore \quad l_e = 1.63 l_u + 3d$$

$$= 1.63(96) + 3(7.5) = 179 \text{ in.}$$

$$R_B = \sqrt{\frac{l_e d}{b^2}} = \sqrt{\frac{179(7.5)}{(5.125)^2}} = 7.15$$

$$K_{bE} = 0.610 \qquad \text{for glulam}$$

$$F_{bE} = \frac{K_{bE} E'_y}{R_B^2} = \frac{0.610(1,600,000)}{(7.15)^2} = 19,091 \text{ psi}$$

$$F_b^* = F_b(C_D)(C_M)(C_t)$$

$$= 2000(1.6)(1.0)(1.0) = 3200 \text{ psi}$$

$$\frac{F_{bE}}{F_b^*} = \frac{19,091}{3200} = 5.97$$

$$\frac{1 + F_{bE}/F_b^*}{1.9} = \frac{1 + 5.97}{1.9} = 3.67$$

$$C_L = \frac{1 + F_{bE}/F_b^*}{1.9} - \sqrt{\left(\frac{1 + F_{bE}/F_b^*}{1.9}\right)^2 - \frac{F_{bE}/F_b^*}{0.95}}$$

$$= 3.67 - \sqrt{(3.67)^2 - 5.97/0.95} = 0.990$$

Volume factor C_V

For DF glulam, $x = 10$.

$$C_V = \left(\frac{21}{L}\right)^{1/x} \left(\frac{12}{d}\right)^{1/x} \left(\frac{5.125}{b}\right)^{1/x} \leq 1.0$$

$$= \left(\frac{21}{16}\right)^{0.1} \left(\frac{12}{7.5}\right)^{0.1} \left(\frac{5.125}{5.125}\right)^{0.1}$$

$$= 1.077 > 1.0$$

$$\therefore \quad C_V = 1.0$$

Neither beam stability nor volume effect has a significant impact on this problem. However, the smaller value of C_L and C_V indicates that stability governs over volume effect.

$$C_L = 0.990$$

$$F_b' = F_b(C_D)(C_M)(C_t)(C_L)$$

$$= 2000(1.6)(1.0)(1.0)(0.990)$$

$$= 3168 \text{ psi}$$

$$\text{Bending stress ratio} = \frac{f_b}{F_b'} = \frac{1830}{3168} = 0.578$$

COMBINED STRESSES:

The bending moment is about the strong axis of the cross section, and the amplification for P-Δ is measured by the column slenderness ratio about the x axis. *Note:* It is a coincidence that the allowable column stress F_c' and the amplification factor for the P-Δ effect are both controlled by $(l_e/d)_x$ in this problem. Recall that F_c' is governed by $(l_e/d)_{\max}$, and the P-Δ effect is controlled by the slenderness ratio for the axis about which the bending moment is applied.

$$\left(\frac{l_e}{d}\right)_{\substack{\text{bending} \\ \text{moment}}} = \left(\frac{l_e}{d}\right)_x = 25.6$$

$$F_{cEx} = \frac{K_{cE}E'}{[(l_e/d)_x]^2} = \frac{0.418(1,600,000)}{(25.6)^2} = 1021 \text{ psi}$$

$$\text{Amplification factor} = \frac{1}{1 - f_c/F_{cEx}} = \frac{1}{1 - 130/1021} = 1.15$$

$$\left(\frac{f_c}{F_c'}\right)^2 + \left(\frac{1}{1 - f_c/F_{cEx}}\right)\frac{f_b}{F_b'} = (0.133)^2 + 1.15(0.578)$$

$$= 0.682 < 1.0 \quad OK$$

> $5\frac{1}{8} \times 7\frac{1}{2}$ axial combination 2 DF glulam with tension laminations (F_{bx} = 2000 psi) is *OK* for combined bending and compression.

NOTE: The critical load combination for the combined load was predetermined to be D + W. If the D + 0.75(L + L$_r$ + W) combination were used, an interaction value of 0.419 would result.

7.16 Design for Minimum Eccentricity

The design procedures for a column with an axial load were covered in detail in Secs. 7.4 and 7.5. A large number of interior columns and some exterior columns qualify as axial-load-carrying members. That is, the applied load is *assumed* to pass directly through the centroid of the column cross section, and, in addition, no transverse bending loads are involved.

Although many columns can theoretically be classified as axial load members, there may be some question about whether the load in practical columns is truly an axial load. In actual construction there may be some misalignment or nonuniform bearing in connections that causes the load to be applied eccentrically.

Some eccentric moment probably develops in columns which are thought to support axial loads only. The magnitude of the eccentric moment, however, is unknown. Many designers simply ignore the possible eccentric moment and design for axial stresses only. This practice may be justified because practical columns typically have square-cut ends. In addition, the ends are attached with connection hardware such that the column end conditions do not exactly resemble the end conditions of an "ideal" pinned-end column. With the restraint provided by practical end conditions, the effective column length is somewhat less than the actual unbraced length. Thus the possible effect of an accidental eccentricity may be compensated by normal field end conditions.

However, Ref. 7.5 states that the possible eccentric moment should not be ignored, and it suggests that columns should be designed for some minimum eccentricity. The minimum eccentricity recommended is similar to the minimum eccentricity formerly required in the design of axially loaded reinforced-concrete columns. In this approach, the moment is taken as the compressive load times an eccentricity of 1 in. or one-tenth the width of the column (0.1d), whichever is larger. The moment is considered independently about both principal axes.

In the design of wood columns, there is no Code requirement to design for a minimum eccentric moment. The suggestion that some designers may provide for an eccentric moment in their column calculations is presented here for information only. Including an eccentric moment in the design of a column is definitely a more conservative design approach. Whether or not eccentricity should be included is left to the judgment of the designer.

7.17   Design Problem: Column with Eccentric Load

Example 7.18 demonstrates the use of the interaction formula for eccentric loads. The load is theoretically an axial load, but the calculations are expanded to include a check for the minimum eccentricity discussed in Sec. 7.16. The same interaction formula would be used in the case of a known eccentricity.

The problem illustrates the significant effect of an eccentricity. Without the eccentricity, the member capacity is simply evaluated by the axial stress ratio of 0.690. However, the combined stress ratio is 0.899 for an eccentricity about the x axis and 0.917 for an eccentricity about the y axis. The combined stress ratios are much closer to the full member capacity, which is associated with a value of 1.0.

This example also demonstrates that F_{by} for a member in the Beams and Stringers size category does not equal F_{bx}. The reduction factor for F_{by} varies with grade.

EXAMPLE 7.18   Column Design for Minimum Eccentricity

The column in Fig. 7.22a is an interior column in a large auditorium. The design roof D + S are theoretically axial loads on the column. Because of the importance of the column, it is desired to provide a conservative design with a minimum eccentricity of 0.1d or 1 in. Bracing conditions are shown. Lumber is Select Structural DF-L, and C_M, C_t, and C_i all equal 1.0.

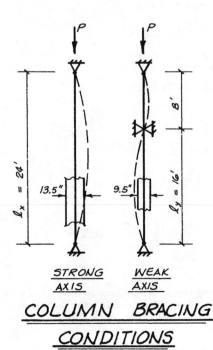

COLUMN BRACING

CONDITIONS

Figure 7.22a Sawn lumber column with different bracing conditions for x and y axes.

$$D = 20 \text{ k}$$

$$\underline{S = 50 \text{ k}}$$

$$P_{TL} = 70 \text{ k} \quad \text{(total load governs over D-}only\text{)}$$

A load duration factor of $C_D = 1.15$ applies throughout the problem.

Try 10 × 14 Sel. Str. DF-L.

The trial size is in the B&S size category. Recall that a member in the Beams and Stringers size category has cross-sectional dimensions of 5 in. or greater and a width that exceeds the thickness by more than 2 in. Design values are obtained from NDS Supplement Table 4D:

DF-L in this size category may be graded under two different sets of lumber grading rules. If any tabulated stresses conflict, use the smaller value (conservative):

$$F_c = 1100 \text{ psi}$$

$$F_{bx} = 1600 \text{ psi*}$$

$$E_x = 1,600,000 \text{ psi}$$

Section properties:

$$b = 9.5 \text{ in.}$$

$$d = 13.5 \text{ in.}$$

$$A = 128.25 \text{ in.}^2$$

$$S_x = 288.6 \text{ in.}^3$$

$$S_y = 203.1 \text{ in.}^3$$

Axial

$$f_c = \frac{P}{A} = \frac{70,000}{128.25} = 546 \text{ psi}$$

$$\left(\frac{l_e}{d}\right)_x = \frac{24 \text{ ft} \times 12 \text{ in./ft}}{13.5 \text{ in.}} = 21.3$$

$$\left(\frac{l_e}{d}\right)_y = \frac{16 \text{ ft} \times 12 \text{ in./ft}}{9.5 \text{ in.}} = 20.2$$

---

*Tabulated values of allowable bending stress for members in the B&S size category are for bending about the *x* axis. When bending is about the weak axis, a size factor is applied as provided in Table 4D of the NDS Supplement. For the Select Structural (Sel. Str.) grade, $C_F = 0.86$ for bending and 1.0 for all other properties.

$E_x = E_y$, and the larger slenderness ratio governs the allowable column stress. Therefore, the strong axis is critical.

$$E' = E(C_M)(C_t)(C_i) = 1,600,000(1.0)(1.0)(1.0)$$

$$= 1,600,000 \text{ psi}$$

For visually graded sawn lumber:

$$K_{cE} = 0.3$$

$$c = 0.8$$

Determine allowable column stress:

$$F_{cE} = \frac{K_{cE}D'}{[(l_e/d)_{max}]^2} = \frac{0.3(1,600,000)}{(21.3)^2} = 1055 \text{ psi}$$

The size factor for compression parallel to grain applies only to Dimension lumber, and C_F defaults to unity for a B&S.

$$F_c^* = F_c(C_D)(C_M)(C_t)(C_F)(C_i)$$

$$= 1100(1.15)(1.0)(1.0)(1.0)(1.0) = 1265 \text{ psi}$$

$$\frac{F_{cE}}{F_c^*} = \frac{1055}{1265} = 0.834$$

$$\frac{1 + F_{cE}/F_c^*}{2c} = \frac{1 + 0.834}{2(0.8)} = 1.146$$

$$C_P = \frac{1 + F_{cE}/F_c^*}{2c} - \sqrt{\left(\frac{1 + F_{cE}/F_c^*}{2c}\right)^2 - \frac{F_{cE}/F_c^*}{c}}$$

$$= 1.146 - \sqrt{(1.146)^2 - 0.834/0.8} = 0.625$$

$$F_c' = F_c(C_D)(C_M)(C_t)(C_P)(C_i)$$

$$= 1100(1.15)(1.0)(1.0)(0.625)(1.0)$$

$$F_c' = 791 \text{ psi} > 546 \qquad OK$$

Alternatively, the axial stress ratio is shown to be less than 1.0:

$$\frac{f_c}{F_c'} = \frac{546}{791} = 0.690 < 1.0$$

The member is adequate for the axial load.

Eccentric Load about Strong Axis

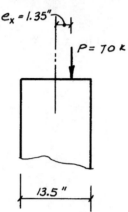

$e_x = 1.35"$

$P = 70k$

$13.5"$

Figure 7.22b Column load applied with eccentricity about x axis.

AXIAL:

The axial stress check is unchanged for this load case.

BENDING:

There are no transverse loads, and the only bending stress is due to the eccentric column force.

$$e_x = 0.1d = 0.1(13.5) = 1.35 \text{ in.} > 1.0$$

$$\text{Ecc. } f_{bx} = \frac{Pe_x}{S_x} = f_c\left(\frac{6e_x}{d_x}\right) = 546\left(\frac{6 \times 1.35}{13.5}\right) = 327 \text{ psi}$$

Size factor

$$C_F = \left(\frac{12}{d}\right)^{1/9} = \left(\frac{12}{13.5}\right)^{1/9} = 0.987$$

Lateral stability

The eccentric moment is about the strong axis of the cross section. Lateral torsional buckling may occur in a plane perpendicular to the plane of bending. Therefore, the unbraced length for lateral stability is 16 ft. Determine l_e in accordance with footnote 1 to NDS Table 3.3.3.

$$l_u = 16 \text{ ft} = 192 \text{ in.}$$

$$\frac{l_u}{d} = \frac{192}{13.5} = 14.2$$

$$7.0 \le 14.2 \le 14.3$$

$$\therefore \quad l_e = 1.63l_u + 3d$$

$$= 1.63(192) + 3(13.5) = 353 \text{ in.}$$

$$R_B = \sqrt{\frac{l_e d}{b^2}} = \sqrt{\frac{353(13.5)}{(9.5)^2}} = 7.27$$

$$K_{bE} = 0.439 \qquad \text{for visually graded sawn lumber}$$

$$F_{bE} = \frac{K_{bE} E_y'}{R_B^2} = \frac{0.439(1,600,000)}{(7.27)^2} = 13,290 \text{ psi}$$

$$F_b^* = F_b(C_D)(C_M)(C_t)(C_F)(C_i)$$

$$= 1600(1.15)(1.0)(1.0)(0.987)(1.0) = 1816 \text{ psi}$$

$$\frac{F_{bE}}{F_b^*} = \frac{13,290}{1816} = 7.318$$

$$\frac{1 + F_{bE}/F_b^*}{1.9} = \frac{1 + 7.318}{1.9} = 4.378$$

$$C_L = \frac{1 + F_{bE}/F_b^*}{1.9} - \sqrt{\left(\frac{1 + F_{bE}/F_b^*}{1.9}\right)^2 - \frac{F_{bE}/F_b^*}{0.95}}$$

$$= 4.378 - \sqrt{(4.378)^2 - 7.318/0.95} = 0.992$$

$$F_{bx}' = F_b(C_D)(C_M)(C_t)(C_L)(C_F)(C_r)(C_i)$$

$$= 1600(1.15)(1.0)(1.0)(0.992)(0.987)(1.0)(1.0)$$

$$= 1802 \text{ psi} > 327 \qquad OK$$

COMBINED STRESSES:

There are two amplification factors for combined stresses when all or part of the bending stress is due to an eccentric load.

Amplification factor for eccentric bending stress

The current check on eccentric bending moment is about the x axis, and the amplification factor is a function of the slenderness ratio for the x axis.

$$\left(\frac{l_e}{d}\right)_x = 21.3$$

The Euler elastic buckling stress was evaluated previously using this slenderness ratio in the axial stress portion of the problem.

$$F_{cEx} = F_{cE} = 1055 \text{ psi}$$

$$\frac{f_c}{F_{cEx}} = \frac{546}{1055} = 0.518$$

$$(\text{Amp Fac})_{ecc} = 1 + 0.234 \left(\frac{f_c}{F_{cEx}}\right) = 1 + 0.234(0.518) = 1.121$$

General P-Δ amplification factor

$$\text{Amp Fac} = \frac{1}{1 - f_c/F_{cEx}} = \frac{1}{1 - 546/1055} = 2.073$$

$$\left(\frac{f_c}{F'_c}\right)^2 + \left(\frac{1}{1 - f_c/F_{cEx}}\right) \frac{f_b + f_c(6e_x/d_x)[1 + 0.234(f_c/F_{cEx})]}{F'_{bx}}$$

$$= (0.690)^2 + (2.073) \left[\frac{0 + 327(1.121)}{1802}\right]$$

$$= 0.899 < 1.0$$

Eccentric load is OK for bending about x axis.

Eccentric Load about Weak Axis

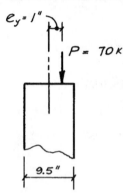

Figure 7.22c Column load applied with eccentricity about y axis.

AXIAL:

The axial stress check remains the same.

BENDING:

The only bending stress is due to the eccentric column force.

$$e_y = 0.1d = 0.1(9.5) = 0.95 \text{ in.} < 1.0$$

$$\therefore \quad e_y = 1.0 \text{ in.}$$

$$\text{Ecc. } f_{by} = \frac{Pe_y}{S_y} = f_c\left(\frac{6e_y}{d_y}\right) = 546\left(\frac{6 \times 1.0}{9.5}\right) = 345 \text{ psi}$$

Determine the allowable bending stress for the y axis. Even with an unbraced length of 24 ft, there is little or no tendency for lateral buckling when the moment is about the y axis. The depth for bending about the y axis is 9.5 in.

$$d = 9.5 < 12$$

$$\therefore \quad C_F = 1.0$$

$$F'_{by} = F_b(C_D)(C_M)(C_t)(C_F)(C_i)$$

$$= 1600(1.15)(1.0)(1.0)(0.86)(1.0)$$

$$= 1582 \text{ psi} > 345 \quad OK$$

COMBINED STRESSES:

Amplification factor for eccentric bending stress

The eccentric bending moment being considered is about the y axis, and the amplification factor is a function of the slenderness ratio for the y axis.

$$\left(\frac{l_e}{d}\right)_y = 20.2$$

$$F_{cEy} = \frac{K_{cE}E'}{[(l_e/d)_y]^2} = \frac{0.3(1,600,000)}{(20.2)^2} = 1175 \text{ psi}$$

$$\left(\frac{f_c}{F'_c}\right)^2 + \frac{f_{by} + f_c(6e_y/d_y)[1 + 0.234(f_c/F_{cEy})]}{F'_{by}[1 - f_c/F_{cEy} - (f_{bx}/F_{bEx})^2]}$$

$$= (0.690)^2 + \frac{0 + 546[6(1.0)/9.5][1 + 0.234(546/1175)]}{1619[1 - 546/1175 - (0/13,255)^2]}$$

$$= 0.917 < 1.0$$

Eccentric load is OK for bending about y axis.

Use   10 × 14   Select Structural   DF-L column.

7.18   References

[7.1] American Forest and Paper Association (AF&PA). 2001. *Allowable Stress Design Manual for Engineered Wood Construction and Supplements and Guidelines,* 2001 ed., AF&PA, Washington DC.

[7.2] American Forest and Paper Association (AF&PA). 2001. *National Design Specification for Wood Construction and Supplement.* 1997 ed., AF&PA, Washington DC.

[7.3] American Institute of Steel Construction (AISC). 2001. *Manual of Steel Construction—Load and Resistance Factor Design,* 3rd ed., AISC, Chicago, IL.

[7.4] Bohnhoff, D.R., Moody, R.C., Verill, S.P., and Shirek, L.F. 1991. "Bending Properties of Reinforced and Unreinforced Spliced Nailed-Laminated Posts," Research Paper FPL-RP-503, Forest Products Laboratory, Forest Service, U.S.D.A., Madison, WI.

[7.5] Gurfinkel, G. 1981. *Wood Engineering,* 2nd ed., Kendall/Hunt Publishing (available through Southern Forest Products Association, Kenner, LA).

[7.6] International Codes Council (ICC). 2003. *International Building Code,* 2003 ed., ICC, Falls Church, VA.

[7.7] Truss Plate Institute (TPI). 2002. National Design Standard for Metal Plate Connected Wood Truss Construction, ANSI/TPI 1-2002, TPI, Madison, WI.

[7.8] Zahn, J.J. 1991. "Biaxial Beam-Column Equation for Wood Members," *Proceedings of Structures Congress '91,* American Society of Civil Engineers, pp. 56–59.

[7.9] Zahn, J.J. 1991. "New Column Design Formula," *Wood Design Focus,* vol. 2, no. 2.

7.19   Problems

Allowable stresses and section properties for the following problems are to be in accordance with the 2001 NDS. Dry service conditions, normal temperatures, and bending about the strong axis apply unless otherwise indicated.

The loads given in a problem are to be applied directly. The load duration factor of 1.6 for problems involving wind or seismic is based on NDS recommendations.

Some problems require the use of a personal computer. Problems that are solved using spreadsheet or equation-solving software can be saved and used as a template for other similar problems. Templates can have many degrees of sophistication. Initially, a template may only be a hand (i.e., calculator) solution worked on a computer. In a simple template of this nature, the user will be required to provide many of the lookup functions for such items as

Tabulated stresses

Lumber dimensions

Duration factor

Wet service factor

Size factor

Volume factor

As the user gains experience with their software, the template can be expanded to perform lookup and decision-making functions that were previously done manually.

Advanced computer programming skills are not required to create effective templates. Valuable templates can be created by designers who normally do only hand solutions. However, some programming techniques are helpful in automating *lookup* and *decision-making* steps.

The first requirement for a template is that it operate correctly (i.e., calculate correct values). Another major requirement is that the input and output be structured in an orderly manner. A sufficient number of intermediate answers should be displayed and labeled so that the solution can be verified by hand.

7.1   A 3 × 8 member in a horizontal diaphragm resists a tension force of 20 k caused by the lateral wind pressure. Lumber is Select Structural DF-L. A single line of 7/8-in.-diameter bolts is used to make the connection of the member to the diaphragm. $C_M = 1.0$, and $C_t = 1.0$, and $C_i = 1.0$.

Find: The allowable axial tension load.

7.2   A 5⅛ × 15 DF axial combination 5 glulam is used as the tension member in a large roof truss. A single row of 1-in.-diameter bolts occurs at the net section of the member. Loads are a combination of dead and snow. Joints are assumed to be pin-connected. MC = 10 percent. $C_t = 1.0$.

Find: *a.* The allowable axial tension load.
  b. Repeat part *a* except that the MC = 15 percent.
  c. Repeat part *a* except that the MC = 18 percent.
  d. Repeat part *a* except that the member is a bending combination 24 F-V8 glulam.

7.3 The truss in Fig. 7.A has a 2 × 4 lower chord of Sel. Str. Spruce-Pine-Fir (South). The loads shown are the result of D = 20 psf and S = 55 psf. There is no reduction of area for fasteners. $C_M = 1.0$, and $C_t = 1.0$, and $C_i = 1.0$. Joints are assumed to be pin-connected.*

Find: Check the tension stress in the member.

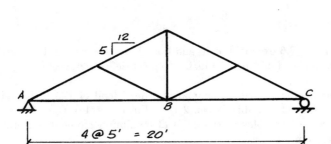

Note: Trusses are spaced
4 ft on center.

Figure 7.A

7.4 Use the hand solution to Prob. 7.1 as a guide to develop a personal computer spreadsheet or equation-solving software template to solve similar problems.
  a. Consider only the specific criteria given in Prob. 7.1.
  b. Expand the template to handle any sawn lumber size. The template is to include a list (i.e., database) of tabulated stresses for all size categories (Dimension lumber, B&S, P&T) of Sel. Str. DF-L.
  c. Expand the database in part *b* to include all stress grades of DF-L from No. 2 through Sel. Str.

7.5 The truss in Fig. 7.A has a 2 × 6 lower chord of No. 1 DF-L. In addition to the loads shown on the sketch, the lower chord supports a ceiling load of 5 psf (20 lb/ft). There is no reduction of member area for fasteners. Joints are assumed to be pin-connected. $C_M = 1.0$, $C_t = 1.0$, and $C_i = 1.0$.

Find: Check combined stresses in the lower chord.

7.6 The truss in Fig. 7.B supports the roof dead load of 16 psf shown in the sketch. Trusses are spaced 24 in o.c., and the roof live load is to be in accordance with

*For trusses with joints which are not pinned (such as toothed metal gusset plates and others), the continuity of the joints must be taken into consideration. For the design of metal-plate-connected trusses, see Ref. 7.7.

the IBC. Lumber is No. 2 DF-L. Fasteners do not reduce the area of the members. Truss joints are assumed to be pin-connected. $C_M = 1.0$, $C_t = 1.0$, and $C_i = 1.0$.

Find: The required member size for the tension (bottom) chord.

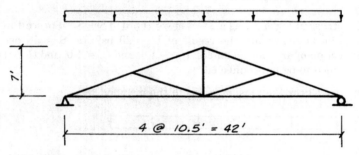

Figure 7.B

7.7   Repeat Prob. 7.6 except that in addition there is a ceiling dead load applied to the bottom chord of 8 psf (16 lb/ft). Neglect deflection.

7.8   The door header in Fig. 7.C supports a dead load of 120 lb/ft and a roof live load of 120 lb/ft. Lumber is No. 2 Hem-Fir. $C_M = 1.0$, $C_t = 1.0$, and $C_i = 1.0$. There are no bolt holes at the point of maximum moment. Lateral stability is not a problem.

Find: *a.* Check the given member size under the following loading conditions:
Vertical loads only
IBC-required combinations of vertical loads and lateral forces
b. Which loading condition is the more severe?

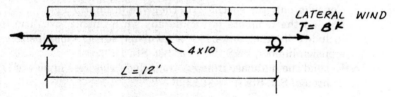

Figure 7.C

7.9   Repeat Prob. 7.8 except the unbraced length is one-half of the span length (that is, $l_u = 0.5L$).

7.10   Use the hand solution to Prob. 7.9 as a guide to develop a personal computer spreadsheet or equation-solving software template to solve similar problems.
a. Consider only the specific criteria given in Prob. 7.9.
b. Expand the template to handle any span length and any unbraced length. The user should be able to choose any trial size of Dimension lumber and any grade of Hem-Fir from No. 2 through Sel. Str.

7.11   A 4 × 4 carries an axial compressive force caused by dead, live, and roof live loads. Lumber is No. 1 DF-L. $C_M = 1.0$, $C_t = 1.0$, and $C_i = 1.0$.

Find: The allowable column load for each load combination if the unbraced length of the member is
a. 3 ft
b. 6 ft
c. 9 ft
d. 12 ft

7.12   Repeat Prob. 7.11 except that the member is a 4 × 6.

7.13   A 6 × 8 carries an axial compressive force caused by dead and snow loads. Lumber is No. 1 DF-L. $C_M = 1.0$, $C_t = 1.0$, and $C_i = 1.0$.

Find: The allowable column load for each load combination if the unbraced length of the member is
a. 5 ft
b. 9 ft
c. 11 ft
d. 15 ft
e. 19 ft

7.14   Use the hand solution to Probs. 7.11 through 7.13 as a guide to develop a personal computer spreadsheet or equation-solving software template to solve similar problems.
a. The user should be able to specify any sawn lumber member size, column length, and tabulated values of F_c and E. Initially limit the template to No. 1 DF-L, and assume that the user will look up and provide the appropriate size factor (if required) for compression.
b. Expand the template to access a database of tabulated stresses for any size and grade of DF-L sawn lumber from No. 2 through Sel. Str. Include provision for the database to furnish the appropriate size factor.

7.15   *Given:*   The glulam column in Fig. 7.D with the following information:

$$D = 20 \text{ k} \qquad L = 90 \text{ k}$$

$$L_r = 40 \text{ k} \qquad L_2 = 10 \text{ ft}$$

$$L_1 = 22 \text{ ft} \qquad L_3 = 12 \text{ ft}$$

The loads are axial forces and the member is axial combination 2 DF glulam without special tension laminations. The column effective length factor is $K_e = 1.0$. $C_M = 1.0$, and $C_t = 1.0$.

Find: Is the column adequate to support the design load?

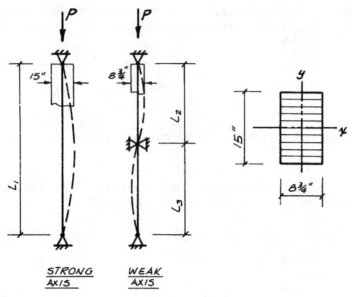

Figure 7.D

7.16   *Given:*   The glulam column in Fig. 7.D with the following information:

$$D = 20 \text{ k} \qquad L = 90 \text{ k}$$

$$L_r = 40 \text{ k} \qquad L_2 = 10 \text{ ft}$$

$$L_1 = 24 \text{ ft} \qquad L_3 = 14 \text{ ft}$$

The loads are axial forces and the member is axial combination 2 DF glulam without special tension laminations. The column effective length factor is $K_e = 1.0$. $C_M = 1.0$, and $C_t = 1.0$.

Find: Is the column adequate to support the design load?

7.17   Use the hand solution to Prob. 7.15 or 7.16 as a guide to develop a personal computer spreadsheet or equation-solving software template to solve similar problems.

   a. Initially the template may be limited to axial combination 2 DF-L, and assume that the user will look up and provide the tabulated properties for the material. The template should handle different loads and unbraced lengths for the x and y axes.

   b. Expand the template to access a database of tabulated stresses for any DF-L glulam combination. Consider either *axial* combinations or *bending* combinations as assigned.

7.18   A sawn lumber column is used to support axial loads of $D = 20 \text{ k}$ and $S = 55$ k. Use No. 1 DF-L. The unbraced length is the same for both the x and y axes of the member. The effective length factor is $K_e = 1.0$ for both axes. $C_M = 1.0$, $C_t = 1.0$ and $C_i = 1.0$.

Find: The minimum column size if the unbraced length is
  a. 8 ft
  b. 10 ft
  c. 14 ft
  d. 18 ft.
  e. 22 ft.
  A computer-based template may be used provided sufficient output is displayed to allow hand checking.

7.19 *Given:* The glulam column in Fig. 7.D with the following information:

$$D = 10 \text{ k} \qquad L = 45 \text{ k}$$

$$L_r = 20 \text{ k} \qquad L_2 = 10 \text{ ft}$$

$$L_1 = 26 \text{ ft} \qquad L_3 = 16 \text{ ft}$$

The minimum eccentricity described in Sec. 7.16 is to be considered. The member is a balanced bending combination 24F-1.8E glulam. The column effective length factor is $K_e = 1.0$. $C_M = 1.0$, and $C_t = 1.0$.

Find: Is the column adequate to support the design loads?

7.20 An 8 × 12 column of No. 1 S-P-F(S) has an unbraced length for buckling about the strong (x) axis of 16 ft and an unbraced length for buckling about the weak (y) axis of 8 ft. $C_M = 1.0$, $C_t = 1.0$, and $C_i = 1.0$.

Find: The allowable axial loads considering D only and D + L.

7.21 A stud wall is to be used as a bearing wall in a wood-frame building. The wall carries axial loads caused by roof dead and live loads. Studs are 2 × 4 Construction-grade Hem-Fir and are located 16 in. o.c. Studs have sheathing attached. $C_M = 1.0$, $C_t = 1.0$, and $C_i = 1.0$.

Find: The allowable load per lineal foot of wall for each load combination if the wall height is
  a. 8 ft
  b. 9 ft
  c. 10 ft

7.22 A stud wall is to be used as a bearing wall in a wood-frame building. The wall carries axial load caused by roof dead and snow loads. Studs are 2 × 6 No. 2 Southern pine and are 24 in. o.c. Studs have sheathing attached. $C_M = 1.0$, $C_t = 1.0$, and $C_i = 1.0$.

Find: The allowable load per lineal foot of wall for each load combination if the wall height is
  a. 10 ft
  b. 14 ft

7.23 *Given:* The exterior column in Fig. 7.E is a 6 × 10 Sel. Str. DF-L. It supports a vertical load due to a girder reaction and a lateral wind force from the horizontal wall framing. The lateral force causes bending about the strong axis of the member, and wall framing provides continuous lateral support about the weak axis. The following values are to be used:

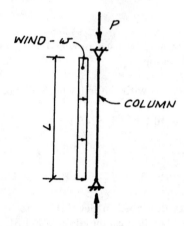

Figure 7.E

$$D = 5 \text{ k} \qquad L = 16 \text{ ft}$$
$$S = 15 \text{ k} \qquad C_M = 1.0$$
$$W = 200 \text{ lb/ft} \qquad C_t = 1.0$$
$$C_i = 1.0$$

Find: Check the column for combined stresses.

7.24   Repeat Prob. 7.23 except that the following values are to be used:

$$D = 5 \text{ k} \qquad L = 21 \text{ ft}$$

$$S = 15 \text{ k} \qquad C_M = 1.0$$

$$W = 100 \text{ lb/ft} \qquad C_t = 1.0$$

$$C_i = 1.0$$

7.25   The truss in Fig. 7.A has a 2 × 10 top chord of No. 2 Hem-Fir. The top of the truss is fully supported along its length by roof sheathing. There is no reduction of area for fasteners. $C_M = 1.0$, $C_t = 1.0$, and $C_i = 1.0$. Joints are assumed to be pin-connected.

Find: Check combined stresses in the top chord. A computer-based template may be used provided sufficient output is displayed to allow hand checking.

7.26   A truss is similar to the one shown in Fig. 7.A except the span is 36 ft and D = 10 psf and S = 20 psf. the top chord is a 2 × 10 of No. 2 Hem-Fir, and it is laterally supported along its length by roof sheathing. There is no reduction of member area for fasteners. $C_M = 1.0$, $C_t = 1.0$, and $C_i = 1.0$. Joints are assumed to be pin-connected.

Find: Check combined stresses in the top chord. A computer-based template may be used provided sufficient output is displayed to allow hand checking.

7.27 A 2 × 6 exterior stud wall is 14 ft tall. Studs are 16 in. o.c. The studs support the following vertical loads per foot of wall:

$$D = 800 \text{ lb/ft}$$

$$L = 800 \text{ lb/ft}$$

$$L_r = 400 \text{ lb/ft}$$

In addition, the wall carries a uniform wind force of 15 psf (horizontal). Lumber is No. 1 DF-L. $C_M = 1.0$, $C_t = 1.0$, and $C_i = 1.0$. Sheathing provides lateral support in the weak direction.

Find: Check the studs, using the IBC-required load combinations. Neglect uplift.

Wood Structural Panels

8.1 Introduction

Plywood, oriented strand board, waferboard, composite panels, and structural particleboard, collectively referred as *wood structural panels,* are widely used building materials with a variety of structural and nonstructural applications. Some of the major structural uses include

1. Roof, floor, and wall sheathing
2. Horizontal and vertical (shearwall) diaphragms
3. Structural components
   a. Lumber-and-plywood beams
   b. Stressed-skin panels
   c. Curved panels
   d. Folded plates
   e. Sandwich panels
4. Gusset plates
   a. Trusses
   b. Rigid frame connections
5. Preservative–treated wood foundation systems
6. Concrete formwork

Numerous other uses of wood structural panels can be cited, including a large number of industrial, commercial, and architectural applications.

As far as the types of buildings covered in this text are concerned, the first two items in the above list are of primary interest. The relatively high allowable loads and the ease with which panels can be installed have made wood structural panels widely accepted for use in these applications. The other topics listed above are beyond the scope of this text. Information on these and other subjects is available from the APA—The Engineered Wood Association.

This chapter will essentially serve as a turning point from the design of the vertical-load-carrying system (beams and columns) to the design of the lateral-force-resisting system (horizontal diaphragms and shearwalls). Wood structural panels provide this transition because it is often used as a structural element in *both* systems.

In the vertical system, structural panels function as the *sheathing* material. As such, it directly supports the roof and floor loads and distributes these loads to the framing system. See Example 8.1. Wall sheathing, in a similar manner, distributes the normal wind force to the studs in the wall. In the lateral-force-resisting system (LFRS), wood structural panels serve as the shear-resisting element.

EXAMPLE 8.1   Wood Structural Panels Used as Sheathing

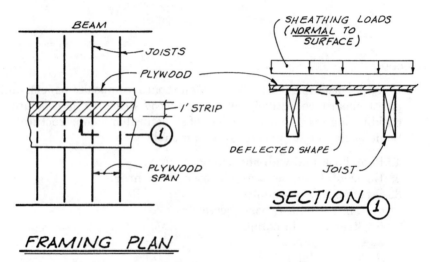

FRAMING PLAN

Figure 8.1 Plywood used to span between framing.

The most popular forms of wood structural panels used for floor or roof sheathing are plywood and oriented strand board (OSB). The term *sheathing load,* as used in this book, refers to loads that are normal to the surface of the sheathing. See Fig. 8.1. Sheathing loads for floors and roofs include dead load and live load (or snow). For walls, the wind force is the sheathing load. Typical sheathing applications use panels continuous over two or more spans. For common joist spacings and typical loads, *design aids* have been developed so that the required sheathing can be chosen without having to perform beam design calculations. A number of these design aids are included in the ASD Manual's Supplement on Wood Structural Panels (Ref. 8.1).

The required *thickness* of the panel is often determined by *sheathing-type loads* (loads normal to the surface of the plywood). On the other hand, the *nailing* requirements for the panel are determined by the *unit shears* in the horizontal or vertical diaphragm. When the shears are high, the required thickness of panel may be governed by the diaphragm unit shears instead of by the sheathing loads.

It should be noted that the required thickness for roof, floor, and wall sheathing are determined using the provisions presented in Chap. 9 of the 2001 NDS, and *design aids* provided in the ASD Manual's Supplement on Wood Structural Panels (Ref. 8.1). Prior to the 2001 NDS, provisions for engineering design with wood structural panels were not included in the NDS whatsoever. Requirements for wood structural panels were determined from design aids provided in Code tables (e.g., Ref. 8.15) or APA literature (e.g., Ref. 8.14). The 2001 NDS does limit its scope to include plywood, oriented strand board, and composite panels. For design with other types of panels, design aids may be available through APA—The Engineered Wood Association.

It is important to realize that the basis for wood structural panel design aids are beam calculations or concentrated load considerations, whichever are more critical. The need may arise for beam calculations if the design aids are found not to cover a particular situation. However, structural calculations for wood structural panels are usually necessary only for the design of a structural-type component (e.g., a lumber-and-plywood beam or stressed-skin panel).

This chapter introduces wood structural panel properties and grades and reviews the procedures used to determine the required thickness and grade of panels for *sheathing applications.* Some of the design aids for determining sheathing requirements are included, but the calculation of stresses in wood panels is beyond the scope of this text. However, the basic structural behavior of structural panels is explained, and some of the unique design aspects of wood panels are introduced. Understanding these basic principles is necessary for the proper use of wood structural panels.

At this time traditional plywood constitutes the majority of total wood structural panel production, but other panel products continue to increase market share, especially oriented strand board (OSB). However, plywood is still the standard by which the other panel products are judged. Because of its wider use in structural applications, plywood and its use as a sheathing material are covered first (Secs. 8.3 to 8.7), and other panel types, including new-generation panels, are introduced in Sec. 8.8. Chapter 9 continues with an introduction to *diaphragm design,* and Chap. 10 covers *shearwalls.* There the calculations necessary for the design of the LFRS are treated in considerable detail.

8.2 Panel Dimensions and Installation Recommendations

The standard size of wood structural panels is 4 ft × 8 ft. Certain manufacturers are capable of producing longer panels, such as 9 ft, 10 ft or 12 ft, but the standard 4 ft × 8 ft dimensions should be assumed in design unless the availability of other sizes is known.

Plywood and the other panel products are dimensionally quite stable. However, some change in dimensions can be expected under varying moisture conditions, especially during the early stages of construction when the material is adjusting to local atmospheric conditions. For this reason, installation in-structions for many roof, floor, and wall sheathing applications recommend a clearance between panel edges and panel ends. See Example 8.2.

EXAMPLE 8.2   Panel Installation Clearances

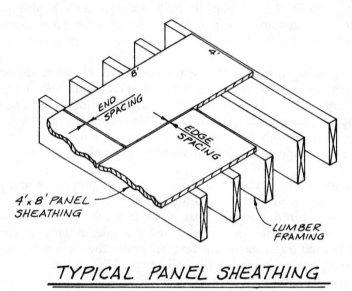

TYPICAL PANEL SHEATHING

Figure 8.2 Clearance between wood structural panels.

Many panel sheathing applications for roofs, floors, and walls recommend an *edge* and *end spacing* of ⅛ in. to permit panel movement with changes in moisture content. Other spacing provisions may apply, depending on the type of panel, application, and moisture content conditions. Refer to the ASD Wood Structural Panel Supplement or the APA's publication *Design/Construction Guide—Residential and Commercial* (Ref. 8.10) for specific recommendations.

The tolerances for panel length and width depend on the panel type. Typical tolerances are +0, −¹⁄₁₆ in., and +0, −⅛ in. Some panel grade stamps include the term *sized for spacing,* and in this case the larger tolerance (+0, −⅛ in.) applies.

The installation clearance recommendations explain the negative tolerance on panel dimensions. By having the panel dimension slightly less than the stated size, the clearances between panels can be provided while maintaining the basic 4-ft module that the use of a wood structural panel naturally implies.

Another installation recommendation is aimed at avoiding *nail popping.* This is a problem that principally affects floors, and it occurs when the sheath-ing is nailed into green supporting beams. As the lumber supports dry, the members shrink and the nails appear to "pop" upward through the sheathing. This can cause problems with finish flooring (especially vinyl resilient flooring and similar products). Squeaks in floors may also develop.

Popping can be minimized by proper nailing procedures. Nails should be driven flush with the surface of the sheathing if the supporting beams are dry. If the supports are green, the nails should be "set" below the surface of the sheathing and the nail holes should not be filled. Squeaks can also be reduced by field-gluing the panels to the supporting beams. For additional information, contact the APA.

Wood structural panels are available in a number of standard thicknesses ranging from ¼ to 1⅛ in. The tolerances for thickness vary depending on the thickness and surface condition of the panel. Panels with veneer faces may have several different surface conditions including unsanded, touch-sanded, sanded, overlaid, and others. See the appropriate specification for thickness tolerances.

8.3   Plywood Makeup

A plywood panel is made up of a number of veneers (thin sheets or pieces of wood). Veneer is obtained by rotating *peeler logs* (approximately 8½-ft long) in a lathe. A continuous veneer is obtained as the log is forced into a long knife. The log is simply unwound or "peeled." See Example 8.3. The veneer is then clipped to the proper size, dried to a low moisture content (2 to 5 percent), and graded according to quality.

EXAMPLE 8.3   Fabrication of Veneer

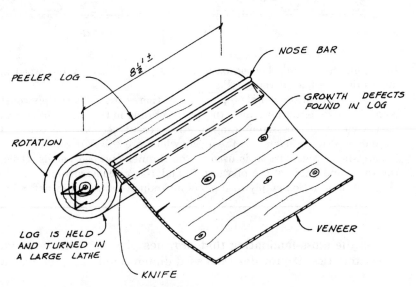

Figure 8.3 Cutting veneer from peeler log.

The log is rotary-cut or peeled into a continuous sheet of *veneer.* Thicknesses range between ¹⁄₁₆ and ⁵⁄₁₆ in. As with sawn lumber, the veneer is graded visually by observing the size and number of defects. Most veneers may be repaired or patched to improve their grade. Veneer grades are discussed in Sec. 8.5.

The veneer is spread with glue and cross-laminated (adjacent layers have the wood grain at right angles) into a plywood panel with an *odd* number of *layers.* See Example 8.4. The panel is then cured under pressure in a large hydraulic press. The glue bond obtained in this process is stronger than the wood in the plies. After curing, the panels are trimmed and finished (e.g., sanded) if necessary. Finally the appropriate grade-trademark is stamped on the panel.

EXAMPLE 8.4   Plywood Cross-Laminated Construction

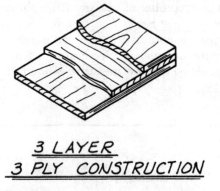

3 LAYER
3 PLY CONSTRUCTION

Figure 8.4*a*

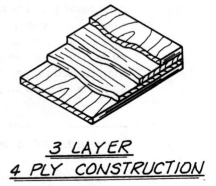

3 LAYER
4 PLY CONSTRUCTION

Figure 8.4*b*

In its simplest form, plywood consists of 3 plies. Each ply has wood grain at right angles to the adjacent veneer (Fig. 8.4*a*).

An extension of the simple 3-ply construction is the 3-layer 4-ply construction (Fig. 8.4*b*). The two center plies have the grain running in the same direction. However, the basic concept of cross-laminating is still present because the two center plies are viewed as a single layer. It is the *layers* which are *cross-laminated.*

Three-layer construction is used in the thinner plywood panels. Depending on the thickness and grade of the plywood, 5- and 7-layer constructions are also fabricated. Detailed information on plywood panel makeup is contained in Ref. 8.18.

It is the cross-laminating that provides plywood with its unique strength characteristics. It provides increased dimensional stability over wood that is not cross-laminated. Cracking and splitting are reduced, and fasteners, such as nails and staples, can be placed close to the edge without a reduction in load capacity.

In summary, *veneer* is the thin sheet of wood obtained from the peeler log. When *veneer* is used in the construction of plywood, it becomes a *ply*. The cross-laminated pieces of wood in a plywood panel are known as *layers*. A layer is often simply an individual ply, but it can consist of more than one ply.

The direction of the grain in a finished panel must be clearly understood. See Example 8.5. The names assigned to the various layers in the makeup of a plywood panel are

1. *Face*—outside ply. If the outside plies are of different veneer quality, the face is the better veneer grade.
2. *Back*—the other outside ply.
3. *Crossband*—inner layer(s) placed at right angles to the face and back plies.
4. *Center*—inner layer(s) parallel with outer plies.

EXAMPLE 8.5   Direction of Grain

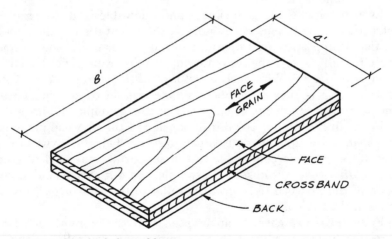

Figure 8.5 Standard plywood layup.

In standard plywood construction, the face and back plies have the grain running parallel to the 8-ft dimension of the panel. Crossbands are inner plies that have the grain at right angles to the face and back (i.e., parallel to the 4-ft dimension). If a panel has more than three layers, some inner plies (centers) will have grain that is parallel to the face and back.

When the stress in plywood is parallel to the 8-ft dimension, there are more "effective" plies (i.e., there are more plies with grain parallel to the stress *and* they are placed farther from the neutral axis of the panel). The designer should be aware that different section properties are involved, depending on how the panel is turned. This is important even if stress calculations are not performed.

If structural calculations are required, the cross-laminations in plywood make stress analysis somewhat more involved. Wood is stronger parallel to the grain than perpendicular to the grain. This is especially true in tension, where wood has little strength across the grain; it is also true in compression but to a lesser extent. In addition, wood is much stiffer parallel to the grain than perpendicular to the grain. The modulus of elasticity across the grain is approximately $\frac{1}{35}$ of the modulus of elasticity parallel to the grain.

Because of the differences in strength and stiffness, the *plies that have the grain parallel to the stress are much more effective* than those that have the grain perpendicular to the stress. In addition, the odd number of layers used in plywood construction causes further differences in strength properties for one direction (say, parallel to the 8-ft dimension) compared with the section properties for the other direction (parallel to the 4-ft dimension). Thus, two sets of cross-sectional properties apply to plywood. One set is used for stresses parallel to the 8-ft dimension, and the other is used for stresses parallel to the 4-ft dimension.

Even if the sheathing thickness and allowable load are read from a table (structural calculations not required), the orientation of the panel and its directional properties are important to the proper use of the plywood. To illustrate the effects of panel orientation, two different panel layouts are considered. See Fig. 8.6. With each panel layout, the corresponding 1-ft-wide beam cross section is shown. The bending stresses in these beams are parallel to the span. For simplicity, the plywood in this example is 3-layer construction.

In the first example, the 8-ft dimension of the plywood panel is parallel to the span (sheathing spans between joists). When the plywood is turned this way (face grain perpendicular to the supports), it is said to be used in the *strong direction.* In the second example, the 4-ft dimension is parallel to the span of the plywood (face grain parallel to the supports). Here the panel is used in the *weak direction.*

From these sketches it can be seen that the cross section for the strong direction has more plies with the grain running parallel to the span. In addition, these plies are located a larger distance from the neutral axis. These two facts explain why the effective cross-sectional properties are larger for plywood oriented with the long dimension of the panel perpendicular to the supports.

8.4 Species Groups for Plywood

A large number of species of wood can be used to manufacture plywood. See Fig. 8.7. The various species are assigned, according to strength and stiffness, to one of five different groups. Group 1 species have the highest-strength char-

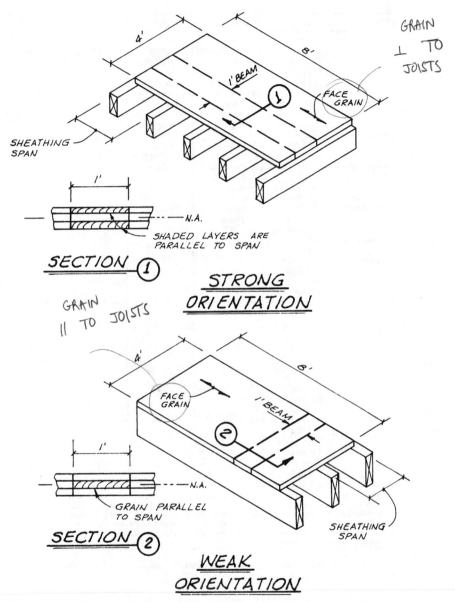

Figure 8.6 Strong and weak directions of plywood. Wood grain that is *parallel* to the span and stress is *more effective* than wood grain that is perpendicular.

acteristics, and Group 5 species have the lowest-strength properties. Allowable stresses have been determined for species groups 1 to 4, and plywood made up of these species can be used in structural applications. Group 5 has not been assigned allowable stresses.

The specifications for the fabrication of plywood allow the mixing of various species of wood in a given plywood panel. This practice allows the more complete usage of raw materials. If it should become necessary to perform stress

Group 1	Group 2	Group 3	Group 4	Group 5
Apitong	Cedar, Port	Alder, Red	Aspen	Basswood
Beech,	Orford	Birch, Paper	Bigtooth	Poplar,
American	Cypress	Cedar, Alaska	Quaking	Balsam
Birch	Douglas	Fir,	Cativo	
Sweet	Fir 2 (a)	Subalpine	Cedar	
Yellow	Fir	Hemlock,	Incense	
Douglas	Balsam	Eastern	Western	
Fir 1 (a)	California	Maple,	Red	
Kapur	Red	Bigleaf	Cottonwood	
Keruing	Grand	Pine	Eastern	
Larch,	Noble	Jack	Black	
Western	Pacific	Lodgepole	(Western	
Maple, Sugar	Silver	Ponderosa	Poplar)	
Pine	White	Spruce	Pine	
Caribbean	Hemlock,	Redwood	Eastern	
Ocote	Western	Spruce	White	
Pine, South.	Lauan	Engelmann	Sugar	
Loblolly	Almon	White		
Longleaf	Bagtikan			
Shortleaf	Mayapis			
Slash	Red			
Tanoak	Tangile			
	White			
	Maple, Black			
	Mengkulang			
	Meranti,			
	Red (b)			
	Mersawa			
	Pine			
	Pond			
	Red			
	Virginia			
	Western			
	White			
	Spruce			
	Black			
	Red			
	Sitka			
	Sweetgum			
	Tamarack			
	Yellow-			
	Poplar			

(a) Douglas Fir from trees grown in the states of Washington, Oregon, California, Idaho, Montana, Wyoming, and the Canadian Provinces of Alberta and British Columbia shall be classed as Douglas Fir No. 1. Douglas Fir from trees grown in the states of Nevada, Utah, Colorado, Arizona and New Mexico shall be classed as Douglas Fir No. 2.
(b) Red Meranti shall be limited to species having a specific gravity of 0.41 or more based on green volume and oven dry weight.

Figure 8.7 Species of wood used in plywood. (*Courtesy of APA—The Engineered Wood Association.*)

calculations, the allowable stresses for plywood calculations have been simplified for use in design. This is accomplished by providing allowable values based on the species group of the face and back plies. The species group of the outer plies is included in the grade stamp of certain grades of plywood (Sec. 8.7). Tabulated section properties are calculated for Group 4 inner plies (the weakest species group allowed in structural plywood). The assumption of Group 4 inner plies is made regardless of the actual makeup. Allowable stresses and cross-sectional properties are given in the APA publication *Plywood Design Specification* (PDS, Ref. 8.14).

Although plywood grades have not yet been covered, it should be noted that some grade modifications can be added to the *sheathing grades* which limit the species used in the plywood. For example, the term STRUCTURAL I can be added to certain plywood grades to provide increased strength properties. The addition of the STRUCTURAL I designation restricts all veneers in the plywood to Group 1 species. Thus the greatest section properties apply to plywood with this designation, because the inner plies of Group 1 species (rather than Group 4 species) are used in calculations.

Besides limiting the species of wood used in the manufacture of plywood, the STRUCTURAL I designation requires the use of exterior glue and provides further restrictions on layup, knot sizes, and repairs over the same grades without the designation. STRUCTURAL I should be added to the plywood grade specification when the increased strength is required, particularly in shear or cross-panel properties (parallel to the 4-ft dimension).

Before the methods for determining the required plywood grade and thickness for sheathing applications can be reviewed, some additional topics should be covered. These include veneer grades, exposure durability classifications, and plywood grades.

8.5 Veneer Grades

The method of producing the veneers which are used to construct a plywood panel was described in Sec. 8.3. Before a panel is manufactured, the individual veneers are graded according to quality. The *grade of the veneers* is one of the factors that determine the *grade of the panel*.

The six basic veneer grades are designated by a letter name:

N	Special-order "natural finish" veneer. Not used in ordinary structural applications.
A	Smooth, paintable surface. Solid-surface veneer without knots, but may contain a limited number (18 in a 4 ft × 8 ft veneer) of neatly made repairs.
B	Solid-surface veneer. May contain knots up to 1 in. in width across the grain if they are both sound and tight-fitting. May contain an unlimited number of repairs.
C-plugged	An improved grade of C veneer which meets more stringent limitations on defects than the normal C veneer. For example, open defects such as knotholes may not exceed ¼ in. by ½ in. Further restrictions apply.
C	May contain open knotholes up to 1 in. in width across the grain and occasional knotholes up to 1½ in. across the grain. This is the minimum-grade veneer allowed in exterior-type plywood.
D	Allows open knotholes up to 2½ in. in width across the grain and occasional knotholes up to 3 in. across the grain. This veneer grade is not allowed in exterior-type plywood.

The veneer grades in this list are given in order of decreasing quality. Detailed descriptions of the growth characteristics, defects, and patching provisions for each veneer grade can be found in Ref. 8.18. Although A and B veneer grades have better surface qualities than C and D veneers, on a structural basis A and C grades are more similar. Likewise, B and D grades are similar, structurally speaking. The reason for these structural similarities is that C veneers can be upgraded through patching and other repairs to qualify as A veneers. See Example 8.6. On the other hand, a B veneer grade can be obtained by upgrading a D veneer. The result is more unbroken wood fiber with A and C veneers.

Except for the special "Marine" exterior grade of plywood, A- and B-grade veneers are used only for face and back veneers. They may be used for the inner plies, but, in general, C and D veneers will be the grades used for the inner plies. It should be noted that D veneers represent a large percentage of the total veneer production, and their use, where appropriate, constitutes an efficient use of materials.

EXAMPLE 8.6   Veneer Grades and Repairs

A veneers are smooth and firm and free from knots, pitch pockets, open splits, and other open defects. A-grade veneers can be obtained by upgrading (repairing) C-grade veneers.

Another upgraded C veneer is C-plugged veneer. Although it has fewer open defects than C, it does not qualify as an A veneer.

B veneers are solid and free from open defects with some minor exceptions. B veneers can be obtained by upgrading (repairing) D-grade veneers. A and B veneers have similar surface qualities, but A and C are structurally similar. Likewise, B and D grades have similar strength properties.

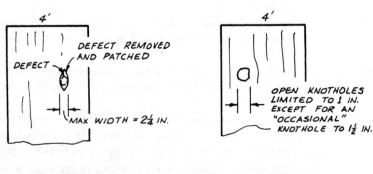

Figure 8.8a   A- and C-grade veneers are structurally similar.

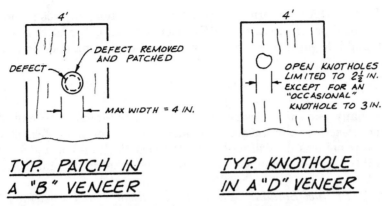

Figure 8.8b B- and D-grade veneers are structurally similar.

8.6 Exposure Durability Classifications

Two exposure durability classifications apply to wood structural panels. One applies to plywood fabricated under PS 1 (Ref. 8.18), and the other applies to panels (both all-veneer plywood and other panels) that are performance-rated. See Sec. 8.7 for additional information on PS 1 and performance-rated panels.

For plywood manufactured under PS 1 the exposure durability classifications are: Exterior and Interior. *Exterior* plywood is glued with an insoluble "waterproof" glue *and* is constructed with a minimum of C-grade veneers. It will retain its glue bond when repeatedly wetted and dried. Exterior plywood is required when it will be permanently exposed to the weather or when the moisture content in use will exceed 18 percent, either continuously or in repeated cycles. In high-moisture-content conditions, the use of preservative–treated panels should be considered.

Interior plywood may be used if it is not exposed to the weather and if the MC in service does not continuously or repeatedly exceed 18 percent. Interior plywood can be manufactured with exterior, intermediate, or interior glue, but it is generally available with exterior glue. Thus, *plywood manufactured with exterior glue is not necessarily classified as Exterior plywood.* If a plywood panel contains a D-grade veneer, it cannot qualify as an Exterior panel even if it is manufactured with exterior glue. The reason for this veneer grade restriction is that the knotholes allowed in the D veneer grade are so large that the glue bond, even with exterior glue, may not stand up under continuous exposure to the weather. Such exposure may result in localized delamination in the area of the knothole.

Interior plywood with any glue type is intended for use in interior (protected) applications. Interior plywood bonded with exterior glue is known as Exposure 1 and is intended for use where exposure to moisture due to long construction delays may occur.

Under APA's performance-rated panel system, plywood and other wood structural panel products are Exterior, Exposure 1, and Exposure 2. For additional information on durability application recommendations, see Ref. 8.10.

8.7 Plywood Grades

For many years the specifications covering the manufacture of plywood have been prescriptive in nature. This means that a method of constructing a plywood panel was fully described by the specification. For a given grade of plywood the species group, veneer grades, and other important factors were specified.

U.S. Product Standard PS 1—Construction and Industrial Plywood (Ref. 8.18) covers the manufacture of traditional *all-veneer* panels known as plywood. For many years PS 1 was a prescriptive-only specification. Although PS 1 still contains prescriptive requirements (a recipe for manufacturing plywood), it now also contains requirements for plywood that can be qualified on the basis of performance tests.

The concept of a *performance standard* was adapted to manufacturing of wood structural panels because a prescriptive type of specification did not lend itself to the development of some of the newer panel products (Sec. 8.8). These panels can be manufactured in a number of different ways using a variety of raw materials. Rather than prescribing how a panel product is to be constructed, a performance standard specifies what the product must do, e.g., load-carrying capability, dimensional stability, and ability to perform satisfactorily in the presence of moisture.

Although performance rating was developed for these newer panel products, it was noted that plywood can also be performance-rated. The performance rating of traditional all-veneer plywood has resulted in the development of newer, thinner thicknesses. For example, $^{15}/_{32}$ in. now replaces $\frac{1}{2}$ in., $^{19}/_{32}$ in. replaces $\frac{5}{8}$ in., and $^{23}/_{32}$ in. replaces $\frac{3}{4}$ in. For more information see APA's publication *Performance Rated Panels* (Ref. 8.9).

There are a large number of grades of plywood. Several examples are given here, but for a comprehensive summary of plywood grades and their appropriate uses, the reader is referred to APA's publication *Design / Construction Guide—Residential and Commercial* (Ref. 8.10) and Ref. 8.6.

Each plywood panel is stamped with a grade-trademark which allows it to be fully identified. A *sanded panel* will have an A- or B-veneer-grade face ply. The back ply may be an A, B, C, or D veneer. The grade-trademark on a sanded plywood panel will include the following:

1. Veneer grade of the face and back

2. Minimum species group (highest species group number from Fig. 8.7) of the outer plies

3. Exposure durability classification

These items essentially identify the plywood. See Example 8.7. Other information included in the stamp indicates the product standard, the manufacturer's mill number (000 shown), and the abbreviation of the "qualified inspection and testing agency." The APA—The Engineered Wood Association is the agency that provides this quality assurance for most of the plywood manufacturers in the United States.

EXAMPLE 8.7   Sanded Plywood Panel

Figure 8.9 Sanded plywood panel. (*Courtesy of APA—The Engineered Wood Association.*)

A typical grade-trademark for a *sanded panel* is shown in Fig. 8.9. Outer plies of A- and B-grade veneers will be sanded. C-plugged will be touch-sanded. Others will be unsanded. For the given example only one side of the panel will be fully sanded.

The sanding operation improves the surface condition of the panel, but in doing so it reduces the thickness of the outer veneers by a measurable amount. In fact, the relative thickness of the layers is reduced by such an amount that different cross-sectional properties are used in strength calculations for sanded, touch-sanded, and unsanded panels. See Tables 3.1 and 3.2 in the ASD Wood Structural Panels Supplement.

Although a sanded plywood grade can be used in a structural application, it is normally not used because of cost. Plywood used in structural applications is often covered with a finish material, and a less expensive plywood grade may be used.

The plywood *sheathing grades* are normally used for roof, floor, and wall sheathing. These are

C-C

C-D

Note that C-C is Exterior-type plywood, and C-D is generally available with exterior glue which qualifies it as Exposure 1. Where added strength is required, these grades can be upgraded by adding STRUCTURAL I to the designation:

C-C STR I

C-D STR I

This grade modification affects both allowable stresses and effective section properties for the plywood.

A sheathing grade of plywood has several different items in the grade-trademark compared with a sanded panel, including the panel thickness and span rating. See Example 8.8.

The *span rating* on sheathing panels is a set of two numbers. The number on the left in the span rating is the maximum recommended span in inches when the plywood is used as roof sheathing. The second number is the maximum recommended span in inches when the plywood is used as subflooring. For example, a span rating of 48/24 indicates that the panel can be used to span 48 in. in a roof system and 24 in. as floor sheathing. In both roof and floor applications it is assumed the panel will be continuous over two or more spans. The purpose of the span rating is to allow the selection of a proper plywood panel for sheathing applications without the need for structural calculations. Allowable roof and floor sheathing loads are covered in later sections of this chapter.

The use of the span rating to *directly* determine the allowable span for a given panel requires that the plywood be oriented in the strong direction (i.e., the long dimension of the panel perpendicular to the supports). In addition, certain plywood *edge support* requirements must be satisfied in order to apply the span rating without a reduction in allowable span.

EXAMPLE 8.8   Plywood Sheathing Grade

Figure 8.10   A sheathing panel (such as C-C) is unsanded. (*Courtesy of APA—The Engineered Wood Association.*)

A typical grade-trademark for a *sheathing grade* of plywood is shown in Fig. 8.10. The stamp indicates the panel is APA Rated Sheathing, *and* it conforms to the Product

Standard PS 1. Because it conforms to PS 1, this is an all-veneer (i.e., plywood) panel. Some APA Rated Sheathing is not traditional all-veneer plywood. These wood structural panels are usually manufactured with some form of reconstituted wood product (Sec. 8.8), and the Product Standard PS 1 will not be referenced in the grade-trademark of these other panels.

The example in Fig. 8.10 also indicates that the plywood is C-C Exterior with the STRUCTURAL I upgrade. The panel is $^{23}/_{32}$ in. thick, and it has a *span rating* of 48/24. Other information includes the manufacturer's mill number (000 shown), and the panel conforms to the *performance standard* PRP-108 recognized in NER-108 (Ref. 8.16).

The ASD Wood Structural Panels Supplement provides design information for plywood, as well as oriented strand board and composite panels. Tables 2.1 and 2.2 in this Supplement provide general information on the grades, thickness, and span ratings for various panel types. The fact to note at this time is that a given span rating may be found on panels of *different* thicknesses. The various thicknesses that comprise a given span rating are listed in Table 5.2 of the ASD Wood Structural Panels Supplement. This table provides both the predominate thickness and alternative thicknesses for each span rating. Table 5.1 provides the section properties for each thickness. The span rating theoretically accounts for the equivalent strength of panels fabricated from different species of wood. Thus, the same span rating may be found on a thin panel that is fabricated from a strong species of wood *and* a thicker panel that is manufactured from a weaker species. Practically speaking, however, plywood for a given span rating is usually constructed so that the thinner (or thinnest) of the panel thicknesses possible will be generally available.

This fact is significant because of the dual function of plywood in many buildings. The minimum *span rating* should be specified for sheathing loads, *and* the minimum *plywood thickness* as governed by lateral diaphragm design (or the minimum thickness compatible with the span rating) should be specified. Thus both span rating and panel thickness are required in specifying a sheathing grade of plywood.

To summarize, panels can be manufactured with different thicknesses for a given span rating. Generally speaking, the thickness that is *available* is the smaller of those listed for a given span rating. If the minimum span rating and minimum thickness are specified, then a panel with a larger span rating and/or thickness may properly be used in the field.

8.8 Other Wood Structural Panels

Reference to wood structural panels in addition to plywood has been made in previous sections of this chapter. These other panel products include composite panels, waferboard, oriented strand board, and structural particleboard. APA

Rated Sheathing and APA Rated Sturd-I-Floor panels include plywood and all the others mentioned. These are recognized by the IBC under *performance standards* such as U.S. Product Standard PS-2 (Ref. 8.17) and NER-108 (Ref. 8.16). Alternatively certain wood structural panels may be produced under a prescriptive type of specification which defines panel mechanical properties such as ANSI A208.1 (Ref. 8.2). These panels can be used for structural roof, floor, and wall sheathing applications. In addition, they can be used to resist lateral forces in horizontal diaphragms and shearwalls.

Wood structural panels that are not all-veneer plywood usually involve some form of reconstituted wood product. A brief description of these panels is given here. See Fig. 8.11.

Oriented strand board (OSB) is a nonveneer panel manufactured from reconstituted wood strands or wafers. The strand-like or wafer-like wood particles are compressed and bonded with phenolic resin. As the name implies, the wood strands or wafers are directionally oriented. The wood fibers are arranged in perpendicular layers (usually three to five) and are thus cross-laminated in much the same manner as plywood. Oriented strand board is manufactured from both softwood species and hardwood species, as well as mixed species.

Oriented strand board was first produced in the early 1980s and has grown to become a major part of the structural wood products industry. Today, OSB is considered in many applications to be interchangeable with plywood. As a

Composite
Panels of reconstituted wood cores bonded between veneer face and back piles.

Waferboard
Panels of compressed wafer-like particles or flakes randomly oriented.

Oriented Strand Board (OSB)
Panels of compressed strand-like or wafer-like particles arranged in layers oriented at right angles to one another.

Structural Particleboard
Panels made of small particles usually arranged in layers by particle size, but not usually oriented.

Figure 8.11 APA Rated Sheathing. Of the four types of sheathing other than plywood, oriented strand board (OSB) is the most widely used. (*Courtesy of APA—The Engineered Wood Association.*)

simple example of this, refer to Table 3.1 from the ASD Wood Structural Panels Supplement where bending stiffness and strength capacities are provided for given span ratings and thicknesses, not for panel type. Table 2.2 in this same Supplement shows that OSB can be used for any span rating, just as plywood can be used for any span rating.

Composite panels (or COMPLY) are recognized by the NDS, along with plywood and OSB, as wood structural panels. Composite panels have a veneer face and back and a reconstituted core. They are typically produced with five layers, such that the center layer is also wood veneer. The reconstituted core is formed from low-quality wood fiber, often from a recycled source.

Waferboard is a nonveneer panel manufactured from reconstituted wood wafers. These wafer-like wood particles or flakes are compressed and bonded with phenolic resin. The wafers may vary in size and thickness, and the direction of the grain in the flakes is usually randomly oriented. The wafers may also be arranged in layers according to size and thickness. Waferboard is the predecessor to OSB, but is not recognized by the NDS as a wood structural panel. APA—The Engineered Wood Association maintains design information regarding use of waferboard.

Structural particleboard is another panel product not recognized by the NDS. It is a nonveneer panel manufactured from small particles (as opposed to larger wafers or strands) bonded with resins under heat and pressure. One primary disadvantage of structural particleboard is its susceptibility to moisture and dimensional instability.

FRP (fiber reinforced plastic) plywood is the latest addition to the suite of wood structural panel products. Fiber reinforced plastic (FRP) sheets are bonded to plywood panels. Currently, the application of FRP to wood panels is limited to plywood, since the bonding of the FRP overlay requires a reasonably smooth surface to adhere and avoid delamination. The application of FRPs to plywood is similar in theory to the use of FRPs in glulam (see Example 6.35). FRP plywood has been used in situations requiring long-lasting durability performance against wearing and weathering. Additional information regarding FRP plywood may be obtained from the APA—The Engineered Wood Association.

As noted, some wood structural panels that are not plywood may involve the use of veneers. For example, composite panels have outer layers that are veneers and an inner layer that is a reconstituted wood core. Other nonplywood panels are completely nonveneer products.

A typical grade-trademark for a nonveneer performance-rated panel includes a number of the same items found in a plywood sheathing stamp. See Example 8.9. However, the grade-trademark found on panels that are not all-veneer plywood does not contain reference to PS 1, and nonveneer panels will not have veneer grades (e.g., C-D) shown in the stamp. Panel grades covered in PS 2 include Sheathing, Structural I Sheathing, and Single Floor. These designations also appear in the grade-trademarks when panels conform to this standard.

EXAMPLE 8.9 Nonveneer Sheathing Grade

APA
THE ENGINEERED WOOD ASSOCIATION

RATED SHEATHING
32/16 15/32 INCH
SIZED FOR SPACING
EXPOSURE 1
— 000 —
PRP-108 HUD-UM-40C

APA
THE ENGINEERED WOOD ASSOCIATION

RATED SHEATHING
32/16 15/32 INCH
SIZED FOR SPACING
EXPOSURE 1
— 000 —
STRUCTURAL I RATED
DIAPHRAGMS-SHEAR WALLS
PANELIZED ROOFS
PS 2-92 SHEATHING
PRP-108 HUD-UM-40C

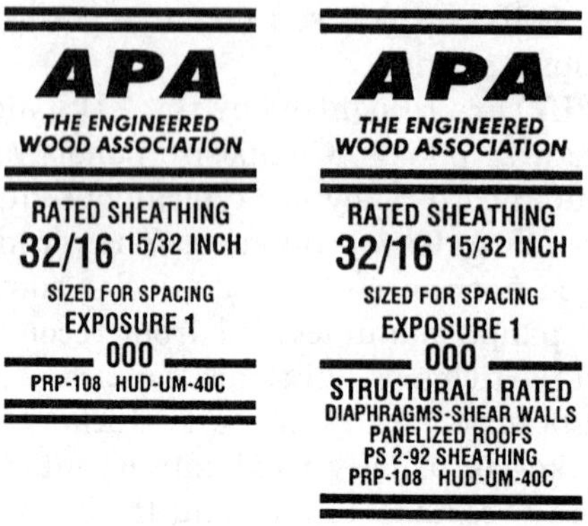

Figure 8.12 Nonveneer sheathing grade. (*Courtesy of APA— The Engineered Wood Association.*)

A typical grade-trademark for a *nonveneer* panel is shown in Fig. 8.12. The stamp indicates that the panel is APA Rated Sheathing, and it has a span rating of 32/16 (Sec. 8.7). In addition, the thickness ($^{15}\!/_{32}$ in.) and durability classification (Exposure 1) are shown.

Other information includes the manufacturer's mill number (000 shown), and the panel is "sized for spacing" (Sec. 8.2). The panel conforms to APA's performance standard PRP108, recognized in NER-108 (Ref. 8.16).

8.9 Roof Sheathing

Wood structural panels account for much of the wood roof sheathing used in the United States. These materials are assumed to be continuous over two or more spans. Plywood, OSB and other panels with directional properties are normally used in the strong direction (long dimension of the panel perpendicular to the supports). However, in panelized roof systems (Sec. 3.2) the panels are often turned in the weak direction for sheathing loads. In this latter case, a thicker panel may be required, but this type of construction results in a savings in labor. In addition, higher diaphragm shears can be carried with the increased panel thickness.

The *span rating* described in Sec. 8.7 appears in the grade-trademark of both traditional plywood and APA's performance-rated sheathing. Recall that plywood, OSB, and other wood structural panels may be performance-rated.

Table 3.1 in the ASD Wood Structural Panels Supplement provides bending stiffness and strength capacities for wood structural panels. These design values can be used to determine the appropriate product (span rating and/or thickness). However, traditionally plywood and other wood structural panels have been viewed more like a proprietary product and thus are selected by evaluating "allowable loads" versus actual design values. Tables 7.1, 7.2, and 7.3 in the Wood Structural Panels Supplement provide allowable uniform load on *sheathing, single floor,* and *sanded panels,* respectively. Note that these tables all assume normal load duration, $C_D = 1.0$. For other durations of load, the allowable loads governed by bending and shear are determined by multiplying the tabulated values by the appropriate load duration factor. The allowable loads governed by deflection would not be adjusted for load duraiton.

The span rating can be used *directly* to determine the sheathing requirements for panels used in the strong direction under typical roof live loading conditions. When used directly, the actual span agrees with the roof span in the two-number span rating. For larger roof loads, the span rating can be used *indirectly* to determine panel sheathing requirements. See Example 8.10.

EXAMPLE 8.10   Determination of Panel Sheathing Requirements

The span rating can be used directly to specify sheathing. When used directly the actual span, or support spacing, agrees with the roof span rating (the first number in the two-number span rating).

The span rating can also be used indirectly to determine sheathing requirements. In this case the roof span in the span rating (the first number in the two-number span rating) will not agree with the actual span or support spacing. To illustrate this, consider a roof system with a sheathing span of 24 in., a D + S load of 85 psf, and an $L/240$ deflection limit. Determine the minimum span-rated sheathing.

Noting that allowable load tables in the ASD Wood Structural Panels Supplement, as with most NDS tables, are presented for normal load duration, $C_D = 1.0$. For the D + S load combination, $C_D = 1.15$. To use Table 7.1 for a duration other than the normal duration, simply multiply the values in the table by the appropriate load duration factor, in this case 1.15. Since in this case the load is given, the load can be divided by the load duration factor and compared to the tabulated values in the table. Therefore, the design allowable load, converted to a normal load basis, is 85/1.15 = 74 psf.

The panel with a span rating of 40/20 is determined to meet the demand. The 40/20 placed on 24-in. center-to-center supports allows 146 psf > 85 psf considering the $L/240$ deflection limit, 130 psf > 74 psf considering bending, and 182 psf > 74 psf considering shear. In this case, the span rating of 40/20 does not relate to the actual span (24 in.).

The maximum allowable spans defined by the span rating assume that the edges parallel to the supports are supported in some fashion. This is referred to as panels *with edge support*. Typically, panel clips are used to provide this edge support, but lumber blocking or another mechanism may be used. Panels *without edge support* have reduced maximum allowable spans. The maximum allowable spans for panels *with* and *without* edge support are provided in Table 6.2 of the ASD Wood Structural Panels Supplement. See Example 8.11.

Panel edge support is intended to limit differential movement between adjacent panel edges. Consequently if some form of edge support is not provided, a thicker panel or a reduced spacing of supports will be required.

A point about the support of panel edges should be emphasized. The use of tongue-and-groove (T&G) edges or panel clips is accepted as an alternative to lumber blocking for sheathing loads only. If blocking is required for diaphragm action (Chap. 9), panel clips or T&G edges (except 1⅛-in.-thick 2-4-1 plywood with stapled T&G edges) cannot be substituted for lumber blocking.

EXAMPLE 8.11   Roof Sheathing Edge Support Requirements

Alternative forms of support for panel edge that is perpendicular to roof framing (Fig. 8.13).

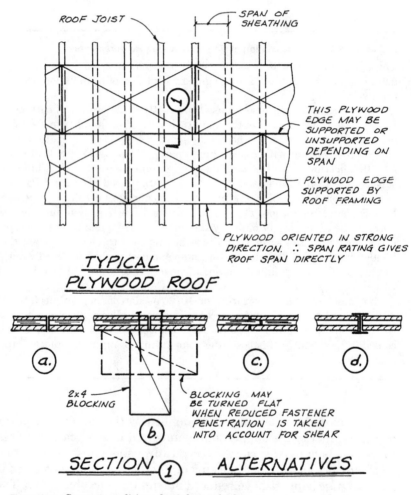

Figure 8.13 Support conditions for 8-ft panel edge.

    a. Unsupported edge. In most cases the NDS requires closer roof joist spacing than given by the *span rating* when the 8-ft panel edges are not supported. Note that panel thickness may be increased as an alternative to providing support to all edges (or reducing the roof beam spacing).

    b. Lumber blocking. Cut and fitted between roof joists.

c. *Tongue-and-groove (T&G) edges.*

d. *Panel clips.* Metal H-shaped clips placed between plywood edges. The IBC, in footnote f of Table 2304.7(3), requires one panel clip midway between supports when the span is less than 48 in. o.c. Two equally spaced clips are required between supports 48 in. o.c.

Use of lumber blocking, T&G edges, and panel clips constitutes edge support for vertical loads. T&G edges and panel clips cannot be used in place of blocking if blocking is required for diaphragm action. Exception: 1⅛-in.-thick plywood with properly stapled T&G edges qualifies as a blocked diaphragm. Diaphragms are covered in Chap. 9.

It was noted at the beginning of this section that panels may be oriented in the weak direction in panelized roof systems. The span rating does not apply directly when panels are used in this manner. However, the design tables in the ASD Wood Structural Panels Supplement include values for panels placed such that the strength axis is parallel to the supports.

8.10   Design Problem: Roof Sheathing

A common roof sheathing problem involves supports that are spaced 24 in. o.c. See Example 8.12. This building is located in a non-snow load area. Consequently the sheathing is designed for a roof live load of 20 psf (because the sheathing spans only 24 in., there is no tributary area reduction for roof live load).

Part 1 of the example considers panels oriented in the strong direction, and several alternatives are suggested.

In part 2, plywood sheathing requirements for a panelized roof are considered. In this case the number of plies used in the construction of the plywood panels is significant. If 5-ply construction is used, the effective section properties in the weak direction are larger. In 4-ply (3-layer) construction, the section properties are smaller, and either stronger wood (STRUCTURAL I) or a thicker panel is required.

EXAMPLE 8.12   Roof Sheathing with a 24-in. Span

Loads

$$D = \;\; 8 \text{ psf}$$

$$\underline{L_r = 20 \text{ psf}} \quad \text{(no snow load)}$$

$$TL = 28 \text{ psf}$$

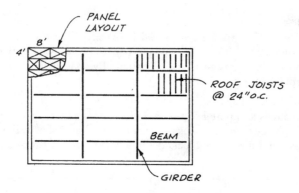

ROOF FRAMING PLAN

Figure 8.14 Panels turned in strong direction.

Part 1

For the roof layout shown, determine the panel sheathing rquirements.

Panel sheathing grades apropriate for this application are Sheathing EXP 1 and STRUCTURAL I Sheathing EXP 1. Both grades are intended for roof applications. The STRUCTURAL I designation means the panel is made with only exterior glue and Group 1 species. STRUCTURAL I should be used when the added shear capacity, and cross-panel strength is necessary for lateral forces. Note that OSB and plywood are both available for these grades. See Table 2.1 in the ASD Wood Structural Panels Supplement.

In Fig. 8.14 the panels are oriented in the strong direction. From Table 7.1 in the ASD Wood Structural Panels Supplement, the required span rating is 24/0. The tabulated allowable uniform loads are as follows:

Load governed by:	Allowable uniform load:	
$L/360$ deflection limit	26 psf (not applicable)	
$L/240$ deflection limit	39 psf > 20 psf (live load deflection)	*OK*
$L/180$ deflection limit	52 psf > 28 psf (total load deflection)	*OK*
Bending strength	52 psf > 28 psf	*OK*
Shear strength	116 psf > 28 psf	*OK*

ASD Wood Structural Panels Supplement Table 5.2 indicates that 24/0 panels may be either ⅜-, ⁷⁄₁₆-, ¹⁵⁄₃₂-, or ½-in. thick, with the ⅜-in. panel being the predominate nominal thickness.

Table 6.2 in the ASD Wood Structural Panels Supplement indicates that with un-blocked edges the maximum allowable span for 24/0 panels is 20 in. Therefore, edge support may be provided by

1. Blocking
2. T&G edges (available in plywood thicknesses ¹⁵⁄₃₂ in. and greater)
3. H-shaped metal clips (panel clips)

See Fig. 8.13.

Although ⅜-in. panels qualify for a 24/0 span rating, ⁷⁄₁₆ in. (or ½-in.) panels are often used to span 24 in. for the roof in this type of building. Also from Table 5.2, ⁷⁄₁₆-

in. panels qualify for a span rating of 24/16. If panels with a span rating of 24/16 are used, the maximum unsupported edge length is 24 in., which is equal to the actual span of 24 in. Therefore, edge support is not required for this alternative. When roofing is to be guaranteed by a performance bond, the roofing manufacturer should be consulted for minimum thickness and edge support requirements.

Summary

The *minimum* panel requirement for this building is ⅜-in. sheathing with a span rating of 24/0 and edges supported.

An alternative plywood sheathing is ⁷⁄₁₆-in. sheathing with a span rating of 24/16 without edge support.

Blocking may be required for either choice for lateral forces (diaphragm action). See Chap. 9.

NOTE: The construction of the panel itself was not specified. That is, any wood structural panel meeting the span rating could be used in this application, including plywood or OSB.

Part 2

In the above building, assume that wood structural panels are is to be used in a panelized roof. Panels will be turned 90 degrees to that shown in Fig. 8.14, so that the long dimension of the panel is parallel to the supports (joists). Here the span rating cannot be used directly because panels are oriented in the weak direction. From Table 7.1 in the ASD Wood Structural Panels Supplement, the required span rating is 48/24. The tabulated allowable uniform loads are as follows:

Load governed by:	Allowable uniform load:	
$L/360$ deflection limit	16 psf (not applicable)	
$L/240$ deflection limit	24 psf > 20 psf (live load deflection)	*OK*
$L/180$ deflection limit	33 psf > 28 psf (total load deflection)	*OK*
Bending strength	38 psf > 28 psf	*OK*
Shear strength	213 psf > 28 psf	*OK*

ASD Wood Structural Panels Supplement Table 5.2 indicates that 48/24 panels may be either ²³⁄₃₂-, ¾-, or ⅞-in. thick, with the ²³⁄₃₂-in. panel being the predominate nominal thickness.

8.11   Floor Sheathing

Wood structural panels are used in floor construction in two ways. One system involves *two layers* of panels, and the other system involves a *single layer.*

The terms used to refer to these different panel layers are:

1. *Subfloor*—the bottom layer in a two-layer system

2. *Underlayment*—the top layer in a two-layer system

3. *Combined subfloor-underlayment*—a single-layer system

A finish floor covering such as vinyl sheet or tile, ceramic tile, hardwood, or carpeting is normally provided. See Ref. 8.10 for specific installation recommendations.

In the two-layer system, the subfloor is the basic structural sheathing material. See Example 8.13. It may be either a sheathing grade of plywood or a nonveneer panel. Recall the two-number span rating from Sec. 8.7. For panels with this span rating the second number is the recommended span in inches when the panel is used as a subfloor. Allowable floor live loads depend on the type of panel.

EXAMPLE 8.13   Two-Layer Floor Construction

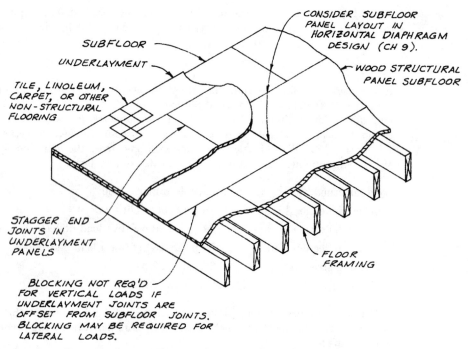

Figure 8.15 Floor construction using a separate subfloor and underlayment.

UNDERLAYMENT-grade panels have C-plugged face veneer and special C-grade inner-ply construction to resist indentations. Typical underlayment thickness is $\frac{1}{4}$ in. for remodeling or use over a panel subfloor and $\frac{3}{8}$ to $\frac{1}{2}$ in. for use over a lumber subfloor or new construction.

When finish flooring has some structural capacity, the underlayment layer is not required. Wood strip flooring and lightweight concrete are examples of finish flooring which do not require the use of underlayment over the subfloor.

As stated for roof sheathing, the ASD Wood Structural Panels Supplement Table 3.1 provides the basic bending stiffness and strength capacities for sheathing. Table 7.1 in the ASD Panels Supplement provides allowable uniform loads for sheathing at various spans or center-to-center spacing of the framing supporting the floor sheathing.

It should be noted that typical wood structural panel applications for floor sheathing are not controlled by *uniform* load criteria, but instead are based on deflection under concentrated loads and how the floor *feels* to passing traffic. These and other subjective criteria relate to user acceptance of floor sheathing. For additional information see Refs. 8.4 and 8.11.

In subfloor construction, panels must be used in the strong direction and must be continuous over two or more spans. Differential movement between adjacent unsupported panel edges is limited by one of the following:

1. Tongue-and-groove edges

2. Blocking

3. ¼-in. underlayment with panel edges offset over the subfloor

4. 1½ in. of lightweight concrete over the subfloor

5. Finish floor of ¾-in. wood strips

As discussed for roof sheathing, the span rating of sheathing can be used *directly* to determine the panel sheathing requirements. For larger loads, the span rating can be used *indirectly* to determine panel sheathing requirements. See Example 8.10.

In two-layer floor construction, the top layer is a grade of panel known as UNDERLAYMENT. The underlayment layer lies under the finish floor covering and on top of the subfloor. It is typically ¼- to ½-in. thick, and its purpose is to provide a solid surface for the direct application of nonstructural floor finishes. UNDERLAYMENT-grade panels are touch-sanded to provide a reasonably smooth surface to support the non-structural finish floor.

Single-layer floor construction is sometimes known as combination subfloor-underlayment because one layer serves both functions. Single-layer floor systems may use thicker grades of UNDERLAYMENT and C-C Plugged Exterior plywood, composite panels, or some form of nonveneer panel. APA's performance-rated panels for single-layer floors are known as Sturd-I-Floor and include plywood, composite, and nonveneer panels. The *span rating* for panels intended to be used in single-layer floor systems is composed of a single number. Here the span rating is the recommended maximum floor span in inches. See Example 8.14.

The basic bending stiffness and strength capacities for single-layer floor panels are given in Table 3.1 of the ASD Wood Structural Panels Supplement. Table 7.2 in the Panels Supplement provides allowable uniform loads for single-layer floor panels at various spans or center-to-center spacing of the framing supporting panels. Also, as with any span-rated product, the span rating of single-layer floor panels can be used *indirectly* to determine panel sheathing requirements for spans other than the rated spans. See example 8.10.

EXAMPLE 8.14 **Single-Layer Floor Panels**

Figure 8.16a Typical grade-trademarks for subfloor-underlayment. (*Courtesy of APA—The Engineered Wood Association.*)

Figure 8.16b Typical grade-trademarks for nonveneer single-layer floor panel. (*Courtesy of APA—The Engineered Wood Association.*)

A typical grade-trademark for a panel combination subfloor-underlayment is shown in Fig. 8.16a. In this example the panel can be identified as plywood because PS 1 is referenced. It is an UNDERLAYMENT grade of plywood that is an interior type with exterior glue (Exposure 1). The panel is APA Rated Sturd-I-Floor, which is also qualified under the performance standard PRP-108. The span rating is 48 in. o.c., and the panel thickness is $1\frac{1}{8}$ in. Other information in the stamp includes the manufacturer's mill number (000 shown), the panel sized for spacing ($+0$, $-\frac{1}{8}$ in. tolerance on panel dimensions), and it has T&G edges.

A typical grade-trademark for a nonveneer single-layer floor panel is shown in Fig. 8.16b. Notice that the grade-trademark does not reference PS 1. Like the panel grade stamp in Fig. 8.16a, this panel is performance-rated under PRP-108. It has a span rating of 20 o.c., and the panel is $\frac{19}{32}$ in. thick. It has a durability rating of Exposure 1. The panel is sized for spacing and has T&G edges. By taking into account the T&G edges and the fact that it is sized for spacing, the panel has a net width of $47\frac{1}{2}$ in.

Single-layer floor systems can be installed with nails. However, the APA glued floor system increases floor stiffness and reduces squeaks due to nail popping (Sec. 8.2). This system uses a combination of field gluing and nailing of floor panels to framing members. Table 6.1 of the ASD Wood Structural Panels Supplement provides recommended minimum nail sizes, types, and spacings for various panel applications. For additional information see Ref. 8.10.

In addition to sheathing and single-layer floor panels, the NDS references generic sanded Exterior plywood panels that are not span-rated. The basic bending stiffness and strength capacities for sanded plywood panels are given for each thickness in Table 3.1 of the ASD Wood Structural Panels Supplement. Table 7.3 in the Panels Supplement provides allowable uniform loads for sanded panels at various spans or center-to-center spacing of the framing supporting the panels.

8.12 Design Problem: Floor Sheathing

In this example a typical floor sheathing problem for an office building is considered. See Example 8.15. The floor utilizes a two-layer floor system with a separate subfloor and underlayment. The subfloor is chosen from the sheathing grades using the two-number span rating described in Sec. 8.7. A ¼-in. plywood UNDERLAYMENT-grade panel is used over the subfloor. If the joints of the underlayment are staggered with respect to the joints in the subfloor, no special edge support is required for the subfloor panels.

A single-layer panel floor could be used as an alternative. Plywood combination subfloor-underlayment (rather than a nonveneer panel) is recommended in Ref. 8.10 when the finish floor is a resilient nontextile flooring or adhered carpet without pad. The span rating for a combination subfloor-underlayment panel consists of a single number in the grade-trademark.

EXAMPLE 8.15   Floor Sheathing with 16-in. Span

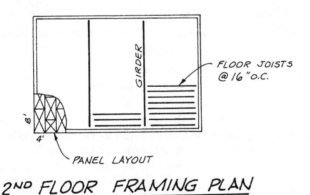

Figure 8.17 Floor construction requires panels in strong direction.

Loads:

$$\text{Floor dead load} = 12 \text{ psf}$$

$$\text{Partition dead load} = 20 \text{ psf}$$

$$\underline{\text{Floor live load} = 50 \text{ psf}}$$

$$\text{TL} = 82 \text{ psf}$$

For the floor layout, determine the sheathing requirements for vertical loads, assuming a separate plywood subfloor and underlayment construction. A resilient-tile finish floor will be used.

Panel sheathing grades appropriate for this application are Sheathing EXP 1 and STRUCTURAL I Sheathing EXP 1. Both grades are intended for floor applications. The STRUCTURAL I designation means the panel is made with only exterior glue and Group 1 species. STRUCTURAL I should be used when the added shear capacity and cross-panel strength is necessary for lateral forces. Note that OSB and plywood are both available for these grades. See Table 2.1 in the ASD Wood Structural Panels Supplement.

From the floor framing plan, the plywood is oriented in the strong direction over two or more spans. Therefore the span rating applies.

$$\text{Joist spacing} = \text{plywood span} = 16 \text{ in.}$$

$$\text{Req'd span rating} = \text{roof span/floor span}$$

$$= 24/16 \text{ or } 32/16$$

Alternatively, the floor sheathing may be selected by comparing the design loads with the allowable loads in Table 7.1 in the ASD Wood Structural Panels Supplement. Assuming a live load deflection limit of $L/360$ and a total load deflection limit of $L/240$, it is noted that the 24/0 span-rated panel nearly satisfies the design requirements:

Load governed by:	Allowable uniform load:
$L/360$ deflection limit	98 psf > 50 psf (live load deflection)
$L/240$ deflection limit	147 psf > 82 psf (total load deflection)
$L/180$ deflection limit	Not applicable
Bending strength	117 psf > 82 psf
Shear strength	179 psf > 82 psf

The 32/16 span rating is selected for this application since the 32/16 panel is available as a 3-ply, 4-ply, and 5-ply plywood panel as well as OSB, allowing greater flexibility in product choice. Furthermore, Table 5.2 of the Panels Supplement indicates that $^{15}\!/_{32}$-, $\frac{1}{2}$-, $^{19}\!/_{32}$-, or $\frac{5}{8}$-in. thick panels are available for this span rating, with the thinnest thickness being the most common. Therefore, the minimum wood structural panel requirement for the subflooring in this problem is

$$^{15}\!/_{32}\text{-in. } 32/16 \text{ span-rated sheathing}$$

Because of the type of finish floor, $\frac{1}{4}$-in. or thicker UNDERLAYMENT-grade plywood should be installed over the subfloor. Underlayment panel edges should be offset with respect to subfloor edges to minimize differential movement between subfloor panels.

Other possible subfloor choices exist including nonveneer panels. As an alternative, a single-layer floor system can be used.

8.13   Wall Sheathing and Siding

Wood structural panels can be used in wall construction in *two basic* ways. In *one method,* the panels serve a structural purpose only. They are attached directly to the framing and serve as sheathing to distribute the normal wind force to the studs, and they may also function as the basic shear-resisting elements if the wall is a shearwall. See Example 8.16. Finished siding of wood or other material is then attached to the outside of the wall.

Typical sheathing grades of plywood and a variety of nonveneer panels are used when finished siding will cover the sheathing. In wall construction, the long dimension of the panel can be either parallel or perpendicular to the studs (supports). For shearwall action all panel edges must be supported. This is provided by studs in one direction and wall plates or blocking between the studs in the other direction.

Finish siding material can be attached by nailing through the wood structural panel sheathing into the wall framing, or it may be attached by nailing directly into the sheathing. See Ref. 8.10 for specific recommendations.

EXAMPLE 8.16   Wood Structural Panel Sheathing with Separate Siding

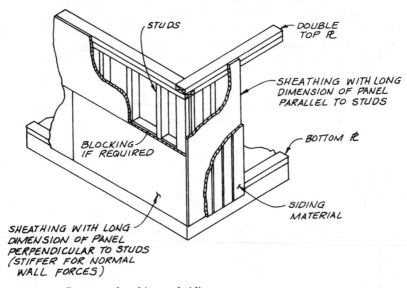

Figure 8.18 Separate sheathing and siding.

In this system the wood structural panels are basically a structural wall element. Wind forces normal to the wall are carried by the sheathing to the wall studs. In some cases the minimum panel requirements are increased if the face grain is not perpendicular to the studs. Finish siding is applied over the sheathing.

If the panel sheathing also functions as a shearwall (lateral forces parallel to wall), panel edges not supported by wall framing must be blocked and nailed (see Chap. 10). *Minimum panel nailing* is 6d common or galvanized box nails at 6 in. o.c. at supported edges and 12 in. o.c. along intermediate supports (studs). See Table 6.1 in the ASD Panels Supplement. Heavier nailing may be required for shearwall action.

In the *second method* of using wood structural panels in wall construction, a single panel layer is applied as combined sheathing-siding. When panels serve as the siding as well as the structural sheathing, a panel siding grade such as APA Rated Siding may be used. See Example 8.17.

Common types of panel siding used for this application are the APA proprietary products known as APA Rated Siding-303. These are exterior plywood panels available with a variety of textured surface finishes. A special type of 303 panel siding is known as Texture 1-11 (T 1-11) and is manufactured in $^{19}\!/_{32}$- and $^{5}\!/_{8}$-in. thicknesses only. It has shiplap edges to aid in weather-tightness and to maintain surface pattern continuity. T 1-11 panels also have $^{3}\!/_{8}$-in.-wide grooves cut into the finished side for decorative purposes. A net panel thickness of $^{3}\!/_{8}$ in. is maintained at the groove. The required thickness of wood structural panels for use as combined sheathing and siding is the same as when used as a structural sheathing only, except as noted in footnote c of Table 4.1A or 4.1B of the ASD Wood Structural Panels *Shear Wall and Diaphragm* Supplement. APA Rated Siding includes a single-number span rating in the grade-trademark which indicates the maximum spacing of studs. Additional information on Rated Siding is available in Ref. 8.10.

EXAMPLE 8.17   Plywood Combined Sheathing-Siding

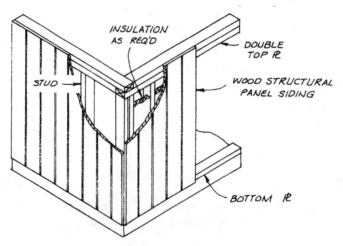

Figure 8.19a Combined sheathing-siding.

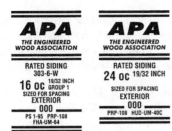

Figure 8.19*b* Typical grade-trademarks for combined sheathing-siding. (*Courtesy of APA—The Engineered Wood Association.*)

In the combined sheathing-siding system, the wood structural panel siding usually has a textured surface finish. These finishes include rough-sawn, brushed, and smooth finish for painting [medium density overlay (MDO)]. In addition to different surface textures, most siding panels are available with grooving for decorative effect.

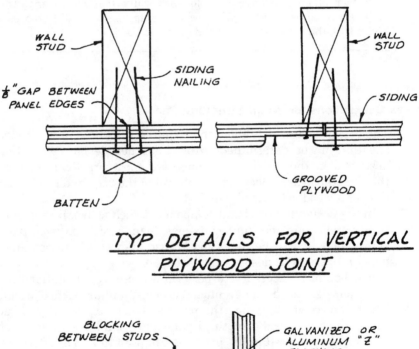

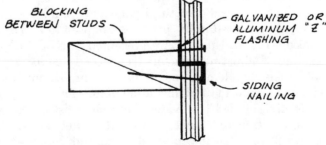

Figure 8.19*c* Panel edge details for combined sheathing-siding.

Typical grade-trademarks for combined sheathing-siding panels are shown in Fig. 8.19*b*. The examples are for APA Rated Siding. A single-number span rating (e.g., 16 or 24 o.c.) indicates maximum recommended stud spacing when the panels run vertically. For additional information including an explanation of siding face grades, see Ref. 8.10.

Wood structural panel siding is usually installed with the 8-ft panel dimension running vertically. However, these panels can also be installed horizontally. Various panel joint details can be used for protection against the weather (see Fig. 8.19*c*).

Nailing requirements for combined sheathing-siding are similar to the nailing for sheathing in the two-layer wall system (Example 8.16). However, hot-dipped galvanized nails are normally used to reduce staining. Casing nails may be used where the presence of a common or box nail head is objectionable. Additional information on plywood nailing for walls is included in Chap. 10. For siding ½-in. thick, or less, 6d nails are used. For thicker siding the nail size is increased to 8d nails.

Heavier nailing may be required for shearwall action. For shearwall design the panel thickness to be used is the *thickness where nailing occurs*. If grooves are nailed (see Fig. 8.19*c*), the net thickness at the groove is used.

8.14   Stress Calculations for Wood Structural Panels

The design aids which allow the required thickness of wood structural panel sheathing to be determined without detailed design calculations have been described in the previous sections. For most practical sheathing problems, these methods are adequate to determine the required grade and panel thickness of a wood structural panel.

In the design of structural components such as box beams with wood structural panel webs and lumber flanges, foam-cores sandwich panels, or stress-skin panels, it will be necessary to use allowable stresses and cross-section panel properties in design calculations.

If it becomes necessary to perform structural calculations for wood structural panels, the designer must become familiar with a number of factors which interrelate to define the panel's structural capacity. Structural design calculations for wood structural panels are similar to that of solid lumber or glulam, as presented in the previous Chaps. 4 to 7. Basic design requirements for wood structural panels are presented in NDS Chap. 9, including the applicability of adjustment factors (see NDS Table 9.3.1). The ASD Wood Structural Panels Supplement, as mentioned throughout this chapter, contains all design value information, including values for the applicable adjustment factors. NDS Table 9.3.1 provides a complete list of the applicable adjustment factors, which include load duration (C_D), wet service (C_M), temperature (C_t), and two factors specific to wood structural panels—grade and construction (C_G) and panel size (C_s).

The grade and construction adjustment factor compensates for different panel constructions (e.g., 3-, 4-, or 5-ply plywood, OSB, or composite panels) as well as different grades within the various panel construction types. See

Panels Supplement Tables 3.1.1, 3.2.1, 3.3.1, and 3.4.1. Note that the design values for wood structural panels have been tabulated with the lowest possible strength value. That is, the grade and construction adjustment factor is defined such that $C_G \geq 1.0$ and can be conservatively assumed to be 1.0. However, in many situations, particularly with STRUCTURAL I panels, C_G is significantly greater than 1.0, and may be as high as 5.2.

The bending and tension strength values tabulated in the ASD Wood Structural Panels Supplement are for panels 24 in. or greater in width, w. The width here is defined as the dimension of the panel perpendicular to the applied stress. The panel size adjustment factor, C_s, reduces the tabulated design values when the width is less than 24 in. See Panels Supplement Table 4.5.

$$w \geq 24 \text{ in.} \qquad C_s = 1.0$$

$$8 \text{ in.} < w < 24 \text{ in.} \qquad C_s = 0.5$$

$$w \leq 8 \text{ in.} \qquad C_s = (8 + w)/32$$

The other adjustment factors for wood structural panels are presented in Sec. 4 of the ASD Wood Structural Panels Supplement.

It will be helpful to review several factors which are unique to plywood structural calculations. These will provide a useful background to the designer even if structural calculations are not required. Although some of these points have been introduced previously, they are briefly summarized here.

Panel section properties. The basic panel cross-section properties for wood structural panels are tabulated in the ASD Wood Structural Panels Supplement Table 5.2. The values in Table 5.2 are based on a *gross* rectangular cross section with a unit 1-ft width. However, since wood structural panels are composite materials, *effective* properties are more appropriate for use in design. Rather than tabulating effective unit stresses, which would then be used with the cross-section properties of Table 5.2, design stiffness and strength capacities are tabulated for wood structural panels. See ASD Panels Supplement Tables 3.1, 3.2, 3.3, and 3.4.

The variables which affect the effective design capacities are

1. Direction of stress

2. Surface condition

3. Species makeup

The *direction of stress* relates to the two-directional behavior of wood structural panels because of their cross-laminated construction. Because of this type of construction, two sets of properties are tabulated. One applies when wood structural panels are stressed parallel to the face grain, and the other applies when they are stressed perpendicular to the face grain.

The *basic surface* conditions for panels are

1. Sanded

2. Touch-sanded

3. Unsanded

It will be recalled from the discussion of the manufacturing of plywood that the relative thickness of the plies is different for these different surface conditions. Thus different properties apply for the three surface conditions.

Finally, the *species makeup* of a panel affects the *effective* properties. The panel grade and construction adjustment factor, C_G, specifically addresses STRUCTURAL I and Marine grades.

Stress calculations.   Prior to the 2001 NDS, provisions for engineering design with wood structural panels was not included in the NDS whatsoever. While specifying wood structural panels with design aids (such as those presented in the previous sections) has not changed significantly, stress calculations have subtly changed. In the past when stress calculations were required, *effective* cross-sectional properties were used with assumed uniform stress properties. The NDS Wood Structural Panels Supplement does *not* tabulate effective cross-sectional properties or unit stresses for panels. Rather, design stiffness and strength capacities are tabulated for wood structural panels. For example, rather than tabulating an effective section modulus, S, and a unit bending stress, F_b, as in the past, the 2001 NDS Panels Supplement tabulates the allowable bending strength capacity, F_bS, per unit width of panel. See ASD Panels Supplement Tables 3.1, 3.2, 3.3, and 3.4.

The use of allowable capacities is similar to ordinary beam design calculations. See Example 8.18.

EXAMPLE 8.18   Plywood Beam Loading and Section Properties

Consider a section of plywood acting as a beam element.

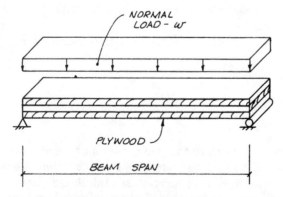

Figure 8.20a Plywood with load normal to surface of panel.

Plywood under Normal (Sheathing) Loads

For loads normal to the surface of the plywood (Fig. 8.20a), design capacities are obtained from the ASD Wood Structural Panels Supplement Table 3.1. The design bending strength, F_bS, and the design bending stiffness, EI, are tabulated for a unit 1-ft width of panel. For shear requirements, see Example 8.19.

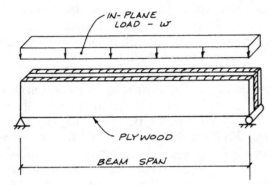

Figure 8.20*b* Plywood with load in plane of panel.

Plywood under In-Plane Loads

Procedures for calculating cross-section properties for plywood loaded in the plane of the panel (Fig. 8.20*b*) are not provided in the NDS or its Supplements. However, procedures are given in the *Plywood Design Specification,* Supplements 2 and 5 (Refs. 8.8 and 8.12). Plywood used in this manner is typically found in fabricated box beams using lumber flanges and plywood webs (Fig. 8.20*c*).

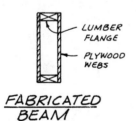

Figure 8.20c Lumber and plywood box beam.

Another factor that is unique to structural calculations for wood structural panels is that there are two different allowable shear capacities. The different allowable shear stresses are a result of the cross-laminations. The type and direction of the loading determine the type of shear involved, and the appropriate allowable shear must be used in checking the stress.

The ASD Wood Structural Panels Supplement refers to these shear stresses as

1. Shear in a plane perpendicular to the plies, or *shear through the thickness* of the plywood

2. Shear *in the plane* of the plies, or *rolling shear*

The first type of shear occurs when the load is in the plane of the panel, as in a diaphragm. See Fig. 8.21*a* in Example 8.19. This same type of stress occurs in fabricated box beams using plywood webs. In the latter case, shear through the thickness of the plywood is the result of flexural (horizontal) shear. The design procedures for fabricated lumber and plywood box beams are covered in PDS, Supplement 2 (Ref. 8.12).

Rolling shear can also be visualized as the horizontal shear in a beam, but in this case the loads are normal to the surface of the panel (as with sheathing loads). See Fig. 8.21*b*. The shear is seen to be "in the plane of the plies" rather than "through the thickness." With this type of loading, the wood fibers that are at right angles to the direction of the stress tend to slide or *roll* past one another. Hence the name *rolling shear*. If the stress is parallel to the face plies, the fibers in the inner crossband(s) are subjected to rolling shear.

The ASD Panels Supplement Tables 3.1 and 3.4 provide different design capacities for the calculation of shear through the thickness and rolling shear. See Example 8.19.

EXAMPLE 8.19   Types of Shear in Plywood

Consider a section of plywood subjected to direct shear.

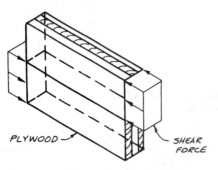

PLYWOOD   SHEAR FORCE

Figure 8.21a Shear through the thickness.

Shear through the Thickness

Shear through the thickness can occur from the type of loading shown in Fig. 8.21*a* (as in diaphragm and shearwall action) or from flexural shear (horizontal shear) caused by the type of loading shown in Fig. 8.20*b*. The stress calculated for these types of loading conditions is based on the *effective thickness for shear.* See ASD Wood Structural Panels Supplement Table 3.4.

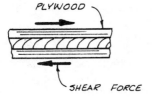

Figure 8.21b Rolling shear is shear in the plane of the plies.

Rolling Shear

Rolling shear occurs in the ply (or plies) that is (are) at right angles to the applied stress (Fig. 8.21b). This type of stress develops when plywood is loaded as shown in Fig. 8.20a. The stress is shear due to bending (horizontal shear) and is calculated from the rolling shear constant Ib/Q. See ASD Panels Supplement Table 3.3. The allowable stress for rolling shear is considerably less than the allowable shear through the thickness.

In Example 8.20, the design capacities for a given wood structural panel are determined. The purpose is to illustrate the use of the ASD Panels Supplement capacity tables and adjustment factors unique to wood structural panels.

EXAMPLE 8.20   Design Capacities of Wood Structural Panels

Determine the design capacities per unit foot of width for a 32/16 span-rated STRUCTURAL I grade 5-ply plywood used under normal load duration and in dry conditions. Assume the panel will be used such that it will have a width of at least 24 in.

Bending Stiffness

Two bending stiffness values are tabulated in the ASD Wood Structural Panels Table 3.1, one for the strength axes being perpendicular to the supports (or framing) and one for the strength axes parallel to the supports. The former is the strong orientation. Panels Supplement Table 3.1.1 provides values for the panel grade and construction adjustment factor.

Strong orientation:

$$EI' = EI(C_M)(C_t)(C_G)$$

$$= (115{,}000)(1.0)(1.0)(1.1)$$

$$= 126{,}500 \text{ lb-in.}^2/\text{ft width}$$

Weak orientation:

$$EI' = EI(C_M)(C_t)(C_G)$$

$$= (8100)(1.0)(1.0)(5.2)$$

$$= 42{,}120 \text{ lb-in.}^2/\text{ft width}$$

Note that as with all tabulated design capacities, these values are based on an assumed 1-ft width of panel.

Bending Strength

As with the stiffness, two bending strength values are tabulated in the ASD Wood Structural Panels Table 3.1, one for the strength axes being perpendicular to the supports (or framing) and one for the strength axes parallel to the supports. The former is the strong orientation.

Strong orientation:

$$F_bS' = F_bS(C_D)(C_M)(C_t)(C_G)(C_s)$$

$$= (370)(1.0)(1.0)(1.0)(1.2)(1.0)$$

$$= 444 \text{ lb-in.}/\text{ft width}$$

Weak orientation:

$$F_bS' = F_bS(C_D)(C_M)(C_t)(C_G)(C_s)$$

$$= (92)(1.0)(1.0)(1.0)(1.7)(1.0)$$

$$= 156 \text{ lb-in.}/\text{ft width}$$

Axial Stiffness

The axial stiffness values are tabulated in the ASD Panels Supplement Table 3.2, with values of C_G provided in Table 3.2.1.

Strong orientation:

$$EA' = EA(C_M)(C_t)(C_G)$$

$$= (4{,}150{,}000)(1.0)(1.0)(1.0)$$

$$= 4{,}150{,}000 \text{ lb/ft width}$$

Weak orientation:

$$EA' = EA(C_M)(C_t)(C_G)$$

$$= (3{,}600{,}000)(1.0)(1.0)(1.0)$$

$$= 3{,}600{,}000 \text{ lb/ft width}$$

Axial Tension

The axial tension capacity values are also tabulated in the ASD Panels Supplement Table 3.2, with values of C_G provided in Table 3.2.1.

Strong orientation:

$$F_t A' = F_t A(C_D)(C_M)(C_t)(C_G)(C_s)$$

$$= (2800)(1.0)(1.0)(1.0)(1.3)(1.0)$$

$$= 3640 \text{ lb/ft width}$$

Weak orientation:

$$F_t A' = F_t A(C_D)(C_M)(C_t)(C_G)(C_s)$$

$$= (1250)(1.0)(1.0)(1.0)(1.3)(1.0)$$

$$= 1625 \text{ lb/ft width}$$

Axial Compression

In addition to the axial stiffness and tension values, axial compression capacity values are also tabulated in the ASD Panels Supplement Table 3.2. Values of C_G are provided in Table 3.2.1.

Strong orientation:

$$F_c A' = F_c A(C_D)(C_M)(C_t)(C_G)$$

$$= (3550)(1.0)(1.0)(1.0)(1.5)$$

$$= 5325 \text{ lb/ft width}$$

Weak orientation:

$$F_c A' = F_c A(C_D)(C_M)(C_t)(C_G)$$

$$= (3100)(1.0)(1.0)(1.0)(1.5)$$

$$= 4650 \text{ lb/ft width}$$

In-Plane Shear

In-plane shear occurs when load is applied normal to the surface of the panels as with floors or roofs. Shear stiffness values are not provided in the ASD Panels Supplement, but shear capacities are tabulated in the Panels Supplement Table 3.3. Two in-plane shear values are tabulated in Table 3.3, one for the strength axes being perpendicular to the supports (or framing) and one for the strength axes parallel to the supports. For all span-rated panels, these two *tabulated* values are equal, but the adjusted values will differ. Values of C_G are provided in Table 3.3.1.

Strong orientation:

$$F_s(Ib/Q)' = F_s(Ib/Q)(C_D)(C_M)(C_t)(C_G)$$

$$= (165)(1.0)(1.0)(1.0)(1.9)$$

$$= 313 \text{ lb/ft width}$$

Weak orientation:

$$F_s(Ib/Q)' = F_s(Ib/Q)(C_D)(C_M)(C_t)(C_G)$$

$$= (165)(1.0)(1.0)(1.0)(1.0)$$

$$= 165 \text{ lb/ft width}$$

Shear Rigidity through the Thickness

Shear through the thickness occurs when load is applied in the plane of the panels, as with shearwalls or diaphragms. Shear rigidity values are provided in the ASD Panels Supplement Table 3.4, with values of C_G provided in Table 3.4.1. While two shear rigidity values are tabulated in Table 3.3—one for the strength axes being perpendicular to the supports (or framing) and one for the strength axes parallel to the supports—neither the tabulated nor the adjusted values differ.

$$G_v t_v' = G_v t_v (C_M)(C_t)(C_G)$$

$$= (27,000)(1.0)(1.0)(1.7)$$

$$= 45,900 \text{ lb/in. of panel depth}$$

Shear Capacity through the Thickness

Shear capacity through the thickness values are also provided in the ASD Panels Supplement Table 3.4, with values of C_G provided in Table 3.4.1. As with the rigidity, two shear capacity values are tabulated in Table 3.3, one for the strength axes being perpendicular to the supports (or framing) and one for the strength axes parallel to the supports, but neither the tabulated nor the adjusted values differ.

$$F_v t_v' = F_v t_v (C_D)(C_M)(C_t)(C_G)$$

$$= (62)(1.0)(1.0)(1.0)(2.0)$$

$$= 124 \text{ lb/in. of shear resisting panel length}$$

Additional discussion of shear through the thickness is provided in Chaps. 9 and 10.

Additional design information on structural calculations for wood structural panels is provided by the APA—The Engineered Wood Association. The primary design reference produced by the APA is the *Plywood Design Specifi-*

cation (PDS, Ref. 8.14). Supplements to the PDS are available (Refs. 8.3, 8.7, 8.8, 8.12, and 8.13) that provide design information and examples for a variety of applications. For performance-rated wood structural panels, section capacities are given in Ref. 8.5.

8.15 References

[8.1] American Forest and Paper Association (AF&PA). 2001. *Allowable Stress Design Manual for Engineered Wood Construction and Supplements and Guidelines,* 2001 ed., AF&PA, Washington DC.

[8.2] American National Standards Institute (ANSI). 1999. "Particleboard, Mat-Formed Wood," ANSI Standard A208.1-99, ANSI, New York, NY.

[8.3] APA—The Engineered Wood Association. 1993. *PDS Supplement #4: Design and Fabrication of Plywood Sandwich Panels,* APA—The Engineered Wood Association, Engineered Wood Systems, Tacoma, WA.

[8.4] APA—The Engineered Wood Association. 1994. *Load-Span Tables for PS 1 Plywood, Technical Note Z802,* APA—The Engineered Wood Association, Engineered Wood Systems, Tacoma, WA.

[8.5] APA—The Engineered Wood Association. 1995. *Design Capacities of APA Performance-Rated Structural-Use Panels, Technical Note N375,* APA—The Engineered Wood Association, Engineered Wood Systems, Tacoma, WA.

[8.6] APA—The Engineered Wood Association. 1995. *Grades and Specifications,* APA—The Engineered Wood Association, Engineered Wood Systems, Tacoma, WA.

[8.7] APA—The Engineered Wood Association. 1995. *PDS Supplement #1: Design and Fabrication of Plywood Curved Panels,* APA—The Engineered Wood Association, Engineered Wood Systems, Tacoma, WA.

[8.8] APA—The Engineered Wood Association. 1995. *PDS Supplement #5: Design and Fabrication of All-Plywood Beams,* APA—The Engineered Wood Association, Engineered Wood Systems, Tacoma, WA.

[8.9] APA—The Engineered Wood Association. 1995. *Performance-Rated Panels,* APA—The Engineered Wood Association, Engineered Wood Systems, Tacoma, WA.

[8.10] APA—The Engineered Wood Association. 1996. *Design/Construction Guide—Residential and Commercial,* APA—The Engineered Wood Association, Engineered Wood Systems, Tacoma, WA.

[8.11] APA—The Engineered Wood Association. 1996. *Load-Span Tables for APA Structural-Use Panels, Technical Note Q225,* APA—The Engineered Wood Association, Engineered Wood Systems, Tacoma, WA.

[8.12] APA—The Engineered Wood Association. 1996. *PDS Supplement #2: Design and Fabrication of Plywood-Lumber Beams,* APA—The Engineered Wood Association, Engineered Wood Systems, Tacoma, WA.

[8.13] APA—The Engineered Wood Association. 1996. *PDS Supplement #3: Design and Fabrication of Plywood Stressed-Skin Panels,* APA—The Engineered Wood Association, Engineered Wood Systems, Tacoma, WA.

[8.14] APA—The Engineered Wood Association. 1997. *Plywood Design Specification (PDS),* APA—The Engineered Wood Association, Engineered Wood Systems, Tacoma, WA.

[8.15] International Codes Council (ICC). 2003. *International Building Code,* 2003 ed., ICC, Falls Church, VA.

[8.16] National Evaluation Service. 2002. "Standards for Structural-Use Panels," Report No., NER-108, International Code Councils (Available from APA—The Engineered Wood Association, Tacoma, WA).

[8.17] National Institute of Standards and Technology (NIST). 1992. *U.S. Product Standard PS 2-92—Performance Standard for Wood-Based Structural-Use Panels,* U.S. Dept. of Commerce, Gaithersburg, MD (reproduced by APA—The Engineered Wood Association, Engineered Wood Systems, Tacoma, WA).

[8.18] National Institute of Standards and Technology (NIST). 1995. *U.S. Product Standard PS 1-95—Construction and Industrial Plywood,* U.S. Dept. of Commerce, Gaithersburg, MD (reproduced by APA—The Engineered Wood Association, Engineered Wood Systems, Tacoma, WA).

8.16   Problems

8.1   What two major types of loading are considered in designing a wood structural panel roof or floor system that also functions as part of the LFRS?

8.2   Regarding the fabrication of plywood panels, distinguish between (1) veneer, (2) plies, and (3) layers.

8.3   In the cross section of plywood panel shown in Fig. 8A, label the names used to describe the 5 plies.

Figure 8.A

5 PLY CONSTRUCTION

8.4   Plywood panels (4 ft × 8 ft) of 5-ply construction (similar to that shown in Fig. 8A) are used to span between roof joists that are spaced 16 in. o.c.

   Find: *a.* Sketch a plan view of the framing and the plywood, showing the plywood oriented in the strong direction.
   b. Sketch a 1-ft-wide cross section of the sheathing. Shade the plies that are effective in supporting the "sheathing" loads.

8.5   Repeat Prob. 8.4 except that the plywood is oriented in the weak direction.

8.6   How are the species of wood used in the fabrication of plywood classified?

8.7   The plywood in Fig. 8A has a grade stamp that indicates *Group* 2.

   Find: *a.* What plies in the panel contain Group 2 species?
   b. If not all plies are of Group 2 species, what is assumed for the others?

8.8   What is the meaning of the term STRUCTURAL I? How is it used?

8.9   What are the veneer grades? Which veneers are the more similar in appearance and surface qualities? Which veneers are structurally more similar? State the reasons for similarities.

8.10   What is the most common type of glue used in the fabrication of plywood (interior or exterior)?

8.11   List and briefly describe the various types of wood structural panels other than plywood.

8.12   What is the difference between a prescriptive specification and a performance standard for the production of wood structural panels?

8.13   Describe the exposure durability classifications for (*a*) plywood manufactured in accordance with the Product Standard PS 1 and (*b*) wood structural panels manufactured under APA's performance standard. (*c*) Which of the durability classifications in parts (*a*) and (*b*) are similar?

8.14 What is the difference between the construction of interior and exterior plywood?

8.15 Briefly describe the *span rating* found in the grade-trademark of the following panels:
a. Sheathing grades
b. APA Rated Sturd-I-Floor
c. APA Rated 303 Siding

8.16 Explain the significance of the following designations that may be found in the grade-trademarks of wood structural panels:
a. PS 1
b. PRP-108

8.17 What ASD Wood Structural Panels Supplement tables provide load/span information for the following?
a. Sheathing
b. Single floor
c. Group 1 sanded panels

8.18 What are the sheathing grades of plywood? Are these sanded, touch-sanded, or unsanded panels? What veneer grades are used for the face and back plies of a sanded plywood panel?

8.19 The spacing of rafters in a roof is 48 in. o.c. Roof dead load = 5 psf. Snow load = 30 psf. Roof sheathing is to be a sheathing grade of plywood, and panels are oriented in the strong direction.

Find: The minimum grade, span rating, thickness, and edge support requirements for the roof sheathing.

8.20 The spacing of joists in a roof is 24 in. o.c. Roof dead load = 8 psf. Snow load = 100 psf. Roof sheathing is to be APA Rated Sheathing (which can be either plywood or a nonveneer panel). Panels are oriented in the strong direction.

Find: The minimum span rating, thickness, and edge support requirements for the roof sheathing.

8.21 The spacing of joists in a floor system is 16 in. o.c., and the design floor (D + L) is 200 psf.

Find: The required panel grade, span rating, thickness, edge support requirements, and panel orientation for the floor sheathing.

8.22 Repeat Prob. 8.21 except that the joist spacing is 24 in. o.c.

8.23 Describe the construction of UNDERLAYMENT-grade plywood. How is it used in floor construction?

8.24 A single-layer floor system is used to support a floor dead load of 10 psf and a floor live load of 75 psf. Panels with a span rating of 16 in. will be used to span between floor joists that are 16 in. o.c.

Find: *a.* Is an APA nonveneer panel able to support these loads, or is a plywood panel necessary?

         b. What are the advantages of field-gluing floor panels to the framing members before they are nailed in place?

         c. What is meant by nail popping, and how is the problem minimized?

8.25   Describe the following types of plywood used on walls (include typical grades):
a. Plywood sheathing
b. Plywood siding
c. Combined sheathing-siding

8.26   Grooved plywood (such as Texture 1-11) is used for combined sheathing-siding on a shearwall. Nailing to the studs is similar to that in Fig. 8.19*c*, the detail for a vertical plywood joint.

Find: What thickness plywood is to be used in the shearwall design calculations?

8.27   Regarding the calculation of stresses in wood structural panels,
a. What is shear through the thickness?
b. What is rolling shear?

8.28   Determine the design capacities per unit foot width for $^{15}\!/_{32}$-in. thick OSB that is 32/16 span-rated. Assume the OSB is used under normal load duration and in dry conditions, and that the panel will be at least 24-in. wide.

8.29   Determine the design capacities per unit foot width for $^{19}\!/_{32}$-in. thick COMPLY that is 16 o.c. span-rated. Assume the COMPLY is used under normal load duration and in dry conditions, and that the panel will be at least 24-in. wide.

Horizontal Diaphragms

9.1 Introduction

The lateral forces that act on conventional wood-frame buildings (bearing wall system) were described in Chap. 2, and the distribution of these forces was covered in Chap. 3. In the typical case, the lateral forces were seen to be carried by the wall framing to the horizontal diaphragms at the top and bottom of the wall sections. A horizontal diaphragm acts as a beam in the plane of a roof or floor that spans between shearwalls. See Fig. 9.1.

The examples in Chap. 3 were basically force calculation and force distribution problems. In addition, the calculation of the unit shear in a diaphragm was illustrated. Although the unit shear is a major factor in a diaphragm design, there are a number of additional items that must be addressed.

The basic design considerations for a horizontal diaphragm are

1. Sheathing thickness
2. Diaphragm nailing
3. Chord design
4. Collector (strut) design
5. Diaphragm deflection
6. Tie and anchorage requirements

The first item is often governed by loads normal to the surface of the sheathing (i.e., by sheathing loads). This subject is covered in Sec. 6.19 for lumber diaphragms and in Chap. 8 for wood structural panel diaphragms. The nailing requirements, on the other hand, are a function of the unit shear. The sheathing thickness and nailing requirements may, however, both be governed by the unit shears *when the shears are large.*

In this chapter the general behavior of a horizontal diaphragm is described, and the functions of the various items mentioned above are explained. This

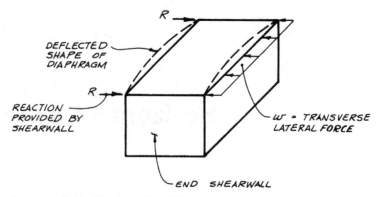

Figure 9.1 Typical horizontal roof diaphragm.

is followed by detailed design considerations for the individual elements. Tie and anchorage requirements are touched upon, but these are treated more systematically in Chap. 15. Shearwalls (vertical diaphragms) are covered in Chap. 10.

Figures illustrating diaphragm design methodologies are generally applicable to both wind and seismic forces. In some cases a figure will note that it specifically uses wind or seismic forces, while in other cases the force is not stated as being one or the other. It should be kept in mind that these calculations will be performed with *allowable stress design (ASD) level forces* for wind and *strength level forces* for seismic.

9.2   Basic Horizontal Diaphragm Action

A horizontal diaphragm can be defined as a large, thin structural element that is loaded in its plane. It is an assemblage of elements which typically includes

1. Roof or floor sheathing

2. Framing members supporting the sheathing

3. Boundary or perimeter members

When properly designed and connected together, this assemblage will function as a horizontal beam that distributes force to the vertical resisting elements in the lateral-force-resisting system (LFRS).

In Sec. 3.3, it was discussed that there are two types of diaphragm analysis that need to be considered: a *rigid diaphragm analysis* when the diaphragm is rigid when compared to the supporting shearwalls, and a *flexible diaphragm analysis* otherwise. This chapter will first introduce flexible diaphragm design, which is the simpler and more common of the two methods. In Sec. 9.11, the code criteria for categorizing diaphragms as *flexible* or *rigid* will be discussed, and a rigid diaphragm analysis will be demonstrated.

The diaphragm must be considered for lateral forces in both transverse and longitudinal directions. See Example 9.1. Like all beams, a horizontal diaphragm must be designed to resist both shear and bending.

In general, a horizontal diaphragm can be thought of as being made up of a shear-resisting element (the roof or floor sheathing) and boundary members. There are two types of boundary members in a horizontal diaphragm (chords and collectors or struts), and the direction of the applied force determines the function of the members (Fig. 9.2). The *chords* are designed to *carry the moment* in the diaphragm.

EXAMPLE 9.1   Horizontal Diaphragm Forces and Boundary Members

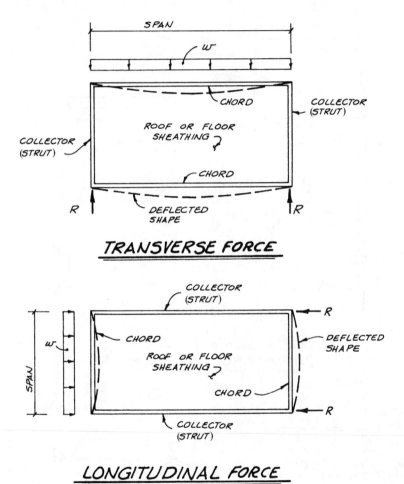

Figure 9.2 Diaphragm boundary members: Chords and Struts.

Diaphragm *boundary members* change functions depending on the direction of the lateral force.

Chords are boundary members that are perpendicular to the applied force. *Collectors* (*struts*) are boundary members that are parallel to the applied force.

An analogy is often drawn between a horizontal diaphragm and a steel wide-flange (W shape) beam. In a steel beam the flanges resist most of the moment, and the web essentially carries the shear. In a horizontal diaphragm, the sheathing corresponds to the web, and the chords are assumed to be the flanges. The chords are designed to carry axial forces created by the moment. These forces are obtained by resolving the internal moment into a couple (tension and compression forces). See Fig. 9.3. The shear is assumed to be carried entirely by the sheathing material. Proper nailing of the sheathing to the framing members is essential for this resistance to develop.

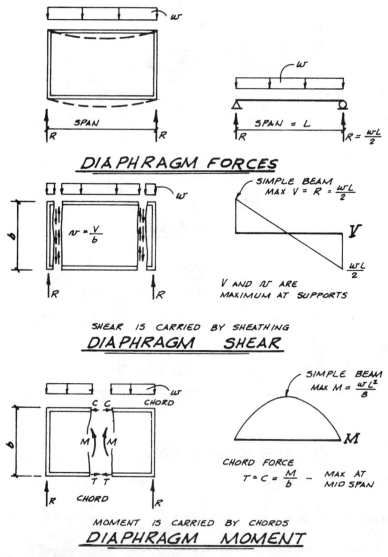

Figure 9.3 Shear and moment resistance in a horizontal diaphragm.

The *collectors (struts)* are designed to transmit the horizontal diaphragm *reactions to the shearwalls*. This becomes a design consideration when the supporting shearwalls are shorter in length than the horizontal diaphragm. See Example 9.2. Essentially, the unsupported horizontal diaphragm unit shear over an opening in a wall must be transmitted to the shearwall elements by the collector (strut).

EXAMPLE 9.2   Function of Collector (Strut)

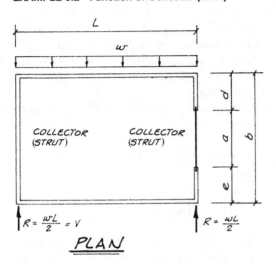

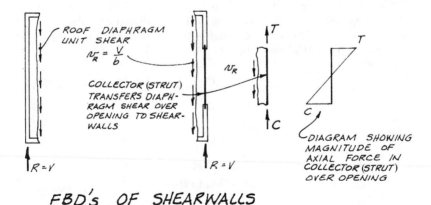

Figure 9.4 Collector (strut) over wall opening.

The left wall has no openings, and roof shear is transferred directly to the wall. The right wall has a large opening and the strut transfers the roof diaphragm shear over the opening to the shearwalls. Here the collector member is assumed to function in both tension and compression. When shearwall lengths d and e are equal, the strut forces T and C are equal. Strut forces for other ratios of d and e are covered in Sec. 9.6.

The *collector* is also commonly known as a *strut* or *drag strut*. These names come from the concept that the collector drags or collects the diaphragm shear into the shearwall. The IBC rquires use of special seismic forces for design of collectors, their splices, and their connections to shearwalls or other vertical resisting elements. These seismic forces, the *maximum seismic load effect* E_m, are found by first looking at the design and detailing requirements of IBC Sec. 1620. IBC Sec. 1620 references ASCE 7 Sec. 9.5.2.6, and Sec. 9.5.2.6.3.1 requires the use of E_m forces for collectors in Seismic Design Categories C and higher. ASCE 7 (Ref. 9.2) exempts structures braced entirely with light-frame shearwalls from this special force level.

The equations for the maximum seismic load effect E_m are given in ASCE 7 equations 9.5.2.7.1-1 and 9.5.2.7.1-2:

$$E_m = \Omega_o Q_E + 0.2 S_{DS} D \qquad\qquad (9.5.2.7.1\text{-}1)$$

and

$$E_m = \Omega_o Q_E - 0.2 S_{DS} D \qquad\qquad (9.5.2.7.1\text{-}2)$$

where Ω_o is the overstrength factor from ASCE 7 Table 9.5.2.2; S_{DS} is the short-period design spectral response acceleration, introduced in Chap. 2; and D is the dead load. The term $\Omega_o Q_E$ is the horizontal component of the seismic force, while $0.2 S_{DS} D$ acts vertically.

This *special seismic force* was first introduced as seismic design force type six in Sec. 3.4. This force is used with the following special seismic load combinations from IBC Sec. 1605.4:

$$1.2\ D + f_1 L + 1.0\ E_m \qquad\qquad (16\text{-}19)$$

$$0.9\ D + 1.0\ E_m \qquad\qquad (16\text{-}20)$$

where D = dead load
$\quad f_1$ = 0.5 or 1.0
$\quad$ L = live load
$\quad E_m$ = maximum seismic load effect

The numbers following each equation are the equation numbers from the 2003 IBC. They are included here to allow the equations to be easily referenced. There are some exceptions to the requirement for this special force level which apply specifically to wood frame construction. These will be discussed later in this chapter.

The special E_m force level and load combinations are used in the Code for a limited number of special elements which are thought to have a major impact on the seismic performance of a building. Besides collectors, the other special elements occurring in wood frame construction are elements providing vertical support under shear walls that occur in an upper floor but do not continue down to the foundation (elements supporting discontinuous shear-walls, discussed in Sec. 16.3). The concept of using the higher E_m force and the special load combinations is that elements which are considered critical to the performance of the building should be designed for the maximum seismic force that is expected in the structure, not the typical IBC design force. Design for this force level will ensure that these critical elements are not the weak link (first item to fail) in the structural system. The Ω_o factor can be found in ASCE 7 Table 9.5.2.2, along with R factors. These special requirements for collectors are illustrated in Sec. 9.7.

9.3 Shear Resistance

The shear-resisting element in the horizontal diaphragm assemblage is the roof or floor sheathing. This can be either lumber sheathing or wood structural panels. The majority of wood horizontal diaphragms use wood structural panel sheathing because of the economy of installation and the relatively high allowable unit shears it provides. Because of its wide acceptance and for simplicity, this chapter deals primarily with wood structural panel diaphragms. Lumber-sheathed diaphragms are covered in Ref. 9.3.

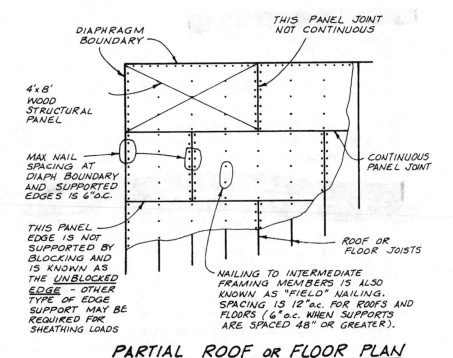

Figure 9.5a Unblocked diaphragm.

For wood structural panel diaphragms, the starting point is the determination of the required panel thickness for sheathing loads, i.e., dead, live, and show loads perpendicular to the plane of the sheathing (Chap. 8). It was previously noted that these loads often determine the final panel thickness. This is especially true for floors because the sheathing loads (due to vertical dead and live loads) are larger and the deflection criteria are more stringent than those for roofs.

The shear capacity for wood structural panel diaphragms is the shear "through the thickness" (Fig. 8.21a). However, in most cases the shear capacity of the panel is not the determining factor. Unit diaphragm shears are usually limited by the *nail capacity* in the wood, rather than the strength of the panel.

The nail spacing required for a horizontal diaphragm may be different at various points in the diaphragm. Because of the importance of the panel nailing requirements, it is necessary for the designer to clearly understand the nailing patterns used in diaphragm construction.

The simplest nailing pattern is found in *unblocked diaphragms*. See Fig. 9.5a. An unblocked diaphragm is one that does *not* have two of the panel edges *supported* by lumber framing. These edges may be completely unsupported, or there may be some other type of edge support such as T&G edges or panel clips (Fig. 8.13). For an *unblocked diaphragm* the *standard nail spacing* requirements are

1. Supported panel edges—6 in. o.c.
2. Along intermediate framing members (known also as *field* nailing)—12 in o.c. (except 6 in. o.c. is required when supports are spaced 48 in. or greater—see the *ASD Wood Structural Panel Shear Wall and Diaphragm Supplement,* Tables 3.1A and 3.1B (Ref. 9.1), footnote b)

When a diaphragm has the panel edges supported with *blocking,* the *minimum* nailing requirements are the same as for the unblocked diaphragms. There are however, more supported edges in a blocked diaphragm. See Fig. 9.5b.

Although the minimum nailing requirements (i.e., maximum nail spacing) are the same for both blocked and unblocked diaphragms, the allowable unit shears are much higher for blocked diaphragms. These higher unit shears are a result of the more positive, direct transfer of stress provided by nailing all four edges of the plywood panel. See Fig. 9.5c.

EXAMPLE 9.3   Wood Structural Panel Diaphragm Nailing

Required nail size is a function of the panel thickness and shear requirements. The panel is shown spanning in the strong direction for normal (sheathing) loads. See Fig. 9.5a for an unblocked diaphragm and Fig. 9.5b for a blocked diaphragm.

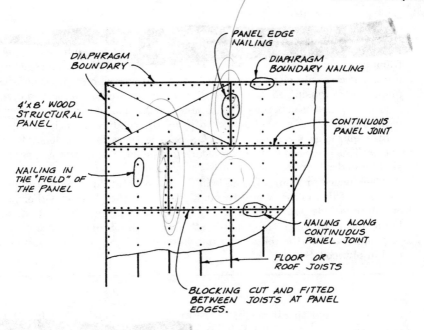

PANEL EDGE NAILING

DIAPHRAGM BOUNDARY NAILING

DIAPHRAGM BOUNDARY

CONTINUOUS PANEL JOINT

4'x 8' WOOD STRUCTURAL PANEL

NAILING IN THE "FIELD" OF THE PANEL

NAILING ALONG CONTINUOUS PANEL JOINT

FLOOR OR ROOF JOISTS

BLOCKING CUT AND FITTED BETWEEN JOISTS AT PANEL EDGES.

PARTIAL ROOF OR FLOOR PLAN

Figure 9.5b Blocked diaphragm.

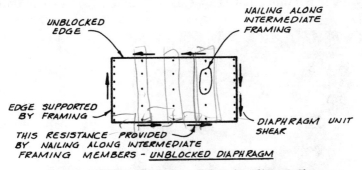

UNBLOCKED EDGE

NAILING ALONG INTERMEDIATE FRAMING

EDGE SUPPORTED BY FRAMING

DIAPHRAGM UNIT SHEAR

THIS RESISTANCE PROVIDED BY NAILING ALONG INTERMEDIATE FRAMING MEMBERS - UNBLOCKED DIAPHRAGM

ISOLATED PANEL FROM FIG 9.5.a

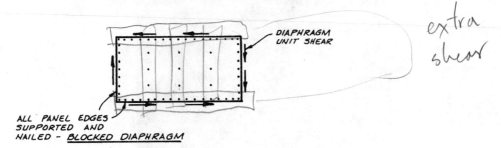

DIAPHRAGM UNIT SHEAR

extra shear

ALL PANEL EDGES SUPPORTED AND NAILED - BLOCKED DIAPHRAGM

ISOLATED PANEL FROM FIG 9.5.b

Figure 9.5c Shear transfer in a wood structural panel diaphragm.

The *direction of the continuous panel joint* and the *direction of the unblocked edge* are used to determine the *load case* for diaphragm design. Different panel layouts and framing arrangements are possible. The continuous panel joint and the unblocked edge are not necessarily along the same line. The diaphragm *load case* (described in Example 9.4) is considered in blocked diaphragms as well as unblocked diaphragms.

From a comparison of the isolated panels in Fig. 9.5c it can be seen that the addition of nailing along the edge of a blocked panel will produce a much stronger diaphragm.

In diaphragms using relatively *thin* panels, there is a tendency for the panel to buckle. This is caused by the edge bearing of panels in adjacent courses as they rotate slightly under high diaphragm forces.

When loaded to failure, the nails in a wood structural panel diaphragm deform, and the *head* of the nail may pull through the face of the plywood. Other possible modes of failure include pulling the nail through the *edge* of the panel and *splitting* the framing members to which the panel is attached. A minimum distance of ⅜ in. from the edge of a panel to the center of a nail is required to develop the design capacity of a nail in plywood. See Fig. 9.5d.

When loaded to ultimate, there are several possible modes of failure that can occur in a plywood diaphragm. Perhaps the most common type of failure is by the nail head pulling through the panel. See Fig. 9.5d.

The 6 in. nail spacing at supported panel edges is used for all *unblocked* diaphragms. For *blocked* diaphragms, however, these spacing provisions are simply the maximum allowed spacing. Much higher allowable unit shears can

nail heads need to be longer

Figure 9.5d Nail deformation at ultimate load. (*APA—The Engineered Wood Association.*)

be obtained by using a decreased nail spacing. Allowable unit shears for both unblocked and blocked diaphragms are given in the *ASD Wood Structural Panel Shear Wall and Diaphragm Supplement,* Tables 3.1A and 3.1B.

The reader is reminded that lateral forces calculated in accordance with IBC equations will be at an *ASD level for wind* and a *strength level for seismic.* Therefore, unit shears due to wind may be compared directly to the ASD allowable shears in the diaphragm Table 3.1A. However, the calculated seismic unit shears will need to be converted to an ASD level before being compared to Table 3.1B values.

Tabulated unit shears for horizontal diaphragms assume that the framing members are Douglas Fir-Larch or Southern Pine. If the diaphragm nailing penetrates into framing members of other species of wood, an adjustment is required that accounts for a reduced nail capacity. Reduction factors for other species of wood are given in footnote a to the allowable shear tables. Tables 3.1A and 3.1B of the ASD Wood Structural Panel Shear Wall and Diaphragm Supplement are tabulated for short-term wind and seismic forces, respectively. Because these represent the common loading conditions for horizontal diaphragms, a load duration factor C_D of 1.6 for wind and seismic is included in the tabulated allowable shears.

At this point, it is important to recognize that the tabulated unit shears for horizontal diaphragms are 40 percent greater for wind (ASD Shear Wall and Diaphragm Supplement, Table 3.1A) than for seismic (Table 3.1B). The increase for allowable unit shear for wind is due to historic design stress levels for diaphragms and a better understanding of wind loading. The IBC (Ref. 9.8) permits the 40 percent increase per Sec. 2306.3.1.

As mentioned, ASD Tables 3.1A and 3.1B are tabulated for short-term (wind and seismic) forces. If a horizontal diaphragm is used to support loads of longer duration, the allowable unit shears will have to be reduced. For example, if the load under consideration was determined to be of 10-year cumulative duration ($C_D = 1.0$), the tabulated allowable shears would be reduced by a factor of 1/1.6. Furthermore, for this situation, the allowable shears from ASD Table 3.1B (seismic) should be used as the base values. That is, the 40 percent increase is specific to wind loading and should not be used for other load durations.

When the design unit shears are large, the required *nail spacing* for *blocked diaphragms* must be carefully interpreted. The nail spacing along the following lines must be considered:

1. Diaphragm boundary

2. Continuous panel joints

3. Other panel edges

4. Intermediate framing members (field nailing)

The field-nailing provisions are always the same: 12 in. o.c. for roofs and floors (except 6-in. o.c. spacing is required when framing is spaced at 48 in. or greater). The nail spacing along the other lines may all be the same (6 in.

o.c. maximum), or some may be different. Allowable unit shears are tabulated for a nail spacing as small as 2 in. o.c.

A great deal of information is incorporated into the ASD Shear Wall and Diaphragm Supplement, Tables 3.1A and 3.1B. Because of the importance of these tables, the remainder of this section deals with a review of Tables 3.1A and 3.1B.

The tables are divided into several main parts. The top parts gives ASD allowable unit shears for diaphragms which have the STRUCTURAL I designation. The values in the bottom half of the tables apply to *all other wood structural panel grades* covered by the NDS.

The latter represents the large majority of wood structural panels, and it includes both sanded and unsanded grades. STRUCTURAL I is stronger (and more expensive) and should be specified when the added strength is required. The ASD allowable shear values for STRUCTURAL I are 10 percent larger than the values for other grades.

In working across the tables it will be noted that the left side is basically descriptive, and it requires little explanation. The two right-hand columns in each table give the allowable unit shears for *unblocked* diaphragms. The central four columns give allowable shears for *blocked* diaphragms.

Throughout the allowable shear tables, reference is made to the *load cases*. Six load cases are defined below in ASD Tables 3.1A and 3.1B. Although somewhat different panel arrangements can be used, all layouts can be classified into one of the six load cases. Because the proper use of unit shear tables depends on the load case, the designer must be able to determine the load case for any layout under consideration.

The *load case* essentially depends on two factors. The *direction of the lateral force* on the diaphragm is compared with the direction of the

1. Continuous panel joint

2. Unblocked edge (if blocking is not provided)

For example, in load case 1 the applied lateral force is perpendicular to the continuous panel joint, and it is also perpendicular to the unblocked edge. Load cases 1 through 4 consider the various combinations of these alternative arrangements. Load cases 5 and 6 have continuous panel joints running in both directions. Load cases 1 through 4 are more common than cases 5 and 6.

The panel layout must be shown on roof and floor framing plans along with the nailing and blocking requirements. It should be noted that the load case is defined by the criteria given above, and it does not depend on the direction of the panel. See Example 9.4.

EXAMPLE 9.4   Diaphragm Load Cases

Determine the *load cases* for two horizontal diaphragms (Fig. 9.6*a* and *b*) in accordance with the ASD Wood Structural Panel Shear Wall and Diaphragm Supplement. Consider both transverse and longitudinal directions.

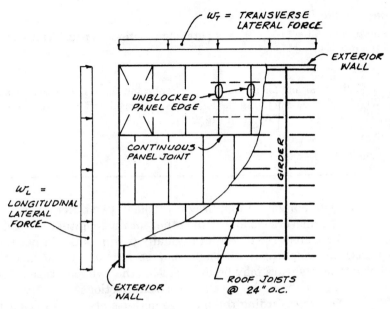

Figure 9.6a Partial roof framing plan.

Panels in Fig. 9.6a are oriented in the strong direction for sheathing loads. In Fig. 9.6b the same building is considered except that the panel layout has been revised. The panels in Fig. 9.6b are oriented in the weak direction for sheathing loads.

Transverse Direction

Transverse force in Fig. 9.6a is perpendicular to continuous panel joint and parallel to unblocked edge.

∴ load case 2

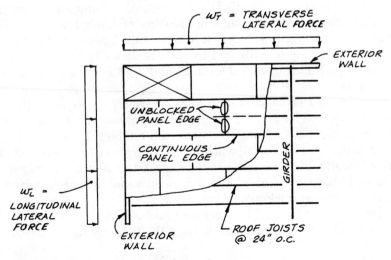

Figure 9.6b Partial roof framing plan.

Longitudinal Direction

Longitudinal force in Fig. 9.6*a* is parallel to continuous panel joint and perpendicular to the unblocked edge.

∴ load case 4

Practically speaking, for an unblocked diaphragm there is load case 1 and *all* others.

When the criteria defining the load case are considered, it is seen that the load cases are the same in Fig. 9.6*b* as in Fig. 9.6*a*:

Transverse—case 2

Longitudinal—case 4

For a given horizontal diaphragm problem, two different load cases apply: one for the transverse load and the other for the longitudinal load. The lateral force in the transverse direction normally produces the larger unit shears in a horizontal diaphragm (this may not be the case for shearwalls), but the allowable unit shears may be different for the two load cases. Thus, the diaphragm shears should be checked in both directions.

Other factors regarding the makeup and use of ASD Tables 3.1A and 3.1B should be noted. For example, the width of the framing members must be considered. Allowable shears for diaphragms using 2-in. nominal framing members are approximately 10 percent less than for diaphragms using wider framing.

The *nail spacing* used for the *diaphragm boundary* on blocked diaphragms is also the nail spacing required along the *continuous panel joint* for load cases 3 and 4. This nail spacing is also to be set at all panel edges for load cases 5 and 6. If load case 1 or 2 is involved, the diaphragm boundary nail spacing is not required at the continuous panel joints, and the somewhat larger nail spacing for "other panel edges" may be used. The selection of the proper nailing specifications is illustrated in several examples later in the chapter.

When the spacing of nails is very close, ASD Tables 3.1A and 3.1B require special precautions to avoid the splitting of lumber framing. For example, under certain circumstances, 3-in. (or wider) nominal framing is required when the spacing of nails is 3 in. or less. Refer to the ASD Shear Wall and Diaphragm Supplement Tables 3.1A and 3.1B footnotes for specific details.

A point should be made about the combination of nail sizes and panel thicknesses in ASD Tables 3.1A and 3.1B. There are times when the nail size and panel thickness for a given design will not agree with the combinations listed in the table. If the allowable shears in the tables are to be used without further justification, the following procedure should be followed: If a thicker panel than that given in the table is used, the allowable shear from the table should be based on the nailing used. No increase in shear is permitted because of the increased thickness.

On the other hand, if a larger nail size is used for a panel thickness given in the table, the allowable shear from the table should be based on the panel thickness. No increase in shear is permitted because of the increased nail size. Thus, without further justification, the combination of panel thickness and nail size given in Tables 3.1A and 3.1B of the ASD Shear Wall and Diaphragm Supplement is "compatible." Increasing one item only does not necessarily provide an increase in allowable unit shear. Reference 9.3 describes a method of calculating allowable diaphragm unit shears by principles of mechanics using the panel shear values and nail strength values.

Another consideration about nail sizes has to do with the penetration of the nail into the framing members. ASD Tables 3.1A and 3.1B indicate that 10d nails require a minimum penetration of $1\frac{5}{8}$ in. into the framing members below the panels. If 2 × 4 blocking is turned flat (dashed lines in Fig. 8.13, alternative *b*), the thickness available for nail penetration is simply the thickness of the 2 × 4 (that is, $1\frac{1}{2}$ in.). This same thickness for nail penetration occurs when light-frame wood trusses use 2 × 4 top chords turned flat. The reduced penetration should be taken into account by reducing the tabulated allowable unit shears in ASD Tables 3.1A and 3.1B. The reduction is figured using a linear interpolation as the ratio of the furnished penetration to the required penetration. For the case of a 10d nail in a 2 × 4 flat, the reduction is $1.5/1.625 = 0.923$ times the tabulated unit diaphragm shear (roughly an 8 percent reduction).

Finally, it should be noted that the other "approved" types of panel edge support (such as T&G edges or panel clips—Fig. 8.13) are substitutes for lumber blocking when considering *sheathing loads* only. They do not qualify as substitutes for blocking for *diaphragm design* (except $1\frac{1}{8}$-in.-thick 2-4-1 panels with properly stapled T&G edges).

9.4 Diaphragm Chords

Once the diaphragm web has been designed (sheathing thickness and nailing), the flanges or chord members must be considered. The determination of the axial forces in the chords is described in Fig. 9.3. The axial force at any point in the chords can be determined by resolving the moment in the diaphragm at that point into a couple (equal and opposite forces separated by a lever arm—the lever arm is the distance between chords):

$$T = C = \frac{M}{b}$$

The tension chord is often the critical member. There are several reasons for this. One is that the allowable stress in compression is often larger than the allowable stress in tension. This assumes that the chord is laterally supported and that column buckling is not a factor. This is usually the case, but the possible effects of column instability in the compression chord must be evaluated.

Another reason that the tension chord may be critical is that the chords are usually not continuous single members for the full length of the building. In order to develop the chord force, the members must be made effective by splicing separate members together. This is less of a problem in compression because the ends of chord members can transmit loads across a splice in end bearing. Tension splices, on the other hand, must be designed.

Because the magnitude of the chord force is calculated from the diaphragm moment, the *magnitude of the chord force follows the shape of the moment diagram*. The design force for any connection splice can be calculated by dividing the moment in the diaphragm at the location of the splice by the distance separating the chords. A simpler, more conservative approach would be to design all diaphragm chord splices for the maximum chord force.

It should also be noted that each chord member must be capable of functioning in either tension or compression. The applied lateral load can change direction and cause tension or compression in either chord.

Some consideration should be given at this point to what elements in a building can serve as the chords for a diaphragm. In a wood-frame building with stud walls, the doubled top plate is usually designed as the chord. See Example 9.5. This type of construction is accepted by contractors and carpenters as standard practice. Although it may have developed through tradition, the concept behind its use is structurally sound.

The top plate members are not continuous in ordinary buildings unless the plan dimensions of the building are very small. Two plate members are used so that the splice in one plate member can be staggered with respect to the splice in the other member. This creates a continuous chord with at least one member being effective at any given point. When chord forces are large, more than two plate members may be required.

In order for the top plate members to act as a chord, they must be adequately connected together. If the chord forces are small, this connection can be made with nails, but if the forces are large, the connection will require the use of wood screws, bolts or steel straps. These are connection design problems, and the procedures given in Chaps. 12 and 13 can be used for the design of these splices. It should also be noted that the chord forces are usually the result of wind or seismic forces, and a C_D of 1.6 applies to the design of wood members and wood connections.

Wood structural panel diaphragms are often used in buildings that have concrete tiltup walls or masonry (concrete block or brick) walls. In these buildings the chord is made up of continuous horizontal reinforcing steel in the masonry or concrete wall. See Example 9.6. If the masonry or concrete is assumed to function in compression only (the usual assumption), the tension chord is critical. The stress in the steel is calculated by dividing the chord force by the cross-sectional area of the horizontal wall steel that is placed at the diaphragm level.

EXAMPLE 9.5   Horizontal Diaphragm Chord—Double Top Plate

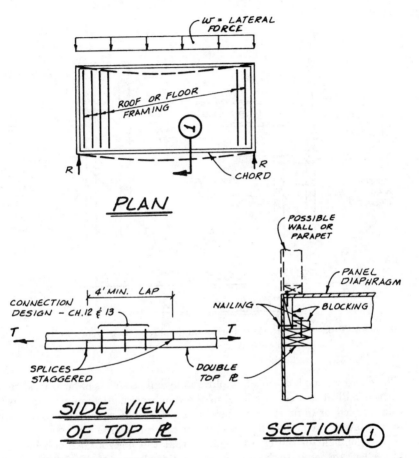

Figure 9.7 Horizontal diaphragm chord in building with wood roof system and woodframe walls. Code minimum lap splice length is 4 ft to form a continuous chord member. Nail or bolt connection must be designed to transmit chord force *T* from one plate member to the other.

The double top plate in a wood-frame wall is often used as the chord for the horizontal diaphragm. Splices are offset so that one member is effective in tension at a splice. Connections for anchoring the horizontal and vertical diaphragms together are covered in Chap. 15.

The double plate in wood walls and the horizontal steel in concrete and masonry walls are probably the two most common elements used as diaphragm chords. However, other building elements can be designed to serve as the chord. As an example of another type of chord member, consider a large window in the front longitudinal wall of a building. See Example 9.7.

EXAMPLE 9.6   Horizontal Diaphragm Chord—Wall Steel

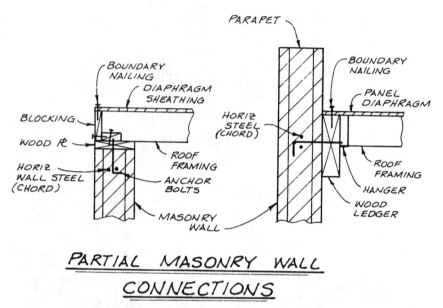

PARTIAL MASONRY WALL

CONNECTIONS

Figure 9.8 Typical connections of horizontal diaphragm to masonry walls.

The chord for the horizontal diaphragm usually consists of horizontal reinforcing steel in the wall at or near the level of the diaphragm. Attempts to design the wood top plate or ledger to function as the chord are usually considered inappropriate because of the larger stiffness of the masonry or concrete walls.

Development of the chord requires that the horizontal diaphragm be adequately attached (anchored) to the shearwalls. In this phase of the design, the spacing of anchor bolts, blocking, and nailing necessary to transfer the forces between these elements are considered. The *details in this sketch are not complete,* and these sketches (Fig. 9.8) are included here to illustrate the diaphragm chords. Anchorage connections are covered in detail in Chap. 15.

The top plate in this longitudinal wall is not continuous. Here the window header supports the roof or floor framing directly, and the header may be designed to function as the chord. The header must be designed for both vertical loads and the appropriate combination of vertical loads and lateral forces. The connection of the header to the shearwall also must be designed. (*Note:* If a cripple stud wall occurs over the header and below the roof framing, a double top plate will be required over the header. In this case the plate can be designed as the chord throughout the length of the wall.)

EXAMPLE 9.7   Header Acting as a Chord

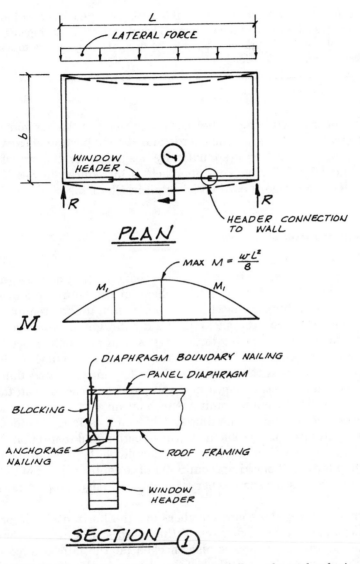

Figure 9.9 The header over an opening in a wall may be used as horizontal diaphragm chord.

Over the window the header serves as the chord. It must be capable of resisting the
maximum chord force in addition to gravity loads. The maximum chord force is

$$T = C = \frac{\text{max. } M}{b}$$

The connection of the header to the wall must be designed for the chord force at that
point:

$$T_1 = C_1 = \frac{M_1}{b}$$

NOTE: For simplicity, the examples in this book determine the chord forces using the dimension b as the width of the building. Theoretically b is the dimension between the centroids of the diaphragm chords, and the designer may choose to use this smaller, more conservative dimension.

The proper functioning of the chord requires that the horizontal diaphragm be effectively anchored to the chords and the supporting shearwalls. *Anchorage* can be provided by systematically designing for the transfer of gravity loads and lateral forces. This approach is introduced in Sec. 10.8, and the design for anchorage is covered in detail in Chap 15.

9.5   Design Problem: Horizontal Roof Diaphragm

In Example 9.8 a wood structural panel roof diaphragm is designed for a one-story building. The maximum unit shear in the transverse direction is the basis for determining the blocking and *maximum* nailing requirements for the diaphragm. However, the shear in the diaphragm is not constant, and it is possible for the nail spacing to be increased in areas of reduced shear. Likewise, it is possible to omit the blocking where the actual shear in the diaphragm is less than the allowable shear for an *unblocked* diaphragm. The locations where these changes in diaphragm construction can take place are easily determined from the unit shear diagram.

The use of changes in nailing and blocking is fairly common, but these changes can introduce problems in construction and inspection. If changes in diaphragm construction are used, they should be clearly shown on the framing plan. In addition, if blocking is omitted in the center portion of the diaphragm, the requirements for some other type of panel edge support must be considered.

In this building the chord members are the horizontal reinforcing bars in the masonry walls, and a check of the chord stress in these bars is shown. The horizontal diaphragm is also checked for the lateral force in the longitudinal direction, and the effect of the change in the diaphragm load case on allowable unit shears is demonstrated.

EXAMPLE 9.8   Wood Structural Panel Roof Diaphragm

The design forces being considered are seismic, and so are calculated at a strength level in accordance with the Code equations. The force notations therefore have a subscript "u" to denote ultimate. Roof framing consists of light-frame wood trusses (top chords on edge) at 24 in. o.c. The panel layout is given (Fig. 9.10a), and the shear and moment in the horizontal diaphragm are summarized below the framing plan. In this example, only the exterior walls are assumed to be shearwalls.

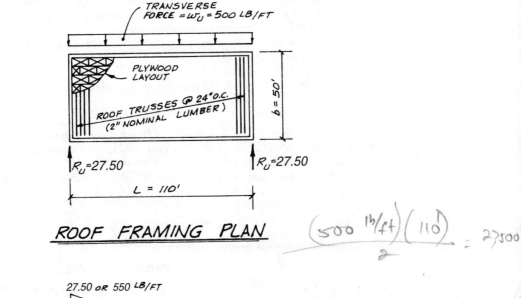

$$\left(\frac{500 \; ^{lb}/_{ft})(110)}{2}\right) = 27{,}500$$

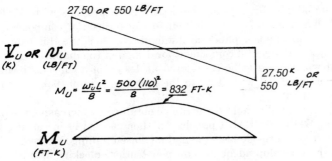

Figure 9.10a Roof plan showing transverse lateral force to horizontal diaphragm. Shear and moment diagrams for the diaphragm are also shown.

1. *Panel sheathing loads.*

Roof Live Load L_r = 20 psf

Roof Dead Load D = 8 psf

Although the roof framing is different, the panel sheathing spans the same distance (24 in.) as the sheathing in Fig. 8.14. The panel choices given in Example 8.12, part 1, apply to the example at hand:

Use ⁷/₁₆-in. Sheathing EXP 1

2. *Diaphragm nailing.* Although this panel is available in STRUCTURAL I, the STRUCTURAL I upgrade will not be used in this problem. If a substantially

higher unit shear were involved *or* if the nailing could be considerably reduced, then STRUCTURAL I should be considered.

Because the 2 × top chord of the light-frame trusses is turned on edge, the 10d nails develop adequate penetration. If the 2 × top chords were turned flat, the reduced nail penetration of 1½ in. would have to be taken into account (Sec. 9.3).

Panel *load case* (transverse lateral force):

The lateral force is *perpendicular to the continuous panel joint and perpendicular to the unblocked edge.*

∴ load case 1

If blocking is not used, the maximum allowable unit shear is

$$\text{Allow. } v = 230 \text{ lb/ft}$$

This comes from *ASD Shear Wall and Diaphragm Supplement* Table 3.1B for $\frac{7}{16}$-in. panels (not STRUCTURAL I) in an unblocked diaphragm, load case 1, 2-in.-nominal framing, and 8d nails. Before comparing this to the calculated roof unit shear, the unit shear needs to be multiplied by ρ and adjusted to an ASD level:

$$v_u = \text{E} = \rho Q_\text{E} = 1.0 \ (550) = 550 \text{ lb/ft}$$

$$v = 0.7\text{E} = 0.7(550) = 385 \text{ lb/ft}$$

The actual shear is greater than the allowable:

$$v = 385 \text{ lb/ft} > 230 \qquad NG$$

A reminder is appropriate here that in accordance with ASCE 7 equations 9.5.2.7-1 and 9.5.2.7-2, in addition to the horizontal component ρQ_E, E has a vertical component of plus and minus $0.2S_{DS}D$. While this example is only addressing the horizontal component for design of the diaphragm, the vertical component would need to be considered in design of a collector beam (Fig. 9.9), in design of wood ledger bolting to the masonry (Fig. 9.10c), and in other similar circumstances.

An increased allowable shear can be obtained by increasing the panel thickness or using STRUCTURAL I panels. Another method of increasing shear capacity used here is to provide blocking and increased nailing (i.e., reduced nail spacing). In order to develop adequate nail penetration, blocking will be turned on edge (not flat). As an alternative, the blocking can be used flat and the tabulated allowable diaphragm shears reduced. From the blocked diaphragm portion of the ASD Shear Wall and Diaphragm Supplement Table 3.1B, the following design is chosen:

> *Use* $^{15}\!/_{32}$-in. Sheathing EXP 1 with 10d common nails
> at 4 in. o.c. boundary*
> 6 in. o.c. all other panel edges
> 12 in. o.c. field
> Blocking required.
> Use 2 × 4 minimum (on edge).

$$\text{Allow. } v = 385 \text{ lb/ft} \cong 385 \text{ lb/ft} \qquad OK$$

---

*Because this diaphragm is load case 1, the 4-in. nail spacing is required at the diaphragm boundary only. The 6-in. nail spacing is used at continuous panels joints as well as other panel edges.

The above nailing and blocking requirements can safely be used throughout the entire diaphragm because these were determined for the maximum unit shear. The variation of diaphragm shear can be shown on a sketch. The *unit shear diagram* is the total shear diagram divided by the diaphragm width (Fig. 9.10*b*).

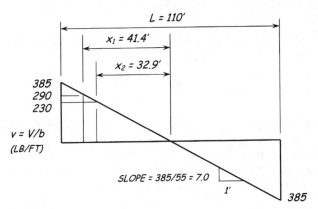

Figure 9.10*b* Unit shear diagram. Similar triangles may be used to determine locations of possible changes in diaphragm construction. Note that both the unit shear forces and the allowable unit shears in this diagram are at an ASD level.

Some designers prefer to take into account the reduced unit shears toward the center of the diaphragm (away from the shearwall reactions). If this is done, the nail spacing can be increased to the Code maximum allowed spacing (that is, 6 in. o.c. at all edges including the diaphragm boundary) where the shear drops off to the corresponding allowable shear for this nailing. This location can be determined by similar triangles. For example,

$$\text{Allow. } v = 290 \text{ lb/ft}$$

when the nail spacing at the diaphragm boundary is 6 in. o.c. and blocking is provided. Then

$$x_1 = \frac{290}{7.00} = 41.4 \text{ ft}$$

This refinement can be carried one step further, and the location can be determined where blocking can be omitted:

$$\text{Allow. } v = 230 \text{ lb/ft} \quad \text{for unblocked diaphragm}$$

Then

$$x_2 = \frac{230}{7.00} = 32.9 \text{ ft}$$

If these changes in diaphragm construction are used, the locations calculated above should be rounded off to some convenient dimensions. These locations and the required types of construction in the various segments of the diaphragm must be clearly shown on the roof framing plan. In this regard, the designer must weigh the saving in labor and materials gained by the changes described above against the possibility that the diaphragm may be constructed improperly in the field. The more variations in nailing and blocking, the greater the chance for error in the field. Increased nail spacing and omission of blocking can represent substantial savings in labor and materials, but inspection becomes increasingly important.

3. *Chord.* The strength level axial force in chord is obtained by resolving the diaphragm moment into a couple:

$$T_u = C_u = \frac{M_u}{b} = \frac{832 \text{ k-ft}}{50 \text{ ft}} = 16.6 \text{ k}$$

Depending on the type of wall system, various items can be designed to function as the *chord* member.

In this example, the walls are made from 8-in. grouted concrete block units, and the horizontal wall steel (two #5 bars) will be checked for chord stresses. (Examples of wood chords are covered elsewhere).

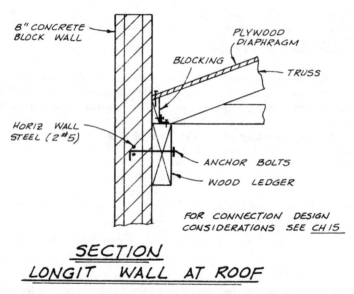

SECTION
LONGIT   WALL   AT ROOF

Figure 9.10c   Horizontal wall steel used as diaphragm chord.

Before calculating the steel stress, the chord force needs to be multiplied by the redundancy/reliability factor ρ and adjusted to an ASD level:

$$T_u = \text{E} = \rho Q_\text{E} = 1.0(16.6) = 16.6 \text{ k}$$

$$T = 0.7\text{E} = 0.7(16.6) = 11.6 \text{ k}$$

Once again a reminder is appropriate that in accordance with ASCE 7 equations 9.5.2.7-1 and 9.5.2.7-2, in addition to the horizontal component ρQ_E, E has a vertical component of plus and minus $0.2S_{DS}D$.

The tension chord is critical, and the stress in the horizontal wall steel can be calculated as the chord force divided by the area of steel. The minimum allowable stress in masonry wall reinforcing steel is 20 ksi under normal loads, and this can be increased by a factor of 1.33 for short-term forces (wind or seismic).

Stress in steel:

$$f_s = \frac{T}{A_s} = \frac{11.6}{2 \times 0.31} = 18.7 \text{ ksi}$$

Allow. $f_s = 20 \times 1.33 = 26.6 \text{ ksi} > 18.7 \quad OK$

The design of other wall-reinforcing steel is a problem in masonry design and is beyond the scope of this book. Steps must be taken to ensure that cross-grain bending (Example 6.1) in the ledger does not occur. The design of anchor bolts and other connections for attaching the horizontal diaphragm to the shearwall is covered in Chap. 15.

4. *Longitudinal lateral force.* The strength level longitudinal force and the corresponding shear and moment diagrams are shown in Fig. 9.10*d*.

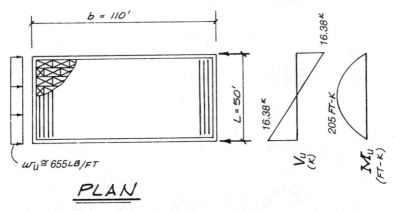

Figure 9.10*d* Roof plan showing longitudinal lateral force to horizontal diaphragm. Shear and moment diagrams are also shown.

The unit shear in a horizontal diaphragm is often critical in the transverse direction. However, the longitudinal shear should be checked, especially if changes in nailing occur. In any event, the longitudinal direction must be analyzed for anchorage forces for connecting the diaphragm and shearwall.

Unit shear:

$$v_u = \frac{V_u}{b} = \frac{16,380}{110} = 149 \text{ lb/ft}$$

The last two steps in determining the seismic force on an element of the primary LFRS are multiplying by ρ and, as part of the basic load combinations, adjusting to an ASD level. In the example in Sec. 3.5, it was shown that $\rho = 1.0$. The adjustments to the diaphragm unit shear are:

$$v_u = \text{E} = \rho Q_\text{E} = 1.0(149) = 149 \text{ lb/ft}$$

$$v = 0.7\text{E} = 0.7(149) = 104 \text{ lb/ft}$$

The load in the longitudinal direction is *parallel* to the *continuous panel joint* and *parallel* to the *unblocked edge*.

∴ load case 3

If the blocking is omitted in the center portion of the diaphragm (up to 35.7 ft on either side of the centerline; see Figure 9.10*b*), the allowable unit shear for an unblocked diaphragm must be used. From Table 3.1B of the *ASD Shear Wall and Diaphragm Supplement,*

$$\text{Allow. } v = 190 \text{ lb/ft} > 104 \qquad OK$$

If the unit shear in the longitudinal direction had exceeded 190 lb/ft, then blocking would have been required for the longitudinal lateral force. When blocking is required for load cases other than cases 1 and 2, the continuous panel joints must be nailed with the same spacing as the diaphragm boundary. The chord strength level force in the transverse walls is less than the chord force in the longitudinal walls:

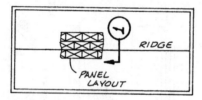

PLAN

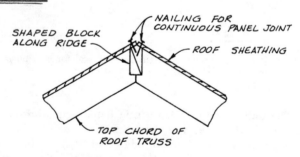

SECTION

Figure 9.10e Horizontal diaphragm detail at ridge.

$$T_u = C_u = \frac{M_u}{b} = \frac{205}{110} = 1.86 \text{ k} < 16.6 \text{ k}$$

∴ two # 5 bars   *OK* for chord around entire building

5. *Nailing along ridge.* The panel sheathing in this building is attached directly to the top chord of the light-frame wood roof trusses (see Fig. 9.10*e*). These trusses prob-

ably have sufficient strength and stiffness to maintain diaphragm continuity at the ridge. However, it is desirable to provide a positive attachment of the sheathing through doubled 2 $\times$ blocking or a shaped 3 $\times$ or 4 $\times$ block.

See Ref. 9.5 for a discussion of diaphragm design considerations at ridges in steeply pitched roofs.

9.6 Distribution of Lateral Forces in a Shearwall

If a shearwall has door or window openings, the total lateral force in the wall is carried by the effective segments in the wall. In wood shearwalls these effective segments are known as *shear panels* (or shearwalls), and in concrete and masonry walls they are referred to as *wall piers*. The designer must understand how to distribute the total force to these resisting elements so that the unit wall shears can be calculated.

The procedures for designing wood shearwalls are given in Chap. 10, but the method for distributing the lateral force is covered at this time. The distribution needs to be understood at this point because it is used to determine the magnitude of the force for the horizontal diaphragm collector (strut). A different procedure is used to distribute the horizontal diaphragm reaction to wood shearwalls and concrete or masonry shearwalls.

In a wood-frame shearwall, the most common approach is to assume that the unit shear is uniform throughout the total length of the wall shear panels. There are some refinements to this approach which will be discussed in Chap. 10. Thus the force in a given panel is in direct proportion to its length. See Example 9.9. The unit shear in one panel is the same as the unit shear in all other panels.

In order for this distribution to be reasonably correct, the wall must be constructed so that the panels function as separate shear-resisting elements. See Chap. 10. In addition, the shear panels are assumed to all have the same height. This is obtained by using continuous full-height studs or posts at each end of the shear panels. The panels must all deflect laterally the same amount. This is accomplished by tying the panels together with the collector (strut).

EXAMPLE 9.9   Distribution of Shear in a Wood Shearwall

In a wood shearwall, the wall is usually assumed to be made up of *separate* shear panels. The shear panels are connected together so that they deflect laterally the same amount at the horizontal diaphragm level. Collector (strut) members provide the connection between the shear panels.

The three shear panels in the wall shown in Fig. 9.11 all have the same height h. This behavior results from the use of double full-height studs or posts at both ends of the shear panels and appropriate shearwall nailing (Chap. 10). Filler panels (wall elements above and below openings which are ignored in the analysis) may be nailed less heavily (i.e., with the Code minimum nailing) to further isolate the shear panels.

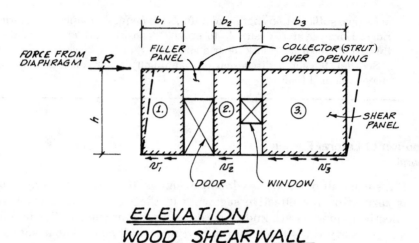

Figure 9.11 Unit shears in a wood-frame shearwall are normally assumed to be uniform.

For the typical wood shearwall, the resistance to the applied lateral force is assumed to be uniform throughout the combined length of the shear panels. Thus

$$v_1 = v_2 = v_3 = \frac{R}{\Sigma b}$$

This approach is appropriate when the aspect ratio (h/b) of each shearwall panel does not exceed 2. Where shearwall panel aspect ratios exceed 2, IBC Sec. 2305.3.3 requires the reduction of the tabulated allowable unit shear values. This topic will be discussed in Chap. 10.

Several types of load resistance develop in a wall subjected to a lateral force. These include bending resistance and shear resistance. For woodframe walls constructed as described above, the shear resistance is the significant form, and the unit shear in this case approaches a uniform distribution. This is only true, however, when the height-to-width (aspect) ratios conform to the Code limitations and the walls are of reasonably similar widths.

In buildings with reinforced-concrete or masonry shearwalls, the distribution of forces to the various elements in the wall differs from the uniform distribution assumed for wood shearwalls. Concrete and masonry walls may have significant combined bending and shear resistance. This combined resistance is measured by the *relative rigidity* of the various piers in the wall. Under a lateral force the piers are forced to undergo the same lateral deformation. The force required to deflect a long pier is larger than the force calculated on the basis of its length only.

Thus, in a concrete or masonry wall, the *unit shear* in a wider pier is greater than the unit shear in a narrower wall segment. The relative rigidity is a function of the height-to-width ration h/b of the pier and not just its width.

The relative rigidity and the question of force distribution are related to the *stiffness* of a structural element: The stiffer (more rigid) the element, the greater the force it "attracts." See Example 9.10. The calculation of relative rigidities for these wall elements is a concrete or masonry design problem and is beyond the scope of this text.

EXAMPLE 9.10   Distribution of Shear in a Concrete or Masonry Shearwall

In a concrete or masonry wall, the "piers" are analyzed to determine their *relative rigidities*. The smaller the height-to-width ratio h/b of a pier, the larger the relative rigidity. The larger the relative rigidity, the greater the percentage of total force R carried by the pier. Pier 2 in the wall shown in Fig. 9.12 has a greater rigidity and a greater unit shear (not just total shear) than pier 1; that is,

$$v_2 > v_1$$

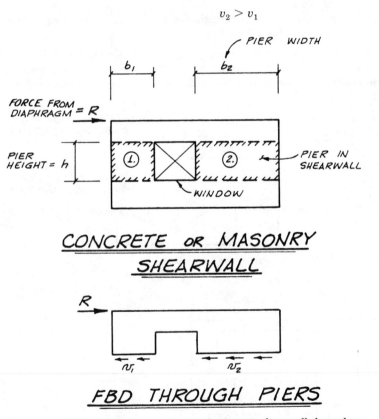

Figure 9.12 Unit shears in a concrete or masonry shearwall depend on relative rigidities of wall piers.

In buildings with concrete or masonry walls, the loads over openings in the walls can be supported in two different ways. In the first system, there is a concrete or masonry wall element over the openings in the wall. See Fig. 9.13a in Example 9.11. This element can be designed as a beam to carry the gravity loads across the opening. Concrete and masonry beams of this type are known as *lintels*. In the second case, a wood *header* may be used to span the opening (Fig. 9.13*b*).

EXAMPLE 9.11   Lintel and Header between Shearwalls

A concrete or masonry lintel, or a wood (or steel) header, can be used to span over an opening in a wall (see Fig. 9.13). The member over the opening must be designed for both vertical loads and lateral forces. Load combinations are

1. Vertical loads $D + (L_r$ or $S)$.
2. Combined vertical and lateral.
   a. For lateral forces parallel to the wall, the force in the header is the collector (strut) force.
   b. For lateral forces perpendicular to the wall, the force in the header is the diaphragm *chord* force.

The chord will be designed using the Code *basic load combinations,* however the collector will need to be designed using the Code *special (overstrength) seismic load combinations,* as discussed in Sec. 9.2. These will be discussed further in Sec. 9.7.

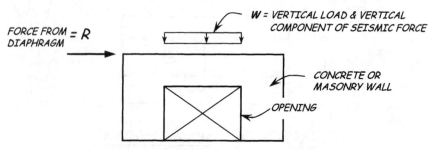

Figure 9.13*a*   Wall with concrete or masonry lintel.

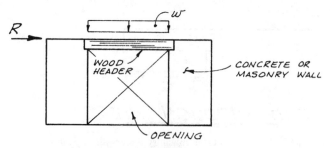

Figure 9.13*b*   Wall with wood header.

A distinction is made between these systems because in the first case the load from the horizontal diaphragm is transmitted to the shearwalls through the lintel, and the diaphragm design is unaffected. In the case of the wood header, however, the unit shear from the horizontal diaphragm must be transmitted through the header/collector (strut) into the shearwalls. The load combinations required for the design of this member are given in Example 9.11.

In summary, the distribution of the *unit shear* in a wall subjected to a lateral force will be as follows:

1. Wood shearwalls—uniform distribution to all shear panels with $h/b \geq 2$

2. Concrete and masonry shearwalls—distribtution is a function of the relative rigidities of the wall piers

For problems in this book involving concrete or masonry walls, it is assumed that the distribution of wall forces and the unit shears are known by previous analysis.

9.7   Collector (Strut) Forces

The perimeter members of a horizontal diaphragm that are parallel to the applied lateral force are the collectors (struts). These members are also known as *drag struts* or *ties,* and their function is illustrated in Fig. 9.4 (Sec. 9.2). The term collector is generally used in this book because it is the term frequently used in the IBC. Essentially the collector pulls or drags the unit shear in an unsupported segment of a horizontal diaphragm (the portion over an opening in a shearwall, for example) into the supporting elements of the shearwall.

The members in a building that serve as the collectors are typically the same members that are used as the *chords* for the lateral force in the perpendicular direction. Thus, the design of the chord and the collector for a given wall may simply involve the design of the same member for different forces. In fact, for a given perimeter member, the chord force is compared with the strut force, and the design is based on the critical force.

The wood design principles for struts and chords are covered in Chap. 7. These members are either tension or compression members, and they may or may not have bending. Bending may result from vertical loads, as in the header in Example 9.11. Where vertical dead loads D are acting, the vertical component of the seismic force $0.2S_{DS}D$ must also be considered in accordance with ASCE 7 equations 9.5.2.7-1 and 9.5.2.7-2, or equations 9.5.2.7.1-1 and 9.5.2.7.1-2 for the maximum earthquake load effect. Connections are also an important part of the collector design. The member itself may be sufficiently strong, but little is gained unless the strut is adequately connected to the supporting shearwall elements. The design of connections is addressed in later chapters, but the calculation of the force in the drag strut is covered now.

The building in Fig. 9.4 has an end shearwall with an opening located in the *center* of the wall. This presents a simple problem for the determination of the collector forces because the total force is shared *equally* by both shear panels. This would be true for the equal-length piers in a concrete or masonry wall also. The force in the collector is maximum at each end of the wall opening. This maximum force is easily calculated as the unit shear in the horizontal diaphragm times one-half the length of the wall opening.

The force in the collector varies linearly throughout the length of the collector. At one end of the opening the collector is in tension, and at the other end it is in compression. Since the lateral force can come from either direction, the ends may be stressed in either tension or compression. The magnitude of the collector force in the two shear panels is not shown in Fig. 9.4. However, the collector force in the top of the shearwall decreases from a maximum at the wall opening to zero at the outside end of the shearwall.

A somewhat more involved problem occurs when an opening is not symmetrically located in the length of the wall. See Example 9.12. Here a greater portion of the total wall load is carried by the longer shear panel. A correspondingly larger percentage of the unsupported diaphragm shear over the opening must be transmitted by the collector to the longer panel.

The magnitude of the force at some point in the collector can be determined in several ways. Two approaches are shown in Example 9.12. The *first method* simply involves summing forces on a free-body diagram (FBD) of the wall. The cut is taken at the point where the collector force is required. The other forces acting on the FBD are the unit shears in the roof diaphragm and the unit shears in the shearwall.

The *second method* provides a plot of the force in the collector along the length of the building. With this diagram the collector force at any point can be readily determined.

The procedure illustrated in Example 9.12 for plotting the collector force can be used for wood or concrete or masonry shearwalls. The only difference is that for masonry or concrete walls, the unit shear in the walls will probably be different.

The two methods presented in Example 9.12 graphically demonstrate the functioning of a collector. These methods are reasonable for the example shown. However, when there are multiple wall openings, some degree of engineering judgment must be exercised in determining the design force for the collector.

Before proceeding with Example 9.12, the discussion of seismic force level for design of collectors needs to be revisited. In Sec. 9.2, it was noted that the IBC considers *collectors* to be critical to the behavior of the lateral force resisting system LFRS. Because of this, a special seismic force level E_m and special seismic load combinations are required for collectors in Seismic Design Category C and higher (ASCE 7 Sec. 9.5.2.7.1). However the Code exempts buildings with a LFRS constructed *entirely with wood or steel light-frame shearwalls* from the special seismic force requirement.

Because Example 9.12 uses wood shearwalls, the special seismic forces and load combinations do not need to be used. If the building in Example 9.12 had masonry walls, the reaction $R = 12.0$ k would have to be multiplied by $\Omega_o = 2.5$. The collector design principles demonstrated in Example 9.12 apply regardless of whether the *basic Code forces* or the E_m forces are used.

The calculation of collector forces is applicable to both wind and seismic design. The notation shown in Example 9.12 is allowable stress design, as would be used for wind forces. For seismic design the forces would generally be at a strength level (and forces and shears would have "u" subscripts included in the notation).

EXAMPLE 9.12   Calculation of Force in Collector

Determine the magnitude of the collector force throughout the length of the front wall in the building shown in Fig. 19.14a. The walls are wood-frame construction. The forces are wind.

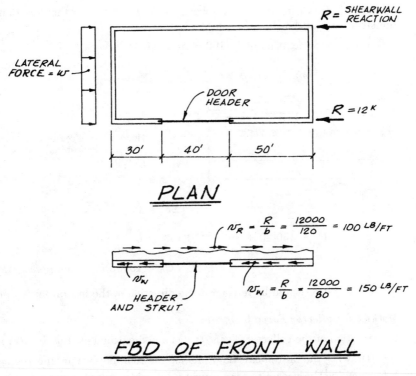

Figure 9.14a   Free body diagram of front longitudinal wall. The forces shown are (1) unit shear in roof diaphragm v_R, (2) unit shear in shearwalls v_W.

Method 1: FBD Approach

The force in the collector at any point can be determined by cutting the member and summing forces. Two examples are shown (Fig. 9.14b and c). One cut is taken through the header, and the other is taken at a point in the wall.

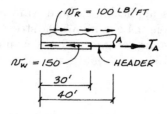

Figure 9.14b Free body diagram of from longitudinal wall taken at point *A*.

a. Collector force at point A (40 ft from left wall):

$$\Sigma F_x = 0$$

$$T_A + 100(40) - 150(30) = 0$$

$$T_A = +500 \text{ lb} \quad \text{(tension)}$$

NOTE: The lateral force can act toward the right or left. If the direction is reversed, the strut force at *A* will be in compression.

b. Collector force at point B (100 ft from left wall):

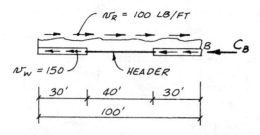

Figure 9.14c Free body diagram of front longitudinal wall taken at point *B*.

$$\Sigma F_x = 0$$

$$C_B + 150(30 + 30) - 100(100) = 0$$

$$C_B = 1000 \text{ lb} \quad \text{(compression)}$$

NOTE: This force would be tension if the direction of the lateral force were reversed.

Method 2: Collector Force Diagram

Construction of the collector (drag strut) force diagram (see Fig. 914*d*) involves

a. Drawing the unit shear diagrams for the roof diaphragm and shearwalls.
b. Drawing the *net* unit shear diagram by subtracting the diagrams in *a*.
c. Constructing the strut force diagram by using the *areas* of the net unit shear diagram as *changes* in the magnitude of the strut force. The strut force is zero at the outside ends of the building.

The relationship between the net unit shear diagram and the collector force diagram is similar to the relationship between the shear and moment diagrams for a beam.

NOTE: Signs of forces are not important because forces can reverse direction.

The values obtained by taking FBDs at points *A* and *B* can be verified by similar triangles on the collector force diagram.

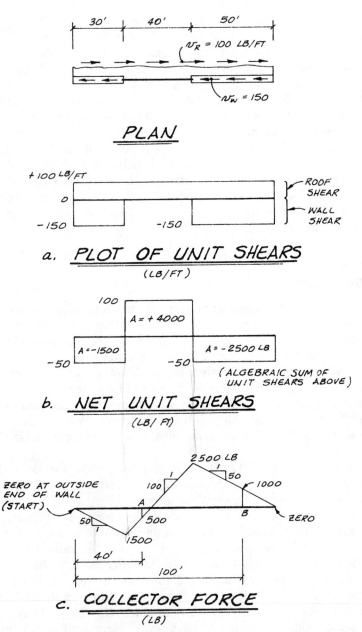

a. **PLOT OF UNIT SHEARS**
(LB/FT)

b. **NET UNIT SHEARS**
(LB/FT)

c. **COLLECTOR FORCE**
(LB)

Figure 9.14d Collector force diagram. The final plot (c. collector force) shows variation in the magnitude of the strut force over the entire length of the front longitudinal wall.

A third approach to collector force calculation for one-story buildings is to determine the maximum force that might develop in a collector regardless of its location along the length of a wall. This force can be taken as the *unit shear in the horizontal diaphragm times the length of the wall opening.* If this is done for the building in Example 9.12, a collector force of $T = 100 \times 40 = 4000$ lb is obtained. Connections at both ends of the strut as well as the member itself would then conservatively be designed for this force.

Although a considerably higher design force is determined in this example (4000 lb versus 2500 and 1500 lb), the method has the advantage of being conservative and easy to apply. When there are multiple openings in a wall, the designer can extend this method to determine a very conservative value for the collector force by taking the unit shear in the horizontal diaphragm times the sum of the lengths of the wall openings. If the force obtained in this manner is overly conservative, a more detailed analysis can be made.

Once the collector forces have been determined, they can be compared with the chord forces at corresponding points along the wall to determine the critical design condition.

9.8   Diaphragm Deflections

The deflection of a horizontal diaphragm has been illustrated in a number of sketches in Chap. 9. This topic is the object of some concern because the walls that are attached to (actually supported by) the horizontal diaphragm are forced to deflect out-of-plane along with the diaphragm. See Example 9.13. If the walls are constructed so they can undergo or "accommodate" these deformations without failure, there is little need for concern. However, a potential problem exists when the deflection is large or when the walls are constructed so rigidly that they can tolerate little deflection. In this case, the diaphragm imposes an out-of-plane deflection on the wall that may cause the wall to be overstressed.

As discussed in Sec. 3.3, there is another reason for calculating the diaphragm deflection, which has to do with classifying the diaphragm as being either *flexible* or *rigid*. This will be discussed further in Sec. 9.11.

There are two methods to evaluate out-of-plane wall displacements resulting from the in-plane deflection of the horizontal diaphragm. The first method that traditionally was the most widely used is basically a rule of thumb, in that no attempt is made to calculate the magnitude of the actual deflection. This method makes use of the span-to-width ratio of the horizontal diaphragm. In this simple approach, the span-to-width ratio L/b is checked to be less than some acceptable limit. For example, the Code sets an upper limit on the span-to-width ratio for blocked wood structural panel horizontal diaphragms of 4.

Thus, this approach assumes that if the *proportions* of the diaphragm are *reasonable,* deflection will not be excessive. Although no numerical value for the deflection of the diaphragm is calculated, this method has traditionally been applied. As a minimum, conformance to the Code required L/b ratio limits should be demonstrated.

EXAMPLE 9.13   Diaphragm Deflections and Span-to-Width Ratios

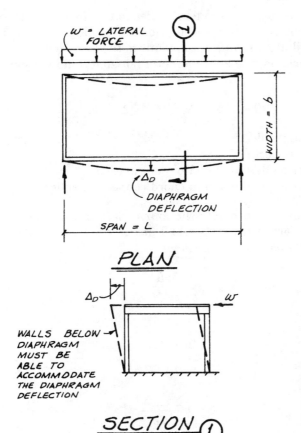

Figure 9.15 Plan and section views showing the deflections caused by lateral force on horizontal diaphragm.

Wood-frame walls are capable of accommodating larger out-of-plane deflections than are more rigid wall systems such as concrete or masonry.

Diaphragm Proportions

The traditional method of controlling deflection was to limit the span-to-width ratio L/b for the diaphragm. For blocked horizontal wood structural panel diaphragms, $L/b \leq 4.0$. For other types of wood diaphragms, see IBC Sec. 2305.2.3.

The current technique of accounting for diaphragm deflections is to calculate a numerical value for the deflection is presented in Sec. 3.3 of the ASD Wood Structural Panel Shear Wall and Diaphragm Supplement and is included in the IBC. The total deflection that occurs in an ordinary beam is made up of two deflection components: bending and shear. However, for most practical beams, the shear deflection is negligible, and only the bending deflection is calculated.

In a horizontal diaphragm, the total deflection is the sum of the deflections caused by a number of factors in addition to bending and shear. All these factors may contribute a significant amount to the total deflection. In the normally accepted formula, the deflection of a horizontal diaphragm is made up of four parts:

$$\text{Total } \Delta = \Delta_b + \Delta_v + \Delta_n + \Delta_c$$

where Δ_b = bending deflection
Δ_v = shear deflection
Δ_n = deflection due to nail slip (deformation)
Δ_c = deflection due to slip in chord connection splices

The equations used to evaluate the four deflection components for a horizontal diaphragm are given in Sec. 3.3 of the *ASD Wood Structural Panel Shear Wall and Diaphragm Supplement:*

$$\Delta_b = \frac{5vL^3}{8EAb}$$

$$\Delta_v = \frac{vL}{4Gt}$$

$$\Delta_n = 0.188Le_n$$

$$\Delta_c = \frac{\Sigma \, (\Delta_c X)}{2b}$$

where A = area of chord cross section, in.^{2}
b = diaphragm width, ft
E = modulus of elasticity of chords, psi
e_n = nail deformation, in. (per Table 3.2 of the *ASD Shear Wall and Diaphragm Supplement*)
G = modulus of rigidity of wood structural panels, psi (see Chap. 8)
L = diaphragm length, ft
t = effective thickness of wood structural panels for shear (see Chap. 8)
v = maximum shear caused by design loads in direction under consideration, lb/ft
$\Sigma(\Delta_c X)$ = sum of the product of individual chord-splice slip values of the diaphragm and its distance to the nearest support, in.-ft

It should be noted that these formulas apply only to *blocked* diaphragms. See Example 9.14. Test data suggests that the deflection of *unblocked* diaphragms is approximately 2.5 times that of a blocked diaphragm. This rule-of-thumb may be used to estimate the deflection of unblocked diaphragms using the above equations.

EXAMPLE 9.14   Calculation of Diaphragm Deflections

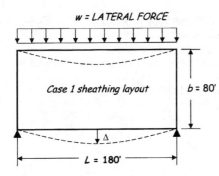

Figure 9.16   Plan of diaphragm.

From a separate analysis, the total lateral force is determined to be 55 k. All framing is assumed to be Douglas fir-Larch and the sheathing to be 32/16 span-rated wood structural panels with 8d common nails. The chords are assumed to be two Douglas fir-Larch No. 1 graded 2 × 8s spliced together.

The following equations are used to evaluate the four deflection components for a horizontal diaphragm:

$$\Delta_b = \frac{5vL^3}{8EAb}$$

$$\Delta_v = \frac{vL}{4Gt}$$

$$\Delta_n = 0.188 L e_n$$

$$\Delta_c = \frac{\Sigma \, (\Delta_c X)}{2b}$$

Considering each deflection component separately:

Bending

$$\Delta_b = \frac{5vL^3}{8EAb}$$

where $v = 55{,}000$ lb/$(2 \times 80$ ft$) = 344$ lb/ft

$\quad L = 180$ ft

$\quad E = 1.7 \times 10^6$ psi for No. 1 DFL 2 × 8

$\quad A = 2(1.5$ in. $\times 7.25$ in.$)$

$\quad\quad = 21.75$ in.^{2}

$\quad b = 80$ ft

$$\Delta_b = \frac{5vL^3}{8EAb} = \frac{5(344)(180)^3}{8(1.7 \times 10^6)(21.75)(80)} = 0.424 \text{ in.}$$

Shear

$$\Delta_v = \frac{vL}{4Gt}$$

Rather than determining G and t separately, the shear rigidity Gt is provided in the ASD Wood Structural Panels Supplement:

$Gt = 27,000$ lb/in. for 32/16 span-rated wood structural panels

$$\Delta_v = \frac{vL}{4Gt} = \frac{(344)(180)}{4(27,000)} = 0.573 \text{ in.}$$

Nail Slip

$$\Delta_n = 0.188Le_n$$

In order to determine the nail deformation e_n, the force per fastener V_n must first be determined. Assuming the 8d nails are spaced at 6 in. o.c. at the edges, two nails per foot are present:

$$V_n = (344 \text{ plf})/(2 \text{ nails per ft})$$

$$= 172 \text{ lb/nail for 8d nails spaced at 6 in. o.c.}$$

It is assumed here that the diaphragm is fabricated and used in the dry condition. From Table 3.2 of the ASD Shear Wall and Diaphragm Supplement:

$$e_n = 1.2(V_n/616)^{3.018}$$

$$= 0.0255 \text{ in.}$$

Note that the nail deformation is increased by 20 percent since STRUCTURAL I was not specified. The diaphragm deflection resulting from nail slip is then

$$\Delta_n = 0.188Le_n = 0.188(180)(0.0255) = 0.863 \text{ in.}$$

Chord Slip

$$\Delta_c = \frac{\Sigma (\Delta_c X)}{2b}$$

The average slip Δ_c in a chord splice connection is assumed to be $\frac{1}{32}$ in., which is half of the $\frac{1}{16}$-in. allowable oversize for the bolt holes. It is assumed that the splice connections are located every 20 ft along a chord. Since there are two chords (tension and compression), the summation can simply be made for one chord and doubled. Therefore,

$$\sum (\Delta_c X) = 2[\tfrac{1}{32}(20 + 40 + 60 + 80 + 80 + 60 + 40 + 20)] = 25 \text{ in.-ft}$$

$$\Delta_c = \frac{\sum (\Delta_c X)}{2b} = \frac{25}{2(80)} = 0.156 \text{ in.}$$

Total Deflection

$$\Delta = 0.424 + 0.573 + 0.863 + 0.156$$

$$\Delta = 2.016 \text{ in.}$$

Considered over a 180 ft length of the diaphragm, this deflection is quite small: $L/1070$. Also a point of consideration is the out-of-plane bending of the wall. Since the base of a wall is essentially a pin connection, the diaphragm deflections cause little bending in the walls. The primary issue, however, is potential vertical instability of the wall caused by this deflection, or the P–Δ effect.

If the deflection of the diaphragm is considered to be excessive, increased nailing or the use of STRUCTURAL I panels are two possible remedies to consider.

One problem with evaluating out-of-plane wall deformations by applying the horizontal diaphragm deflection calculation is that there are no definite criteria against which the calculated deflection must be checked. The simple deflection limits used for ordinary beams (such as $L/180$ and $L/360$) are not applied to diaphragm deflections. These types of limits do not evaluate the out-of-plane effects that the deformations impose on the walls supported by the diaphragm.

Reference 9.6 suggests a possible deflection criterion for diaphragms that support concrete or masonry walls, but generally the evaluation of the calculated deflection is left to the judgment of the designer. It should be noted that the limits on the span-to-width ratio given in IBC Table 2305.2.3 are a requirement of the Code, and they must be satisfied.

In this introduction to diaphragm design, the limits on the span-to-width ratio will be relied upon for controlling deflections. As noted, this is fairly common practice in ordinary building design. Diaphragm deflections are examined in greater detail in Sec. 16.10.

9.9 Diaphragms with Interior Shearwalls

All of the diaphragms considered up to this point have been supported by shearwalls that are the exterior walls of the building. Although it is convenient to introduce horizontal diaphragms by using this basic system, many buildings make use of interior shearwalls in addition to exterior shearwalls. This section will address analysis of buildings with flexible diaphragms and interior shearwalls. Analysis of buildings with rigid diaphragms is covered in Sec. 9.11 and 16.10.

The roof or floor assembly is assumed to act as a number of separate horizontal diaphragms. See Example 9.15. As a result, they are treated as simply supported beams that span between the respective shearwall supports. Thus the *shear in the diaphragm* can be determined using the methods previously covered for buildings with exterior shearwalls. The difference is that the span of the diaphragm is now measured between adjacent parallel walls and not simply between the exterior walls of the building. With these separate diaphragms there will be different unit shears in the diaphragms unless the spans and loads happen to be the same for the various diaphragms.

The *forces to the shearwalls* are calculated as the sum of the reactions from the horizontal diaphragms that are supported by the shearwall. Thus, for an exterior wall the shearwall force is the reaction of one horizontal diaphragm. For an interior shearwall, however, the force to the wall is the sum of the diaphragm reactions on either side of the wall.

The chord force for the diaphragms is determined in the same manner as for a building with exterior shearwalls. The moments in the respective diaphragms are calculated as simple beam moments. Once the chord forces are determined, the chords are designed in accordance with appropriate procedures for the type of member and materials used. The diaphragm design procedure shown in Example 9.15 is equally applicable with wind or seismic forces. The example uses strength level forces, denoted with a "u" subscript. Resulting seismic unit shears must be reduced to an ASD level before they can be compared with Code allowable values.

EXAMPLE 9.15   Flexible Horizontal Diaphragm with Interior Shearwall

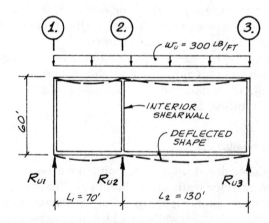

Figure 9.17a Plan view of building with interior and exterior shearwalls. Deflected shapes of assumed separate horizontal diaphragms are shown.

Determine the unit shear in the horizontal diaphragm and shearwalls in the building shown in Fig. 9.17 for the transverse lateral force of 300 lb/ft. Also calculate the diaphragm chord forces. The applied lateral forces used in this example are seismic and are therefore at a strength level.

Horizontal Diaphragm

Flexible horizontal diaphragms are assumed to span between the exterior and interior shearwalls. The continuity at the interior wall is disregarded, and two simply supported diaphragms are analyzed for internal unit shears. See Fig. 9.17b.

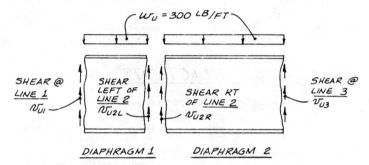

Figure 9.17b Free body diagrams of horizontal diaphragms.

$$V_{u1} = V_{u2L} = \frac{w_u L_1}{2} = \frac{(0.300)70}{2} = 10.5 \text{ k}$$

$$v_{u1} = v_{u2L} = \frac{V_{u1}}{b} = \frac{10,500}{60} = 175 \text{ lb/ft}$$

$$V_{u2R} = V_{u3} = \frac{w_u L_2}{2} = \frac{(0.300)130}{2} = 19.5 \text{ k}$$

$$v_{u2R} = v_{u3} = \frac{V_{u2R}}{b} = \frac{19,500}{60} = 325 \text{ lb/ft}$$

The nailing requirements for each diaphragm can be determined separately on the basis of the respective unit shears. Note that both diaphragms 1 and 2 will have their own nailing requirements, including boundary nailing along line 2. These nailing provisions must be clearly shown on the design plans.

As an alternative to double boundary nailing, it is possible to design the nailing along line 2 for the combined shear from diaphragms 1 and 2:

$$v_{u2} = \frac{10,500 + 19,500}{60} = 500 \text{ lb/ft}$$

Shearwalls

The unit shears in the shearwalls can be determined by calculating the reactions carried by the shearwalls. Shearwall reactions must balance the loads (shears) from the horizontal diaphragm. See Fig. 9.17c.

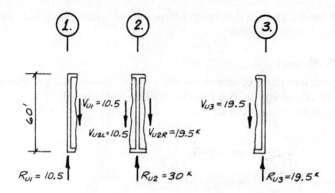

SHEARWALL REACTIONS

Figure 9.17c Plan view showing free body diagrams of the three transverse shearwalls.

Unit wall shears (no openings in walls):

$$v_{u1} = \frac{R}{b} = \frac{10,500}{60} = 175 \text{ lb/ft}$$

$$v_{u2} = \frac{30,000}{60} = 500 \text{ lb/ft}$$

$$v_{u3} = \frac{19,500}{60} = 325 \text{ lb/ft}$$

Alternate shearwall reaction calculation:

Because the diaphragms are assumed to be simply supported, the shearwall reactions can also be determined using the tributary widths to the shearwalls. See Fig. 9.17d.

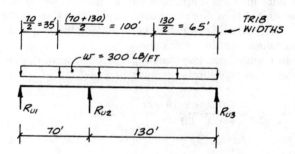

Figure 9.17d Lateral force diagram showing tributary widths to shearwalls (schematic similar to Fig. 9.17a).

$$R_{u1} = (0.300)35 = 10.5 \text{ k}$$

$$R_{u2} = (0.300)100 = 30 \text{ k}$$

$$R_{u3} = (0.300)65 = 19.5 \text{ k}$$

NOTE: The calculated unit shears for the shearwalls are from the horizontal diaphragm reactions. If the lateral force is a seismic force, there will be an additional wall shear generated by the dead load of the shearwall itself (see Chap. 3).

Diaphragm Chord Forces

The distribution of force in the chord follows the shape of the simple beam moment diagram for the diaphragms. See Fig. 9.17e.

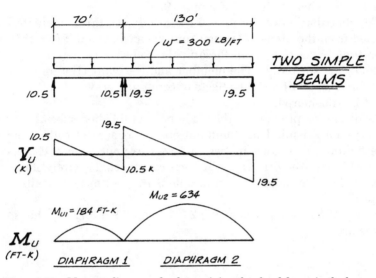

Figure 9.17e Moment diagrams for determining the chord forces in the horizontal diaphragms.

$$T_{u1} = C_{u1} = \frac{M_1}{b} = \frac{184}{60} = 3.1 \text{ k}$$

$$T_{u2} = C_{u2} = \frac{M_2}{b} = \frac{634}{60} = 10.6 \text{ k}$$

The simple span procedure for handling horizontal diaphragms and interior shearwalls illustrated in Example 9.15 is probably the most common approach used to design the general class of buildings covered in this book. However, Ref. 9.6 suggests that the continuity of the horizontal diaphragm at an interior shearwall should not be disregarded. Additional diaphragm research is needed to better understand the interaction of the various components in a lateral-force-resisting system.

Additional design considerations are required for seismic forces on buildings that are *structurally irregular.* An example of an irregularity is an interior shearwall on the second floor of a two-story building that is offset horizontally from the interior shearwall on the first floor. This and other types of seismic irregularities are covered in Chap. 16.

9.10 Interior Shearwalls with Collectors (Struts)

If a horizontal diaphragm is supported by an interior shearwall that is not the full width of the building, or if there are openings in the wall, a collector will be required. As with the collectors in exterior walls (Sec. 9.7), the collector will transfer the unit shear from the unsupported portion of the horizontal diaphragm into the shearwall. In the case of an interior shearwall, the collector force will be the sum of the forces from the horizontal diaphragms on both sides of the shearwall. See Example 9.16.

The distribution of the force in the collector throughout its length can be plotted from the areas of the net unit shear diagram. From this plot it can be seen that the critical axial force in the collector is at the point where it connects to the shearwall. The member and the connection must be designed for this force. In addition, the effects of combined gravity loads and lateral forces must be considered.

As discussed previously, the Code has established special seismic forces E_m and special seismic load combinations for design of collectors (drag struts). The building discussed in Sec. 9.7 was exempt from these special requirements because the primary LFRS was constructed entirely of light-frame wood shearwalls. The building in Example 9.16 has masonry walls, and is not exempt.

From ASCE 7 equations 9.5.2.7.1-1 and 9.5.2.7.1-2, the special seismic forces are:

$$E_m = \Omega_o Q_E + 0.2 S_{DS} D$$

and

$$E_m = \Omega_o Q_E - 0.2 S_{DS} D$$

From IBC Sec. 1605.4, the corresponding *special seismic load combinations* are:

$$1.2D + f_1 L + 1.0\ E_m \tag{16-19}$$

$$0.9D + 1.0\ E_m \tag{16-20}$$

In Example 9.16, the collector force will first be calculated as Q_E and then be multiplied by Ω_o to obtain E_m.

EXAMPLE 9.16   Building with Interior Collector

The building of Example 9.15 is modified in this example so that the interior shearwall is not the full width of the building. Determine the collector force for the interior shearwall. See Fig. 9.18.

Assume:

$$\Omega_o = 2.5 \ \text{(ASCE 7 Table 9.5.2.2)}$$

Also assume that the shears and reactions from Example 9.15 are strength level seismic design forces.

Unit shear in shearwall: $v_{u2} = \dfrac{R_{u2}}{b} = \dfrac{30,000}{40} = 750 \ \text{lb/ft}$

The interior shearwall carries the reactions of the horizontal diaphragms. Part of this force is transferred to the shearwall by the collector. The maximum collector force is 10 k and is obtained from the collector force diagram in Fig. 9.18.

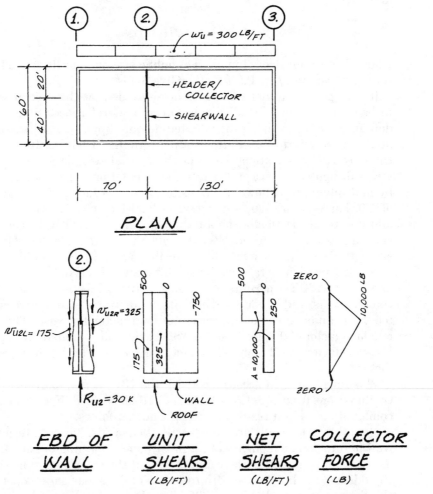

Figure 9.18 Construction of collector (strut) force diagram. The steps involved are similar to those introduced in Fig. 9.14d.

Collector Force Adjustment

$$E_m = \Omega_o Q_E = 2.5(10,000) = 25,000 \text{ lb strength level}$$

The Ω_o multiplier adjusts the collector member force to the estimated maximum earthquake force that can develop. IBC requires use of the special seismic force E_m for both allowable stress and strength design methods:

$$1.2D + f_1 L + E_m \tag{16-19}$$

and

$$0.9D + E_m \tag{16-20}$$

where f_1 is 1.0 for floors in places of public assembly or floor live loads greater than 100 psf, and f_1 is 0.5 for other live loads.

It is important to note that it is not intended for the special seismic forces to be compared to ASD allowable forces. IBC references Sec. 1617.1 for the definition of E_m. Although it is noted as only applicable when the simplified analysis procedure is used, the last paragraph of IBC Sec. 1617.1.1.2 provides the only IBC description of use of the special seismic forces with allowable stress design. This section permits the calculation of a strength level capacity by multiplying the allowable stress or force by a factor of 1.7 and a phi factor of 1.0. For wood design, a duration of load factor is permitted to be used in addition to the 1.7 factor. This pseudo-strength level can be compared to the special seismic force level. Alternatively, the strength can be calculated using strength design provisions and applicable strength design phi factors.

ASCE 7 provisions take a slightly different approach, but result in the same end effect. ASCE 7 permits E_m to be substituted for E in the ASD basic load combinations, and in addition, permits an allowable stress increase of 1.2 (Sec. 9.5.2.7.1) and the use of a duration of load factor for wood. The combination of a load factor of 0.7, for E in the ASD basic load combinations, and the stress increase factor of 1.2 results in a factor of 1.71—virtually the same as the IBC factor.

When designing in accordance with the IBC, it is important that IBC load combinations from equations 16-19 and 16-20 be used. Use of the ASCE 7 load combinations would result in different design forces.

Not all interior walls are necessarily shearwalls. For an interior woodframe wall to function as a shearwall, it must have sheathing of adequate strength, and it must be nailed to the framing so that the shear resistance is developed. In addition, the height-to-width ratio h/b of the shear panel must be less than the limits given in IBC Table 2305.3.3. (Procedures for designing shearwalls are covered in Chap. 10.) the diaphragm must also be connected to the shearwall so that the shears are transferred.

In addition to these strength requirements, some consideration should be given to the relative widths and locations of the various walls. To illustrate this point, consider an interior wall with a width that is very small in comparison with the adjacent parallel wall. See Example 9.17.

Because of the large width b_e of the right exterior wall, it will be a more rigid support for the horizontal diaphragm than the small interior wall. The large rigidity of the exterior wall will "attract" and carry a large portion of the lateral force tributary to *both* walls.

Because of this, it would be better practice to ignore the small interior wall and design the nearby exterior wall for the entire tributary lateral force. The small interior wall should then be sheathed and nailed lightly to ensure that it does not interfere with the diaphragm span.

Attempts can be made to compare the relative rigidities of the various shearwalls. This is similar to the rigidity analysis of the various piers in a single concrete or masonry wall (Sec. 9.6). However, in wood-frame construction the question of whether to consider a wall as an "effective" shearwall is normally a matter of judgment. The relative width of the shearwalls and their proximity to other effective walls are used in making this determination.

EXAMPLE 9.17   Effectiveness of Interior Shearwalls

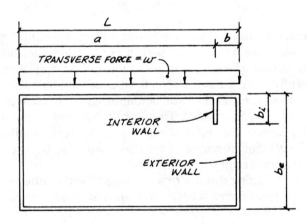

Figure 9.19 Comparison of effective shearwalls.

Using tributary widths, the *calculated* load to the interior shearwall is

$$R_i = \left(\frac{a}{2} + \frac{b}{2} \right) w$$

The load to the right exterior wall would be much smaller:

$$R_e = \left(\frac{b}{2}\right)w$$

Under *actual* loading, however, the load carried by the right exterior wall will approach

$$R_e = \left(\frac{L}{2}\right)w$$

The reason for this is that the large exterior wall is more rigid than the interior wall. Thus, in order to distribute the force from the horizontal diaphragm to the walls using tributary widths, the shearwalls must be *effective*.

The effectiveness of a wall in providing true shearwall action depends on several factors, including

1. Relative lengths of the walls. If b_i and b_e were more nearly equal, the interior wall would be more effective.

2. Relative location of the walls. If b were larger in comparison to a, the interior shearwall would also begin to function as a separate support.

The evaluation of shearwall effectiveness is, to a large extent, a matter of judgment. Walls which are judged to be ineffective (i.e., ignored as shearwalls) should be constructed so that they do not interfere with the assumed action of the diaphragm.

9.11   Diaphragm Flexibility

Horizontal diaphragms can be classified according to their tendency to deflect under load. Concrete floor or roof slabs deflect very little under lateral forces and are generally classified as *rigid diaphragms*. Wood diaphragms, on the other hand, generally deflect more and can be classified as either *flexible* or *rigid diaphragms* depending on the ratio of diaphragm deflection Δ_D (Fig. 3.6c) to shearwall deflection Δ_S (Fig. 3.5c). There are other systems which classify diaphragms into additional categories such as semirigid, semiflexible, and very flexible (Ref. 9.6). For purposes of this text, only the basic classification of rigid and flexible diaphragms is considered.

The purpose of introducing these terms is to describe the difference between the distribution of lateral forces for the two types of diaphragms. In addition to describing the distribution of lateral forces to shearwalls, there is a second reason for introducing the classifications of *flexible* and *rigid* horizontal diaphragms. For Seismic Design Categories C and up, the seismic forces for anchorage of concrete or masonry walls to floors or roofs (ASCE 7 Sec. 9.5.2.6.3.2) are dependent on whether or not the floor or roof diaphragm is flexible.

With a rigid diaphragm there is a torsional moment that must be considered. A *torsional moment* is developed if the centroid of the applied lateral force does not coincide with the center of resistance (center of rigidity) of the supporting shearwalls. See Example 9.18. The effect of the torsional moment is to cause wall shears in addition to those which would be developed if no eccentricity existed. This is essentially a combined stress problem in that shears are the result of direct loading and an eccentric moment.

EXAMPLE 9.18   Comparison of Flexible and Rigid Diaphragms

Basic classification of horizontal diaphragms:

1. Flexible
2. Rigid

Forces to supporting walls are simple beam (tributary) reactions for a *flexible* diaphragm (Fig. 9.20a).

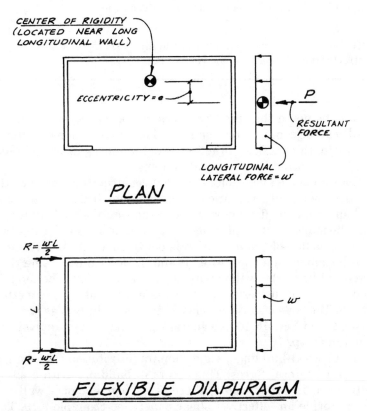

Figure 9.20a   Distribution of lateral force in a flexible diaphragm.

Forces to walls supporting a *rigid* diaphragm R are the sum of the forces due to direct shear proportional to the wall stiffnesses (as if P were applied at the center of rigidity) R_d and eccentric forces R_{ecc} due to the torsional moment $P(e)$ (Fig. 9.20b). These forces are discussed in more detail in Sec. 16.10.

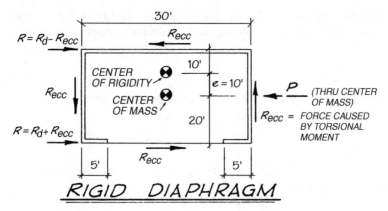

Figure 9.20*b*   Distribution of lateral force in a rigid diaphragm.

NOTE: The force due to rotation adds to the direct shear force in the open front wall and subtracts from the force in the rear wall.

It should be noted that the Code does not permit wood framing to support the dead load of concrete roof or floor slabs (except nonstructural coverings less than 4 in. thick). Thus, practice is that diaphragms of reinforced concrete will not be supported by wood wall systems.

The Code makes no mention as to what construction materials the criterion does or does not apply. Historically, practice has been to categorize all wood frame diaphragms as flexible without evaluating the Code criteria.

Today the majority of wood frame structural designs still categorize all wood diaphragms as flexible, and rotation is not considered. As a result, the lateral force to the supporting shearwalls is simply determined using the respective tributary widths to the walls as illustrated in Sec. 9.9. The force is then distributed to the various resisting elements in the wall by the methods covered in Sec. 9.6. The code criteria for classifying a diaphragm as *flexible* or *rigid* is demonstrated in Sec. 16.10. A general rigid diaphragm analysis is also demonstrated in Chap. 16.

In certain wood buildings, rigid diaphragm rotation may be necessary in order to resist lateral forces. This occurs in buildings with a side that is essentially open. In other words, there is no segment of a wall for sufficient length to provide an "effective" shear panel. See Example 9.19. The only way that a building of this nature can be designed is to take the torsional moment in rotation. The torsional moment is resolved into a couple with the forces acting in the transverse walls.

EXAMPLE 9.19   Rotation in an Open Sided Building

In buildings with no effective shearwall on one side, rotation of the diaphragm can be used to carry the torsional moment into the walls perpendicular to the applied lateral force (not allowed for buildings with concrete or masonry walls). Even in entirely wood-framed buildings, the Code limits the depth of the diaphragm normal to the open side to 25 ft or two-thirds the diaphragm width, whichever is smaller. (For one-story buildings, the width is limited to 25 ft or the full diaphragm width, whichever is smaller.)

Rear wall must carry the entire longitudinal load:

$$R = wL$$

Force to transverse walls due to rotation:

$$\text{Torsion} = T = Re = \frac{wL^2}{2}$$

Resolve torsion into a couple separated by the distance between transverse walls:

$$F_1 = F_2 = \frac{T}{b} = \frac{wL^2}{2b}$$

For more information see the APA publication *Diaphragms* Ref. 9.2.

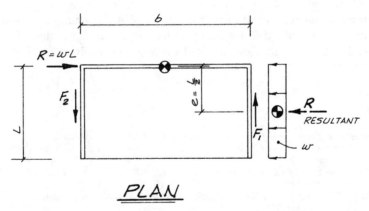

Figure 9.21 Plan of building with open side (i.e., no effective shearwall along one side).

It should be noted that this is not necessarily a desirable design technique. In fact, the *Code does not allow the consideration of rotation of a wood diaphragm in a building with concrete or masonry walls.* The required approach in these types of buildings, and the desirable approach in all cases, is to provide a shearwall along the open side of the building.

The question of rotation brings up an important point. The designer, insofar as practical, should strive for structural *symmetry*. In Fig. 9.21 it is quite unlikely that the design could be modified to have walls of exactly the same length at the front and back of the building. However, the design of a shearwall with an effective length sufficient to develop unit shears of a reasonable magnitude along the open side is recommended. The addition of this amount of symmetry will aid in reducing damage due to rotation.

Problems involving rotation contributed to a number of the failures in residential buildings in the 1971 San Fernando, 1989 Loma Prieta, and 1994 Northridge earthquakes. Reference 9.5 analyzes these failures and recommends construction details to avoid these problems. For additional recommendations on the design of horizontal wood diaphragms see Refs. 9.4, 9.6, and 9.7.

The rigid diaphragm analysis in Example 9.19 was a simple case with a very straight forward solution. A flexible diaphragm analysis would currently be a lot more common in practice. It is recommended that the reader analyze several example buildings using both the flexible and rigid diaphragm analyses to develop a feeling for how much the distribution of shear changes between the two approaches.

While the force distribution to shearwalls will be different in each building studied, the reader will be able to develop an understanding of what type of plan layout results in significant redistribution of forces to shearwalls between the two methods. The layouts resulting in significant differences generally have a substantial distance between their center of mass and center of rigidity. Where this occurs, it is recommended that the diaphragms and shearwalls be examined using a rigid diaphragm analysis. Another approach is to perform both flexible and rigid diaphragm analyses and design each element for the larger of the resulting forces. Rigid diaphragm analysis is discussed further in Chap. 16.

9.12   References

[9.1] American Forest and Paper Association (AF&PA), 2001. *Allowable Stress Design Manual for Engineered Wood Construction and Supplements and Guidelines,* 2001 ed., AF&PA, Washington, DC.

[9.2] American Society of Civil Engineers (ASCE), 2003. *Minimum Design Loads for Buildings and Other Structures (ASCE 7-02),* New York, NY.

[9.3] APA—The Engineered Wood Association, 1990. APA Research Report 138—Plywood Diaphragms, Tacoma, WA.

[9.4] APA—The Engineered Wood Association, 1997. Design/Construction Guide—Diaphragms and Shear Walls, APA, Tacoma, WA.

[9.5] Applied Technology Council (ATC), 1976. A Methodology for Seismic Design and Construction of Single Family Dwellings, ATC, Redwood City, CA.

[9.6] Applied Technology Council (ATC), 1981. Guidelines for Design of Horizontal Wood Diaphragms (ATC-7), Redwood City, CA.

[9.7] Bower, Warren H., 1974. "Lateral Analysis of Plywood Diaphragms," *Journal of the Structural Division,* ASCE, Vol. 100, No. ST4, American Society of Civil Engineers, New York, NY.

[9.8] International Codes Council (ICC), 2003. *International Building Code,* 2003 ed., ICC, Falls Church, VA.

9.13   Problems

For simplicity the following horizontal diaphragm design problems all use wood structural panel sheathing. Sheathing nailing and allowable unit shears are to be taken from the ASD Wood Structural Panel Shear Wall and Diaphragm Supplement.

Allowable stresses and section properties for wood members in the following problems are to be in accordance with the 2001 NDS. Dry-service conditions, normal temperatures, and bending about the strong axis apply unless otherwise noted. Based on recommendations in the NDS, a load duration factor of 1.6 is to be used for problems involving wind or seismic forces. The loads given in a problem are to be used with the IBC *basic load combinations*.

Some problems require the use of a computer equation solving program. Problems that are solved on a computer can be saved and used as a *template* for other similar problems. Templates can have many degrees of sophistication. Initially, a template may only be a hand (i.e., calculator) solution worked on a spreadsheet. In a simple template of this nature, the user will be required to provide many of the lookup functions for such items as

Allowable diaphragm unit shears (lb/ft)

Diaphragm nailing specifications

Tabulated stresses for wood members

As the user gains experience with computer solutions, a template can be expanded to perform many of the lookup and decision-making functions that were previously done manually.

Advanced computer programming skills are not required to create effective templates. Valuable templates can be created by designers who normally do only hand solutions. However, some programming techniques are helpful in automating *lookup* and *decision-making* steps.

The first requirement is that a computer template operate correctly (i.e., calculate correct values). Another major consideration is that the input and output should be structured in an orderly manner. A sufficient number of intermediate answers are to be displayed and labeled so that the solution can be verified by hand.

9.1   *Given:*   The single-story commercial building in Fig. 9.A. The wood structural panel is $^{15}\!/_{32}$-in. thick 32/16 span-rated sheathing unblocked. Nails are 8d common. Critical lateral forces (wind) are

$$w_T = 320 \text{ lb/ft}$$

$$w_L = 320 \text{ lb/ft}$$

Vertical loads are

$$\text{Roof Dead Load D} = 10 \text{ psf}$$

$$\text{Snow S} = 30 \text{ psf}$$

Find: *a.* For the transverse direction:
    1. Roof diaphragm unit shear
    2. Panel load case (state criteria) and allowable unit shear
b. For the longitudinal direction:
    1. Roof diaphragm unit shear
    2. Panel load case (state criteria) and allowable unit shear
c. What is the required nailing?

     d. If blocking is not required, do the edges of the panel need some other type of edge support?

9.2   *Given:*   The single-story commercial building in Fig. 9.A. The wood structural panel is $^{15}/_{32}$-in. thick 32/16 span-rated STR I sheathing. Critical lateral forces (strength level seismic) are

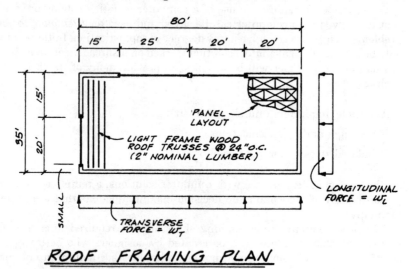

Figure 9.A

$$w_u = w_T = 320 \text{ lb/ft}$$

$$w_u = w_L = 570 \text{ lb/ft}$$

     Find: *a.* Roof diaphragm adjusted unit shear.
                b. Is the IBC span-to-width ratio satisfied? show criteria.
                c. Is blocking required for lateral forces? If it is required, calculate at what point it can be omitted. Show all criteria such as diaphragm load case, allowable unit shears, and nailing requirements.
                d. The maximum adjusted chord force.

9.3   Use the hand solution to Prob. 9.1 or 9.2 as a guide to develop a computer *template* to solve similar problems.

     a. Consider only the specific criteria given in the problems.
     b. Expand the template to handle any diaphragm unit shear covered in the *ASD Shear Wall and Diaphragm Supplement.* In other words, input the values from the ASD Supplement into a computer template, and have the template return the required nail specification and blocking requirements. Macros may be used as part of the template.

9.4   *Given:*   The single-story wood-frame warehouse in Fig. 9.B. Roof sheathing of $^{3}/_{8}$-in. 24/0 span-rated STR I sheathing is adequate for vertical "sheathing" loads. Strength level seismic lateral forces to roof diaphragm are

$$w_u = w_T = 330 \text{ lb/ft}$$

$$w_u = w_L = 570 \text{ lb/ft}$$

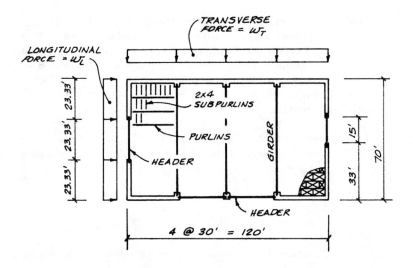

ROOF FRAMING PLAN

Figure 9.B

Find: a. Design the roof diaphragm, considering lateral forces in both directions. Show all criteria including design shears, load cases, allowable shears, and nailing requirements. If blocking is required, determine at what points it can be omitted.

b. Calculate the maximum chord forces for both lateral forces. Also calculate the chord forces at the ends of all headers in the exterior walls.

c. Plot the distribution of the strut force for each of the walls with openings. Compare the strut forces and chord forces to determine the critical loading.

9.5 Use the hand solution to Prob. 9.4 as a guide to expand the computer template from Prob. 9.3. The expanded template should evaluate the diaphragm *chord forces* and *drag strut* forces. In addition to maximum chord forces, the template should allow the user to select various arbitrary locations for the evaluation of chord and strut forces.

9.6 Repeat Prob. 9.4 except that the panel is $^{15}/_{32}$-in. 32/16 span-rated STR I sheathing and the lateral forces are

$$w_u = w_T = 630 \text{ lb/ft}$$

$$w_u = w_L = 790 \text{ lb/ft}$$

9.7 *Given:* The building in Fig. 9.A and the strength level seismic lateral forces are

$$w_u = w_T = 320 \text{ lb/ft}$$

$$w_u = w_L = 570 \text{ lb/ft}$$

Find: a. The maximum adjusted chord forces. Also determine the adjusted chord forces at each end of the wall openings.

   b. Plot the distribution of the strut forces for the two walls that have openings. Compare the magnitude of the adjusted strut forces with the chord forces.

NOTE: A computer template may be used to solve this problem. Verify results with hand solution.

9.8   *Given:*   Assume that the framing plan in Fig. 9.A is to be used for the second-floor framing of a two-story retail sales building. Lateral wind forces are

$$w_T = 450 \text{ lb/ft}$$

$$w_L = 450 \text{ lb/ft}$$

Floor live load L is to be in accordance with IBC. Panel is STR I.

Find: a. Required span-rating and panel thickness.

   b. Required panel nailing. Check unit shears in both transverse and longitudinal directions. Show all criteria in design calculations.

   c. Calculate the maximum chord force and drag force in the building. Assume that three 2 × 6's of No. 1 DF-L are used as the top plate in the building. Check the maximum stress in this member if it serves as the chord and strut. At any point in the plate, two 2 × 6's are continuous (i.e., only one member is spliced at a given location). A single row of 1-in.-diameter bolts is used for the connections. $C_M = 1.0$, and $C_t = 1.0$.

   d. Assuming splices in the chords are spaced every 20 ft, determine the deflection of the diaphragm. Assume the structure is constructed and maintained in the dry condition.

NOTE: Design of the bolts is not part of this problem.

9.9   The one-story building in Fig. 9.C has the following loads:

   Roof dead load D = 12 psf
   Wall dead load D = 150 pcf (normal-weight concrete)
   Wind = 20 psf
   Strength level seismic force = 0.200W
   Tributary wall height to roof diaphragm = 10 ft
   Roof live load L_r in accordance with IBC Table 1607.1

Diaphragm chord consists of two #5 continuous horizontal bars in the wall at the diaphragm level.

Find: a. Required panel span-rating and thickness.

   b. Critical lateral forces.

   c. Required plywood nailing based on the diaphragm unit shears. Show all design criteria. Consider forces in both directions.

   d. Determine the deflection of the diaphragm. Assume the structure is constructed and maintained in the dry condition.

e. Check the stress in the chord reinforcing steel. Allow. $f_s = 20$ ksi $\times$ 1.33.

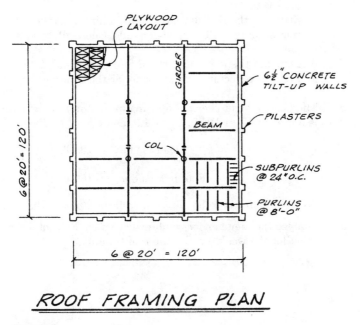

ROOF FRAMING PLAN

Figure 9.C

9.10 *Given:* The one-story building in Fig. 9.D. Roof sheathing is $^{15}/_{32}$-in. 32/16 span-rated panels spanning between roof joists spaced 24 in. o.c. Nails are 10d common. For the transverse strength level seismic force, the panel layout is load case 1.

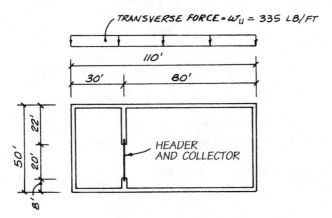

PLAN

Figure 9.D

Find: *a.* Design the roof diaphragm, assuming that only the exterior walls are effective in resisting the lateral force. Show all design criteria including span-width ratio, nailing, and blocking requirements. Omit blocking where possible. Do not consider deflection of the diaphragm.

b. Redesign the roof diaphragm, assuming that the interior *and* exterior walls are effective shearwalls. Show all criteria, except deflection.

c. Compare the maximum chord forces for *a* and *b*.

d. Plot the collector force diagram for the interior shearwall.

9.11 *Given:* The header/strut over the opening in the interior shearwall in Fig. 9.D is a 4 × 14 No. 1 DF-L. $C_M = 1.0$, and $C_t = 1.0$. The vertical load is

$$\text{Dead load } w_D = 100 \text{ lb/ft}$$

$$\text{Roof live load } w_{L_r} = 160 \text{ lb/ft}$$

The axial force is to be determined from a plot of the drag strut force.

Find: Analyze the member for combined stresses. Full lateral support is provided for the weak axis by the roof framing.

10

Shearwalls

10.1 Introduction

Shearwalls comprise the vertical elements in the lateral-force-resisting system (LFRS). They support the horizontal diaphragm and transfer the lateral forces down into the foundation. The procedures for calculating the forces to the shearwalls are covered in Chaps. 3 and 9.

Wood horizontal diaphragms are often used in buildings with masonry or concrete shearwalls as well as buildings with wood-frame shearwalls. However, the design of masonry and concrete shearwalls is beyond the scope of this book, and this chapter deals with the design of wood-frame shearwalls only.

A number of sheathing materials can be used to develop shearwall action in a wood-frame wall. These include

1. Wood structural panels (e.g., plywood and OSB)
2. Gypsum wallboard (drywall)
3. Interior and exterior plaster (stucco)
4. Fiberboard (including fiber-cement panels)
5. Lumber sheathing (diagonal, or horizontal sheathing with diagonal bracing)

There may be other acceptable materials, but these are representative. When the design forces (wall shears) are relatively small, the normal wall covering material may be adequate to develop shearwall action. However, when the unit shears become large, it is necessary to design special sheathing and nailing to develop the required capacity. Wood structural panels provide greater shearwall capacity than the other sheathing types listed above.

10.2 Basic Shearwall Action

Essentially a shearwall cantilevers from the foundation and is subjected to one or more lateral forces. See Example 10.1. As the name implies, the basic form of resistance is that of a shear element (Fig. 3.5c). The concept of "shear panels" in a wood-frame wall leads to the assumption that the lateral force is generally distributed uniformly throughout the total length of all panels (Sec. 9.6). Exceptions will be discussed in Sec. 10.7.

EXAMPLE 10.1 Cantilever Action of a Shearwall

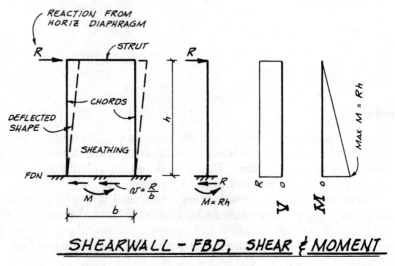

SHEARWALL – FBD, SHEAR & MOMENT

Figure 10.1 Deflected shape of one-story shearwall. Also shown are cantilever loading diagram (schematic), and shear and moment diagrams.

Figure 10.1 shows a typical shear panel for a one-story building. If there are additional stories, there will be additional forces applied to the shearwall at the diaphragm levels. If the lateral force is a seismic force, there will also be an inertial force generated by the mass of the shearwall (Chap. 3).

A variety of sheathing materials can be used to develop shearwall action for resisting lateral forces. The IBC (Ref. 10.13) sets limits on the height-to-width ratios h/b for shear panels in order to develop effective shear (deep-beam) resistance. Limits on h/b vary with different sheathing materials. See IBC Table 2305.3.3.

A number of items should be considered in the design of a shearwall. These include

1. Sheathing thickness

2. Shearwall nailing

3. Chord design (tension and compression)

4. Collector or strut design (tension and compression)

5. Shear panel proportions

6. Anchorage requirements (hold-downs and shear)

7. Deflection

These factors are essentially the same as those required for a horizontal diaphragm, but there are some differences in procedures.

The *sheathing thickness* depends on the type of material used in the wall construction. Loads normal to the surface and the spacing of studs in the wall may determine the thickness of the sheathing, but the unit shear often controls the thickness. In other cases the sheathing thickness may be governed by the required fire rating of a wall. *Shearwall nailing* or stapling requirements are a function of the unit shear in the wall and the materials of construction.

As with a horizontal diaphragm, the *chords* are designed to carry the moment, and chords are required at both ends of a shearwall. The *collector* (or *strut*) for a shearwall is the same member as the collector for the horizontal diaphragm. It is the connecting link between the shearwall and the horizontal diaphragm for loads parallel to the shearwall. Design forces for the collector are covered in detail in Chap. 9.

The *proportions* of a shear panel are measured by its height-to-width ratio h/b. In buildings with two or more stories, the height of the shearwall is the vertical distance between horizontal diaphragms. The IBC sets upper limits on the height-to-width ratio for various wall sheathing materials used as shear panels. Specific values of these limits will be covered in the sections dealing with the various construction materials. Panels are considered as effective shearwall elements if these limits are not exceeded. Shear panels which satisfy the height-to-width ratio criteria are considered by many designers to also have proportions which will not allow the panel to deflect excessively under load (this is similar to the span-to-width ratios for horizontal diaphragms—Sec. 9.8). Nonetheless, IBC provisions for calculating shearwall deflections and related limitations on story drift are Code requirements. Calculation of shearwall deflections is addressed in Sec. 10.12.

10.3   Shearwalls Using Wood Structural Panels

Perhaps the most common wood-frame shearwall is the type that uses wood structural panel sheathing or plywood siding. Today, however, other wood panel products provide additional choices for shearwall construction. Allowable unit shears for various types of wood structural panel shearwalls are given in both IBC Table 2304.6.1 (reproduced in Appendix C of this book) and ASD Wood Structural Panel Shear Wall and Diaphragm Supplement (Ref. 10.1) Tables 4.1A and 4.1B. Included in these tables are design values for shearwalls constructed with STRUCTURAL I, sheathing, and plywood panel siding complying with PS 1, PS 2, or NER-108 (Refs. 10.19 to 10.21). See Secs. 8.7 and 8.8 for a description of these panel products. While the tabulated values in the IBC table and ASD Supplement table are the same, the IBC table includes a number of footnotes that are not included in the ASD Supplement table. The user of the IBC table must adhere to the information in these footnotes. For convenience, this text will refer to the ASD Supplement tables for tabulated allowable unit shears for use in design examples.

Tabulated allowable unit shears for shearwalls assume that the framing members are Douglas Fir-Larch or Southern Pine. If the diaphragm nailing penetrates into framing members of other species of wood, an adjustment that accounts for a reduced nail capacity in lower density species is required. Reduction factors for other species of wood are given in a footnote to the allowable shear tables. In addition it should be noted that the tabulated shearwall values are for short-term (ASD Supplement Table 4.1A for wind or Table 4.1B for seismic) forces.

As with horizontal diaphragms (see Sec. 9.3), it is important to recognize that the tabulated unit shears for shearwalls are 40 percent greater for wind (ASD Shear Wall and Diaphragm Supplement Table 4.1A) than for seismic (Table 4.1B). The increase for the allowable unit shear for wind is due to historic design stress levels for shearwalls and a better understanding of wind loading. The IBC (Ref. 10.13) permits the 40 percent increase per IBC Sec. 2306.3.1.

As mentioned, ASD Supplement Tables 4.1A and 4.1B are tabulated for short-term (wind and seismic) forces. If a horizontal diaphragm is used to support loads of longer duration, the allowable unit shears will have to be reduced (see Sec. 4.15 of this book). For example, if the load under consideration was determined to be of a 10-year cumulative duration ($C_D = 1.0$), the tabulated allowable shears would be reduced by a factor of 1/1.6. Furthermore, for this situation, the allowable shears from ASD Supplement Table 4.1 (seismic) should be used as the base values. That is, the 40 percent increase is specific to wind loading and should not be used for other load durations.

Wood structural panel sheathing is often provided on only one side of a wall, and then finish materials are applied to both sides of the wall. If the design unit shears are higher than can be carried with a single layer, another layer of sheathing can be installed on the other side of the wall. See Fig. 10.2a. The addition of a second layer of sheathing doubles the shear capacity of the wall. The capacity of the wall covering (drywall, plaster, stucco, etc.) is not additive to the shear capacity of the wood structural panel.

Plywood siding and certain code-recognized proprietary structural wood panel sidings can also be used to resist shearwall forces. This siding can be nailed directly to the studs, or it may be installed over a layer of ⅝-in. gypsum sheathing. See Fig. 10.2b. A wall construction that gives a 1-hour fire rating uses 2 × 4 studs at 16 in. o.c. with ⅝-in. Type X gypsum sheathing and minimum ⅜-in. plywood siding on the outside, and ⅝-in. Type X gypsum wallboard on the interior. When grooved plywood panels are used, the thickness to be considered in the shearwall design is the net thickness where the plywood is nailed (i.e., the thickness at groove or shiplap edge of the panel).

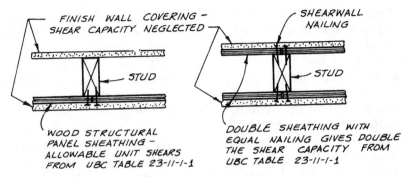

Figure 10.2a Wood structural panel sheathing.

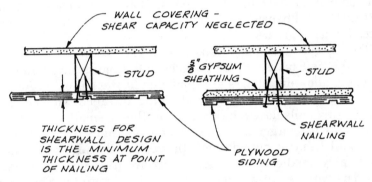

Figure 10.2b Plywood siding.

In a few instances it is theoretically possible to nail certain grooved plywood siding panels through the full thickness of the panel (i.e., away from the grooves or edges). In these cases, thicker wall studs are necessary at the boundary of the panels to provide sufficient panel and framing edge nailing distance. Sample calculations for unit shear values based upon the plywood strength and fastener capacity are given in Ref. 10.28.

EXAMPLE 10.2 Wood Structural Panel Shearwalls

Allowable unit shears for wood structural panel sheathing and plywood siding are given in the ASD Wood Structural Panel Shear Wall and Diaphragm Supplement Tables 4.1A and 4.1B. These tables cover both sheathing or siding attached directly to framing and sheathing or siding applied over ½- or ⅝-in. gypsum sheathing.

Allowable shear values are given for panels nailed with *common* or *galvanized box nails* and *galvanized casing nails*. Finishing nails are not allowed for shearwalls.

As noted, the allowable unit shears and nailing requirements for wood structural panel shearwalls are given in Tables 4.1A and 4.1B of the ASD Wood Structural Panel Shear Wall and Diaphragm Supplement. It will be helpful at this point to review the organization of these tables. The shears given in the right side of the table are for panels installed over ½- or ⅝-in. gypsum sheathing. The allowable shears to the left of these values apply to panels that are nailed directly to the wall framing.

Ultimate load tests of shear panels indicate that failure occurs by the nail head pulling through the panel and/or nails bending and withdrawing from the framing. Common nails and box nails have the same-diameter head and length, and for this reason, tests have shown that panels built with *common nails* and *galvanized box nails* develop approximately the same shear resistance. Allowable unit shears in the top portion of ASD Supplement Tables 4.1A and 4.1B apply to both of these types of nails. The first section applies to STRUCTURAL I wood structural panels grade. The section below this applies to all other panel grades, including plywood panel siding, when nailed with common or galvanized box nails. Note that allowable unit shears for STRUCTURAL I panels are approximately 10 percent higher than comparable values for other grades.

Although galvanized box nails can be used to install plywood siding, the size of the nail head may be objectionable from an appearance standpoint. Therefore, allowable shears are tabulated at the bottom of ASD Supplement Tables 4.1A and 4.1B for plywood siding installed with *galvanized casing nails*. Casing nails have a head diameter that is smaller than that of common or box nails but larger than that of finishing nails. Because of the tendency of the nail head to pull through the siding, the allowable unit shears are considerably smaller when casing nails are used instead of galvanized box nails. The various types of nails are described more fully in Chap. 12.

ASD Shear Wall Supplement Tables 4.1A and 4.1B apply to shear panels installed either horizontally or vertically. In the tables it is also assumed that all panel edges are supported by either the wall studs, plates, or blocking. Footnotes give adjustments to the tabulated shears where appropriate. The tables also give special requirements designed to avoid the splitting of lumber framing when closely spaced nails are used. For example, when the nail spacing is 2 in. o.c., 3-in. nominal (or wider) framing is required at adjoining panel edges.

The additional footnotes found in IBC Table 2306.4.1 require staggering of edge nailing to different framing members when shearwalls are sheathed on both faces; specify 3× studs at abutting panel edges when the design unit shear is above 350 plf; reference requirements for sill plate size, and anchorage and steel plate washers; and they give specifics for nail galvanizing and staple crown width.

IBC Table 2305.3.3 sets an upper limit on h/b of 3.5 for shearwalls resisting wind forces and 2.0 for seismic forces. Per the footnote to IBC Table 2305.3.3, the aspect ratio h/b for seismic forces may be increased up to 3.5 provided the tabulated allowable unit shears are reduced by the multiplier $2b/h$. Regardless, these are the largest height-to-width ratios permitted for wood-frame shear panels. For other sheathing materials the maximum value for h/b is as low as 1.0 (Sec. 10.4).

These limits can be important in the selection of the materials for a shear panel. See Example 10.3. The larger the allowable height-to-width ratio h/b, the shorter the permitted width of a shear panel for a given wall height. Thus if the width available for a shear panel is restricted, the use of a wood structural panel shearwall may be required in order to satisfy the design unit shear within height-to-width limitations (Fig. 10.3).

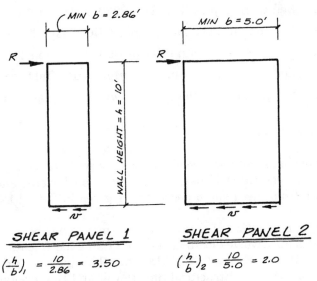

Figure 10.3 Minimum shearwall widths for different height-to-width ratios. Panels with smaller widths are ignored in design.

EXAMPLE 10.3   Height-to-Width Ratios for Wood-Frame Shearwalls

Shearwalls with wood structural panels considering wind have an upper limit on h/b of 3.5. Thus, for a 10-ft-high shearwall, the minimum width is 2.86 ft (shear panel 1). Shearwalls with wood structural panels for seismic forces have an upper limit on h/b of 2.0 without reduction in tabulated values. Here a 10-ft high shearwall must have a width of at least 5.0 ft (shear panel 2).

Some common wall coverings can be used in shearwall design. For example, gypsum board and plaster walls have a limit of h/b of 2.0 with blocking provided at all panel edges. If blocking is not provided, the limit on h/b reduces to 1.5. Unit shear values for gypsum board and plaster are provided in IBC Table 2306.4.5. See also Sec. 10.4.

The discussion to this point has centered on one of the primary design tables for shearwalls, namely, Tables 4.1A and 4.1B in the ASD Wood Structural Panel Shear Wall and Diaphragm Supplement. This Supplement is new to the NDS suite of documents, but tables such as Tables 4.1A and 4.1B have been available through APA—The Engineered Wood Association for many years. Additionally, the APA tables have been included in the model building codes (e.g., the Uniform Building Code) that preceded the IBC (see Chap. 1) for many years as well. However, there are a number of potential shearwall applications involving structural panels that are not covered in Tables 4.1A and 4.1B. In addition, some points in the tables need confirmation and clarification. Some of these applications and needs include:

1. Unblocked shearwalls
2. Panels installed with staples (IBC Table 2306.4.1)
3. Panels attached to light-gage metal studs (IBC Chap. 22)
4. Shearwalls with panels on both sides of wall (IBC Table 2306.4.1)
5. Panels installed over ⅝-in. gypsum sheathing (IBC Table 2306.4.1)
6. Increased stud spacing and the need for wider framing members to avoid splitting
7. Monotonic versus cyclic load

A number of shearwall tests, conducted by the APA—The Engineered Wood Association, address these issues. The results, conclusions, and recommendations drawn from these and other tests are given in Refs. 10.23, 10.24, and 10.28.

Traditionally, evaluation of shearwall performance has been based on *monotonic testing*. This involves loading the top of an experimental shearwall as shown in Fig. 10.1 of Example 10.1. Design tables such as Tables 4.1A and 4.1B from the ASD Wood Structural Shear Wall and Diaphragm Supplement are based on data from such tests. The issue of shearwall performance under *cyclic (reversed) loading* versus shearwall performance under *monotonic (non-reversed) loading* is of recent concern. While monotonic loading, or at least non-reversed loading, may be appropriate for evaluating shearwall performance under lateral loads caused by wind, ground motion during a seismic event results in cyclic (reversed) forces on the shearwall.

ASTM Standard E2126 (Ref. 10.4) has been established for cyclic load testing of wood-frame shearwalls. As the research and design communities sort out the many loading protocols currently being used, this standard is a logical location for the established consensus. Cyclic loading protocols currently being used include: the sequential phased displacement (SPD) protocol (Ref. 10.27), the International Standards Organization (ISO) protocol (Ref. 10.15), and the CUREE Protocols (Ref. 10.16). This latter set of protocols was developed as part of a recent comprehensive research program on the seismic performance of wood-frame buildings, coordinated by the Consortium of Universities for Research in Earthquake Engineering (CUREE). Testing comparing the effect of various loading protocols can be found in Ref. 10.10. From available testing results, it can be observed that wood-structural panel shearwall-component performance at peak capacity, and deflection at peak capacity, can range from very close to monotonic results for some protocols, to significantly below monotonic results for others. The first steps have been taken toward developing a rational basis for the loading protocol in the CUREE program, but continuing development is required. Additional issues, influencing shearwall testing results, require consideration—including the presence of finish materials and the details at the test wall boundaries. Reference 10.6 provides discussion of the significant amount of shearwall test data made recently available.

The original basis for the shearwall tables, now in the ASD Supplement and IBC, was for calculating allowable unit shears for each combination of sheathing and nailing, based on allowable nail loads. This was supplemented by checking the factors of safety between design and peak capacities for a limited number of sheathing and nailing combinations—using monotonic, non-reversed loading protocols. In the future, when the issue of appropriate loading protocol is addressed, tabulated allowable shear values may need to be revised. Shearwall performance under cyclic loading is also influenced by hold-down performance. If the shearwall is allowed to slip or lift up from the foundation, its performance is drastically reduced. See Refs. 10.6, 10.23, and 10.26 for further details on cyclic testing and performance of wood shearwalls.

In another research project, the APA studied the composite action of shearwalls that make use of glue, in addition to mechanical fasteners, for the attachment of sheathing materials. The use of glue increases strength and stiffness. However, the mode of failure switched to rolling shear in the panel or shear along the glueline. These modes of failure are non-ductile, so the energy absorption in seismic loading is minimal. See Ref. 10.30 for a summary of these results.

10.4 Other Sheathing Materials

It was mentioned at the beginning of this chapter that wall materials other than wood structural panels can be used to develop shearwall action. If the design shears are high, wood structural panel sheathing may be required. However, if the design forces are relatively low, other wall coverings may have sufficient strength to carry the shears directly.

The IBC has made several significant changes to seismic bracing of light-frame buildings that use sheathing materials other than wood structural panels. First, ASCE 7 Table 9.5.2.2 (referenced by IBC Sec. 1617.4) assigns a much lower R-factor to light-frame wall systems when sheathed with other than wood structural panel sheathing. The R-factors are 2 for a bearing wall system and 2.5 for a building frame system; as compared to 6.5 and 7, respectively, using wood structural panel sheathing. Second, ASCE 7 Table 9.5.2.2 limits this system to a maximum building height of 35 feet in Seismic Design Category D, and prohibits the system in Seismic Design Categories E and F. Finally, the IBC has limited the sheathing material permitted for this system in Seismic Design Category D to diagonal lumber sheathing.

For these light-frame systems, common bracing materials are stucco (exterior plaster), gypsum lath and plaster (interior plaster), gypsum wallboard (drywall), particleboard, fiberboard, and diagonal lumber sheathing. Additional proprietary bracing materials are becoming commonly used; however, rather than being incorporated directly into the building code, these materials are evaluated on a case-by-case basis by the evaluation services arms of the code development groups.

Allowable unit shears, construction requirements, and other provisions are found in IBC Sec. 2306.4. Lumber sheathed shearwalls, addressed in Sec. 2306.4.2, refer to the IBC diaphragm section for allowable unit shears and construction requirements. No Seismic Design Category limits are placed on lumber sheathed diaphragms, other than the system limits, which prohibit use in Seismic Design Categories E and F. A shearwall aspect ratio maximum limit of 2 can be found in IBC Table 2305.3.3.

In the past, lumber sheathing was used extensively for wood-frame shearwalls. Lumber sheathing can be applied horizontally (known as *straight sheathing*), but is relatively weak and flexible. Diagonally applied sheathing is considerably stronger and stiffer because of its triangulated (truss) action. See Fig. 10.4. Shearwalls using lumber sheathing have largely been replaced with wood structural panel shearwalls. Because of their relatively limited use, the design of lumber-sheathed walls is not covered in this text. A treatment of lumber shearwalls is given in Ref. 10.29.

Particleboard and fiberboard shearwalls are addressed in IBC Secs. 2306.4.3 and 2306.4.4, respectively, with allowable unit shears given in Tables 2306.4.3 and 2308.9.3(4). Use of both materials for shearwalls is prohibited in Seismic Design Categories D, E, and F. For particleboard shearwalls, IBC Table 2305.3.3 indicates that the aspect ratio limits are the same as for wood structural panel shearwalls. For fiberboard shearwalls, an aspect ratio maximum limit of 1.5 is applicable. Fiberboard wall sheathing is often also used for insulation.

IBC Sec. 2306.4.5 addresses use of stucco, interior plaster over gypsum lath, and gypsum wallboard (drywall). Allowable unit shears, given in Table 2306.4.5, range from 60 to 250 plf. Both blocked and unblocked sheathing panel edges are addressed. Footnote a of Table 2306.4.5 indicates that the values are for wind and seismic loading in Seismic Design Categories A, B,

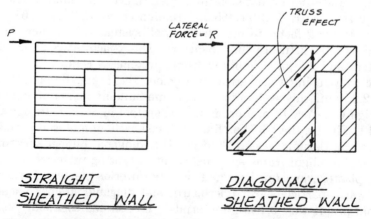

Figure 10.4 Lumber-sheathed shearwalls.

and C. No provision is given for use of these materials in required shearwalls in Seismic Design Category D. Recent cyclic testing of shearwall components has included gypsum wallboard and stucco, alone and in combination with other materials. See Refs. 10.10, 10.18, and 10.22.

10.5 Bracing in Wood-Frame Walls

The Code requires that most buildngs and other structures have an "engineered design" for both vertical loads and lateral forces. *Engineered design* generally involves comparing a calculated demand to an allowable stress or force. Based on a history of generally acceptable performance, however, conventional (prescriptive) light-frame construction is also recognized in the Code. In this method, members and fastenings are selected from tables rather than being calculated. While some of the tables have a basis in calculations, others are based only on historic construction practice. Two sets of conventional construction provisions are now a part of the International Code Council codes. Conventional light-frame construction provisions are included in Sec. 2308 of the IBC. In addition, prescriptive provisions for one- and two-family dwellings can be found in the International Residential Code (IRC) (Ref. 10.14). The IRC provisions go well beyond light-frame construction, to include concrete and masonry wall construction. While the IRC applies to residential occupancies, the only occupancy limit placed on the IBC provisions is the exclusion of Seismic Use Group III (primarily essential facilities). The IBC provisions do, however, place restrictive story limits on nonresidential occupancies. IBC Sec. 2308 also refers to the American Forest and Paper Association (AF&PA) Woodframe Construction Manual for One and Two Family Dwellings (Ref. 10.3) as an alternative standard for prescriptive construction. Use of either the AF&PA Woodframe Construction Manual or the SBCCI Standard for Hurricane-Resistant Residential Construction (Ref. 10.25) is required when the basic wind speed exceeds 100 mph. This text will discuss the IBC conventional light-frame provisions.

Lateral force resistance is ensured in conventional light-frame buildings by requiring some form of wall bracing. As is true for engineering design provisions, the prescriptive seismic design provisions have varying scope limits and bracing material limits based on Seismic Design Category. IBC Sec. 2308.9.3 defines eight wall bracing materials including let-in braces, diagonal lumber sheathing, wood structural panel sheathing, fiberboard sheathing, gypsum wallboard sheathing, particleboard sheathing, and hardboard siding. Use of these bracing materials for Seismic Design Categories A, B, and C is specified in Table 2308.9.3(1). Let-in braces are prohibited in Seismic Design Category C and in the bottom story of three-story buildings in Seismic Design Categories A and B.

Diagonal let-in braces have often been used in the past for bracing of conventional light-frame buildings. See Fig. 10.5. Experience in the 1971 San Fernando earthquake demonstrated the inadequacy of let-in braces. Many of these braces either failed in tension or pulled out of the bottom plate (Ref.

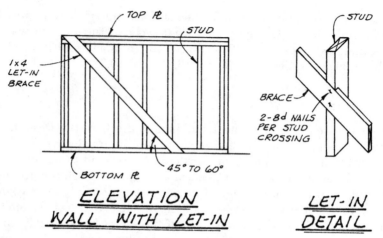

Figure 10.5 Let-in bracing.

10.11). If these braces are used in buildings that require structural calculations, they are essentially ignored for design purposes.

Bracing requirements for Seismic Design Categories D and E are found in IBC Table 2308.12.4, where type and length of bracing is a function of the design spectral response acceleration S_{DS} and the number of stories above the story being braced. Choice of bracing *material* for Categories D and E is more restrictive than for Seismic Design Category C.

It should be understood that the bracing provided in these IBC tables are prescriptive requirements. In other words, tradition and experience indicate that if a building follows these prescribed forms of bracing, reasonable performance can be expected. Because of this, structural design is not required for a building that qualifies as conventional light-frame construction. On the other hand, if a building clearly does not satisfy the criteria for conventional construction, an engineered lateral force design is required. Both the IBC and IRC provisions allow specific nonconventional portions of the building to be engineered, while the balance of the building is designed using the conventional construction provisions. With this mix of design methods, the designer needs to consider the ability of the engineered portions to behave in a manner complimentary to the conventional construction behavior.

An engineered lateral force design requires more than a nailing schedule for a shearwall. The shearwall is part of a system that also includes horizontal diaphragms, collectors (struts), anchorage connections, and an overturning analysis. An engineered lateral force design requires a clearly identified load path for the transfer of lateral forces.

Conventional light-frame construction, on the other hand, does not involve detailed consideration of the load path. Tables of minimum fastening requirements are used for interconnection of members. Positive overturning anchorage is seldom provided, although use of anchor bolts indirectly provides limited uplift resistance. The lesser requirements for detailing and interconnection are very significant characteristics of conventional light-frame construction.

Reference 10.12 identifies an interesting development in the language of the construction industry that has taken place in recent years. Contractors and builders are generally aware of the need for shearwalls in wood-frame construction, and the phrases *shear the house* and *shear the building* have evolved. Here the word *shear* has become a verb that implies adding plywood sheathing to the walls of a building, perhaps throughout the entire structure.

For a building that does not meet the conventional wall bracing provisions, a proposal to "shear the entire building" may be made to a building official. This offer is usually made to obtain a building permit without the added time and expense required to obtain an appropriate lateral-force design. *Shearing* the structure by arbitrarily adding heavily nailed sheathing is not a substitute for proper design of the *lateral-force-resisting system* (LFRS). See Ref. 10.12 for additional discussion.

10.6   Shearwall Chord Members

The vertical members at both ends of a shear panel are the chords of the vertical diaphragm. See Example 10.4. As with a horizontal diaphragm, the chords are designed to carry the entire moment. The moment at the base of the shear panel is resolved into a couple which creates axial forces in the end posts.

Some designers analyze the chords in a shearwall for forces created by the *gross overturning moment* in the wall. This is a conservative approach in designing for *tension* because it neglects the reduction in moment caused by the dead load framing into the wall. For a definition of gross, net, and design net overturning moments, see Example 2.9 in Sec. 2.10.

Even if the dead-load-resisting moment is considered, the critical stress may be the tension at the net section. Reductions in the cross-sectional area often occur at the base of the tension chord for hold-down connections.

EXAMPLE 10.4   Shearwall Chord Members

For a wood structural panel shearwall with an aspect ratio (h/b) not greater than 2, the unit shear is generally assumed to be uniform throughout. For a lateral wind force, the unit shear is the same at the top and bottom of the two shear panels. The gross overturning moment (OM) is calculated at the base of each shear panel:

$$M_1 = v \times b_1 \times h$$

$$M_2 = v \times b_2 \times h$$

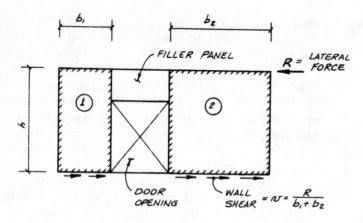

SHEARWALL WITH TWO PANELS

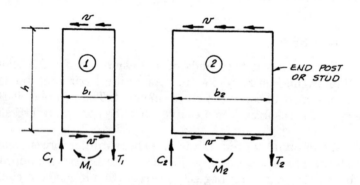

FBD's OF SHEAR PANELS

Figure 10.6 Elevation view of wall with two shearwall panels. Below are free body diagrams of shear panels. The moment at the base of a shear panel is resolved into a *couple*. The couple forms two concentrated chord forces: one in tension and the other in compression.

Chord forces are calculated by dividing gross OMs by the length of the shear panels:

$$T_1 = C_1 = \frac{M_1}{b_1} = vh$$

$$T_2 = C_2 = \frac{M_2}{b_2} = vh$$

Thus, considering the effects of the *gross OM*, the chord forces for the two panels are equal:

$$T_1 = C_1 = T_2 = C_2 = vh$$

A somewhat modified (but similar) procedure is used to calculate the gross chord forces when additional lateral forces are involved. These forces can be generated by additional stories (horizontal diaphragms) or, in the case of seismic forces, a lateral force due to the mass of the shearwall itself (Fig. 10.7d).

Chord forces for the two panels are *not equal* if the resisting dead load moments are considered. When the effects of the dead load are taken into account, larger chord forces will occur in the shorter-width shear panel.

Shearwall chord forces can be analyzed by using the expressions developed in this example or by using the basic statics approach described in Example 2.9 (Sec. 2.10). The chord forces should be in agreement regardless of the method used.

In the *compression* chord, the gross overturning force alone may not represent the critical design force. *Gravity loads* may be carried by the compression chord *in addition to the force due to overturning,* and, if applicable, the vertical component of the seismic force, $0.2S_{DS}D$. If the chord members bear on the sill plate, compression perpendicular to the grain of the wall plates may be a consideration. With a column serving as a chord, a bearing plate may be used to reduce the bearing stress. Lateral forces to a shearwall may come from either direction, and chords must be designed for both tension and compression.

Tests conducted by APA demonstrated that compression perpendicular to grain on wall plates under a shearwall chord can, in fact, be a limiting design consideration. The tests were conducted on heavily loaded shearwalls that had sheathing on both sides of the wall. See Refs. 10.23, 10.24, and 10.28 for details.

The usual assumption is that the shear is uniform throughout the width of a wood-frame shearwall. This common assumption is used throughout this book. However, one reference indicates that a nonuniform distribution of shear may develop, depending on how shearwall overturning is handled. See Ref. 10.7 for details about this alternative approach.

In order to ensure that the panels in a shearwall function as assumed (i.e., separate panels of equal height), double full-height studs or posts should be used for the chords at the ends of the shear panels (Example 9.9 in Sec. 9.6). These tend to emphasize the vertical continuity of the panels. Recent testing (Ref. 10.9) has indicated that the sheathing and nailing of wall areas, neglected in the structural calculations (above and below windows and above doors), can significantly reduce wall deflection and anchorage forces. Where practical, it is recommended that these areas be sheathed and fastened in a manner similar to the portions designated as shearwalls.

10.7 Design Problem: Shearwall

In Example 10.5 the exterior transverse walls of a rectangular building are designed for a seismic force. The building meets the conditions of many "typical" buildings as outlined in Example 3.5 in Sec. 3.4. The *strength-level* seismic base shear coefficient for a building with wood bearing walls and wood structural panel sheathing is assumed as 0.200 (see Example 3.5 in Sec. 3.4).

There is a roof overhang on the left end of the building which develops a smaller, uniformly distributed, lateral seismic force than the main portion of the roof diaphragm. The difference in these two is the seismic force generated by the two longitudinal walls (i.e., the walls perpendicular to the direction of the earthquake force). The reaction of the left shearwall is made up of the cantilevered roof overhang plus the tributary force of the main roof diaphragm.

The total shear at the midheight of the transverse walls is made up of the tributary roof diaphragm reactions plus the seismic force due to the dead load of the top half of the shearwall.

In this example a wood structural panel shearwall is used for the left wall. In the right wall, two shear panels are designed using the same type of sheathing as the left wall, but the nailing is increased to meet the higher shear requirements.

The maximum chord force occurs in the walls with the higher unit shears. Calculations indicate that one 2 × 8 stud is sufficient to resist the tension chord force, but two 2 × 8 studs or a 4 × 8 post are required to carry the compression chord force. In order to function together, the two studs must be effectively connected together and attached to the sheathing.

EXAMPLE 10.5   Shearwall Design

Design the two transverse shearwalls in the building shown in Fig. 10.7a. Studs are 2 × 8s of No. 2 DF-L, and walls are stucco over wood structural panels. The wall dead load D is 20 psf. The *strength-level* seismic force coefficient is 0.200g. The diaphragm uniform forces are 364 plf and 280 plf for the main diaphragm and overhang, respectively. MC ≤ 19 percent, $C_M = 1.0$, $C_t = 1.0$.

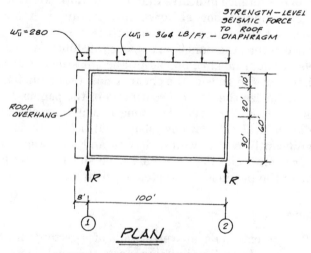

Figure 10.7a Plan view of building showing *strength-level* lateral seismic force to the horizontal diaphragm.

Diaphragm Reactions

Strength-level forces to shearwalls from horizontal diaphragm are calculated on a tributary-width basis.

Wall 1: R_u = reaction from overhang + reaction from main roof

$$= (280 \times 8) + \left(364 \times \frac{100}{2}\right) = 20.44 \text{ k}$$

Wall 2: $R_u = 364 \times \dfrac{100}{2} = 18.2 \text{ k}$

Shearwall 1

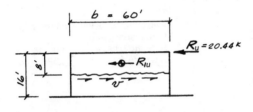

ELEV. WALL 1

Figure 10.7*b* Elevation of left transverse wall (shearwall 1). Seismic force shown is at *strength level.*

Calculate wall seismic force, using the dead load of the *top half* of the wall (Chap. 3). Using the top half of the wall follows dynamic theory (Fig. 2.14*b*). Some designers use *entire* wall dead load (conservative).

$$R_{1u} = 0.200\text{W} = 0.200(20 \text{ psf} \times 8 \times 60) = 1.92 \text{ k}$$

The total *strength-level* design seismic shear force is

$$V_u = R_u + R_{1u} = 20.44 + 1.92 = 22.36 \text{ k}$$

$$v_u = \frac{V_u}{b} = \frac{22,360}{60} = 373 \text{ lb/ft}$$

Since this is at the *strength level,* it must be converted for use in ASD by multiplying by 0.7. Assuming a redundancy ratio $\rho = 1$:

$$v = 0.7v_u = 0.7(373) = 261 \text{ lb/ft}$$

NOTE: The seismic coefficient of 0.200 used in this problem is for wood structural panel shearwalls. The only other shearwall sheathing material that would be permitted in Seismic Design Category D is diagonal lumber sheathing. For diagonal lumber sheathing the R-factor would be reduced from 6.5 to 2, resulting in a seismic force coefficient of 0.65. Because the use of diagonal lumber sheathing is not common, this example will use wood structural panel sheathing.

From ASD Shear Wall Table 4.1B, ⅜-in. sheathing with 6d common nails

Allow. $v = 300$ lb/ft > 261 *OK*

> *Use* ⅜-in. wood structural panel sheathing with
>   6d common nails at
>   4-in. o.c. panel edges
>   12-in. o.c. field
>   Blocking required

Shearwall 2

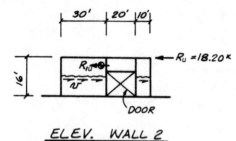

ELEV. WALL 2

Figure 10.7c Elevation of right transverse wall (shearwall 2). Seismic force shown is at *strength level.*

The total *strength-level* design seismic shear force is conservatively calculated using R_1 from shearwall 1:

$$V_u = R_u + R_{1u} = 18.20 + 1.92 = 20.12 \text{ k}$$

$$v_u = \frac{V_u}{b} = \frac{20{,}120}{30 + 10} = 503 \text{ lb/ft}$$

Since this unit shear is at the strength level, it must be converted for use in ASD by multiplying by 0.7

$$v = 0.7(503) = 352 \text{ lb/ft}$$

> *Use*   wood structural panel similar to wall 1 except
>   6d common nails at
>   3-in. o.c. panel edges
>   12-in o.c. field
>   Blocking required

Allow. $v = 390 > 352$ lb/ft   *OK*

NOTE: The unit shear of 352 plf is just above the 350 plf limit when IBC Table 2306.4.1, footnote i reqires use of 3× studs at abutting sheathing panels. Engineering judgment may be exercised here, since the 350 plf limit is exceeded by less that 1 percent.

Tension Chord

The chord stresses are critical in wall 2 (Fig. 10.7d) because of the higher wall shears. The tension chord can be checked conservatively using the gross OM:

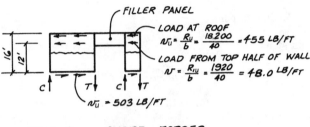

WALL 2 - CHORD FORCES

Figure 10.7d Free body diagrams of panels in shearwall 2. Seismic forces and unit shears shown at the *strength level.*

The forces and unit shears shown in Fig. 10.7d are the *strength level.* To design the tension chord, the unit shears must be converted for use in ASD by multiplying by 0.7. Assume $\rho = 1.0$

Load at top of shearwall (roof): $v = 0.7(v_u) = 0.7(455) = 319$ lb/ft

Load from top half of shearwall: $v = 0.7(v_u) = 0.7(48) = 33.6$ lb/ft

$$T = C = vh = 319(16) + (33.6)12 = 5.51 \text{ k} \quad \text{(using gross OM)}$$

Check the tension stress in one 2 × 8 stud:

$$\text{Net area} = A_n = 1\tfrac{1}{2}(7\tfrac{1}{4} - 1) = 9.38 \text{ in.}^2$$

$$f_t = \frac{T}{A_n} = \frac{5510}{9.38} = 587 \text{ psi}$$

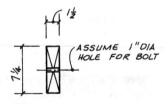

Figure 10.7e Net area of one study.

Lumber is No. 2 DF-L. A 2 × 8 is in the Dimension lumber size category, and tabulated stresses and size factors are from NDS Supplement Table 4A:

$$F_t = 575 \text{ psi} \qquad C_F = 1.2 \qquad \text{for } F_t \text{ in a No. 2 } 2 \times 8$$

$$F_c = 1350 \text{ psi} \qquad C_F = 1.05 \qquad \text{for } F_c \text{ in a No. 2 } 2 \times 8$$

$$F_{c\perp} = 625 \text{ psi}$$

$$E = 1,600,000 \text{ psi}$$

Allowable tension stress:

$$F_t' = F_t(C_D)(C_m)(C_t)(C_F)$$

$$= 575(1.6)(1.0)(1.0)(1.2)$$

$$= 1104 \text{ psi} > 598 \qquad OK$$

> One 2×8 is adequate for the tension chord of the shearwall.

The stress in the tension chord shown above is based on the gross chord force of 5.5 k. If the tension chord had been overstressed using the gross OM, the resisting dead load moment could be considered. However, if consideration is given to the dead load D, and seismic forces control design, the vertical component of seismic forces must also be considered. In order to properly consider the combination of vertical loads and horizontal forces, ASCE 7 equations 9.5.2.7-1 and 9.5.2.7-2 are required for the seismic force:

$$E = \rho Q_E + 0.2 S_{DS} D$$

and

$$E = \rho Q_E - 0.2 S_{DS} D$$

These equations for seismic force need to be combined with IBC basic load combinations 16-10 and 16-12 from IBC Sec. 1605.3.1:

$$D + (W \text{ or } 0.7E) + L + (L_r \text{ or } S \text{ or } R)$$

and

$$0.6D + 0.7E$$

When the ASD basic load combinations are used, the force being calculated becomes an ASD force. This will be clarified by use of the variables F, T, and C, rather than F_u, T_u, and C_u. This is not consistent with the approach recommended in the balance of this text, but it cannot be avoided when combining different load types. In this example, loads L, S, and R are taken as zero. Because the last equation has the least dead load acting, it will control design of the tension chord for shearwall uplift:

$$T = 0.6D - 0.7(\rho Q_E) - 0.7(0.2 S_{DS} D)$$

Assuming $S_{DS} = 1.3g$ and $\rho = 1.0$ gives the following:

$$T = 0.6D - 0.7(1.0)Q_E - 0.7(0.2)(1.3)D$$

$$= 0.6D - 0.7Q_E - 0.18D$$

$$= 0.42D - 0.7Q_E$$

Some thought has to be given to the correct signs for the tension force. The goal is to combine the tension uplift from the shearwall (given a negative sign for tension) with the minimum possible counteracting dead load. For that intent, the shearwall uplift and dead load are in opposing directions. Even with the vertical seismic component, inclusion of the dead load will still be reducing the net tension force due to $0.7Q_E$. As a result, it is still conservative to ignore the effect of dead load. Using a similar procedure, the seismic force combination for compression can be solved as:

$$C = 1.18D + 0.7Q_E + L_r$$

In this case, the compression force is increased when the full load combination is considered. It is not conservative to ignore the effect of vertical loads for compression design.

Although not required in this problem, the reduced tension chord force* will be illustrated for comparison. Two approaches can be used to evaluate the effect of the resisting moment:

1. Basic statics using free-body diagrams (refer to Example 2.9 in Sec. 2.10).
2. Subtraction of the dead load tributary to the tension chord from the gross chord force

The second method is illustrated in this example. However, both methods give the same results, and it is suggested that the reader apply both procedures and compare the results. The second method is quick and easy to apply to a one-story shearwall. In the case of a multistory shearwall, the second method requires careful attention to be applied properly, and the first method may be easier to use until experience is gained.

For a given unit shear v, the shorter-width panel in a given shearwall (10 ft in this example) will be critical. A wall dead load of 20 psf was given in the problem statement. In addition, assume that the wall is a bearing wall that supports a portion of the roof. Say 12-ft-long roof joists (not shown) frame into the right end wall, and the roof dead load = 15 psf. For the 10-ft-long shearwall, the tributary dead load to the tension chord is one-half the shear panel length (5 ft).

$$\text{Roof } D = 15 \text{ psf} \times (12/2) \text{ ft} = 90 \text{ lb/ft}$$

$$\underline{\text{Wall } D = 20 \text{ psf} \times 16 \text{ ft} = 320 \text{ lb/ft}}$$

$$\text{Total } D = 410 \text{ lb/ft of wall*}$$

Using the just calculated ASD load combination for members critical in tension (uplift), $D = 410 \text{ lb/ft} \times (5 \text{ ft tributary length}) = 2050 \text{ lb}$, and $Q_E = 5510 \text{ lb}$, ASD level tension chord force is calculated as:

---

*Reference 10.7 indicates that a nonuniform wall shear may occur if this reduced force is considered.

$$T = 0.42D - 0.7Q_E$$

$$T = 0.42(2050) - 0.7(5510)$$

$$T = 861 - 3857 = -2996 \text{ lb, use 3000 lb}$$

Here the dead load and uplift (overturning) force are acting in opposite directions, thus the plus sign for the dead load and the minus sign for the seismic force.* In this case the change in the chord force due to dead load is about 20 percent of the gross uplift force. It is important to note that proper application of the ASD load combinations require that the seismic forces be converted to an ASD level at this step. This is made clear by using the symbol T, rather than T_u. Converting the forces a second time could result in significant undersizing of members or connections.

NOTE: A single 2× stud may not provide adequate thickness for fastening to the anchor brackets (hold-downs). Conventional construction uses either double 2× studs or a single 4× stud for all shearwall chords to provide greater anchorage thickness. However, compression may control the design and require either double 2× studs or a single 4× stud.

Compression Chord

Using the just calculated ASD load combination for members critical in compression (downward reaction), D = 410 lb/ft × (5 ft tributary length) = 2050 lb, Q_E = 5510 lb, and L_r = 20 psf × (12 ft/2)(5 ft tributary length) − 600 lb.

$$C = 1.18D + 0.7Q_E + L_r$$

$$C = 1.18(2050) + 0.7(5510) + 600$$

$$T = 2419 + 3857 + 600 = 6880 \text{ lb}$$

Here all force components are acting in the same direction (downward), so all have a positive sign.

Check the column allowable load of one stud. Sheathing provides lateral support about the weak axis.

$$\left(\frac{l_e}{d}\right)_y = 0$$

$$\left(\frac{l_e}{d}\right)_x = \frac{16 \text{ ft} \times 12 \text{ in./ft}}{7.25} = 26.5$$

$$E' = E(C_M)(C_t) = 1,600,000(1.0)(1.0)$$

$$= 1,600,000 \text{ psi}$$

For visually-graded sawn lumber used as a column, the values of K_{cE} and c are

---

*Roof live load is omitted when overturning is checked.

$$K_{cE} = 0.3$$

$$c = 0.8$$

$$F_{cE} = \frac{K_{cE}E'}{(l_c/d)^2} = \frac{0.3(1,600.00)}{(26.5)^2} = 684 \text{ psi}$$

$$F_c' = F_c(C_D)(C_M)(C_t)(C_F)$$

$$= 1350(1.6)(1.0)(1.0)(1.05) = 2268 \text{ psi}$$

$$\frac{F_{cE}}{F_c^*} = \frac{684}{2268} = 0.302$$

$$\frac{1 + F_{cE}/F_c^*}{2c} = \frac{1 + 0.302}{2(0.8)} = 0.813$$

$$C_P = \frac{1 + F_{cE}/F_c^*}{2c} - \sqrt{\left(\frac{1 + F_{cE}/F_c^*}{2c}\right)^2 - \frac{F_{cE}/F_c^*}{c}}$$

$$= 0.813 - \sqrt{(0.813)^2 - 0.302/0.8} = 0.281$$

$$F_c' = F_c(C_D)(C_M)(C_t)(C_F)(C_P)$$

$$= 1350(1.6)(1.0)(1.0)(1.05)(0.281)$$

$$= 637 \text{ psi}$$

Check bearing perpendicular to the grain on the wall plate: A stud serving as a shearwall chord occurs at the end of the wall plate. Therefore, $C_b = 1.0$.

$$F_{c\perp}' = F_{c\perp}(C_M)(C_t)(C_b)$$

$$= 625(1.0)(1.0)(1.0)$$

$$= 625 < 637$$

$$\therefore F_{c\perp}' \text{ governs}$$

$$\text{Allow. } P = F_c'A = 0.625(1.5 \times 7.25) = 6.8 \text{ k} \cong 6.88 \text{ k} \qquad OK$$

In this case, the use of one 2× stud would result in an overstress of about 1 percent. This would generally be considered within the certainty with which the forces can be calculated, and therefore would be acceptable. For a variety of reasons, relating to detailing, it is often desirable to have a double 2× or a 4× post at the ends of shearwalls; even when a 2× stud can be shown to be adequate by calculation.

NOTE: The column calculations given above for the compression chord are very conservative. The maximum compressive force at the base of the wall was used for the design load. However, the base of the chord is attached at the floor level, and column buckling is prevented at that point. Some reduced axial column force occurs at the critical location for buckling.

Check the compressive stress at the base of the chord, using the net area (Fig. 10.7e). Column buckling does not occur.

$$f_c = \frac{P}{A_n} = \frac{7560}{2(9.38)} = 403 \text{ psi}$$

$$F'_c = F_c(C_D)(C_M)(C_t)(C_F)(C_P)$$

$$= 1350(1.6)(1.0)(1.0)(1.05)(1.0)$$

$$= 2268 \text{ psi} \geqslant 403 \quad OK$$

> *Use* two 2 × 8 studs of No. 2 DF-L for all shearwall chords

If the length of the shorter shearwall panel in Wall 2, Fig. 10.7c, had been 4.6 ft instead of 10 ft, the aspect ratio would be 16 ft/4.6 ft = 3.5. Wall panels with this aspect ratio are still permitted to be designed as shearwalls; however, in accordance with IBC Table 2305.3.3, footnote a, use of a reduced allowable unit shear is required. The reduction factor is 2b/h = 2(4.6)/16 = 0.58. The unit shear reduction is based on observed larger deflections in high aspect ratio shearwalls.

When unit shears are calculated in walls combining longer and shorter length panels, one approach is to calculate an effective length for wall panels with aspect ratios greater than 2. The effective length of the 4.6 ft panel would be reduced to 4.6(0.58) = 2.7 ft. The unit shear would then be calculated using the full long panel length of 30 ft plus the effective short panel length of 2.7 ft. The sheathing, nailing, and other fastenings should be the same for both panels, and be based on the unit shear calculated using the effective length.

The load duration factor C_D = 1.6 used in Example 10.5 for checking the tension and compression chords is in accordance with the recommendations of the NDS (Ref. 10.2) and IBC (Ref. 10.9) for wind and seismic forces.

10.8 Perforated Shearwall Design Method

The traditional shearwall design approach used in Example 10.5 is termed the *segmented shearwall design method*. When door or window openings are present, such as wall line 2, only full-height *segments* of walls are assumed effective in resisting wind and seismic forces. A relatively new approach to designing shearwalls with door and window openings is the *perforated shearwall design method*. See IBC Sec. 2305.3.7.2.2. The primary advantage of the method is a reduction in the number of required hold-downs. The segmented approach requires each end of each segmented wall (each chord) to be anchored to the foundation with a hold-down bracket. For example, shearwall 2 in Example 10.5 would require four hold-downs using the segmented shearwall method. The perforated shearwall method, however, considers the entire wall to act as a single unit. By using the perforated shearwall method, only two hold-downs to the foundation would be required, one at each end of the wall (i.e., at the building corners), for this same wall. Example 10.6 illustrates the use of the perforated shearwall design method on shearwall 2 in Example 10.5. See Refs. 10.3, 10.8, and 10.17 for additional information on the method.

EXAMPLE 10.6   Perforated Shearwall Design

Consider the right transverse wall (shearwall 2) of the building shown in Fig. 10.7a which was designed using the *segmented shearwall approach*. The wall is shown in Fig. 10.8a. Compare the allowable shearwall capacity of the *segmented* wall to that using the *perforated shearwall approach*.

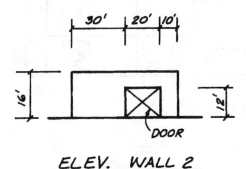

Figure 10.8a Elevation of shearwall 2 (see Example 10.5).

Traditional Shearwall Strength

The shearwall was designed with ⅜-in. wood structural panel sheathing and 6d common nails at 3-in. o.c. along panel edges and 12-in. o.c. in the field. Blocking was required. The tabulated allowable shearwall strength from ASD Shear Wall Supplement Table 4.1B is

$$v = 390 \text{ lb/ft}$$

and the allowable capacity is then

$$V_{\text{segmented}} = vb = 390(30 + 10) = 15.6 \text{ k}$$

Perforated Shearwall Strength

The allowable capacity using the perforated shearwall method $V_{\text{perforated}}$ is determined by multiplying the tabulated shear capacity of each full-height segment $(vb)_{fh}$ by an opening adjustment factor C_o

$$V_{\text{perforated}} = C_o(vb)_{fh}$$

NOTE: Shear is not uniformly distributed along the full length of a perforated shear-wall. Attachment of the diaphragm to the shearwall or shearwall to the foundation is valid only in areas with full height sheathing, such as is done in the segmented approach. Therefore, it is inappropriate to calculate a unit shear strength v using the perforated approach.

The opening adjustment factor C_o is tabulated with values dependent on the ratio of unrestrained opening height to the total wall height h_o/h, and the ratio of wall length with full-height sheathing to the total wall length b_{fh}/b. See Figs. 10.8b and 10.8c, and IBC Table 2305.3.7.2. Additional background and derivation of the factor C_o is covered in Refs. 10.8 and 10.17.

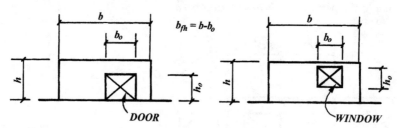

Figure 10.8b Wall and opening dimensions.

	h_o/h				
b_{fh}/b	1/3	1/2	2/3	5/6	1.0
			C_o		
0.1	1.00	0.69	0.53	0.43	0.36
0.2	1.00	0.71	0.56	0.45	0.38
0.3	1.00	0.74	0.59	0.49	0.42
0.4	1.00	0.77	0.63	0.53	0.45
0.5	1.00	0.80	0.67	0.57	0.50
0.6	1.00	0.83	0.71	0.63	0.56
0.7	1.00	0.87	0.77	0.69	0.63
0.8	1.00	0.91	0.83	0.77	0.71
0.9	1.00	0.95	0.91	0.87	0.83
1.0	1.00	1.00	1.00	1.00	1.00

Figure 10.8c Tabulated opening adjustment factor C_o as a function of the ratio of unrestrained opening height to the total wall height h_o/h, and the ratio of wall length with full-height sheathing to the total wall length b_{fh}/b. Use linear interpolation between tabulated values. See also IBC Table 2305.3.7.2.

From Fig. 10.8a

$$\frac{b_{fh}}{b} = \frac{30 + 10}{60} = 0.66$$

If a door opening of $h_o = 12$ ft is assumed, then $h_o/h = 3/4$. Using liner interpolation between $b_{fh}/b = 0.6$ and 0.7, and between $h_o/h = 2/3$ and 5/6, then $C_o = 0.706$.

The tabulated allowable shearwall strength for a full-height segment from ASD Shear Wall Table 4.1B is

$$v = 390 \text{ lb/ft}$$

The allowable shearwall capacity for the perforated wall is

$$V_{\text{perforated}} = C_o v b_{fh} = 0.706(390)(40) = 11.0 \text{ k}$$

as compared to 15.6 k from the segmented approach.

If the door height were changed from $h_o = 12$ ft to $h_o = 8$ ft, the capacity determined using the segmented approach would not change. However, using the perforated approach, the opening adjustment factor C_o increases to $C_o = 0.854$ and the capacity of the wall increases to $V = (0.854)(390)(40) = 13.3$ k.

The allowable shear capacity using the perforated shearwall method will always be less than that of an identical wall designed using the segmented shearwall method. The main advantage of the perforated approach is not in design capacity, but rather the perforated shearwall method requires a reduced number of hold-downs. The segmented approach requires each end of each segmented wall (each chord) to be anchored to the foundation with a hold-down bracket. In this example of shearwall, four hold-downs would be required. The perforated approach considers the entire wall to act as a single unit. Using the perforated shearwall method, only two hold-downs to the foundation are required—one at each end of the wall (i.e., at the building corners). Using the perforated approach, the loss in allowable shear capacity due to reduced restraint against overturning is offset by the economic advantage of reducing the number of hold-downs.

The perforated shearwall design method is new and only recently adopted into the Code. The main advantage of the perforated shearwall design method is not in design capacity or simplicity of design, but rather that it requires a reduced number of hold-downs as compared to the segmented shearwall design method.

The perforated approach considers the entire wall to act as a single unit. Using the perforated shearwall design method requires the use of only two hold-downs to the foundation—one at each end of the wall (i.e., at the building corners). The segmented shearwall design method requires each end of each full-height segmented wall to be anchored to the foundation with a hold-down bracket. The shearwall in Example 10.6 had a single opening and four hold-downs would be required using the segmented approach versus two hold-downs using the perforated approach. There is a corresponding loss in allowable shear capacity using the perforated shearwall design method due to reduced restraint against overturning. This reduced capacity is offset by the economic advantage of reducing the number of hold-downs.

Requirements for distributed anchorage along the bottom of perforated shearwalls are different than for segmented shearwalls. Rather than providing shear anchorage for an average unit shear calculated as V/b_{fh}, the average unit shear is further multiplied by $1/C_o$. For the example problem configuration with the door opening, this would result in the average unit shear being multiplied by 1.42. This adjustment is made because, in the perforated shear-wall method, shearwall segments at the ends of walls that have hold-down brackets can develop the same capacities as segmented shearwalls, while other segments that do not have hold-down brackets develop somewhat lesser capacities. Because of this variation, the unit shear at the base of the wall will be higher in some locations than in others. Locations of highest unit shears will generally be different when the seismic force direction reverses. For this reason, an amplified unit shear is required for base anchorage throughout the perforated shearwall length.

In addition, between the hold-down anchors, the wall base must be fastened for a unit uplift force equal to the amplified unit shear force. This fastening can be provided by the anchor bolts at the foundation level. At an upper framed floor level, this fastening could be sill plate nails in withdrawal, tie-down straps, or a number of other devices. See the commentary to the NEHRP Provisions (Ref. 10.5) for additional information regarding anchorage of perforated shearwalls.

10.9 Anchorage Considerations

Six design considerations for shearwalls were listed in Sec. 10.2. The first five of these items have been covered (forces for designing the collector were covered in Chap. 9). The last item (anchorage) is considered now. Generally speaking, anchorage refers to the tying together of the main elements so that the building will function as a unit in resisting the design loads. Although gravity loads need to be considered, the term *anchorage* emphasizes the design for lateral forces. This includes uplift caused by overturning moments.

Anchorage can be systematically provided by considering the transfer of the following loads:

1. Vertical (gravity) loads

2. Lateral forces parallel to a wall

3. Lateral forces perpendicular to a wall

If this systematic design approach is followed where the horizontal diaphragm connects to the various shearwalls, and where the shearwalls are attached to the foundation, the building will naturally be tied together. See Example 10.7.

The remainder of this chapter provides an introduction to anchorage considerations. The transfer of gravity loads is first summarized. For lateral forces, it is convenient to separate the discussion of anchorage into two parts. The transfer of lateral forces at the *base of the shearwall* is fairly direct, and this chapter concludes with a consideration of these anchorage problems. The anchorage of the *horizontal diaphragm to the shearwall* is somewhat more complicated, and this problem is covered in detail in Chap. 15.

The question of anchorage is basically a connection design problem. The scope of the discussion in this chapter is limited to the calculation of the forces involved in the connections. Once the magnitude of the forces is known, the techniques of Chaps. 12 through 14 can be used to complete the problem.

EXAMPLE 10.7   Basic Anchorage Criteria

The process of tying the building together can be approached systematically by considering loads and forces in the three principal directions of the building. The critical locations are where the horizontal diaphragms connect to the vertical diaphragms and shearwalls tie into the foundation. See Fig. 10.9.

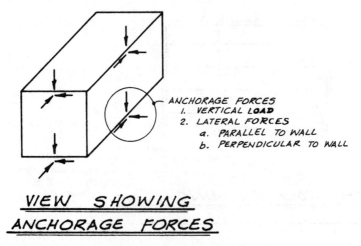

VIEW SHOWING
ANCHORAGE FORCES

Figure 10.9 Anchorage forces in three principal directions.

10.10   Vertical (Gravity) Loads

As noted, the term *anchorage* implies connection design for lateral forces. However, this can be viewed as part of an overall connection design process. In general connection design, the first loads that come to mind are the gravity loads. These start at the roof level and work their way down through the structure. The "flow" of these forces is fairly easy to visualize. See Example 10.8.

The magnitude of the loads required for the design of these connections can usually be obtained from the beam and column design calculations. The hardware used for these connections is reviewed in Chap. 14.

EXAMPLE 10.8   Connections for Gravity Loads

Design for gravity loads involves the progressive transfer of loads from their origin at the lightest framing member, through the structure, and eventually into the foundation. Consider the "load path" for the transfer of gravity loads in the roof framing plan (Fig. 10.10). This requires the design of the following connections:

a. Joist connections to roof beam and bearing wall

b. Roof beam connections to girder and bearing wall

c. Girder connections to column and header

d. Header connection to column

e. Column connection to footing, and stud connection to wall plate (these connections are not shown on the sketch)

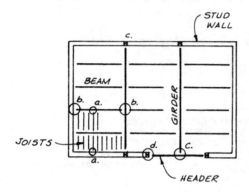

Figure 10.10 Connections for the transfer of gravity loads.

10.11 Lateral Forces Parallel to a Wall

A shearwall resists lateral forces that are parallel to the wall. In doing so, it cantilevers from the foundation, and both a shear and a moment are developed at the base of the wall. Anchorage for the shear and moment is provided by separate connections.

The connections for resisting the *moment* will be considered first. It will be recalled that the moment is carried by the chords, and the chord forces are obtained by resolving the moment into a couple. See Fig. 10.11. The chord forces may be calculated from the gross overturning moment, or the resisting moment provided by the dead load may be considered. Both procedures are illustrated in Example 10.5 (Sec. 10.7).

For connection design, the tension chord is generally of main concern. The tension chord force and the connection at the base of the tension chord will become important for

1. Tall shear panels (i.e., when h is large)

2. Narrow panels (i.e., when b is small)

3. Small resisting dead loads

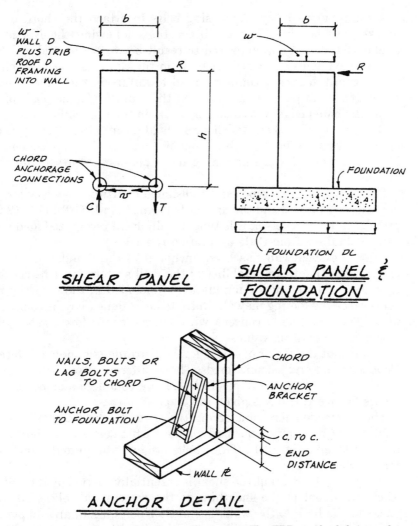

Figure 10.11 Anchorage for shearwall moment. The FBD on the left is used to design the anchorage connection of the shear panel to the foundation. The FBD on the right is used to check overall moment stability. The anchor bracket detail involves a connection between the bracket and the shearwall chord, and between the bracket and the foundation. Fastener spacing requirements (end distance and center-to-center) are important. See Chaps. 12 and 13 for details.

When these conditions exist, a large tension uplift force develops at the base of the wall. If these conditions do not exist, there may be no uplift force at the base of the wall.

The tension tie for the shearwall anchorage is often made with an engineered or prefabricated metal bracket. These brackets are attached to the chord member with nails, bolts, or lag bolts, and to the foundation with an anchor bolt. The number and size of nails, bolts, or lag bolts, and the required embedment length for the anchor bolt, depend on the design force.

A recent move away from using bolts to attach the chord to the anchor bracket has occurred due to observations following the Northridge earthquake. Significant slip appeared to result between the stud and bracket due to oversized holes and wood crushing in the chord around the bolts. This slip in the chord-to-anchor connection led to failure in the nailed connection between the wood panel sheathing and the stud. An "unzipping" of the nailing along the panel edge was caused by a slip in the connection between the chord and the anchor bracket. Additionally, hold-down posts often were observed to split out from the hole to the bottom of the chord, leaving the bolt ineffective. Nails and lag bolts are becoming more popular to attach the chord to the anchor bracket.

Connections of this type can be designed by using the wood design methods of Chaps. 12 and 13 and the anchor bolt values in IBC Table 1912.2. However, prefabricated brackets of this type usually have recognized load values which can be obtained from Code evaluation reports.

After the shearwall has been anchored to the foundation for the tension chord force, the overall stability of the wall system must be checked. In this case the dead load of the foundation is included in the resisting moment. In some cases the footing may also have to be designed for the compressive chord force. This becomes a concern with a large chord force, especially in areas with low soil-bearing values.

Recall that the factor of safety for overturning is handled differently by the Code for wind and seismic forces. These differences were discussed briefly in Example 2.9 and in Example 10.5. A more complete analysis of the requirements for overturning is given in Chap. 16.

The connection for transferring the *shear* to the foundation is considered now. This connection is normally made with a series of anchor bolts. See Example 10.9 and Fig. 10.12. These bolts are in addition to the anchor bolts used for the chord forces.

It will be recalled that the shear is essentially carried by the wall sheathing. The attachment of the sheathing to the bottom wall plate will transfer the shear to the base of the wall. The anchor bolts, then, are simply designed to transfer the shear from the bottom plate into the foundation.

Although the design of bolted connections is covered in a later chapter, the basic design procedure is briefly described here. The strength of the anchor bolt is determined as the bolt capacity parallel to the grain in the wood *or* by the strength of the anchor bolt in the concrete footing. The capacity of the bolt in the wood plate is covered in Chap. 13, and the capacity of the anchor bolt in the concrete footing is given in IBC Table 1912.2. The smaller of these two values is used to determine the size and number of anchor bolts.

The anchor bolt values in IBC Table 1912.2 are based on a minimum required embedment length of bolt into the concrete footing. In addition, use of the full tabulated design values requires a minimum edge distance from the side of the footing to the center of the anchor bolt. For all bolt diameters, design values are tabulated for an edge distance on 6 bolt diameters. Also tabulated are design values for ½-, ⅝-, and ¾-in. bolts with an edge distance of 10 bolt diameters. However, the required edge distance can be reduced 50 percent if the allowable loads are reduced 50 percent. Linear interpolation may be used for intermediate edge distances. Edge distance and other bolt spacings based on the requirements of the wood member are given in Chap. 13.

Based on observations of foundation sill splitting in laboratory shearwall tests and in buildings following the Northridge earthquake, requirements for thicker foundation sills and for steel plate washers on anchor bolts have been introduced in Seismic Design Categories D, E, and F. These requirements can be found in IBC Sec. 2305.10.3. The steel plate washers are required in all shearwalls in Seismic Design Categories D, E, and F. The 3× foundation sills are required for ASD design shears greater than 350 plf. Both the thicker sill and the plate washer help to delay or avoid the cross-grain splitting of the sill plate that can occur when the sheathing nailing, at the uplifting end of the wall, pulls up on the foundation sill.

Because the bottom wall plate is supported directly on concrete, there is a potential problem with termites and decay. If a hazard exists, the Code requires that protection be provided. This is accomplished by using lumber for the sill plate which has been preservative-treated or wood that has a natural resistance to decay (e.g., foundation redwood). An introduction to pressure-treated lumber is given in Sec. 4.9.

EXAMPLE 10.9   Anchorage for Shearwall Shear

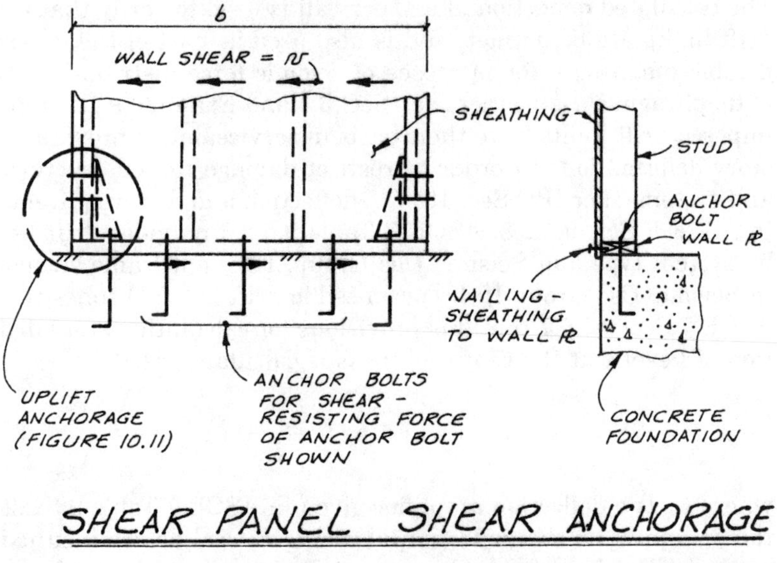

SHEAR PANEL SHEAR ANCHORAGE

Figure 10.12 A separate set of anchor bolts is provided for shear transfer. These anchor bolts are in addition to those provided for the chord uplift connections.

The anchor bolt requirement can be determined by first assuming a size of anchor bolt and determining the allowable load per bolt. If the allowable load per anchor bolt is Z' (parallel-to-grain design value), and the total lateral force parallel to the shear panel is the unit shear times the panel width ($v \times b$), the required number of anchor bolts is

$$N = \frac{v \times b}{Z'}$$

The average spacing is approximately

$$\text{Spacing} = \frac{b}{N} = \frac{bZ'}{vb} = \frac{Z'}{v}$$

(This spacing is approximate because starting and ending anchor bolts must be set in from the panel ends a sufficient distance to clear the chords and tie-down brackets.)

The Code minimum anchor bolt requirement for wood-frame walls is given in IBC Sec. 2308.6 as ½-in. diameter anchor bolts at 6 ft-0 in. o.c. Additionally, a properly sized nut and washer must be used on each anchor bolt.

A minimum of two bolts or anchor straps are required per wall plate, and one bolt or strap is required within 12 in. of, and at least 4 in. from, the end of each plate piece. Anchor bolts for shearwalls are usually larger and more closely spaced than the Code minimum.

10.12   Shearwall Deflections

The calculated deflection of a shearwall is used to verify that Code maximum drift limits are being met, and is also used as part of the determination of a flexible diaphragm for purposes of seismic force distribution. For discussion of diaphragm classification, see Sec. 3.3 and Example 9.18 in Sec. 9.11. Code-imposed drift limits have their basis in serviceability, providing some limit on story deformations in order to restrict damage to nonstructural components and systems. Per IBC Sec. 1617.3, deflection and drift requirements are given in ASCE 7 Sec. 9.5.2.8. ASCE 7 Table 9.5.2.8 provides drift limits that vary by system type and Seismic Use Group. For light-frame shearwall buildings in Seismic Use Group I, the permissible drift is 0.020 times the story height. ASCE 7 Sec. 9.5.5.7 contains provisions for calculating story drift. The deflection of Level x at the center of mass is calculated as:

$$\delta_x = \frac{C_d \delta_{xe}}{I}$$

where C_d is a deflection amplifier given in ASCE 7 Table 9.5.2.2 (the R-factor table), and δ_{xe} is the deflection at *strength level* determined using an elastic analysis. For light-frame bearing wall systems with wood structural panel sheathing, C_d is 4.0, and δ_{xe} is the shearwall deflection introduced in Chap. 3 as Δ_s. See Fig. 10.13.

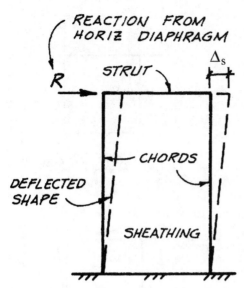

Figure 10.13 Deflection of a shearwall is termed *story drift* Δ_s.

In the past, the most critical story drift at any location in a story was limited to the Code allowable drift. In contrast, the IBC only requires investigation of the drift at the center of mass of any story, unless the building has a torsional irregularity. While this is reasonably accomplished for a rigid diaphragm structure, it is difficult to apply to a structure that uses a flexible diaphragm distribution. For flexible diaphragm structures, one approach is to continue limiting each individual shearwall drift to the Code maximum allowable.

The ASD Wood Structural Panel Shear Wall and Diaphragm Supplement provides a method for calculating the deflection of a shearwall. The method accounts for bending and shear in the wall assembly, as well as nail deformation and anchorage slip (see Fig. 10.14):

$$\Delta_s = \Delta_b + \Delta_v + \Delta_n + \Delta_a$$

where Δ_b = bending deflection of the shearwall
$\quad\quad \Delta_v$ = shear deflection of the shearwall
$\quad\quad \Delta_n$ = deflection of the shearwall due to nail slip (deformation)
$\quad\quad \Delta_a$ = deflection of the shearwall due to anchorage slip and rotation

A shearwall acts similar to a cantilevered I-beam. The chords resist the moment analogous to the flanges of an I-beam, and axial elongation and shortening of the chords causes deflection. Therefore, the bending rigidity of a shearwall is related to the axial rigidity of the chords EA. The *bending deflection* of a shearwall is then determined from

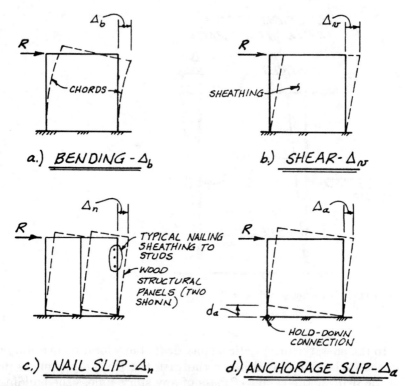

a.) _BENDING - Δ_b_

b.) _SHEAR - Δ_ν_

c.) _NAIL SLIP - Δ_n_

d.) _ANCHORAGE SLIP - Δ_a_

Figure 10.14 Deflection components of a shearwall. (a) _Bending deflection Δ_b._ Shearwall chords carry the moment and act like the flanges of a deep I-beam. The bending deflection of the shearwall is related to the axial shortening and elongation of the compression and tension chords (flanges). (b) _Shear deflection Δ_ν._ Wall sheathing carries the shear in a shearwall and act like the web of a deep I-beam. The shear deflection shown of the shearwall is that of a pure shear element. (c) _Nail slip deflection Δ_n._ Slip in the nailed connections between the sheathing and framing allows individual sheathing panels to move relative to the framing and relative to each other. This slip reduces the effectiveness of the sheathing to resist shear and results in additional shearwall deflection. (d) _Anchorage slip deflection Δ_a._ Slip in the anchorage detail will allow rigid body rotation of the shearwall and deflection Δ_a. Slip in the anchorage can occur both between the chord and the anchor bracket, and between the anchor bracket and the foundation.

$$\Delta_b = \frac{8vh^3}{EAb}$$

where v = shear force at the top of the wall (lb/ft)
$\quad\quad h$ = height of the wall (ft)
$\quad\quad b$ = width of the wall (ft)
$\quad\quad E$ = modulus of elasticity of the chord (psi)
$\quad\quad A$ = cross-sectional area of chord (in.^{2})

While the chords resist the bending moment, the sheathing in a shearwall acts like the web of an I-beam and resists the applied shear. The _shear deflection_ is then related to the shear rigidity Gt of the sheathing:

$$\Delta_v = \frac{vh}{Gt}$$

where G = modulus of rigidity of the wood structural panel sheathing (psi)
t = effective thickness of the wood structural panel sheathing (in.)

If the sheathing were rigidly attached to the studs *and if* the shearwall were rigidly attached to the foundation, the bending and shear terms Δ_b and Δ_v would account for *all* the shearwall deflection. Neither of these conditions exists in typical wood shearwalls, and therefore the two additional deflection terms Δ_n and Δ_a must be accounted for in design. The prediction of these deflections relies on experimental data and empirical relationships.

Slip in the nailed connections between the sheathing and framing members reduces the effectiveness of the sheathing in resisting shear by allowing individual sheathing panels to slip relative to each other. The deflection due to *nail slip* Δ_n is estimated from the relationship

$$\Delta_n = 0.75 h e_n$$

where e_n = nail deformation (in.).

The nail deformation e_n is the slip resulting between the stud and the sheathing at the design load. It is dependent on fastener type, minimum penetration, maximum fastener load V_n, and whether green or dried (seasoned) lumber is used at the time of fabrication. Whether the lumber is initially green or dry, it is assumed that the shearwall will be dry in service and the lumber will dry in place (dry use).

The nail deformation e_n can be determined from the equations listed in ASD Shear Wall Supplement Table 3.2:

6d common nail (1¼-in. minimum penetration, 180 lb maximum fastener load):

Green lumber (dry use)   *dry (seasoned) lumber (dry use)*

$$e_n = \left(\frac{V_n}{434}\right)^{2.314} \qquad\qquad e_n = \left(\frac{V_n}{456}\right)^{3.144}$$

8d common nail (1⅜-in. minimum penetration, 220 lb maximum fastener load):

Green lumber (dry use)   *dry (seasoned) lumber (dry use)*

$$e_n = \left(\frac{V_n}{857}\right)^{1.869} \qquad\qquad e_n = \left(\frac{V_n}{616}\right)^{3.018}$$

10d common nail (1½-in. minimum penetration, 260 lb maximum fastener load):

Green lumber (dry use)   *dry (seasoned) lumber (dry use)*

$$e_n = \left(\frac{V_n}{977}\right)^{1.894}$$ $$e_n = \left(\frac{V_n}{769}\right)^{3.276}$$

14-ga staple (1 to 2 in. minimum penetration, 140 lb maximum fastener load):

Green lumber (dry use)   *dry (seasoned) lumber (dry use)*

$$e_n = \left(\frac{V_n}{902}\right)^{1.464}$$ $$e_n = \left(\frac{V_n}{596}\right)^{1.999}$$

14-ga staple (2 in. minimum penetration, 170 lb maximum fastener load):

Green lumber (dry use)   *dry (seasoned) lumber (dry use)*

$$e_n = \left(\frac{V_n}{674}\right)^{1.873}$$ $$e_n = \left(\frac{V_n}{461}\right)^{2.776}$$

The fastener load V_n is determined by dividing the maximum unit shear force v by the number of nails per foot at the interior panel edges. The equations listed are for Structural I wood structural panels. If the sheathing is not Structural I, then the resulting nail deformation e_n should be increased by 20 percent.

Slip or rotation in the anchorage detail permits rigid body rotation of the shearwall. The calculation of the deflection of the shearwall due to slip or rotation in the anchorage detail Δ_a is similar to the calculation of the shearwall deflection due to nail slip. A design slip is estimated for the connection and the deflection of the shearwall is proportioned by the height-to-width ratio of the wall

$$\Delta_a = \frac{h}{b} d_a$$

where d_a = anchorage slip (in.).

The anchorage slip d_a is the total slip between the chord and the anchor bracket and the anchor bracket and the foundation at design load. For many prefabricated anchor brackets, manufacturers provide information regarding slip. For the most part, the anchor slip values provided by manufacturers include the deflection measure between the anchor bolt and the strap, or bracket, at location of fastening to the wood post. Not included is the slip due to deformation of fasteners to the post. For purposes of the examples in this text, a vertical anchor slip of $\frac{1}{8}$ in. will be used. This includes approximately $\frac{1}{16}$ in. for anchor bolt and bracket deformation and $\frac{1}{16}$ in. for deformation of the fasteners to the post. Alternately, data from shearwall testing could be used to determine anchor slip.

In summary, the total story drift of a shearwall is calculated from

$$\Delta_s = \Delta_b + \Delta_v + \Delta_n + \Delta_a = \frac{8vh^3}{EAb} + \frac{vh}{Gt} + 0.75he_n + \frac{h}{b}d_a$$

The use of this equation for the calculation of the shearwall deflection is illustrated in Example 10.10. The shearwall for the building shown in Fig. 10.7 was designed in Example 10.5. In Example 10.10, the deflection or story drift of the left transverse wall (shearwall 1) is calculated. From this, the maximum inelastic response displacement is determined and compared to the IBC limit.

EXAMPLE 10.10   Shearwall Deflection

Determine the deflection Δ_s of the left transverse wall (shearwall 1) of the building shown in Example 10.5. The shearwall is shown in Fig. 10.15 and is 60-ft long and 16-ft high. Two No. 2 DF-L 2 × 8 studs are used as the shearwall chords and ⅜-in. wood structural panel sheathing is used with 6d common nails at 3-in. o.c. along panel edges and 12-in. o.c. in the field.

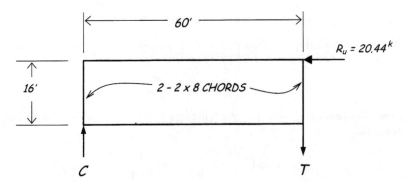

Figure 10.15   Elevation of shearwall. Seismic force shown is at *strength level.*

NOTE: IBC Sec. 1617.3 requires all shearwall deflections to be calculated using earthquake forces without reduction. That is, *strength-level* seismic forces are used to determine shearwall displacements. Considering the load combination D + 0.7E + L + (L$_r$ or S or R), the 0.7 reduction factor shall not be used. No live load, snow load, or rain load is specified, thus the load combination reduces to D + E.

From Example 10.5:

strength-level shear force at top of wall

$$v_u = \frac{V_u}{b} = \frac{20{,}440}{60} = 341 \text{ lb/ft}$$

chord properties

$$E = 1{,}600{,}000 \text{ psi}$$

$$A = 2(1.5)(7.25) = 21.75 \text{ in.}^2$$

To determine the shear rigidity, the span-rating of the sheathing is required. The span-rating for the $\frac{3}{8}$-in. sheathing was not specified in Example 10.5; however, ASD Wood Structural Panel Supplement Table 5.2 indicates 24/0 is the only span-rating available for this thickness. From ASD Wood Structural Panel Supplement Table 3.4, the rigidity is given as

$$G_v t_v = 25,000 \text{ lb/in.}$$

Note that exterior glue is assumed for the sheathing, which is common.

To calculate the nail deformation e_n, the shear force per nail V_n must be determined by dividing the *strength-level* unit shear force v_u by the number of nails per foot at the interior panel edges. The *strength-level* unit shear is determined using load combination D + E. For this calculation, the seismic force including the entire wall dead load is added to the shear at the top of the wall v_u (see Example 10.5). The wall dead load D = 20 psf. The unit shear of the wall is determined using the seismic base shear coefficient 0.002W (see Example 10.5):

$$v_{\text{wall}} = 0.200W = 0.200Dh = 0.200(20 \text{ psf})(16 \text{ ft}) = 64 \text{ lb/ft}$$

The total unit shear is

$$v = 341 + 64 = 405 \text{ lb/ft}$$

With a nail spacing of 3 in. o.c., 4 nails per ft are used along the panel edges.

$$(V_n)_u = \frac{405}{4} = 101.25 \text{ lb/nail}$$

Since the sheathing used is not Structural I, the value of e_n must be increased by 20 percent.

$$e_n = 1.20 \left(\frac{(V_n)_u}{456} \right)^{3.144} = 1.20 \left(\frac{101.25}{456} \right)^{3.144} = 0.0106$$

The story drift Δ_s can now be determined

$$\Delta_s = \Delta_b + \Delta_v + \Delta_n + \Delta_a$$

$$= \frac{8v_u h^3}{EAb} + \frac{v_u h}{Gt} + 0.75he_n + \frac{h}{b} d_a$$

$$= \frac{8(405)(16)^3}{(1,600,00)(21.75)(60)} + \frac{(405)(16)}{(25,000)} + 0.75(16)(0.0106) + \frac{16}{60}(0.125)$$

$$= 0.0064 + 0.26 + 0.13 + 0.033$$

$$\Delta_s = 0.43 \text{ in.}$$

According to IBC Sec. 1617.3, the calculated shearwall displacement is multiplied by a *deflection amplification factor* C_d from IBC Table 1617.6.2. The deflection amplification factor for wood-frame walls is $C_d = 4$, thus

$$\Delta = C_d \Delta_s = (4)(0.43) = 1.72 \text{ in.}$$

The IBC limit for shearwall displacement is 0.025 times the shearwall height.

$$\Delta_{\text{limit}} = 0.020(16 \times 12) = 3.84 \text{ in.} > 1.72 \text{ in.} \quad OK$$

10.13 Lateral Forces Perpendicular to a Wall

In addition to being designed for the shearwall forces covered previously in this chapter, walls must be designed for forces that are normal to the wall. The wall must be adequately anchored to the foundation and the roof diaphragm to resist design forces normal to the wall due to tributary wind and seismic forces. See Example 10.11.

EXAMPLE 10.11   Force Perpendicular to a Wall

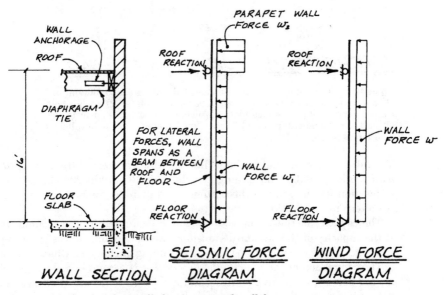

Figure 10.16 Section thru wall showing normal wall forces.

The procedures for determining wind and seismic forces were presented in Chap. 2. See Example 2.17. As illustrated in Fig. 10.16, the distribution of normal wall forces resulting from wind pressures and seismic forces differ significantly. From an analysis of the forces acting normal to the wall, the main wall system may be designed as well as the anchorage to the footing (floor reaction).

Anchorage considerations at the horizontal diaphragm level are discussed in Chap. 15.

In the case of the lateral force *parallel* to the wall, the sheathing is assumed to transfer the force into the bottom plate of the wall. However, for the lateral force *normal* to the wall, the sheathing cannot generally be relied upon to transfer these forces. Connections should be designed to transfer the reaction from the stud to the bottom plate and then from the plate to the foundation. See Example 10.12 and Fig. 10.17.

EXAMPLE 10.12   Anchorage for Perpendicular to Wall Force

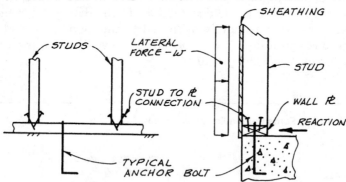

ANCHORAGE FOR NORMAL LOADS

Figure 10.17 Two connections transfer normal wall forces to foundation: (1) nail connection between stud and bottom wall plate, (2) anchor bolt connection of wall plate to foundation. Anchor bolt design is governed by either the perpendicular to wall forces shown above or by the parallel to wall forces in Fig. 10.12.

The lateral force normal to the wall is usually transferred from the *stud to the bottom wall plate* by nails. The nailing schedule in IBC Table 2304.9.1 gives the standard connection for the stud to the bottom (sole) plate as four 8d common or box toenails. An alternate connection is two 16d end nails.

The capacity of these connections can be evaluated by the techniques of Chap. 12. If the capacity of the standard stud-to-plate connections is less than the applied load, some other type of connection (such as a framing anchor; see Chap. 14) will be required.

The connection of the *plate to the foundation* is accomplished with anchor bolts. The anchor bolt requirement can be determined by first assuming a size of anchor bolt and determining the allowable load per bolt.

If the allowable load per anchor bolt is Z' (perpendicular-to-grain design value), and the total foundation reaction (Example 10.11) is the reaction per foot R times the width of the wall b, then the required number of anchor bolts is

$$N = \frac{R \times b}{Z'}$$

This number must be compared with the required number of bolts determined for anchoring the parallel-to-wall forces (Example 10.9). The larger number of bolts controls.

The standard method of connecting wall studs to the bottom plate is with toenails or nails driven through the bottom plate into the end of the stud. The latter connection is often used when a wall panel is preframed and then lifted into place. Methods for evaluating the capacity of nailed connections are given in Chap. 12.

Experience during the 1971 San Fernando earthquake with conventional wood-frame houses proved the standard connection between wall studs and the bottom plate to be inadequate in many cases resulting in wall uplift at corners. Reference 10.11 was developed as a guide to home designers in an attempt to reduce these and other types of failures. This reference recommends the following anchorage for conventional wood-frame walls:

1. Framing anchors (Chap. 14) are to be used to fasten studs to the bottom wall plate at exterior corners. A ½-in.-diameter anchor bolt is to be installed within 2¾ in. of the corner studs (in addition to other anchor bolts). This is essentially an uplift connection.

2. In areas of high seismic risk (e.g., Seismic Design Category D), the following additional anchorage is suggested. Framing anchors are to be used to connect the first two studs to the plate in each wall adjacent to the exterior corners.

3. Studs in the first story of two-story constructions are to be attached to the sole plate with framing anchors at 4-ft. intervals along the wall. This is required for all wall coverings except plywood.

For additional information see Ref. 10.11.

The force in the bottom plate is transferred to the foundation with anchor bolts. Because lateral forces parallel and perpendicular to the wall are not considered simultaneously, the same anchor bolts can be used for both of these forces.

The procedure for determining the required number of bolts is much the same as for parallel-to-wall loads. Here the strength of the anchor bolt is determined as the load capacity of the bolt perpendicular to the grain in the wood plate *or* by the strength of the anchor bolt in the concrete footing. The smaller of these two values is used as the allowable load per anchor bolt.

10.14   References

[10.1] American Forest and Paper Association (AF&PA). 2001. *Allowable Stress Design Manual for Engineered Wood Construction and Supplements and Guidelines,* 2001 ed., AF&PA, Washington, DC.

[10.2] American Forest and Paper Association (AF&PA). 2001. *National Design Specification for Wood Construction and Supplement,* 2001 ed., AF&PA, Washington, DC.

[10.3] American Forest and Paper Association (AF&PA). 2001. *Wood Frame Construction Manual for One- and Two-Family Dwellings,* 2001 ed., AF&PA, Washington, DC.

[10.4] American Society for Testing and Materials (ASTM). 2002. "Standard Test Methods for Cyclic (Reversed) Load Test for Shear Resistance of Framed Walls for Buildings" (ASTM E2126-02), ASTM, West Conshohocken, PA.

[10.5] Building Seismic Safety Council (BSSC). 2000. *NEHRP Recommended Provisions for Seismic Regulations for New Buldings and Other Structures,* 2000 ed., BSSC, Washington, DC.

[10.6] Cobeen, K., Russell, J. E. and Dolan, J. D. 2003. *Recommendations for Earthquake Resistance in the Design and Construction of Woodframe Buildings* (CUREE Publication No. W-30), Consortium of Universities for Research in Earthquake Engineering (CUREE) Richmond, CA.

[10.7] Diekmann, Edward F. 1998. "Diaphragms and Shearwalls," *Wood Engineering and Construction Handbook,* 3rd ed., K. F. Faherty and T. G. Williamson, eds., McGraw-Hill, New York, NY.

[10.8] Douglas, Bradford K., and Sugiyama, H. 1995. *Perforated Shearwall Design Approach.* American Forest and Paper Association (AF&PA), Washington, DC.

[10.9] Fischer, David et al. 2001. *Shake Table Tests of a Two-Story Woodframe House* (CUREE Publication No. W-06), Consortium of Universities for Research in Earthquake Engineering (CUREE), Richmond, CA.

[10.10] Gatto, Kip, and Chia-Ming Uang. 2002. *Cyclic Response of Woodframe Shearwalls: Loading Protocol and Rate of Loading Effects* (CUREE Publication No. W-13), Consortium of Universities for Research in Earthquake Engineering (CUREE), Richmond, CA.

[10.11] Goers, Ralph W. 1976. ATC-4, *A Methodology for Seismic Design and Construction of Single Family Dwellings,* Applied Technology Council, Redwood City, CA.

[10.12] Henry, John R. 1990. "Adequate Bracing Versus Lateral Design: A New Look at an Old Problem," *Building Standards,* September–October 1990, International Conference of Building Officials (ICBO), Whittier, CA.

[10.13] International Code Council (ICC). 2003. *International Building Code,* 2003 ed., ICC, Country Club Hills, IL.

[10.14] International Code Council (ICC). 2003. *International Residential Code,* 2003 ed., ICC, Country Club Hills, IL.

[10.15] International Organization for Standardization (ISO). 1998. "Timber Structures—Joints Made with Mechanical Fasteners—Quasi Static Reversed Cyclic Test Method" (ISO/TC 165 WD 16670), ISO, Secretariat—Standards Council of Canada, Ottawa, Canada.

[10.16] Krawinkler, H., Parisi, F., Ibarra, L., Ayoub, A., and Medina, R. 2001. *Development of a Testing Protocol for Woodframe Structures* (CUREE Publication No. W-02), Consortium of Universities for Research in Earthquake Engineering (CUREE), Richmond, CA.

[10.17] Line, Philip, and Douglas, Bradford K. 1996. *Perforated Shearwall Design Method,* International Wood Engineering Conference, New Orleans, LA, pp. 2-345–2-349.

[10.18] McMullin, Kurt M., and Merrick, Dan. 2002. *Seismic Performance of Gypsum Walls: Experimental Test Program* (CUREE Publication No. W-15), Consortium of Universities for Research in Earthquake Engineering (CUREE), Richmond, CA.

[10.19] National Evaluation Service Committee. 1997. "Standards for Structural-Use Panels," Report No, NER-108, Council of American Building Officials (Available from APA—The Engineered Wood Association, Tacoma, WA).

[10.20] National Institute of Standards and Technology (NIST). 1992. *U.S. Product Standard PS 2-92—Performance Standard for Wood-Based Structural-Use Panels,* U.S. Dept. of Commerce, Gaithersburg, MD (reproduced by APA—The Engineered Wood Association, Engineered Wood Systems, Tacoma, WA).

[10.21] National Institute of Standards and Technology (NIST). 1995. *U.S. Product Standard PS 1-95—Construction and Industrial Plywood,* U.S. Dept. of Commerce, Gaithersburg, MD (reproduced by APA—The Engineered Wood Association, Engineered Wood Systems, Tacoma, WA).

[10.22] Pardoen, G., Waltman, A., Kazanjy, R., Freund, E., and Hamilton, C. 2003. *Testing and Analysis of One-Story and Two-Story Shear Walls Under Cyclic Loading* (CUREE Publication No. W-25), Consortium of Universities for Research in Earthquake Engineering (CUREE), Richmond, CA.

[10.23] Rose, John D. 1998. *Laboratory Report 158—Preliminary Testing of Wood Structural Panel Shear Walls Under Cyclic (Reversed) Loading,* APA—The Engineered Wood Association, Engineered Wood Systems, Tacoma, WA.

[10.24] Rose, J.D., and Keith, E.D. 1996. *Laboratory Report 157—Wood Structural Panel Shear Walls with Gypsum Wallboard and Window / Door Openings,* APA—The Engineered Wood Association, Engineered Wood Systems, Tacoma, WA.

[10.25] Southern Building Code Congress International (SBCCI). 1999. *Standard for Hurricane Resistant Construction* (SSTD 10-99), SBCCI, Birmingham, AL.

[10.26] Structural Engineers Association of Southern California (SEAOSC). 2001. *Report of Light-Frames Walls With Wood-Sheathed Shear Panels.* CoLA-UCI Light Frame Test Committee Subcommittee of Research Committee and Department of Civil & Environmental Engineering, University of California, Irvine, CA.

[10.27] Structural Engineers Association of Southern California (SEAOSC). 1997. *Standard Method of Cyclic (Reversed) Load Tests for Shear Resistance of Framed Walls for Buildings.* SEAOSC, Whittier, CA.

[10.28] Tissel, John R. 1996. *Laboratory Report 154—Structural Panel Shear Walls,* APA—The Engineered Wood Association, Engineered Wood Systems, Tacoma, WA.

[10.29] Tissel, John R., and Elliott, James R. 1997. *Research Report 138—Plywood Diaphragms,* APA—The Engineered Wood Awssociation, Engineered Wood Systems, Tacoma, WA.

[10.30] Tissel, John R., and Rose, John D. 1988. *Research Report 151—Plywood End Shear Walls in Mobile Homes,* APA—The Engineered Wood Association, Engineered Wood Systems, Tacoma, WA.

10.15   Problems

The following shearwall problems all involve some type of wood structural panel. Nailing and allowable unit shears are taken from the ASD Wood Structural Panel Shear Wall and Diaphragm Supplement Tables 4.1A and 4.1B. Allowable stresses and section properties for wood members in the following problems are to be in accordance with the NDS. Dry-service conditions, normal temperatures, and bending about the strong axis apply unless otherwise noted. Based on recommendations in the NDS, a load duration factor of 1.6 is to be used for problems involving wind or seismic forces.

Some problems require the use of spreadsheet or equation-solving software. Problems that are solved using such software can be saved and used as a *template* for other similar problems. Templates can have many degrees of sophistication. Initially, a template may only be a hand (i.e., calculator) solution worked using spreadsheet or equation-solving software. In a simple template of this nature, the user will be required to provide many of the lookup functions for such items as

Allowable shearwall unit shears (lb/ft)

Shearwall nailing specifications

Tabulated stresses for wood members

As the user gains experience with their software, a template can be expanded to perform many of the lookup and decision-making functions that were previously done manually.

Advanced computer programming skills are not required to create effective templates. Valuable templates can be created by designers who normally do only hand solutions. However, some programming techniques are helpful in automating *lookup* and *decision-making* steps.

The first requirement is that a template operate correctly (i.e., calculate correct values). Another major consideration is that the input and output be structured in an orderly manner. A sufficient number of intermediate answers are to be displayed and labeled in a template so that the solution can be verified by hand.

10.1   *Given:*   The single-story wood-frame building in Fig. 10.A. The critical lateral wind force in the longitudinal direction is given. Sheathing is $^{15}/_{32}$-in. STR I with 8d common nails.

Find: a. The unit shear in the shear panel along line 1.

b. The required nailing for the shearwall along line 1.

c. The net maximum uplift force at the base of the wall along line 1. Consider the resisting moment provided by the weight of the wall (assume no roof load frames into longitudinal walls). Wall dead load D = 20 psf.

d. Are the proportions of the shear panel in the above wall within the limits given in the IBC?

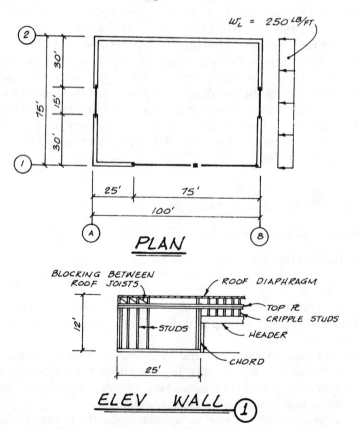

Figure 10.A

10.2 *Given:* The single-story wood-frame building in Fig. 10.A. The critical lateral wind force in the longitudinal direction is given. Wood structural panel sheathing is designed to function as a shearwall. Studs are spaced 16 in. o.c.

Find: a. The minimum required thickness of sheathing, assuming that the face grain may run either horizontally or vertically. (Sheathing is used under a nonstructural wall covering.)
b. The unit shear in the shear panel along line 1.
c. Using the panel sheathing thickness determined in *a*, select the required spacing of 6d common nails for the shearwall along line 1. Indicate blocking requirements (if any). Check the proportions of the shear panel.
d. The maximum chord force in the shear panel along line 1, neglecting any resisting moment. Is a doubled 2 × 6 stud adequate to resist this chord force? Lumber is No. 2 DF-L. A single row of ¾-in.-diameter bolts is used for the anchorage connection at the base of the wall. $C_M = 1.0$, and $C_t = 1.0$.
e. The uplift force for the design of the connection at the base of the tension chord. Consider the resisting moment provided by the wall and the roof dead load supported by the wall. The wall supports a 5-ft tributary width of roof dead load. Roof D = 10 psf. Wall D = 15 psf.

10.3 *Given:* The single-story wood-frame building in Fig. 10.A. The *transverse* lateral wind force (not shown in the sketch) is 375 lb/ft. The dead load of the roof and the wall is 250 lb/ft along the wall. Texture 1-11 (Sec. 8.13) plywood siding is applied directly to studs spaced 16 in o.c.

Find: a. Unit shear in the roof diaphragm and in the shearwalls along lines *A* and *B*.
b. The required nailing for the plywood siding to function as a shear panel. Use galvanized casing nails. Show all criteria. Check the proportions of the shear panel.
c. The net design uplift force at the base of the shear panel chords.

10.4 *Given:* The wood-frame building in Fig. 10.A. The critical lateral wind force in the longitudinal direction is given. The exterior wall finish is stucco (⅞-in. portland cement plaster over woven wire mesh).

Find: Determine if the longitudinal wall along line 2 is capable of functioning as a shear wall for the given lateral force. Specify nailing. Check diaphragm proportions. Show all criteria.

10.5 Use the hand solution to Prob. 10.1, 10.2, or 10.3 as a guide to develop a computer *template* to solve similar problems.
a. Consider only the specific criteria given in the problems.
b. Expand the template to handle any unit shear covered in ASD Shear Wall Supplement Tables 4.1A and 4.1B. In other words, input the values from the tables into a spreadsheet table *or* database and have the template return the required nail specification and blocking requirements.

10.6   *Given:*   The one-story wood-frame building in Fig. 10.B. The critical lateral force in the longitudinal direction is the given *strength-level* (unfactored or not reduced) seismic force. The roof overhang has a higher lateral force because of the larger dead load of the plastered eave. Neglect the seismic force generated by the dead load of the wall itself. Wall sheathing is $^{15}/_{32}$-in. STR I with 10d common or galvanized box nails. $C_M = 1.0$, and $C_t = 1.0$.

      Find: *a.* Design the shear panels in the wall along line 1. Show all criteria including nailing, blocking, and shear panel proportions.

                 b. Check shear panel chords using two 2 × 6s of No. 2 DF-L. Consider the resisting moment provided by a wall dead load of 20 psf. Assume no roof dead load frames into wall. A single row of $^{3}/_{4}$-in.-diameter bolts is used to anchor the chord to the foundation.

                 c. Plot the distribution of the strut force along line 1. Check the stresses caused by the maximum strut force if a double 2 × 6 wall plate serves as the collector (strut). A single row of $^{7}/_{8}$-in.-diameter bolts is used for the splices. Lumber is No. 2 DF-L.

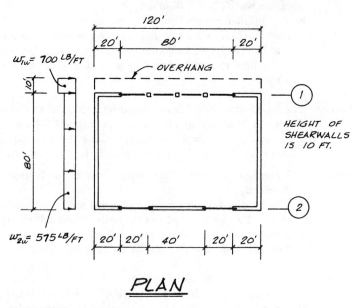

PLAN

Figure 10.B

10.7   *Given:*   The one-story building in Fig. 10.B. The longitudinal *strength-level* (unfactored or not reduced) seismic force to the roof diaphragm is given. Texture 1-11 (Sec. 8.13) plywood siding is applied over $^{5}/_{8}$-in. gypsum sheathing for the wall along line 2. In addition to the load shown, consider the seismic force generated by the dead load of the top half of the wall. Assume wall D = 15 psf throughout and a seismic coefficient of 0.200.

      Find: *a.* Design the shear panels in the wall along line 2, using galvanized box nails. Show all criteria.

     b. Check the shear panels chords, using two 2 × 4s of No. 2 DF-L. Neglect any resisting moment. A single row of ⅝-in.-diameter bolts is used to anchor the chord to the foundation. $C_M = 1.0$ and $C_t = 1.0$.

     c. Plot the distribution of the strut force along line 2. Check the stresses caused by the maximum strut force if a double 2 × 4 wall plate serves as the strut. A single row of ⅝-in.-diameter bolts is used for the splices. Lumber is No. 2 DF-L.

10.8   *Given:*   The one-story wood-frame building in Fig. 10.B. Roof dead load D = 10 psf. Wall dead load D = 18 psf (assume constant throughout). Wind = 20 psf. Seismic coefficient = 0.200.

     Find: *a.* Calculate the critical lateral force in the transverse direction.

          b. The exterior wall covering for the transverse walls is stucco (⅞-in. portland cement plaster on expanded metal lath). Determine if these walls can function adequately as shearwalls. Consider the shear at the midheight of the wall. Show all criteria and give construction requirements.

10.9   *Given:*   The one-story wood-frame building in Chap. 9, Fig. 9.D. The transverse lateral wind force is $w_r = 355$ lb/ft. The height of the building is 10 ft.

     Find: *a.* The unit shears in the three transverse shearwalls.

          b. Design the three transverse shearwalls, using ⅜-in. wood structural panel sheathing and 8d common or galvanized box nails. Studs are 16 in. o.c. Show all criteria including shearwall proportions.

          c. Calculate the chord forces for each of the three transverse shearwalls. Because the layout of the roof framing is not completely given, calculate the chord forces, using the gross overturning moments.

10.10 *Given:*   The elevation of the two-story wood-frame shearwall in Fig. 10.C. The wind force reactions from the roof diaphragm and second-floor diaphragm are shown in the sketch. Vertical loads carried by the wall are also shown in the sketch (dead load values include the weight of the wall). Wall sheathing is ¹⁵⁄₃₂-in. STR I with 8d common or galvanized box nails.

     Find: *a.* The required nailing for the shear panel. Give full specifications.

          b. The required tie (collector strut) forces for the connections at *A*, *B*, *C*, and *D*.

          c. The net design uplift force for the tension chord at the first- and second-floor levels.

          d. The maximum force in the compression chord at the first- and second-floor levels. Consider the effects of both gravity loads and lateral forces. Assume that the header is supported by the shear panel chord.

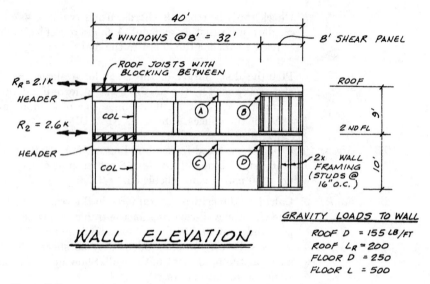

GRAVITY LOADS TO WALL

ROOF D = 155 LB/FT
ROOF L_R = 200
FLOOR D = 250
FLOOR L = 500

WALL   ELEVATION

Figure 10.C

e. If the allowable shear load on one anchor bolt is 1500 lb, determine the number and approximate spacing of anchor bolts necessary to transfer the shear from bottom wall plate to the foundation. These bolts are in addition to those required for uplift.

10.11 _Given:_ The wood-frame shearwall in Fig. 10.D. The lateral wind force from the roof diaphragm to the end shear panel is given. Gravity loads are also shown on the sketch. The dead load includes both roof and wall dead load. The shear panel has $^{15}\!/_{32}$-in. sheathing.

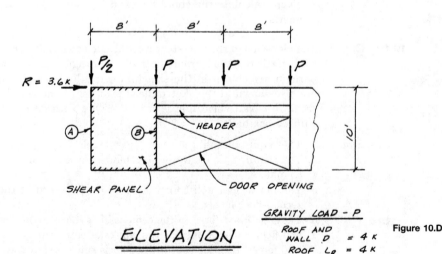

GRAVITY LOAD – P

ROOF AND
WALL   D   = 4 K
ROOF L_R = 4 K

Figure 10.D

ELEVATION

Find: *a.* Determine the required nailing for the plywood.
b. Calculate the net uplift force for the shear panel chord. Does this occur at column *A* or *B*?
c. Determine the maximum compressive force in the shear panel chord. Consider the effects of gravity loads. Assume that the header is supported by the shear panel chord. What is the critical load combination? Specify which chord is critical.
d. Assume that the allowable shear load per anchor bolt is 1000 lb. Determine the required number and approximate spacing of anchor bolts necessary to transfer the wall shear from the bottom plate to the foundation. These bolts are in addition to those required for uplift.

10.12 *Given:* The elevation of the two-story wood-frame shearwall in Fig. 10.E. Vertical loads and lateral wind forces are shown in the sketch. Wall framing is 2 in. nominal. Studs are 16 in. o.c. Wall sheathing is $^{15}/_{32}$-in. STR I with 8d common or galvanized box nails.

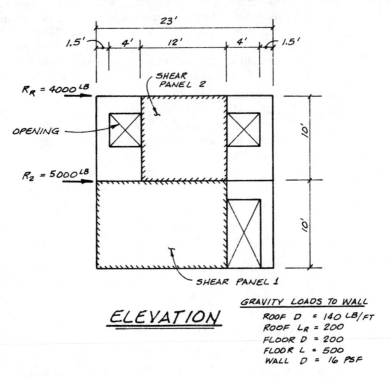

ELEVATION

GRAVITY LOADS TO WALL
ROOF D = 140 LB/FT
ROOF L_R = 200
FLOOR D = 200
FLOOR L = 500
WALL D = 16 PSF

Figure 10.E

Find: *a.* The required nailing for the first- and second-story shear panels.
b. The net uplift force at the base of shear panels 1 and 2.
c. The maximum compressive force at the end of the first-floor shear panel. State the critical load combination.
d. Assume that the allowable shear load of an anchor bolt parallel to the wall is 1500 lb. Determine the required number and approxi-

mate spacing of anchor bolts necessary to transfer the shear from the wall into the foundation. These bolts are in addition to those required for uplift.

10.13 Repeat Prob. 10.12 except that shear panel 1 is 12 ft long and lies directly below shear panel 2.

10.14 *Given:* The two-story shearwall system shown in Fig. 10.E. Wall framing is 2 in. nominal. Studs are 16 in. o.c. Wall sheathing is $^{15}\!/_{32}$-in. STR I with 8d common or galvanized box nails with nailing 6 in. o.c. along edges and 12 in. o.c. in the field. The second story windows are 4 ft high, and the first floor door is 8 ft high. The loads shown in Fig. 10.e are not applicable to this problem. Assume seismic forces govern.

　　　　　Find: The allowable lateral loads R_R and R_2 using the perforated shearwall design method.

10.15 *Given:* The one-story wood-frame building shown in Fig. 10.B. The *strength-level* (unfactored or not reduced) seismic force in the longitudinal direction is shown. Wall framing is No. 2 grade DF-L 2 × 8s, 16 in. o.c. The chords are 2-2 × 8 nailed together. Wall sheathing is $^{15}\!/_{32}$-in. STR I with 8d galvanized box nails 3 in. o.c. at the edges and 12 in. o.c. in the field. Wall dead load D = 20 psf. A low-slip anchorage detail is used with $d_a = 0.125$ in.

　　　　　Find: a. The story drift Δ_s of wall line 1.
　　　　　　　　 b. The story drift Δ_s of wall line 2.
　　　　　　　　 c. Compare the results of parts a and b to the IBC limit.

Wood Connections—Background

11.1 Introduction

The 1991 NDS (Ref. 11.1) introduced an entirely new method for evaluating the strength of laterally loaded connections. The new method is known as the *yield limit model for dowel-type fasteners,* and it represents the adoption of an engineering mechanics approach to the design of wood connections. It replaces the old empirical method of predicting the capacity of many common wood fasteners.

Although the equation format of the NDS provides a well organized design specification, the move to an engineering mechanics approach for fasteners has introduced some rather complicated equations for connection design. The designer, however, has the option of applying the formulas (which is best done on a computer) or using a set of fastener capacity tables that are based on the yield limit equations.

Chapter 11 starts by introducing the common types of fasteners used in typical wood connections, and the notation system for connection design is described. Greater detail about the connections and the required adjustment factors is included in the chapters that deal with specific fasteners. The chapter concludes with a review of the general concepts of the yield limit theory. Important strength considerations are included along with a summary of the various possible yield modes that are considered in this approach to connection design.

11.2 Types of Fasteners and Connections

There are a wide variety of fasteners and many different types of joint details that can be used in wood connections. When structural members are attached with fasteners or some other type of hardware, the joint is said to be a *mechanical connection.*

Mechanical connections are distinguished from connections made with *adhesives.* Adhesives are normally used in a controlled environment, such as a glulam, plywood, structural composite lumber, or wood I-joist manufacturing plant under the control of a formal quality assurance program. In some cases adhesives are used in the field in combination with nails for attaching sheathing to framing members. However, since adhesives and mechanical fasteners exhibit different levels of connection rigidity, increases in connection capacity are not typically associated with their combined usage. The connections covered in this discussion are limited to mechanical connections.

The design of connections using the following types of fasteners is covered in this book:

1. Nails

2. Bolts

3. Lag bolts or lag screws

4. Split ring and shear plate connectors

Some background material is introduced here, but much more detailed descriptions of the fasteners and design requirements for connections are given in Chaps. 12 and 13.

Connections are generally classified according to the direction of loading. *Shear connections* have the load applied *perpendicular* to the length of the fastener. See Example 11.1. These connections are further classified as to the number of shear planes. The most common applications are *single shear* and *double shear,* but additional shear planes are possible. According to the NDS (Sec. 11.3.8), connections with more than two shear planes are to be analyzed by evaluating each shear plane as a single-shear connection. The design value for the weakest shear plane is then multiplied by the number of shear planes to obtain the total connection capacity.

EXAMPLE 11.1   Typical Wood Connections Subjected to Shear

The most common fasteners used in wood connections are nails, bolts, and lag bolts. In shear connections the load acts perpendicular to the length of the fasteners. One or more shear planes occur.

Split ring and shear plate connectors are also used in shear-type connections to provide additional bearing area on the wood members for added load capacity.

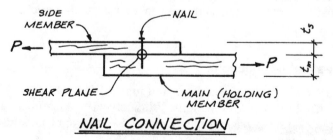

Figure 11.1a Single-shear (two-member) wood-to-wood *nail connection.* A similar connection can be made with a steel side plate (wood-to-metal).

The symbol Z represents the nominal fastener load capacity for these types of connections. *Shear* connections are also known as *laterally loaded* connections. The other basic type of connections in wood design is withdrawal connections (Example 11.2).

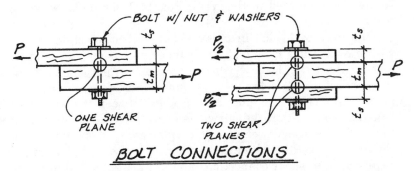

Figure 11.1b Single-shear (two-member) and double-shear (three-member) *bolt connections.* Wood-to-wood connections are shown, but connections with steel side plates are common (wood-to-metal).

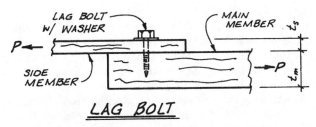

Figure 11.1c Single-shear (two-member) wood-to-wood *lag bolt connection.* Another common application is with a steel side plate (wood-to-metal).

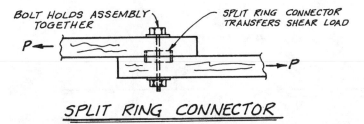

Figure 11.1d *Split ring and shear plate connectors* fit into precut grooves. A split ring connector (shown in sketch) is used only for a wood-to-wood connection. Shear plate connectors can be used for either wood-to-wood *or* wood-to-metal connections. See Fig. 13.26 for photographs of split ring and shear plate connectors in partially assembled joints.

The second major type of loading in a wood connection has the applied load *parallel* to the length of the fastener, and the fastener is loaded in tension. When nails and lag bolts are subjected to this type of loading, the concern is that the fastener may pull out of the wood member. *Withdrawal loading* is the term applied to this situation (see Example 11.2). Nominal withdrawal values are tabulated in the NDS (Ref. 11.3) for nails, wood screws and lag bolts.

Bolts may be loaded in tension, but such connections are not subject to withdrawal. The strength of the fastener in tension is usually not critical, but technically it should be verified. Compression perpendicular to grain under the head of a bolt or lag bolt is a normal design consideration. A washer or plate of sufficient size must be provided between the wood member and the fastener head so that the allowable bearing stress $F'_{c\perp}$ is not exceeded. Refer to Sec. 6.8 for a review of allowable compressive stress perpendicular to grain. Finally, it should be noted that split ring and shear plate connectors are basically used in shear connections and are not subjected to tension.

EXAMPLE 11.2   Typical Wood Connections Loaded in Withdrawal and Tension

Withdrawal loading attempts to pull a nail or lag bolt (lag screw) out of the main (holding) wood member. The symbol W represents the nominal fastener load capacity for these types of connections. Note that the main member is the member that receives the pointed end of the fastener. In a withdrawal connection the load is parallel to the length of the fastener, and the fastener is stressed in tension. Bolts can be loaded in tension, but they are not subject to withdrawal.

A washer of adequate size must be provided under the head and nut of a bolt loaded in tension, and under the head of a lag bolt loaded in withdrawal, so that allowable compressive stress perpendicular to the grain is not exceeded.

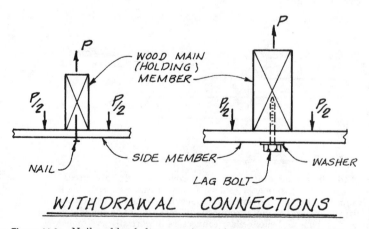

Figure 11.2a Nail and lag bolt connections subject to withdrawal loading.

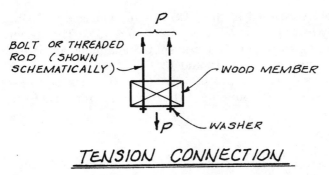

TENSION CONNECTION

Figure 11.2b Bolts used in tension.

The notation system for fastener design is based on the two types of loading just described. The following symbols are used:

Z = nominal design value for single fastener subjected to lateral *shear* load

W = nominal design value for single fastener subjected to *withdrawal* load

The term *nominal design value* refers to the basic load capacity for a fastener as defined by a table or equation in the NDS. As with all tabulated properties for wood design, the nominal fastener design value applies to a specific set of conditions. These idealized conditions include such things as loads of *normal* (10-year) *duration*, lumber that is *initially dry* and *remains dry* in service, no sustained exposure to high *temperature*, and so on. The idealized conditions for connections are similar to the conditions that apply to the tabulated stresses for a wood member such as a beam. Thus, the nominal fastener design value is simply the starting point for determining the allowable load capacity for a particular structural application.

Adjustment factors for load duration, moisture content conditions, temperature, and a number of other design considerations may be required to convert the nominal design value to an *allowable design value* for a single fastener. A prime is added to the symbol to indicate that the appropriate adjustments have been applied:

Z' = Z × (product of applicable adjustment factors for shear connection)

W' = W × (product of applicable adjustment factors for withdrawal connection)

It should be noted that the NDS provides design criteria for a number of fasteners in addition to those listed above. The appropriate adjustment factors for all fasteners covered in the NDS are summarized in NDS Table 10.3.1, *Applicability of Adjustment Factors for Connections.* The specific adjustment factors for nails, bolts, lag bolts, and split ring and shear plate connectors are reviewed in detail in Chaps. 12 and 13.

The fasteners covered in this book are those that are most likely to be found in a typical wood-frame building. Other fasteners included in the NDS are

1. Wood screws

2. Metal plate connectors

3. Timber rivets

4. Drift bolts and drift pins

5. Spike grids

Wood screws are similar to lag bolts, but wood screws are smaller in diameter, have a slightly tapered shank, and do not have a hex (or square) bolt head. Wood screws are inserted with a screwdriver or screw gun, but lag bolts are installed with a wrench. Wood screws have allowable design values for lateral load comparable to nails. Due to the gripping action of threads, wood screws have significantly larger allowable design values for withdrawal load than nails. They have some application in manufactured housing, but are not often used in conventional wood-frame construction. For larger design loads, lag bolts are the more likely choice.

Metal plate connectors are frequently found in wood buildings that use prefabricated light-frame wood trusses. A photograph of metal plate connected trusses was given in Fig. 6.33*b.* Metal plate connectors are light-gage metal plates that connect the truss members together. Metal plate connectors are also known as *nail plates* or *truss plates.* Truss plates are typically produced by pressing or stamping sheet metal to form teeth or barbs which protrude from the side of the plate. See Fig. 11.3. The protrusions serve as the "nails" for the truss plate and become embedded in the wood members during fabrication of a wood truss. Alternatively, conventional nails can be driven through holes punched (without protrusions) in nail plates.

The load capacity of a particular type of truss plate is determined by testing, and the load rating is usually given in pounds per square inch of contact area between the metal plate and the wood member. When a truss is analyzed, the forces to be transferred between the wood members are determined. The required size of metal plate connector is then obtained by dividing the force by the allowable load per square inch for the truss plate. The overall size and configuration of a metal plate connector must provide the necessary contact areas for each member in the joint. Truss plates can be found in many wood buildings, but they can be viewed as a specialty item. The design of metal plate connectors is usually handled by a truss manufacturer, and fabrication is typically restricted to a manufacturing plant operating under a formal quality control program. See Ref. 11.12 for information.

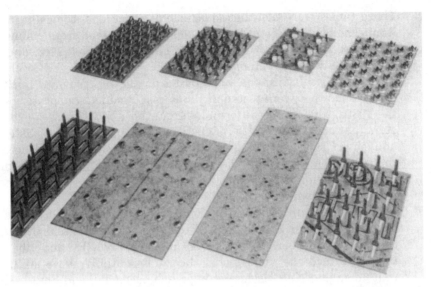

Figure 11.3 Metal plate connectors. (*Courtesy of the FPL.*)

Timber rivets (also called *glulam rivets*) are a relatively new type of fastener that was introduced in the 1997 edition of the NDS. The reader should not confuse a timber rivet with a rivet in steel construction. They do not look anything alike. Timber rivets are hardened steel fasteners shaped somewhat like a nail, but with a countersunk head and a rectangular-shaped cross section with rounded corners. The cross-sectional dimensions are approximately ⅛ in. by ¼ in., and rivet lengths range from 1½ in. to 3½ in. Timber rivets are driven with a hammer through predrilled holes in steel side plates. They are installed with the longer (¼ in.) cross-sectional dimension parallel-to-grain to prevent splitting in wood members. Their primary application is for resisting very large loads in multiple rivet, heavy glulam timber connections.

Drift bolts and *drift pins* are unthreaded steel rods that are driven into a hole bored through one wood member and into the adjacent member. *Spike grids* are intended for use in wood pile and pole construction. Drift pins and spike grids are also considered specialty fasteners. A number of fasteners are used in wood construction in addition to those included in the NDS. Information on a variety of fasteners may be obtained from the *Wood Handbook* (Ref. 11.9), the *TCM* (Ref. 11.5), the International Staple, Nail and Tool Association (see Ref. 11.10), Ref. 11.7, and catalogs of connection hardware suppliers. Some of the hardware used in conjunction with wood fasteners is covered in Chap. 14.

11.3 Yield Model for Laterally Loaded Fasteners

In the past the capacity of laterally loaded nails, bolts, and lag bolts was obtained from fastener capacity tables given in earlier editions of the NDS.

These tables were also included in building codes, typically in abbreviated form. The load values in these tables were obtained from empirical formulas which are summarized in the NDS Commentary (Ref. 11.2) and various editions of the *Wood Handbook* (Ref. 11.9).

The empirical formulas were based on the *proportional limit* strength of connections that were tested at the Forest Products Laboratory (FPL) during the 1930s. The empirical formulas from the *Wood Handbook* contain parameters that are important to connection performance such as fastener diameter and length. However, the empirical expressions simply reproduced load values that were obtained from laboratory testing programs. These formulas cannot be derived using principles of engineering mechanics and may not be appropriate for all connections.

In the 1940s an approach to connection analysis was developed in Europe that was based on the yielding of the various elements in the connection. In the 1980s a number of papers were published by FPL and other researchers that confirmed the yield model theory. Essentially these studies concluded that an engineering mechanics approach based on yield limit theory is appropriate for analyzing *dowel-type fasteners* in wood connections (Refs. 11.8 and 11.11).

Traditionally a dowel is thought to be a circular rod of wood or metal that is placed in a hole which has been drilled in a wood member or members. However, in the context of the NDS, the term *dowel-type fastener* simply refers to fasteners with a generally cylindrical shape such as bolts, screws, and nails.

The idea is that these fasteners act similar to a dowel as they bear against the grain in a wood member under a lateral load. The term *lateral load* here refers to a load that is perpendicular to the length of the fastener, i.e., a *shear connection*. The load may or may not include the effects of lateral wind or earthquake forces. The yield limit model applies only to shear connections and not to fasteners loaded in withdrawal. Withdrawal connections are still covered by empirical formulas provided in the NDS.

In utilizing the yield limit theory, the various yield modes that can occur in a given type of connection are analyzed. A load capacity is computed for each of the various modes. The yield limit is then taken as the smallest of these load capacities. This critical load defines the *yield mechanism* for the connection. It should be noted that the yield limit equations for Z in the NDS include appropriate factors of safety.

An engineering mechanics approach to connection design, such as the yield limit model, has both advantages and disadvantages. The principal advantage is that the designer can mathematically analyze a connection of practically any configuration. Thus, the designer is not limited to joint configurations and details that are given in the NDS or Code tables. In the past, if a connection did not match the details in the table, the designer was forced either to revise the connection or to extrapolate values from a table that may not have been appropriate.

In addition, the yield limit model is able to take into account the strength of different fastener materials. In the older empirical method, a certain fastener strength was simply presumed. However, the yield theory directly incorporates the yield strength of a fastener into the analysis. A final advantage is that the yield limit model can be used to evaluate the strength of a connection involving wood members of different species, or a connection between a wood member and another material.

On the negative side, the application of the yield model to connection design can involve the use of some rather cumbersome equations. The number of equations varies depending on whether the fastener is in single shear or double shear. However, regardless of the particular conditions, application of the yield limit model in equation form requires the consideration of a number of possible yield modes. The equations can be lengthy and awkward to solve by hand on a calculator. Fortunately the yield equations are easy to program and solve on a computer using any of the popular spreadsheet or equation-solving software.

In keeping with the *equation format,* the NDS contains a complete set of yield limit equations for the analysis of dowel-type fasteners. However, the NDS takes a balanced approach to this problem. It recognizes that some designers may not need or want the sophistication provided by the formal yield limit equations. Thus, for the designer who does only an occasional wood connection design, or for the designer who simply does not want to become involved with the yield equations, the NDS also provides fastener capacity tables for various connection configurations.

The NDS fastener capacity tables are based on the yield limit equations, but are simplified from the general theory by reducing the number of parameters used to obtain load capacity. For example, the tables apply to connections with all members from the same species of wood, and they cover only a limited number of species (specific gravities). The tables for bolts and lag bolts also handle only parallel- and perpendicular-to-grain loadings. Therefore, the tables are simple, conservative, and relatively easy to use. They apply to a certain set of common but limited conditions.

The fastener capacity tables are limited in scope, and the real advantage of the yield limit model comes in the automated evaluation of the yield equations. In this way all of the details affecting a particular connection may be taken into account. In addition, there are some fastener problems that *require* use of the yield limit equations. For example, the NDS bolt tables cover only parallel- and perpendicular-to-grain loading, and the equations will be necessary to evaluate loading at some other angle to grain.

A programmed solution to the yield equations using a spreadsheet or equation-solving software can be easier than looking up a value in the NDS fastener tables. Solving the yield limit equations is, in fact, an elementary task on a spreadsheet. After the equations have been solved once, a *template* is created which can be used for other similar problems. The template can also be expanded to include some of the other design provisions such as the adjustment factors for wood connections.

The remaining portions of this chapter summarize several of the important properties that affect the behavior of dowel-type connections. This is given in order to develop a general understanding of the rationale behind the theory, without attempting to derive the lengthy equations. The possible yield modes are also reviewed for typical two-member and three-member connections.

11.4   Factors Affecting Strength in Yield Model

It has been stated that the yield limit model is appropriate for the design of nails, bolts, lag bolts, and wood screws used in shear connections. For reasons given in Sec. 11.2, the design of wood screw connections is not covered in this book. Split ring and shear plate connectors may not behave as dowel-type fasteners, and because of their rather limited use, they are addressed only briefly (Chap. 13).

The yield limit model for dowel-type fasteners is based upon engineering mechanics, and it uses *connection geometry* and *material properties* to evaluate strength. The primary factors used to compute the nominal design value Z include

1. Fastener diameter D
2. Dowel bearing length l
3. Fastener bending yield strength F_{yb}
4. Dowel bearing strength of wood member F_e
   a. Specific gravity (dry) of wood member G
   b. Angle of load to grain of wood member θ
   c. Relative size of fastener (large or small diameter)
5. In the case of metal side plates, the dowel bearing strength F_e of the metal

The notation in the above list is usually expanded by adding subscripts m and s to the appropriate terms to indicate whether the symbols apply to the main member or the side member. For example, l_m, F_{em}, G_m, and θ_m refer to the dowel bearing length, dowel bearing strength, specific gravity, and angle of load to grain for the *main member*. Likewise, l_s, F_{es}, G_s, and θ_s represent the corresponding properties for the *side member*. *Member thicknesses* (t_m and t_s) were labeled in Fig. 11.1. For bolted connections, the dowel bearing length l and member thickness t are identical. However for nail, wood screw, and lag bolt connections the dowel bearing length in the main member is typically less than the main member thickness ($l_m < t_m$). In a two-member bolted connection, the main member is the thicker member; and in a three-member connection, the main member is the center member. In a two-member nail or lag bolt connection, the member holding the pointed end of the fastener is the main member. The other members in the connections are the side members.

The bending *yield strength of the fastener F_{yb}* is a property which is used to predict the load capacity of a mechanism that involves the formation of a plastic hinge (Sec. 11.6) in the fastener. The NDS provides typical values of bending yield strength for various types of fasteners. These values may be used in the event that material properties for a specific fastener are not available. Values of F_{yb} for the fasteners given in the NDS are representative of those found in the field and are generally thought to be conservative. See NDS Appendix I for additional information.

The *dowel bearing strength* F_e is the strength property of the members in a connection that resists *embedding* of a dowel. Designers familiar with structural steel can best relate this property to the allowable bearing stress F_p on a steel plate in a bolted connection.

In wood design, the value of F_e depends on a number of different factors. The formulas for dowel bearing strength are given in Sec. 11.5. It is important to understand that the members in a connection can have different dowel bearing strengths. Because F_e depends on the specific gravity of the wood, one way to have different dowel bearing strengths in the same connection is to have main and side members from different species.

The angle of load to grain θ also may cause F_e to be different for the main and side members. Several typical uses of bolts and lag bolts in shear-type connections were illustrated in Example 11.1 (Sec. 11.2). The connections in this earlier example were made up of only wood members, and in each case the load was applied parallel to the length of the member. The load, therefore, was parallel to the grain, and the angle of load to grain was zero ($\theta = 0$ degrees).

The wood members in a connection may have different angles of load to grain. For example, the load from a fastener can cause bearing parallel to the grain in one member and bearing perpendicular to the grain in another member. Other connections may involve loading at an angle θ other than 0 or 90 degrees to the grain. See Fig. 11.4a in Example 11.3.

The dowel bearing strength for a wood member is greatest when the load is parallel to the grain and weakest for a load perpendicular to the grain. For intermediate angles, the bearing strength lies between the parallel- and perpendicular-to-grain values. (*Exception:* small-diameter fasteners have the same dowel bearing strength regardless of angle of load to grain—see Sec. 11.5.)

In other common connections, metal side members are used. These can be full-size structural-steel members, but often the side plates are some form of metal connection hardware. See Fig. 11.4b. The dowel bearing strength of a steel plate is much greater than the dowel bearing strength of a wood member.

EXAMPLE 11.3   Examples of Different Dowel Bearing Strengths in the Same Connection

Different dowel bearing strengths can result from different angles of load to grain or the use of side members of different materials.

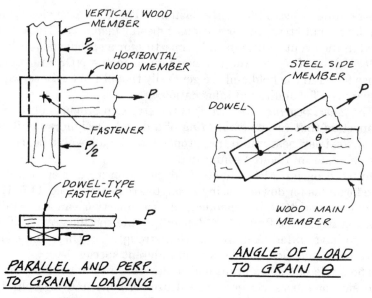

Figure 11.4a Depending on fastener size, the angle of load to grain may have an effect on the dowel bearing strength of a wood member. In the sketch on the left, the load transmitted by the fastener is parallel to the grain in the horizontal wood member and perpendicular to the grain in the vertical member. On the right, the load is at an angle θ to the direction of the grain in the wood member.

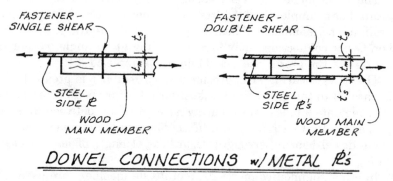

Figure 11.4b Dowel-type fasteners are often used in wood connections with steel side plates. Single-shear and double-shear applications are common in wood-to-metal connections. The dowel bearing strength F_e for a steel plate is obviously much larger than F_e for a wood member.

If a metal side member is used in a connection, the plate is to be designed in accordance with recognized engineering practice. For example, a steel side plate should meet the requirements of the American Institute of Steel Construction (Ref. 11.4) or the American Iron and Steel Institute (Ref. 11.6). In addition, a multiple of the *ultimate tensile strength* of the steel F_u is used to define the dowel bearing strength of metal plates for use in the yield limit equations for the wood connection.

The major factors that affect the nominal design value Z of an individual dowel-type fastener according to the yield limit model have been briefly introduced. Some additional information regarding dowel bearing strength and the bending of fasteners in a connection is given in the following sections.

Recall that there are a number of other adjustments that may affect the allowable load Z'. These additional factors may apply to all types of connections (e.g., load duration factor C_D), or the factors may apply only to specific types of fasteners (e.g., penetration depth factor C_d for split ring and shear plate connectors). Adjustment factors are covered in subsequent chapters dealing with specific fastener types.

11.5   Dowel Bearing Strength

The *property of wood* that affects the nominal design value Z of a nail, bolt, or lag bolt is known as *dowel bearing strength*. In much of the research literature, this property is termed *embedding strength,* hence the notation F_e. It is related to the crushing strength of the wood member under loading from a dowel subjected to a shear load. As noted in the previous section, the dowel bearing strength varies with the *specific gravity* of the wood and the relative size of the dowel. For large-diameter fasteners, the *angle of load to grain* also affects the dowel bearing strength.

Values of specific gravity G for the various combinations of lumber species are given in NDS Table 11.3.2A. Dowel bearing strengths F_e are given in NDS Table 11.3.2 for various values of G, and in NDS Table 11.3.2B for plywood and OSB panels. Dowel bearing strengths for structural composite lumber products must be obtained from the manufacturer's technical literature or Code evaluation reports. In practice the dowel bearing strengths for lumber, glulam, or panel products in a connection problem will typically be obtained directly from NDS Tables 11.3.2 and 11.3.2B. This will undoubtedly be the case for a hand solution of the yield limit equations. In contrast, if the equations for Z are programmed on the computer or solved on a spreadsheet, the *formulas for F_e* given in Example 11.4 may be useful.

Tests have shown that the dowel bearing strength for *large-diameter* fasteners depends on the angle of load to grain. Consequently, there is one formula for dowel bearing strength for parallel-to-grain loading ($F_{e\parallel}$) and another for perpendicular-to-grain loading ($F_{e\perp}$). For intermediate angles of load to grain, the Hankinson formula is used to obtain a dowel bearing strength ($F_{e\theta}$) that lies between $F_{e\parallel}$ and $F_{e\perp}$. The Hankinson formula was first introduced in Sec. 6.8 in a form for evaluating the allowable bearing stress in a wood member. The same expression, with a simple change in notation, is appropriate for determining the dowel bearing strength for fastener design.

Small-diameter dowels do not exhibit the directional strength characteristics of large fasteners, and a single expression defines dowel bearing strength regardless of the direction of loading. The shank diameter separating the categories of large dowels and small dowels is ¼ in. Large-diameter fasteners (such as bolts and lag bolts) are installed in drilled holes. In contrast, small dowel-type fasteners (such as nails) are usually driven into the wood with a hammer or pneumatic gun.

EXAMPLE 11.4   Formulas for Dowel Bearing Strength

Numerical values of dowel bearing strength are listed in the NDS:

Sawn lumber and glulam	NDS Table 11.3.2
Plywood and OSB	NDS Table 11.3.2B

The values in the NDS tables are based on the following expressions:

LARGE DOWELS ($D \geq 0.25$ in.):

Parallel-to-grain loading:
$$F_{e\parallel} = 11,200(G)$$

Perpendicular-to-grain loading:
$$F_{e\perp} = 6100(G^{1.45})D^{-0.5}$$

The Hankinson formula gives F_e at an angle of load to grain θ

$$F_{e\theta} = \frac{F_{e\parallel}F_{e\perp}}{F_{e\parallel}\sin^2\theta + F_{e\perp}\cos^2\theta}$$

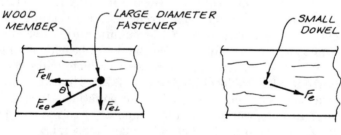

$F_{e\parallel}$ – PARALLEL TO GRAIN
$F_{e\perp}$ – PERPENDICULAR TO GRAIN
$F_{e\theta}$ – ANGLE TO GRAIN θ

F_e – ANY ANGLE TO GRAIN

LARGE DOWELS

SMALL DOWELS

Figure 11.5 Definition of dowel bearing strength depends on relative size of fastener. The dowel bearing strength F_e for *large*-diameter fasteners is a function of angle of load to grain. Specific gravity and diameter also are important. The F_e for *small*-diameter fasteners depends only on specific gravity.

SMALL DOWELS ($D < 0.25$ in.):

All angles of load to grain: $F_e = 16,600\, G^{1.84}$

where F_e = dowel bearing strength; subscripts $\parallel$, $\perp$, and θ indicate strengths parallel to grain, perpendicular to grain, and at an angle to grain θ

$\quad\quad$ G = specific gravity (dry) of wood member

$\quad\quad$ D = diameter of dowel-type fastener

The yield limit model is capable of handling a wide variety of wood connections including those in which the main and side members have different dowel bearing strengths. Several situations are described in Sec. 11.4 where F_{em} and F_{es} are not equal. On the other hand, in a number of design conditions the dowel bearing strength will be the same for the main and side members.

The formulas for dowel bearing strength in Example 11.4 are provided in the footnotes to NDS Table 11.3.2. Again, these formulas may be more convenient than the tables for F_e if the yield limit equations for Z are programmed on a computer. The dowel bearing strength formulas are evaluated for several fasteners in Example 11.5 to simply confirm the NDS tables.

EXAMPLE 11.5   NDS Dowel Bearing Values

Use the appropriate formulas from Example 11.4 to evaluate the dowel bearing strength values in NDS Table 11.3.2 for the following fasteners. Note that the values for F_e in the NDS tables are rounded to the nearest 50 psi.

LARGE DOWELS

Bolts:

a. $\frac{5}{8}$-in.-diameter bolt in DF-L with load applied parallel to grain ($\theta = 0$ degrees) Values of specific gravity are available in NDS Table 11.3.2A. For DF-L the specific gravity is 0.50.

$$F_{e\parallel} = 11{,}200(G) = 11{,}200 \times 0.50 = 5600 \text{ psi}$$

This value agrees with $F_{e\parallel}$ in NDS Table 11.3.2.

b. $\frac{3}{4}$-in.-diameter bolt in DF-L with load applied perpendicular to grain ($\theta = 90$ degrees)

$$F_{e\perp} = 6100(G)^{1.45}D^{-0.5}$$

$$= 6100(0.50)^{1.45}(0.75)^{-0.5}$$

$$= 2578 \text{ psi} \approx 2600 \quad\quad \text{from NDS Table 11.3.2}$$

SMALL DOWELS

Nails: 16d common nail at any angle of load to grain in DF-L

$$F_e = 16{,}600(G)^{1.84} = 16{,}600(0.50)^{1.84}$$

$$= 4637 \text{ psi} \approx 4650 \text{ psi} \quad\quad \text{from NDS Table 11.3.2}$$

In the case of a connection between a wood member and a metal side plate, it is necessary to have the equivalent of an embedding strength value for the metal member. Steel design standards specify bearing strengths of 2.4 times the ultimate tensile strength ($F_p = 2.4F_u$) for hot-rolled steel plates (Ref. 11.4) and 2.2 times the ultimate tensile strength ($F_p = 2.2F_u$) for cold-formed steel plates (Ref. 11.6). As noted in NDS Appendix I, steel bearing strengths F_p are reduced by a factor of 1.6 to obtain dowel bearing strengths F_e that can be used directly in NDS *yield limit equations*. The 1.6 reduction factor permits F_e values for steel to be used in conjunction with applicable load duration factors for wood members in a connection. Thus, the dowel bearing strengths for steel members in connections are $F_e = 1.5F_u$ for hot-rolled steel plates and $F_e = 1.375F_u$ for cold-formed steel plates. In general, steel plates with thickness less than ¼ in. are assumed to be cold-formed steel, and plates with thickness greater than or equal to ¼ in. are assumed to be hot-rolled steel. For wood members in connections, F_e is defined as in Example 11.4.

This section concludes with a brief discussion of the dowel bearing capacity of a fastener in a *glulam* beam and in *wood structural panel* sheathing. It will be recalled from Chap. 5 that higher-quality laminating stock is located at the outer tension and compression zones in a glulam bending member, and lower-quality material is placed in the inner core. In fact, the manufacturing specifications permit the mixing of different species of wood in the same member (i.e., the inner core may be from a weaker species than the outer laminations). Although mixing different species groups is permitted, most current practice is to manufacture a glulam member with material from a single species group.

If a glulam beam is produced from a single species group, a single specific gravity applies to the member. Values of G can be obtained from the NDS tables that were previously described. Dowel bearing strength for the glulam may be computed using the expressions given in Example 11.4, or the value of F_e may be read from NDS Table 11.3.2.

In the event that a glulam is manufactured with material from more than one species group, there will be a different dowel bearing strength for each species group. See Example 11.6. A conservative approach to connection design is to use F_e based on the specific gravity for the inner core laminations. To take advantage of a larger F_e for the outer laminations, the designer will have to investigate the layup of a given member. In other words, to use the dowel bearing strength for the outer laminations, the designer must be assured that the fastener, or group of fasteners, is located entirely within the laminations with the higher F_e.

EXAMPLE 11.6   Dowel Bearing Strength in Glulam Beams

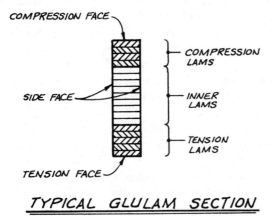

COMPRESSION FACE

COMPRESSION LAMS

SIDE FACE

INNER LAMS

TENSION LAMS

TENSION FACE

TYPICAL GLULAM SECTION

Figure 11.6 Distribution of laminations in a glulam follows bending stress distribution.

Most glulam beams produced today are manufactured with laminating stock from a single species group, and F_e is a function of a single specific gravity. In a member with outer laminations from one species group and inner laminations from another, two different specific gravities apply. To use the larger F_e, the fasteners must be located entirely within the laminations from the denser species group.

Wood structural panels such as plywood and oriented strand board (OSB) may include plies from a single species group or from multiple species groups. Preliminary research (Ref. 11.8) indicates that when all plies are from the same species group, the dowel bearing strength for a small-diameter fastener ($D < 0.25$ in.) in plywood can be obtained from the formula given in Example 11.4. However, often the species make-up of plywood or OSB panels is not known by designers. According to NDS Table 11.3.2B, $F_e = 4650$ psi and $G = 0.5$ can be conservatively used for all grades of OSB and for STRUCTURAL I or Marine grades of plywood. For all other grades of plywood, $F_e = 3350$ psi and $G = 0.42$ should be used. These F_e values are based on dowel bearing tests for nails in panel products conducted by APA—The Engineered Wood Association. Although not explicitly stated in the NDS, previous guidance from APA—The Engineered Wood Association indicates that $G = 0.5$ may be conservatively used to calculate $F_{e\parallel}$ and $F_{e\perp}$ values for larger diameter fasteners ($D \geq 0.25$ in.).

11.6   Plastic Hinge in Fastener

Dowel bearing strength is the primary factor that measures the strength of the wood members in a connection. In addition to this, the strength of the fastener can affect the yield limit mechanism of a connection. The mode of yielding in a fastener is related to the formation of one or more *plastic hinges*. Designers who are familiar with the behavior of structural-steel members already understand the concept of a plastic hinge, but it is reviewed here briefly.

At normal temperatures, structural steel behaves in a ductile manner. In other words, the material can undergo large deformations before rupture occurs. This behavior is best seen on a stress-strain diagram for steel. See Fig. 11.7. The horizontal yield plateau is characteristic of the material. The ability of steel to undergo large deformations without rupture is essential to the formation of a plastic hinge.

In structural-steel design, a plastic hinge is usually introduced by examining the bending stress diagram in a beam as the member is loaded through several stages. See Example 11.7. In order for a plastic hinge to form, the beam must be braced so that buckling does not occur.

Consider what happens to the member as the load is increased to the point where the bending stress reaches yield F_y at the outside fibers. The beam does not collapse simply because the outer fibers in the member reach F_y. In fact, the member can carry additional load beyond this point as inner portions of the cross section reach yield. Because of the ductile nature of the material, this behavior continues until the entire section yields.

The deformation of the outside fibers of the beam takes place over a relatively small portion of the beam length. On either side of this localized yielding, the beam remains relatively straight. This behavior leads to the concept of a plastic hinge.

A plastic hinge corresponds to a rectangular bending stress block in which all of the fibers have reached the yield stress. By definition, a real hinge is a structural element that does not transmit moment. In contrast, a plastic hinge develops the plastic moment capacity M_p of the cross section. However, once M_p is reached at the point of local maximum moment, additional bending moment cannot be developed by the member. Thus, a plastic hinge functions similar to a real hinge as far as *additional* loading is concerned.

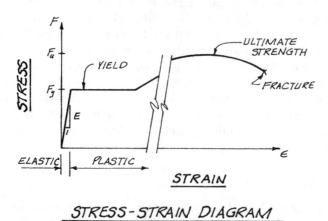

Figure 11.7 Typical stress-strain diagram for a ductile steel.

EXAMPLE 11.7   Formation of a Plastic Hinge in a Steel Beam

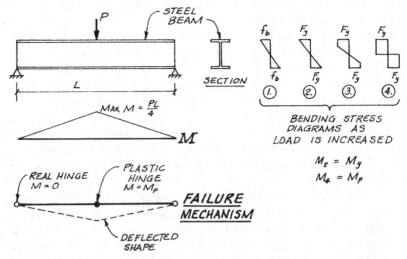

Figure 11.8 Formation of a plastic hinge (P.H.) in a structural steel beam. In a simply supported beam, the formation of a single plastic hinge causes a failure or collapse mechanism to develop.

The loading diagram for the W-shape beam has a concentrated load at the midspan. Several bending stress diagrams are given to show various stages as the load is increased. The steps in the formation of a plastic hinge are as follows:

1. Small load P causes a small bending stress f_b. The member is elastic, and the stress diagram is triangular with zero stress at the neutral axis and maximum stress at the outside fibers. The maximum bending stress is below the yield point.

2. Larger bending load causes the maximum bending stress to reach yield at the outside fibers ($f_b = F_y$). The moment in the beam that corresponds to this stress diagram is the yield moment M_y, and the load is known as the yield load P_y.

3. As the concentrated load is increased, the beam does not collapse (assuming that buckling is prevented). Because of the ductile nature of the material, the outer fibers simply continue to deform with no increase in stress. However, the inner fibers increase in stress until yielding occurs. This pattern of increasing the stress of the inner fibers continues until all of the fibers eventually yield.

4. When all fibers of the cross section have reached F_y, a *plastic hinge* has formed. The moment at the cross section is known as the plastic moment M_p, and the cross section at the point of maximum moment is incapable of carrying any additional moment. Thus, the plastic hinge functions as a real hinge as far as additional loading is concerned.

When a single plastic hinge forms in a simply supported beam, a failure mechanism develops. The plastic hinge at the midspan causes the formation of an unstable linkage (three hinges on a straight line) or failure mechanism. Note, in the sketch of the linkage, a real hinge is shown as an open circle, and the plastic hinge is shown as a solid circle. The load that causes a collapse mechanism to form is known as the *ultimate* or *collapse load,* and it is given the symbol P_u.

The cross-sectional property that measures the plastic moment capacity of a member is known as the plastic section modulus Z. The plastic moment capacity is the product of the plastic section modulus and the yield stress ($M_p = ZF_y$). This relationship is analogous to the one for maximum elastic moment, which is the elastic section modulus times the yield stress ($M_y = SF_y$). Additional information on plastic section modulus can be found in books on steel structures.

Now that the concept of a plastic hinge has been introduced, the idea of a failure mechanism can be addressed in a little more detail. In a steel beam, a failure mechanism is a structural form that cannot support a load. It is characterized by large, unbounded displacements. In a beam or similar structure, the failure mechanism is associated with the collapse load.

The method of structural analysis that deals with the determination of collapse mechanisms is known as *plastic analysis*. Plastic analysis involves adding a sufficient number of plastic hinges to an initially stable structure to form a failure mechanism. The most elementary form of a collapse mechanism for a steel beam is one with three hinges on a straight line. Once this form develops, the structure is unstable. One of the objectives in plastic analysis is to determine the collapse load P_u, which is the smallest load that will cause a failure mechanism to develop.

The linkage mechanism in Fig. 11.8 was created by the formation of a single plastic hinge in the beam. Whether or not a single plastic hinge causes the development of a collapse mechanism depends on the support conditions for the member. In a simply supported beam, only one plastic hinge is needed to produce a failure mechanism. However, for other support conditions, additional plastic hinges may be required. See Example 11.8. The beam in this example is fixed at one end and pinned at the other, and the development of a collapse mechanism occurs in several stages.

The first plastic hinge logically occurs at the point of maximum moment. However, the first plastic hinge does not produce a failure mechanism. For additional loading (beyond the loading required to cause the first plastic hinge), the beam appears to be a member that is pinned at both ends. There is a real hinge at one end of the member and a plastic hinge at the other end. A collapse mechanism with three hinges in line requires the formation of a second plastic hinge. The second plastic hinge develops under the concentrated load.

EXAMPLE 11.8   Formation of Two Plastic Hinges in a Steel Beam

A plastic hinge first develops at the fixed end, which is the point of maximum moment. However, a failure mechanism does not occur until an increased load causes a second plastic hinge under the concentrated load. The linkage created by the second plastic hinge has three hinges on a straight line: one real hinge and two plastic hinges.

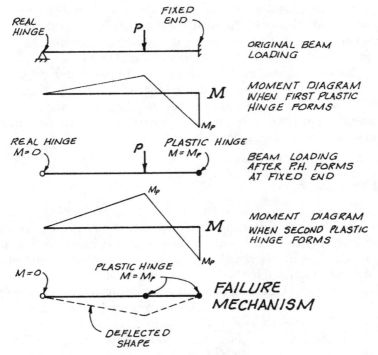

REAL
HINGE

FIXED
END

P

ORIGINAL BEAM
LOADING

M

MOMENT DIAGRAM
WHEN FIRST PLASTIC
HINGE FORMS

M_P

REAL HINGE
$M = 0$

P

PLASTIC HINGE
$M = M_P$

BEAM LOADING
AFTER P.H. FORMS
AT FIXED END

M_P

M

MOMENT DIAGRAM
WHEN SECOND PLASTIC
HINGE FORMS

M_P

$M = 0$

PLASTIC HINGE
$M = M_P$

M_P

**FAILURE
MECHANISM**

DEFLECTED
SHAPE

Figure 11.9 Steel beam fixed at one end and pinned at the other. Two plastic hinges are required to develop a mechanism.

The formation of a plastic hinge in a structural-steel member and the development of a failure mechanism have been discussed here for orientation and review. There are some major differences between the behavior of a dowel-type steel fastener used in a wood connection and the structural-steel beams just reviewed. One difference is that the cross section of a fastener is circular. However, the configuration of the cross section does not change the concept of a plastic hinge.

Another difference is that the mechanism in a wood connection is not that of a structural-steel beam. The behavior of a dowel-type fastener is more complex. In addition, there is a reserve capacity in a wood connection; hence the term *yield limit* is applied to wood connections rather than *collapse load*.

However, the elementary beam examples are useful to define the concept of a plastic hinge and a mechanism. At this point it should simply be understood that dowel-type fasteners (bolts, lag bolts, and nails) of a ductile metal can develop one or more plastic hinges as a *yield mechanism* is formed in a wood connection.

The NDS contains a complete set of yield limit equations to evaluate the nominal load capacity of the various yield mechanisms. Some of the mechanisms involve crushing of the wood fibers (exceeding the dowel bearing strength), and others are the result of a plastic hinge forming in the fastener combined with crushing of the wood fibers. The designer need not be fully conversant with the derivation of the yield limit equations, but an understanding of the various yield modes is important. In particular, ductile yield modes that involve the formation of plastic hinges in fasteners (e.g., Mode III and Mode IV) may be preferred for resisting earthquake and other dynamic loads. A general summary of these modes is given in Sec. 11.7 for single-shear and double-shear wood connections.

11.7   Yield Limit Mechanisms

Several properties critical to the performance of dowel-type fasteners have been introduced in this chapter. These include connection geometry, dowel bearing strength of the members, and a plastic hinge in a fastener. These factors are used in the *yield limit equations* for laterally loaded nails, bolts, and lag bolts to evaluate the nominal design value Z for a single fastener.

To summarize the general procedure, there are several possible yield modes for dowel-type fasteners. An equation is provided in the NDS to evaluate the load capacity for each mode. The nominal design value Z for a particular fastener is then taken as the smallest value considering all of the appropriate modes.

The remaining portion of this chapter describes the various yield limit modes that can occur in shear-type connections. The yield modes are common to all dowel-type connections, and the yield limit mechanisms described here will be referenced in subsequent chapters. Some modes that apply to single-shear (two-member) connections do not occur in double-shear (three-member) connections because of the symmetric loading that takes place in double-shear connections. See Example 11.9. The yield limit equations are given in Example 11.10 for single-shear and double-shear connections. The yield limit modes are as follows:

Mode I: Dowel bearing failure under uniform bearing.   The bearing stress for Mode I yielding is uniform because in this yield mechanism the fastener does not rotate or bend. The dowel bearing strength of a wood member is simply exceeded under uniform bearing. The possibility of Mode I yielding must be considered in each wood member in the connection. Consequently, a Mode I mechanism is further classified as Mode I_m if the dowel bearing strength is exceeded in the main member. If the overstress occurs in the side member, the mechanism is Mode I_s.

The need to check for a bearing yield mechanism in each member stems from the fact that member thickness and dowel bearing strength can be different for the main and side members (i.e., the products of $l_m F_{em}$ and $l_s F_{es}$ may not be equal). Different strengths can be the result of mixing different species of wood in a connection or of having a different angle of load to grain for the main and side members.

It should be recalled that there is also a definition of dowel bearing strength for steel members as well as wood members. The F_e for a steel member is a necessary parameter for the general yield limit equations.

Mode II: Dowel bearing failure under nonuniform bearing. The second principal yield mechanism also causes the dowel bearing strength to be exceeded. In this case, the fastener remains straight, but it undergoes rigid body rotation. Rotation causes a nonuniform bearing stress, and the wood fibers are crushed at the outside face of each member. Because of symmetry, Mode II yielding does not occur in a double-shear connection.

Mode III: Plastic hinge located near each shear plane. In a Mode III mechanism, the dowel bends and a plastic hinge forms in the fastener. In a single-shear connection, the plastic hinge can occur within the side member or the main member, near the shear plane. There are two yield equations associated with Mode III, and the mechanisms are further classified as Mode III_m and Mode III_s.

If the plastic hinge occurs inside one member, there is a corresponding crushing of the wood fibers in the other member. This is one reason why the yield mechanisms in a wood connection are more complex than those for the steel beams described in Sec. 11.6. The mechanism is classified as Mode III_m when the dowel bearing strength is exceeded in the main member. Likewise, Mode III_s indicates a crushing of the wood fibers in the side member. The notation emphasizes the member in which the dowel bearing strength is exceeded, rather than the member which contains the plastic hinge.

In a double-shear connection, two plastic hinges may form (one near each shear plane) with crushing of the wood fibers in the side members. This is known as a Mode III_s mechanism. A Mode III_m mechanism does not occur in double-shear connections.

Mode IV: Two plastic hinges near each shear plane. The final mechanism covered by the yield model involves the formation of two plastic hinges at each shear plane. Mode IV yielding can occur in both single-shear and double-shear connections.

The yield limit model for fasteners thus covers four general modes of yielding. In a single-shear connection, Mode I and Mode III can occur in either the main member or side member, and these bring the total possible number of yield mechanisms to six. As noted, there is one yield equation for each yield mode. Because of symmetry there are only four yield mechanisms for a double-shear connection.

The NDS Commentary (Ref. 11.2) provides additional information on the historical development of the yield limit equations and a comprehensive summary of the supporting research.

EXAMPLE 11.9   Yield Modes

Dowel-type fasteners can yield in several modes. See Fig. 11.10. These modes are also sketched in *NDS Appendix I*.

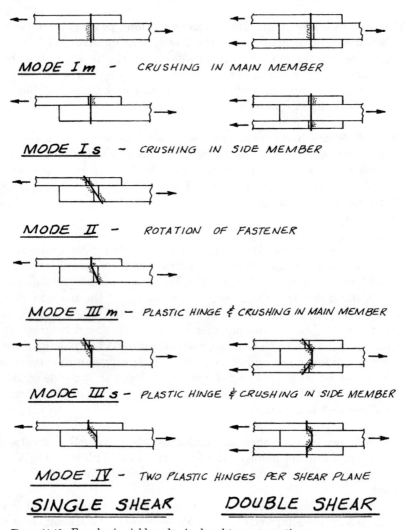

Figure 11.10 Four basic yield modes in dowel-type connections.

To fully cover the four basic yield modes, there are a total of six yield limit equations for single-shear connections. The two additional equations are required because Mode I and Mode III mechanisms can occur in either the main or side members. In double-shear connections, Mode II and Mode III$_m$ mechanisms do not occur, and three-member connections are fully covered by four yield limit equations.

EXAMPLE 11.10   **Yield Limit Equations**

Single-Shear Connections

Mode I_m (NDS equation 11.3-1):

$$Z = \frac{Dl_m F_{em}}{R_d}$$

Mode I_s (NDS equation 11.3-2):

$$Z = \frac{Dl_s F_{es}}{R_d}$$

Mode II (NDS equation 11.3-3):

$$Z = \frac{k_1 Dl_s F_{es}}{R_d}$$

Mode III_m (NDS equation 11.3-4):

$$Z = \frac{k_2 Dl_m F_{em}}{(1 + 2R_e)R_d}$$

Mode III_s (NDS equation 11.3-5):

$$Z = \frac{k_3 Dl_s F_{em}}{(2 + R_e)R_d}$$

Mode IV (NDS equation 11.3-6):

$$Z = \frac{D^2}{R_d} \sqrt{\frac{2F_{em}F_{yb}}{3(1 + R_e)}}$$

Double-Shear Connections

Model I_m (NDS equation 11.3-7):

$$Z = \frac{Dl_m F_{em}}{R_d}$$

Mode I_s (NDS equation 11.3-8):

$$Z = \frac{2Dl_s F_{es}}{R_d}$$

Mode III_s (NDS equation 11.3-9):

$$Z = \frac{2k_3 Dl_s F_{em}}{(2 + R_e)R_d}$$

Model IV (NDS equation 11.3-10):

$$Z = \frac{2D^2}{R_d} \sqrt{\frac{2F_{em}F_{yb}}{3(1 + R_e)}}$$

where Z = nominal design value for a single fastener connection (Z is taken as the smallest value from all applicable yield limit equations), lb

$$k_1 = \frac{\sqrt{R_e + 2R_e^2(1 + R_t + R_t^2) + R_t^2R_e^3} - R_e(1 + R_t)}{(1 + R_e)}$$

$$k_2 = -1 + \sqrt{2(1 + R_e) + \frac{2F_{yb}(1 + 2R_e)D^2}{3F_{em}l_m^2}}$$

$$k_3 = -1 + \sqrt{\frac{2(1 + R_e)}{R_e} + \frac{2F_{yb}(2 + R_e)D^2}{3F_{em}l_s^2}}$$

$$R_e = \frac{F_{em}}{F_{es}}$$

$$R_t = \frac{l_m}{l_s}$$

D = fastener diameter, in.
l_m = dowel bearing length in main member, in.
l_s = dowel bearing length in side member, in.
F_{em} = dowel bearing strength of main member, psi
F_{es} = dowel bearing strength of side member, psi
F_{yb} = bending yield strength of fastener, psi
R_d = coefficient specified in NDS Table 11.3.1B for reducing connection yield capacity to an allowable design value

As discussed in Chaps. 12 and 13, the reduction coefficient R_d ranges from 2.2 to 5.0 depending on the fastener diameter, yield mode, and angle between direction of load and direction of grain orientation in a wood member.

In practice, the designer will establish a trial configuration for a particular wood connection. This defines the geometry and strength parameters for use in determining the nominal design value Z for a single fastener. This nominal strength is the smallest load capacity obtained from all of the appropriate yield equations for the connection. The nominal strength is then subject to the other adjustments that account for the number of fasteners, duration of load, moisture content, and so on. The allowable fastener load with *all* adjustments taken into account is given the symbol Z'. Once the allowable load has been determined, a decision can be made about the trial joint configuration.

The basic concept of the yield model approach to wood connection design has been introduced in Chap. 11. With this background, the reader can explore the detailed design provisions for the common fasteners used in wood connections. In spite of the rather technical nature of the general yield theory, it should be recalled that the designer is not forced to obtain load capacities for all connections by evaluating the yield limit equations. The NDS contains a number of fastener capacity tables that allow the designer to read directly a fastener value for many common applications. The tables are based on the appropriate yield limit equations and a simplified set of criteria. The intent of the NDS is to provide the occasional user with an alternative to programming the equations or to solving the equations by hand. Although the tables provide appropriate design values for many connections, the tables do not cover all loading conditions, and use of the yield limit equations cannot be avoided entirely. It has been noted previously that the yield limit equations are easy to solve on a modern spreadsheet program, and a spreadsheet template may involve less effort than reading the NDS tables.

11.8 References

[11.1]   American Forest and Paper Association (AF&PA). 1991. *National Design Specification for Wood Construction and Supplement,* 1991 ed., AF&PA, Washington, DC.

[11.2]   American Forest and Paper Association (AF&PA). 1999. *Commentary on the National Design Specification for Wood Construction,* 1997 ed., AF&PA, Washington, DC.

[11.3]   American Forest and Paper Association (AF&PA). 2001. *National Design Specification for Wood Construction and Supplement.* 2001 ed., AF&PA, Washington, DC.

[11.4]   American Institute of Steel Construction (AISC). 1989. *Manual of Steel Construction,* 9th ed., AISC, Chicago, IL.

[11.5]   American Institute of Timber Construction (AITC). 1994. *Timber Construction Manual,* 4th ed., AITC, Englewood, CO.

[11.6]   American Iron and Steel Institute (AISI). 1996. *Cold-Formed Steel Design Manual.* 1996 ed., AISI, Washington, DC.

[11.7]   American Society of Civil Engineers (ASCE). 1996. *Mechanical Connections in Wood Structures,* Manual on Engineering Practice No. 84, ASCE, New York, NY.

[11.8]   Aune, P., and Patton-Mallory, M. 1986. "Lateral Load-Bearing Capacity of Nailed Joints Based on Yield Theory," Research Papers FPL 469 and 470, Forest Products Laboratory, Forest Service, U.S.D.A., Madison, WI.

[11.9]   Forest Products Laboratory (FPL). 1999. *Wood Handbook: Wood as an Engineering Material, General Technical Report 113,* FPL, Forest Service, U.S.D.A., Madison, WI.

[11.10] National Evaluation Service Inc. 1997. "Power-Driven Staples and Nails for Use in All Types of Building Construction," Report No. NER-272, International Codes Council (Available from ISANTA, Chicago, IL).

[11.11] Patton-Mallory, M. 1989. "Yield Theory of Bolted Connections Compared with Current U.S. Design Criteria," *Proceedings of the 2nd Pacific Timber Engineering Conference,* University of Auckland, New Zealand.

[11.12] Truss Plate Institute (TPI). 2002. *National Design Standard for Metal Plate Connected Wood Truss Construction,* ANSI/TPI 1-2002, TPI, Madison, WI.

11.9 Problems

11.1 Define the following terms and briefly describe how the terms relate to wood connections:

 a. Dowel-type fastener

 b. Plastic hinge

 c. Dowel bearing strength

 d. Yield mechanism

 e. Angle of load to grain

11.2 Sketch the yield mechanisms that may occur in a single-shear (two-member) connection that uses a dowel-type fastener. Along with the sketch, give a brief description of each mode.

11.3 Sketch the yield mechanisms that may occur in a double-shear (three-member) dowel-type connection. Give a brief description of each mode.

11.4 Describe the general concept behind the yield model theory of connection analysis.

11.5 Define the following symbols as used in the yield model for dowel-type fasteners: D, l_s, F_{em}, F_u, G_s, θ_s, F_{yb}.

11.6 Use the appropriate expressions from Example 11.4 to compute the dowel bearing strengths for the following bolt connections. Compare the results with the values in NDS Table 11.3.2.
a. ½-in.-diameter bolt in Hem-Fir with load applied parallel to grain
b. ½-in.-diameter bolt in Hem-Fir with load applied perpendicular to grain
c. ½-in.-diameter bolt in Hem-Fir with angle of load to grain of 40 degrees
d. 1-in.-diameter bolt in DF-L with load applied parallel to grain
e. 1-in.-diameter bolt in DF-L with load applied perpendicular to grain
f. 1-in.-diameter bolt in DF-L with angle of load to grain of 60 degrees

11.7 Use the appropriate expressions from Example 11.4 to compute the dowel bearing strengths for the following lag bolt (lag screw) connections. Compare the results with the values in NDS Table 11.3.2.
a. ⅝-in.-diameter lag bolt in Southern Pine with load applied parallel to grain
b. ⅝-in.-diameter lag bolt in Southern Pine with load applied perpendicular to grain
c. ⅝-in.-diameter lag bolt in Southern Pine with angle of load to grain of 55 degrees
d. ⅝-in.-diameter lag bolt in DF-L with load applied parallel to grain
e. ⅝-in.-diameter lag bolt in DF-L with load applied perpendicular to grain
f. ⅝-in.-diameter lag bolt in DF-L with angle of load to grain of 45 degrees

11.8 Use the appropriate expressions from Example 11.4 to compute the dowel bearing strengths for the following nail connections. Compare the results with the values in NDS Table 11.3.2.
a. 8d nail in Southern Pine with load applied parallel to grain
b. 8d nail in Southern Pine with load applied perpendicular to grain
c. 8d nail in Southern Pine with angle of load to grain of 30 degrees
d. 10d nail in DF-L with load applied parallel to grain
e. 10d nail in DF-L with load applied perpendicular to grain
f. 10d nail in DF-L with angle of load to grain of 45 degrees

Nailed and Stapled Connections

12.1 Introduction

The general use of nails in the construction of horizontal diaphragms and shearwalls was addressed in previous chapters. In addition, the two basic forms of nailed connections in structural applications (*shear* and *withdrawal*) were described in the introductory chapter on wood connections. It is recommended that the reader review the joint configurations, nomenclature, and yield modes for these connections in Chap. 11 before proceeding with the detailed design provisions given here.

The usual practice in the United States is to design nailed connections when the loads to be transmitted are relatively small, and to use other types of fasteners (e.g., bolts) for joints with larger loads. This is the general approach to connection design covered in this book. However, this contrasts with recent practice of some designers in Europe and New Zealand to also use nails in connections subjected to large forces. Obviously connections with heavy loads require the use of a large number of nails, and special fabrication practices are necessary. The preference for these types of connections is based on the expected ductile behavior of nailed joints and a corresponding concern about the possible lack of ductility in certain bolted connections, especially those involving large-diameter bolts.

The basic concept of a nail attaching one member to another member has undergone little change, but many developments have occurred in the configuration of nails and in the methods of installation. For example, Ref. 12.14 lists over 235 definitions and descriptions relating to the use of nails. The majority of these nails are fasteners for special applications, and the structural designer deals with a relatively limited number of nail types. Power-driving equipment has contributed substantially to the advances in this basic type of wood connection.

The different types of nails are distinguished by the following characteristics:

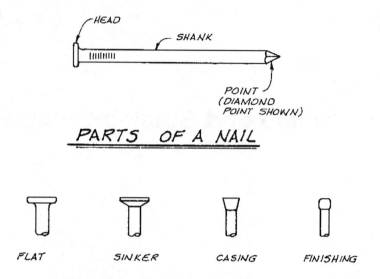

Figure 12.1 Definitions of nail terms.

1. Nail head

2. Shank

3. Nail point

4. Material type

5. Surface condition

The first three items have to do with the configuration of the nail. See Fig. 12.1, in which four typical nail heads are shown along with the widely used diamond-shaped nail point. Many other types of nails are described in Refs. 12.7, 12.9, and 12.14.

This chapter introduces the main types of nails used in structural applications. The yield limit equations for laterally loaded nails are summarized, and the modification factors for determining allowable fastener design values are reviewed. Design procedures for the various types of nailed connections are outlined.

12.2 Types of Nails

The NDS (Ref. 12.3) provides structural design guidelines for the following basic types of nails:

1. Common nails and spikes

2. Box nails

3. Sinker nails

4. Threaded hardened-steel nails

The sizes of common, box, and sinker nails are specified by the pennyweight of the nail, abbreviated d. The pennyweight for different types of nails specifies the length, shank diameter, and head size of the nail. The sizes of the nails listed above are given in NDS Appendix L.

Common, box, and sinker nails are all fabricated from low-carbon steel wire with a plain shank and a diamond point. Common nails and box nails have a flat head, while sinker nails have a countersunk head as shown in Fig. 12.1. For a given pennyweight common nails and box nails have the same length, while sinker nails are ⅛ in. or ¼ in. shorter. All three types of nails have different diameters for a given pennyweight. Threaded nails are typically fabricated from high-carbon steel wire and have an annularly or helically threaded shank, a flat head, and a diamond point (some low-carbon threaded steel nails have sinker heads). See Fig. 12.2.

Common nails represent the basic structural nail. Design values for common wire nails are for a nail surface condition that is described as *bright*. This term is used to describe nails with a natural bare-metal finish, and design values for this surface condition can be considered conservative for other surface conditions. For a given pennyweight, the diameter of common wire nails is larger than the diameter of box nails, and allowable design values are correspondingly higher. Because of their larger diameter, common nails have less tendency to bend when driven manually.

Box nails and *sinker nails* are also widely used. They are available in several different surface conditions including bright, galvanized (zinc-coated), and cement-coated. Design values for nails in the bright condition may conservatively be used for other surface conditions. Coated nails are discussed below. The wire diameter of box nails and sinker nails is smaller than that for common nails, and the allowable design values are correspondingly smaller. Box nails and sinker nails also have a greater tendency to bend during driving.

Common spikes are similar in form to common wire nails except that they have a larger diameter. Because of their larger diameter, spikes have higher allowable design values, but the spacing requirements between fasteners are greater in order to avoid splitting of the wood. Predrilling of holes may also be necessary to avoid splitting.

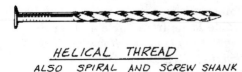

HELICAL THREAD
ALSO SPIRAL AND SCREW SHANK

ANNULAR THREAD
ALSO RING SHANK

Figure 12.2 Threaded hardened-steel nails.

Threaded hardened-steel nails are often made from high-carbon steel and are heat-treated and tempered to provide greater strength. These nails are especially useful when the lumber moisture content will vary during the service life of the structure. The threaded shank provides greater withdrawal capacity in these circumstances. In addition, the higher yield strength values associated with high-carbon steel provide improved lateral (shear) resistance.

Design values for threaded hardened-steel nails are based on the wire diameter, except that for annularly threaded nails with threads on the shear plane, design values should be based on the root diameter of the nail.

The structural designer may encounter some types of *nails not covered in the NDS*. These may include

1. Finishing nails

2. Casing nails

3. Cooler nails

4. Zinc-coated nails

5. Cement-coated nails

Finishing nails are slender, bright, low-carbon steel nails with a finishing head and a diamond point. These nails are used for finish or trim work, and they may be used for the installation of nonstructural paneling. However, because of the small nail head, finishing nails are not allowed for the attachment of structural sheathing or siding (Sec. 10.3).

Casing nails are slender, low-carbon steel nails with a casing head and a diamond point. These are available with a zinc coating (galvanizing) to reduce staining of wood exposed to the weather. Galvanized casing nails can be used for the structural attachment of plywood siding when the appearance of a flat nail head is considered objectionable. Because of the smaller diameter of the nail head, allowable plywood shearwall values for casing nails are smaller than for common or galvanized box nails (Sec. 10.3).

Cooler nails are slender, low-carbon steel nails with a flat head and diamond point, and are usually cement-coated. The head diameter is the same as or smaller than that of a common wire nail of the same length. When gypsum wallboard (drywall) of the required thickness is properly installed with the Code-required size and number of cooler nails, a wall may be designed as a shear panel using the allowable loads from Table 4.3B of the *Special Design Provisions for Wind and Seismic Supplement, Allowable Stress Design Manual for Engineered Wood Construction* (Ref. 12.1) (Sec. 10.4).

It was noted earlier that nails can be obtained with a number of different surface conditions. The bright condition was previously described, and several others are introduced in the remainder of this section.

Zinc-coated (galvanized) nails are intended primarily for use where corrosion and staining resistance are important factors in performance and appearance. In addition, resistance to withdrawal may be increased if the zinc coating is evenly distributed. However, extreme irregularities of the coating may actually reduce withdrawal resistance. It should also be pointed out that the use of galvanized nails to avoid staining is not fully effective. Where staining should be completely avoided for architectural purposes, stainless steel nails can be used. The expense of stainless steel nails is much greater than that of galvanized nails.

Cement-coated nails are coated by tumbling or submerging in a resin or shellac (cement is not used). Resistance to withdrawal is increased because of the larger friction between the nail and the wood. However, there are substantial variations in the uniformity of cement coatings, and much of the coating may be removed in the driving of the nail. These variables cause large differences in the relative resistance to withdrawal for cement-coated nails. The increased resistance to withdrawal for these types of nails is, in most cases, only temporary.

12.3   Power-Driven Nails and Staples

Major advances in the installation of driven fasteners such as nails and staples have been brought about by the development of pneumatic, electric, and mechanical guns. The use of such power equipment greatly increases the speed with which fasteners can be installed. The installation of a nail or other fastener has been reduced to the pull of a trigger and can be accomplished in a fraction of a second. See Fig. 12.3.

Conventional round flathead nails can be installed with power equipment. To accomplish this, the nails are assembled in clips or coils that are fed automatically into the driving gun. The round head on these types of fasteners prevents the shanks of the nails from coming in close contact with adjacent

Figure 12.3 Typical power fastening equipment. On the left, nailing through bottom wall plate with Senco SN325+ framing nailer. On the right, stapling plywood sheathing with Senco M series stapler. (*Courtesy of Senco Products, Inc.*)

nails, and therefore, a relatively large clip is required for a given number of nails.

The size of the clip can be reduced by modifying the shape of the conventional nail. A number of different configurations have been used to make more compact fastener clips. *Wire staples, T nails,* and *modified roundhead nails* are all in current use. See Fig. 12.4. Fasteners such as staples are often made from relatively thin wire. The tendency of this type of fastener to split the wood is reduced. In addition, standard power-driven nails are often thinner and shorter than corresponding pennyweight sizes of hand-driven nails.

Testing programs have resulted in Code recognition of a number of power-driven fasteners for use in structural design. For example, the International Staple, Nail and Tool Association has structural design values published in NER 272 (Ref. 12.13). National Evaluation Reports (NERs) are published jointly by the three model building code agencies under the name of the National Evaluation Service. Recognition for other power-driven fasteners has also been obtained.

The literature giving design information is extensive and is not reproduced in this text. The remainder of this chapter deals specifically with the basic types of nails covered in the NDS. The designer, however, should be able to locate *equivalent* power-driven fasteners in the references.

It should be pointed out that the setting of power-driving guns is an important factor in the performance of a fastener. Care should be taken that the driving gun does not cause the head of a fastener to penetrate below the surface of the wood. Fasteners improperly installed will have a substantially lower load-carrying capacity. Precautions should also be taken in the manual driving of nails, and the head of a nail should not be "set" deeply below the

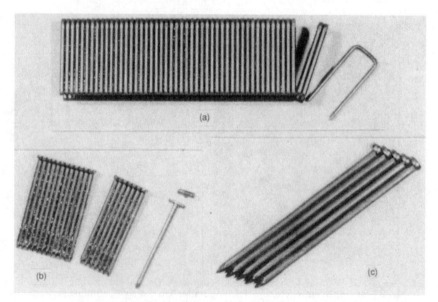

Figure 12.4 Typical power-driven fasteners. (*a*) Wire staples; (*b*) T nails; (*c*) modified roundhead nails. (*Photograph by Mike Hausmann.*)

surface of the wood. Because of the rapidity and consistency obtained with power-driving equipment, the installation setting is of particular importance with power-driven fasteners. The type of member will affect the setting of the gun. For example, when plywood roof sheathing is installed on a panelized roof, the gun should be adjusted for nailing into the different supporting members (2×'s, 4×'s, and glulams).

12.4   Yield Limit Equations for Nails

A laterally loaded nail connection is usually a *single-shear* connection between a wood or steel *side member* and a wood *main member.* See Example 12.1. Recall from Chap. 11 that the dowel bearing strength F_e for a given species of wood is constant for a small-diameter fastener regardless of the angle of load to grain. Therefore, the angle of load to grain θ is not a concern in the design of nailed connections. Furthermore, if the wood members in a connection are *both from the same species,* the dowel bearing strengths are the same ($F_{em} = F_{es}$). The load tables for wood-to-wood connections in the NDS are limited to this simplified condition.

The dowel bearing strength for a nail connection is a function of the specific gravity of the wood (Example 11.4 in Sec. 11.5). Different values of F_e occur when the main member and side member are from different species. Connections involving mixed species are easily accommodated in the yield limit equations by using the corresponding dowel bearing strengths for the main and side members.

Mixing different species of wood in a nailed connection is a common occurrence. For example, plywood manufactured from Douglas Fir (specific gravity of 0.50) may be nailed to Hem-Fir sawn-lumber framing (specific gravity of 0.43). The respective dowel bearing strengths in this connection are $F_{es} = 4650$ psi for the plywood side member and $F_{em} = 3500$ psi for the framing or main member (see Example 12.5 later in this chapter).

If a nailed connection uses a metal side plate, the dowel bearing strengths for the main member and side member are also different. The bearing strength of a cold-formed steel member may conservatively be taken as 1.375 times the ultimate tensile strength of the steel side plate $F_e = 1.375F_u$.

EXAMPLE 12.1   Single-Shear Nail Connections

The most common nailed connections are single-shear (two-member) joints with a wood main member and either a wood or a cold-formed steel side member. The nail passes through the *side member* and into the main member. The *main member* or *holding member* receives the pointed end of the fastener. The *penetration* is the length of the nail in the main member. The penetration is computed as the length of the nail minus the thickness of the side member. If the nail extends beyond the main member, the penetration is taken as the thickness of the main member:

$$\text{Penetration } p = l - t_s \leq t_m$$

The *dowel bearing length* of the nail in the main member is used in yield limit equations to determine the nominal design value Z for a nail connection. According to NDS 11.3.4 the dowel bearing length in the main member l_m is equivalent to the nail penetration when the nail penetration is at least 10 times the shank diameter of the nail, $p \geq 10D$. Alternatively, the dowel bearing length in the main member is taken as the penetration minus the length of the tapered tip of the nail when $6D \leq p < 10D$. According to the AF&PA *Technical Report 12—General Dowel Equations for Calculating Lateral Connection Values*, the length of tapered tip for nails is typically in the range of $1.3D$ to $2D$ (see Ref. 12.4). Thus the dowel bearing length in the main member can be determined by:

$$l_m = \begin{cases} p \leq t_m & \text{when } p \geq 10D \\ p - 2D \leq t_m & \text{when } 6D \leq p < 10D \end{cases}$$

$$\therefore l_m = \begin{cases} l - t_s \leq t_m & \text{when } p \geq 10D \\ l - t_s - 2D \leq t_m & \text{when } 6D \leq p < 10D \end{cases}$$

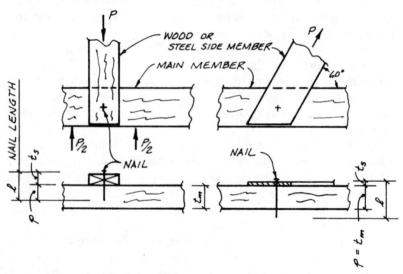

Figure 12.5a Wood-to-wood and wood-to-metal connections.

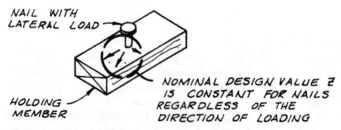

Figure 12.5b Angle of load to grain does not affect nominal strength of laterally loaded nail.

The *penetration* of a nail is defined as the distance that the nail extends into the main member. In order to use the nominal design value Z of a nail according to the yield limit equations, the penetration must be a minimum of 6 times the shank diameter of the nail, $p \geq 6D$. The *dowel bearing length* of a nail in the main member is defined as the portion of the nail penetration that is used in yield limit equations to determine the nominal design value Z for a nail connection. The dowel bearing length l_m is defined in Example 12.1 for different ranges of nail penetration in the main member.

There are a number of possible nail connections in addition to those shown in Fig. 12.5a. For example, nails can be used in three-member connections, and the NDS provides the appropriate method for obtaining the nominal design value Z for joints in double shear. Another type of connection occurs between a plywood panel and light-gage metal studs. In this case the wood panel becomes the side member, and the sheet-metal stud is the main member because it receives the point of the fastener. Fasteners for this application may be nails, pins, or screws. These and other types of connections can be analyzed using yield limit theory. The reader should contact APA—The Engineered Wood Association for information on this latter type of connection.

The purpose of this book is to provide a general introduction to nailed connections and the yield limit model, and it is not feasible to address all possible types of nailed connections. Only the basic *single-shear* connections in Fig. 12.5a are covered in this book. Six possible yield modes are to be considered, and the yield limit equations from the NDS for these modes are given in Example 12.2.

As an alternative to using the yield limit equations, nominal design values for nails for a number of common applications are available in the NDS tables. These tables cover single-shear connections for three basic types of nails: common, box, and sinker nails. The fastener capacity tables are based on the yield limit equations with nail penetration of $p = 10D$ in the main member, and are thus directly applicable for connections with $p \geq 10D$. For nail connections with $6D \leq p < 10D$ a footnote to the NDS tables provides a conservative reduction factor of $p/10D$ to account for the reduced nail penetration in the main member. The reduction factor may be used to adjust tabulated design values, in lieu of calculating a reduced connection capacity directly from yield limit equations. The NDS tables cover *wood-to-wood* connections (for the case that both members are of the same species), *wood-to-metal* connections, and *wood-to-panel* connections. See NDS Tables 11N to 11R.

The design value for many nailed connection problems is covered by the NDS tables. However, once a program is developed or a spreadsheet template is created, it will often be more convenient to obtain nail design values from the computer. In addition, some problems will not be covered by the NDS tables, and application of the yield limit equations will be *required*. The examples in Sec. 12.5 illustrate the step-by-step solution of the yield equations.

EXAMPLE 12.2   Yield Limit Equations for Nail in Single Shear

The basic strength of a single dowel-type fastener subjected to a lateral load is known as the *nominal design value* and is given the symbol Z. The nominal design value is taken as the smallest load capacity obtained by evaluating all of the yield limit equations for a given type of connection. The *allowable design value* Z' for a single fastener may then be obtained by multiplying the nominal design value by the appropriate adjustment factors described in Sec. 12.6.

Six yield modes apply to a single-shear nail connection, as given in NDS Tables 11.3.1A and 11.3.1B. The yield modes are sketched in Fig. 11.10 (Sec. 11.7).

Mode I_m (NDS equation 11.3-1):

$$Z = \frac{Dl_m F_{em}}{K_D}$$

Mode I_s (NDS equation 11.3-2):

$$Z = \frac{Dl_s F_{es}}{K_D}$$

Mode II (NDS equation 11.3-3):

$$Z = \frac{k_1 Dl_s F_{es}}{K_D}$$

Mode III_m (NDS equation 11.3-4):

$$Z = \frac{k_2 Dl_m F_{em}}{(1 + 2R_e)K_D}$$

Mode III_s (NDS equation 11.3-5):

$$Z = \frac{k_3 Dl_s F_{em}}{(2 + R_e)K_D}$$

Mode IV (NDS equation 11.3-6):

$$Z = \frac{D^2}{K_D}\sqrt{\frac{2F_{em}F_{yb}}{3(1 + R_e)}}$$

where Z = nominal design value for nail (Z is to be taken as smallest value from six yield limit equations), lb

$$k_1 = \frac{\sqrt{R_e + 2R_e^2(1 + R_t + R_t^2) + R_t^2 R_e^3} - R_e(1 + R_t)}{(1 + R_e)}$$

$$k_2 = -1 + \sqrt{2(1 + R_e) + \frac{2F_{yb}(1 + 2R_e)D^2}{3F_{em}l_m^2}}$$

$$k_3 = -1 + \sqrt{\frac{2(1 + R_e)}{R_e} + \frac{2F_{yb}(2 + R_e)D^2}{3F_{em}l_s^2}}$$

$$R_e = \frac{F_{em}}{F_{es}}$$

$$p = l - t_s \le t_m$$

$\qquad$ = penetration of nail

l_m = dowel bearing length of nail in main member (member holding point), in.

t_s = thickness of side member, in.

t_m = thickness of main member, in.

l_s = dowel bearing length of nail in side member, in.

l = length of nail, in.

F_{es} = dowel bearing strength of side member, psi

F_{em} = dowel bearing strength of main (holding) member, psi

F_{yb} = bending yield strength of fastener, psi

D = diameter of nail or spike, in.

K_D = reduction coefficient for fasteners with $D < 0.25$ in.

$$= \begin{cases} 2.2 & \text{for } D \le 0.17 \text{ in.} \\ 10D + 0.5 & \text{for } 0.17 \text{ in.} < D < 0.25 \text{ in.} \end{cases}$$

Values of F_e may be obtained from NDS Tables 11.3.2 or 11.3.2B, or they may be computed using the equation given in Sec. 11.5 of this book.

The yield limit equations in the NDS can be used to obtain allowable design values for any fastener with a known bending yield strength F_{yb}. It will be recalled that the bending yield strength of a dowel-type fastener is one of the major factors relating to the formation of a plastic hinge in a connection (Sec. 11.6). The term *bending yield strength* indicates that the yield stress is obtained from a bending test of the nail rather than a tension test. See Fig. 12.6. The bending yield stress is obtained from an *offset* taken parallel to the initial

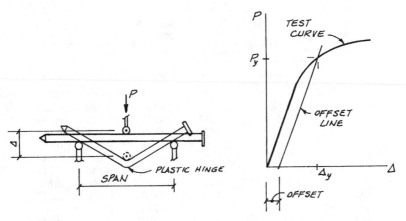

Figure 12.6 Test used to determine bending yield strength of a nail. An offset to the load deflection curve is used to obtain P_y and Δ_y from which the bending yield strength F_{yb} is computed.

slope of the load deflection curve. The nail is loaded as a simply supported beam with a concentrated load at the midspan (Ref. 12.8).

To provide guidance to the designer regarding the strength of nails generally available in the marketplace, a limited survey of box and common nails was conducted (NDS Appendix I). The survey found that bending yield strength is higher for smaller-diameter nails. The apparent reason for this increase in F_{yb} is the additional work-hardening of the material that takes place in the manufacturing of small-diameter nails.

Based on the results of the nail survey, the following formula for the bending yield strength of low-carbon steel nails and spikes (Ref. 12.12) was developed:

$$F_{yb} = 130.4 - 213.9D$$

According to NDS Appendix I, the bending yield strength is assumed to be 30 percent higher for threaded hardened-steel nails:

$$F_{yb} = 1.3(130.4 - 213.9D)$$

In these equations, D is the diameter of the fastener (in.), and F_{yb} is the bending yield strength (ksi). Note that F_{yb} in other equations is in psi. The 30 percent increase for threaded hardened-steel nails is a soft conversion based on previous NDS design values. The diameters and lengths of common, box, and sinker nails are summarized for comparison in Fig. 12.7a. These dimensions are also available in NDS Appendix L. The NDS fastener capacity tables for nails are based on the above formulas, but values of F_{yb} in NDS Tables 11N to 11R were conservatively *truncated*. Formula values and truncated values of F_{yb} are compared in Fig. 12.7b. The truncated values of F_{yb} are listed in the footnotes to the NDS tables.

When connection design values are obtained from yield limit equations or from NDS tables, the design process is referred to as *engineered construction*. It has not been common practice for manufacturers to identify the bending yield strength of fasteners used in wood construction. However, the primary

Sizes of Common, Box, and Sinker Nails

	Length in.		Wire diameter, in.		
Pennyweight	Common and Box nails	Sinker nails	Common nails	Box nails	Sinker nails
6d	2	1.875	0.113	0.099	0.092
7d	2.25	2.125	0.113	0.099	0.099
8d	2.5	2.375	0.131	0.113	0.113
10d	3	2.875	0.148	0.128	0.120
12d	3.25	3.125	0.148	0.128	0.135
16d	3.5	3.25	0.162	0.135	0.148
20d	4	3.75	0.192	0.148	0.177
30d	4.5	4.25	0.207	0.148	0.192
40d	5	4.75	0.225	0.162	0.207
50d	5.5	—	0.244	—	—
60d	6	5.75	0.263	—	0.244

Figure 12.7a Table of nail sizes.

Common, Box, and Sinker Nails

Nail diameter, in.	F_{yb} from formula, ksi	Truncated F_{yb} used in NDS tables, ksi
0.099	109	100
0.113	106	100
0.120	105	100
0.128	103	100
0.131	102	100
0.135	102	100
0.148	99	90
0.162	96	90
0.177	93	90
0.192	89	80
0.207	86	80
0.225	82	80
0.244	78	70
0.263	74	70

Figure 12.7b Bending yield strengths for nails used to develop nominal design values Z in NDS Tables 11N to 11R. Values of F_{yb} may be verified in the footnotes to NDS tables.

specification for nails and spikes in the United States (Ref. 12.9) now includes supplementary provisions that enable purchasers to specify *engineered construction nails* with minimum bending yield strengths. Since specification of bending yield strength for fasteners is a relatively new practice, it is recommended that designers either specify *engineered construction nails* or identify the minimum F_{yb} for each type of nail used in the design.

Several numerical examples of nail connections using the yield limit equations are given in the next section.

12.5 Applications of Yield Limit Equations

Example 12.3 evaluates the nominal design value of a 16d common nail in a connection between two sawn lumber members. The smallest value from the six yield equations is obtained from NDS equation 11.3-6, which indicates that a Mode IV yield governs. This type of yield mode has two plastic hinges in the fastener. See Fig. 11.10 (Sec. 11.7) for a sketch of the yield mode.

Because both members are the same species, the design value of the connection can be verified using NDS Table 11N.

EXAMPLE 12.3   Yield Limit Equations for Single-Shear Wood-to-Wood Nail Connection

Determine the nominal design value for the nail in Fig. 12.8. Both pieces of lumber are DF-L, and the nail is a 16d common nail. The angle of load to grain does not affect the problem and is not specified.

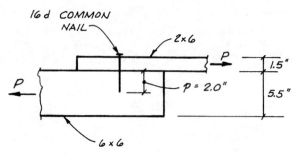

Figure 12.8 Single-shear wood-to-wood connection.

Dimensions of nail are obtained from the table in Fig. 12.7a or from NDS Appendix L:

$$D = 0.162 \text{ in.}$$

$$l = 3.5 \text{ in.}$$

Penetration

$$l_s = t_s = 1.5 \text{ in.}$$

$$p = l - t_s \le t_m$$

$$= 3.5 - 1.5 = 2.0 \text{ in.} < 5.5 \text{ in.}$$

$$10D = 10(0.162) = 1.62 \text{ in.} < 2.0 \text{ in.}$$

$$\therefore l_m = p = l - t_s = 2.0 \text{ in.}$$

The bending yield strength of a 16d common nail is assumed to be

$$F_{yb} = 90 \text{ ksi} = 90,000 \text{ psi}$$

This value can be obtained from the table in Fig. 12.7b or from NDS Table 11N. Specific gravity is the same for both main and side members and is obtained from NDS Table 11.3.2A:

$$G_m = G_s = 0.50$$

Dowel bearing strength can be computed or read from NDS Table 11.3.2:

$$F_{em} = F_{es} = 4650 \text{ psi}$$

Coefficients for use in yield equations:

$$K_D = 2.2$$

$$R_t = \frac{l_m}{l_s} = \frac{2}{1.5} = 1.33$$

$$R_e = \frac{F_{em}}{F_{es}} = \frac{4650}{4650} = 1.0$$

$$1 + R_e = 1 + 1.0 = 2.0$$

$$1 + 2R_e = 1 + 2(1.0) = 3.0$$

$$2 + R_e = 2 + 1.0 = 3.0$$

$$k_1 = \frac{\sqrt{R_e + 2R_e^2(1 + R_t + R_t^2) + R_t^2 R_e^3} - R_e(1 + R_t)}{(1 + R_e)}$$

$$= \frac{\sqrt{1.0 + 2(1.0^2)(1 + 1.33 + 1.33^2) + (1.33^2)(1.0^3)} - 1.0(1 + 1.33)}{2.0}$$

$$= 0.4916$$

$$k_2 = -1 + \sqrt{2(1 + R_e) + \frac{2F_{yb}(1 + 2R_e)D^2}{3F_{em}l_m^2}}$$

$$= -1 + \sqrt{2(2.0) + \frac{2(90,000)(3.0)(0.162^2)}{3(4650)(2.0^2)}}$$

$$= 1.0625$$

$$k_3 = -1 + \sqrt{\frac{2(1 + R_e)}{R_e} + \frac{2F_{yb}(2 + R_e)D^2}{3F_{em}l_s^2}}$$

$$= -1 + \sqrt{\frac{2(2.0)}{1.0} + \frac{2(90,000)(3.0)(0.162^2)}{3(4650)(1.5^2)}}$$

$$= 1.1099$$

Yield limit equations:

Mode I_m (NDS equation 11.3-1):

$$Z = \frac{Dl_m F_{em}}{K_D} = \frac{0.162(2.0)(4650)}{2.2} = 685 \text{ lb}$$

Mode I_s (NDS equation 11.3-2):

$$Z = \frac{Dl_sF_{es}}{K_D} = \frac{0.162(1.5)(4650)}{2.2} = 514 \text{ lb}$$

Mode II (NDS equation 11.3-3):

$$Z = \frac{k_1Dl_sF_{es}}{K_D} = \frac{0.4916(0.162)(1.5)(4650)}{2.2} = 253 \text{ lb}$$

Mode III_m (NDS equation 11.3-4):

$$Z = \frac{k_2Dl_mF_{em}}{(1 + 2R_e)K_D} = \frac{1.0625(0.162)(2.0)(4650)}{3.0(2.2)} = 243 \text{ lb}$$

Mode III_s (NDS equation 11.3-5):

$$Z = \frac{k_3Dl_sF_{em}}{(2 + R_e)K_D} = \frac{1.1099(0.162)(1.5)(4650)}{3.0(2.2)} = 190 \text{ lb}$$

Mode IV (NDS equation 11.3-6):

$$Z = \frac{D^2}{K_D} \sqrt{\frac{2F_{em}F_{yb}}{3(1 + R_e)}}$$

$$= \frac{(0.162)^2}{2.2} \sqrt{\frac{2(4650)(90,000)}{3(2.0)}} = 141 \text{ lb}$$

Nominal design value is selected as the smallest value from the yield equations:

$$\boxed{Z = 141 \text{ lb}}$$

This agrees with the value listed in NDS Table 11N.

The second example of the yield limit equations uses an 8d box nail to connect a metal side plate to a sawn lumber main member of Southern Pine. See Example 12.4. The dowel bearing strength for a cold-formed steel plate may conservatively be taken as 1.375 times the ultimate tensile strength of the steel. In this problem a Mode III_s yield mechanism controls which indicates that a single plastic hinge forms in the nail.

EXAMPLE 12.4   Yield Limit Equations for Single-Shear Wood-to-Metal Nail Connection

Determine the nominal design value for the nail in Fig. 12.9. The side member is a 12-gage plate of ASTM 653 grade 33 steel, and the main member is a 4 × 4 of Southern

Pine. The fastener is an 8d box nail. The angle of load to grain does not affect nailed connections and is not given.

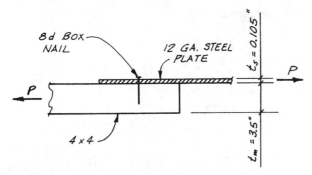

Figure 12.9 Single-shear wood-to-metal connection.

The dimensions of the nail are obtained from the table in Fig. 12.7a or NDS Appendix L:

$$D = 0.113 \text{ in.}$$

$$l = 2.5 \text{ in.}$$

Penetration:

$l_s = t_s = 0.105$ in. (12-gage plate thickness from Ref. 12.5 or NDS Table 11N)

$$p = l - t_s \leq t_m$$

$$= 2.5 - 0.105 = 2.40 \text{ in.} < 3.5 \text{ in.}$$

$$10D = 10(0.113) = 1.13 \text{ in.} < 2.40 \text{ in.}$$

$$\therefore l_m = p = l - t_s = 2.40 \text{ in.}$$

Bending yield strength of an 8d box nail is assumed to be

$$F_{yb} = 100,000 \text{ psi}$$

Specific gravity for Southern Pine is obtained from NDS Table 11.3.2A:

$$G_m = 0.55$$

Dowel bearing strength for Southern Pine can be computed or read from NDS Table 11.3.2:

$$F_{em} = 5550 \text{ psi}$$

Dowel bearing strength of steel side member:

$$F_{es} = 1.375F_u = 61{,}850 \text{ psi}$$

Coefficients for yield equations:

$$K_D = 2.2$$

$$R_t = \frac{l_m}{l_s} = \frac{2.4}{0.105} = 22.86$$

$$R_e = \frac{F_{em}}{F_{es}} = \frac{5550}{61{,}850} = 0.0897$$

$$1 + R_e = 1 + 0.0897 = 1.0897$$

$$1 + 2R_e = 1 + 2(0.0897) = 1.1795$$

$$2 + R_e = 2 + 0.0897 = 2.0897$$

$$k_1 = \frac{\sqrt{R_e + 2R_e^2(1 + R_t + R_t^2) + R_t^2 R_e^3} - R_e(1 + R_t)}{(1 + R_e)}$$

$$= \frac{\sqrt{0.0897 + 2(0.0897^2)(1 + 22.86 + 22.86^2) + (22.86^2)(0.0897^3)} - 0.0897(1 + 22.86)}{1.0897}$$

$$= 0.8287$$

$$k_2 = -1 + \sqrt{2(1 + R_e) + \frac{2F_{yb}(1 + 2R_e)D^2}{3F_{em}l_m^2}}$$

$$= -1 + \sqrt{2(1.0897) + \frac{2(100{,}000)(1.1795)(0.113^2)}{3(5550)(2.40^2)}}$$

$$= 0.4869$$

$$k_3 = -1 + \sqrt{\frac{2(1 + R_e)}{R_e} + \frac{2F_{yb}(2 + R_e)D^2}{3F_{em}l_s^2}}$$

$$= -1 + \sqrt{\frac{2(1.0897)}{0.0897} + \frac{2(100{,}000)(2.0897)(0.113^2)}{3(5550)(0.105^2)}}$$

$$= 6.3049$$

Yield limit equations:

Mode I_m (NDS equation 11.3-1):

$$Z = \frac{Dl_m F_{em}}{K_D} = \frac{0.113(2.4)(5550)}{2.2} = 684 \text{ lb}$$

Mode I_s (NDS equation 11.3-2):

$$Z = \frac{Dl_s F_{es}}{K_D} = \frac{0.113(0.105)(61,850)}{2.2} = 334 \text{ lb}$$

Mode II (NDS equation 11.3-3):

$$Z = \frac{k_1 Dl_s F_{es}}{K_D} = \frac{0.8287(0.113)(0.105)(61,850)}{2.2} = 276 \text{ lb}$$

Mode III_m (NDS equation 11.3-4):

$$Z = \frac{k_2 Dl_m F_{em}}{(1 + 2R_e)K_D} = \frac{0.4869(0.113)(2.40)(5550)}{1.1795(2.2)} = 282 \text{ lb}$$

Mode III_s (NDS equation 11.3-5):

$$Z = \frac{k_3 Dl_s F_{em}}{(2 + R_e)K_D} = \frac{6.3049(0.113)(0.105)(5550)}{2.0897(2.2)} = 90 \text{ lb}$$

Mode IV (NDS equation 11.3-6):

$$Z = \frac{D^2}{K_D} \sqrt{\frac{2F_{em}F_{yb}}{3(1 + R_e)}}$$

$$= \frac{(0.113)^2}{2.2} \sqrt{\frac{2(5550)(100,000)}{3(1.0897)}} = 107 \text{ lb}$$

Nominal design value is selected as the smallest value from the yield equations:

$$\boxed{Z = 90 \text{ lb}}$$

This agrees with the value listed in NDS Table 11P.

The first two nail examples were given to clarify the use of the yield formulas and to verify the NDS tables. The concept behind the tables is to cover many of the common nail applications. However, the yield limit equations allow the designer to evaluate a much wider variety of connections. The final nail problem in this section requires use of the yield equations because the specific conditions are not addressed by the tables.

The 10d common nail in Example 12.5 is driven through a ½-in. thickness of Douglas Fir plywood into a sawn lumber framing member of Hem-Fir. Because different species are involved, there are different dowel bearing strengths for the main and the side members. The yield mechanism is Mode III_s because NDS equation 11.3-5 produces the smallest nominal design value.

EXAMPLE 12.5   Yield Limit Equations for Single-Shear Plywood-to-Lumber Nail Connection

Determine the nominal design value for the nail in Fig. 12.10. The side member is ½-in. STR I DF plywood panel, and the main member is a 2 × 6 piece of Hem-Fir. The fastener is a 10d common nail. The angle of load to grain does not affect nailed connections and is not given.

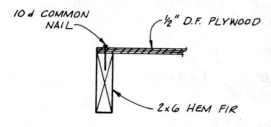

Figure   12.10 Single-shear   plywood-to-lumber connection.

The dimensions of the nail are obtained from the table in Fig. 12.7a or NDS Appendix L:

$$D = 0.148 \text{ in.}$$

$$l = 3.0 \text{ in.}$$

Penetration: The nail is driven into the edge of the 2 × 6, and $t_m = 5.5$.

$$l_s = t_s = 0.5 \text{ in.}$$

$$p = l - t_s \le t_m$$

$$= 3.0 - 0.5 = 2.5 \text{ in.} < 5.5 \text{ in.}$$

$$10D = 10\,(0.148) = 1.48 \text{ in.} < 2.5 \text{ in.}$$

$$\therefore l_m = p = l - t_s = 2.5 \text{ in.}$$

Bending yield strength of a 10d common nail is assumed to be

$$F_{yb} = 90{,}000 \text{ psi}$$

Specific gravity for Hem-Fir is obtained from NDS Table 11.3.2A:

$$G_m = 0.43$$

Dowel bearing strength for Hem-Fir can be computed or read from NDS Table 11.3.2:

$$F_{em} = 3500 \text{ psi}$$

Because plywood is STR I, the panel contains veneer from all Group I species, and the specific gravity for DF can be used to determine dowel bearing strength. Values are obtained from NDS Table 11.3.2B:

$$G_s = 0.50$$

$$F_{es} = 4650 \text{ psi}$$

Coefficients for yield equations:

$$K_D = 2.2$$

$$R_t = \frac{l_m}{l_s} = \frac{2.5}{0.5} = 5.0$$

$$R_e = \frac{F_{em}}{F_{es}} = \frac{3500}{4650} = 0.7527$$

$$1 + R_e = 1 + 0.7527 = 1.7527$$

$$1 + 2R_e = 1 + 2(0.7527) = 2.5054$$

$$2 + R_e = 2 + 0.7527 = 2.7527$$

$$k_1 = \frac{\sqrt{R_e + 2R_e^2(1 + R_t + R_t^2) + R_t^2 R_e^3} - R_e(1 + R_t)}{(1 + R_e)}$$

$$= \frac{\sqrt{0.7527 + 2(0.7527^2)(1 + 5.0 + 5.0^2) + (5.0^2)(0.7527^3)} - 0.7527(1 + 5.0)}{1.7527}$$

$$= 1.3156$$

$$k_2 = -1 + \sqrt{2(1 + R_e) + \frac{2F_{yb}(1 + 2R_e)D^2}{3F_{em}l_m^2}}$$

$$= -1 + \sqrt{2(1.7527) + \frac{2(90,000)(2.5054)(0.148^2)}{3(3500)(2.5^2)}}$$

$$= 0.9120$$

$$k_3 = -1 + \sqrt{\frac{2(1 + R_e)}{R_e} + \frac{2F_{yb}(2 + R_e)D^2}{3F_{em}l_s^2}}$$

$$= -1 + \sqrt{\frac{2(1.7527)}{0.7527} + \frac{2(90,000)(2.7527)(0.148^2)}{3(3500)(0.5^2)}}$$

$$= 1.9651$$

Yield limit equations:

Mode I_m (NDS equation 11.3-1):

$$Z = \frac{Dl_m F_{em}}{K_D} = \frac{0.148(2.5)(3500)}{2.2} = 589 \text{ lb}$$

Mode I_s (NDS equation 11.3-2):

$$Z = \frac{Dl_s F_{es}}{K_D} = \frac{0.148(0.5)(4650)}{2.2} = 156 \text{ lb}$$

Mode II (NDS equation 11.3-3):

$$Z = \frac{k_1 Dl_s F_{es}}{K_D} = \frac{1.3156(0.148)(0.5)(4650)}{2.2} = 206 \text{ lb}$$

Mode III_m (NDS equation 11.3-4):

$$Z = \frac{k_2 Dl_m F_{em}}{(1 + 2R_e)K_D} = \frac{0.9120(0.148)(2.5)(3500)}{2.5054(2.2)} = 214 \text{ lb}$$

Mode III_s (NDS equation 11.3-5):

$$Z = \frac{k_3 Dl_s F_{em}}{(2 + R_e)K_D} = \frac{1.9651(0.148)(0.5)(3500)}{2.7527(2.2)} = 84 \text{ lb}$$

Mode IV (NDS equation 11.3-6):

$$Z = \frac{D^2}{K_D} \sqrt{\frac{2F_{em}F_{yb}}{3(1 + R_e)}}$$

$$= \frac{(0.148)^2}{2.2} \sqrt{\frac{2(3500)(90,000)}{3(1.7527)}} = 109 \text{ lb}$$

Nominal design value is selected as the smallest value from the yield equations:

$$\boxed{Z = 84 \text{ lb}}$$

Whether the NDS tables or the yield limit equations are used, the determination of the nominal design value Z is the first step in developing the allowable design value Z' for a single nail. Other important strength properties and the required adjustment factors that may be necessary to obtain the allowable design value for a single fastener are covered in the next section.

12.6   Adjustment Factors for Laterally Loaded Nails

The yield limit equations for laterally loaded nailed connections were reviewed in Sec. 12.5. As a reminder, *nominal design value* Z is the notation assigned to the load capacity of a single fastener as obtained from the *yield equations*. In order to be used in the design of a particular connection, the nominal design value must be converted to an *allowable design value* Z'.

This is done by multiplying the nominal value by a series of adjustment factors. Depending on the circumstances, the adjustment factors could have a major or minor effect on the allowable design value. The adjustment factors required in the design of connections are summarized in NDS Table 10.3.1. If the particular design conditions agree with the conditions for the nominal design value, the adjustment factors simply default to unity.

The base conditions associated with the nominal design value of a nail connection are as follows:

1. Load is normal (10-year) duration.

2. Wood is initially dry at the time of fabrication of a connection and remains dry in service.

3. Temperature range is normal.

4. Penetration of nail is at least 6 fastener diameters.

5. Nail is driven into the side grain of the main (holding) member.

6. Nail is not part of diaphragm or shearwall nailing (Chaps. 9 and 10).

7. Nail is not driven as a toenail.

If a nailed connection is used under exactly these conditions, the adjustment factors are all unity and $Z' = Z$.

For a laterally loaded nailed connection that does not satisfy the base conditions described above, the following adjustment factors may be required:

$$Z' = Z(C_D C_M C_t C_{eg} C_{di} C_{tn})$$

where C_D = load duration factor (Sec. 4.15)
   C_M = wet service factor (Sec. 4.14)
   C_t = temperature factor (Sec. 4.19)
   C_{eg} = end grain factor
   C_{di} = diaphragm factor
   C_{tn} = toenail factor

The total allowable design value for a connection using two or more nails is the sum of the individual allowable design values Z'. This summation is appropriate if the nails are all of the same type and size because each fastener will have the same yield mode. The design capacity for the connection is, then, the number of nails N times the allowable design value for one fastener:

$$\text{Allow. } P = N(Z')$$

A number of adjustment factors have been described in previous sections, and these are reviewed only briefly. Other adjustments apply only to nailed connections and require greater explanation.

Load duration factor C_D.   For most loading conditions, the load duration factors for connections are the same as those applied to allowable stresses in the design of structural members such as beams, tension members, and columns.

However, there is one exception to this general rule which relates to *impact loads*. Although impact loads are not part of the usual design loads for many wood structures, they are required in certain applications. Impact loads are applied suddenly and last for a very short period of time (2.0 seconds or less load duration). See Fig. 4.13 in Sec. 4.15. An example of impact loading is the increased force caused by the starting or stopping of a crane as it lifts or lowers a load. In the design of a wood *member* under impact loading, the designer may increase the allowable stress by a factor of $C_D = 2.0$.

Most laboratory tests of connections are for load durations of 5–10 minutes, providing ample support for using $C_D = 1.6$ under short term loads. However, there is very little laboratory test data for connections under impact load conditions. Therefore, in the design of a wood *connection* for impact loading the NDS conservatively limits the load duration factor to $C_D = 1.6$.

Wet service factor C_M. The concept of an allowable stress adjustment based on moisture content was introduced in Sec. 4.14. It will be recalled that the wet-service factors for member design vary with the type of stresses. Different adjustments also apply in connection design depending on the type of fastener and the moisture content of the wood at several stages. The moisture content is considered (1) at the time of fabrication of the joint and (2) in service. The nominal design value applies to nailed connections in which the wood is initially dry and remains dry in service. For connection design the term *dry* refers to wood that has a moisture content of 19 percent or less.

If the moisture content exceeds the dry limits, either at the time of fabrication *or* in service, a reduction in load capacity is required for common nails, box nails, sinker nails, and common wire spikes. For laterally loaded connections $C_M = 0.7$ for these conditions. A reduction in design value for high moisture content is not required for threaded hardened-steel nails. Values of C_M are listed in NDS Table 10.3.3, *Wet Service Factors, C_M, for Connections.*

Temperature factor C_t. If wood is used in an application with sustained exposure to higher-than-normal temperature conditions, the allowable stresses for member design and the allowable design values for connections are to be reduced by an appropriate temperature factor C_t. Refer to Sec. 4.19 for a brief review of the temperature factor. Numerical values of C_t for connections are obtained from NDS Table 10.3.4. The temperature range for most wood-frame buildings does not require adjustment of design values, and C_t can usually be set equal to one.

Nail penetration. The penetration of a nail was defined in Fig. 12.5a as the distance that the nail extends into the main or holding member. The dowel bearing length of a nail in the main member l_m was defined as a function of nail penetration p in Example 12.1. To obtain any design value Z for a nail, the nail penetration in the main member must be at least 6 times the diameter of the nail ($p \geq 6D$). In previous editions of the NDS, only four yield limit equations (rather than six) were specified for nailed connections, and a penetration depth factor was provided for nail penetration in the range $6D \leq p \leq 12D$. However, all six yield limit equations are specified for nailed connections in the 2001 NDS. Thus, the penetration depth factor is no longer required since dowel bearing length in the main member l_m is explicitly addressed in the yield limit equations.

End grain factor C_{eg}. The nominal design value applies to a laterally loaded connection with the nail driven into the *side grain* of the main member. Recall that the main member receives the pointed end of the fastener. In this basic shear connection, the load is perpendicular to the length of the fastener, and the axis of the nail is perpendicular to the wood fibers in the holding member. This is the strongest and most desirable type of nailed connection. See Fig. 12.11a in Example 12.6. When a nail is driven into the side grain, the end grain factor does not apply or may be set equal to one.

In another type of single-shear connection, the nail penetrates into the *end grain* of the holding member (Fig. 12.11b). The load is still perpendicular to the length of the nail, but the fastener axis is now parallel to the grain. This is a much weaker type of connection. To determine the reduced allowable design value, the nominal design value is multiplied by the end grain factor $C_{eg} = 0.67$.

EXAMPLE 12.6   Comparison of Side Grain and End Grain Nailing

Lateral Resistance in Side Grain

Two examples of nails used in the *basic* shear connection are shown in Fig. 12.11a. The basic connection has the nail driven into the *side grain* (i.e., perpendicular to the grain) of the holding member, and the load is applied perpendicular to the length of the nail. For nails driven into the side grain, $C_{eg} = 1.0$.

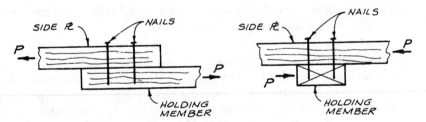

Figure 12.11a   Nails in side grain.

Lateral Resistance in End Grain

The second main type of lateral load connection has the nail driven into the end grain (i.e., parallel to the grain) of the holding member, and the load is applied perpendicular to the length of the nail. The front and side views of an end grain connection are shown in Fig. 12.11b with two different sets of lateral loads. For nails driven into the end grain, $C_{eg} = 0.67$.

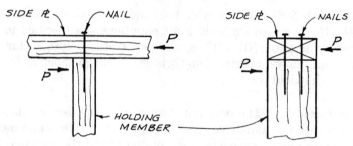

Figure 12.11b Nails in end grain.

Diaphragm factor C_{di}. Design of nailing for horizontal diaphragms and shear-walls using standard tables was covered in detail in Chaps. 9 and 10. The allowable unit shears in Tables 3.1A, 3.1B, 4.1A, and 4.2B of the *Wood Structural Panel Shear Wall and Diaphragm Supplement, Allowable Stress Design Manual for Engineered Wood Construction* (Ref. 12.1) are the result of diaphragm tests conducted by APA. As an alternative to designing diaphragms and shearwalls with these tables, the designer may compute the allowable unit shear. The allowable shear will be controlled by the shear capacity of the plywood panel *or* by the lateral design value of the fasteners.

Tests on diaphragms and shearwalls have demonstrated that the capacity of nails used in diaphragms is greater than the capacity of nails used in other structural connections. The diaphragm factor C_{di} is used to adjust the nominal design value Z to an allowable design value for use in diaphragm and shear-wall design. Note that the diaphragm factor is only to be used for nails that attach plywood sheathing to framing members; C_{di} does not apply to other connections that may simply be a part of the diaphragm. For example, C_{di} is not used in the design of diaphragm chord splices or drag strut connections made with nails. See Ref. 12.15 for an example of how to compute the allowable unit shear for a diaphragm using C_{di} and other appropriate factors. See also the note in Example 15.4 (Sec. 15.3).

Toenail factor C_{tn}. In many cases it is not possible to nail directly through the side member into the holding member. In these circumstances toenails may be used. See Example 12.7. Toenails are nails that are driven at an angle of 30 degrees to the side member and are started approximately one-third of the nail length from the intersection of the two members.

Anyone who has ever driven a toenail realizes that the installation details shown in Fig. 12.12a are only approximated in the field. However, the geometry given allows the designer to evaluate the dowel bearing length in the side member l_s and the penetration $p_L = l_m$ of the toenail into the holding member for use in the yield limit equations. Penetration requirements for a laterally loaded toenail are the same as those for a conventional shear connection. Common applications of toenails are for stud to wall plate, beam to wall plate, and blocking to plate connections.

The capacity of a toenail is obtained by multiplying the nominal design value Z by the toenail factor C_{tn}. The toenail factor is $C_{tn} = 0.83$ for a laterally loaded toenail, and a different toenail factor applies to withdrawal connections (Sec. 12.12). The toenail factor does not apply to conventional laterally loaded connections. In other words, C_{tn} defaults to unity ($C_{tn} = 1.0$) for the ordinary single-shear connections in Fig. 12.11a.

Finally, it will be noted that some authors draw a distinction between the types of connections shown in Fig. 12.12a and b. The difference has to do with the direction of the grain in the side member. For example, assume that the vertical member in Fig. 12.12a is a wall stud. The toenail, therefore, goes through the *end grain* of the side member. In contrast, the toenails in Fig. 12.12b run through the *side grain* of the beams and the blocking.

EXAMPLE 12.7   Toenail Connection

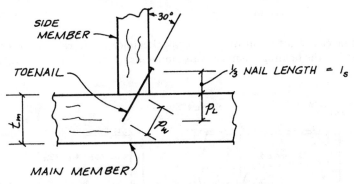

Figure 12.12a Toenail installation details and geometry.

Dimensions for toenails for use in yield limit equations:

l = length of nail

l_s = dowel bearing length in side member for toenail design

= projected length of toenail in side member

= one-third nail length

= $l/3$

p_L = penetration of toenail for *laterally loaded* connection

= projected length of toenail in main (holding) member

= $l \cos 30 - l_s \le t_m$

l_m = dowel bearing length of toenail in main member

$$= \begin{cases} p_L & \text{when } p_L \geq 10D \\ p_L - \text{projected length of tapered tip} & \text{when } 6D \leq p_L < 10D \end{cases}$$

$$= \begin{cases} l \cos 30 - l_s \leq t_m & \text{when } p_L \geq 10D \\ l \cos 30 - l_s - 2D \cos 30 \leq t_m & \text{when } 6D \leq p_L < 10D \end{cases}$$

There are two different definitions of penetration for toenails. The penetration p_L of a toenail in a laterally loaded connection is the *projected* length of the toenail in the main member. The definition of penetration p_W for a toenail in a *withdrawal* connection changes to the actual length of the nail in the main member. Although the yield limit equations do not apply to withdrawal connections, the two definitions of nail penetration are shown in Fig. 12.12a for comparison.

p_W = penetration of toenail for a connection *loaded in withdrawal*

= total length of nail in main member

$$= l - \frac{l/3}{\cos 30} \leq \frac{t_m}{\cos 30}$$

The capacity of a toenail in a withdrawal connection is related to the surface area of the nail in the holding member. Nail connections loaded in withdrawal are addressed in Sec. 12.12.

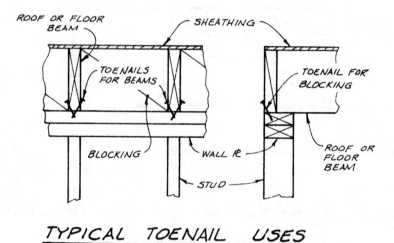

TYPICAL TOENAIL USES

Figure 12.12b Typical applications of toenails. On the left, toenails connect beams to top plate of stud wall. On the right, toenails attach blocking to wall plate.

When this difference is recognized, the connection in Fig. 12.12a is referred to as a *toenail,* and the connections in Fig. 12.12b are termed *slant nails.*

According to Ref. 12.10, toenails require a reduction in design value ($C_{tn} = 0.83$), and slant nails do not ($C_{tn} = 1.0$).

There is a problem with semantics here, because field personnel refer to both types of installations as toenails. Perhaps to avoid confusion, the NDS does not recognize the difference between the two applications, and $C_{tn} = 0.83$ is currently used for both.

Several numerical examples will now be given to demonstrate the practical design of nails in single-shear connections. In many cases the nominal design values for the nails can be determined from either the yield limit equations or the NDS nail tables.

12.7 Design Problem: Nail Connection for Knee Brace

The type of brace used in this example is typical for a number of unenclosed structures including carports, sheds, and patios. See Example 12.8. The design load is the lateral seismic force. It is assumed to be carried equally by all of the columns, and the reaction at the base of the column is determined first.

The horizontal component of force in the brace is then calculated using the free-body diagram of a column. After the horizontal component is known, the axial force in the brace can be evaluated. The force in the brace is used to determine the required number of nails to attach the brace. The same connection will be used at the top and bottom of the brace.

The nominal design value Z for the nail is obtained from a previous example, and the problem essentially deals with the determination of the allowable design value Z'.

EXAMPLE 12.8   Knee Brace Connection

The carport shown in Fig. 12.13a uses 2 × 6 knee braces to resist the longitudinal seismic force. Determine the number of 16d common nails required for the connection of the brace to the 4 × 4 post. Material is Southern Pine lumber that is dry at the time of construction. Normal temperatures apply.

Force to one row of braces:

$$R = \frac{wL}{2} = 76 \left(\frac{22}{2}\right) = 836 \text{ lb}$$

Assume the force is shared equally by all braces.

$$\Sigma M_0 = 0$$

$$3H - 209(10) = 0$$

$$H = 697 \text{ lb}$$

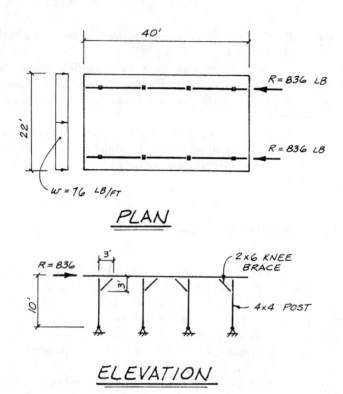

PLAN

ELEVATION

Figure 12.13a Lateral force in longitudinal direction.

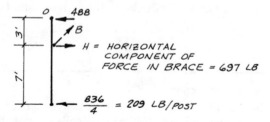

FBD OF COLUMN

Figure 12.13b FBD to determine axial force B in brace.

$$B = \sqrt{2}H = \sqrt{2}(697)$$

$$= 985 \text{ lb axial force in knee brace}$$

$$= \text{force on nailed connection}$$

The nominal design value for a 16d common nail in Southern Pine can be evaluated using the yield equations (Sec. 12.4), or it can be obtained from NDS Table 11N. Nominal design value from NDS Table 11N

$$Z = 154 \text{ lb/nail}$$

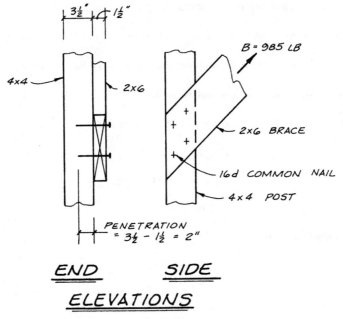

Figure 12.13c Nail connection of brace to column.

Adjustment Factors

Penetration

According to the footnotes for NDS Table 11N, the tabulated nominal design values Z are applicable for connections with nail penetration of $p \geq 10D$

$$10D = 10(0.162) = 1.62 \text{ in.} < 2.0 \text{ in.} \qquad \text{OK}$$

Moisture content

Because the building is "unenclosed," the brace connection may be exposed to the weather, and the severity of this exposure must be judged by the designer. Assume that a reduction for high moisture content is deemed appropriate, and the wet service factor C_M is obtained from NDS Table 10.3.3.

$$C_M = 0.7$$

Load duration

The load duration factor recommended in the NDS for seismic forces is $C_D = 1.6$.

Other adjustment factors

All other adjustment factors for allowable nail capacity do not apply to the given problem, and each can be set equal to unity:

$C_t = 1.0$ because normal temperature range is assumed

$C_{eg} = 1.0$ because nails are driven into side grain of holding member

$C_{di} = 1.0$ because connection is not part of nailing for diaphragm or shearwall

$C_{tn} = 1.0$ because nails are not toenailed

Allowable load for 16d common nail in Southern Pine:

$$Z' = Z(C_D C_M C_t C_{eg} C_{di} C_{tn})$$

$$= 154(1.6)(0.7)(1.0)(1.0)(1.0)(1.0)$$

$$= 172 \text{ lb/nail}$$

Required number of nails:

$$N = \frac{B}{Z'} = \frac{985}{172} = 5.73$$

> *Use*   six 16d common nails each end of knee
> brace for high-moisture conditions.*

If the reduction for wet service is not required, $C_M = 1.0$. The revised connection is

$$Z' = 154(1.6)(1.0) = 246 \text{ lb/nail}$$

$$N = \frac{985}{246} = 4.00$$

> *Use*   four 16d common nails each end of knee
> brace if moisture is not a concern.

For an unenclosed building of this nature, the designer must evaluate the exposure of the connections and the moisture content of the members. This example illustrates the effects of different assumed moisture conditions. The load duration factor of $C_D = 1.6$ for seismic forces was also illustrated.

Other adjustment factors for converting Z to Z' default to 1.0 in this example. The reasons for the defaults to unity are listed in this first nail example, but in subsequent problems the adjustments are set equal to 1.0 without detailed explanation.

---

*The use of preservative pressure-treated wood should be considered (Sec. 4.9).

The number of nails that can be accommodated in a connection may be limited by the spacing requirements (joint details). The layout of this connection is considered in Sec. 12.14.

12.8 Design Problem: Top Plate Splice

In Example 12.9, a splice in the double top plate of a wood-frame wall is designed. The forces in the plate are caused by the horizontal diaphragm action of the roof. The longitudinal wall plate is designed for the *chord force* (produced by the transverse lateral force) or the *collector (strut) force* (produced by the longitudinal lateral force). The larger of the two forces is used for the connection design. In this building the chord force controls, but the drag force could have been the critical force if the opening in the wall had been somewhat longer than 14 ft.

In buildings with larger lateral forces, the magnitude of the chord or drag force may be too great for a nailed connection. In this case, bolts or some form of connection hardware could be used for the plate splice.

At any splice, either the top or bottom member of the chord will be continuous. If a check of the stress in the chord were shown, the tension force would be divided by the cross-sectional area of *one* of the plate members. This example concentrates on the design of the connection only, and the stress in the chord/drag-strut member is not included.

An interesting point about the yield limit equations is highlighted in this example. The results of three of the six required equations are the same as those in a previous nail problem (Example 12.3). However, the values given by the Mode I_m, Mode II, and Mode III_m equations are considerably less in the current example. The lower values of Z are the result of a reduced nail penetration caused by a main member that is only 1.5 in. thick. In Example 12.9, the values of Mode I_m and Mode I_s are the same because the dowel bearing lengths in the main and side members are equal. The values of Mode III_m and Mode III_s are also identical for the same reason.

The designer needs to be aware of the subtle way this problem changed from the earlier example. The outcome was the same (that is, $Z = 141$ lb in both examples), but the results could have been different.

It should be noted that the tables in the NDS *do not* take into account the thickness of the main member. In other words, the tables assume that $l_m = p = 10D$ and the nail does not protrude out of the holding member. This does not affect the result of Example 12.9 because the Mode I_m, Mode II, and Mode III_m yield mechanisms do not control. Consequently the nominal design value agrees with the value listed in NDS Table 11N. However, this may not always be the case.

EXAMPLE 12.9 Top Plate Splice

The building shown in Fig. 12.14a has stud walls with a double 2× top plate around the entire building. Lateral seismic forces to the horizontal diaphragm are given.

Splices in the double top plate for the *longitudinal wall* are to be designed for the horizontal diaphragm chord force or drag force (whichever is larger). Lumber is dry DF-L that will remain dry in service. Normal temperatures apply. Connection is to be made with 16d common nails.

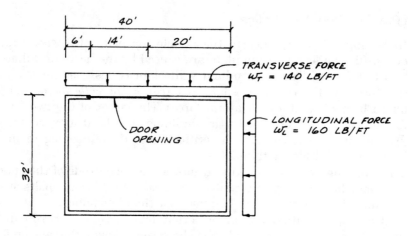

Figure 12.14a   Lateral forces to roof diaphragm. Consider force in each direction separately.

Transverse Lateral Force

Top plate of wall serves as diaphragm chord.
    Moment in diaphragm:

$$M = \frac{wL^2}{8} = \frac{140(40)^2}{8} = 28{,}000 \text{ ft-lb}$$

Chord force:

$$T = C = \frac{M}{b} = \frac{28{,}000}{32} = 875 \text{ lb}$$

For a discussion of horizontal diaphragm chords and collectors (struts), see Chap. 9.

Longitudinal Lateral Force

Wall plate serves as drag strut (Fig. 12.14b).
    Shearwall reaction:

$$R = \frac{wL}{2} = \frac{160(32)}{2} = 2560 \text{ lb}$$

Wall with opening has the critical collector (strut) force.
    Roof diaphragm unit shear:

$$v_r = \frac{R}{b} = \frac{2560}{40} = 64 \text{ lb/ft}$$

Unit shear in shearwall:

$$v_r = \frac{R}{b} = \frac{2560}{6 + 20} = 98.5 \text{ lb/ft}$$

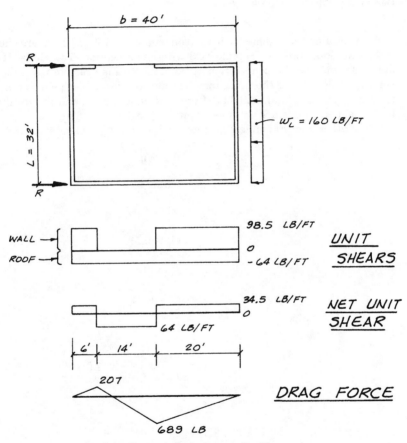

Figure 12.14b Force in collector (strut) along wall with opening.

Splice Connection

Max. chord force = 875 lb   governs

Max. drag force = 689 lb

If the Code minimum lap of 4 ft is used for the top plate splice, the connection within the lap must transmit the full chord force.

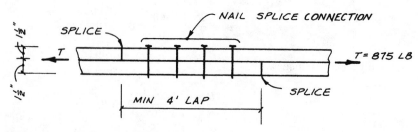

Figure 12.14c   Splice in top wall plate.

The yield limit equations for a 16d common nail in Fig. 12.14c are the same as the yield limit equations in Example 12.3 with one exception: The Mode III_m yield mechanism depends on the penetration of the nail. In this problem the nail protrudes beyond the thickness of the main member, and the furnished penetration is less. NDS equations 11.3-1, 11.3-3, and 11.3-4 are reevaluated here, and the results of the other three yield equations are obtained from the previous example.

Information from Example 12.3 that applies to the problem at hand:

$$D = 0.162 \text{ in.}$$

$$l = 3.5 \text{ in.}$$

$$F_{yb} = 90{,}000 \text{ psi}$$

$$G_m = G_s = 0.50$$

$$F_{em} = F_{es} = 4650 \text{ psi}$$

$$l_s = t_s = 1.5 \text{ in.}$$

$$K_D = 2.2$$

$$R_e = \frac{F_{em}}{F_{es}} = \frac{4650}{4650} = 1.0$$

$$1 + R_e = 2.0$$

$$1 + 2R_e = 3.0$$

$$2 + R_e = 3.0$$

Reduced nail penetration:

$$p = l - t_s \le t_m$$

$$= 3.5 - 1.5 = 2.0 > 1.5$$

$$\therefore p = 1.5 \text{ in.}$$

$$10D = 10(0.162) = 1.62 \text{ in.} > 1.5 \text{ in.}$$

$$l_m = l - t_s - 2D \le t_m$$

$$= 3.5 - 1.5 - 2(0.162) = 1.68 \text{ in.} > 1.5 \text{ in.}$$

$$\therefore l_m = 1.5 \text{ in.}$$

Revised values of R_t, k_1, and k_2 with dowel bearing length of 1.5 in.:

$$R_t = \frac{l_m}{l_s} = \frac{1.5}{1.5} = 1.0$$

$$k_1 = \frac{\sqrt{R_e + 2R_e^2(1 + R_t + R_t^2) + R_t^2 R_e^3} - R_e(1 + R_t)}{(1 + R_e)}$$

$$= \frac{\sqrt{1.0 + 2(1.0^2)(1 + 1.0 + 1.0^2) + (1.0^2)(1.0^3)} - 1.0(1 + 1.0)}{2.0}$$

$$= 0.4142$$

$$k_2 = -1 + \sqrt{2(1 + R_e) + \frac{2F_{yb}(1 + 2R_e)D^2}{3F_{em}l_m^2}}$$

$$= -1 + \sqrt{2(2.0) + \frac{2(90,000)(3.0)(0.162^2)}{3(4650)(1.5^2)}}$$

$$= 1.1099$$

Mode I$_m$ (NDS equation 11.3-1):

$$Z = \frac{Dl_m F_{em}}{K_D} = \frac{0.162(1.5)(4650)}{2.2} = 514 \text{ lb}$$

Mode II (NDS equation 11.3-3):

$$Z = \frac{k_1 Dl_s F_{es}}{K_D} = \frac{0.4142(0.162)(1.5)(4650)}{2.2} = 213 \text{ lb}$$

Mode III$_m$ (NDS equation 11.3-4):

$$Z = \frac{k_2 Dl_m F_{em}}{(1 + 2R_e)K_D} = \frac{1.1099(0.162)(1.5)(4650)}{3.0(2.2)} = 190 \text{ lb}$$

These are less than the values for Mode I$_m$, Mode II, and Mode III$_m$ for the connection in Example 12.3.

Results of other yield equations from Example 12.3:

Mode I$_s$ (NDS equation 11.3-2)

$$Z = 514 \text{ lb}$$

Mode III_s (NDS equation 11.3-5)

$$Z = 190 \text{ lb}$$

Mode IV (NDS equation 11.3-6)

$$Z = 141 \text{ lb}$$

Refer to Fig. 11.10 in Sec. 11.7 for sketches of the yield modes. Notice that the same value of Z is obtained for Mode I_m and Mode I_s for the top plate splice. Similarly, the same value of Z is obtained for Mode III_m and Mode III_s. This is to be expected because of the equal thicknesses of the main and side members. The length of the nail is the same in both members.

The nominal design value is taken as the smallest value from the yield formulas:

$$Z = 141 \text{ lb}$$

Adjustment factors for this problem are all unity except the load duration factor. For seismic forces the NDS recommends

$$C_D = 1.6$$

Required penetration to use value of Z:

$$6D = 6(0.162) = 0.97 < 1.5 \text{ in.} \qquad OK$$

Allowable design value for a 16d common nail in DF-L:

$$Z' = Z(C_D C_M C_t C_{eg} C_{di} C_{tn})$$

$$= 141(1.6)(1.0)(1.0)(1.0)(1.0)(1.0)$$

$$= 225 \text{ lb/nail}$$

Required number of nails:

$$N = \frac{T}{Z'} = \frac{875}{225} = 3.88$$

Use four 16d common nails between splice points.

Although it was not done in this example, the designer can specify the location of the splices in the top plate. If these locations are clearly shown on the plans, the design forces (chord and collector loads) that occur at these points may be used for connection design. In this example the *maximum* forces were used, and consequently the top plate splices may occur at any point along the length of the wall.

12.9 Design Problem: Shearwall Chord Tie

The requirements for anchoring a shearwall panel to the foundation were discussed in Chap. 10. A similar problem occurs in the attachment of a second-story shearwall to the supporting first-story shearwall. The same basic forces must be considered.

In Example 12.10 the tension chord of the second-story wall is spliced to the chord of the first-story wall. This splice is required because in typical *platform* construction the studs are not continuous from the roof level to the foundation. The first-story studs stop at the double wall plate that supports the second-floor framing. Separate second-story studs are then constructed on top of the second-floor platform.

The purpose of the splice, then, is to develop a continuous shearwall chord from the roof level to the foundation. Splices of this type are referred to as *continuity ties* or *splices*. Various types of connections can be used for these ties. In this example the connection consists of a steel strap connected to the first- and second-story chords with 8d box nails. For larger forces, bolts or lag screws can be used in place of nails. Calculations to verify the tension capacity of the strap are not shown in this example, but they would normally be required in design. In addition, the layout and spacing requirements of the nails in the connection should be considered.

Chord forces can be calculated using either the gross overturning moment or the design overturning moment. Both are illustrated, and the choice of which method to use in a particular situation is left to the judgment of the designer.

EXAMPLE 12.10 Shearwall Chord Tie

Design the connection tie between the first- and second-story shearwall chords for the right end wall (Fig. 12.15a). A 12-gage ASTM A653 grade 33 steel strap with 8d box nails is to be used. Lumber is dry Southern Pine that will remain dry in service. Normal temperatures apply. Wind force reactions from the horizontal roof and second-floor diaphragms are given.

Chord Force

The chord forces at the second-story level are caused by the lateral forces at the roof level. See Fig. 12.15b.

From the gross overturning moment OM (refer to Example 10.4),

$$\text{Gross } T = C = vh = (150)(12) = 1800 \text{ lb} \quad \text{(conservative)}$$

From the design OM (refer to Example 2.9),

$$\text{Design OM} = \text{gross OM} - 0.6(\text{dead load RM})$$

$$= (150 \times 10)(12) - 0.6(140 \times 10)(5)$$

$$= 13.8 \text{ ft-k}$$

$$\text{Net } T = \frac{M}{b} = \frac{13,800}{10} = 1380 \text{ lb}$$

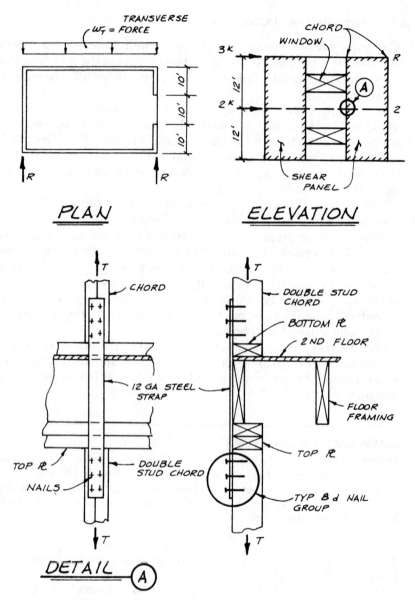

Figure 12.15a Shearwall chord connection.

Alternate calculation

$$\text{Design } T = \text{gross } T - 0.6(\text{trib. DL})$$

$$= 1800 - 0.6(140 \times 5) = 1380 \text{ lb}$$

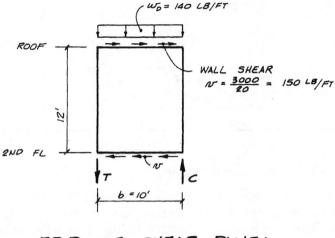

FBD OF SHEAR PANEL

Figure 12.15b Shearwall overturning.

Nail Connection

The nominal design value for an 8d box nail installed through a metal side plate into a Southern Pine main member was computed using the yield model formulas in Example 12.4. The same value applies to the problem at hand.

$$Z = 90 \text{ lb}$$

Load duration factor for wind is

$$C_D = 1.6$$

All other adjustment factors are unity for this problem, and the allowable design value is

$$Z' = Z(C_D) = 90(1.6) = 144 \text{ lb/nail}$$

For the gross overturning moment:

$$N = \frac{1800}{144} = 12.5$$

> *Use* thirteen 8d box nails each end of steel strap into multiple studs.

For the design overturning moment:

$$N = \frac{1380}{144} = 9.6$$

> *Use*   ten 8d box nails each end of steel
> strap to multiple studs.

12.10   Design Problem: Laterally Loaded Toenail

The example to illustrate the design capacity of a toenail connection is the Code attachment of a stud to the bottom plate of a wall. See Example 12.11. The nailing schedule in IBC Table 2304.9.1 lists four 8d common toenails as one of the acceptable methods of attaching a stud to the sole plate. Design forces and special considerations for this attachment in areas of high seismic risk were discussed in Sec. 10.13.

The force on the given connection is wind perpendicular to the wall. The penetration of the nail and the thickness of the side member for use in the yield limit equations are determined using the geometry from Fig. 12.12a (Sec. 12.6). Recall that l_s and l_m are to be taken as the *projected* lengths of the nail in the side and main members.

The geometry thus established is not directly covered by the NDS fastener capacity tables, and the yield limit expressions are evaluated in the example. This would normally be done on the computer, and the full numerical example is given for illustration.

EXAMPLE 12.11   Capacity of Toenail Connection

Determine the allowable wind force reaction at the base of a wood-frame shearwall which uses the standard Code toenailing of four 8d common nails (Fig. 12.16). Lumber is dry DF-L that will remain dry in service. Normal temperatures apply.

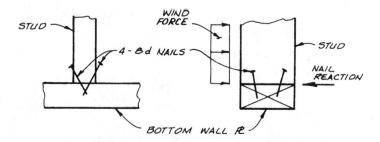

TYPICAL   STUD   TO   ℞   CONNECTION

Figure 12.16 Toenail connection at bottom of stud subject to lateral wind force.

Geometry:

Dimensions of an 8d common nail are obtained from the table in Fig. 12.7*a* or NDS Appendix L:

$$D = 0.131 \text{ in.}$$

$$l = 2.5 \text{ in.}$$

For a toenail, the thickness of the side member and penetration are determined using the geometry given in Fig. 12.12*a*. The bottom wall plate is a nominal 2×, and $t_m = 1.5$ in.

$$l_s = \frac{l}{3} = \frac{2.5}{3} = 0.83 \text{ in.}$$

$$p_L = l \cos 30 - l_s \le t_m$$

$$= 2.5 \cos 30 - 0.83 \le 1.5$$

$$\therefore p_L = 1.33 \text{ in.}$$

$$10D = 10(0.131) = 1.31 \text{ in.} < 1.33 \text{ in.}$$

$$\therefore l_m = p_L = 1.33 \text{ in.}$$

Coefficients for yield equations:

Bending yield strength of an 8d common nail is obtained from the table in Fig. 12.7*b* or NDS Table I1:

$$F_{yb} = 100{,}000 \text{ psi}$$

Specific gravity and dowel bearing strength are the same for both the main member and side member and are obtained from NDS Tables 11.3.2A and 11.3.2:

$$G_m = G_s = 0.50$$

$$F_{em} = F_{es} = 4650 \text{ psi}$$

Coefficients for use in yield equations:

$$K_D = 2.2$$

$$R_e = \frac{F_{em}}{F_{es}} = \frac{4650}{4650} = 1.0$$

$$1 + R_e = 2.0$$

$$1 + 2R_e = 3.0$$

$$2 + R_e = 3.0$$

$$R_t = \frac{l_m}{l_s} = \frac{1.33}{0.83} = 1.5981$$

$$k_1 = \frac{\sqrt{R_e + 2R_e^2(1 + R_t + R_t^2) + R_t^2R_e^3} - R_e(1 + R_t)}{(1 + R_e)}$$

$$= \frac{\sqrt{1.0 + 2(1.0^2)(1 + 1.5981 + 1.5981^2) + (1.5981^2)(1.0^3)} - 1.0(1 + 1.5981)}{2.0}$$

$$= 0.5623$$

$$k_2 = -1 + \sqrt{2(1 + R_e) + \frac{2F_{yb}(1 + 2R_e)D^2}{3F_{em}l_m^2}}$$

$$= -1 + \sqrt{2(2.0) + \frac{2(100{,}000)(3.0)(0.131^2)}{3(4650)(1.33^2)}}$$

$$= 1.1015$$

$$k_3 = -1 + \sqrt{\frac{2(1 + R_e)}{R_e} + \frac{2F_{yb}(2 + R_e)D^2}{3F_{em}l_s^2}}$$

$$= -1 + \sqrt{\frac{2(2.0)}{1.0} + \frac{2(100{,}000)(3.0)(0.131^2)}{3(4650)(0.83^2)}}$$

$$= 1.2501$$

Yield equations:

Mode I_m (NDS equation 11.3-1):

$$Z = \frac{Dl_mF_{em}}{K_D} = \frac{0.131(1.33)(4650)}{2.2} = 369 \text{ lb}$$

Mode I_s (NDS equation 11.3-2):

$$Z = \frac{Dl_sF_{es}}{K_D} = \frac{0.131(0.83)(4650)}{2.2} = 231 \text{ lb}$$

Mode II (NDS equation 11.3-3):

$$Z = \frac{k_1Dl_sF_{es}}{K_D} = \frac{0.5623(0.131)(0.83)(4650)}{2.2} = 130 \text{ lb}$$

Mode III_m (NDS equation 11.3-4):

$$Z = \frac{k_2Dl_mF_{em}}{(1 + 2R_e)K_D} = \frac{1.1015(0.131)(1.33)(4650)}{3.0(2.2)} = 135 \text{ lb}$$

Mode III$_s$ (NDS equation 11.3-5):

$$Z = \frac{k_3 D l_s F_{em}}{(2 + R_e)K_D} = \frac{1.250(0.131)(0.83)(4650)}{3.0(2.2)} = 96 \text{ lb}$$

Mode IV (NDS equation 11.3-6):

$$Z = \frac{D^2}{K_D}\sqrt{\frac{2F_{em}F_{yb}}{3(1 + R_e)}}$$

$$= \frac{(0.131)^2}{2.2}\sqrt{\frac{2(4650)(100,000)}{3(2.0)}} = 97 \text{ lb}$$

Nominal design value:

Nominal design value is taken as the smallest value from the yield formulas:

$$Z = 96 \text{ lb}$$

Adjustment factors:

Load duration factor for wind is

$$C_D = 1.6$$

Limit for penetration:

$$6D = 6(0.113) = 0.68 \text{ in.}$$

$$0.68 \text{ in.} < 1.33 \text{ in.} \quad \text{OK}$$

Toenail factor for laterally loaded toenails:

$$C_{tn} = 0.83$$

All other adjustments do not apply to this problem and are set equal to unity.

Allowable design value:

$$Z' = Z(C_D C_{tn})$$

$$= 96(1.6)(0.83) = 128 \text{ lb/nail}$$

Allowable reaction:

Allowable wind reaction at base of stud:

$$R = N \times Z' = 4(128) = \boxed{511 \text{ lb/stud}}$$

12.11   Design Problem: Laterally Loaded Connection in End Grain

The final example of a shear connection involves nailing a member into the end grain of a roof beam. See Example 12.12. This is not a very desirable type of connection, but it is frequently seen in the field and serves as an illustration of when the end grain factor C_{eg} is required. There are a number of alternative connections, with nails driven into the side grain, that can be used in place of the given connection. These alternative connections are generally preferred and become necessary for larger loads. The alternative connections may involve some type of metal hardware (Chap. 14).

EXAMPLE 12.12   Nail into End Grain of Roof Beam

Determine the allowable load that can be transmitted by the nail connection in Fig. 12.17. Lumber is DF-L that is initially dry and remains dry in service. Normal temperatures apply. The connection is made with two 16d common nails into the end grain of the main member. Loads are (D + S).

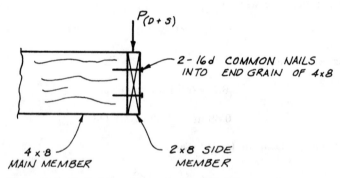

Figure 12.17 Nails driven into end grain of holding member.

Nominal design value:

The nominal design value in this example is the same as the nail in Example 12.3. The member sizes are different, but the dimensions and other factors affecting the nominal design value are the same in the two problems. Thus, from Example 12.3

$$Z = 141 \text{ lb}$$

Allowable design value:

Load duration factor for (D + S):

$$C_D = 1.15$$

End grain factor:

$$C_{eg} = 0.67$$

Other adjustment factors have default values of unity.

$$Z' = Z(C_D C_{eg}) = 141(1.15)(0.67)$$

$$= 109 \text{ lb/nail}$$

Allowable load on connection:

$$P_{(D+S)} = N \times Z' = 2 \times 109$$

$$= \boxed{218 \text{ lb} \quad \text{(dry service)}}$$

NOTE: If the connection is exposed to the weather or if high-humidity conditions exist, the wet service factor will apply, and the allowable load becomes

$$Z' = Z(C_D C_M C_{eg}) = 141(1.15)(0.7)(0.67)$$

$$= 76 \text{ lb/nail}$$

$$P_{(D+S)} = 2 \times 76 = \boxed{152 \text{ lb} \quad \text{(wet service)}}$$

12.12 Nail Withdrawal Connections

The concept of a fastener loaded in withdrawal was introduced in Chap. 11 (Fig. 11.2). In the *basic withdrawal connection,* the nail passes through the side member and into the *side grain* of the main member. In this connection the load is parallel to the length of the nail, and the load attempts to pull the nail out of the holding member. Generally speaking, connections loaded in withdrawal are weaker and less desirable than the same connections subjected to shear loading. The NDS provides tabulated design values for fasteners subjected to "pullout," but the usual recommendation is that withdrawal loading of unthreaded fasteners should be avoided (in favor of other more positive types of connections) where practical.

The yield limit model for dowel-type fasteners applies only to shear connections, and the design values for nails in withdrawal are based on empirical test results. The *nominal design value* for the basic withdrawal connection is given the symbol W. In addition to nail size and type, the withdrawal capacity depends on the specific gravity of the wood and the penetration p of the nail into the holding member.

NDS Table 11.2C gives the nominal withdrawal design value *per inch of penetration* for common nails, box nails, common wire spikes, and threaded hardened steel nails. If the value from the table is given the symbol w, the nominal design value for a single nail is

$$W = w \times p$$

As explained previously, the nominal design value W represents the load capacity for a nail used under a set of base conditions, similar to the base conditions for laterally loaded connections:

1. Load is normal (10-year) duration.
2. Wood is initially dry at the time of fabrication of a connection, and it remains dry in service.
3. Temperature range is normal.
4. Nail is driven into the side grain of the main (holding) member.
5. Nail is not driven as a toenail.

To obtain the *allowable design value* W' for another set of conditions, the nominal design value is multiplied by the appropriate adjustment factors:

$$W' = W(C_D C_M C_t C_{tn})$$

where C_D = load duration factor (Sec. 4.15)
$\quad\quad C_M$ = wet service factor (Sec. 4.14)
$\quad\quad C_t$ = temperature factor (Sec. 4.19)
$\quad\quad C_{tn}$ = toenail factor

If more than one nail of the same size and type is used, the allowable withdrawal design value for the connection is the allowable withdrawal design value for one fastener W' times the number of nails N:

$$P = N(W')$$

Perhaps the most basic point to understand about nails loaded in withdrawal has to do with the direction of the grain of the holding member. See Example 12.13. The nominal design values apply only to connections with the nail penetrating into the *side grain* of the holding member (Fig. 12.18a). This has been referred to as the *basic* withdrawal connection. The other possible arrangement is to have the nail penetrate into the *end grain* of the holding member (Fig. 12.18b).

Nails that are driven into the end grain have very low capacities and exhibit considerable variation in withdrawal values. Consequently, the NDS does not provide withdrawal design values for nails driven into the end grain. This point could be emphasized by introducing an end grain factor of $C_{eg} = 0$ into the formula for allowable withdrawal design values.

EXAMPLE 12.13   Comparison of Withdrawal Connections

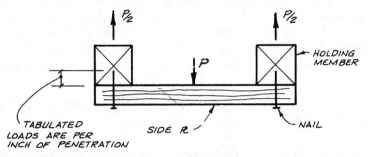

Figure 12.18a Basic withdrawal connection.

Withdrawal from Side Grain

The *basic* withdrawal connection has the nail driven into the *side grain* (i.e., perpendicular to the grain) of the holding member, and the load is parallel to the length of the nail (Fig. 12.18a).

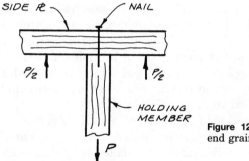

Figure 12.18b Withdrawal from end grain—not allowed.

Withdrawal from End Grain—Not Permitted

The second example is a withdrawal connection with the nail driven into the *end grain* (i.e., parallel to the grain) of the holding member. The load is again parallel to the length of the nail (Fig. 12.18b). This is an inherently weak connection, and it is not permitted in design (i.e., it has no allowable design value).

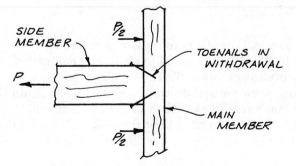

Figure 12.18c Toenail connection withdrawal from side grain.

Toenail Loaded in Withdrawal

A *toenail* connection may be loaded in withdrawal as long as the nails penetrate the *side grain* of the holding member. For withdrawal loading, the toenail factor is $C_{tn} = 0.67$ (compared with 0.83 for toenails subjected to shear loading). For a toenail in withdrawal, NDS 11.5.4 indicates that the wet service factor need not be applied. In other words, C_M may be set equal to 1.0 for toenails loaded in withdrawal.

A brief summary of the adjustment factors for withdrawal connections is given. This is followed by two numerical examples.

Load duration factor C_D. The load duration factor C_D for nail connections subject to withdrawal is the same as that applied in the design of laterally loaded nail connections as well as structural members such as beams, tension members, and columns. Refer to Secs. 12.6 and 4.15 for a review of load duration and the numerical values for C_D.

Wet service factor C_M. The concept of an adjustment factor based on moisture content was introduced in Sec. 4.14. In most cases, nails loaded in withdrawal are more negatively affected by variations in moisture content than are other connections. As with shear connections, the moisture content of the wood is considered (1) at the time of fabrication of the joint and (2) in service. The nominal design value applies to withdrawal connections in which the wood is initially dry and remains dry in service. For other moisture content conditions, values of C_M are obtained from NDS Table 10.3.3.

The performance of threaded hardened steel nails is better than other types of nails, and $C_M = 1.0$ for all moisture content conditions. The NDS also specifies $C_M = 1.0$ for all toenail connections loaded in withdrawal, regardless of nail type.

Temperature factor C_t. If wood is used at higher-than-normal temperatures for an extended period of time, the nominal design values for connections are to be reduced by the temperature factor C_t. Numerical values of C_t are obtained from NDS Table 10.3.4. It should be noted that the temperature range in most wood-frame buildings does not require a reduction in fastener load capacity, and C_t is usually taken as 1.0.

Toenail factor C_{tn}. The installation of toenails was covered earlier in this chapter. The geometry in Fig. 12.12a is used to evaluate penetration in determining the nominal withdrawal value W. The nominal design value is then reduced by the toenail factor for withdrawal $C_{tn} = 0.67$. Note again that the wet service factor does not apply to toenails loaded in withdrawal ($C_M = 1.0$).

Because withdrawal connections are not the preferred connection, only two brief examples are given. The first is a withdrawal connection with the nail driven perpendicular to the member and into the side grain. See Example 12.14. The second problem demonstrates the procedure for a toenail connection loaded in withdrawal. See Example 12.15.

EXAMPLE 12.14   Basic Withdrawal Connection

Assume that the hanger in Fig. 12.18a is used to suspend a dead load P. The side member is a 2 × 8, and the holding members are 6 × 6s. Lumber is Western Hemlock that is initially green, and seasoning occurs in service. Normal temperatures apply. Six 16d box nails are used at both ends of the side member. Determine the allowable load P.

Nail diameter and length are obtained from the table in Fig. 12.7a or from NDS Appendix L:

$$D = 0.135 \text{ in.}$$

$$l = 3.5 \text{ in.}$$

Specific gravity is obtained from NDS Table 11.3.2A:

$$G = 0.47$$

Tabulated withdrawal value from NDS Table 11.2C:

$$w = 28 \text{ lb/in.}$$

Penetration:

$$p = l - t_s \leq t_m$$

$$= 3.5 - 1.5 \leq 5.5$$

$$\therefore p = 2.0 \text{ in.}$$

Nominal design value:

$$W = p \times w = 2.0(28) = 56 \text{ lb/nail}$$

Allowable design value:

$$W' = W(C_D C_M C_t C_{tn})$$

$$= 56(0.9)(0.25)(1.0)(1.0) = 12.6 \text{ lb/nail}$$

The load is applied symmetrically, and the total load capacity is

$$\text{Allow. } P = N(W') = 12(12.6) = \boxed{151 \text{ lb}}$$

EXAMPLE 12.15   Toenail Withdrawal Connection

Assume that the connection in Fig. 12.18c has three 8d common nails. Both members are 2-in. nominal DF-L that is initially dry and remains dry in service. Normal temperatures apply. Determine the allowable wind force P.

Nail diameter and length are obtained from the table in Fig. 12.7a or NDS Appendix L:

$$D = 0.131 \text{ in.}$$

$$l = 2.5 \text{ in.}$$

Penetration for a toenail in a withdrawal connection (Fig. 12.12a):

$$p_W = l - \frac{l/3}{\cos 30} \leq \frac{t_m}{\cos 30}$$

$$= 2.5 - \frac{2.5}{3 \cos 30} \leq \frac{1.5}{\cos 30}$$

$$= 1.54 < 1.73$$

$$\therefore p_W = 1.54 \text{ in.}$$

Specific gravity from NDS Table 11.3.2A:

$$G = 0.50$$

Tabulated withdrawal value from NDS Table 11.2C:

$$w = 32 \text{ lb/in.}$$

Nominal design value:

$$W = p_W \times w = 1.54(32) = 49.3 \text{ lb/nail}$$

Allowable design value:

$$W' = W(C_D C_t C_{tn})$$

$$= 49.3(1.6)(1.0)(0.67) = 53 \text{ lb/nail}$$

Total load capacity:

$$\text{Allow. } P = N(W') = 3(53) = \boxed{159 \text{ lb}}$$

12.13   Combined Lateral and Withdrawal Loads

In addition to the direct withdrawal problem, there are occasions when nails and spikes are subjected to combined lateral and withdrawal loading. See Fig.

12.19. The resultant load acts at an angle α to the surface of the main member. This situation can occur for nails that attach roof sheathing to framing members on low-slope roofs. The nail withdrawal loading is due to wind uplift on the roof sheathing, while the simultaneous lateral loading is caused by transfer of wind loads from the walls into the roof diaphragm (Sec. 9.2). For this type of loading the NDS includes a provision for calculating an allowable resultant design value Z'_α which must equal or exceed the total design load.

The first step in computing the allowable resultant design value is to evaluate the nominal design value Z of the nail or spike for *shear* loading using the appropriate yield limit equations. Then the nominal *withdrawal* design value W is computed using the tabulated value from NDS Table 11.2C and the penetration depth of the nail in the main member ($W = w \times p$). The necessary adjustment factors are applied to the nominal values to obtain allowable values Z' and W'. The interaction between these values is taken into account by the following formula to give the allowable resultant:

$$Z'_\alpha = \frac{Z'W'}{Z' \sin \alpha + W' \cos \alpha}$$

This interaction equation is similar to the Hankinson formula (Sec. 11.5 and 13.16), differing only in the magnitude of the exponents for the trigonometric terms in the denominator.

12.14   Spacing Requirements

The spacing requirements for nail connections given in NDS 11.1.5.6 are rather general. Nail spacing "shall be sufficient to prevent splitting of the wood."

In the past, the layout of nails in a connection has typically been left to the discretion of the field personnel fabricating the connection. However, because an increasingly large number of nails are used in some structural connections,

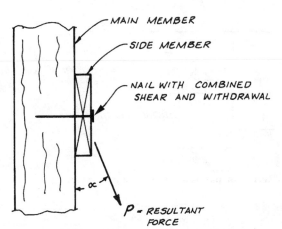

Figure 12.19 Nail subjected to combined lateral and withdrawal loading.

the designer may need to address the layout of fasteners in certain cases. Judgment must be used in the detailing of nail connections because there is not a recognized standard for the placement of nails. The NDS Commentary (Ref. 12.2) provides partial guidance in Table C12.4-1.

In order to discuss the spacing of mechanical fasteners, it is necessary to define some terms that are used in detailing wood connections. See Fig. 12.20. These definitions apply to nails as well as other types of fasteners, but there are additional clarifications required for bolts and other connectors.

End distance is the distance, measured parallel to the grain, from the center of the nearest fastener to the end grain of a square-cut piece of lumber. *Edge distance* is the distance, measured perpendicular to the grain, from the center of the nearest fastener to the edge of the member. *Center-to-center distance* (also known as *nail spacing*) is the distance from the center of one fastener to the center of an adjacent fastener in a row. A *row* of fasteners is defined as two or more fasteners in a line parallel to the direction of load, and *row spacing* is the perpendicular distance between rows of fasteners.

In most cases, nails are simply driven into wood members without drilling. However, preboring nail holes may be used to avoid splitting the wood. The NDS limits the diameter of a pilot hole to a maximum of 90 percent of the nail diameter for wood with a specific gravity greater than 0.6, and 75 percent of the nail diameter for less dense wood ($G \leq 0.6$). The full lateral design value Z' and withdrawal design value W' apply to connections with or without prebored holes. Preboring is an effective method of avoiding the splitting of wood members.

There are many connection detailing questions that are left unanswered under the NDS Commentary criteria. Although the NDS Commentary addresses a wide range of connection configurations, there are no guidelines given for center-to-center nail spacing when loading is at an angle to grain, nor for end distance requirements when loads are applied perpendicular to grain.

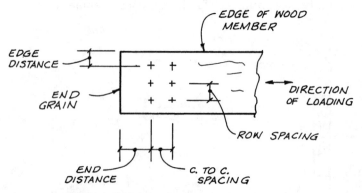

Figure 12.20 Definition of terms for nail spacing. Drawing shows *three rows* of nails.

Another problem occurs in members that do not have square-cut ends. See Example 12.16. As the angle θ in Fig. 12.22 becomes small, the perpendicular distance from the end cut on the diagonal member to the center of the nail becomes small if the end distance (parallel to the grain) is kept constant. A similar problem occurs with the edge distance for the vertical member. For an end cut that is not square, the TCM (Ref. 12.6) recommends that the *perpendicular distance from the end cut* should not be less than the *edge distance*. Similar logic may reasonably be applied to the perpendicular distance from the nail to the edge of the vertical member. It should be noted that the recommendation just cited from the TCM is a general recommendation for fasteners, and the TCM does not provide specific spacing criteria for nail connections. For nails, the TCM indicates that "unusual" splitting is to be avoided.

EXAMPLE 12.16 Nail Spacing

Determine the nail spacing for the knee brace connection designed in Example 12.8 (Sec. 12.7). See Fig. 12.22 for a scale drawing. Apply the NDS Commentary criteria to the extent possible. Can size 16d common nails be accommodated?

NOTE: As the angle θ becomes small, the perpendicular distance from the nail to the end cut may become too small. Therefore, use the general recommendation from AITC that the perpendicular distance from the end cut be not less than the edge distance. The edge distance guidelines will also be applied to the edge distance for the vertical member, even though the load is not parallel to the axis of the vertical member.

	Connection configuration		
	Wood side members *without* prebored holes	Wood side members *with* prebored holes **OR** steel side plates *without* prebored holes	Steel side plates *with* prebored holes
center-to-center spacing			
parallel to grain	15D	10D	5D
perpendicular to grain	10D	5D	2.5D
end distance parallel to grain			
tension loading	15D	10D	5D
compression loading	10D	5D	3D
edge distance	2.5D	2.5D	2.5D
row spacing			
nails staggered in adjacent rows	2.5D	2.5D	2.5D
nails *in-line* in adjacent rows	5D	3D	2.5D

Figure 12.21 Recommended minimum nail spacings based on NDS Commentary provisions.

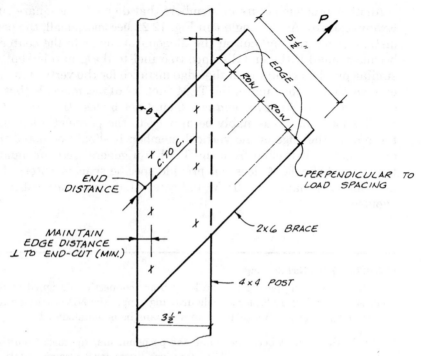

Figure 12.22 Detailing of nail connection for brace.

For connections without prebored holes, the NDS Commentary specifies

for the knee brace:

$$\text{Min. end distance} = 15D = 15 \times 0.162 = 2.43 \text{ in.}$$

$$\text{Min. spacing (c. to c. distance)} = 15D = 2.43 \text{ in.}$$

$$\text{Min. row spacing} = 5D = 5 \times 0.162 = 0.81 \text{ in.}$$

for the 4 × 4 post:

$$\text{Edge distance} = 2.5D = 2.5 \times 0.162 = 0.41 \text{ in.}$$

In order to accommodate these spacings, the post width must be:

$$(2 \times 2.43 \text{ in.} \times \cos \theta) + (2 \times 0.41 \text{ in.}) = (2 \times 2.43 \times 0.707) + (2 \times 0.41) = 4.26 \text{ in.}$$

Since the post is only 3.5 in. wide, a six nail pattern cannot be accommodated without prebored holes. However, if nail holes are prebored the spacing and end distance requirements for the knee brace reduce to:

$$\text{Min. end distance} = 10D = 10 \times 0.162 = 1.62 \text{ in.}$$

$$\text{Min. spacing (c. to c. distance)} = 10D = 1.62 \text{ in.}$$

$$\text{Min. row spacing} = 3D = 3 \times 0.162 = 0.49 \text{ in.}$$

The minimum required post width is reduced to:

$$(2 \times 1.62 \text{ in.} \times \cos \theta) + (2 \times 0.41 \text{ in.}) = (2 \times 1.62 \times 0.707) + (2 \times 0.41) = 3.11 \text{ in.}$$

Therefore a six nail pattern with prebored holes meets the NDS Commentary requirements for wood-to-wood connections.

A scale drawing is helpful in determining the spacing requirements for a connection. Because the spacing provisions perpendicular to the load are not specified, the total number of nails that can be used in a limited area is not fully defined.

The spacings given in Example 12.16 represent a mix of very limited criteria for nail connections from the NDS Commentary and the TCM:

1. Distance perpendicular to the end cut equal to the edge distance comes from the general spacing recommendations in the TCM.
2. Recommended spacings, end distances and edge distances for nailed connections with either wood or steel side members and with or without prebored holes are provided in the NDS Commentary.

Clearly there are a number of possible interpretations of nail spacing criteria. Furthermore, other sources offer much different spacing recommendations. For example, without prebored holes, Ref. 12.10 recommends an end distance of 20D and other spacings from this reference are generally larger than those cited.

At this point the reader should understand that certain connections will require careful consideration of nail spacing, and that considerable judgment is required in detailing connections. Nail patterns should be chosen by the designer with a basic understanding of the material and the tendency of wood to split if adequate spacing is not provided.

Definitive criteria for spacing in a nail connection does not exist at this time. The criteria included in the NDS Commentary represent committee consensus, but additional research is necessary to confirm spacing criteria.

12.15   Nailing Schedule

IBC Table 2304.9.1 is a *nailing schedule* which gives the minimum nailing that is to be used for a number of common connections in wood-frame construction. The table covers such items as the nailing of beams (joists and rafters) to the top plate of a wall. For this connection three 8d toenails are specified (Fig. 12.12b). The required nailing for numerous other connections is covered by this schedule. Copies of a number of IBC tables are included in Appendix C of this book.

The requirements given by the IBC should be regarded as a *minimum*. They apply to conventional wood-frame construction as well as construction that has been structurally designed. A larger number of nails (or some other type of connection hardware) may be required when structural calculations are performed. Special consideration should be given to the attachment of the various elements in the lateral-force-resisting system.

12.16 References

[12.1] American Forest and Paper Association (AF&PA). 2001. *Allowable Stress Design Manual for Engineered Wood Construction and Supplements and Guidelines,* 2001 ed., AF&PA, Washington, DC.

[12.2] American Forest and Paper Association (AF&PA). 1999. *Commentary on the National Design Specification for Wood Construction,* 1997 ed., AF&PA, Washington DC.

[12.3] American Forest and Paper Association (AF&PA). 2001. *National Design Specification for Wood Construction and Supplement.* 2001 ed., AF&PA, Association, Washington DC.

[12.4] American Forest and Paper Association (AF&PA). 1999. *Technical Report 12—General Dowel Equations for Calculating Lateral Connection Values,* AF&PA, Washington, DC.

[12.5] American Institute of Steel Construction (AISC). 1989. *Manual of Steel Construction,* 9th ed., AISC, Chicago, IL.

[12.6] American Institute of Timber Construction (AITC). 1994. *Timber Construction Manual,* 4th ed., AITC, Englewood, CO.

[12.7] American Society for Testing and Materials (ASTM). 2003. "Standard Terminology of Nails for Use with Wood and Wood-Base Materials," ASTM F547-01, *Annual Book of Standards, Vol. 01.08,* ASTM, Philadelphia, PA.

[12.8] American Society for Testing and Materials (ASTM). 2003. "Standard Test Method for Determining Bending Yield Moment of Nails," ASTM F1575-01, *Annual Book of Standards, Vol. 01.08,* ASTM, Philadelphia, PA.

[12.9] American Society for Testing and Materials (ASTM). 2003. "Standard Specification for Driven Fasteners: Nails, Spikes, and Staples," ASTM F1667-02, *Annual Book of Standards, Vol. 01.08,* ASTM, Philadelphia, PA.

[12.10] Hoyle, R.J., and Woeste, F.E. 1989. *Wood Technology in the Design of Structures,* 5th ed., Iowa State University Press, Ames, IA.

[12.11] International Codes Council (ICC). 2003. *International Building Code,* 2003 ed., ICC, Falls Church, VA.

[12.12] McLain, T.E. 1991. "Engineered Wood Connections and the 1991 National Design Specification," *Wood Design Focus,* Vol. 2, No. 2.

[12.13] National Evaluation Service Inc. 1997. "Power-Driven Staples and Nails for Use in All Types of Building Construction," Report No. NER-272, International Codes Council (Available from ISANTA, Chicago, IL).

[12.14] Stern, E.G. 1967. "Nails—Definitions and Sizes, a Handbook for Nail Users," Report No. 61, Virginia Tech, Blacksburg, VA

[12.15] Tissell, J.R., and Elliot, J.R. 1997. *Plywood Diaphragms, Research Report 138,* APA—The Engineered Wood Association, Engineered Wood Systems, Tacoma, WA.

12.17 Problems

Many of the nail problems in this section require the *nominal design value Z* and the *allowable design value Z'* as either final or intermediate answers. These values should be identified as part of the general solution along with the information specifically requested.

The nominal design values for a given nail may be obtained by solving the yield limit equations or by using the NDS nail tables. Tables are to be used only if they apply directly to the design conditions stated in the problem and are to be referenced in the solution. Yield limit equations may be evaluated by either calculator or computer. If the computer is used, output must be sufficient so that all steps in the solution can be verified.

In all problems normal temperatures are assumed, and $C_t = 1.0$. All adjustment factors are to be referenced to the appropriate NDS table or section so that the source of the value may be verified.

Some problems require the use of a spreadsheet or equation-solving software. Problems that are solved on a spreadsheet can be saved and used as a *template* for other similar problems. Templates can have many degrees of sophistication. Initially, a template may only be a hand (i.e., calculator) solution worked on a spreadsheet. In a simple template of this nature, the user will be required to provide many of the lookup functions for such items as

1. Nail diameter and length

2. Bending yield strength of nail

3. Specific gravity and dowel bearing strength of wood member

4. Dowel bearing strength of metal side member

As the user gains experience with spreadsheets, a template can be expanded to perform many of the lookup and decision-making functions that were previously done manually.

Advanced programming skills are not required to create effective spreadsheets templates. Valuable templates can be created by designers who normally do only hand solutions. However, some programming techniques are helpful in automating *lookup* and *decision-making* functions.

The first requirement is that a spreadsheet operate correctly (i.e., calculate correct values). Another major consideration is that the input and output must be structured in an orderly manner. A sufficient number of intermediate answers should be displayed and labeled in a template so that the solution can be verified by hand.

12.1 Use spreadsheet or equation-solving software to set up a template for the solution of the yield limit equations for a laterally loaded nail. Consider a single-shear wood-to-wood connection, using the following variables for input:

Nail diameter D

Nail length l

Thickness of side member t_s

Specific gravity of the side member G_s

Thickness of main member t_m

Specific gravity of main member G_m

The template is to do the following:

a. Compute the dowel bearing strength for the main and side members F_{es} and F_{em}.

b. Determine the dowel bearing length of the nail in the main member l_m.

c. Evaluate the coefficients for use in the yield limit equations.

d. Solve the four yield limit equations and select the smallest value as the *nominal design value Z*.

12.2 Expand or modify the template from Prob. 12.1 to handle a single-shear connection with a steel side plate. Input remains the same except the ultimate tensile strength of the metal side plate F_u replaces the specific gravity of the side member G_s.

12.3   Expand the template from Prob. 12.1 to handle laterally loaded *toenail* connections *in addition to* conventional shear connections. Input is to essentially remain the same as Prob. 12.1, but some method of distinguishing between a toenail and a nail not driven as a toenail is required.

One approach is to leave the thickness of the side member blank (i.e., $t_s = 0$) to indicate a toenail connection. The spreadsheet would then evaluate l_s for the toenail using the geometry in Fig. 12.12a.

12.4   Expand the template in Prob. 12.1, 12.2, or 12.3 (as assigned) to convert the *nominal design value Z* to an *allowable design value Z'*. The input must be expanded to include the appropriate adjustment factors. Include, as a minimum, the adjustments for:

a. Load duration factor C_D

b. Wet service factor C_M

c. End grain factor C_{eg}

d. Toenail factor C_{tn} (applies only if expanding Prob. 12.3)

12.5   The shed in Fig. 12.A has double 2×6 knee braces on each column to resist the lateral force. Lumber is Southern Pine. The braces are to be connected to the columns with 16d common nails. Lateral force to roof diaphragm w_L is 67 lb/ft. Heights: $h_1 = 9$ ft, $h_2 = 2$ ft.

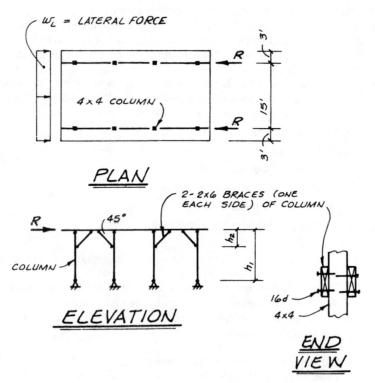

Figure 12.A

Find: a. The required number of nails for the brace connection, assuming that $C_M = 1.0$.

b. The required number of nails for the brace connection, assuming that the connection is exposed to the weather.

c. How would *a* and *b* change if threaded hardened-steel nails with $D = 0.148$ in. and $l = 3.5$ in. were used?

12.6 The shed in Fig. 12.A has double 2 × 6 knee braces on each column to resist the lateral force. Lumber is Redwood (open grain). The braces are to be connected to the columns with 16d common nails. Lateral force to roof diaphragm is $w_L = 46$ lb/ft. Heights: $h_1 = 12$ ft, $h_2 = 4$ ft.

Find: a. The required number of nails for the brace connection, assuming that $C_M = 1.0$.

b. The required number of nails for the brace connection, assuming that the connection is exposed to the weather.

c. How would *a* and *b* change if threaded hardened-steel nails with $D = 0.148$ in. and $l = 3.5$ in. were used?

12.7 Repeat Prob. 12.6 except that 16d box nails are to be used.

12.8 The single-story wood-frame building in Fig. 12.B has a double 2 × 4 top wall plate of Southern Pine. This top plate serves as the chord and drag strut along line 1. Loads are $w_T = 134$ lb/ft and $w_L = 223$ lb/ft. $C_M = 1.0$.

Find: a. The maximum chord force along line 1.

b. The maximum drag force along line 1.

c. Determine the number of 16d common nails necessary to splice the top plate for the maximum force determined in *a* and *b*.

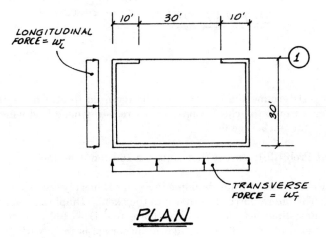

PLAN

Figure 12.B

12.9   Repeat Prob. 12.8 except that 16d box nails are to be used.

12.10 The single-story wood-frame building in Fig. 12.B has a double 2 × 4 top wall plate of Douglas Fir-Larch. The top plate serves as the chord and drag strut along line 1. Loads are $w_T = 100$ lb/ft and $w_L = 250$ lb/ft. $C_M = 1.0$.

   Find: a. The maximum chord force along line 1.
   b. The maximum drag force along line 1.
   c. Determine the number of 16d common nails necessary to splice the top plate for the maximum force determined in *a* and *b*.

12.11 Repeat Prob. 12.10 except that 16d box nails are to be used.

12.12 *Given:* The elevation of the shearwall in Fig. 12.C. The lateral wind force and the resisting dead load of the shearwall are shown on the sketch. The uplift force at the bottom of the wall is anchored to the foundation by a metal strap embedded in the concrete foundation. The 10-gage ASTM A653 grade 33 steel strap is attached to the shearwall chord (two 2 × 4s) with 10d common nails. Lumber is Hem-Fir. $C_M = 1.0$. Assume that the steel strap is adequate.

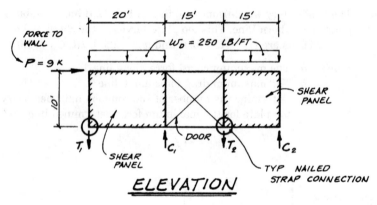

Figure 12.C

   Find: Determine the number of nails required to attach the metal anchor strap to the chord. Consider the resisting dead load when calculating the anchorage force.

12.13 Repeat Prob. 12.12 except that nails are 16d common nails.

12.14 The two-story wood-frame building in Fig. 12.D has plywood shear panels along line 1. The lateral forces are shown in the sketch. Diaphragm span length $L = 30$ ft. Resisting dead loads to line 1 are roof D = 100 lb/ft, floor D = 120 lb/ft, and wall D = 10 psf. Tie-down anchorage is to be with 12-gage ASTM A653 grade 33 steel straps and 16d common nails. Studs are 2 × 6 of Spruce-Pine-Fir (South). Assume that the metal strap is adequate. $C_M = 1.0$.

Find: The number of nails required for the anchorage connections at A and B.

12.15   Repeat Prob. 12.14 except that the diaphragm span length is $L = 40$ ft.

12.16   The attachment of a roof diaphragm to the supporting shearwall is shown in Fig. 12.E. The shear transferred to the wall is $v = 230$ lb/ft. Toenails in the blocking are used to transfer the shear from the roof diaphragm into the double top plate. Lumber is DF-L. $C_M = 1.0$.

Find: The required number of 8d common toenails per block if roof beams are 16 in o.c.

12.17   Repeat Prob. 12.16 except $v = 180$ lb/ft and nails are 8d box toenails.

12.18   The studs in the wood-frame wall in Fig. 12.F span between the foundation and the horizontal diaphragm. Roof beams are anchored to the wall top plate with 8d box toenails. Lumber is Hem-Fir. Roof beams are spaced 24 in o.c. Wind $= 20$ psf. Wall height $= 10$ ft. $C_M = 1.0$.

Find: The required number of toenails per roof beam.

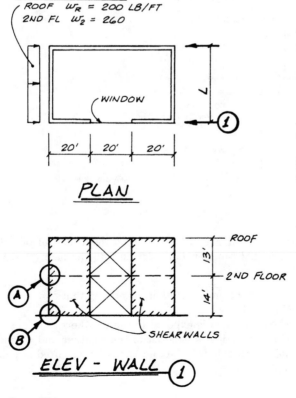

<u>DIAPHRAGM FORCES</u>
ROOF $w_R = 200$ LB/FT
2ND FL $w_2 = 260$

WINDOW

20'   20'   20'

<u>PLAN</u>

ROOF
13'
2ND FLOOR
14'
SHEARWALLS

<u>ELEV - WALL</u> ①

Figure 12.D

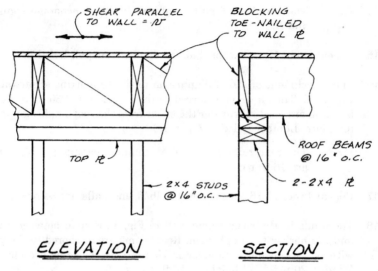

Figure 12.E

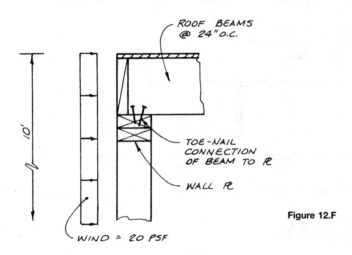

Figure 12.F

12.19 The support shown in Fig. 12.G carries a dead load of 500 lb. Lumber is
DF-L. Nails are 20d common nails driven into the side grain of the holding
member.

Find: The required number of nails, assuming that
  a. $C_M = 1.0$
  b. The connection is exposed to the weather.
  c. The connection is exposed to the weather and threaded hardened
       nails with $D = 0.177$ in. and $l = 4$ in. are used in place of common
       nails.

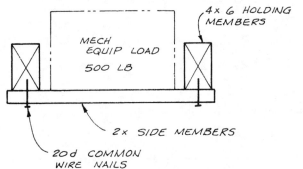

MECH
EQUIP LOAD
500 LB

4 x 6 HOLDING
MEMBERS

2 x SIDE MEMBERS

20 d COMMON
WIRE NAILS

Figure 12.G

12.20 Determine the IBC minimum required nailing for the following connections.
Give nail size, type, and required number.

a. Joist to sill or girder, toenail

b. Bridging to joist, toenail each end

c. Sole plate to joist or blocking, face nail

d. Top plate to stud, end nail

e. Stud to sole plate, toenail

f. Double studs, face nail

g. Doubled top plates, face nail

h. Top plates, laps, and intersections; face nail

i. Ceiling joists to plate, toenail

j. Continuous header to stud, toenail

k. Ceiling joists to parallel rafters, face nail

l. Rafter to plate, toenail

Bolts, Lag Bolts, and Other Connectors

13.1 Introduction

Once the vertical-load-supporting system and lateral-force-resisting system of a building have been designed, attention is turned to the design of the connections. The importance of connection design has been emphasized in previous chapters, and the methods used to calculate the forces on several types of connections have been illustrated.

When the design forces are relatively small, the connections may often be made with nails (Chap. 12). However, bolts, lag bolts, and other connectors are normally used for larger loads. There are several additional reasons why nailed connections were covered in a separate chapter. Recall from Chap. 11 that the design value for a small-diameter fastener, such as a nail, is independent of the angle of load to grain. The angle of load to grain, however, is a major consideration in the design of large-diameter fasteners. The other principal reason is that there are a number of adjustment factors that apply to bolts, lag bolts, and other connectors which do *not* apply to nails.

Connections made with larger-diameter fasteners often involve the use of some form of structural steel hardware. For many common connections this hardware may be available prefabricated from manufacturers or suppliers. Other connections, however, will require the fabrication of special hardware. Prefabricated hardware, when available, is often more economical than made-to-order hardware. Chapter 14 continues the subject of connection design by giving examples of common types of hardware. These examples are accompanied by comments about both good and poor connection layout practices.

The types of fasteners covered in this chapter include

1. Bolts
2. Lag bolts or lag screws

3. Other connectors
   a. Split rings
   b. Shear plates

Examples of several connections using these fasteners were given in the introductory chapter on connections. It is recommended that the reader review the joint configurations, nomenclature, and yield modes that are summarized in Chap. 11 before proceeding with the detailed design provisions given here. Because of their greater use, the majority of the coverage in Chap. 13 is devoted to bolted connections. This is followed by lag bolts. Split ring and shear plate connectors are often viewed as specialty connections, and for this reason their use is covered only briefly.

13.2   Bolt Connections

Most bolts are used in laterally loaded dowel-type connections. These are further described by giving the number of shear planes and the types of structural members connected. For example, the most common connections are either *single-shear* (two-member) or *double-shear* (three-member) joints. See Example 13.1. Connections can be *wood-to-wood* or *wood-to-metal*. A single-shear connection also occurs when a wood member is attached by anchor bolts to a concrete foundation or a concrete or masonry wall. These can be referred to as *wood-to-concrete* or *wood-to-masonry* connections, but they are often simply termed *anchor bolt* connections.

EXAMPLE 13.1   Bolts in Shear Connections

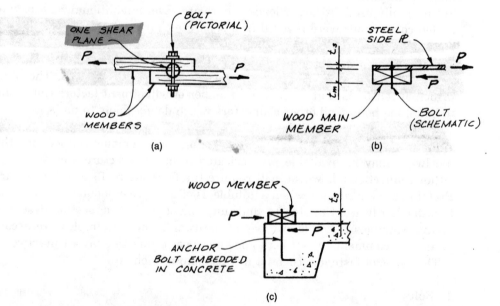

(a)

(b)

(c)

Figure 13.1 Single-shear connections. (*a*) Wood-to-wood. (*b*) Wood-to-metal. (*c*) Wood-to-concrete anchor bolt.

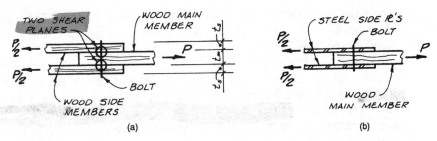

Figure 13.2 Double-shear connections. (*a*) Wood-to-wood. (*b*) Wood-to-metal.

In shear connections the load is perpendicular to the axis of the bolt. The angle of load to grain in the wood can be zero (load parallel to grain), 90 degrees (load perpendicular to grain), or at some intermediate angle θ. The angle of load to grain affects the design value of bolted connections.

Bolted connections can have more than two shear planes (i.e., more than three members). For multiple shear planes, the NDS (Ref. 13.1) states that the design value is to be evaluated by first determining the single-shear capacity for each shear plane. The nominal design value for the connection is determined by multiplying the value for the weakest shear plane by the number of shear planes. In this book, only the common cases of single- and double-shear bolt connections are covered.

Washers of adequate size are to be provided between a wood member and the bolt head, and between the wood member and the nut. Washers can be circular or square. The use of washers applies to bolts in shear as well as to those loaded in tension. For bolts in tension (Fig. 11.2*b* in Sec. 11.2), the size of washer needs to be such that the bearing stress in the wood member does not exceed the allowable compressive stress perpendicular to grain $F'_{c\perp}$.

The size of the washer is not particularly critical for bolts loaded in shear, and the washer simply protects the wood as the nut is tightened. The nut should be snugged tight, but care should be taken so that the nut is not installed too tight. In other words, the washer should not be embedded in the wood member by overtightening the nut. On the other hand, if a wood member seasons in service, the width of the wood member may shrink so that the nut may need to be retightened after a period of time.

The quality of bolts used in wood connections has not been a major concern in the past. However, with the move to a design procedure based on the yield model for dowel-type fasteners, there is a need to have recognized mechanical properties for bolts. Recall that some yield modes are based on the formation of one or more plastic hinges in the fastener, and this requires bolts with predictable bending yield strengths F_{yb} and ductility characteristics.

Designers familiar with steel know that the trend in steel structures is to move away from the use of low-strength A307 bolts and toward the use of high-strength A325 bolts. However, this is not the case in wood structures, and A307 bolts are often used in wood connections. The current problem with specifying A307 bolts is that the ASTM standard for these bolts provides only an ultimate tensile strength F_u, and there is no minimum F_y covered by the specification (see Ref. 13.8).

Without a published value for yield strength, there is a conceptual problem in using the yield limit model with ASTM A307 bolts. As a practical matter, A307 bolts have performed satisfactorily in the past, and these fasteners must have a reasonable F_y. The tables in the NDS are based on $F_{yb} = 45$ ksi, which is generally thought to be a conservative value of bending yield strength for A307 bolts.

As an alternative to specifying A307 bolts, the designer may choose to use another bolt specification such as SAE J429, *Mechanical and Material Requirements for Externally Threaded Fasteners* (Ref. 13.11). This specification provides a more complete set of strength criteria for bolts including F_y. Note that F_y is the yield strength determined from a tension test. See NDS Appendix I for an approximate relationship for bending yield strength F_{yb} as a function of tension yield F_y and tension ultimate F_u.

Bolts should be installed in properly drilled holes, and members in a connection should be aligned so that only light tapping is required to insert the bolts. If the bolt holes are too small, an excessive amount of driving will be required to install the bolts, and splitting of the wood may be induced. Obviously, splitting will greatly reduce the shear capacity of a bolted connection. On the other hand, if bolt holes are too large, nonuniform bearing stresses may occur. The NDS specifies that bolt holes are to be a minimum of $\frac{1}{32}$ in. to a maximum of $\frac{1}{16}$ in. larger than the bolt diameter.

The manner in which the holes are bored in the wood also has an effect on load capacity. A bolt hole with a smooth surface develops a higher load with less deformation than one that has a rough surface. See Fig. 13.3. Smooth surfaces are obtained with sharp drill bits, a proper drill speed, and a slow rate of feed. See the *Wood Handbook* (Ref. 13.9) for more information.

A related problem concerning fabrication tolerances can occur for A307 bolts in wood construction. Manufacturers of A307 bolts are permitted to provide *reduced diameter body* bolts having unthreaded shank diameter approximately 9 percent smaller than the *outside-to-outside* thread diameter (major diameter) of the bolts. This effectively doubles the gap between the bolt shank and the inner surface of the bolt hole when reduced diameter body bolts are installed in connections with properly drilled holes. The use of reduced diameter body bolts may lead to excessive displacement at connections, and can also preclude effective load distribution among individual fasteners in multiple-bolt connections. In order to avoid these problems, it is recommended that designers specify *full diameter body* A307 bolts for connections where overall displacement or load distribution to all fasteners in a group is critical.

Bolt sizes used in wood connections typically range from $\frac{1}{2}$ through 1 in. in diameter, in increments of $\frac{1}{8}$ in. In the past, large bolts ($1\frac{1}{4}$- and $1\frac{1}{2}$-in. diameter) were also used. However, for a number of years AITC expressed concern about the use of large-diameter bolts in wood members (see Ref. 13.4).

Figure 13.3 Fabrication technique affects the performance of bolt connections. The bolt hole on the left has a rough surface as the result of torn wood fibers due to improper drilling. The hole on the right is smooth and will make a better connection. Both holes were made with the same twist drill bit. The smooth hole was produced by using a higher drill speed (rpm) and a lower rate of feed (in./min) through the piece. (*Courtesy of the FPL.*)

The reasons cited for this concern are the difficulty in accurately drilling larger holes and the difficulty in aligning large-diameter bolts in a multiple-bolt connection. There is some opinion that the greater stiffness of large-diameter bolts may not allow uniform bearing on all fasteners in a connection. If there is nonuniform bearing in a connection as the result of inaccuracies in fabrication (Sec. 7.2), small-diameter bolts may be flexible enough to allow a redistribution of stresses in the connection so that the bolts become more uniformly loaded. With large-diameter bolts there is concern that splitting of the wood member may occur before a redistribution of stress to less highly loaded bolts can occur. For these reasons, the NDS effectively restricts the use of large diameter bolts in wood connections by only providing dowel bearing strengths (F_e) and tabulated design values (Z) for bolts 1 in. and less in diameter. Additional research should clarify the effect of fabrication tolerances on the strength of large- and small-diameter bolts.

13.3 Bolt Yield Limit Equations for Single Shear

The yield limit equations for bolts in single shear are covered in this section. The value obtained from the yield limit equations is the *nominal design value* Z for a single bolt. The adjustment factors necessary to convert the nominal value to an *allowable design value* Z' are covered in Sec. 13.5.

Many of the coefficients needed to evaluate the yield limit equations have been previously defined. Recall that for a large-diameter fastener, such as a bolt, the dowel bearing strength may be different for the main and side members even if the members are from the same species. This occurs when the angle of load to grain is different for the main and side members ($\theta_m \neq \theta_s$).

Recall also from Sec. 11.5 that the dowel bearing strength parallel to grain $F_{e\parallel}$ depends only on the specific gravity of the wood. However, dowel bearing strength perpendicular to grain $F_{e\perp}$ is a function of both specific gravity and bolt diameter. $F_{e\parallel}$ and $F_{e\perp}$ can be computed given the specific gravity of the wood and the fastener diameter, or values can be obtained from NDS Table 11.3.2, *Dowel Bearing Strengths*. If needed, the dowel bearing strength at an angle of load to grain $F_{e\theta}$ can be determined using the Hankinson formula. The dowel bearing strength of a metal member may be taken equal to 1.5 times the ultimate tensile strength for hot-rolled steel plates ($F_e = 1.5F_u$) or 1.375 times the ultimate tensile strength for cold-formed steel plates ($F_e = 1.375F_u$) (see Sec. 11.5). Finally, in wood-to-concrete connections the NDS recommends a dowel bearing strength (F_e) of 6000 psi. This value is generally considered appropriate for both concrete and masonry (see Sec. 15.4).

Yield limit equations from the NDS for a *bolt* in *single shear* are given in Example 13.2. This type of connection requires six yield limit equations because any of the six possible yield modes (Fig. 11.10 in Sec. 11.7) may control. The equations require the use of several coefficients that are, in turn, defined by some rather lengthy expressions. Again, the use of computer spreadsheet or equation-solving software greatly aids in the solution of the yield limit equations.

EXAMPLE 13.2   Bolt Yield Limit Equations for Single Shear

The following yield limit equations apply to the three types of single-shear bolt connections that were illustrated in Fig. 13.1. Yield modes are sketched in Fig. 11.10 (Sec. 11.7).

The nominal design value for one bolt in a single-shear connection between two members is the smallest value obtained from the following six equations.

Mode I_m (NDS equation 11.3-1):

$$Z = \frac{Dl_m F_{em}}{4K_\theta}$$

Mode I_s (NDS equation 11.3-2):

$$Z = \frac{Dl_s F_{es}}{4K_\theta}$$

Mode II (NDS equation 11.3-3):

$$Z = \frac{k_1 Dl_s F_{es}}{3.6K_\theta}$$

Mode III_m (NDS equation 11.3-4):

$$Z = \frac{k_2 Dl_m F_{em}}{3.2(1 + 2R_e)K_\theta}$$

Mode III$_s$ (NDS equation 11.3-5):

$$Z = \frac{k_3 D l_s F_{em}}{3.2(2 + R_e)K_\theta}$$

Mode IV (NDS equation 11.3-6):

$$Z = \frac{D^2}{3.2K_\theta} \sqrt{\frac{2F_{em}F_{yb}}{3(1 + R_e)}}$$

where Z = nominal design value for bolt in single shear (Z is to be taken as the smallest value from six yield limit equations), lb

$$k_1 = \frac{\sqrt{R_e + 2R_e^2(1 + R_t + R_t^2) + R_t^2 R_e^3} - R_e(1 + R_t)}{(1 + R_e)}$$

$$k_2 = -1 + \sqrt{2(1 + R_e) + \frac{2F_{yb}(1 + 2R_e)D^2}{3F_{em}l_m^2}}$$

$$k_3 = -1 + \sqrt{\frac{2(1 + R_e)}{R_e} + \frac{2F_{yb}(2 + R_e)D^2}{3F_{em}l_s^2}}$$

$$R_e = \frac{F_{em}}{F_{es}}$$

$$R_t = \frac{l_m}{l_s}$$

$$K_\theta = 1 + \frac{\theta}{360}$$

D = bolt diameter, in.
l_m = dowel bearing length in main (thicker) member, in.
l_s = dowel bearing length in side (thinner) member, in.
F_{em} = dowel bearing strength of main member, psi:
$$= \begin{cases} F_{e\parallel} & \text{for load parallel to grain} \\ F_{e\perp} & \text{for load perpendicular to grain} \\ F_{e\theta} & \text{for load at angle to grain } \theta \text{ (see Hankinson formula below)} \end{cases}$$
  For a concrete or masonry main member, F_{em} = 6000 psi
F_{es} = dowel bearing strength of side member, psi:
$$= \begin{cases} F_{e\parallel} & \text{for load parallel to grain} \\ F_{e\perp} & \text{for load perpendicular to grain} \\ F_{e\theta} & \text{for load at angle to grain } \theta \text{ (see Hankinson formula below)} \end{cases}$$
  For a hot-rolled steel side member, $F_{es} = 1.5F_u$
  For a cold-formed steel side member, $F_{es} = 1.375F_u$
F_{yb} = bending yield strength of bolt, psi
θ = maximum angle of load to grain ($0 \le \theta \le 90$ degrees) for any member in connection

Dowel bearing strength at an angle of load to grain θ is given by the Hankinson formula:

$$F_{e\theta} = \frac{F_{e\parallel}F_{e\perp}}{F_{e\parallel} \sin^2 \theta + F_{e\perp} \cos^2 \theta}$$

In wood-to-metal connections the designer must ensure that the bearing capacity of the steel side member is not exceeded in accordance with recognized steel design practice. In wood-to-concrete or wood-to-masonry connections the designer must ensure that the bearing capacity of the concrete or masonry is not exceeded in accordance with recognized concrete and masonry design practice. This can be done for concrete in IBC Table 1912.2 (Ref. 13.10). See Example 15.6 (Sec. 15.4) for two sample anchor bolt problems.

As an alternative to solving the yield limit equations, the nominal design values for a number of common *single-shear* connections are available in NDS Tables 11A to 11E.

For *wood-to-wood* connections, four nominal design values are given in the NDS tables. Four values are listed because the equations used to generate the tables produce a different bolt capacity depending on the *angle of load to grain* for the members in the connection. Values of Z are given for the following conditions:

1. Load *parallel* to grain in *both* main member and side member. The notation in the NDS tables for this condition is $Z_{\parallel}$.

2. Load *perpendicular* to grain in *side* member and parallel to grain in main member, $Z_{s\perp}$.

3. Load *perpendicular* to grain in *main* member and parallel to grain in side member, $Z_{m\perp}$.

4. Load *perpendicular* to grain in *both* main member and side member, $Z_{\perp}$.

For *wood-to-steel, wood-to-concrete,* and *wood-to-masonry* connections, the NDS tables provide two values of Z based on the angle of load to grain for the wood member:

1. Load parallel to grain in wood member, $Z_{\parallel}$.

2. Load perpendicular to grain in wood member, $Z_{\perp}$.

Metal side plates are assumed to be ¼-in.-thick ASTM A36 steel in NDS Tables 11B and 11D.

The NDS tables may be helpful in providing nominal bolt design values for a number of loading conditions without the need to evaluate the yield limit formulas. However, the NDS tables are restricted to parallel- and perpendicular-to-grain loadings, and the yield limit equations must be evaluated for connections with an intermediate angle of load to grain ($0 < \theta < 90$ degrees).

Two examples are given to illustrate the application of the yield limit equations. The first problem verifies a bolt value in one of the NDS tables. See Example 13.3. The load is perpendicular to the grain in the main member and parallel to the grain in the side member. Both wood pieces are of the same species, but there are different dowel bearing strengths for the two members because of the angle of load to grain. The problem is not a complete design because the conditions necessary to convert the nominal design value to an allowable design value are not given. The example is simply intended to illustrate the yield limit equations and the NDS tables.

EXAMPLE 13.3 Nominal Design Value: Single-Shear Bolt in Wood-to-Wood Connection

The horizontal load in the connection shown in Fig. 13.4 is parallel to the grain in the horizontal member and perpendicular to the grain in the vertical member. Lumber is DF-L. Determine the nominal design value of the single ¾-in. full-diameter body A307 bolt.

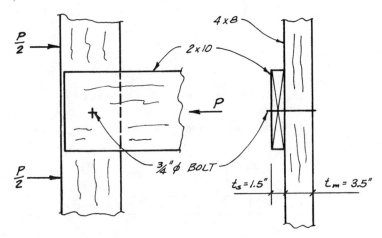

Figure 13.4 Single-shear wood-to-wood connection.

Summary of Known Values

Summarize the information necessary to evaluate the yield limit equations. Bolt diameter, member thicknesses, and angles of load to grain are obtained from the sketch. Dowel bearing strengths can be computed using the formulas from Example 11.4 (Sec. 11.5), or they can be read from NDS Table 11.3.2. Specific gravity is given in NDS Table 11.3.2A.

$$D = 0.75 \text{ in.}$$

$$F_{yb} = 45 \text{ ksi} \qquad \text{(assumed for A307 bolt)}$$

$$l_m = t_m = 3.5 \text{ in.}$$

$$\theta_m = 90 \text{ degrees}$$

$$F_{em} = 2600 \text{ psi} \qquad (F_{e\perp} \text{ for ¾-in. bolt in DF-L})$$

$$l_s = t_s = 1.5 \text{ in.}$$

$$\theta_s = 0 \text{ degrees}$$

$$F_{es} = 5600 \text{ psi} \qquad (F_{e\parallel} \text{ for bolt in DF-L})$$

$$G_m = G_s = 0.50$$

Coefficients for Yield Limit Equations

$$R_e = \frac{F_{em}}{F_{es}} = \frac{2600}{5600} = 0.464$$

$$1 + R_e = 1 + 0.464 = 1.464$$

$$1 + 2R_e = 1 + 2(0.464) = 1.929$$

$$2 + R_e = 2 + 0.464 = 2.464$$

$$R_t = \frac{l_m}{l_s} = \frac{3.5}{1.5} = 2.333$$

$$1 + R_t = 1 + 2.333 = 3.333$$

$$1 + R_t + R_t^2 = 3.333 + (2.333)^2 = 8.778$$

$$\theta = 90 \text{ degrees} \qquad (\textit{Note: } \theta \text{ is the larger of } \theta_m \text{ and } \theta_s.)$$

$$K_\theta = 1 + \frac{\theta}{360} = 1 + \frac{90}{360} = 1.25$$

$$k_1 = \frac{\sqrt{R_e + 2R_e^2(1 + R_t + R_t^2) + R_t^2 R_e^3} - R_e(1 + R_t)}{(1 + R_e)}$$

$$= \frac{\sqrt{0.464 + 2(0.464^2)(8.778) + (2.333)^2(0.464^3)} - 0.464(3.333)}{1.464}$$

$$= 0.4383$$

$$k_2 = -1 + \sqrt{2(1 + R_e) + \frac{2F_{yb}(1 + 2R_e)D^2}{3F_{em}l_m^2}}$$

$$= -1 + \sqrt{2(1.464) + \frac{2(45,000)(1.929)(0.75^2)}{3(2600)(3.5^2)}}$$

$$= 0.9876$$

$$k_3 = -1 + \sqrt{\frac{2(1 + R_e)}{R_e} + \frac{2F_{yb}(2 + R_e)D^2}{3F_{em}l_s^2}}$$

$$= -1 + \sqrt{\frac{2(1.464)}{0.464} + \frac{2(45,000)(2.464)(0.75^2)}{3(2600)(1.5^2)}}$$

$$= 2.663$$

Yield Limit Equations

Mode I_m (NDS equation 11.3-1):

$$Z = \frac{Dl_m F_{em}}{4K_\theta} = \frac{0.75(3.5)(2600)}{4(1.25)} = 1365 \text{ lb}$$

Mode I_s (NDS equation 11.3-2):

$$Z = \frac{Dl_sF_{es}}{4K_\theta} = \frac{0.75(1.5)(5600)}{4(1.25)} = 1260 \text{ lb}$$

Mode II (NDS equation 11.3-3):

$$Z = \frac{k_1Dl_sF_{es}}{3.6K_\theta} = \frac{0.4383(0.75)(1.5)(5600)}{3.6(1.25)} = 614 \text{ lb}$$

Mode III_m (NDS equation 11.3-4):

$$Z = \frac{k_2Dl_mF_{em}}{3.2(1 + 2R_e)K_\theta} = \frac{0.9876(0.75)(3.5)(2600)}{3.2(1.929)(1.25)} = 874 \text{ lb}$$

Mode III_s (NDS equation 11.3-5):

$$Z = \frac{k_3Dl_sF_{em}}{3.2(2 + R_e)K_\theta} = \frac{2.663(0.75)(1.5)(2600)}{3.2(2.464)(1.25)} = 790 \text{ lb}$$

Mode IV (NDS equation 11.3-6):

$$Z = \frac{D^2}{3.2K_\theta}\sqrt{\frac{2F_{em}F_{yb}}{3(1 + R_e)}}$$

$$= \frac{0.75^2}{3.2(1.25)}\sqrt{\frac{2(2600)(45,000)}{3(1.464)}} = 1026 \text{ lb}$$

The nominal design value is selected as the smallest value from the yield limit equations:

$$\boxed{Z = 614 \text{ lb}}$$

Compare with value from NDS Table 11A, *Bolt Design Values (Z) for Single Shear (two member) Connections.* Enter table with t_m = 3.5, t_s = 1.5, D = 0.75, and G = 0.50:

$$Z_{m\perp} = 610 \text{ lb} \approx 614 \text{ lb} \qquad OK$$

The second example is a bolt in single shear that is not covered by the NDS tables, and the yield limit equations must be used. The connection is between a 4× wood member and a ½-in.-thick hot-rolled angle. The angle is A36 steel, and the minimum tensile strength is F_u = 58 ksi. Example 13.4 has an angle of load to grain of 45 degrees, and $F_{e\theta}$ is computed with the Hankinson formula.

EXAMPLE 13.4   Nominal Design Value: Single-Shear Bolt in Wood-to-Metal Connection
The bolt through steel angle in Fig. 13.5 provides a load to the horizontal wood member at an angle of 45 degrees to the grain. The angle is A36 steel, and the wood member

is Hem-Fir. Determine the nominal design value of a single ⅝-in. full-diameter body A307 bolt.

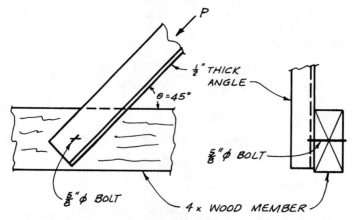

Figure 13.5 Single-shear wood-to-steel connection.

Summary of Known Values

Dowel bearing strength parallel and perpendicular to grain in Hem-Fir can be computed using the formulas in Example 11.4 (Sec. 11.5), or they can be obtained from NDS Table 11.3.2.

$$D = 0.625 \text{ in.}$$

$$F_{yb} = 45 \text{ ksi} \quad \text{(assumed for A307 bolt)}$$

$$l_m = t_m = 3.5 \text{ in.}$$

$$\theta_m = 45 \text{ degrees}$$

$$F_{em\parallel} = 4800 \text{ psi}$$

$$F_{em\perp} = 2250 \text{ psi} \quad (F_{e\perp} \text{ for ⅝-in. bolt in Hem-Fir})$$

$$l_s = t_s = 0.5 \text{ in.}$$

$$F_{es} = 1.5F_u = 87{,}000 \text{ psi} \quad \text{(for A36 steel)}$$

Coefficients for Yield Limit Equations

$$F_{e\theta} = \frac{F_{e\parallel}F_{e\perp}}{F_{e\parallel}\sin^2\theta + F_{e\perp}\cos^2\theta}$$

$$= \frac{4800(2250)}{4800\sin^2 45 + 2250\cos^2 45} = 3064 \text{ psi}$$

$$R_e = \frac{F_{em}}{F_{es}} = \frac{3064}{87{,}000} = 0.03522$$

$$1 + R_e = 1 + 0.03522 = 1.035$$

$$1 + 2R_e = 1 + 2(0.03522) = 1.070$$

$$2 + R_e = 2 + 0.03522 = 2.035$$

$$R_t = \frac{l_m}{l_s} = \frac{3.5}{0.5} = 7.0$$

$$1 + R_t = 1 + 7.0 = 8.0$$

$$1 + R_t + R_t^2 = 8.0 + 7.0^2 = 57.0$$

$$\theta = 45 \text{ degrees}$$

$$K_\theta = 1 + \frac{\theta}{360} = 1 + \frac{45}{360} = 1.125$$

$$k_1 = \frac{\sqrt{R_e + 2R_e^2(1 + R_t + R_t^2) + R_t^2 R_e^3} - R_e(1 + R_t)}{(1 + R_e)}$$

$$= \frac{\sqrt{0.03522 + 2(0.03522^2)(57.0) + 7.0^2(0.03522^3)} - 0.03522(8.0)}{1.035}$$

$$= 0.1367$$

$$k_2 = -1 + \sqrt{2(1 + R_e) + \frac{2F_{yb}(1 + 2R_e)D^2}{3F_{em}l_m^2}}$$

$$= -1 + \sqrt{2(1.035) + \frac{2(45,000)(1.070)(0.625^2)}{3(3064)(3.5^2)}}$$

$$= 0.5507$$

$$k_3 = -1 + \sqrt{\frac{2(1 + R_e)}{R_e} + \frac{2F_{yb}(2 + R_e)D^2}{3F_{em}l_s^2}}$$

$$= -1 + \sqrt{\frac{2(1.035)}{0.03522} + \frac{2(45,000)(2.035)(0.625^2)}{3(3064)(0.5^2)}}$$

$$= 8.483$$

Yield Limit Equations

Mode I_m (NDS equation 11.3-1):

$$Z = \frac{Dl_m F_{em}}{4K_\theta} = \frac{0.625(3.5)(3064)}{4(1.125)} = 1489 \text{ lb}$$

Mode I_s (NDS equation 11.3-2):

$$Z = \frac{Dl_s F_{es}}{4K_\theta} = \frac{0.625(0.5)(87,000)}{4(1.125)} = 6042 \text{ lb}$$

Mode II (NDS equation 11.3-3):

$$Z = \frac{k_1 D l_s F_{es}}{3.6 K_\theta} = \frac{0.1367(0.625)(0.5)(87,000)}{3.6(1.125)} = 917 \text{ lb}$$

Mode III$_m$ (NDS equation 11.3-4):

$$Z = \frac{k_2 D l_m F_{em}}{3.2(1 + 2R_e)K_\theta} = \frac{0.5507(0.625)(3.5)(3064)}{3.2(1.070)(1.125)} = 958 \text{ lb}$$

Mode III$_s$ (NDS equation 11.3-5):

$$Z = \frac{k_3 D l_s F_{em}}{3.2(2 + R_e)K_\theta} = \frac{8.483(0.625)(0.5)(3064)}{3.2(2.035)(1.125)} = 1109 \text{ lb}$$

Mode IV (NDS equation 11.3-6):

$$Z = \frac{D^2}{3.2K_\theta} \sqrt{\frac{2F_{em}F_{yb}}{3(1 + R_e)}}$$

$$= \frac{0.625^2}{3.2(1.125)} \sqrt{\frac{2(3064)(45,000)}{3(1.035)}} = 1022 \text{ lb}$$

The nominal design value is selected as the smallest value from the yield limit equations:

$$\boxed{Z = 917 \text{ lb}}$$

13.4 Bolt Yield Limit Equations for Double Shear

The yield limit equations and several numerical examples for bolts in single-shear connections were given in the previous section. In addition to this application, bolts are frequently used in double shear. From a review of the various yield modes in Fig. 11.10 (Sec. 11.7), it should be clear that there are a different set of yield limit equations for double- and single-shear connections.

The yield limit equations for bolt connections in double shear are given in Example 13.5. Because of symmetric behavior, only four yield equations are required for a double-shear wood-to-wood connection. In addition, the steel side members must be designed in accordance with recognized steel design practice (Ref. 13.3).

EXAMPLE 13.5 Yield Limit Equations for Bolts in Double Shear

The following yield limit equations from the NDS apply to the two common types of double-shear connections illustrated in Fig. 13.2. The yield modes are sketched in Fig. 11.10.

Wood-to-Wood Connection

The nominal design value for one bolt in a double-shear connection between three wood members is the smallest value obtained from the following four equations. The thicknesses of the *side* members are assumed to be equal. See the NDS for the case of unequal thickness.

Mode I_m (NDS equation 11.3-7):

$$Z = \frac{D l_m F_{em}}{4 K_\theta}$$

Mode I_s (NDS equation 11.3-8):

$$Z = \frac{2 D l_s F_{es}}{4 K_\theta}$$

Mode III_s (NDS equation 11.3-9):

$$Z = \frac{2 k_3 D l_s F_{em}}{3.2 (2 + R_e) K_\theta}$$

Mode IV (NDS equation 11.3-10):

$$Z = \frac{2 D^2}{3.2 K_\theta} \sqrt{\frac{2 F_{em} F_{yb}}{3(1 + R_e)}}$$

where Z = nominal design value for bolt in double shear (Z is to be taken as smallest value from four yield limit equations), lb

$$k_3 = -1 + \sqrt{\frac{2(1 + R_e)}{R_e} + \frac{2 F_{yb}(2 + R_e) D^2}{3 F_{em} l_s^2}}$$

$$R_e = \frac{F_{em}}{F_{es}}$$

$$K_\theta = 1 + \frac{\theta}{360}$$

D = bolt diameter, in.
l_m = dowel bearing length in main (center) member, in.
l_s = dowel bearing length in one of the side members, in.
F_{em} = dowel bearing strength of main (center) member, psi:

$$= \begin{cases} F_{e\parallel} & \text{for load parallel to grain} \\ F_{e\perp} & \text{for load perpendicular to grain} \\ F_{e\theta} & \text{for load at angle to grain } \theta \text{ (see Hankinson formula below)} \end{cases}$$

F_{es} = dowel bearing strength of side member, psi:

$$= \begin{cases} F_{e\parallel} & \text{for load parallel to grain} \\ F_{e\perp} & \text{for load perpendicular to grain} \\ F_{e\theta} & \text{for load at angle to grain } \theta \text{ (see Hankinson formula below)} \end{cases}$$

(For hot-rolled steel side members, $F_{es} = 1.5 F_u$.)
F_{yb} = bending yield strength of bolt, psi
θ = maximum angle of load to grain ($0 \le \theta \le 90$ degrees) for any member in connection

Dowel bearing strength at an angle of load to grain θ is given by the Hankinson formula:

$$F_{e\theta} = \frac{F_{e\|}F_{e\perp}}{F_{e\|}\sin^2\theta + F_{e\perp}\cos^2\theta}$$

Wood-to-Metal Connections

For one bolt in a double-shear connection between a *wood main member* and *two steel side plates,* the nominal design value is taken as the smallest value obtained from the four yield limit equations given above for a double-shear connection. In addition, the designer must ensure that the bearing capacity of the steel side members is not exceeded in accordance with accepted steel design practice.

A less frequently encountered double-shear connection involves a *steel main (center) member* and *two wood side members.* In this situation a Mode I_m mechanism represents a bearing failure in the main steel member. The yield limit equations are evaluated to determine the nominal design value for the bolt. In this case also, the designer must ensure that the bearing capacity of the steel member is adequate in accordance with recognized steel design practice.

As with single shear, the NDS bolt tables may be helpful in providing nominal design values for a number of loading conditions without the need to evaluate the yield limit formulas. Values of Z for double-shear connections are given in NDS Tables 11F to 11I. Again, the NDS tables are limited to parallel- and perpendicular-to-grain loading only, and the yield limit equations are required for connections with an intermediate angle of load to grain $(0 < \theta < 90$ degrees).

One example is given to illustrate the yield limit equations for double shear and to gain familiarity with the NDS bolt tables. See Example 13.6. Although the example deals with a connection that can be read from an NDS table, a computer solution on a spreadsheet or other program will normally be used for the cases not covered by a table. The example illustrates the yield limit formulas, and the adjustment factors necessary to convert the nominal design value to an allowable bolt capacity are covered in the next section.

EXAMPLE 13.6   Nominal Design Value: Double-Shear Bolt in Wood-to-Metal Connection

Determine the nominal design value of the single $\frac{7}{8}$-in. full-diameter body A307 bolt in Fig. 13.6. The load in the connection is parallel to the grain in the 6× wood member, and $\frac{1}{4}$-in.-thick steel side plates are used. The wood member is DF-L, and the side plates are A36 steel with $F_u = 58$ ksi.

Summary of Known Values

Dowel bearing strength of the wood member can be computed using the formulas in Example 11.4 (Sec. 11.5), or it can be read from NDS Table 11.3.2. Specific gravity is also available in NDS Table 11.3.2A.

$$D = 0.875 \text{ in.}$$

$$F_{yb} = 45 \text{ ksi} \quad \text{(assumed for A307 bolt)}$$

$$l_m = t_m = 5.5 \text{ in.}$$

$$\theta_m = 0 \text{ degrees}$$

$$F_{em} = 5600 \text{ psi} \quad (F_{e\parallel} \text{ for DF-L})$$

$$G_m = 0.50$$

$$l_s = t_s = 0.25 \text{ in.}$$

$$F_{es} = 1.5F_u = 87,000 \text{ psi} \quad \text{(for A36 steel)}$$

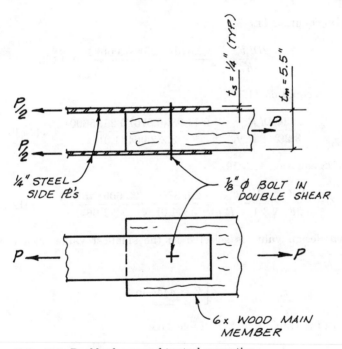

Figure 13.6 Double-shear wood-to-steel connection.

Coefficients for Yield Limit Equations

$$R_e = \frac{F_{em}}{F_{es}} = \frac{5600}{87,000} = 0.06437$$

$$1 + R_e = 1 + 0.06437 = 1.064$$

$$2 + R_e = 2 + 0.06437 = 2.064$$

$$\theta = 0 \text{ degrees}$$

$$K_\theta = 1 + \frac{\theta}{360} = 1 + \frac{0}{360} = 1.0$$

$$k_3 = -1 + \sqrt{\frac{2(1 + R_e)}{R_e} + \frac{2F_{yb}(2 + R_e)D^2}{3F_{em}l_s^2}}$$

$$= -1 + \sqrt{\frac{2(1.064)}{0.06437} + \frac{2(45,000)(2.064)(0.875^2)}{3(5600)(0.25^2)}}$$

$$= 11.98$$

Yield Limit Equations

Mode I_m (NDS equation 11.3-7):

$$Z = \frac{Dl_m F_{em}}{4K_\theta} = \frac{0.875(5.5)(5600)}{4(1.0)} = 6783 \text{ lb}$$

Mode I_s (NDS equation 11.3-8):

$$Z = \frac{2Dl_s F_{es}}{4K_\theta} = \frac{2(0.875)(0.25)(87,000)}{4(1.0)} = 9516 \text{ lb}$$

Mode III_s (NDS equation 11.3-9):

$$Z = \frac{2k_3 Dl_s F_{em}}{3.2(2 + R_e)K_\theta} = \frac{2(11.98)(0.875)(0.25)(5600)}{3.2(2.064)(1.0)} = 4444 \text{ lb}$$

Mode IV (NDS equation 11.3-10):

$$Z = \frac{2D^2}{3.2K_\theta}\sqrt{\frac{2F_{em}F_{yb}}{3(1 + R_e)}} = \frac{2(0.875^2)}{3.2(1.0)}\sqrt{\frac{2(5600)(45,000)}{3(1.064)}} = 6012 \text{ lb}$$

The nominal design value is selected as the smallest value from the yield limit equations:

$$\boxed{Z = 4444 \text{ lb}}$$

Compare with the value from NDS Table 11G:

$$Z_\parallel = 4440 \approx 4444 \text{ lb} \qquad OK$$

13.5 Adjustment Factors for Bolts

The yield limit equations for single- and double-shear bolt connections were reviewed in previous sections. The term *nominal design value* is assigned to the capacity of a single bolt as obtained from the yield limit equations. For use in the design of a particular connection, the nominal design value Z must be converted to an *allowable design value* Z'. As with nails, this is done by multiplying the nominal value by a series of adjustment factors.

The adjustments convert the base conditions for a bolt to the design conditions for a given problem. Depending on the circumstances, the adjustment factors could have a major or minor effect on the allowable design value. The adjustment factors required in the design of connections are summarized in NDS Table 10.3.1. If the particular design conditions agree with the conditions for the nominal design value, the adjustment factors simply default to unity.

The base conditions associated with the nominal design value are as follows:

1. Load is normal (10-year) duration.
2. Wood is initially dry at the time of fabrication of a connection and remains dry in service.
3. Temperature range is normal.
4. There is one bolt in a line parallel to the applied load (i.e., only one bolt per row).
5. Spacing provisions for bolts (geometry requirements for the connection) are satisfied.

If a bolt connection is used under exactly these conditions, the adjustment factors are all unity, and $Z' = Z$.

For a laterally loaded bolt connection that does not satisfy the base conditions described above, the following adjustment factors may be required:

$$Z' = Z(C_D C_M C_t C_g C_\Delta)$$

where C_D = load duration factor (Sec. 4.15)
C_M = wet service factor (Sec. 4.14)
C_t = temperature factor (Sec. 4.19)
C_g = group action factor
C_Δ = geometry factor

The total allowable capacity for a connection using two or more bolts is the sum of the individual allowable design values Z'. This summation is appropriate if the bolts are all the same type and size, because each fastener will have the same yield mode. The capacity for the connection is the number of bolts N times the allowable design value for one fastener:

$$\text{Allow. } P = N(Z')$$

A number of adjustment factors were described previously and are mentioned only in summary. Other adjustments require greater explanation. Some of these new factors will also be required for other connectors such as lag bolts, split rings, and shear plates.

Load duration factor C_D. The load duration factor C_D for bolted connections is the same as that applied to nailed connections. Refer to Sec. 12.6 for a review of load duration and the numerical values for C_D.

Wet service factor C_M. The concept of an allowable stress adjustment based on moisture content was introduced in Sec. 4.14. However, the wet service factors for member design vary with the type of stress. Different adjustments also apply in connection design depending on the type of fastener and the moisture content of the wood at several stages. The moisture content is considered (1) at the time of fabrication of the joint and (2) in service. The nominal design value applies to a bolt connection in which the wood is initially dry and remains dry in service. For connection design, the term *dry* refers to wood that has a moisture content of 19 percent or less. For these conditions, there is no reduction in the design value of a bolt for moisture content, and $C_M = 1.0$. Other moisture content conditions may require significant reductions of the nominal design value (that is, $C_M < 1.0$).

The proper design of bolted connections requires that the effects of changes in moisture content be considered in the design of a wood structure. In the past, problems were created in certain bolted connections by designers who were not familiar with the unique characteristics of wood.

It is necessary for the designer to understand the material if it is to be used properly. Wood undergoes dimensional changes as the moisture content varies. Shrinkage (or swelling) is caused by changes in moisture content below the fiber saturation point (Sec. 4.7). The large majority of the dimensional change takes place across the grain, and very little occurs parallel to the grain. These volumetric changes require that the *arrangement of bolts* in a connection be taken into account *in addition to the moisture content conditions* mentioned above.

Problems involving volume changes develop when bolts are held rigidly in position. Perhaps the best illustration is in a connection that uses a metal side plate and parallel rows of bolts. See Example 13.7. The bolts are essentially fixed in position by the metal side plates (the only movement permitted is that allowed by the clearance between the bolts and the holes). If the lumber is at an initially high moisture content and is allowed to season in place after fabrication of the connection, serious splits and cracks can develop.

The shrinkage that occurs is primarily across the grain. Wood is very weak in tension perpendicular to grain, and the result will be the development of cracking parallel to the grain. Obviously this type of cracking greatly reduces the strength of a connection.

The problem of cross-grain shrinkage can often be eliminated by proper detailing of a connection. This can be accomplished by providing for the unrestricted movement of the bolts across the grain. For example, in the case of the steel side plate, separate side members for the two rows of bolts can be used. These separate plates will simply move with the bolts as the moisture content changes, and cross-grain tension and the associated cracking will be avoided. Another approach is to use slotted holes in the steel side plates so that movement is not restricted. Other examples of proper connection detailing to avoid splitting are given in Chap. 14.

EXAMPLE 13.7   Cross-Grain Cracking

Figure 13.7a shows a critical connection for cross-grain cracking. As the wood loses moisture, cross-grain shrinkage occurs, and the bolts are rigidly held in position by the steel plates. The lumber becomes stressed in tension across the grain (a weak state of stress for wood).

Wet Service Factors (C_M)

In the NDS the following definitions apply to connections:

"Wet" condition—MC > 19 percent

"Dry" condition—MC ≤ 19 percent

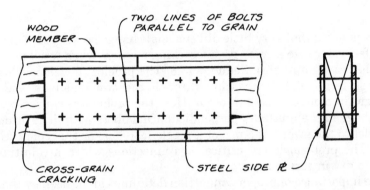

Figure 13.7a   Restrained shrinkage causes tension stress perpendicular to grain.

For the type of connection in Fig. 13.7a, the following values of C_M are specified in NDS Table 10.3.3:

1. When the lumber is initially dry and remains dry in service, C_M = 1.0.
2. When the lumber used to fabricate a connection is initially wet and seasons in place to a dry condition, the allowable design value is reduced by 60 percent (i.e., C_M = 0.40).

The cross-grain cracking problem can be minimized by proper joint detailing. The use of separate side plates for each line of bolts allows the free movement of the bolts across the grain. Slotted holes in steel side plates can also be used. With proper detailing C_M can be taken as 1.0 for bolt connections in dry service conditions, or 0.7 for bolt connections in wet service conditions. See Fig. 13.7b.

NDS spacing criteria (see below) limit the spacing between outer rows of bolts to a maximum of 5 in. unless separate side plates are used. If separate plates are not used, the 5-in. limit on the spacing across the grain provides a limited width across the grain in which shrinkage can occur.

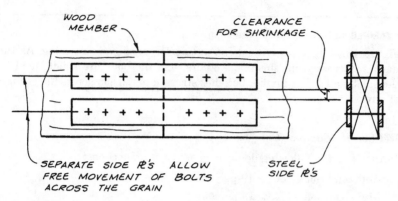

Figure 13.7b Proper connection detailing eliminates restraint across the grain.

It was noted that very little dimensional change occurs parallel to the grain. This fact can cause a cross-grain cracking problem when wood members frame together at right angles. See Fig. 13.8. The problem develops when there are multiple rows of bolts passing through the two members and significant changes in moisture content occur. Here the problem is caused by cross-grain shrinkage in the mutually perpendicular members with little or no shrinkage parallel to the grain. The use of a single fastener for this connection is desirable. The problem is not critical in connections which are initially dry and remain dry in service.

The importance of proper connection detailing can be seen by the magnitude of the wet service factor required when severe cross-grain cracking can de-

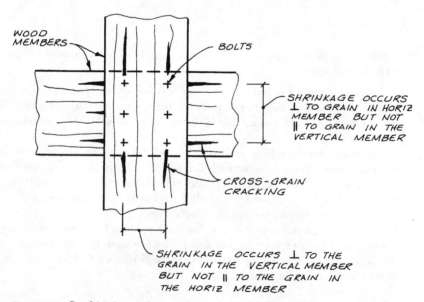

Figure 13.8 Cracking in wood-to-wood connection.

velop. Example 13.7 indicates that a C_M of 0.40 is required when these connections are fabricated with wood members that are initially wet and season
to a dry condition in service. A summary of the C_M values for wood connections
is given in NDS Table 10.3.3, *Wet Service Factors, C_M, for Connections*.

It should be noted that $C_M = 1.0$ for connections where cross-grain cracking
is prevented (e.g., connections that are initially dry and remain dry). The NDS
also indicates that $C_M = 0.7$ for connections with proper detailing when the
lumber is wet at the time of fabrication and will season in place to a dry
service condition. Examples of proper connection detailing include:

1. One fastener only

2. Two or more fasteners in a single line parallel to the grain (no shrinkage
   parallel to grain)

3. Fasteners in two or more lines parallel to the grain with separate side
   plates for each line (movement not restricted)

Note there can be a fairly large difference between the initial and final
moisture contents in the *dry* range, and shrinkage can occur (Sec. 4.7). This
is especially true in dry climate zones. In view of the concern about the effects
of cross-grain shrinkage in critical connections, the designer may want to
make reductions of allowable stress when large differences in moisture content are expected even in the dry range.

As noted earlier, however, the most desirable solution is to detail the connection so that shrinkage cannot induce cross-grain cracking in the first place.
Providing for unrestricted movement is a key factor. Recommended connection
detailing practices with this intention are given in Chap. 14.

Temperature factor C_t. If wood is used in an application with sustained exposure to higher-than-normal temperature conditions, the allowable design
values for connections should be reduced by an appropriate temperature factor
C_t. Refer to Sec. 4.19 for a brief review of the temperature factor. Numerical
values of C_t for connections are given in NDS Table 10.3.4. The temperature
range for most wood-frame buildings does not require adjustment of bolt values, and C_t is often unity.

Group action factor C_g. Research has shown that each bolt in a row does not
carry an equal portion of the total load on a connection. This is also true for
other large-diameter fasteners such as lag bolts, split rings, and shear plate
connectors. To account for this behavior, the nominal design value is reduced
based on the number of fasteners in a row. A *row* is considered to be a *group
of fasteners in a line parallel* to the direction of *loading*. The term *group action
factor C_g* is the reduction factor that accounts for the nonuniform loading of
fasteners. The value of C_g decreases as the number of fasteners in a row increases. The group action factor does not apply to small-diameter fasteners
such as nails.

The NDS defines the group action factor with an equation and provides
tables of C_g values for a limited set of conditions. The formula for the group

action factor is rather cumbersome to solve by hand on a calculator, and it is preferable to solve the expression on the computer. The definition of group action factor in equation form allows the designer to handle a much wider variety of loading conditions and joint details. See Example 13.8.

The primary consideration in determining C_g is the number of fasteners in a row n. The number of fasteners in a row n should not be confused with the number of fasteners in the connection N. If there is only one row of fasteners, then n and N are equal.

In addition to the number of fasteners in a row, the group action factor depends on

Ratio of the stiffnesses of members being connected

Diameter of fastener

Spacing of fasteners in a row

Load/slip modulus for the fastener

Although net area is used to evaluate member stresses at a connection, the distribution of forces to the fasteners is a function of relative stiffness. For an axial force member, the stiffness of a member is the product of *gross* cross-sectional area and modulus of elasticity.

EXAMPLE 13.8   Formula for Group Action Factor C_g

The nominal design value for a fastener Z is reduced by the group action factor C_g. The value of C_g decreases as the number of fasteners in a row increases.

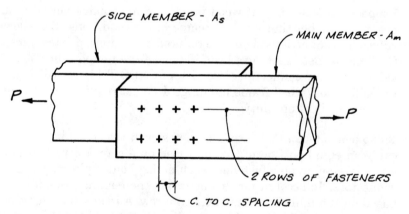

Figure 13.9 Definition of terms for C_g. Group action factor primarily depends on the number of fasteners in a row. For sample connection: Number of fasteners in row = n = 4; number of fasteners in connection = N = 8.

The number of fasteners in a row n is used to determine the group action factor C_g. The total number of fasteners N is used to determine the total allowable capacity for the connection

$$\text{Allow. } P = N(Z')$$

The formula for C_g is the same for bolts, lag bolts, split rings, and shear plates. The differences in these fasteners are reflected in the load/slip modulus γ for the connection. The group action factor is given by the following formula:

$$C_g = \left[\frac{m(1 - m^{2n})}{n[(1 + R_{EA}m^n)(1 + m) - 1 + m^{2n}]} \right] \left[\frac{1 + R_{EA}}{1 - m} \right]$$

where n = number of fasteners in a row

R_{EA} = smaller ratio of member stiffnesses

$\phantom{R_{EA}}$ = lesser of $\dfrac{E_s A_s}{E_m A_m}$ or $\dfrac{E_m A_m}{E_s A_s}$

E_m = modulus of elasticity of main member, psi

A_m = gross cross-sectional area of main member, in.^{2}

E_s = modulus of elasticity of side member, psi

A_s = gross cross-sectional area of side member, in.^{2} (*Note:* If there is more than one side member, A_s is the sum of the gross cross-sectional areas of the side members.)

$m = u - \sqrt{u^2 - 1}$

$u = 1 + \gamma \left(\dfrac{s}{2} \right) \left[\dfrac{1}{E_m A_m} + \dfrac{1}{E_s A_s} \right]$

s = center-to-center spacing between adjacent fasteners in a row, in.

γ = load/slip modulus for a connection, lb/in.

 = $180,000(D^{1.5})$ for dowel-type fasteners such as bolts or lag bolts in a wood-to-wood connection

 = $270,000(D^{1.5})$ for dowel-type fasteners such as bolts or lag bolts in a wood-to-metal connection

 = $500,000$ for 4-in.-diameter split ring or shear plate connectors

 = $400,000$ for 2½-in.-diameter split ring or 2⅝-in.-diameter shear plate connectors

D = diameter of fastener, in.

As an alternative to the equation for group action factor, the NDS provides expanded tables for C_g that may be used for certain applications. Again, the concept in the NDS is to define terms, to the extent possible, with general formulas that can be readily programmed, *and* to provide tables for use by the designer who does only an occasional wood design problem. Numerical values of C_g are given in NDS Tables 10.3.6A to 10.3.6D.

Geometry factor C_Δ. The NDS provides spacing requirements for bolts. These are typically given as a number of bolt diameters D (e.g., $4D$ is the required end distance for a wood member in compression). In many cases the NDS provides two sets of spacings. The larger ones can be viewed as *base dimensions* that are required in order to use the full *nominal design value* for the fastener. The *geometry factor C_Δ* is the multiplier that is used to reduce the nominal design value Z if the spacings furnished in a given connection are less than the base dimensions. Thus, if the base spacing requirements for a connection are all satisfied, the geometry factor defaults to unity ($C_\Delta = 1.0$). Note that C_Δ does not exceed 1.0 if additional space is provided.

The second set of spacings given in the NDS is the *minimum permitted spacings*. For example, the minimum permitted end distance for a member in compression is $2D$. An end distance less than this minimum is unacceptable. The geometry factor could be viewed as zero ($C_\Delta = 0$) if the furnished spacings are less than the minimums.

A practical application of the geometry factor occurs when the furnished spacing lies between the base dimension and the minimum spacing. The geometry factor is a simple straight-line reduction. See Example 13.9. If there is more than one spacing that does not satisfy the basic spacing requirements, C_Δ is governed by the worst case. In other words, the smallest C_Δ for a given connection is to be used for all fasteners in a connection. However, the reductions are not cumulative, and several geometry factors need not be multiplied together.

EXAMPLE 13.9   Geometry Factor C_Δ

The geometry factor C_Δ is used to reduce the nominal design value for a bolt if the base dimensions are not satisfied. As an illustration, determine the geometry factor for the connection shown in Fig. 13.10 considering end distance only. Assume that all other base dimensions are provided.

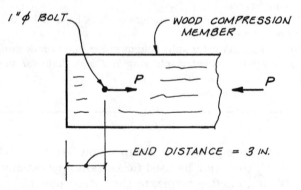

Figure 13.10 End distance for compression member.

From NDS Table 11.5.1B, the end distances for a member loaded in axial compression are as follows:

$$\text{Base end distance} = 4D = 4(1) = 4 \text{ in.} \qquad \text{(end distance required to obtain } full \text{ design value)}$$

$$\text{Minimum end distance} = 2D = 2(1) = 2 \text{ in.} \qquad \text{(end distance required to obtain } any \text{ design value)}$$

The end distance provided in Fig. 13.10 is between the base dimension of 4 in. and the minimum of 2 in. The geometry factor is the ratio of the furnished end distance to the end distance required to obtain the full design value

$$C_\Delta = \frac{\text{furnished end distance}}{\text{base end distance}} \leq 1.0$$

$$= \frac{3 \text{ in.}}{4 \text{ in.}} = 0.75$$

A geometry factor is to be calculated for each spacing that does not satisfy the base dimensions. However, geometry factors are not cumulative, and if more than one geometry factor is calculated, the *smallest* C_Δ is used for *all* fasteners in a connection. The other geometry factors are disregarded.

Although the NDS allows the bolt spacings to be reduced below the base dimensions (provided the allowable design value is reduced using the appropriate C_Δ), it is considered good detailing practice to provide the full base spacings whenever possible.

Now that the concept of the geometry factor has been introduced, the spacing requirements for bolted connections can be considered in detail. The dimensions for locating bolts in a connection are given to the centerline of the bolts. The detailing of a bolted connection requires consideration of

1. End distance

2. Edge distance

3. Center-to-center (c. to c.) spacing of bolts in a row

4. Row spacing

Bolted connections are somewhat more complicated than nailed connections because the direction of loading in relation to the direction of the grain must be considered. Other factors such as bolt slenderness l/D and the sense of force in the member (tension or compression) may also affect the spacing requirements. The *bolt slenderness ratio* is defined as the *smaller* value of

$$\frac{l_m}{D} = \frac{\text{bearing length of bolt in main member}}{\text{bolt diameter}}$$

$$\frac{l_s}{D} = \frac{\text{total bearing length of bolt in side member(s)}}{\text{bolt diameter}}$$

If there is more than one side member, l_s is taken as the combined thickness of the side members.

Because of the complexity, the spacing requirements are best summarized on a sketch. The NDS spacing criteria for *parallel-to-grain loading* are given in Example 13.10. the spacing requirements to obtain the full design value ($C_\Delta = 1.0$) are shown first, followed by the minimum permitted spacings in parentheses.

Note that there are different end distances for members stressed in tension and compression. For tension members it further depends on whether the member is a softwood or hardwood species. The 5-in. limit on the distance between outer rows of bolts was addressed previously (see Wet Service Factor above).

EXAMPLE 13.10   Bolt Spacing Requirements: Parallel-to-Grain Loading

Parallel-to-grain loading spacing requirements (Fig. 13.11) are as follows:

1. *End distance*
   a. Members in tension
      Softwood species = $7D$ ($3.5D$ minimum)
      Hardwood species = $5D$ ($2.5D$ minimum)
   b. Members in compression = $4D$ ($2D$ minimum)
2. *Edge distance:* bolt slenderness = l/D
   a. For $l/D \le 6.0$: edge distance = $1.5D$
   b. For $l/D > 6.0$: edge distance = $1.5D$ or $0.5 \times$ row spacing, whichever is larger
3. *Center-to-center* spacing = $s = 4D$ ($3D$ minimum)
4. *Row spacing* = $1.5D$. Row spacing between outer rows of bolts shall not exceed 5 in. unless separate side plates are used for each row.

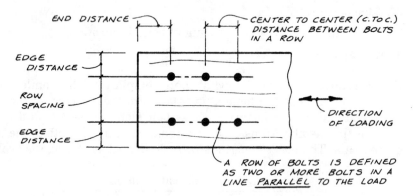

PARALLEL TO GRAIN LOADING

Figure 13.11 Bolt spacing requirements: parallel to grain loading.

The NDS spacing criteria for *perpendicular-to-grain loading* are given in Example 13.11. Again, the spacing requirements to obtain the full design value ($C_\Delta = 1.0$) are shown first, followed by the minimum permitted spacings in parentheses. Note that there are different edge distances for the *loaded edge* and the *unloaded edge*. The bolt can be viewed as delivering a load to the wood member. The loaded edge is the edge of the piece toward which the load is acting (i.e., the edge in the direction of bolt displacement for each member). The unloaded edge is the other side of the member.

EXAMPLE 13.11   Bolt Spacing Requirements: Perpendicular-to-Grain Loading

Perpendicular-to-grain loading spacing requirements (Fig. 13.12) are as follows:

1. *End distance = 4D   (2D minimum)*
2. *Edge distance*
   a. Loaded edge = 4D
   b. Unloaded edge = 1.5D
3. *Center-to-center* spacing to obtain full design value is governed by the spacing requirements of the member(s) to which the bolts are attached (e.g., steel side plates* or wood member loaded parallel to grain).   (3D minimum)
4. *Row spacing*
   a. For $l/D \leq 2.0$: row spacing = 2.5D.
   b. For $l/D \geq 6.0$: row spacing = 5D.
   c. For $2.0 < l/D < 6.0$: interpolate between 2.5D and 5D.

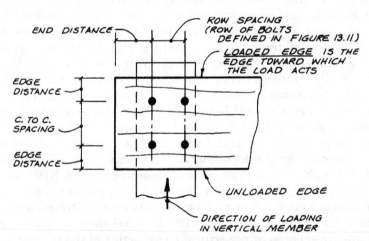

PERPENDICULAR TO GRAIN LOADING

Figure 13.12   Bolt spacing requirements: perpendicular to grain loading.

NOTE: Shear in the member must be checked in accordance with NDS Sec. 3.4.3.

---

*The center-to-center and edge spacing requirements for steel plates are given in Ref. 13.3.

Specific requirements for the spacing of bolts in a connection where the load is at some angle other than 0 or 90 degrees to the direction of the grain are not given in the NDS. It is stated, however, that "the gravity axis of each member shall pass through the center of resistance of the group of fasteners to insure uniform stress in the main member and a uniform distribution of load to all fasteners." In addition to providing this symmetry, the spacing provisions for both parallel and perpendicular loading can be used as a guide in detailing the connection.

This concludes the basic summary of adjustment factors for bolted connections. Several numerical examples are provided in the following sections to illustrate the design procedures for bolts.

13.6   Tension and Shear Stresses at a Multiple Fastener Connection

The 2001 NDS introduced specific procedures for evaluating stresses at multiple-fastener connections in wood tension members loaded parallel to grain. The new provisions in NDS Appendix E (Ref. 13.1) are based on recent tests of multiple-fastener connections with bolts spaced closely together. These tests indicate that connections with rows of closely spaced bolts may fail before reaching the connection yield capacity predicted by the yield limit equations, even when the predicted yield capacity is reduced by the appropriate group action factor C_g and geometry factor C_Δ. The failures observed at multiple-bolt connections in wood tension members involve combinations of shear failure and tension failure in the wood, and are analogous to *shear rupture* failure at bolted connections in steel tension members (see Ref. 13.3). Thus, the procedures in NDS Appendix E constitute a final check of connection capacity in wood tension members after the required number of fasteners has been determined based on the yield limit equations, and a fastener pattern has been established based on NDS spacing requirements.

The new provisions in NDS Appendix E for multiple-fastener connections include three design checks of connection capacity in wood tension members. The first is a check of *net section tension* as specified in NDS Secs. 3.1.2 and 3.8.1. This provision addresses tensile stress on the net cross-sectional area of a tension member. Net section tension capacity is determined based on the allowable tension design value F'_t for the wood member, the number of fasteners that are placed at the critical cross section of the tension member, and the projected area of each fastener in the cross section. This design check was described in Chap. 7.

The second design check addresses failure of the wood member in shear along the perimeters of each row of bolts. The term for this failure mode is *row tear-out,* and it involves the formation of *shear plugs* along each line of bolts oriented parallel to the longitudinal axis of the tension member (see Fig. 13.13). Row tear-out capacity is based on the allowable shear design value F'_v for the wood member, the number of fasteners in each row n_i, and the critical distance between fasteners in a row s_{crit}. Row tear-out failure is assumed to occur whenever shear failure initiates between any two fasteners in a row, or between the last fastener in a row and the end of the tension member. Thus, the critical distance between fasteners in a row s_{crit} is defined as the lesser of the minimum spacing between fasteners in a row or the distance from the last fastener in a row to the end of the tension member. For glued laminated timber members the nominal shear design value F_v must be multiplied by a reduction factor of 0.8 when evaluating shear stresses at a connection (see footnotes to Tables 5A and 5B in the NDS Supplement, Ref. 13.1).

Figure 13.13 Row tear-out failure in the main member for a multiple-bolt connection. Shear plug failures occur between the last bolt in a row and the end of the glulam member.

The third design check specified in NDS Appendix E addresses *group tear-out,* which is a combination of net section tension and row tear-out. Group tear-out involves failure of the wood member around the perimeter of the entire group of fasteners. Thus, group tear-out capacity is based on the net section tension capacity of the cross section between outer rows of fasteners, plus the shear capacity along the perimeter of the outer rows of fasteners.

When the connection capacity is governed by row tear-out or group tear-out, the connection capacity can be increased by increasing the spacing between fasteners. However, designers should avoid spacing outer rows of fasteners more than 5 in. apart unless separate side plates are provided, as illustrated in Example 13.7 (Sec. 13.5).

The 2001 NDS provisions for row tear-out and group tear-out are based on tests of a limited number of multiple-bolt connection configurations. It is assumed that these provisions are equally applicable to all large diameter fasteners including split ring and shear plate connectors, bolts, and lag bolts that penetrate the entire thickness of a wood member. See Example 13.12.

EXAMPLE 13.12   Tension and Shear Stresses at a Multiple-Fastener Connection

Tension and shear stresses at a multiple-fastener connection can lead to three possible failure modes in wood tension members loaded parallel to grain. In some cases these failure modes may occur prior to achieving the connection yield mode predicted from yield limit equations. The symbol Z' is used to designate the allowable capacity at a connection, and subscripts are employed to indicate whether the allowable capacity is based on consideration of net section tension Z'_{NT}, row tear-out Z'_{RT}, or group tear-out Z'_{GT}. When evaluating tension and shear stresses at connections, the allowable capacity is not multiplied by the number of fasteners in the connection N since the fasteners in critical regions of the connection are directly considered when calculating

Z'_{NT}, Z'_{RT}, and Z'_{GT}. Net section tension failure was discussed in Chap. 7. Row tear-out and group tear-out failure modes are illustrated in Fig. 13.14.

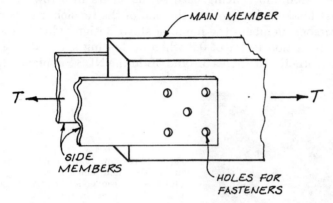

Figure 13.14a Multiple fastener connection.

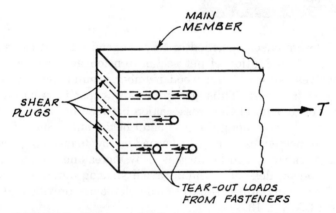

Figure 13.14b Row tear-out failure along each row of fasteners; dashed lines indicate locations of shear failure in wood member.

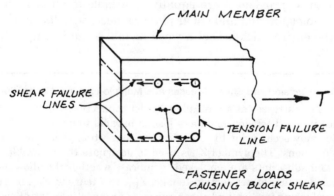

Figure 13.14c Group tear-out failure around the perimeter of fastener group; dashed lines indicate locations of tension failure and shear failure in wood member.

Net section tension (NDS equation E.2-1):

$$Z'_{NT} = F'_t A_n$$

Row tear-out (NDS equations E.3-2 and E.3-3):

$$Z'_{RT} = \sum_{i=1}^{n_{row}} Z'_{RT_i} = \sum_{i=1}^{n_{row}} n_i F'_v t s_{crit}$$

Group tear-out (NDS equation E.4-1):

$$Z'_{GT} = \frac{Z'_{RT-1}}{2} + \frac{Z'_{RT-n}}{2} + F'_t A_{group-net}$$

where F'_t = allowable tension design value parallel to grain, psi (see Chap. 7)

F'_v = allowable shear design value, psi (see Chap. 6)

A_n = net cross-sectional area of tension member, in.^{2} (see Chap. 7)

$A_{group-net}$ = net cross-sectional area of tension member between outer rows of fasteners, in.^{2}

t = thickness of tension member, in.

s_{crit} = critical spacing between fasteners in row i, in.

= the lesser of the minimum spacing between fasteners in row i, or the distance from the last fastener in row i to the end of the tension member

n_i = number of fasteners in row i

n_{row} = number of rows of fasteners in a fastener group

Z'_{RT_i} = row tear-out capacity for any (ith) row of fasteners in a fastener group, lb

Z'_{RT-1} = row tear-out capacity for the first row of fasteners in a fastener group, lb

Z'_{RT-n} = row tear-out capacity for the last (nth) row of fasteners in a fastener group, lb

As discussed in Sec. 13.5, a row of fasteners is defined as a group of fasteners in a line parallel to the direction of loading. The formula for calculating row tear-out capacity Z'_{RT} is based on the assumption that shear failure for each row of fasteners occurs along two lines, one failure line on each side of the row and extending from the first fastener in the row to the end of the wood member. Since failure of the entire row may occur when shear failure initiates between any two fasteners in the row, the critical distance between fasteners in a row s_{crit} is used to calculate row tear-out capacity. If the shear stress due to applied load were uniformly distributed between fasteners in a row, the tear-out capacity for each row of fasteners would be estimated by:

$$Z'_{RT_i} = 2(n_i F'_v t s_{crit})$$

However, experimental test data indicates that the actual row tear-out capacity is approximately half the magnitude calculated using this formula. Therefore, a triangular stress distribution is assumed to occur between adjacent fasteners in a row. This means that the shear stress due to applied load in a multiple-fastener connection is maximum immediately adjacent to a fastener and decreases linearly to zero near the next fastener in the row. Since the actual row tear-out capacity is reached when F'_v is exceeded at any location along a row of fasteners, the tear-out capacity for each row of fasteners is thus:

$$Z'_{RT_i} = \tfrac{1}{2}[2(n_i F'_v t s_{crit})] = n_i F'_v t s_{crit}$$

The formula for group tear-out is based on net section tension failure of the area between outer rows of fasteners $A_{group-net}$ combined with shear failure along the perimeter of the outer rows of bolts. Thus the row tear-out capacities for each of the outer rows of fasteners (Z'_{RT-1} and Z'_{RT-n}) are divided by 2 since only one shear failure line occurs at each row. According to NDS Sec. 3.1.2, when fasteners are arranged in a staggered pattern the net section tension area $A_{group-net}$ is typically determined assuming staggered fasteners occur at the same cross section.

13.7 Design Problem: Multiple-Bolt Tension Connection

The multiple-fastener connection at the end of a glulam beam illustrates the importance of comparing connection capacity based on yield limit equations with capacity based on tension and shear stresses in a wood tension member at a connection. See Example 13.13. Even though the NDS base dimensions are met or exceeded in this connection ($C_\Delta = 1.0$), the capacity at the connection is governed by group tear-out and row tear-out capacities. Both group tear-out and row tear-out are associated with shear failure in the wood member along rows of bolts. The group tear-out and row tear-out capacities can be increased by increasing bolt spacing in the connection.

EXAMPLE 13.13 Multiple-Bolt Tension Connection

Determine the allowable capacity of the multiple-bolt double-shear tension connection at the end of the 24F-V4 Douglas Fir glulam member (from the 24F-1.8E Stress Class) shown in Fig. 13.15a. Two ¼-in.-thick A36 steel side plates are attached to the 5⅛ × 12 glulam member using six 1-in.-diameter A307 bolts in a grid pattern. Each steel

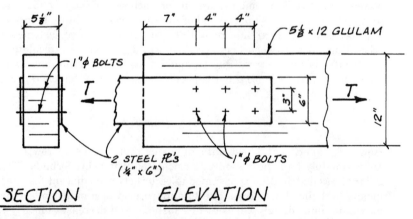

Figure 13.15a Six-bolt tension connection; double-shear wood-to-steel.

side plate is 6 in. wide. The spacing between rows of bolts is 3 in., the spacing between bolts in a row is 4 in., and the distance from the last bolt in each row to the end of the glulam member is 7 in. The axial tension force in the glulam member is a seismic force. The glulam member is initially dry and will remain dry in service. The temperature range is normal.

The nominal design value Z for a single-bolt connection can be determined from the yield limit equations for double-shear wood-to-metal connections. Alternatively, $Z_\parallel$ can be obtained from NDS Table 11I. The following parameters apply for this connection:

$$D = 1.0 \text{ in.}$$

$$F_{yb} = 45 \text{ ksi}$$

$$t_m = 5.125 \text{ in.}$$

$$\theta_m = 0 \text{ degrees}$$

$$G_m = 0.50$$

$$F_{em} = 5600 \text{ psi} \qquad (F_{e\parallel} \text{ for bolt in DF-L})$$

$$t_s = 0.25 \text{ in.}$$

$$F_{es} = 1.5F_u = 1.5(58{,}000) = 87{,}000 \text{ psi} \qquad (\text{for A36 steel plate})$$

Nominal design value from NDS Table 11I:

$$Z_\parallel = 5720 \text{ lb}$$

The reader may wish to verify this nominal design value by applying the yield limit equations and observing that Mode III$_s$ governs.

Allowable connection capacity based on bolt yield limit equations:

$$n = 3 \text{ bolts in each row}$$

$$C_D = 1.6 \qquad \text{for loading due to wind or seismic}$$

$$C_M = 1.0 \qquad \text{given}$$

$$C_t = 1.0 \qquad \text{given}$$

The NDS base dimensions and formulas for the geometry factor C_Δ are summarized in Examples 13.9 and 13.10 (Sec. 13.5). In this problem:

End distance = 7 in. = 7D   (tension member)
c.-to-c. spacing between bolts in a row = s = 4 in. = 4D
Spacing between rows = 3 in. > 1.5D
Edge distance = 4.5 in. > 1.5D   (for l_m/D = 5.125)

$$C_\Delta = 1.0 \qquad (\text{since all NDS base dimensions are met or exceeded})$$

The formula and coefficients for the group action factor C_g are summarized in Example 13.8 (Sec. 13.5):

$$\gamma = \text{load/slip modulus for wood-to-metal bolt connection}$$

$$= 270,000(D^{1.5}) = 270,000(1.0)^{1.5} = 270,000 \text{ lb/in.}$$

$$E_m = 1,700,000 \text{ psi} \quad (E_{\text{axial}} \text{ from NDS Supplement Table 5A})$$

$$A_m = (5.125)(12) = 61.5 \text{ in.}^2$$

$$E_m A_m = 1,700,000(61.5) = 104,550,000$$

$$E_s = 29,000,000 \text{ psi}$$

$$A_s = (2)(0.25)(6.0) = 3.0 \text{ in.}^2 \quad (\text{for two steel side plates})$$

$$E_s A_s = 29,000,000(3.0) = 87,000,000$$

$$R_{EA} = \frac{E_s A_s}{E_m A_m} = \frac{87,000,000}{104,550,000} = 0.8321 \quad (\text{since } E_s A_s < E_m A_m)$$

$$1 + R_{EA} = 1 + 0.8321 = 1.832$$

$$u = 1 + \gamma \left(\frac{s}{2}\right) \left[\frac{1}{E_m A_m} + \frac{1}{E_s A_s}\right]$$

$$= 1 + 270,000 \left(\frac{4.0}{2}\right) \left[\frac{1}{104,550,000} + \frac{1}{87,000,000}\right]$$

$$= 1.011$$

$$m = u - \sqrt{u^2 - 1}$$

$$= 1.011 - \sqrt{1.011^2 - 1} = 0.8601$$

$$1 + m = 1 + 0.8601 = 1.860$$

$$1 - m = 1 - 0.8601 = 0.1399$$

$$m^{2n} = 0.8601^{2(3)} = 0.4049$$

$$1 + R_{EA} m^n = 1 + 0.8321(0.8601)^3 = 1.530$$

$$C_g = \left[\frac{m(1 - m^{2n})}{n[(1 + R_{EA}m^n)(1 + m) - 1 + m^{2n}]}\right] \left[\frac{1 + R_{EA}}{1 - m}\right]$$

$$= \left[\frac{0.8601(1 - 0.4049)}{3[(1.530)(1.860) - 1 + 0.4049]}\right] \left[\frac{1.832}{0.1399}\right]$$

$$= 0.993 \quad (\text{compare with } C_g \approx 0.99 \text{ from NDS Table 10.3.6C})$$

$$Allow.\ P = N(Z') = N(Z)(C_D C_M C_t C_g C_\Delta)$$

$$= (5720)(1.6)(1.0)(1.0)(0.993)(1.0)$$

$$= 54{,}540 \text{ lb}$$

Allowable connection capacity based on tension and shear stresses in the glulam member:

The formulas for net section tension Z'_{NT}, row tear-out Z'_{RT}, and group tear-out Z'_{GT} are summarized in Example 13.12 (Sec. 13.6).

Since the bolts penetrate the wide face of the 24F-V4 glulam member, select the nominal shear design value for bending about the strong x-axis and multiply by a reduction factor of 0.8 in accordance with NDS Supplement Table 5A:

$$F_v = (0.8)240 \text{ psi} = 192 \text{ psi}$$

$$F'_v = F_v(C_D C_M C_t)$$

$$= 192(1.6)(1.0)(1.0)$$

$$= 307 \text{ psi}$$

$$F'_t = F_t(C_D C_M C_t)$$

$$= 1100(1.6)(1.0)(1.0)$$

$$= 1760 \text{ psi}$$

Net section tension (NDS equation E.2-1):

$$A_n = 5.125[12 - 2(1.0 + \tfrac{1}{16})] = 50.6 \text{ in.}^2 \qquad \text{(see Fig. 13.15}b\text{)}$$

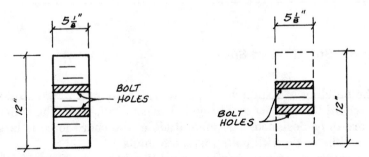

Figure 13.15b Net section of glulam member (left); portion of net area between rows of bolts $(A_{\text{group-net}})$ for group tear-out (right).

The $\frac{1}{16}$ in. was added to the bolt diameter to account for drilling oversize holes in accordance with NDS Sec. 11.1.2.

$$Z'_{\text{NT}} = F'_t(A_n) = 1760(50.6) = 89{,}070 \text{ lb} > 54{,}540 \text{ lb}$$

Row tear-out (NDS equations E.3-2 and E.3-3):

$$s_{crit} = s = 4.0 \text{ in.}$$

$$Z'_{RT-1} = Z'_{RT-2} = nF'_v t s_{crit} = 3(307)(5.125)(4.0) = 18{,}890 \text{ lb}$$

$$Z'_{RT} = \sum_{i=1}^{n_{row}} Z'_{RT_i} = 18{,}890 + 18{,}890 = 37{,}780 \text{ lb} < 54{,}540 \text{ lb}$$

Group tear-out (NDS equation E.4-1):

$$A_{group-net} = 5.125[3 - 2(\tfrac{1}{2})(1.0 + \tfrac{1}{16})] = 9.93 \text{ in.}^2 \qquad (\text{see Fig. 13.15}b)$$

$$Z'_{GT} = \frac{Z'_{RT-1}}{2} + \frac{Z'_{RT-2}}{2} + F'_t A_{group-net}$$

$$Z'_{GT} = \frac{18{,}890}{2} + \frac{18{,}890}{2} + 1760(9.93) = 36{,}370 \text{ lb} < 54{,}540 \text{ lb}$$

> The allowable capacity is 36,370 lb due to group tear-out at the connection.

The net section tension capacity Z'_{NT} at the connection is much larger than the capacity based on yield limit equations for the bolts. However, the group tear-out capacity Z'_{GT} at the connection is approximately 67 percent of the capacity based on yield limit equations for the bolts. The row tear-out capacity Z'_{RT} is only marginally higher than Z'_{GT}. Z'_{GT} could be increased by increasing the spacing between rows of bolts (thus increasing $A_{group-net}$), as well as the spacing between bolts in a row (thus increasing s_{crit}). Similarly, Z'_{RT} could be increased by increasing the spacing between bolts in a row.

13.8   Design Problem: Bolted Chord Splice for Diaphragm

For the building in Example 13.14, the top plate of a 2 × 6 stud wall serves as the chord for the roof diaphragm. The connections for the splice in the top plate are to be designed. The magnitude of the chord force is large, and the splice will be made with bolts instead of nails.

The location of the splices in the chord should be considered. In this building the chord is made up of fairly long members (say 20 ft), and the splices are offset by half the length of the members (10 ft in this example). At some point between the splices, the chord force is shared equally by the two members. Therefore, the connections at each end of a chord member may be designed for one-half of the total chord force.

This is in contrast to the nailed chord splice designed in Example 12.9 (Sec. 12.8). If the lap of the members in the chord is kept to a minimum, the entire chord force must be transmitted by the single connection within the lap. Either method of splicing the chord may be used in practice.

The connection is designed by first assuming a bolt size and determining the nominal design value for the fastener. In this problem Z can be found using the yield limit equations or the NDS bolt tables. Except for t_m and θ_m, the current example is very similar to Example 13.3. Therefore, the determination of the nominal value Z is not shown and is left to the reader as an exercise.

With Z known, the approximate number of bolts is estimated, and the allowable design value for the connection is determined and compared with the design load. As with many practical problems, a number of the possible adjustment factors default to unity. However, the load duration factor and the group action factor are required to obtain Z'.

Part 1 of the example follows the recommended practice of providing the full base dimensions for the end distance and center-to-center spacing. Part 2 is given for illustrative purposes only, to show how the geometry factor is applied should the base dimensions not be provided.

EXAMPLE 13.14   Bolt Splice for Tension Chord

A large wood-frame warehouse uses 2×6 stud walls. A double 2×6 top plate functions as the chord of the horizontal roof diaphragm. See Fig. 13.16a. Determine the number and spacing of ¾-in. full-diameter body bolts for the splice connection in the top plate. The bolts are in a single row. Lumber is Select Structural DF-L. The chord force at the connection is the result of seismic loading and has been determined to be 9.3 k. The splice is to be designed for one-half of this load. The stress in the member is also to be evaluated. $C_m = 1.0$, and $C_t = 1.0$.

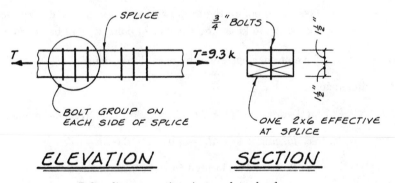

Figure 13.16a   Bolt splice connections in top plate chord.

Bolt Design—Part 1

In part 1, spacings are to be determined in accordance with the base dimensions to obtain the full design value for the bolt. The following parameters are known about the connection. The nominal design value for the bolt can be determined from the yield limit equations for a single-shear wood-to-wood connection. As an alternative to the equations, Z can be obtained from NDS Table 11A. Evaluation of Z is left as an exercise for the reader.

$$D = 0.75 \text{ in.}$$

$$F_{yb} = 45 \text{ ksi} \qquad \text{(assumed for A307 bolts)}$$

$$t_m = 1.5 \text{ in.}$$

$$\theta_m = 0 \text{ degrees}$$

$$F_{em} = 5600 \text{ psi} \qquad (F_{e\|} \text{ for bolt in DF-L})$$

$$t_s = 1.5 \text{ in.}$$

$$\theta_s = 0 \text{ degrees}$$

$$F_{es} = 5600 \text{ psi} \qquad (F_{e\|} \text{ for bolt in DF-L})$$

$$G_m = G_s = 0.50$$

Nominal design value:

$Z = 725 \text{ lb}$ (Mode II yield mechanism given by NDS equation 11.3-3. This compares with $Z = 720$ lb in NDS Table 11A.)

Adjustment factors:

$C_D = 1.6$ for lateral loading due to wind or seismic

$C_M = 1.0$ given

$C_t = 1.0$ given

$C_g \approx 0.95$ (assumed) before the group action factor can be determined, it is necessary to know the number of bolts. C_g will be verified later.

$C_\Delta = 1.0$ for bolt spacings given by base dimensions.

Base dimensions:

End distance $= 7D = 7(0.75) = 5.25$ in.   (tension member)

c.-to-c. spacing $= s = 4D = 4(0.75) = 3.0$ in.

Bolts will be located in the center of the 2 × 6s.

Edge distance $= 0.5(5.5) = 2.75$ in.

$l/D = 1.5/0.75 = 2.0 < 6.0$

Min. edge distance $= 1.5D = 1.5(0.75)$

$= 1.125 \text{ in.} < 2.75 \text{ in.} \qquad OK$

Estimate number of bolts:

$$\text{Load on one bolt group} = \frac{T}{2} = \frac{9300}{2} = 4650 \text{ lb}$$

$$\text{Approx. } Z' = Z(C_D C_M C_t C_g C_\Delta)$$

$$= 725(1.6)(1.0)(1.0)(0.95)(1.0)$$

$$= 1102 \text{ lb/bolt}$$

$$N \approx \frac{4650}{1102} = 4.22 \quad \textit{say} \quad 5 \text{ bolts}$$

Allowable design value:
Verify group action factor by evaluating the equation for C_g for five bolts in a row.

$$\gamma = \text{load/slip modulus for wood-to-wood bolt connection}$$

$$= 180,000(D^{1.5}) = 180,000(0.75^{1.5}) = 116,900 \text{ lb/in.}$$

$$E_m = E_s = 1,900,000 \text{ psi}$$

$$A_m = A_s = 8.25$$

$$R_{EA} = 1.0$$

NOTE: R_{EA} is the minimum ratio of member stiffnesses [i.e., the smaller of $E_m A_m/(E_s A_s)$ and $E_s A_s/(E_m A_m)$]. In this example, the member stiffnesses are equal, hence $R_{EA} = 1.0$.

$$u = 1 + \gamma \left(\frac{s}{2}\right) \left[\frac{1}{E_m A_m} + \frac{1}{E_s A_s}\right]$$

$$= 1 + 116,900 \left(\frac{3.0}{2}\right) \left[\frac{1}{1,900,000(8.25)} + \frac{1}{1,900,000(8.25)}\right]$$

$$= 1.022$$

$$m = u - \sqrt{u^2 - 1}$$

$$= 1.022 - \sqrt{1.022^2 - 1} = 0.8097$$

$$1 + m = 1 + 0.8097 = 1.810$$

$$1 - m = 1 - 0.8097 = 0.1903$$

$$m^{2n} = 0.8097^{2(5)} = 0.1211$$

$$1 + R_{EA} = 1 + 1.0 = 2.0$$

$$1 + R_{EA} m^n = 1 + 1.0(0.8097^5) = 1.348$$

$$C_g = \left[\frac{m(1 - m^{2n})}{n[(1 + R_{EA}m^n)(1 + m) - 1 + m^{2n}]} \right] \left[\frac{1 + R_{EA}}{1 - m} \right]$$

$$= \left[\frac{0.8097(1 - 0.1211)}{5[(1.348)(1.810) - 1 + 0.1211]} \right] \left[\frac{2.0}{0.190} \right]$$

$$= 0.958$$

Allowable capacity for one splice connection:

$$\text{Allow. } P = N(Z') = N(Z)(C_D C_M C_t C_g C_\Delta)$$

$$= 5(725)(1.6)(1.0)(1.0)(0.958)(1.0)$$

$$= 5556 \text{ lb} > 4650 \text{ lb} \quad OK$$

Use   five ¾-in.-diameter bolts each side of splice.

NOTE: The spacing of bolts must be shown on the design plans.

Bolt Design—Part 2

The bolt spacings given in part 1 are the preferred spacings even though the allowable capacity is greater than the applied load. For illustration, determine the effect on the capacity of the connection if reduced spacings are used.

Reduced spacings and geometry factor:

$$\text{Reduced end distance} = 5.0 \text{ in.} < 7D = 5.25 \text{ in.} \qquad \text{(base)}$$

$$> 3.5D = 2.625 \text{ in.} \qquad \text{(minimum)}$$

$$\therefore \text{ End } C_\Delta = \frac{5.0}{5.25} = 0.952$$

$$\text{Reduced c.-to-c. spacing} = s = 2.5 \text{ in.} < 4D = 3.0 \text{ in.} \qquad \text{(base)}$$

$$> 2D = 1.5 \text{ in.} \qquad \text{(minimum)}$$

$$\therefore \text{ Center-to-center } C_\Delta = \frac{2.5}{3.00} = 0.833$$

A single geometry factor is used for the entire connection, and the smaller value applies:

$$C_\Delta = 0.833$$

Revised group action factor:

$$u = 1 - \gamma \left(\frac{s}{2} \right) \left[\frac{1}{E_m A_m} + \frac{1}{E_s A_s} \right]$$

$$= 1 - 116,900 \left(\frac{2.5}{2} \right) \left[\frac{1}{1,900,000(8.25)} + \frac{1}{1,900,000(8.25)} \right]$$

$$= 1.019$$

$$m = u - \sqrt{u^2 - 1}$$
$$= 1.019 - \sqrt{1.019^2 - 1} = 0.8246$$

$$1 + m = 1 + 0.8246 = 1.825$$
$$1 - m = 1 - 0.8246 = 0.1754$$

$$m^{2n} = 0.8246^{2(5)} = 0.1454$$

$$1 + R_{EA} = 1 + 1.0 = 2.0$$

$$1 + R_{EA}m^n = 1 + 1.0(0.8246^5) = 1.381$$

$$C_g = \left[\frac{m(1 - m^{2n})}{n[(1 + R_{EA}m^n)(1 + m) - 1 + m^{2n}]}\right]\left[\frac{1 + R_{EA}}{1 - m}\right]$$

$$= \left[\frac{0.8246(1 - 0.1454)}{5[(1.381)(1.825) - 1 + 0.1454]}\right]\left[\frac{2.0}{0.1754}\right]$$

$$= 0.965$$

Allowable capacity for one splice connection:

$$\text{Allow. } P = N(Z') = 5(725)(1.6)(0.965)(0.833)$$
$$= 5(932)$$
$$= 4662 \text{ lb} > 4650 \text{ lb} \qquad OK$$

Although the allowable capacity is greater than the applied load, the connection capacity has been substantially reduced (4662 lb versus 5556 lb) as a result of bolt spacing.

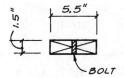

Figure 13.16b Net section at splice.

Allowable connection capacity based on tension and shear stresses in one 2 × 6 member:

The formulas for net section tension Z'_{NT} and row tear-out Z'_{RT} are summarized in Example 13.12 (Sec. 13.6).

Net section tension (NDS equation E.2-1):

$$A_n = 1.5[5.5 - (\tfrac{3}{4} + \tfrac{1}{16})] = 7.03 \text{ in.}^2$$

The $\tfrac{1}{16}$ in. was added to the bolt diameter to account for drilling oversize holes in accordance with NDS Sec. 11.1.2.

$$f_t = \frac{T}{A_n} = \frac{9300}{7.03} = 1323 \text{ psi}$$

$$F_t' = F_t(C_D)(C_M)(C_t)(C_F)$$

$$= 1000(1.6)(1.0)(1.0)(1.3)$$

$$= 2080 \text{ psi} > 1323 \text{ psi} \qquad OK$$

Row tear-out (NDS equations E.3-2 and E.3-3):

$$n = \text{ five bolts in a single row}$$

$$s_{\text{crit}} = s = 2.5 \text{ in.}$$

$$F_v' = F_v(C_D)(C_M)(C_t)$$

$$= 180(1.6)(1.0)(1.0)$$

$$= 288 \text{ psi}$$

$$Z_{\text{RT}}' = nF_v'ts_{\text{crit}}$$

$$= 5(288)(1.5)(2.5) = 5400 \text{ lb} > 4650 \text{ lb} \qquad OK$$

> *Use* double 2×6 Sel. Str. DF-L for top plate.

13.9   Shear Stresses in a Beam at a Connection

The evaluation of shear stresses in a beam is covered in Chap. 6. When a beam is supported by a connection with bolts, lag bolts, split rings, or shear plates, there are additional considerations for horizontal shear (NDS Sec. 3.4.3). The computed shear stress is determined using the *effective depth* d_e, which is taken as the depth of the member minus the distance from the unloaded edge of the member to the *center* of the nearest bolt or lag bolt. For timber connectors d_e is taken as the depth of the beam minus the distance from the unloaded edge to the *edge* of the nearest connector. See Example 13.15.

It should be clear that the tendency of a beam to develop a crack parallel to the grain will be affected by the magnitude of the shear stress in the member. In addition, the tendency of the member to split will be affected by the proximity of the connection to the end of the beam. Consequently, when the connection is less than 5 times the depth of the beam d from the end of the member, the computed shear stress is increased by the square of the ratio of the depth to the effective depth $(d/d_e)^2$. When the connection is greater than $5d$ from the end of the member, the splitting tendency is reduced, and the shear stress is computed without the $(d/d_e)^2$ increase. The allowable shear design value is as defined in Chap. 6, except that for glued laminated timber beams the nominal shear design value F_v must be multiplied by a reduction factor of 0.8 when evaluating shear stresses at a connection (see footnotes to Tables 5A and 5B in the NDS Supplement, Ref. 13.1).

EXAMPLE 13.15   Shear Stresses at a Connection

The shear in a beam supported by fasteners is checked using the effective depth d_e (Fig. 13.17). For bolts and lag bolts, d_e is the depth of the member minus the distance from the unloaded edge to the center of the nearest fastener. For split ring and shear plate connectors, d_e is the depth of the member minus the distance from the unloaded edge to the nearest edge of a connector. The loaded and unloaded edges are determined by examining the *force on the beam* at the fastener. The loaded edge is the edge toward which the force acts. The unloaded edge is the opposite edge.

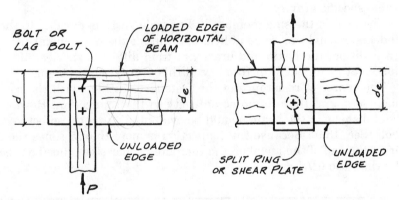

Figure 13.17 Definition of effective depth d_e for evaluation of shear in a beam supported by fasteners.

The total shear V in the beam at a connection is computed in the usual manner (Chap. 6). The NDS provides an equation for the design shear V_r' which must be greater than or equal to the calculated shear force on the member V when the connection is located less than $5d$ from the end of the member:

$$V_r' = \frac{2}{3} F_v' b d_e \left(\frac{d_e}{d}\right)^2 \geq V$$

The approach outlined here, where the design shear is compared to the applied shear force, is not typical in allowable stress design (ASD). However, it is quite common when the load and resistance factor design (LRFD) method is used. For ASD it is more common to compare the allowable shear design value to the stress in the member caused by the shear force. In this case the above equation can be rewritten in terms of stresses. Depending on the location of the connection, the actual shear stress is then compared with the allowable shear design value as follows:

When the connection is less than $5d$ from the end of the member,

$$f_v = \frac{3V}{2bd_e}\left(\frac{d}{d_e}\right)^2 \leq F'_v$$

When the connection is more than $5d$ from the end of the member,

$$f_v = \frac{3V}{2bd_e} \leq F'_v$$

For a discussion of the allowable shear design value F'_v, see Chap. 6.

13.10   Design Problem: Bolt Connection for Diagonal Brace

This final bolt example involves a connection that must be evaluated with the yield limit equations because the NDS bolt tables cover only parallel- and perpendicular-to-grain loading, and they handle only wood members with the same specific gravity.

The load in the side member in Example 13.16 is parallel to the grain ($\theta = 0$ degrees), and the angle of load to grain for the main member is 45 degrees. In addition, the beam and brace are from different species. Many of the adjustment factors default to unity in this connection problem.

The center of resistance of the fastener coincides with the centroid of the members, and there is no eccentricity in the connection. However, the horizontal shear in the beam should be evaluated using the effective depth to the bolt (Sec. 13.9). Because the connection is more than 5 times the beam depth from the end of the member, the computed shear stress need not be multiplied by the ratio $(d/d_e)^2$.

EXAMPLE 13.16   Bolt Connection for Diagonal Brace

The beam in Fig. 13.18a supports a dead load of 575 lb at its free end and is connected to the diagonal brace with a single ¾-in. full-diameter body bolt. The horizontal beam is a 24F-V4 DF glulam (from the 24F-1.8E Stress Class), and the diagonal brace is No. 1 Hem-Fir sawn lumber. Wood is initially dry and remains dry in service. Normal temperature conditions apply.

Determine if the bolt is adequate, and check the shear stress in the beam at the connection. Assume that other stresses in the members are OK.

Solve for force in brace.

$$\Sigma M_c = 0$$

$$6T_v = 575(10)$$

$$T_v = 958 \text{ lb}$$

$$T = \sqrt{2}T_v = \sqrt{2}(958) = 1355 \text{ lb}$$

NOTE: Segment BC of the beam is subjected to combined axial and bending stresses (Chap. 7).

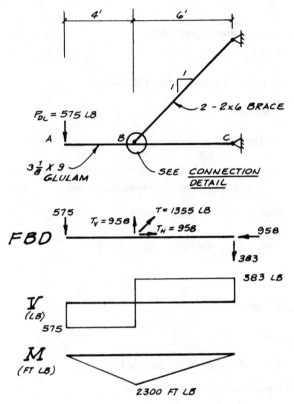

Figure 13.18a Member AC supported by diagonal brace at B.

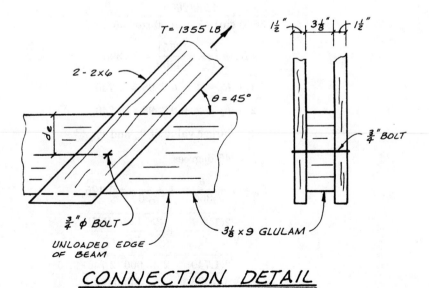

CONNECTION DETAIL

Figure 13.18b Bolt connection at B: double shear wood-to-wood.

Bolt Capacity

Summary of values for yield limit equations:

$$D = 0.75 \text{ in.}$$

$$F_{yb} = 45 \text{ ksi} \qquad \text{(assumed for A307 bolts)}$$

$$t_m = 3.125 \text{ in.}$$

$$\theta_m = 45 \text{ degrees}$$

$$G_m = 0.50 \qquad \text{(dowel bearing strength may be computed}$$
$$\text{using } G \text{ or read from NDS Table 11.3.2.)}$$

$$F_{e\parallel} = 5600 \text{ psi} \qquad \text{for bolt in DF-L}$$

$$F_{e\perp} = 2600 \text{ psi} \qquad \text{for } \tfrac{3}{4}\text{-in. bolt in DF-L}$$

$$t_s = 1.5 \text{ in.}$$

$$\theta_s = 0 \text{ degrees}$$

$$G_s = 0.43$$

$$F_{es} = F_{e\parallel} = 4800 \text{ psi} \qquad \text{for bolt in Hem-Fir}$$

Coefficients for yield equations:

$$F_{em} = F_{e\theta} = \frac{F_{e\parallel}F_{e\perp}}{F_{e\parallel}\sin^2\theta + F_{e\perp}\cos^2\theta}$$

$$= \frac{5600(2600)}{5600\sin^2 45 + 2600\cos^2 45} = 3551 \text{ psi}$$

$$R_e = \frac{F_{em}}{F_{es}} = \frac{3551}{4800} = 0.7398$$

$$1 + R_e = 1 + 0.7398 = 1.740$$

$$2 + R_e = 2 + 0.7398 = 2.740$$

$$\theta = \text{larger value of } \theta_m \text{ and } \theta_s$$

$$= 45 \text{ degrees}$$

$$K_\theta = 1 + \frac{\theta}{360} = 1 + \frac{45}{360} = 1.125$$

$$k_3 = -1 + \sqrt{\frac{2(1 + R_e)}{R_e} + \frac{2F_{yb}(2 + R_e)D^2}{3F_{em}l_s^2}}$$

$$= -1 + \sqrt{\frac{2(1.740)}{0.7398} + \frac{2(45,000)(2.740)(0.75^2)}{3(3551)(1.5^2)}}$$

$$= 2.239$$

Yield limit equations:
Mode I_m (NDS equation 11.3-7):

$$Z = \frac{Dl_m F_{em}}{4K_\theta} = \frac{0.75(3.125)(3551)}{4(1.125)} = 1850 \text{ lb}$$

Mode I_s (NDS equation 11.3-8):

$$Z = \frac{2Dl_s F_{es}}{4K_\theta} = \frac{2(0.75)(1.5)(4800)}{4(1.125)} = 2400 \text{ lb}$$

Mode III_s (NDS equation 11.3-9):

$$Z = \frac{2k_3 Dl_s F_{em}}{3.2(2 + R_e)K_\theta}$$

$$= \frac{2(2.239)(0.75)(1.5)(3551)}{3.2(2.740)(1.125)} = 1814 \text{ lb}$$

Mode IV (NDS equation 11.3-10):

$$Z = \frac{2D^2}{3.2K_\theta} \sqrt{\frac{2F_{em} F_{yb}}{3(1 + R_e)}}$$

$$= \frac{2(0.75^2)}{3.2(1.125)} \sqrt{\frac{2(3551)(45,000)}{3(1.740)}} = 2445 \text{ lb}$$

Nominal design value:

$$Z = 1814 \text{ lb}$$

Allowable design value:

The bolt is located on the center lines of the two members.
The required edge distance to the loaded edge of the horizontal member is $4D$.

$$4D = 4(0.75) = 3.0$$

$$\text{Edge distance} = \frac{9}{2} = 4.5 > 3.0 \quad OK$$

Assume that all other dimensions in the connection (including end distance for the tension member) will be chosen to satisfy the base dimension spacings for the connection in accordance with Sec. 13.5. Thus $C_\Delta = 1.0$.

$$Z' = Z(C_D C_M C_t C_g C_\Delta)$$

$$= 1814(0.9)(1.0)(1.0)(1.0)(1.0)$$

$$= 1633 \text{ lb} > 1355 \text{ lb} \quad OK$$

| ¾-in.-diameter bolt | OK |

Shear Stress

The shear stress in the $3\frac{1}{8} \times 9$ glulam beam is not critical, but calculations will be shown to demonstrate the criteria given in Sec. 13.9. Compare *5 times the beam depth* with the distance of the connection from the end of the beam.

$$5d = 5 \times 9 = 45 \text{ in.} < 48 \text{ in.}$$

The joint is more than 5 times the depth from the end. The computed stress need not be multiplied by $(d/d_e)^2$:

$$f_v = \frac{3V}{2bd_e} = \frac{3(575)}{2(3.125)(9.0/2)} = 61.3 \text{ psi}$$

Since the bolts penetrate the wide face of the 24F-V4 glulam beam, select the nominal shear design value for bending about the strong x-axis and multiply by a reduction factor of 0.8 in accordance with NDS Supplement Table 5A:

$$F_v = (0.8)240 = 192 \text{ psi}$$

$$F'_v = F_v(C_D C_M C_t)$$

$$= 192(0.9)(1.0)(1.0)$$

$$= 172.9 \text{ psi} > 61.3 \text{ psi} \quad OK$$

Shear in beam is OK.

It should be noted that shear at the connection in the glulam beam was OK since the location of the connection was greater than $5d$ from the end of the beam. If the connection had been located closer than $5d$ to the end of the beam, then the shear at the connection would not have been OK since:

$$f_v = \frac{3V}{2bd_e}\left(\frac{d}{d_e}\right)^2 = \frac{3(575)}{2(3.125)(4.5)}\left(\frac{9}{4.5}\right)^2 = 245.3 \text{ psi}$$

$$F'_v = 172.9 \text{ psi} < 245.3 \text{ psi} \quad NG$$

Possible solutions, when $F'_v < f_v$ at a connection, include increasing d_e by locating the bolt below (or above) the neutral axis or increasing bd_e by selecting a larger glulam beam.

13.11   Lag Bolt Connections

Lag bolts are relatively large-diameter fasteners that have a *wood screw* thread and a square or hexagonal bolt head. See Fig. 13.19a. The NDS and the TCM (Refs. 13.1 and 13.6) refer to these fasteners as lag screws. The term *lag bolt* implies a large-diameter fastener with a bolt head, and the term *lag screw* indicates a fastener with a wood screw thread. Both refer to the same type of fastener, and the term *lag bolt* is generally used in this book.

A distinction should be made between large-diameter lag bolts (lag screws) and similar small-diameter fasteners known as *wood screws*. Recall from Chap. 11 that the dowel bearing strength is different for large- and small-diameter fasteners. For lag bolts $F_{e\perp}$ is a function of diameter. The methods of installation for lag bolts and wood screws are also different. Because of their large size, lag bolts are installed with a wrench, and wood screws are usually installed with a screwdriver or screw gun. The NDS covers the design of both lag bolts and wood screws, but only lag bolts are addressed in this text.

Lag bolts are used when an excessively long bolt would be required to make a connection or when one side of a through-bolted connection may be inaccessible. Lag bolts can be used in shear connections (Fig. 13.19*b*) or withdrawal connections (Fig. 13.19*c*). Both applications are used in practice, and the designer should be familiar with the design procedures for both types of problems.

EXAMPLE 13.17   Lag Bolt Connections

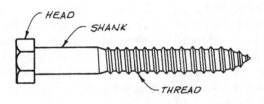

TYPICAL LAG BOLT

Figure 13.19*a* Lag bolt. Large-diameter fastener with wood screw thread and bolt head. Also known as lag screw.

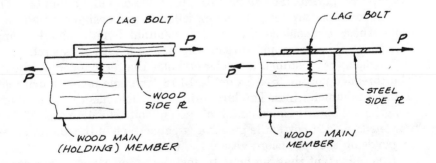

SHEAR TYPE CONNECTIONS

Figure 13.19*b* Lag bolts in single-shear connections. Wood-to-wood and wood-to-metal connections are the common applications.

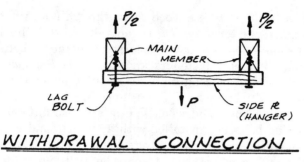

WITHDRAWAL   CONNECTION

Figure 13.19c Lag bolts in withdrawal connection.

Lag bolts are installed in special prebored holes that accommodate the shank diameter and provide holding material for the thread. Washers or steel plates are required under the fastener head (similar to bolts).

As with the majority of other fasteners used in wood connections, there is not a currently recognized national standard that defines mechanical properties for lag bolts. However, lag bolts are typically manufactured from the same wire stock as bolts, and $F_{yb} = 45$ ksi appears to be conservative for commonly available lag bolts ⅜ in. or greater in diameter. For smaller diameters, the bending yield strength is assumed to increase because of work-hardening (similar to nails): $F_{yb} = 70$ ksi for ¼-in. diameter, and $F_{yb} = 60$ ksi for 5/16-in. diameter. It is felt that these values can reasonably be used in the yield limit equations until a standard is developed. In the meantime, the designer should list the bending yield strength as part of the design specifications.

The configuration of lag bolts is given in NDS Appendix L *Typical Dimensions for Standard Hex Lag Screws*. This table provides a number of important dimensions that are used in the installation and design calculations for a lag bolt. These dimensions include the nominal length, shank diameter and length, root diameter and thread length, and length of tapered tip.

Lag bolts are installed in prebored holes that accommodate the shank and the threaded portion. See Example 13.18. The hole involves drilling with two different-diameter bits. The larger-diameter hole has the same diameter and length as the unthreaded shank of the lag bolt. The diameter of the lead hole for the threaded portion is given as a percentage of the shank diameter that depends on the specific gravity of the wood.

It is important that lag bolts be installed properly with a wrench. Driving with a hammer is an unacceptable method of installation. Soap or another type of lubricant should be used on the lag bolt or in the pilot hole to facilitate installation and to prevent damage to the fastener.

EXAMPLE 13.18   Installation of Lag Bolt

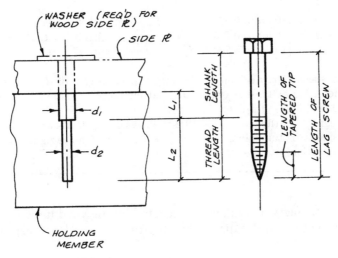

Figure 13.20*a*   Hole sizes.

A lag bolt is installed in a pilot hole. The dimensions of the pilot hole in the main or holding member are as follows:

For the shank:

$$L_1 = \text{shank length} - \text{side plate thickness} - \text{washer thickness}$$

$$d_1 = \text{shank diameter}$$

Figure 13.20*b*   Photographs of lag bolt threads. (*Courtesy of the FPL.*)

For the threaded portion:

$$L_2 \geq \text{thread length}$$

$$d_2 = \begin{cases} (0.65 \text{ to } 0.85) \times d_1 & \text{for wood with } G > 0.6 \\ (0.60 \text{ to } 0.75) \times d_1 & \text{for wood with } 0.5 < G \leq 0.6 \\ (0.40 \text{ to } 0.70) \times d_1 & \text{for wood with } G \leq 0.5 \end{cases}$$

For a given specific-gravity range, the larger percentages of d_1 should be used for larger-diameter lag bolts. It is important that lag bolts be installed by turning with a wrench into lead holes of proper size.

The photograph on the left in Fig. 13.20b shows deep, clean-cut threads of a lag bolt that was properly installed in the correct size lead hole. On the right, the threads have shallow penetration into the wood because of an oversize pilot hole.

The shear capacity of a lag bolt is covered by the yield limit equations (Sec. 13.12). The withdrawal capacity is empirically based on test results (Sec. 13.15).

13.12 Yield Limit Equations for Lag Bolts

The yield limit equations for lag bolts in single-shear (two-member) connections are identical to the yield limit equations for bolts in single shear, as summarized in Example 13.2. Lag bolts are not generally used in double-shear applications. The fastener diameter D in the yield limit equations is typically taken as the root diameter of the threads D_r for most lag bolt connections since lag bolt threads are frequently located near the shear plane where plastic hinges form in the fastener (see Sec. 11.7). Alternatively, the unthreaded shank diameter may be used for D in the yield limit equations if the moment and bearing behavior at the threaded region of the lag bolt are explicitly addressed in a more detailed analysis (see NDS Sec. 11.3.6 and Ref. 13.2).

According to NDS Sec. 11.3.4, the dowel bearing length in the main member l_m is defined in two ways based on the total length of the lag bolt in the main member (the member that receives the threaded end of the lag bolt). If the lag bolt penetrates a distance of $10D$ or greater into the main (holding) member, then the dowel bearing length l_m is taken as the total length of the lag bolt in the main member. However, if the lag bolt penetrates a distance less than $10D$ into the main member, the dowel bearing length l_m is taken as the total length of the lag bolt in the main member minus the length of the tapered tip E of the lag bolt (see NDS Appendix L). This is similar to the definition of dowel bearing length in the main member for nails. See Example 12.1. The dowel bearing strength F_e can be computed given the specific gravity of the wood, or values can be obtained from NDS Table 11.3.2, *Dowel Bearing Strengths*.

As an alternative to solving the yield limit equations, the nominal design value for a lag bolt used in certain common applications may be read from

NDS Tables 11J and 11K. Tabulated design values are based on the root diameter of the lag bolt threads D_r and lag bolt penetration in the main member of $p = 8D$. Thus, the NDS table values are directly applicable for connections with $p \geq 8D$. For lag bolt connections with $4D \leq p < 8D$ a footnote to the NDS tables provides a conservative reduction factor of $p/8D$ to account for the reduced lag bolt penetration in the main member. The reduction factor may be used to adjust tabulated design values in lieu of calculating a reduced connection capacity directly from yield limit equations. Design values are given for parallel- and perpendicular-to-grain loading. The notation is similar to that used in the bolt tables:

1. Load *parallel* to grain in *both* the main member and the side member, $Z_\parallel$.

2. Load *perpendicular* to grain in the *side* member and parallel to grain in the main member, $Z_{s\perp}$.

3. Load *perpendicular* to grain in the *main* member and parallel to grain in the side member, $Z_{m\perp}$.

4. Load *perpendicular* to grain in *both* the main member and the side member, $Z_\perp$.

When the load is at an intermediate angle of load to grain, the yield limit equations must be evaluated.

One example is given to illustrate the application of the yield limit equations for lag bolts. See Example 13.19. The connection is between a steel side plate and a glulam beam.

EXAMPLE 13.19   Lag Bolt Nominal Design Value: Single-Shear Wood-to-Metal Connection

Determine the nominal design value of a ⅝-in.-diameter by 6-in.-long lag bolt used to connect a ¼-in.-thick metal side plate of A36 steel and a DF-L glulam beam. Load is parallel to grain in the glulam.

Summary of Known Values

Dowel bearing strength can be computed (Sec. 11.5) given the specific gravity of the glulam beam, or it can be read as $F_{e\parallel}$ from NDS Table 11.3.2.

$L = 6$ in.

$D_r = 0.471$ in.   [for a ⅝-in.-diameter lag bolt (see NDS Appendix L)]

$E = {}^{13}\!/_{32}$ in. $= 0.406$ in.   (see NDS Appendix L)

$F_{yb} = 45$ ksi   (assumed for ⅝-in.-diameter lag bolt)

$\theta_m = 0$ degrees

$G_m = 0.50$

$$F_{em} = 5600 \text{ psi} \qquad (F_{e\parallel} \text{ for lag bolt in DF-L})$$

$$l_s = t_s = 0.25 \text{ in.}$$

$$10D = 10(0.625) = 6.25 \text{ in.}$$

Total lag bolt length in main member $= L - t_s = 6 - 0.25 = 5.75 \text{ in.} < 6.25 \text{ in.}$

$$\therefore l_m = L - t_s - E = 6 - 0.25 - 0.406 = 5.34 \text{ in.}$$

$$F_{es} = 1.5 \, F_u = 87{,}000 \text{ psi} \qquad (\text{for A36 steel plate})$$

Coefficients for Yield Limit Equations

$$R_e = \frac{F_{em}}{F_{es}} = \frac{5600}{87{,}000} = 0.06437$$

$$1 + R_e = 1 + 0.06437 = 1.064$$

$$2 + R_e = 2 + 0.06437 = 2.064$$

$$1 + 2R_e = 1.129$$

$$R_t = \frac{l_m}{l_s} = \frac{5.34}{0.25} = 21.4$$

$$1 + R_t = 1 + 21.4 = 22.4$$

$$1 + R_t + R_t^2 = 22.4 + 21.4^2 = 479.3$$

$$\theta = 0 \text{ degrees}$$

$$K_\theta = 1 + \frac{\theta}{360} = 1 + \frac{0}{360} = 1.0$$

$$k_1 = \frac{\sqrt{R_e + 2R_e^2(1 + R_t + R_t^2) + R_t^2 R_e^3} - R_e(1 + R_t)}{(1 + R_e)}$$

$$= \frac{\sqrt{0.06437 + 2(0.06437^2)(479.3) + (21.4^2)(0.06437^3)} - 0.06437(22.4)}{1.064}$$

$$= 0.5626$$

$$k_2 = -1 + \sqrt{2(1 + R_e) + \frac{2F_{yb}(1 + 2R_e)D^2}{3F_{em}l_m^2}}$$

$$= -1 + \sqrt{2(1.064) + \frac{2(45,000)(1.129)(0.471^2)}{3(5600)(5.34^2)}}$$

$$= 0.4750$$

$$k_3 = -1 + \sqrt{\frac{2(1 + R_e)}{R_e} + \frac{2F_{yb}(2 + R_e)D^2}{3F_{em}l_s^2}}$$

$$= -1 + \sqrt{\frac{2(1.064)}{0.06437} + \frac{2(45,000)(2.064)(0.471^2)}{3(5600)(0.25^2)}}$$

$$= 7.504$$

Yield Limit Equations

Mode I_m (NDS equation 11.3-1):

$$Z = \frac{Dl_mF_{em}}{4K_\theta} = \frac{0.471(5.34)(5600)}{4(1.0)} = 3524 \text{ lb}$$

Mode I_s (NDS equation 11.3-2):

$$Z = \frac{Dl_sF_{es}}{4K_\theta} = \frac{0.471(0.25)(87,000)}{4(1.0)} = 2561 \text{ lb}$$

Mode II (NDS equation 11.3-3):

$$Z = \frac{k_1Dl_sF_{es}}{3.6K_\theta} = \frac{0.5626(0.471)(0.25)(87,000)}{3.6(1.0)} = 1602 \text{ lb}$$

Mode III_m (NDS equation 11.3-4):

$$Z = \frac{k_2Dl_mF_{em}}{3.2(1 + 2R_e)K_\theta} = \frac{0.4750(0.471)(5.34)(5600)}{3.2(1.129)(1.0)} = 1854 \text{ lb}$$

Mode III_s (NDS equation 11.3-5):

$$Z = \frac{k_3Dl_sF_{em}}{3.2(2 + R_e)K_\theta} = \frac{7.504(0.471)(0.25)(5600)}{3.2(2.064)(1.0)} = 749 \text{ lb}$$

Mode IV (NDS equation 11.3-6):

$$Z = \frac{D^2}{3.2K_\theta}\sqrt{\frac{2F_{em}F_{yb}}{3(1 + R_e)}} = \frac{0.471^2}{3.2(1.0)}\sqrt{\frac{2(5600)(45,000)}{3(1.064)}} = 871 \text{ lb}$$

The nominal design value is selected as the smaller value from the yield formulas:

$$\boxed{Z = 749 \text{ lb}}$$

Compare with values from NDS Table 11K:

$$Z = 750 \text{ lb} \approx 749 \text{ lb}$$

13.13   Adjustment Factors for Lag Bolts in Shear Connections

The yield limit equations for lag bolts in single shear were discussed in Sec. 13.12. The *nominal design value Z* represents the capacity of a single lag bolt or lag screw as obtained from the yield equations. In the design of a given connection, the nominal design value is converted to an *allowable design value Z'* by a series of adjustment factors.

The adjustment factors convert the base conditions to the conditions for a specific design situation. The adjustment factors required in the design of connections are summarized in NDS Table 10.3.1.

The base conditions associated with the nominal design value are as follows:

1. Load is normal (10-year) duration.
2. Wood is initially dry at the time of fabrication of a connection and remains dry in service.
3. Temperature range is normal.
4. There is one lag bolt in a line parallel to the applied load (i.e., only one lag bolt per row).
5. Spacing provisions for lag bolts (geometry requirements for the connection) are satisfied.
6. Penetration of lag bolt in the holding member (not including the tapered tip) is a minimum of 4 times the shank diameter (that is, $p \geq 4D$).
7. Lag bolt penetrates the side grain (not the end grain) of the main (holding) member.

If a lag bolt is used in a connection under exactly these conditions, the adjustment factors all default to unity, and $Z' = Z$.

For a shear connection that does not satisfy the base conditions described above, the following adjustment factors may be required:

$$Z' = Z(C_D C_M C_t C_g C_\Delta C_{eg})$$

where C_D = load duration factor (Sec. 4.15)
C_M = wet service factor (Sec. 4.14)
C_t = temperature factor (Sec. 4.19)

C_g = group action factor
C_Δ = geometry factor
C_{eg} = end grain factor

The total allowable capacity for a connection using two or more lag bolts is the sum of the individual allowable design values Z'. This summation is appropriate if the lag bolts are all the same type and size because each fastener will have the same yield mode. The allowable capacity for the connection is the number of lag bolts N times the allowable design value for one fastener:

$$\text{Allow. } P = N(Z')$$

A number of adjustment factors are described elsewhere in this book and are mentioned only in summary. Some of these were covered in Chap. 4, and others, which apply to both lag bolts and bolts, were summarized in Sec. 13.5. Only the adjustment factors that are unique to lag bolts are covered in detail.

Load duration factor C_D. The load duration factor for a lag bolt connection is the same as that applied to the design of other fasteners. Refer to Sec. 12.6 for a review of load duration and the numerical values for C_D.

Wet service factor C_M. The concept of an adjustment factor based on moisture content was introduced in Sec. 4.14. The discussion regarding possible damage in a wood connection due to cross-grain volume changes applies to both lag bolts and bolts. See Sec. 13.5 for a discussion of this potentially serious problem. The numerical values of C_M are also the same for lag bolts and bolts and are given in NDS Table 10.3.3, *Wet Service Factors, C_M, for Connections.*

The first objective in good design practice is to keep the connection dry where possible. Second, the connection should be detailed so that cross-grain cracking does not occur. Wood is very weak in tension perpendicular to grain, and the conditions that cause tension stress across the grain can be eliminated, or minimized, by proper detailing. For example, separate steel side plates for rows of lag bolts parallel to the grain will allow volume changes to occur without locking in stresses.

Temperature factor C_t. If wood is used in an application with high temperatures, the allowable design value for a lag bolt is to be reduced by the temperature factor C_t. Refer to Sec. 4.19 for a brief review of the temperature factor. Numerical values of C_t for connections are given in NDS Table 10.3.4. The temperature range for most wood-frame buildings does not require a reduction of design values, and the temperature factor is often taken as unity.

Group action factor C_g. In a connection that has more than one lag bolt in a row, it is necessary to reduce the nominal design value by a multiplier known as the group action factor C_g. The group action factor primarily depends on the number of fasteners in a row. The group action factor for a lag bolt connection is the same as for a bolt connection. Refer to Sec. 13.5 for a discussion and definition of C_g. Note that C_g does not exceed unity ($C_g \leq 1.0$).

Geometry factor C_Δ.   End distance, edge distance, center-to-center spacing, and row spacing requirements for lag bolts are the same as for bolts. Refer to Sec. 13.5 for a discussion of spacing requirements. Specific criteria are given in Example 13.10 for parallel-to-grain loading and in Example 13.11 for perpendicular-to-grain loading.

Examples 13.10 and 13.11 summarize the base dimensions for fastener spacing. It is recommended that the base dimensions be used in designing and detailing a connection whenever practical. However, if conditions do not permit the use of the full base dimensions, it is possible to reduce certain spacings with a corresponding reduction in allowable design value. The minimum spacing limits are also given in these examples.

The geometry factor C_Δ is the multiplier that is used to reduce the nominal design value Z if the furnished spacings are less than the base dimensions. If the base requirements for a connection are all satisfied, the geometry factor defaults to unity ($C_\Delta = 1.0$). For other conditions, the geometry factor for lag bolts is the same as for bolts. Refer to Sec. 13.5 for the definition of C_Δ for reduced spacings. Note that C_Δ does not exceed unity ($C_\Delta \leq 1.0$).

Lag bolt penetration.   The penetration of a lag bolt is defined as the distance that the fastener extends into the main (holding) member minus the length of the tapered tip. Dimensions of typical lag bolts are given in NDS Appendix L, and the length of tapered tip E is included in this table. Thus, the penetration of a lag bolt can be computed as the length of the lag bolt minus the thickness of the side member minus the washer thickness (if applicable) minus the length of the tapered end:

$$p = L - t_s - t_{\text{washer}} - E$$

To obtain any design value Z for a lag bolt connection, the penetration must be at least 4 times the diameter of the lag bolt ($p \geq 4D$). In previous editions of the NDS only three yield limit equations (rather than six) were specified for lag bolt connections, and a penetration depth factor was provided for lag bolt penetration in the range $4D \leq p \leq 8D$. However, all six yield limit equations are specified for lag bolt connections in the 2001 NDS. Thus, the penetration depth factor is no longer required since the dowel bearing length of the lag bolt in the main member l_m is explicitly addressed in the yield limit equations.

End grain factor C_{eg}.   The nominal design value Z applies to a laterally loaded shear connection with the lag bolt installed into the *side grain* of the main member. Again, the main member is the member that receives the threaded end of the fastener. See Fig. 13.21a. This is the strongest and most desirable type of lag bolt connection. When a lag bolt is installed in the side grain, the end-grain factor does not apply, or it can be viewed as having a default value of unity ($C_{eg} = 1.0$).

EXAMPLE 13.20 Comparison of Lag Bolts in Side Grain and End Grain

Lateral Resistance in Side Grain

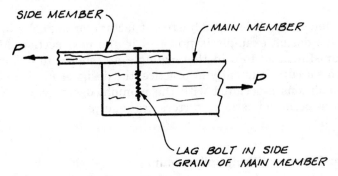

Figure 13.21*a* Lag bolt in side grain.

The basic shear connection has the lag bolt installed in the side grain of the main member (Fig. 13.21*a*). The full nominal design value applies to this type of loading, and the end grain factor is $C_{eg} = 1.0$.

Lateral Resistance in End Grain

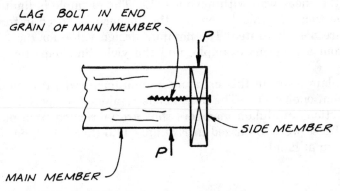

Figure 13.21*b* Lag bolt in end grain.

A second type of shear connection has the lag bolt installed in the end grain of the holding member (Fig. 13.21*b*). In other words, the lag bolt is installed parallel to the grain of the main member. The connection still qualifies as a laterally loaded connection because the load is perpendicular to the axis of the fastener. For lag bolts installed in the end grain, $C_{eg} = 0.67$. For a lag bolt in the end grain, $F_{em} = F_{e\perp}$.

In another type of shear connection, the lag bolt is installed in the end grain of the holding member (Fig. 13.21b). This is a much weaker and less desirable connection. However, laterally loaded connections of this type are permitted, but the nominal design value is reduced by an end grain factor of $C_{eg} = 0.67$. A similar factor is used in connections with nails driven into the end grain (Sec. 12.6).

This concludes the review of adjustment factors for lag bolts in shear connections. One design example is provided in the next section. Many of the possible adjustment factors in this example are shown to default to unity. Although this occurs frequently in practice, the designer needs to be aware of the base conditions associated with the nominal design value. When other circumstances occur, adjustment factors are required.

13.14   Design Problem: Collector (Strut) Splice with Lag Bolts

This example makes use of lag bolts to attach a glulam beam to a concrete shearwall. The load is the result of a wind or seismic force on the horizontal diaphragm. See Example 13.21. The force on the connection is obtained from the collector (strut) force diagram which is given in the problem. For a review of collector (strut) force diagrams, refer to Sec. 9.10.

The force is transferred from the glulam beam by lag bolts in a shear connection to a steel plate. A second connection transfers the load from the steel plate to the shear wall with anchor bolts. The example is limited to the lag bolt connection, and calculations for the anchor bolts used in the second connection are not illustrated. The nominal design value for the lag bolt is obtained from a previous example, and the yield limit equations are not repeated.

The glulam beam in this example is manufactured with laminations that are all from one species. The designer is reminded of special considerations for connections in glulam members that are fabricated from more than one species. These were discussed in Example 11.6 (Sec. 11.5), but do not affect the problem at hand.

EXAMPLE 13.21   Splice with Lag Bolts

The roof beam in the building in Fig. 13.22 serves as the collector (strut) for two horizontal diaphragms. The maximum collector force occurs at the connection of the roof beam and the concrete shearwall, and the magnitude of the load is obtained from the collector (strut) force diagram.

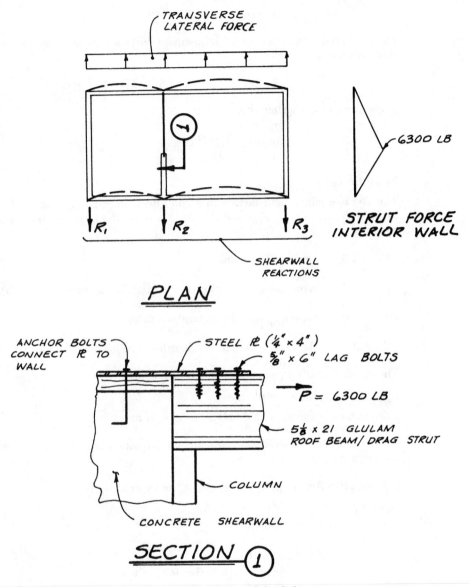

Figure 13.22 Collector (strut) connection made with lag bolts.

Determine the number of ⅝-in.-diameter, × 6-in. lag bolts necessary to connect the steel splice plate to the Douglas Fir (24F-V4) glulam beam (from the 24F-1.8E Stress Class). The nominal design value for the lag bolt is obtained from Example 13.19. Assume that the splice plate and the connection to the wall are adequate. A single row of lag bolts is to be used, and the base dimensions for lag bolt spacings are to be summarized. $C_M = 1.0$, and $C_t = 1.0$.

Trial Design

The nominal design value for a ⅝-in.-diameter 6-in. lag bolt from Example 13.19 is governed by Mode III$_s$ yield behavior:

$$Z = 749 \text{ lb}$$

For an initial trial, assume all adjustment factors default to unity except the load duration factor:

$$Z' \approx Z(C_D) = 749(1.6) = 1198 \text{ lb/lag bolt}$$

Approximate number of bolts

$$N = \frac{P}{Z'} = \frac{6300}{1198} = 5.26 \qquad say \quad 6 \text{ lag bolts}$$

Final Design

The possible adjustment factors that can affect the capacity of a lag bolt connection include C_D, C_M, C_t, C_g, C_Δ, and C_{eg}. At this point the following factors are known or assumed:

$C_D = 1.6$ for wind or seismic

$C_M = 1.0$ given (assumes connection is initially dry and remains dry in service)

$C_t = 1.0$ assumes normal temperature range

$C_{eg} = 1.0$ lag bolts installed in side grain (top face) of glulam

The remaining adjustments will now be considered.

Penetration depth:
   Thickness of the metal side plate $t_s = \frac{1}{4}$ in.

A washer is not required since the steel side plate separates the lag bolt head from the wood member.

From NDS Appendix L, the length of the tapered tip is

$$E = {}^{13}\!/_{32} \text{ in.}$$

$$p = L - t_s - E$$

$$= 6 - 0.25 - 0.406$$

$$= 5.34 \text{ in.}$$

Minimum penetration to develop the full design value is

$$p = 4D = 4(0.625) = 2.5 \text{ in.} < 5.34 \text{ in.} \qquad OK$$

Geometry factor:
   Connection is a single row of lag bolts with load parallel to grain. Spacing requirements for lag bolts are the same as for bolts. For parallel-to-grain loading, spacing requirements are summarized in Example 13.10. If there is adequate room, it is recommended that base dimensions be used in detailing a connection.
   Base dimensions for a single row of bolts in a tension member (softwood species):

End distance $= 7D = 7(0.625) = 4.375$ in.   *use*   5 in.

c.-to-c. spacing $= s = 4D = 4(0.625) = 2.5$ in.   *use*   3 in.

Edge distance $= 1.5D = 1.5(0.625) = 0.94$ in.

The row of lag bolts will be installed along the centerline of the beam.

$$\text{Edge distance} = \frac{5.125}{2} = 2.56 \text{ in.} > 0.94 \quad OK$$

The furnished spacings all exceed the base dimensions.

$$\therefore C_\Delta = 1.0$$

Group action factor:

The formula and coefficients for the group action factor are summarized in Example 13.8 (Sec. 13.5).

Load/slip modulus $\gamma = 270,000(D^{1.5})$ for lag bolts in wood-to-metal connection

$$= 270,000(0.625^{1.5}) = 133,400 \text{ lb/in.}$$

Member stiffnesses:

$$E_m = E_{\text{axial}} = 1,700,000 \text{ psi} \quad \text{(NDS Supplement Table 5A)}$$

$$A_m = 5.125 \times 21 = 107.6 \text{ in.}^2$$

$$E_m A_m = 1,700,000(107.6) = 183,000,000$$

$$E_s = E_{\text{steel}} = 29,000,000 \text{ psi}$$

$$A_s = 0.25 \times 4 = 1.0 \text{ in.}^2$$

$$E_s A_s = 29,000,000(1.0) = 29,000,000$$

Spacing of fasteners in a row

$$s = \text{c.-to-c. spacing} = 3 \text{ in.}$$

Coefficients for C_g:

$$n = \text{number of lag bolts in row} = 6$$

$$u = 1 + \gamma \left(\frac{s}{2}\right) \left[\frac{1}{E_m A_m} + \frac{1}{E_s A_s}\right]$$

$$= 1 + 133,400 \left(\frac{3.0}{2}\right) \left[\frac{1}{183,000,000} + \frac{1}{29,000,000}\right]$$

$$= 1.008$$

$$m = u - \sqrt{u^2 - 1}$$

$$= 1.008 - \sqrt{1.008^2 - 1} = 0.8813$$

$$1 + m = 1 + 0.8813 = 1.881$$

$$1 - m = 1 - 0.8813 = 0.1187$$

$$m^{2n} = 0.8813^{2(6)} = 0.2195$$

$$\frac{E_m A_m}{E_s A_s} = \frac{183,000,000}{29,000,000} = 6.309$$

$$\frac{E_s A_s}{E_m A_m} = \frac{1}{6.309} = 0.1585$$

$$R_{EA} = \text{smaller of ratios of member stiffnesses}$$

$$= 0.1585$$

$$1 + R_{EA} = 1 + 0.1585 = 1.159$$

$$1 + R_{EA} m^n = 1 + 0.1585(0.8813)^6 = 1.074$$

$$C_g = \left[\frac{m(1 - m^{2n})}{n[(1 + R_{EA}m^n)(1 + m) - 1 + m^{2n}]} \right] \left[\frac{1 + R_{EA}}{1 - m} \right]$$

$$= \left[\frac{0.8813(1 - 0.2195)}{6[(1.074)(1.881) - 1 + 0.2195]} \right] \left[\frac{1.159}{0.1187} \right]$$

$$= 0.902$$

Allowable design value:

$$Z' = Z(C_D C_M C_t C_g C_\Delta C_{eg})$$

$$= 1145(1.6)(1.0)(1.0)(0.902)(1.0)(1.0)$$

$$= 1081 \text{ lb}$$

$$\text{Allow. } P = N(Z') = 6(1081) = 6485 \text{ lb} > 6300 \text{ lb} \qquad \textit{OK}$$

> *Use*   6 lag bolts in a single row
> $\frac{5}{8}$-in. diameter $\times$ 6 in. long.
> End distance = 5 in.
> c.-to-c. spacing = 3 in.

13.15   Lag Bolts in Withdrawal

The concept of a lag bolt being subject to a withdrawal load was introduced in Chap. 11 (Fig. 11.2a). Withdrawal loading attempts to pull the fastener out of the holding member.

The yield limit equations for dowel-type fasteners apply only to shear-type connections, and the yield limit theory does not apply to lag bolts in withdrawal. The withdrawal values for lag bolts are based on an empirical formula summarized in the NDS.

The *nominal design value* for the basic withdrawal connection is given the symbol W. The basic withdrawal connection is with the lag bolt installed in the side grain of the holding member. NDS Table 11.2A gives the nominal withdrawal design value *per inch of thread penetration* into the main member.

If the value from the table is given the symbol w and the *thread* penetration for withdrawal is given the symbol p_w, the nominal design value for a single lag bolt is

$$W = w \times p_w$$

The penetration p for a shear connection was explained in Sec. 13.13. For shear, the penetration is taken as the length of the lag bolt in the main (holding) member minus the length of the tapered end. In other words, in a shear connection, a portion of the unthreaded shank may be included in the penetration length. However, the effective penetration for withdrawal loading p_w depends on the effective thread length in the holding member. See Fig. 13.23. NDS Appendix L gives the length of the threaded portion of the lag bolt T, the length of the tapered tip E, and the length of the full-diameter threads $T - E$. Assuming that the threads are all within the main member, the effective penetration for withdrawal is

$$p_w = T - E$$

The nominal design value W represents the capacity for a lag bolt used under a set of base conditions. The base conditions associated with the nominal design value are as follows:

1. Load is normal (10-year) duration.

2. Wood is initially dry at the time of fabrication of connection and remains dry in service.

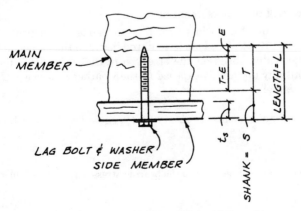

Figure 13.23 Effective thread length for withdrawal.

3. Temperature range is normal.

4. Lag bolt is installed in the side grain of the holding member.

To obtain the allowable design value W' for a different set of conditions, the nominal design value is multiplied by adjustment factors:

$$W' = W(C_D)(C_M)(C_t)(C_{eg})$$

where C_D = load duration factor (Sec. 4.15)
  C_M = wet service factor (Sec. 4.14)
  C_t = temperature factor (Sec. 4.19)
  C_{eg} = end grain factor

A discussion of adjustment factors for lag bolts in *shear* connections was given in Sec. 13.13. The majority of this material also applies to lag bolts in *withdrawal* connections.

However, the end grain factor is different for a withdrawal connection. The default condition for C_{eg} = 1.0 still applies to the base condition of a lag bolt installed into the *side grain* of the main member. But for a withdrawal connection into the *end grain* of the holding member, the end grain factor is C_{eg} = 0.75 instead of the 0.67 value used for shear connections. Connections into the end grain should be avoided where possible. Although the loading shown is not withdrawal, Fig. 13.21*a* and *b* illustrates lag bolts installed in the side grain and end grain.

The total allowable withdrawal capacity for a connection is the number of lag bolts N times the allowable design value for one fastener:

$$\text{Allow. } P = N(W')$$

An example of a simple lag bolt withdrawal connection is given in Example 13.22.

EXAMPLE 13.22   Lag Bolt Withdrawal Connection

The load P is an equipment dead load which is suspended from two Southern Pine supporting beams (Fig. 13.24). The load is centered between the two beams. Determine the allowable design value based on the strength of the ½ in.-diameter × 6 in. lag bolts. Lumber is initially dry and is exposed to the weather in service. C_t = 1.0.
  Nominal design value:

Length of lag bolt in main member = $L - t_s - t^*_{\text{washer}}$ = 6 − 1.5 − 0.125 = 4.375 in.

---

  *Washer must be sufficiently large to keep the bearing stress perpendicular to grain below the allowable ($f_{c\perp} \leq F'_{c\perp}$).

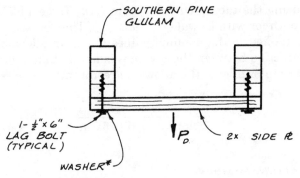

Figure 13.24 Withdrawal loading on lag bolts.

Thread length $= T = 3\frac{1}{2}$ in.   (NDS Appendix L)

All of the threads are inside the main member.

Effective penetration for withdrawal $p_w = T - E = 3\frac{3}{16}$ in.

Specific gravity for Southern Pine:

$$G = 0.55$$

From NDS Table 11.2A

$$w = 437 \text{ lb/in.}$$

$$W = w(p_w) = 437(3\tfrac{3}{16}) = 1393 \text{ lb/lag bolt}$$

Allowable design value:

$$W' = W(C_D)(C_M)(C_t)(C'_{eg})$$

$$= 1393(0.9)(0.7)(1.0)(1.0)$$

$$= 878 \text{ lb}$$

$$\text{Allow. } P = 2(878) = 1756 \text{ lb}$$

13.16 Combined Lateral and Withdrawal Loads

In addition to the direct withdrawal problem, there are occasions when lag bolts are subjected to combined lateral and withdrawal loading. See Fig. 13.25. The resultant load acts at an angle α to the surface of the main member. The practice for this type of loading is to determine an allowable resultant design value Z'_α which must equal or exceed the total design load.

The first step in computing the allowable design value is to evaluate the nominal design value Z of the lag bolt for *shear* loading using the appropriate yield limit equations. For the next step, the nominal *withdrawal* design value

W is computed using the tabulated value from NDS Table 11.2A and the effective thread length for withdrawal ($W = w \times p_w$). The necessary adjustment factors are then applied to the nominal values to obtain allowable values Z' and W'. The interaction between these values is finally taken into account by the Hankinson formula, to give the allowable resultant:

$$Z'_\alpha = \frac{Z'W'}{Z' \sin^2 \alpha + W' \cos^2 \alpha}$$

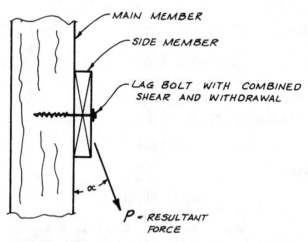

Figure 13.25 Lag bolt subjected to combined lateral and withdrawal loading.

13.17 Split Ring and Shear Plate Connectors

Split rings and shear plates are connectors that are installed in precut grooves in wood members. Split ring and shear plate connectors provide a large bearing surface to resist shearing-type forces in a wood connection. A bolt or lag bolt is required through the center of the split ring or shear plate to hold the assembly together. A sketch of the load transfer was shown in the introductory chapter on connections in Fig. 11.1*d*.

Split rings are only for wood-to-wood connections because the steel ring fits into a groove cut into the mating surfaces of the members being connected. See Fig. 13.26*a*. The steel ring is "split" to provide simultaneous bearing on the inner core and the outer surface of the groove.

Shear plates can be used for wood-to-metal connections because the shear plate is flush with the surface of the wood. Shear plates may also be used for wood-to-wood connections, but a shear plate is required in each wood member. Both wood-to-wood and wood-to-steel connections are shown in Fig. 13.26*b*. Shear plate wood-to-wood connections are easier to assemble or disassemble than split-ring connections and may be desirable where ease of erection is important.

EXAMPLE 13.23   Split Ring and Shear Plate Connectors

Split rings are for wood-to-wood connections only. Figure 13.26a is an unassembled joint showing the split ring installed in the center member and the precut groove in one of the side members ready for installation. Also shown are the bolt, washer, and nut required to hold the assembly together. In a three-member joint, a second split ring is required at the mating surface between the center member and the other side member.

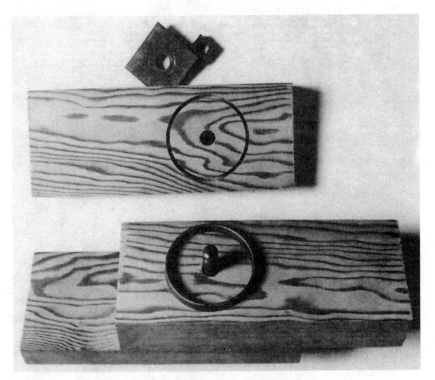

Figure 13.26a Split ring connection. (*Courtesy of the FPL.*)

Joint A at the top of Fig. 13.26b shows a wood-to-wood connection ready for assembly. A shear plate is required in each mating face of the wood members. Therefore, in the three-member joint illustrated, a total of four shear plates is required. Shear plates have the advantage of creating a surface that is flush with the surface of the wood member. Although more connectors are required, joints made with shear plates are easier to assemble than a similar split ring connection. Joint B is a three-member wood-to-steel connection. One shear plate is required at each interface between the wood and steel members.

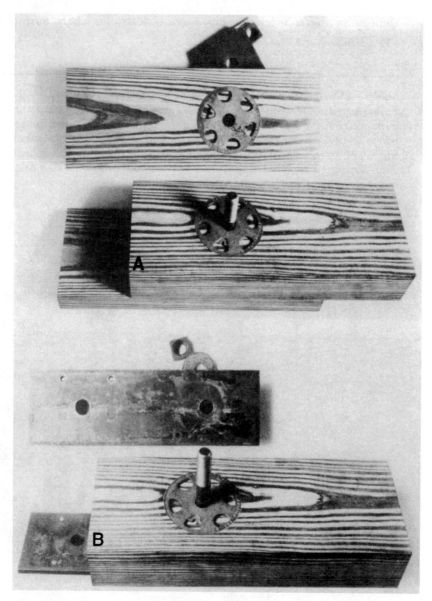

Figure 13.26b Shear plate connections. (*Courtesy of the FPL.*)

The allowable design values for split rings and shear plates are higher than those for bolts or lag bolts. However, split-ring and shear plate connectors require special fabrication equipment, and their use is limited by fabrication costs. These types of connectors may be used where relatively large loads must be transferred in a fairly limited amount of space, but they are much less common than simple bolted connections. Split ring and shear plate connectors are more widely used in glulam arches and heavy timber trusses than in shearwall-type buildings. See Example 13.24.

EXAMPLE 13.24   Typical Arch Connection Using Timber Connectors

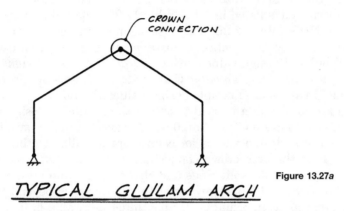

CROWN CONNECTION

TYPICAL GLULAM ARCH

Figure 13.27a

Arch connections are shown in Fig. 13.27a and b to illustrate how shear plates can be used in the end grain of a member. Allowable design values for timber connectors are given in NDS Tables 12.2A and 12.2B. Similar tables are available in the TCM. Information on the design of arches can also be found in the TCM.

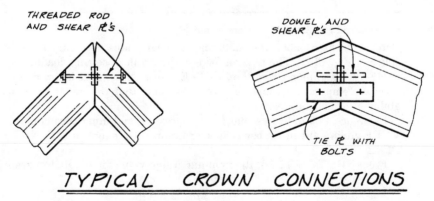

THREADED ROD AND SHEAR ℞'s

DOWEL AND SHEAR ℞'s

TIE ℞ WITH BOLTS

TYPICAL   CROWN CONNECTIONS

Figure 13.27b

For arches with steep slopes, two shear plates can be used with a threaded rod. Washers are counterbored into the arch. For flat arches, two shear plates can be used on a dowel for vertical reactions. The tie plate keeps the arch from separating.

Split rings are available in 2½- and 4-in. diameters, and shear plates are available in 2⅝- and 4-in. diameters. Other pertinent dimensions are given in NDS Appendix K. For the design of split rings and shear plates, commercial lumber species are divided into four groups (A to D) in NDS Table 12A. Information on the species groupings for timber connectors in glulam members is available in the TCM (Ref. 13.6) and Ref. 13.5.

Design values for split rings and shear plates are not covered by the yield limit theory that is applied to nails, bolts, and lag bolts. Nominal design values for split rings are listed in NDS Table 12.2A, and values for shear plates are found in NDS Table 12.2B. The basic connection for both split rings and shear plates is with the connector installed in the side grain of the wood member. Tabulated design values include *parallel-to-grain design value P* and *perpendicular-to-grain design value Q*. Adjustment factors are provided to convert tabulated values to allowable design values P' and Q'.

The Hankinson formula is used to obtain the allowable design value N' at an angle of load to grain other than 0 or 90 degrees. Further adjustments are used for obtaining design values for connectors installed in the end grain of a member (as in the arch crown connection in Fig. 13.27b). Spacing requirements are important for split rings and shear plates, and design criteria are given in the NDS. A detailed review of this material is beyond the scope of this book. Additional information for obtaining allowable design values and connector spacings is available in Ref. 13.6.

A brief example will illustrate the use of the NDS tables for timber connectors. See Example 13.25. The connection involves the design of shear plates for the splice of a chord in a horizontal diaphragm.

EXAMPLE 13.25   Shear Plate Connection

Determine the number of 4-in.-diameter shear plates necessary to develop the chord splice in Fig. 13.28. The tension force is the result of seismic loading. The member is a 24F-V4 DF glulam (from the 24F-1.8E Stress Class) that is dry and remains dry in service. Temperature range is normal. Assume that the metal splice plates and the glulams* are adequate.

Reference 13.5 indicates that the compression and inner core laminations of a 24F-V4 DF glulam are in timber connector group B (this agrees with NDS Table 12A for DF-L).

From NDS Table 12.2B, the nominal design value for parallel-to-grain loading is

$$P = 4320 \text{ lb}$$

---

*For tension calculations, the net area of a wood member with shear plates must allow for the projected area of the connectors in addition to the projected area of the bolt hole. See NDS Section 3.1.2 and Appendix K.

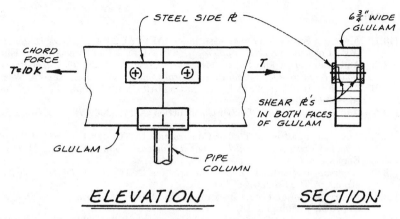

Figure 13.28 Diaphragm chord splice made with shear plates.

The following adjustment factors apply to this problem:

$C_D = 1.6$ wind or seismic force

$C_M = 1.0$ moisture content factor

$C_t = 1.0$ normal temperature range

$C_g = 1.0$ one shear plate in a row

$C_\Delta = 1.0$ spacings will exceed base dimensions

$C_{st} = 1.11$ steel side plate factor for parallel-to-grain loading in species group B (NDS Table 12.2.4)

Allowable design value for one shear plate:

$$P' = P(C_D)(C_M)(C_t)(C_g)(C_\Delta)(C_{st})$$

$$= 4320(1.6)(1.0)(1.0)(1.0)(1.0)(1.11)$$

$$= 7672 \text{ lb}$$

There is a shear plate in each face of the glulam.

$$\text{Allow. } T = 2(7672) = 15,300 \text{ lb} > 10,000 \text{ lb} \quad OK$$

> *Use* total of four 4-in.-diameter shear
> plates (two each side of splice).

13.18   References

[13.1]   American Forest and Paper Association (AF&PA). 2001. *National Design Specification for Wood Construction and Supplement.* 2001 ed., AF&PA, Association, Washington DC.

[13.2]   American Forest and Paper Association (AF&PA). 1999. *Technical Report 12—General Dowel Equations for Calculating Lateral Connection Values,* AF&PA, Washington, DC.

[13.3]   American Institute of Steel Construction (AISC). 1989. *Manual of Steel Construction,* 9th ed., AISC, Chicago, IL.

[13.4]   American Institute of Timber Construction (AITC). 2002. *Bolts in Glued Laminated Timber, AITC Technical Note 8,* AITC, Englewood, CO.

[13.5]   American Institute of Timber Construction (AITC). 2001. *DESIGN Standard Specifications for Structural Glued Laminated Timber of Softwood Species,* AITC 117-2001, AITC, Englewood, CO.

[13.6]   American Institute of Timber Construction (AITC). 1994. *Timber Construction Manual,* 4th ed., AITC, Englewood, CO.

[13.7]   American Iron and Steel Institute (AISI). 1996. *Cold-Formed Steel Design Manual.* 1996 ed., AISI, Washington DC.

[13.8]   American Society for Testing and Materials (ASTM). 2003. "Standard Specification for Carbon Steel Bolts and Studs, 60,000 psi Tensile Strength," ASTM A307-00, *Annual Book of Standards,* Vol. 01.08, ASTM, Philadelphia, PA.

[13.9]   Forest Products Laboratory (FPL). 1999. *Wood Handbook: Wood as an Engineering Material, General Technical Report 113,* FPL, Forest Service, U.S.D.A., Madison, WI.

[13.10]   International Codes Council (ICC). 2003. *International Building Code,* 2003 ed., ICC. Falls Church, VA.

[13.11]   Society of Automotive Engineers (SAE). 1999. "Mechanical and Material Requirements for Externally Threaded Fasteners," SAE J 429, SAE, Warrendale, PA.

13.19   Problems

Problems 13.1 through 13.9 involve work with a computer spreadsheet or equation-solving software. The first requirement is for a spreadsheet to operate correctly, (i.e., to compute correct values). Another major consideration is the format of input and output. A sufficient number of intermediate results should be displayed so that the computer solution can be conveniently verified by hand.

It will be necessary to test a number of different connection configurations to verify that the equation for each yield limit mode is programmed correctly. Save each template for use on other connection problems.

Many of the problems require the nominal design value Z and the allowable design value Z' as either final or intermediate answers. These values are to be identified as part of the general solution along with the information specifically requested.

Nominal design values may be obtained by solving the yield limit equations or by using the NDS tables. Tables are to be used only if they apply directly to the design conditions stated in the problem and are to be referenced in the solution. Yield limit equations may be evaluated by either calculator or computer.

In all problems, normal temperatures are assumed, and $C_t = 1.0$. All adjustment factors are to be referenced to the appropriate NDS table or section so that the source of the values may be verified.

13.1   Use a computer spreadsheet or equation-solving software to solve the yield limit equations for a bolt in a single-shear wood-to-wood connection. Input the following:

Diameter of bolt D (in.)

Bending yield strength of bolt F_{yb} (ksi)

Thickness of main member t_m (in.)

Specific gravity of main member G_m

Angle of load to grain for main member θ_m (degrees)

Thickness of side member t_s (in.)

Specific gravity of side member G_s

Angle of load to grain for side member θ_s (degrees)

The software template is to do the following:

a. Compute the dowel bearing strengths parallel ($F_{e\parallel}$), perpendicular ($F_{e\perp}$), and at angle θ ($F_{e\theta}$) to the grain for both the main and side members.

b. Evaluate the coefficients for use in the yield limit equations: R_e, R_t, θ, K_θ, k_1, k_2, and k_3. Other intermediate answers may be provided to aid in checking.

c. Solve the six yield limit equations, and select the smallest value as the *nominal design value Z*.

13.2 Expand or modify the template from Prob. 13.1 to handle a bolt in a single-shear connection with a steel side plate (wood-to-metal connection). Input remains the same, except G_s and θ_s are not required, and the ultimate tensile strength of the metal side plate F_u is an input.

13.3 Expand or modify the template from Prob. 13.1 to handle a single-shear connection with an anchor bolt (wood-to-concrete connection). Input remains the same, except G_m and θ_m are not required, and F_e for the concrete main member is an input.

NOTE: If desired, this template may be expanded to include a database or table of nominal design values for the capacity of the anchor bolt in the concrete (IBC Table 1912.2).

13.4 Expand the template in Prob. 13.1, 13.2, or 13.3 (as assigned) to convert the *nominal design value Z* to an *allowable design value Z'*. Input must be expanded to include the appropriate adjustment factors. Include, as a minimum, adjustments for

a. Load duration factor C_D

b. Wet service factor C_M

c. Group action factor C_g

d. Geometry factor C_Δ

NOTE: C_D and C_M may be input by the user, but C_g and C_Δ should be generated by the template with a minimum of input. Necessary input is to be determined as part of the problem. Macros may be used if desired but are not required.

13.5 Use a computer spreadsheet or equation-solving software to solve the yield limit equations for a double-shear wood-to-wood connection, using the following variables for input:

Bolt diameter D (in.)

Bending yield strength of bolt F_{yb} (ksi)

Thickness of main member t_m (in.)

Specific gravity of main member G_m

Angle of load to grain for main member θ_m (degrees)

Thickness of side member t_s (in.)

Specific gravity of side member G_s

Angle of load to grain for side member θ_s (degrees)

The software template is to do the following:

a. Compute the dowel bearing strengths parallel ($F_{e\parallel}$), perpendicular ($F_{e\perp}$), and at angle θ ($F_{e\theta}$) to the grain for both the main and side members.

b. Evaluate the coefficients for use in the yield limit equations: R_e, R_t, θ, K_θ, and k_3. Other intermediate answers may be provided to aid in checking.

c. Solve the four yield limit equations, and select the smallest value as the *nominal design value Z*.

13.6   Expand or modify the template from Prob. 13.5 to handle a bolt in a double-shear connection with steel side plates (wood-to-metal connection). Input remains the same, except G_s and θ_s are not required, and the ultimate tensile strength of the metal side plate F_u is an input.

13.7   Expand the template in Prob. 13.5 or 13.6 (as assigned) to convert the *nominal design value Z* to an *allowable design value Z'*. Input is to be expanded to include the appropriate adjustment factors. Include as a minimum:

Load duration factor C_D

Wet service factor C_M

Group action factor C_g

Geometry factor C_Δ

NOTE: C_D and C_M may be input by the user, but C_g and C_Δ should be generated by the template with a minimum of input. Necessary input is to be determined as part of the problem. Macros may be used if desired but are not required.

13.8   Use a computer spreadsheet or equation-solving software to set up the solution of the yield equations for a single-shear lag bolt connection. Consider both *wood-to-wood* and *wood-to-steel* connections, using the following variables for input:

Diameter of lag bolt D (in.)

Length of lag bolt L (in.)

Bending yield strength of lag bolt F_{yb} (ksi)

Specific gravity of main member G_m

Angle of load to grain for main member θ_m (degrees)

Thickness of side member t_s (in.)

Specific gravity of side member G_s

Angle of load to grain for side member θ_s (degrees)

The software template is to do the following:

a. Compute the dowel bearing strengths parallel ($F_{e\parallel}$), perpendicular ($F_{e\perp}$), and at angle θ ($F_{e\theta}$) to the grain for both the main and side members.

b. Evaluate the coefficients for use in the yield limit equations: R_e, R_t, θ, K_θ, k_1, k_2, and k_3. Other intermediate answers may be provided to aid in checking.

c. Solve the six yield limit equations for a *wood-to-wood* connection or a *wood-to-steel* connection. Select the smallest value as the *nominal design value Z*.

13.9   Expand the template in Prob. 13.8 to convert the *nominal design value Z* to an *allowable design value Z'*. Input must be expanded to include the appropriate adjustment factors. Include as a minimum:

Load duration factor C_D

Wet service factor C_M

Group action factor C_g

Geometry factor C_Δ

End grain factor C_{eg}

NOTE: C_D, C_M, and C_{eg} may be input by the user, but C_g and C_Δ should be generated by the template with a minimum of input. Necessary input is to be determined as part of the problem. Macros may be used if desired but are not required.

13.10   The connection in Fig. 13.A uses a single row of five ¾-in.-diameter bolts. Lumber is No. 1 Hem-Fir. The load is the result of (D + W). $C_M = 1.0$.

Find: a. Allowable design value for the bolt connection.
b. Allowable tension in the wood members.
c. Indicate the base spacing requirements for the bolts.
d. Row tear-out capacity for the connection.

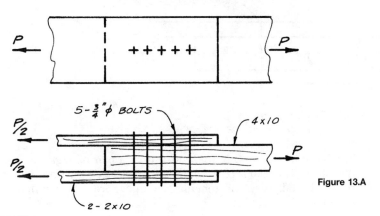

Figure 13.A

13.11   A connection similar to the connection in Fig. 13.A has a single row of *six* (instead of five) ¾-in.-diameter bolts. Lumber is No. 1 DF-L. Load is the result of (D + S). $C_M = 1.0$.

Find: a. Allowable design value for the bolt connection.
b. Allowable tension in the wood members.
c. Indicate the base spacing requirements for the bolts.
d. Row tear-out capacity for the connection.

13.12   Repeat Prob. 13.10 except that the lumber is green at the time of fabrication and it later seasons in service.

13.13   Repeat Prob. 13.11 except that the lumber is green at the time of fabrication and it later seasons in service.

13.14   Repeat Prob. 13.10 except that the connection is exposed to the weather.

13.15   Repeat Prob. 13.11 except that the connection is exposed to the weather.

13.16   The connection in Fig. 13.B uses side plates of A36 steel and four 1-in. bolts on each side of the splice. Bolt spacing is $4D$ and end distance is $7D$. Lumber is No. 1 DF-L. The force is seismic.

   Find: Allowable design value for the bolts if:
       a. $C_M = 1.0$.
       b. Green lumber is used that later seasons in service.
       c. Connection is exposed to the weather.
       d. Lumber is above the FSP in service.

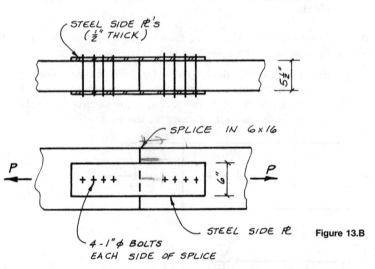

Figure 13.B

13.17   The beam in Fig. 13.C is suspended from a pair of angles (2 L 4 × 3 × ⁵⁄₁₆ of A36 steel). The load is (D + L). Lumber is Select Structural Spruce-Pine-Fir (S) that is initially dry and remains dry in service. Assume that the angles are adequate.

   Find: *a.* Determine if the bolts are adequate.
       b. Check the shear stress in the beam at the connection.
       c. Are the base spacing requirements satisfied?

13.18   Repeat Prob. 13.17 except the hanger is not vertical. It forms an angle of 60 degrees to the horizontal upward to the right.

13.19   Repeat Prob. 13.17 except that the lumber is initially at the FSP and it seasons in service.

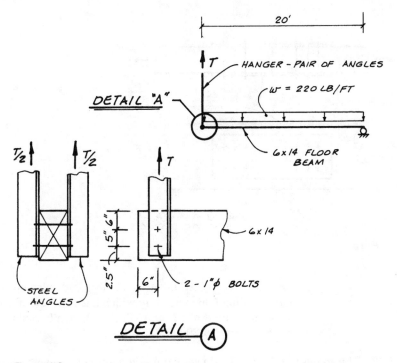

Figure 13.C

13.20 The connection in Fig. 13.D uses two rows of ⅞-in.-diameter bolts with three fasteners in each row. The load carried by the connection is (D + L). The wood member is a 6¾ × 16.5 DF glulam with $G = 0.50$. It is initially dry (at or near the EMC) and remains dry in service.

Find: a. Allowable connection design value based on the capacity of the bolts. Show the base dimension spacing requirements on a sketch of the connection.
   b. Check the stresses in the main member and side plates for the load in (a). Assume A36 steel plate which has an allowable tensile stress on the gross area of 22 ksi ($0.6F_y$) and an allowable tensile stress on the net area of 29 ksi ($0.5F_u$).

13.21 The connection in Fig. 13.D uses two rows of ⅞-in.-diameter bolts with *five* (instead of three) fasteners in each row. Load carried by the connection is (D + S). The wood member is a 6 × 14 Select Structural DF-L that is initially wet, and it later seasons in place.

Find: a. Allowable design value for the connection, based on the strength of the bolts.
   b. Allowable design value for the connection if separate side plates are used for the two rows of bolts. Show the base dimension spacing requirements for the bolts in the connection.

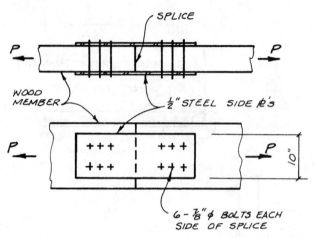

Figure 13.D

    c. Check the stresses in the main member and side plates for the load in (*b*). Assume A36 steel plate which has an allowable tensile stress on the gross area of 22 ksi ($0.6F_y$) and an allowable tensile stress on the net area of 29 ksi ($0.5F_u$).

13.22 The connection in Fig. 13.E uses ¾-in.-diameter bolts in single shear. Lumber is DF-L that is initially dry (near the EMC) and it remains dry in service. The load is caused by (D + L).

    Find: *a.* The allowable design value for the connection, based on the strength of the bolts.

              b. Check tension and shear stresses at the connection in each member and select bolt spacings and end distances to provide the allowable design value determined in part *a*. Show the spacing requirements for the bolts on a sketch of the connection.

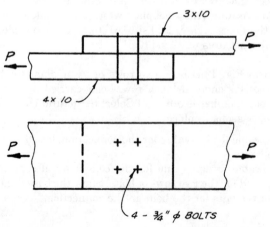

Figure 13.E

13.23 The connection in Fig. 13.F uses 1-in.-diameter bolts and serves as a tie for lateral wind forces. Lumber is DF-L that is initially dry (near the EMC) and it remains dry in service.

Find: a. Allowable design value for the connection based on the strength of the bolts.
 b. Check tension and shear stresses at the connection in the side members. Check shear stresses at the connection in the main member assuming the connection is located at least $5d$ from the end of the member. Select bolt spacing, end distance, and edge distances to provide the allowable design value determined in part *a*. Show the spacing requirements for the bolts on a sketch of the connection.
 c. If initially green lumber is used, what will be the effect on the allowable design value?

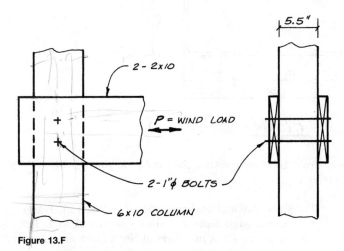

Figure 13.F

13.24 The connection in Fig. 13.G uses one 1-in.-diameter bolt that carries a lateral wind force. Lumber is Southern Pine that is dry ($C_M = 1.0$). Assume that the members are adequate.

Find: Allowable design value for the connection.

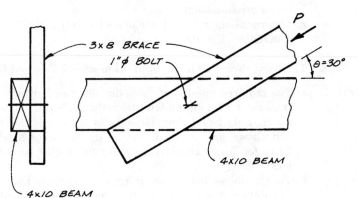

Figure 13.G

13.25 The building in Fig. 13.H uses 2 × 6 wall framing of No. 1 Hem-Fir. The double top wall plate serves as the horizontal diaphragm chord and drag strut. The top plate is fabricated from 20-ft-long lumber, and splices occur at 10-ft intervals. Design lateral forces are given. $C_M = 1.0$.

Find: Design and detail the connection splice in the top plate of wall 1, using ¾-in.-diameter bolts.

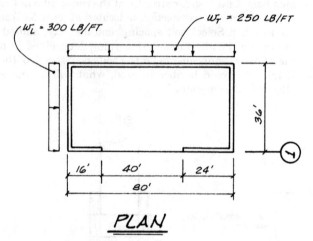

Figure 13.H

13.26 Repeat Prob. 13.25 except that the lumber is No. 1 DF-L and the bolts are ⅞ in. diameter.

13.27 *Given:*   The shearwall and anchorage detail in Fig. 13.I. The gravity loads and the lateral wind force are shown in the sketch. The chords in the shearwall are 4 × 4 members of No. 2 DF-South. Assume the bracket is adequate. $C_M = 1.0$.

Find: *a.* Number of ¾-in.-diameter bolts necessary to anchor the chord to the hold-down bracket. Assume an area of 2.5 in.² for the bracket. Sketch the base dimension spacing requirements for the bolts.
   b. Check the tension and shear stresses of the chord at the connection (bolts are in a single row).
   c. Determine the required size of anchor bolt to tie the bracket to the foundation. Use IBC Table 1912.2.

13.28 *Given:*   The wood shearwall in Fig. 13.I. Lumber is DF-L, and $C_M = 1.0$.

Find: *a.* The required number of ¾-in.-diameter anchor bolts necessary to transfer the 4-k lateral force from the 2× bottom plate to the 2000-psi concrete foundation. These bolts are in addition to the anchor bolt shown in the sketch for the uplift anchor bracket. See Sec. 10.11 for design considerations for the anchor bolts for this problem.
   b. For the anchor bolts found in part *a*, determine the total allowable foundation reaction (Figs. 10.16 and 10.17) perpendicular to the wall for short-term (wind or seismic) forces.

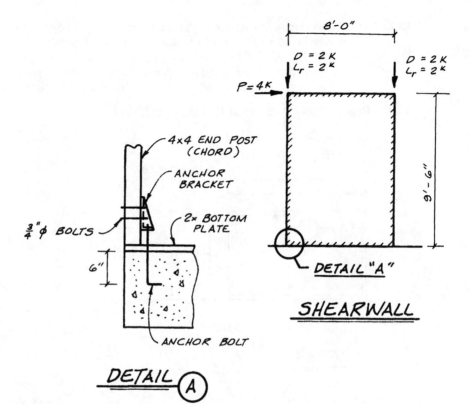

SHEARWALL

DETAIL (A)

Figure 13.I

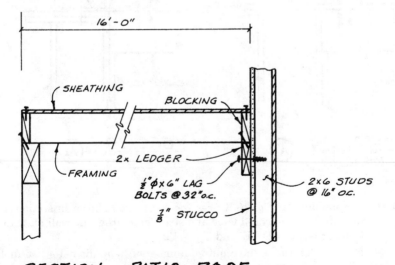

SECTION - PATIO ROOF

Figure 13.J

13.29   The connection of the ledger to the stud wall in Fig. 13.J uses ½-in.-diameter × 6-in. lag bolts. Patio D = 6 psf, and L_r = 10 psf. Lumber is DF-L.

Find: a. Determine if the connection is adequate to carry the design loads if the lumber is initially dry and remains dry in service.

b. Is the connection adequate if the lumber is exposed to the weather?

13.30   Repeat Prob. 13.29 except that the L_r is 20 psf.

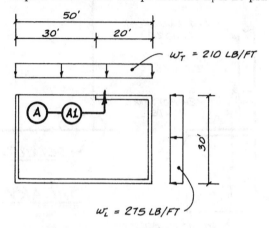

PLAN

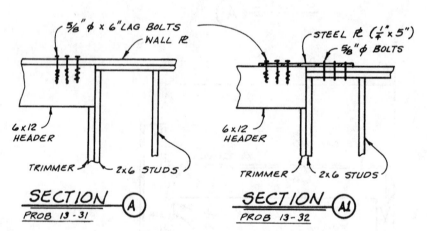

SECTION (A)
PROB 13-31

SECTION (A1)
PROB 13-32

Figure 13.K

13.31   The building in Fig. 13.K uses the connection shown in section A to transfer the lateral force (chord force and strut force) from the wall plate to the header. Lumber is No. 1 DF-L, and C_M = 1.0.

Find: Determine the required number of ⅝-in.-diameter × 6-in. lag bolts for the connection.

13.32   Repeat Prob. 13.31 except that the lateral force transfer is made with the connection shown in section A1. In addition to the number of lag bolts, determine the required number of ⅝-in.-diameter bolts (A307) necessary to connect the steel splice plate to the double 2× top plate.

Chapter 14

Connection
Hardware

14.1 Introduction

Chapters 12 and 13 covered the basic fasteners that are used in the construction of wood buildings. In addition to these fasteners, most wood connections involve some type of metal connection hardware. This hardware may be a simple tension strap, but considerably more complicated devices are often used.

This chapter introduces some typical connection hardware details. A critique of the connection accompanies a number of the details, and particular attention is given to the effects of cross-grain shrinkage. These principles should be given special attention if it is necessary to design, detail, and fabricate special connection hardware. An example of several connection hardware calculations is given after the general review.

In a number of cases, prefabricated connection hardware may be available for common connections. In any event, the basic principles of connection design should be understood by the designer. The chapter concludes with a review of some typical prefabricated connection hardware.

14.2 Connection Details

The hardware used in a connection must itself be capable of supporting the design loads, and the structural steel design principles of Refs. 14.2 and 14.5 should be used to determine the required plate thickness and weld sizes. Some of these principles are demonstrated in examples, but a comprehensive review of steel design is beyond the scope of this book.

Some suggested details for common wood connections are included in Refs. 14.3, 14.4, and 14.6, with an emphasis on avoiding cross-grain cracking caused by changes in moisture content. These details are aimed at minimizing restraint in connections that will cause splitting. Where restraint does occur, the distance across the grain between the restraining elements should be kept small so that the total shrinkage between these points is minimized. Where possible, the connections should be designed to accommodate the shrinkage (and swelling) of the wood without initiating built-in stresses.

The majority of the details in this section are taken or adapted from Ref. 14.3 by permission of AITC. The sketches serve as an introduction to the types of connections that are commonly used in wood construction. See Example 14.1. Both good and poor connection design practices are illustrated. The details were developed specifically for glulam construction, where members and loads are often large. The principles are valid for both glulam and sawn lumber of any size, but the importance of these factors increases with increasing member size.

The connection details in Example 14.1 are not to scale and are to be used as a guide only. Actual designs will require complete drawings with dimensions of the connection and hardware. The designs must take into consideration the spacing requirements for bolts (Examples 13.10 and 13.11 in Sec. 13.5) or the other types of fasteners. Examples of the following connections are included in Example 14.1:

1. Typical beam-to-column connections
2. Beam-to-girder saddle connection
3. Beam face hanger connection
4. Beam face clip connection
5. Cantilever beam hinge connection
6. Beam connection for uplift
7. Beam connection to continuous column
8. Beam with notch in tension side
9. Beam with end notch
10. Inclined beam—lower support detail
11. Inclined beam—upper support detail
12. Suspending multiple loads from beam
13. Suspending isolated loads from beam
14. Truss heel connection
15. Additional truss joint considerations
16. Moisture protection at column base

References 14.3, 14.4, and 14.6 cover a number of connections in addition to these. However, the details given here illustrate the basic principles involved in proper connection design. A numerical example is provided in the following section for the design of a beam-to-column connection. This is followed by some additional considerations for beam hinge connections.

EXAMPLE 14.1   Connection Details*

Beam-to-Column Connections

If compression perpendicular to grain $F_{c\perp}$ at the beam reaction is satisfactory, the beam may be supported directly on a column. A T-bracket can be used to tie the various members together (Fig. 14.1a).

The bottom face of the beam bears on the column. If cross-grain shrinkage occurs in length a, cracking or damage in the region of the T-bracket will result. The problem can be minimized by keeping a small.

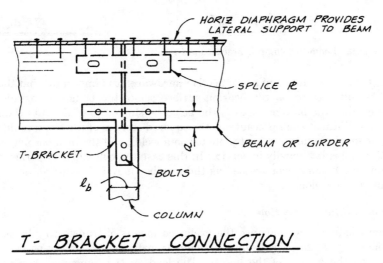

T- BRACKET   CONNECTION

Figure 14.1a   Beam-to-column connection with T-bracket.

For larger loads the T-bracket will be replaced with a U-bracket (Fig. 14.1b). The U-bracket can provide a bearing length l_b which is greater than the width of the column. The spacing requirements for the fasteners must also be considered in determining the length of the bracket.

If a separate splice plate is used near the top of the beam (Fig. 14.1a), the effects of the end rotation and beam separation should be considered. Bolts in tightly fitting holes can restrict the movement of the end of the beam, and splitting may be induced. Slotted holes can reduce this problem. If the beam carries an axial drag or chord force, the T-bracket or U-bracket can be designed to transmit the lateral force.

U-brackets (Fig. 14.1b) may be fabricated from bent plates when the bracket does not cantilever a long distance beyond the width of the column. For longer U-brackets, the vertical plates will probably be welded to a thicker base plate. Calculations for a U-bracket connection are given in Sec. 14.3.

---

*Adapted from Ref. 14.3, courtesy of AITC.

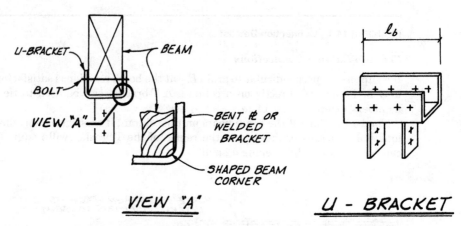

Figure 14.1b Beam-to-column U-bracket hardware.

In U-brackets and similar connection hardware, it is important that the wood member be fully seated on the bearing surface *before* the bolts are installed. This may require that the bottom edges of the beam be shaped to conform to the inside of the bracket. Without such precautions the squared corners of the beam could initially rest on the inside radius of a bent plate (or on a weld), and the member may not fully seat until it is loaded heavily in service. In this case the bolt capacity perpendicular to the grain may become overstressed as the member moves downward, and a split in the beam may develop.

Beam Saddle Connection

Beam saddle connections are the preferred means of transferring beam reactions to girders (Fig. 14.2a). The beam reaction is transferred by bearing perpendicular to the grain on the *bottom* of the hanger. The load on the hanger is then transferred by bearing perpendicular to the grain through the *top* of the saddle to the girder. This type of connection is recommended for larger loads.

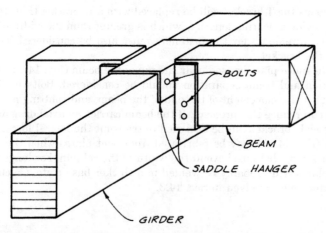

Figure 14.2a Beam hanger with saddle feature providing bearing on the *bottom* of beam and on the *top* of girder.

The tops of beams are shown higher than the girder to allow for shrinkage in beams without affecting roof sheathing or flooring. The idea is to have the tops of beams and girders at the same level after shrinkage occurs. For shrinkage calculations see Example 4.3 (Sec. 4.7).

Note in Fig. 14.2a that the bolts are located *near* the bearing points to minimize cross-grain effects. Figure 14.2b illustrates the possible effects of cross-grain shrinkage if the bolts are located *improperly*. The bolt in the beam is located some distance away from the initial point of bearing (section 1). After the beam is in service for some time, it loses moisture and shrinks across the grain. The bolt is held rigidly in position by the hanger bracket. Under a relatively small applied load, the bolt capacity may be large enough to support the beam reaction, and the bottom face of the member may shrink *up* from its intended bearing point (section 2).

However, when the member finally becomes heavily loaded, the bearing capacity of the bolt perpendicular to the grain may be exceeded, and a split at the end of the member will develop as the beam seats again on the hanger (section 3). The end split not only reduces the capacity of the member, but it also renders the bolt ineffective in resisting lateral (wind or seismic) forces that may create an axial force in the beam. Improperly located bolts as shown in Fig. 14.2b should be *avoided*.

Although the bolts should be located near the bearing face of the member, care should be taken to avoid boring holes in the high-quality outer lamination in bending glulams. This is especially important when bolts occur in the tension zone of the member.

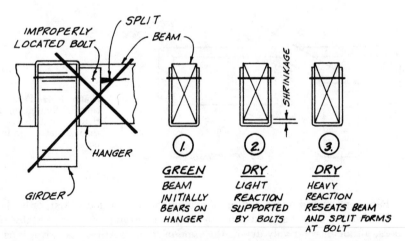

Figure 14.2b *Avoid* improper bolt location in beam hanger. Three stages are identified that lead to the formation of a split. Proper detailing will avoid splits by placing bolts closer to the beam bearing surface.

Beam Face Hanger Connection

When the reaction of the beam or purlin is relatively small, the hanger can be bolted to the face of the girder (Fig. 14.3). The bolts in the main supporting beam or girder should be placed in the upper half of the member, but not too close to the top of the beam where extreme fiber-bending stresses are maximized. This is especially important in the tension zone which occurs on the top side of cantilever girders.

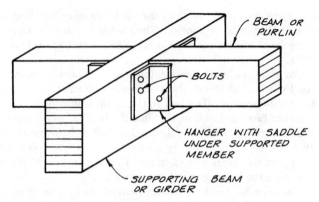

Figure 14.3 Beam hanger with saddle under supported beam. Beam reaction is transferred to girder through fasteners instead of bearing as in Fig. 14.2a.

Beam Face Clip Connection

Face connections without the saddle feature shown in Fig. 14.3 may be used for light loads (Fig. 14.4a). However, construction is more difficult than with the saddle because the beams must be held in place while the bolts are installed. The connection at the end of the supported beam must be checked for shear in accordance with the procedures outlined in Sec. 13.9.

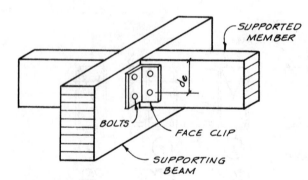

Figure 14.4a Face clip connection is limited to light loads. Beam reaction is transferred through fasteners in both members.

Face clips should be limited to connections with light loads. Larger loads will require a long row of bolts perpendicular to the grain through steel side plates. Glulam timbers, although relatively dry at the time of manufacture, may shrink as the EMC is reached in service. The problems associated with cross-grain tension should be avoided by not using the type of connection in Fig. 14.4b.

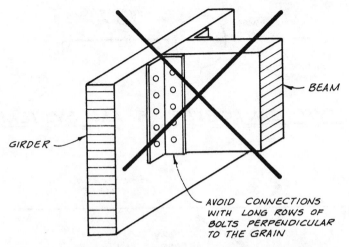

Figure 14.4b *Avoid* long rows of bolts perpendicular to grain through steel side plates. Changes in moisture content will result in cross-grain volume changes, and cross-grain tension and splitting may occur. The saddle connection in Fig. 14.2*a* is the preferred method of transferring heavy loads.

Cantilever Beam Hinge Connection

PROBLEM

For vertical loads the *suspended* beam in Fig. 14.5*a* bears on the *bottom* of the hinge connector, and the hinge connector in turn bears on the *top* of the *supporting* (cantilever) member. Cantilever systems may be subjected to lateral forces as well as gravity bending loads. Some designers have used a tension connector with bolts lined up in both members near the bottom of the beam. The problem with this arrangement is that as the wood shrinks, the supporting member will permit the saddle to move downward. The bolts in the cantilever, however, will restrain this movement, and a split may occur. The revised detail (Fig. 14.5*b*) avoids this problem.

SUGGESTED REVISION

The connection in Fig. 14.5*b* does not interfere with possible shrinkage in the members and cracking is avoided. It also allows a slight rotation in the joint. A positive tension tie for lateral forces can be provided with a separate tension strap at the middepth of the joint. This tension strap may also be an integral part of the hanger if the holes in the tension tie are slotted vertically in deep members. Also see Sec. 14.4.

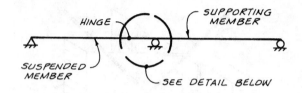

TYP CANTILEVER BEAM SYSTEM

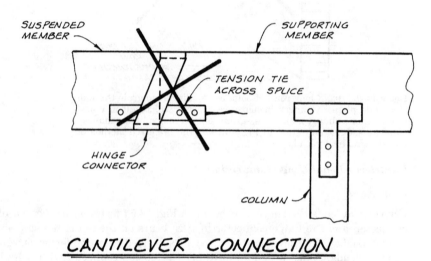

CANTILEVER CONNECTION

Figure 14.5a *Avoid* cantilever beam hinge connections with improperly located tension tie which causes cracking as cross-grain shrinkage occurs.

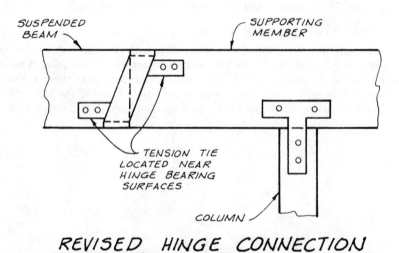

REVISED HINGE CONNECTION

Figure 14.5b *Revised* cantilever beam hinge connection with tension ties properly located near top and bottom bearing surfaces.

Beam Connection for Uplift

In this example, gravity loads are transferred by bearing perpendicular to grain. Adequate fastenings must also be provided to carry horizontal and uplift forces. See Fig. 14.6. These loads are usually of a transient nature and are of short duration. The fastener is usually placed toward the lower edge of the member. The distance d_e must be no less than the required perpendicular-to-grain edge distance for the type of fastener used. It must also be large enough so that the shear stresses caused by uplift (calculated in accordance with the methods of Sec. 13.9) are not excessive. The end distance e must also be adequate for the type of fastener used.

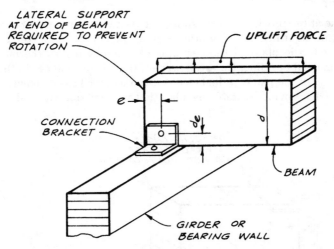

Figure 14.6 Gravity loads are transferred by bearing. Connection hardware may be necessary for uplift and lateral forces. Proper detailing considers fastener spacing requirements.

Beam Connection to Continuous Column

PROBLEM

Connections of the type shown in Fig. 14.7a should not be used on large beams or girders because tension perpendicular to the grain and horizontal shear at the lower fastener tend to cause splitting of the member.

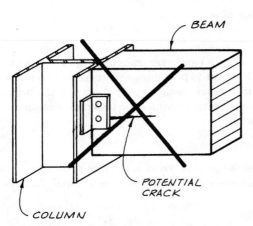

Figure 14.7a Avoid transferring large reactions at the end of a beam with fasteners.

SUGGESTED REVISION

When the design is such that the bending member does not rest on the top of a column, wall, or pilaster, but frames into another member, several methods can be used to support the ends. The preferred method is to transfer the end reaction by bearing perpendicular to grain. See Fig. 14.7b.

1. Vertical beam reaction is transferred by bearing perpendicular to grain on the beam seat angle.

2. Positive connection is made between the beam and the column with fasteners in the tie clip.

3. Clip angle at top provides lateral stability to beam, but it is not attached to the top of the beam with fasteners. In other words, the clip angle prevents rotation perpendicular to the plane of the beam, and it thereby braces the member against lateral torsional buckling. However, because there is no fastener in the beam at the top, rotation in the plane of the beam is accommodated as the member is loaded in bending. This provides a stable member that will not split the end.

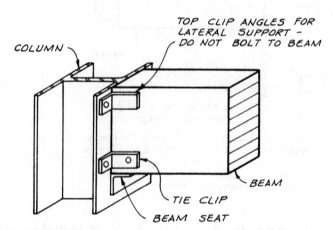

Figure 14.7b *Revised* connection transfers beam reaction by bearing on the bottom of the member.

Beam with Notch in Tension Side

PROBLEM

The detail in Fig. 14.8 is critical in cantilever framing (negative moment at support). The top *tension* fibers have been cut to provide space for recessed connection hardware or for the passage of conduit or other elements over the top of the beam. This is particularly serious in glulam construction because the tension laminations are critical in the performance of the structure.

SUGGESTED REVISION

Notches in the tension zone are serious stress raisers and should be avoided. Revise the detail in Fig. 14.8 so that the cut in the beam is eliminated. Also see Sec. 6.2.

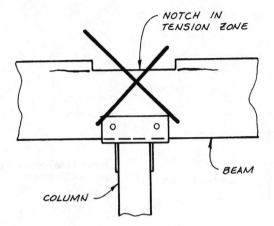

Figure 14.8 *Avoid* notching the tension side of a wood member.

Beam with End Notch

PROBLEM

An abrupt notch in the end of a wood member creates two problems (Fig. 14.9a). One is that the effective shear strength of the member is reduced because of the smaller depth d_n and stress concentration. In addition, the exposure of end grain in the notch will permit a more rapid migration of moisture in the lower portion of the member, and a split may develop.

For horizontal shear calculations at notched end of a beam, see Sec. 6.5.

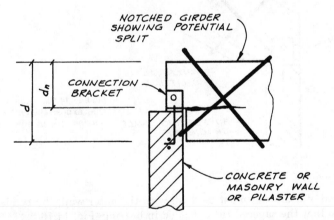

Figure 14.9a *Avoid* square cut notches.

SUGGESTED REVISION

Where the height of the top of a beam is limited, the beam seat should be lowered to a pilaster or specially designed seat in the wall (Fig. 14.9b). A lateral support connection at the top of the girder is required for stability. There is no fastener between the wood member and lateral support hardware. Therefore, the lateral support connection prevents rotation perpendicular to the plane of the girder, but it allows rotation of the end of the member for in-plane loading.

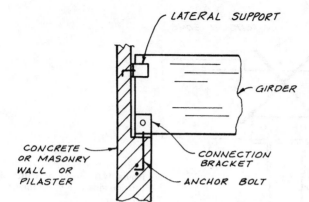

Figure 14.9b *Revised* beam support without notch.

Inclined Beam-Lower Support Details

PROBLEM

The condition in Fig. 14.10*a* is similar to the problem in Fig. 14.9*a*, but it may not be as evident. The shear strength of the end of the member is reduced, and the exposed end grain may result in splitting.

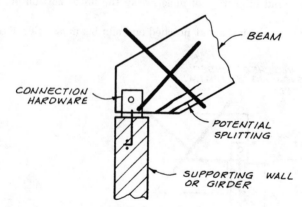

Figure 14.10*a* *Avoid* end cuts that expose the end grain on the *tension* side of a beam.

SUGGESTED REVISION

Where the end of the beam must be flush with the outer wall, the beam seat should be lowered so that the tapered cut is loaded in bearing (Fig. 14.10*b*).

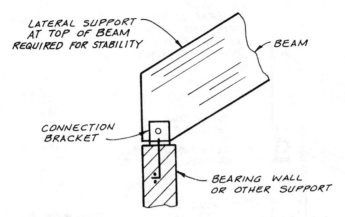

Figure 14.10*b* *Revised* taper cut is detailed so that the end grain is stressed in compression.

Inclined Beam-Upper Support Details

PROBLEM

The bird's mouth cut in Fig. 14.11*a* is commonly used in rafters to obtain a horizontal bearing surface. In major, heavily loaded beams, large cuts of this nature substantially reduce the effective depth of the member at the notch and should be avoided. Stress concentrations also cause high shear stress at the notch and may cause splitting.

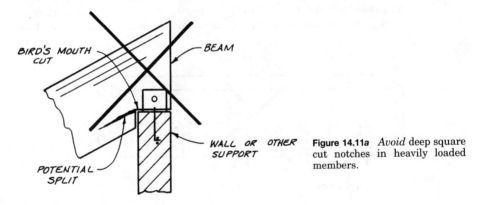

Figure 14.11*a* *Avoid* deep square cut notches in heavily loaded members.

SUGGESTED REVISION

A welded bracket with a sloping seat can be used to avoid notching the member (Fig. 14.11*b*). The bracket must be properly designed to support the beam reaction. Eccentricity in the connection should be avoided.

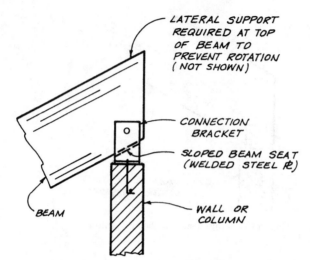

Figure 14.11b *Revised* beam support without end notch.

Suspending Multiple Loads from Beam

PROBLEM

Suspending multiple loads below the neutral axis (NA) of a wood member is not recommended (Fig. 14.12a). Tension stresses perpendicular to the grain tend to concentrate at the fastenings. Wood is weak in cross-grain tension, and tension perpendicular to grain can interact with horizontal shear to cause splitting. Hanger loads may be mechanical or electric equipment suspended below the beam, or they may be reactions from joists, purlins, or other members framing into the side of the beam.

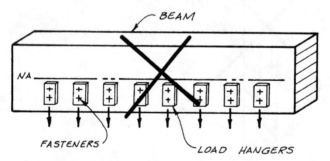

Figure 14.12a *Avoid* suspending multiple loads below the neutral axis.

SUGGESTED REVISION 1

It is better design practice to locate the fasteners above the neutral axis, as shown in Fig. 14.12b. Where possible, the loads should be transferred by bearing (Fig. 14.12c).

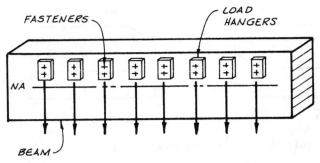

Figure 14.12b *Revised* detail with loads suspended above the neutral axis.

SUGGESTED REVISION 2

The preferred method is to suspend loads from the top of a beam or girder, as in Fig. 14.12c. This approach transfers the loads by bearing on the top of the beam. It avoids tension perpendicular to grain and stress concentrations from large loads being transferred by fasteners. Transferring loads by compression perpendicular to grain is generally the preferred method of load transfer in all cases, but it becomes increasingly important as the magnitude of the load increases.

Note that this method was recommended previously for the transfer of beam reactions to a girder in Fig. 14.2a. The saddle transfers the load by bearing rather than by fasteners.

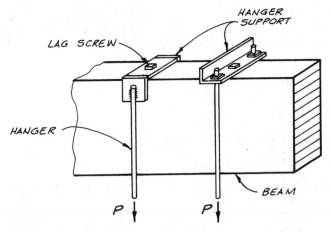

Figure 14.12c *Revised* detail with loads transferred by bearing (compression perpendicular to grain).

Suspending Isolated Loads from Beam

The connections shown in Fig. 14.13 are for isolated loads. Recommendations for suspending isolated loads are similar to those for multiple loads.

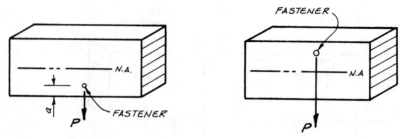

Figure 14.13 Isolated loads suspended from a beam.

An isolated light load may be suspended below the neutral axis of a beam. However, the load should be limited to one that can be transferred with small fasteners (e.g., small conduit loads). The distance a must equal or exceed the required edge distance, and Ref. 14.3 recommends a minimum of 6 in. Shear stresses in the member must be checked according to Sec. 13.9. It is good practice to locate fasteners above the neutral axis. Even better practice is to detail the connection as shown in Fig. 14.12c.

Truss Heel Connection

PROBLEM

The truss heel connection in Fig. 14.14a has structural steel gusset plates with bolts or shear plates for fasteners. There are two problems with this detail. First, the connection involves an eccentricity because the forces at the joint do not intersect at the center of the group of fasteners. Second, the steel gusset plates hold the fasteners rigidly in position, and this can cause the wood truss members to split as the joint attempts to rotate when the truss is loaded.

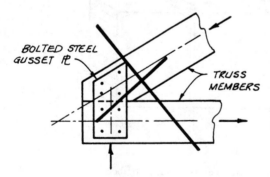

Figure 14.14a *Avoid* eccentric forces and rigid truss connections.

SUGGESTED REVISION

A truss connection is shown in Fig. 14.14b in which the force in the top chord is taken by bearing. This connection has the advantage of concentric forces. The single fastener in the top chord member and the clearance provided between the chord members allow free joint rotation as the truss is loaded. Cross-grain shrinkage in the bottom chord of the truss can be eliminated by the proper placement of bolts in vertically slotted holes.

See the TCM (Ref. 14.4) for a numerical example of a truss design.

NOTE: For the design of light-frame wood trusses using light-gage metal plates, see Ref. 14.8.

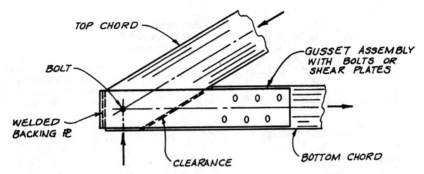

Figure 14.14b *Revised* connection detail with concentric forces. This detail also accommodates rotation between truss members.

Additional Truss Joint Considerations

PROBLEM

When the centerlines of members (and correspondingly the axial forces in the truss) do not intersect at a common point, considerable shear and moment may result in the bottom chord (Fig. 14.15a). When these stresses are combined with a high tension stress in the member, failure may occur.

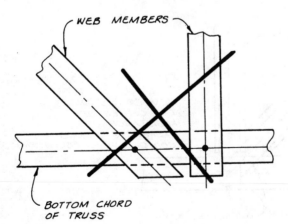

Figure 14.15a *Avoid* truss connections with eccentric forces.

PROBLEM

The welded steel gusset plate in Fig. 14.15b has a rigidity problem similar to the solid gusset plate in Fig. 14.14a. The welded plates are held in position, and splitting of the wood members may occur as the truss attempts to deform under load.

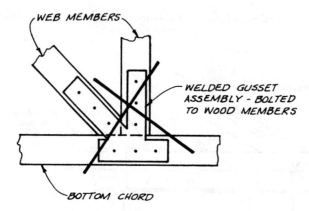

Figure 14.15b *Avoid* rigid truss connections which do not allow rotation between truss members. Prying and splitting may occur.

SUGGESTED REVISION

For trusses with single-piece members, the chords and web members should be in the same plane, and separate straps or gusset plates should be used for the connection. See Fig. 14.15c. The separate gusset plates should accommodate truss deflection and joint rotation without splitting of the wood members.

Steel gusset plates must be properly designed for the applied tension or compression forces. Shims should be provided between the steel plates and the wood members where necessary to maintain gusset plate alignment. For example, shims should be used between the vertical gusset plate and the vertical wood member to account for the thickness of the diagonal gusset plate. Without these shims the vertical gusset plate may become damaged as the bolts are tightened.

See the TCM (Ref. 14.4) for a numerical example of a truss design.

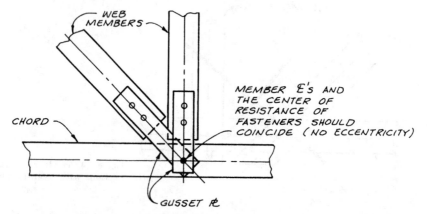

Figure 14.15c *Revised* connection detail with separate gusset plates. Fastener arrangement allows joint rotation without prying or splitting.

Moisture Protection at Column Base

PROBLEM

Some designers try to conceal the base of a column or an arch by placing concrete around the connection (Fig. 14.16a). Moisture may migrate into the lower portion of the wood and cause decay.

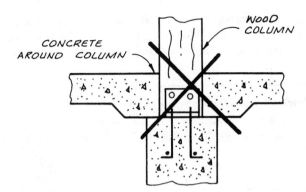

Figure 14.16a *Avoid* joint details that entrap moisture and promote decay.

SUGGESTED REVISION

Detail the column base at the top of the floor level (Fig. 14.16b). Note that the base of the column is separated from concrete by a bottom bearing plate. Where subjected to splashing water, untreated columns should be supported by piers projecting up at least 1 in. (IBC 2304.11.2.6; Ref. 14.7) above the finished floor.

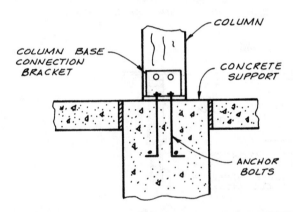

Figure 14.16b *Revised* connection detail does not entrap moisture.

14.3 Design Problem: Beam-to-Column Connection

Example 14.2 illustrates a typical beam-to-column connection. The beam is $5\frac{1}{8} \times 27$ glulam, and the column is a 4-in. steel tube. Steel tubes and steel pipe columns are commonly used to support wood beams.

The design of this connection starts with a consideration of the required bearing length for the wood beam. Calculations of this nature were introduced in Chap. 6. The required overall length of the U-bracket for bearing stresses is 8 in. In later calculations this length is shown to be less than the length required to satisfy the bolt spacing criteria.

The example concludes with a check of the bending stresses in the U-bracket. Because of the small cantilever length, a bent plate is used for the bracket. For longer cantilever lengths, a welded U-bracket will be required.

For special connection hardware, complete details of the bracket should be included on the structural plans. However, U-brackets of the type covered in this example are generally available prefabricated from commercial sources.

EXAMPLE 14.2   U-Bracket Column Cap

Design a U-bracket to connect two $5\frac{1}{8} \times 27$ glulam beams to the 4-in. steel tube column. Gravity roof loads are given in Fig. 14.17a. There is no uplift force on the beam-to-column connection.

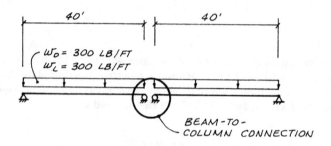

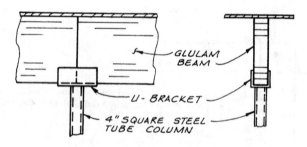

Figure 14.17a Beam to column connection.

The lateral seismic force causes an axial force of 6 k in the glulam beams. Glulam is 24F-V4 Douglas Fir (from the 24F-1.8E Stress Class) which is initially dry and remains dry in service. Temperature range is normal.

Vertical Load

Reaction from one beam:

$$R = \frac{wL}{2} = \frac{600(40)}{2} = 12{,}000 \text{ lb}$$

Allowable bearing stress:

$$F'_{c\perp} = F_{c\perp}(C_M C_t C_b) = 650(1.0)(1.0)(1.0)$$

$$= 650 \text{ psi}$$

$$\text{Req'd bearing length} = \frac{R}{b \times F'_{c\perp}} = \frac{12{,}000}{5.125 \times 650}$$

$$= 3.6 \text{ in./beam}$$

Minimum length of bracket = $2(3.6) = 7.2$ in.   *say* 8 in. minimum

Lateral Force

Assume that the lateral force will be transferred from one glulam to the other through the U-bracket made from ¼-in.-thick A36 steel. Use ¾-in.-diameter bolts in double shear with steel side plates.

$$D = 0.75 \text{ in.}$$

$$F_{yb} = 45 \text{ ksi} \quad \text{(assumed for A307 bolt)}$$

$$l_m = 5.125 \text{ in.}$$

$$\theta_m = 0 \text{ degrees}$$

$$F_{em} = 5600 \text{ psi} \quad (F_e \text{ for DF-L})$$

$$l_s = 0.25 \text{ in.}$$

$$F_{es} = 87{,}000 \text{ psi} = 1.5\,F_u \quad \text{for A36 steel}$$

The nominal design value for the bolt may be computed as the smallest value from the four yield limit equations for a double-shear wood-to-metal connection (Examples 13.5 and 13.6 in Sec. 13.4). However, the given conditions are covered directly by an NDS table, and Z is obtained from NDS Table 11I as

$$Z = 3340 \text{ lb}$$

The reader may wish to verify this value by applying the yield limit equations and observing that Mode IIIs governs.

Allowable design value:
    Load duration factor C_D from the NDS (Ref. 14.1) is 1.6 for seismic forces. Both C_m and C_t are 1.0. A row is defined as a line of fasteners *parallel* to the direction of loading. Assume $C_g = 1.0$. The geometry factor C_Δ will also be unity because the base dimensions for bolt spacing will be provided as a minimum. Therefore,

$$Z' = Z(C_D C_M C_t C_g C_\Delta)$$

$$= 3340(1.6 \times 1.0 \times 1.0 \times 1.0 \times 1.0)$$

$$= 5344 \text{ lb/bolt}$$

$$\text{Req'd number of bolts} = \frac{6000}{5344} \approx 1.12$$

> *Use* two ¾-in.-diameter bolts each side of splice.

Although there are two bolts on each side of the splice, two rows will be provided with one bolt in each row.

$$\therefore C_g = 1.0 \quad OK$$

Connection Details

A number of possible connection details can be used. One possible design is included in Fig. 14.17*b*.

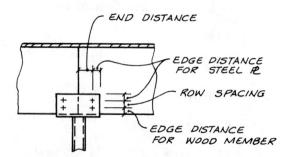

Figure 14.17*b* Connection details.

BOLT SPACINGS

Lateral force is carried by the U-bracket in tension, and vertical load is carried by bearing. Spacing requirements are governed by the lateral force (parallel-to-grain loading criteria—Example 13.10 in Sec. 13.5).

HORIZONTAL SPACING

$$\text{End distance} = 7D = 7(\tfrac{3}{4}) = 5\tfrac{1}{4} \text{ in.}$$

$$\text{Steel plate edge distance} = 1\tfrac{1}{4} \text{ in.} \quad \text{(minimum from Ref. 14.2)}$$

$$\text{Total bracket length} = (5\tfrac{1}{4} + 1\tfrac{1}{2})2 = 13\tfrac{1}{2} \text{ in.}$$

$$13\tfrac{1}{2} > 8 \text{ in.}$$

Base dimension spacing provisions govern over bearing ($f_{c\perp}$).

VERTICAL SPACING

$$\text{Steel plate edge distance} = 1\tfrac{1}{4} \text{ in.} \qquad say \ 1\tfrac{1}{2} \text{ in.}$$

$$\text{Row spacing} = 1\tfrac{1}{2}D = 1\tfrac{1}{2}(\tfrac{3}{4}) = 1\tfrac{1}{8} \text{ in.} \qquad say \ 1\tfrac{1}{2} \text{ in.}$$

$$\text{Bolt slenderness} = \frac{l}{D} = \frac{5.125}{0.75} = 6.83 > 6.0$$

$$\therefore \text{Edge distance*} = 1\tfrac{1}{2}D = 1\tfrac{1}{8} \text{ in.}$$

$$or \qquad \tfrac{1}{2} \text{ row spacing} = \tfrac{3}{4} \text{ in.} \qquad \text{(the larger governs)}$$

$$Say \qquad 1\tfrac{1}{2} \text{ in.}$$

$$\text{Minimum bracket height} = 3(1\tfrac{1}{2}) = 4\tfrac{1}{2} \text{ in.}$$

PROPORTIONS OF U-BRACKET

As a rule of thumb, a bracket height of one-third the length is reasonable:

$$\tfrac{1}{3} \times 13.5 = 4.5 \text{ in.} \qquad \text{(agrees with vertical bolt spacing requirements)}$$

U-Bracket Stresses

The plan and elevation of the connection detail are shown in Fig. 14.17c. The bracket essentially cantilevers from the face of the column. [If a round column (e.g., a pipe column) is used, the cantilever length is figured from the face of an equivalent square column of equal area.]

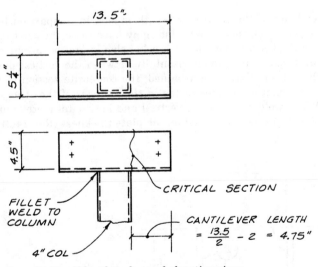

Figure 14.17c U-bracket plan and elevation views.

---

*Uplift force was given as zero. If uplift occurs, the edge distance is $4D$ as governed by perpendicular-to-grain criteria (Example 13.11 in Sec. 13.5) or possibly greater if governed by horizontal shear criteria (Sec. 13.9).

The width of the U-bracket is determined by adding a clearance to the width of the beam to allow installation of the member. Reference 14.3 recommends a maximum clearance of ¼ in. In this example a clearance of ⅛ in. is used (inside bracket width = 5⅛ + ⅛ = 5¼ in.). See uniform bearing considerations in Fig. 14.1*b*.

At this point it can be determined whether a bent plate bracket or welded plate bracket should be used (Fig. 14.17*d*). If the cantilever length of the bracket is small in comparison to the distance between the vertical legs, a bracket of uniform thickness throughout can be used. This can be obtained by folding a plate into the desired shape. The bending stresses in the bracket caused by the cantilever action must be evaluated.

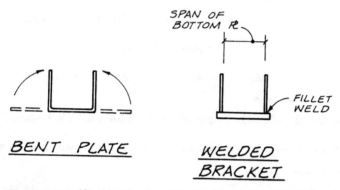

Figure 14.17*d* Alternative ways of forming U-bracket. For lighter loads a folded (bent) plate may be used. Larger loads may require a welded bracket with a thicker bottom plate capable of spanning between the vertical legs.

On the other hand, if the cantilever length is long in comparison to the distance between the vertical legs, the bottom plate may have to be thicker than the vertical legs. In this case the bottom plate of the bracket should first be designed for the bending stresses developed as it spans horizontally between the vertical plates. After the thickness of the bottom plate is determined, the composite section of the U-bracket can be checked for cantilever stresses (as in the analysis of the stresses in the bent plate bracket). The weld between the vertical and horizontal plates must also be designed. The minimum weld size based on the plate thickness (Ref. 14.2) often governs the weld size.

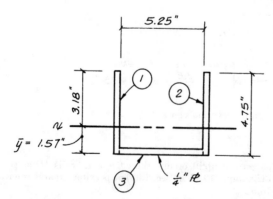

Figure 14.17*e* Dimensions for computing section properties of U-bracket.

In the problem at hand, a bent plate bracket will be used because the cantilever length of 4.75 in. (Fig. 14.17c) is less than the distance between vertical plates (5¼ in.). Check cantilever bending stresses using ¼-in. plate (Fig. 14.17e).

Section properties are conveniently calculated in a table from the formula

$$I_x = \Sigma I_g + \Sigma A d^2$$

Element	A	y	Ay	$I_g = \dfrac{bh^3}{12}$	d	Ad^2
1	4.75 × ¼ = 1.19	2.38	2.82	2.23	0.80	0.76
2	1.19	2.38	2.82	2.23	0.80	0.76
3	5.25 × ¼ = 1.31	0.125	0.16	0.01	1.45	2.76
	3.69		5.80	4.47		4.28

$$\bar{y} = \frac{Ay}{A} = \frac{5.80}{3.69} = 1.57 \text{ in.}$$

$$I_x = \Sigma I_g + \Sigma A d^2 = 4.47 + 4.28 = 8.75 \text{ in.}^4$$

Uniform load on bracket:

$$w = \frac{R}{A} = \frac{12{,}000 \times 2}{5.125 \times 13.5} = 347 \text{ psi}$$

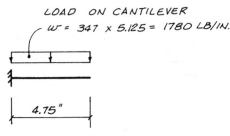

LOAD ON CANTILEVER
w = 347 × 5.125 = 1780 LB/IN.

4.75"

Figure 14.17f

$$M = \frac{wL^2}{2} = \frac{1780(4.75)^2}{2} = 20 \text{ in.-k}$$

$$f_b = \frac{Mc}{I} = \frac{20(3.18)}{8.75} = 7.28 \text{ ksi}$$

For A36 steel plate,

$$F_b = 22 \text{ ksi* } > 7.28 \qquad OK$$

> *Use*   ¼-in.-thick plate U-bracket.

---

*The width-to-thickness ratios of plate elements in the compression zone are less than the limits given in Sec. B5. of Ref. 14.2.

The U-bracket must be welded to the top of the column. Without design loads, the weld size is governed by plate thickness requirements (Ref. 14.2).

14.4 Cantilever Beam Hinge Connection

The type of connection used for an internal hinge in a wood cantilever beam system was introduced in Fig. 14.5. There the detailing of the connection was considered to minimize the effects of shrinkage. In the design of this type of connection, the equilibrium of the hinge must also be considered. See Fig. 14.18. The saddle hinge with tension ties located near the bearing surfaces is recommended for gravity loads. The tension ties are required to balance the eccentric moment generated by the vertical reactions.

Once the dimensions of the hinge connection have been established, the force on the tension ties can be calculated. This force is used to determine the number and size of bolts in the tension ties as well as the size and length of the weld between the tension tie and the main portion of the hinge connector.

Other considerations include spacing requirements for the bolts in the connection. The top and bottom bearing plates for the hinge must be capable of spanning horizontally between the vertical plates, and the welds between the vertical and horizontal plates must be designed.

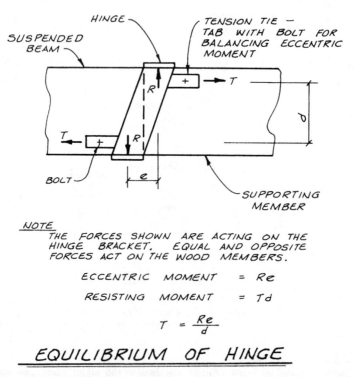

Figure 14.18 Saddle hinge with tension ties for gravity loads.

In many cases there will be a tension force at the hinge due to an axial force in the beam. This is in addition to the tension load caused by gravity loads. This typically occurs in buildings with concrete and masonry walls under seismic forces. Here the beams are designed as continuous cross ties between horizontal diaphragm chords (Chap. 15). A tension force can also result from diaphragm chord or drag strut action.

The addition of a lateral force to the hinge connection problem can be handled in several ways. One solution is to provide a *welded* splice across the connection. See Fig. 14.19.

In the example shown, the seismic ties are located at the midheight of the hinge connector. Vertically slotted holes are required in the welded seismic ties to allow for cross-grain shrinkage. Note that the bolts must be located properly in the vertical slots to accommodate the movement between the bolts and the points of bearing in the connection. In addition, the nuts on these bolts should be hand-tightened only to permit movement.

In a different approach, *separate* tension ties for the seismic force are provided across the hinge in a typical double-shear bolt connection. This connection is similar to Fig. 14.19 except the tie plates are continuous across the hinge and are not welded to the hinge connector. In this case the use of vertically slotted holes is not required because the separate splice plates will not restrain movement across the grain. Shims may be used to isolate the seismic tension tie from the hinge connector.

Other solutions to the seismic force problem are possible, but the two described are representative.

14.5 Prefabricated Connection Hardware

A number of manufacturers produce prefabricated connection hardware for use in wood construction. Catalogs are available from manufacturers or suppliers which show the configurations and dimensions of the hardware. In ad-

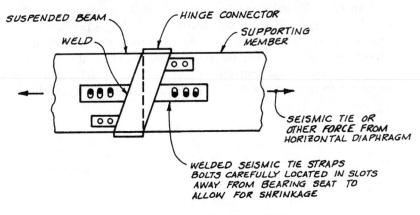

Figure 14.19 Saddle hinge with seismic tie.

dition, the catalogs generally indicate allowable design loads for connections using the hardware. However, the designer should verify that the hardware and the allowable loads are recognized by the local building code authority.

In many cases prefabricated hardware will be the most economical solution to a connection problem. This is especially true if the hardware is a "shelf item." For example, light-gage steel joist or purlin hangers (similar in configuration to the beam face hanger in Fig. 14.3) are usually available as a stock item. Allowable loads for some hardware have been established through testing, and design values may be higher than can be justified by stress calculations. Obviously this type of hardware represents an efficient use of materials.

For connections involving larger loads, the appropriate hardware may be available from a manufacturer. However, the designer should have the ability to apply the fastener design principles of the previous chapters *and* basic structural steel design principles (Ref. 14.2) to develop appropriate connection details for a particular job. In some cases it may be more economical to fabricate special hardware, *or* there may not be suitable prefabricated hardware for a certain application.

Many manufacturers offer a number of variations of the hardware illustrated in Example 14.1. The fittings may range from light-gage steel to ¼-in. and thicker steel plate. Allowable loads vary correspondingly. Several other types of connection hardware are generally available from manufacturers. See Example 14.3. These include

1. Framing anchors

2. Diaphragm-to-wall anchors

3. Tie straps

Some hardware that has been illustrated in previous chapters is also available commercially. For example, shearwall tie-down brackets (Fig. 10.11) are produced by a number of manufacturers. Metal knee braces (Fig. 6.26*b*) and metal bridging (mentioned in Example 6.7) are also available.

With some knowledge of structural steel design and the background provided in Chaps. 11 to 14, the designer should be prepared to develop appropriate connections for the transfer of both vertical loads and lateral forces. Chapters 15 and 16 conclude the discussion of the lateral force design problem with final anchorage considerations and an introduction to some advanced problems in diaphragm design, structural irregularity, and overturning.

EXAMPLE 14.3   Connection Hardware

Framing Anchors

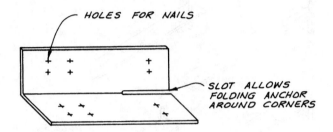

Figure 14.20a

Framing anchors (Fig. 14.20a) are light-gage steel connection brackets that are usually installed with nails. They have a variety of uses, especially where conventional nail connections alone are inadequate. Framing anchors generally can be bent and nailed in a number of different ways (Fig. 14.20b). Allowable loads depend on the final configuration of the anchor and the direction of loading (different allowable loads apply to loadings A through E).

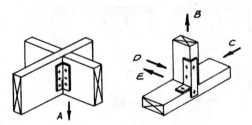

Figure 14.20b Framing anchor applications.

Diaphragm-to-Wall Anchors

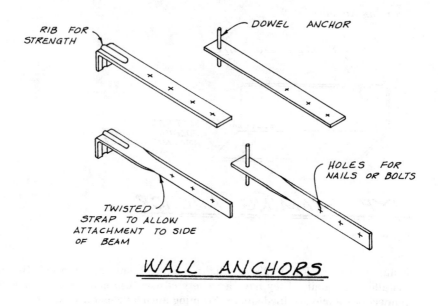

RIB FOR STRENGTH

DOWEL ANCHOR

TWISTED STRAP TO ALLOW ATTACHMENT TO SIDE OF BEAM

HOLES FOR NAILS OR BOLTS

WALL ANCHORS

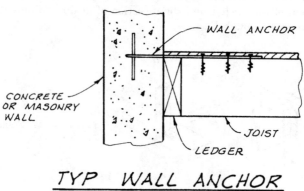

WALL ANCHOR

CONCRETE OR MASONRY WALL

JOIST

LEDGER

TYP WALL ANCHOR APPLICATION

Figure 14.21 Wall anchors are used to tie a concrete or masonry wall to a horizontal diaphragm.

Diaphragm-to-wall anchors (Fig. 14.21) are known commercially by various names (purlin anchors, joist anchors, strap anchors, etc.). These anchors are used to tie concrete or masonry walls to horizontal diaphragms (Chap. 15). Their purpose is to prevent the ledger from being stressed in cross-grain bending (Example 6.1).

Note that the connection in Fig. 14.21 is only part of the total wall-to-diaphragm connection. It carries the lateral force that acts perpendicular to the wall. The anchor-bolt-to-ledger connection in Fig. 6.1*b* is used for vertical loads and lateral forces parallel to the wall. See Example 15.6 (Sec. 15.4) for additional information.

Tie Straps

TIE STRAP

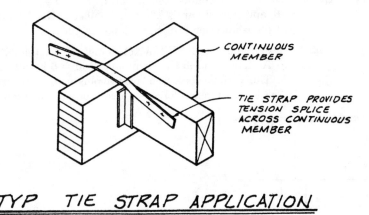

TYP TIE STRAP APPLICATION

Figure 14.22

Tie straps (Fig. 14.22) are typically used to provide a connection across an intervening member. Ties of this type may be required if the beams are part of a horizontal diaphragm (Chap. 15).

14.6 References

[14.1] American Forest and Paper Association (AF&PA). 2001. *National Design Specification for Wood Construction and Supplement.* 2001 ed., AF&PA, Washington DC.

[14.2] American Institute of Steel Construction (AISC). 1989. *Manual of Steel Construction,* 9th ed., AISC, Chicago, IL.

[14.3] American Institute of Timber Construction (AITC). 1994. *Typical Construction Details, AITC 104-84,* AITC, Englewood, CO.

[14.4] American Institute of Timber Construction (AITC). 1994. *Timber Construction Manual,* 4th ed., AITC, Englewood, CO.

[14.5] American Iron and Steel Institute (AISI). 1996. *Cold-Formed Steel Design Manual.* 1996 ed., AISI, Washington DC.

[14.6] APA—The Engineered Wood Association. 1999. *Glulam Connection Details, EWS Form T300,* APA—The Engineered Wood Association, Engineered Wood Systems, Tacoma, WA.

[14.7] International Codes Council (ICC). 2003. *International Building Code,* 2003 ed., ICC, Falls Church, VA.

[14.8] Truss Plate Institute (TPI). 2002. *National Design Standard for Metal Plate Connected Wood Truss Construction,* ANSI/TPI 1-2002, TPI, Madison, WI.

14.7   Problems

The design of connection hardware usually involves the use of structural steel design principles as well as wood design calculations. For the following problems, structural steel is A36, and design methods and allowable stresses should conform to the specifications contained in Ref. 14.2. Design of wood members and fasteners shall be in accordance with the 2001 NDS. $C_t = 1.0$.

14.1 *Given:* The beam-to-column connection in Fig. 14.A. The two beams are 6 × 16, and the column is a 6 × 6. Lumber is No. 1 DF-L. $C_M = 1.0$. Bolts in the connection are ⅞-in.-diameter A307.

   Find: Design and detail a U-bracket connection for the following loads:
   a. An axial load in the column of dead = 5 k and snow = 15 k.
   b. The vertical load from part *a* plus a tension force in the beam of 8 k. The tension force is a result of a seismic force.

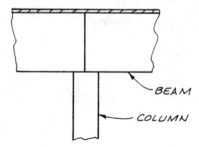

BEAM

COLUMN

Figure 14.A

14.2 Repeat Prob. 14.1 except that the lumber is Select Structural Hem-Fir.

14.3 *Given:* The beam-to-column connection in Fig. 14.A. The two beams are 6¾ × 31.5 24F-V4 DF glulam (from the 24F-1.8E Stress Class). $C_M = 1.0$. Bolts in the connection are ¾-in. diameter A307. The column is an 8-in. standard steel pipe column.

   Find: Design and detail a U-bracket connection for the following loads:
   a. A total axial load in the column of (D + S) = 70 k.
   b. The vertical load from *a* plus a tension force in the beam of 6 k. The tension force is a result of a seismic force.

NOTE: The cantilever length of the U-bracket is to be determined using the dimensions of an effective square column. The effective square column has an area equal to the area of a circle with a diameter of 8.625 in. (the outside diameter of an 8-in. standard pipe). Design the bottom plate of the U-bracket for bending between the vertical side plate supports.

14.4   *Given:*   The beam-to-girder connection in Fig. 14.B. The members are 20F-V3 DF glulam (from the 20F-1.5E Stress Class). The girder is a $6\frac{3}{4} \times 30$, and the beams are $3\frac{1}{8} \times 15$. $C_M = 1.0$.

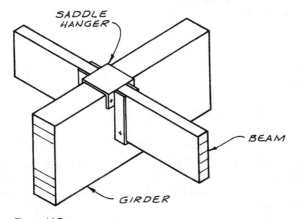

SADDLE HANGER

BEAM

GIRDER

Figure 14.B

Find: Design and detail the connection hardware if each beam transmits a reaction of 4.5 k to the girder. Load is (D + L$_r$).

14.5   *Given:*   The beam-to-girder connection in Fig. 14.C. The girder is a $5\frac{3}{4} \times 25.5$ 16F-V6 DF glulam (from the 16F-1.3E Stress Class), and the beam is a 4×14 sawn member No. 1 DF-L. Bolts are $\frac{3}{4}$-in. diameter A307. $C_M = 1.0$.

Find: Design and detail the connection hardware if the beam reaction on the girder is 2000 lb. Load is (D + L$_r$).

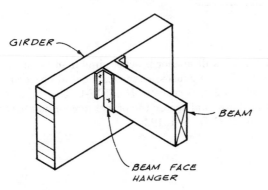

GIRDER

BEAM

BEAM FACE HANGER

Figure 14.C

14.6 *Given:* The hinge connector in Fig. 14.D. The girders are $6\frac{3}{4} \times 31.5$ 20F-V3 DF glulam (from the 20F-1.5E Stress Class). $C_M = 1.0$. Bolts are $\frac{7}{8}$-in. diameter A307.

Find: Design and detail the hinge connector for the following loads:
a. A vertical load of 16 k. Load is (D + S).
b. The vertical load from part a plus a tension force in the girder of 8 k.

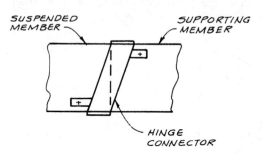

Figure 14.D

14.7 *Given:* The hinge connector in Fig. 14.E. The members are 24F-V1 SP glulams (from the 24F-1.7E Stress Class). The supporting member is 5×24.75, and the suspended member is a 5×28.875. $C_M = 1.0$. Bolts are $\frac{3}{4}$-in. diameter A307.

Find: Design and detail the hinge connector for the following loads:
a. A vertical load of 12 k. Load is (D + L_r).
b. The vertical load from part a plus a tension force in the girder of 4 k.

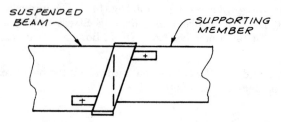

Figure 14.E

14.8 Design and detail a shearwall anchor bracket similar to the one shown in Fig. 10.11. The lateral wind force causes a tension force in the chord of 4 k. The chord is composed of two 2×6 studs of No.1 DF-L. Use $\frac{3}{4}$-in. A307 bolts for the connection to the chord, and choose the proper anchor bolt for the connection to the foundation from IBC Table 1912.2.

Diaphragm-to-Shearwall Anchorage

15.1 Introduction

Anchorage was defined previously as the tying together of the major elements of a building with an emphasis on the transfer of lateral (wind and earthquake) forces. A systematic approach to anchorage involves a consideration of load transfer in the three principal directions of the building.

Chapter 10 illustrated the anchorage requirements at the base of a wood-frame shearwall. This chapter continues the anchorage problem with a detailed analysis of the connection between the horizontal diaphragm and the shearwalls. Additional considerations for shearwall anchorage and overturning are addressed in Chap. 16. Several typical anchorage details are analyzed including connections to wood-frame shearwalls and concrete or masonry shearwalls.

Following the subject of diaphragm-to-shearwall anchorage, the subdiaphragm concept is introduced. This design technique was developed to ensure the integrity of a horizontal plywood diaphragm that supports seismic forces generated by *concrete* or *masonry walls*. Its purpose is to satisfy the Code requirement that "continuous ties" be provided to distribute these larger seismic forces into the diaphragm. Referenced in Sec. 1620 of the IBC, ASCE 7 Sec. 9.5.2.6.3.2 includes subdiaphragm requirements for Seismic Design Categories C and higher. This is also an anchorage problem, but it is unique to buildings with concrete or masonry walls.

15.2 Anchorage Summary

Horizontal diaphragm anchorage refers to the design of connections between the horizontal diaphragm and the vertical elements of the building. These elements may support the horizontal diaphragm, or they may transfer a force to the diaphragm.

A systematic approach to anchorage is described in Example 10.6 (Sec. 10.8). This simply involves designing the connections between the various resisting elements for the following loads and forces:

1. Vertical loads
2. Lateral forces parallel to a wall
3. Lateral forces perpendicular to a wall

In designing for these gravity loads and lateral forces, it is important that a *continuous path* be developed. The path must take the load or force from its source, through the connections, and into the supporting element.

The "path" in the progressive transfer of *vertical loads* is described in Example 10.7 (Sec. 10.9) and is not repeated here. Many of the connections for vertical loads can be made with the hardware described in Chap. 14.

There are several types of *lateral forces parallel* to a wall. See Example 15.1. These include

1. Basic unit shear transfer
2. Collector (strut) force
3. Horizontal diaphragm chord force

The methods used to calculate collector (strut) forces and horizontal diaphragm chord forces were given in Chap. 9, and several connection designs were demonstrated in Chaps. 12 and 13. These forces are listed here simply to complete the anchorage design summary. However, the connection for the transfer of the unit shear has not been discussed previously, and it requires some additional consideration. The connections used to transfer this shear are different for different types of walls. Typical details of these connections and the calculations involved are illustrated in the following sections.

EXAMPLE 15.1   Anchorage Forces Parallel to Wall

Consider the anchorage forces *parallel* to the right end wall in the building shown in Fig. 15.1a. Three different forces are involved.

Unit Shear Transfer

The unit shear in the roof diaphragm v_R (Fig. 15.1a) must be transferred to the supporting elements (collector strut and shearwall).

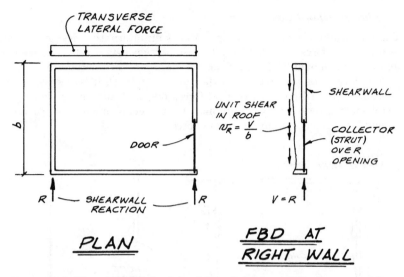

Figure 15.1a Plan view showing lateral force to horizontal diaphragm, and free body diagram cut a small distance away from right shearwall and collector (strut).

Drag Strut Connection to Shearwall

The force in the collector (strut) must be transferred to the shearwall (Fig. 15.1*b*). The magnitude of the *strength level* collector (strut) force may be read from the collector (strut) force diagram. In addition to this horizontal force, vertical forces due to dead and live loads need to be considered in design of the collector connection to the shearwall. Further, in Seismic Design Categories C and higher, where the structure is not entirely of light-frame construction, ASCE 7 Sec. 9.5.2.6.1 requires that the collector (strut) and its connection to the wall be capable of resisting the special (amplified) seismic forces of ASCE 7 Sec. 9.5.2.7.1.

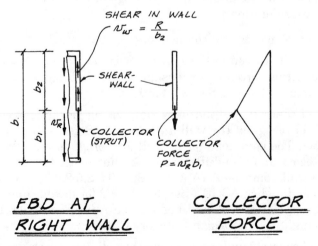

Figure 15.1b *Collector (strut) force* caused by lateral force to diaphragm in the transverse direction.

Diaphragm Chord Connection

The diaphragm chord force is Fig. 15.1c is an alternative design criterion to the collector (strut) force in Fig. 15.1b. The connection between the member over the door and the wall should be designed for the *chord force* or the *collector (strut) force*, whichever is larger. As was true with the collector (strut), it is necessary to combine the horizontal chord force with vertical forces due to dead and live loads. Unlike the collector (strut), there is no requirement for the chord member or connection to be designed for special (amplified) seismic forces.

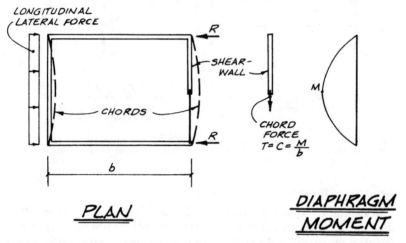

Figure 15.1c *Diaphragm chord force* caused by lateral force to diaphragm in the longitudinal direction.

There are also several types of *lateral forces perpendicular* to a wall that must be considered in designing the anchorage of the horizontal diaphragm. These include consideration for

1. Wind force on elements and components.

2. Minimum load path connections in all Seismic Design Categories using seismic connection forces F_p for design of the wall anchorage to the supporting structure (ASCE 7 Sec. 9.6.2.6.1.1).

3. Anchorage of concrete and masonry walls using seismic forces F_p normal to the wall for design of the wall anchorage in Seismic Design Categories B and higher. The equations for F_p wall anchorage forces vary by Seismic Design Category and also differ from those introduced in Sec. 2.15 for design of the wall component (ASCE 7 Sec. 9.5.2.6.2.8 for Seismic Design Category B and higher, ASCE 7 Sec. 9.5.2.6.3.2 for flexible diaphragms and Seismic Design Category C and higher, and ASCE 7 Sec. 9.6.1.3 for rigid diaphragm and Seismic Design Category C and higher).

4. Anchorage of concrete and masonry walls in all Seismic Design Categories, using a minimum *strength level* horizontal force of not less than 280 lb/ft (ASCE 7 Sec. 9.5.2.6.1.2).

For walls other than concrete and masonry, it is recommended (but not required) that the anchorage forces of ASCE 7 Sec. 9.5.2.6.2.8 also be used in design. While these anchorage forces are not likely to control over wind for most light-frame walls, heavy finishes such as masonry veneer can cause them to control.

The design considerations listed above are alternate design forces, and anchorage must be designed for the maximum value. These forces act normal to the wall (either inward or outward) and attempt to separate the wall from the horizontal diaphragm. See Example 15.2.

Similar forces were introduced in Examples 2.16 (Sec. 2.15) and 10.11 (Sec. 10.13). In the sketches in these examples, the force labeled *roof reaction* corresponds to the anchorage force being discussed for wind forces and is similar to the anchorage force being discussed for seismic forces.

In wood-frame buildings, wind usually controls the design of anchorage for perpendicular-to-wall forces. For masonry and concrete walls, however, the seismic force F_p often governs because of the large dead load of these walls. The Code minimum force (e.g., 280 lb/ft) for these types of walls ensures a minimum connection regardless of the calculated seismic forces.

EXAMPLE 15.2   Anchorage Forces Perpendicular to Wall

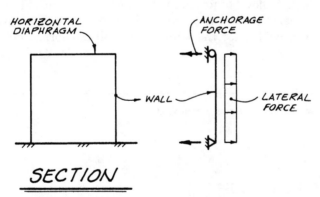

SECTION

Figure 15.2 Schematic of one-story building. In most buildings, wall framing is assumed to span vertically between the foundation and the roof diaphragm.

Anchorage for normal-to-wall forces must be designed for the largest of the following:

1. Wind force tributary to horizontal diaphragm.
2. Minimum load path connection forces F_p in all Seismic Design Categories.
3. For anchorage of concrete and masonry walls in Seismic Design Categories B and higher, seismic forces F_p normal to the wall.
4. For anchorage of concrete and masonry walls in all Seismic Design Categories, not less than a minimum strength level horizontal force of 280 lb/ft.

Anchorage must be provided for normal forces acting inward and outward.

A large number of connection details can be used to anchor the horizontal diaphragm to the walls for both parallel and perpendicular forces. The type of wall framing (wood frame, concrete, or masonry) and the size and direction of the framing members for the horizontal diaphragm affect the choice of the anchorage connection. Examples of two typical anchorage details are given in Secs. 15.3 and 15.4.

15.3   Connection Details—Horizontal Diaphragm to Wood-Frame Wall

A typical anchorage connection of a horizontal plywood diaphragm to a wood-frame wall is considered here. Two anchorage details are shown in Example 15.3: one for the transverse wall and one for the longitudinal wall.

In practice, additional details will be required. For example, the connection where the girder ties into the transverse wall must be detailed in the structural plans. The vertical reaction of the girder, and the horizontal diaphragm chord force caused by lateral forces in the longitudinal direction, need to be considered. However, the two connection details given in the sketches are intended to define the problem and be representative of the methods used for anchorage.

As noted previously, the key to the anchorage problem is to provide a continuous *path* for the "flow" of forces in the connection. Each step in this flow is labeled on the sketch, and a corresponding explanation is given. Sample calculations are provided for the transverse wall anchorage detail. See Example 15.4.

EXAMPLE 15.3   Anchorage to Wood-Frame Walls

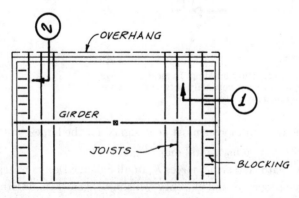

FRAMING   PLAN

Figure 15.3a One story building with wood roof system and wood-frame stud walls.

Different anchorage details are required for the diaphragm attachment to the transverse and longitudinal walls. More complicated roof or floor framing will require additional anchorage details.

The following details shown are representative only, and other properly designed connections are possible. The path for the transfer of anchorage forces is described for each connection.

Transverse Wall Anchorage

Parallel-to-wall force (diaphragm shear)

a. Diaphragm boundary nailing transfers shear into end joist (Fig. 15.3*b*).

b. The shear must be transferred from the end joist into the wall plate. The top plate here serves as the diaphragm chord and strut. Various types of connections can be used for this transfer. Framing anchor is shown. Alternate connections not shown are toenails or additional blocking, as in Example 9.5. The designer should exercise caution when specifying toenails since the toenails may split the blocking. For this reason, IBC Sec. 2305.1.4 limits the allowable shear transferred by toenails to 150 lb/ft in Seismic Design Categories D, E, and F (Ref. 15.5).

c. Shearwall edge nailing transfers shear into wall sheathing.

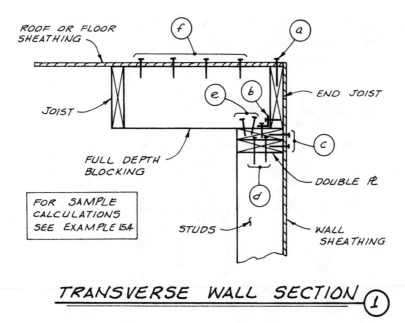

Figure 15.3*b*   Anchorage connections between roof and transverse shearwall.

Perpendicular-to-wall force (wind critical for most wood-frame walls)

d. Connect stud to double plate for reaction at top of stud. Standard connection from IBC Table 2304.9.1 is two 16d nails into end grain of stud. Alternate connection for larger forces is framing anchor.

    e. Full-depth blocking normal to wall is required to prevent the rotation of the joists under this type of force. This blocking is necessary whether or not the horizontal diaphragm is a "blocked" diaphragm. Spacing of blocks depends on lateral force (2 to 4 ft is typical). Connect top plate to block for tributary lateral force. Toenails are shown (not allowed for concrete or masonry walls in Seismic Design Category C and higher per ASCE 7 Sec. 9.5.2.6.3.2). Alternate connection not shown is framing anchor.

    f. Connect block to horizontal diaphragm sheathing for same force considered in *e.* For consistency, use the same size of nails as for diaphragm nailing.

Longitudinal Wall Anchorage

Parallel-to-wall force (diaphragm shear)

    a. Diaphragm boundary nailing transfers shear into blocking (Fig. 15.3c).

    b. Connection of blocking to double plate transfers shear into top plate. Framing anchor is shown. Alternate connections not shown are toenails (limited to 150 lb/ft seismic forces in Seismic Design Categories D, E, and F, per IBC Sec. 2305.1.4) or additional blocking, as in Example 9.5.

    c. Shearwall edge nailing transfers shear from top plate to wall sheathing.

Perpendicular-to-wall force (wind critical for most wood-frame walls)

    d. Same as transverse wall connection *d* (Fig. 15.3b).

    e. Connect top plate to joist for the tributary lateral force. Framing anchor is shown.

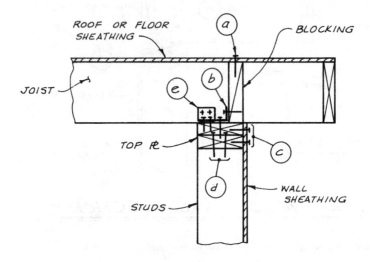

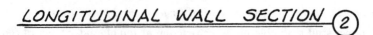

Figure 15.3c Anchorage connections between roof and longitudinal shear-wall.

Alternate connection not shown is with toenails (not allowed for concrete or masonry walls in Seismic Design Category C or higher, per ASCE 7 Sec. 9.5.2.6.3.2). The connection of the joist to the horizontal diaphragm sheathing is provided directly by the diaphragm nailing. Specific nail design similar to the transverse wall connection f is usually not required.

EXAMPLE 15.4   Anchorage of Transverse Wood-Frame Wall

Design the connection for the anchorage of the transverse wall for the building in Example 15.3 (section 1), using the following loads. The letter designations in this example correspond to the designations of the connection details in Fig. 15.3b. Lumber is DF-L, and plywood is ½-in. STR I DF with 10d common nails. $C_M = 1.0$, and $C_t = 1.0$.

Known Information

   Roof unit shear parallel to wall = 205 lb/ft   (ASD wind or seismic)

   Wind force perpendicular to wall = 26 psf   (ASD)

   Seismic $S_{DS} = 1.0g$

   Wall unit weight = 15 psf

   Wall height = 10 ft

Assume allowable load on framing anchor for parallel to wall forces (item b in Fig. 15.3b).

   Allow. V = 410 lb/anchor*   (normal duration)

                = 410(1.6) = 656 lb   (wind or seismic)

Seismic Force Perpendicular to Wall (Strength)

   Minimum connection (ASCE 7 Sec. 9.5.2.6.1.1) = $0.133S_{DS}w$

                        = 0.133(1.0)15 psf = 2 psf

Alternate minimum force (recommended but not required)

   Seismic Design Category B and up (ASCE 7 Sec. 9.5.2.6.2.8) = $0.4S_{DS}w$

                        = 6 psf

Wind force will control design forces for anchorage perpendicular to the wall.

---

*In practice, the capacity of the framing anchor is obtained from the evaluation report giving Code-recognized design values for a proprietary product.

Wall Anchorage Force Perpendicular-to-Wall

Wall anchorage T and C = 26 psf (10 ft wall height/2) = 130 lb/ft

Parallel-to-Wall Forces

a. The nailing for item a in Fig. 15.3b is obtained from Table 3.1A of the *Wood Structural Panel Shear Wall and Diaphragm Supplement, Allowable Stress Design Manual for Engineered Wood Construction* (Ref. 15.1).
   Minimum allowable roof shear = 300 lb/ft > 205 lb/ft   *OK*
   Minimum edge nailing (10d common nails at 6 in. o.c.) will transfer roof shear into end joist.
b. Item b in Fig. 15.3b connects end joist to plate with framing anchors (assume diaphragm width = b = 40 ft):

$$V = vb = 205 \times 40 = 8200 \text{ lb}$$

$$\text{Number of anchors} = \frac{8200}{656} \approx 13$$

$$\text{Spacing of anchors} \approx \frac{40}{13} = 3.08 \text{ ft}$$

Connection b:	*Use* 13 framing anchors total.
	Space anchors at approximately 3 ft-0 in. o.c.

Alternate toenail connection:
   Assume 10d common toenails. The yield limit equations from Chap. 12 for the strength of the toenails will be shown for review. In practice, the yield limit equations will be solved on a spreadsheet template or other computer program. For the remaining nail problems in this example, the nominal design value will be obtained from NDS tables (Ref. 15.3).

Dimensions and bending yield strength of 10d common nail can be obtained from the tables in Fig. 12.7a and b (Sec. 12.4) or from NDS Appendix I and Appendix L.

$$D = 0.148 \text{ in.}$$

$$l = 3.0 \text{ in.}$$

$$F_{yb} = 90 \text{ ksi}$$

For a toenail the thickness of the side member (for use in the yield limit equations) and the penetration are determined using the geometry given in Fig. 12.12a (Sec. 12.6). The thickness of the main member t_m is assumed to be the combined thicknesses of the double 2× members forming the top wall plate.

$$t_m = 2(1.5) = 3.0 \text{ in.}$$

$$l_s = \frac{l}{3} = \frac{3.0}{3} = 1.0 \text{ in.}$$

$$p_L = l \cos(30) - l_s \le t_m$$

$$= 3.0 \cos(30) - 1.0 \le 3.0$$

$$\therefore p_L = 1.60 \text{ in.}$$

$$10D = 10(0.148) = 1.48 \text{ in.} < 1.60 \text{ in.}$$

$$\therefore l_m = p_L = 1.60 \text{ in.}$$

Coefficients for yield limit equations:

The specific gravity and dowel bearing strength are the same for both the main member and the side member and are obtained from NDS Tables 11.3.2A and 11.3.2:

$$G_m = G_s = 0.50$$

$$F_{em} = F_{es} = 4650 \text{ psi}$$

$$K_D = 2.2 \qquad \text{for } D \leq 0.17 \text{ in.}$$

$$R_e = \frac{F_{em}}{F_{es}} = \frac{4650}{4650} = 1.0$$

$$1 + R_e = 1 + 1.0 = 2.0$$

$$1 + 2R_e = 1 + 2(1.0) = 3.0$$

$$2 + R_e = 2 + 1.0 = 3.0$$

$$R_t = \frac{l_m}{l_s} = \frac{1.60}{1.0} = 1.6$$

$$k_1 = \frac{\sqrt{R_e + 2R_e^2(1 + R_t + R_t^2) + R_t^2 R_e^3} - R_e(1 + R_t)}{(1 + R_e)}$$

$$= \frac{\sqrt{1.0 + 2(1.0^2)(1 + 1.6 + 1.6^2) + (1.6^2)(1.0^3)} - 1.0(1 + 1.6)}{2.0}$$

$$= 0.5628$$

$$k_2 = -1 + \sqrt{2(1 + R_e) + \frac{2F_{yb}(1 + 2R_e)D^2}{3F_{em}l_m^2}}$$

$$= -1 + \sqrt{2(2.0) + \frac{2(90,000)(3.0)(0.148)^2}{3(4650)(1.60)^2}}$$

$$= 1.081$$

$$k_3 = -1 + \sqrt{\frac{2(1 + R_e)}{R_e} + \frac{2F_{yb}(2 + R_e)D^2}{3F_{em}l_s^2}}$$

$$= -1 + \sqrt{\frac{2(2.0)}{1.0} + \frac{2(90,000)(3.0)(0.148)^2}{3(4650)(1.0)^2}}$$

$$= 1.202$$

Yield limit equations:

Mode I_m (NDS equation 11.3-1):

$$Z = \frac{Dl_m F_{em}}{K_D} = \frac{0.148(1.6)(4650)}{2.2} = 501 \text{ lb}$$

Mode I_s (NDS equation 11.3-2):

$$Z = \frac{Dl_s F_{es}}{K_D} = \frac{0.148(1.0)(4650)}{2.2} = 313 \text{ lb}$$

Mode II (NDS equation 11.3-3):

$$Z = \frac{k_1 Dl_s F_{es}}{K_D} = \frac{0.5628(0.148)(1.0)(4650)}{2.2} = 176 \text{ lb}$$

Mode III_m (NDS equation 11.3-4):

$$Z = \frac{k_2 Dl_m F_{em}}{K_D(1 + 2R_e)} = \frac{1.081(0.148)(1.60)(4650)}{2.2(3.0)} = 180 \text{ lb}$$

Mode III_s (NDS equation 11.3-5):

$$Z = \frac{k_3 Dl_s F_{em}}{K_D(2 + R_e)} = \frac{1.202(0.148)(1.0)(4650)}{2.2(3.0)} = 125 \text{ lb}$$

Mode IV (NDS equation 11.3-6):

$$Z = \frac{D^2}{K_D} \sqrt{\frac{2F_{em}F_{yb}}{3(1 + R_e)}}$$

$$= \frac{(0.148)^2}{2.2} \sqrt{\frac{2(4650)(90,000)}{3(2.0)}} = 118 \text{ lb}$$

Nominal design value:

The nominal design value is taken as the smallest value from the yield formulas:

$$Z = 118 \text{ lb}$$

Adjustment factors:

Load duration factor from the NDS for wind:

$$C_D = 1.6$$

Limit for penetration:

$$6D = 6(0.148) = 0.89 \text{ in.} \quad \text{(minimum penetration)}$$

$$0.89 \text{ in.} < 1.60 \text{ in.} \quad OK$$

Toenail factor for laterally loaded toenails:

$$C_{tn} = 0.83$$

All other adjustment factors default to unity in this problem.

Allowable design value:

$$Z' = Z(C_D C_M C_t C_{eg} C_{di} C_{tn})$$

$$= 118(1.6)(1.0)(1.0)(1.0)(1.0)(0.83)$$

$$= 157 \text{ lb/nail}$$

$$\text{Req'd spacing} = \frac{157 \text{ lb/nail}}{205 \text{ lb/ft}} = 0.764 \text{ ft} = 9.2 \text{ in.} \quad say \quad 9 \text{ in. o.c.}$$

> Alternate connection *b*:   *Use*   10d common toenails
> at 9 in. o.c.

From IBC Table 2304.9.1, the alternate connection *b* using toenails requires a minimum 8d common or box nail at 6 in. o.c. The above connection exceeds this minimum.

c. Plywood shearwall edge nailing transfers shear from wall plate into shearwall (Table 4.1A of the *Wood Structural Panel Shear Wall and Diaphragm Supplement, Allowable Stress Design Manual for Engineered Wood Construction*) (Ref. 15.1).

Perpendicular-to-Wall Forces

d. Check connection of stud to wall plate for separation force. Assume two 16d common nails through wall plate into end grain of stud (IBC Table 2304.9.1).

Lumber is DF-L, and the thickness of the side member is the thickness of one of the wall plates ($t_s = 1\frac{1}{2}$ in.). The nominal design value for a 16d nail could be obtained from the yield limit equations. However, the plate-to-stud connection is covered directly in NDS Table 11N. The reader may wish to verify the results of the yield limit equations as an exercise.

$$Z = 141 \text{ lb}$$

The following adjustments apply:

$$C_D = 1.6 \quad \text{for wind}$$
$$C_{eg} = 0.67 \quad \text{Nail is driven into end grain of stud.}$$

Check nail penetration:

$$p = l - t_s = 3.5 - 1.5 = 2.0 \text{ in.}$$

$$10D = 10(0.162) = 1.62 \text{ in.} < 2.0 \text{ in.}$$

$$\therefore l_m = p = 2.0 \text{ in.}$$

All other adjustment factors are unity.

$$Z' = Z(C_D C_M C_t C_{eg} C_{di} C_{tn})$$

$$= 141(1.6)(1.0)(1.0)(0.67)(1.0)(1.0)$$

$$= 151 \text{ lb/nail}$$

Allowable load for stud connection:

$$\text{Allow. } P = N(Z') = 2(151) = 302 \text{ lb/stud}$$

Load on one stud connection is the force per foot of wall times the spacing of the studs.

$$P = 130 \text{ lb/ft} \times 1.33 \text{ ft}$$

$$= 173 \text{ lb/stud} < 302 \qquad OK$$

e. Assume that blocking is spaced 24 in. o.c.

$$\text{Load per block} = 130 \text{ lb/ft} \times 2 \text{ ft} = 260 \text{ lb}$$

Capacity of 10d common toenails:

$$Z' = 157 \text{ lb/nail} \qquad (\text{from } b)$$

$$\text{Number of toenails} = \frac{260 \text{ lb/block}}{157 \text{ lb/nail}} \approx 2$$

> *Use* two 10d common toenails per block. Blocks at 24 in. o.c.

f. Attach blocking to roof sheathing.

$$\text{Load per block} = 260 \text{ lb} \quad (\text{from } e)$$

For the general diaphragm nailing 10d common nails in ½-in. STR I DF plywood are used. Determine the number of 10d nails required to transmit the load from the block into the roof diaphragm.

The yield limit equations could be evaluated for a 10d common nail through ½-in. DF plywood into DF-L blocking. However, a conservative nominal design value will be obtained from NDS Table 11Q for $\frac{7}{16}$-in. DF plyowod:

$$Z = 87 \text{ lb/nail}$$

The following adjustments apply:

$$C_D = 1.6 \qquad \text{for wind}$$

$$C_{di} = 1.1 \qquad \text{Nail is into plywood diaphragm.}$$

Check nail penetration:

$$p = l - t_s = 3.0 - 0.5 = 2.5 \text{ in.}$$

$$10D = 10(0.148) = 1.48 \text{ in.} < 2.5 \text{ in.}$$

$$\therefore l_m = p = 2.5 \text{ in.}$$

Other adjustment factors default to unity.

$$Z' = Z(C_D C_M C_t C_{eg} C_{di} C_{tn})$$

$$= 87(1.6)(1.0)(1.0)(1.0)(1.1)(1.0)$$

$$= 153 \text{ lb/nail}$$

NOTE: Reference 15.8 recommends a 10 percent reduction in nail value when diaphragm nailing is driven into narrow (2-in. nominal) framing. The 0.9 multiplying factor in the next step is to account for this reduction. Depending on materials, other adjustments may apply. Contact APA for the latest information regarding allowable nail design values for diaphragms using the yield limit equations.

$$\text{Modified } Z' = 153(0.9) = 138 \text{ lb/nail}$$

$$\text{Number of nails} = \frac{260 \text{ lb/block}}{138 \text{ lb/nail}} \approx 2$$

Length of block will depend on spacing of roof joists (assume joists are 24 in. o.c.). If nails are spaced 6 in. o.c.,

$$N = \frac{24}{6} \approx 4 \text{ nails/block} > 2 \qquad OK$$

> *Use* 10d common nails at 6 in. o.c.
> through plywood into blocking.

15.4 Connection Details—Horizontal Diaphragm to Concrete or Masonry Walls

The anchorage of concrete tilt-up and masonry (concrete block and brick walls) to horizontal plywood diaphragms has been the subject of considerable discussion in the design profession. The use of plywood diaphragms in buildings with concrete or masonry walls is very common, especially in one-story commercial and industrial buildings. Some failures of these connections occurred in recent earthquakes (Ref. 15.4), and revised design criteria have been included in the Code to strengthen these attachments.

The Code has the following requirements for the anchorage of concrete and masonry walls for seismic forces:

In Seismic Design Categories B and higher (ASCE 7 Sec. 9.5.2.6.2.8):

1. Interconnection of the wall elements and connection to supporting framing systems shall have sufficient ductility, rotational capacity, or sufficient strength to resist shrinkage, thermal changes, and differential foundation settlement when combined with seismic forces.
2. Where the anchor spacing exceeds 4 ft, walls must be designed to resist bending between anchors.

In addition, in Seismic Design Categories C and higher (ASCE 7 Sec. 9.5.2.6.3.2):

3. Diaphragms shall be provided with continuous ties or struts between diaphragm chords to distribute these anchorage forces into the diaphragms. Added chords may be used to form subdiaphragms to transmit the anchorage forces to the continuous cross ties. The maximum length-to-width ratio of the structural diaphragm shall be 2½ to 1. In wood diaphragms the continuous ties shall be in addition to the diaphragm sheathing.
4. Wall anchorage connections shall extend into the diaphragm a distance to develop the force transfer into the diaphragm.
5. Wall anchorage shall not be accomplished by use of toenails or nails subject to withdrawal nor shall wood ledgers be used in cross-grain bending or cross-grain tension.
6. The strength design forces for steel elements of the wall anchorage system, other than anchor bolts and reinforcing steel, shall be 1.4 times the forces otherwise required by this section.
7. Diaphragm to wall anchorage using embedded straps shall be attached to, hooked around, the reinforcing steel or otherwise terminated so as to effectively transfer forces to the reinforcing steel.
8. When elements of the wall anchorage system are loaded eccentrically, or are not perpendicular to the wall, the system shall be designed to resist all components of the forces induced by the eccentricity.
9. When pilasters are present in the wall, the anchorage force at the pilasters shall be calculated considering the additional load transferred to the pilasters. However, the minimum unit anchorage force at a floor or roof shall not be reduced.

Two different anchorage problems are considered in this section, and the path used to transfer the forces is described. The first anchorage detail involves a *ledger connection* to a wall with a parapet. See Example 15.5. The vertical load must be transferred in addition to the lateral forces. The anchor bolts are designed to carry both the vertical load and the lateral force parallel to the wall. This requires that two different load combinations be checked, and the capacity of the bolts in the wall and in the ledger must be determined.

A special anchorage device is provided to transmit the lateral force normal to the wall into the framing members. This is required to prevent cross-grain bending in the ledger. A numerical example is provided for the anchorage of a masonry wall with a ledger. See Example 15.6.

EXAMPLE 15.5   Typical Ledger Anchorage to Masonry Wall

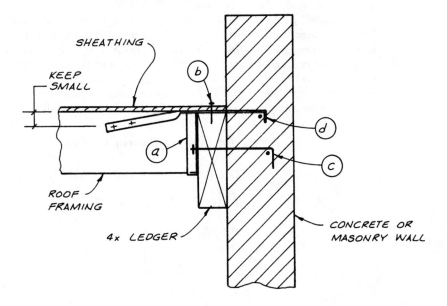

TYPICAL LEDGER CONNECTION

Figure 15.4a Anchorage connections to masonry wall with ledger.

a. Vertical loads (D + [L$_r$ or S]) are transferred from roof framing members to ledger with prefabricated metal hangers.

b. Horizontal diaphragm unit shear parallel to the wall is transferred to ledger by diaphragm boundary nailing.

c. The vertical load and unit shear from *a* and *b* are transferred from the ledger to the wall by the anchor bolts. The strength of the anchor bolts is governed by the capacity of the bolts in the

1. *Wall*

    Concrete walls—IBC Table 1912.2

    Masonry walls—IBC Sec. 2109.7.3

2. *Wood ledger* with bolt in single shear (Chap. 13)

    The load cases in IBC Sec. 1605.3.1 can be simplified to the following to determine the bolt ASD level loading (floor live L does not occur, and rain load R is not being considered in design)

    (*a*) D + L + (L$_r$ or S)

    (*b*) D + (W or 0.7E) + (L$_r$ or S)

    (*c*) 0.6D + W

    (*d*) 0.6D + 0.7E

    Except for equation *a*, both horizontal and vertical forces are included in the load combination equations. Vertical forces will act perpendicular to the ledger grain and horizontal forces will act parallel to the ledger grain. It must be remembered that E has both horizontal and vertical components:

$$E = \rho Q_E \pm 0.2D$$

This will be applicable to load combinations b and d.

In addition, for load combinations with two or more variable loads, IBC Sec. 1605.3.1.1 permits the variable loads to be multiplied by 0.75. This will be applicable to load combination b. When the factor of 0.75 is used, however, it is also necessary to check load combination b without the 0.75 but with only one transient load or force, and design for the most critical condition. This will result in load combinations $b1$ and $b2$ for this example.

In this case equations b and d could be modified to the following:

($b1$)
$$D + 0.75 (0.7(\rho Q_E + 0.2D)) + 0.75(L_r)$$
$$= D + 0.525\rho Q_E + 0.11D + 0.75L_r$$
$$= 1.11D + 0.525\rho Q_E + 0.75L_r$$

($b2$)
$$D + 0.7(\rho Q_E + 0.2D)$$
$$= D + 0.7\rho Q_E + 0.14D$$
$$= 1.14 D + 0.7\rho Q_E$$

(d)
$$0.6D + 0.7\rho Q_E - 0.7(0.2D) = 0.46D + 0.7\rho Q_E$$

It will always be conservative, however, to use equation b without the permitted factor of 0.75, allowing equation b to take a simpler form:

(b)
$$D + 0.7(\rho Q_E + 0.2D) + L_r$$
$$= D + 0.7\rho Q_E + 0.7(0.2D) + L_r$$
$$= 1.14D + 0.7\rho Q_E + L_r$$

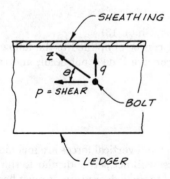

Figure 15.4b Anchor bolt connection between ledger and masonry wall. Vectors shown are resisting forces provided by anchor bolt under gravity load and lateral force parallel to wall.

d. Lateral forces perpendicular to the wall are carried by diaphragm-to-wall anchors (Fig. 14.21 in Sec. 14.5) or similar-type hardware. If the strap is bent down to allow bolting through the framing member, the fold-down distance should be kept small to minimize eccentricity. However, the required edge distance for the fasteners must be provided. The anchors are embedded in the wall, hooked around horizontal wall steel, and nailed, bolted, or lag bolted to the roof framing. The purpose of this separate anchorage connection for normal wall forces is to avoid cross-grain bending in the wood ledger (Example 6.1 in Sec. 6.2).

Subdiaphragm Problem

Once the perpendicular-to-wall force has been transferred into the framing members, it is carried into the sheathing. Because these normal wall forces may be large for concrete and masonry walls, it is important that the framing members extend back into the diaphragm (i.e., away from the wall) in order to make this transfer. The concept of subdiaphragms (Sec. 15.5) was developed to ensure that the framing is sufficiently anchored into the diaphragm.

EXAMPLE 15.6   Anchorage of Concrete Wall with Ledger

Design the connections for the ledger anchorage in Example 15.5. Assume Seismic Design Category D, with $S_{DS} = 1.0g$, and assume seismic controls over wind forces. Assume snow load S is not applicable at this building location. The letter designations in this example correspond to the designations of the connection details in Fig. 15.4*a*. The following information is known.

Loads

Roof dead load D = 60 lb/ft   (along wall)

Roof live load L_r = 100 lb/ft

Roof diaphragm shear = 420 lb/ft   (parallel to wall)

∴ Use Code minimum separation force of 280 lb/ft normal to wall.

Construction

½-in. STR I plywood roof sheathing is used.

Roof framing members are 2 × 6 at 24 in. o.c.

Ledger is 4 × 12 DF-L that is dry and remains dry in service. Normal temperature conditions apply.

Walls are 6-in. concrete = 75 psf

Wall height is 12 ft to roof with a one foot parapet above

Wall tributary height to diaphragm is (12/2) + 1 = 7 ft

Wall Anchorage Force (Perpendicular to wall)

First the wall weight tributary to the diaphragm needs to be calculated

$$W = (75 \text{ psf})(7 \text{ ft tributary}) = 525 \text{ lb/ft}$$

Three anchorage forces will be checked to determine the controlling anchorage force for this building. The three forces were introduced in Sec. 15.2 but will be considered in greater detail here.

1. Minimum load path connections (ASCE 7 Sec. 9.5.2.6.1.1)

$$F_p = 0.133 S_{DS} w_c \geq 0.05 w_c$$

$$= 0.133(1.00)(525 \text{ lb/ft}) = 0.133 w_c = 70 \text{ lb/ft}$$

2a. For Seismic Design Categories B and higher (ASCE 7 Sec. 9.5.2.6.2.8)

$$F_p = 0.4 S_{DS} I w_c \geq 400 S_{DS} I$$

$$= 0.4(1.0)(1.0)(525 \text{ lb/ft}) \geq 400(1.0)(1.0)$$

$$= 210 \text{ lb/ft} \geq 400 \text{ lb/ft}$$

$$= 400 \text{ lb/ft}$$

2b. For Seismic Design Categories C and higher (ASCE 7 Sec. 9.5.6.3.2)

$$F_p = 0.8 S_{DS} I w_c = 0.8(1.0)(1.0)(525 \text{ lb/ft}) = 420 \text{ lb/ft}$$

3. Minimum strength level force (ASCE 7 Sec. 9.5.2.6.1.2)

$$F_p \geq 280 \text{ lb/ft}$$

From all of these force requirements, the controlling *strength level* design force for wall anchorage is 420 lb/ft.

Load Combinations

From the load combinations introduced in Example 15.5, and because only dead load D, roof live load L_r and seismic forces E are being considered, the following load combinations need to be considered:

a. $D + L_r$
b1. $1.11D + 0.525\rho Q_E + 0.75 L_r$
b2. $1.14D + 0.7\rho Q_E$
d. $0.46D + 0.7\rho Q_E$

Of these load combination a involves only vertical forces and will use a different duration of load factor C_D of 1.25. Load combinations b and d involve both horizontal and vertical forces and will use a duration of load factor C_D of 1.6.

Anchorage

a. Roof beams are 2 ft o.c. Reaction of roof beam on ledger $D + L_r$:

$$R = (60 + 100)2 = 320 \text{ lb}$$

Choose prefabricated metal hanger from manufacturer with Code-recognized capacity greater than 320 lb.

b. From Table 3.1B of the *Wood Structural Panel Shear Wall and Diaphgram Supplement, Allowable Stress Design Manual for Engineered Wood Construction* (Ref. 15.1) for ½- (or ¹⁵⁄₃₂)-in. STR I plywood, allow. v = 320 lb/ft (>300) for a *blocked* diaphragm with 2-in. nominal framing and 10d common nails at:

6-in. o.c. edges

12-in. o.c. field

c. Assume ¾-in. anchor bolts at 32 in. o.c. For this anchor bolt spacing, compare the *actual* load and *allowable* load for two cases:

1. Vertical load D + L$_r$
2. Vertical load and lateral force parallel to wall

(1) *Vertical load combination:*

$$D + L_r = (60 + 100) \left(\frac{32}{12}\right) = 427 \text{ lb/bolt}$$

Check bolt capacity in concrete wall with f'_c = 2500 psi (IBC Table 1912.2).

$$B_v = 2250 \text{ lb/bolt} > 427 \text{ lb/bolt} \quad OK$$

Check bolt capacity in ledger.

The capacity of a wood-to-concrete connection is defined in NDS Table 11E. The nominal design value for a wood-to-concrete connection is taken as the single-shear capacity for a connection with the following definitions:

l_m = dowel bearing length in the main member for use in yield limit equations is taken equal to the embedment length of the anchor bolt in the masonry.

F_{em} = dowel bearing strength of main member for use in yield limit equations is taken equal to the dowel bearing strength of the concrete or masonry. The NDS recommends 6000 psi for $f'_c \geq 2000$ psi.

Application of the yield limit equations from Chap. 13 can be used to determine the anchor bolt capacity. In practice, however, NDS Table 11E will be used which provides bolt design values for single shear connections with sawn lumber anchored to concrete. If the particular case is not addressed in the NDS Table, the yield limit equations must be solved. Such is the case when $f'_c < 2000$ psi and the dowel bearing strength of the concrete or masonry is not defined in the NDS. The dowel bearing strength of 6000 psi used in the NDS was derived from several research studies involving the dowel bearing strength of concrete and lateral connections. The research indicated dowel bearing strengths on the order of 2 to 5 times the compressive strength of concrete, and the NDS initially considered adopting a dowel bearing strength for concrete and masonry equal to three times the compressive strength ($3 \times f'_c$ or f'_m). However, in consideration of past practice which allowed 6150 psi for dowel bearing strength of concrete and masonry along with the research studies, the NDS adopted a dowel bearing strength of 6000 psi for the purpose of tabulating wood-to-concrete connection values. Although data to support the use of a specific dowel bearing strength when $f'_c < 2000$ psi is currently not available, past practice of using dowel bearing strengths of 6150 psi resulted in satisfactory connection performance regardless of concrete strength (see Ref. 15.2). Therefore, AF&PA currently recommends the use of a dowel bearing strength for concrete equal to 6000 psi for all strengths of concrete, and Table 11E can be assumed valid for any concrete strength.

For vertical (gravity) loads, the load is perpendicular to grain:

$$D = 0.75 \text{ in.}$$

$$F_{yb} = 45 \text{ ksi} \quad \text{(assumed for A307 bolt)}$$

$$l_s = t_s = 3.5 \text{ in.}$$

$$\theta_s = 90 \text{ degrees}$$

$$F_{es} = 2600 \text{ psi} \quad (F_{e\perp} \text{ for } \tfrac{3}{4}\text{-in. bolt in DF-L from NDS Table 11.3.2})$$

For an anchor bolt problem:

$$l_m = t_m = 6.0 \text{ in.} \quad \text{(assumed embedment)}$$

$$F_{em} = 6000 \text{ psi}$$

Coefficients for yield limit equations:

$$R_e = \frac{F_{em}}{F_{es}} = \frac{6000}{2600} = 2.31$$

$$1 + R_e = 1 + 2.31 = 3.31$$

$$1 + 2R_e = 1 + 2(2.31) = 5.62$$

$$2 + R_e = 2 + 2.31 = 4.31$$

$$R_t = \frac{l_m}{l_s} = \frac{6.0}{3.5} = 1.71$$

$$1 + R_t = 1 + 1.71 = 2.71$$

$$1 + R_t + R_t^2 = 2.71 + 1.71^2 = 5.63$$

$$\theta_{max} = 90 \text{ degrees}$$

$$K_\theta = 1 + \frac{\theta_{max}}{360} = 1 + \frac{90}{360} = 1.25$$

$$k_1 = \frac{\sqrt{R_e + 2R_e^2(1 + R_t + R_t^2) + R_t^2 R_e^3} - R_e(1 + R_t)}{(1 + R_e)}$$

$$= \frac{\sqrt{2.31 + 2(2.31^2)(5.63) + (1.71^2)(2.31^3)} - 2.31(2.71)}{3.31}$$

$$= 1.1062$$

$$k_2 = -1 + \sqrt{2(1 + R_e) + \frac{2F_{yb}(1 + 2R_e)D^2}{3F_{em}l_m^2}}$$

$$= -1 + \sqrt{2(3.31) + \frac{2(45,000)(3.31)(0.75^2)}{3(6000)(6.0^2)}}$$

$$= 1.657$$

$$k_3 = -1 + \sqrt{\frac{2(1 + R_e)}{R_e} + \frac{2F_{yb}(2 + R_e)D^2}{3F_{em}l_s^2}}$$

$$= -1 + \sqrt{\frac{2(3.31)}{2.31} + \frac{2(45,000)(4.31)(0.75^2)}{3(6000)(3.5^2)}}$$

$$= 0.9635$$

Yield limit equations:

Mode I_m (NDS equation 11.3-1):

$$Z = \frac{Dl_m F_{em}}{4K_\theta} = \frac{0.75(6.0)(6000)}{4(1.25)} = 5400 \text{ lb}$$

Mode I_s (NDS equation 11.3-2):

$$Z = \frac{Dl_s F_{es}}{4K_\theta} = \frac{0.75(3.5)(2600)}{4(1.25)} = 1365 \text{ lb}$$

Mode II (NDS equation 11.3-3):

$$Z = \frac{k_1 Dl_s F_{es}}{3.6K_\theta} = \frac{1.1062(0.75)(3.5)(2600)}{3.6(1.25)} = 1678 \text{ lb}$$

Mode III_m (NDS equation 11.3-4):

$$Z = \frac{k_2 Dl_m F_{em}}{3.2(1 + 2R_e)K_\theta} = \frac{1.657(0.75)(6.0)(6000)}{3.2(5.62)(1.25)} = 1990 \text{ lb}$$

Mode III_s (NDS equation 11.3-5):

$$Z = \frac{k_3 Dl_s F_{em}}{3.2(2 + R_e)K_\theta} = \frac{0.9635(0.75)(3.5)(6000)}{3.2(4.31)(1.25)} = 880 \text{ lb}$$

Mode IV (NDS equation 11.3-6):

$$Z = \frac{D^2}{3.2K_\theta}\sqrt{\frac{2F_{em}F_{yb}}{3(1+R_e)}}$$

$$= \frac{0.75^2}{3.2(1.25)}\sqrt{\frac{2(6000)(45,000)}{3(3.31)}} = 1037 \text{ lb}$$

Nominal design value is selected as the smallest value from the yield limit formulas:

$$Z = 880 \text{ lb}$$

NOTE: The value of Z determined here agrees with the $Z_\perp$ design value tabulated in NDS Table 11E.

Load duration factor for the combination $(D + L_r)$ is $C_D = 1.25$. With a minimum edge distance for the loaded edge of $4D = 4(0.75) = 3.0$ in., the full bolt capacity can be used (i.e., $C_\Delta = 1.0$). Other adjustment factors are unity for this problem.

$$Z' = Z(C_D C_M C_t C_g C_\Delta)$$

$$= 880(1.25)(1.0)(1.0)(1.0)(1.0)$$

$$= 1100 \text{ lb/bolt} > 427 \text{ lb/bolt} \quad OK$$

(2) *Vertical loads plus lateral force parallel to wall:*

Recall the loads considered in this example include dead, roof live and seismic. The applicable ASD load combinations are

b1. $1.11D + 0.525\rho Q_E + 0.75L_r$
b2. $1.14D + 0.7\rho Q_E$
d. $0.46D + 0.7\rho Q_E$

The variable ρ will be assumed to be 1.0. It is not readily obvious which of these load combinations governs thus all three will be checked. Considering first $0.46D + 0.7Q_E$:

Perpendicular-to-grain component q:

$$q = 0.46D = 0.46(60)\left(\frac{32}{12}\right) = 74 \text{ lb/bolt}$$

Parallel-to-grain component p:

$$p = \text{lateral force} = 0.7(420)\left(\frac{32}{12}\right) = 784 \text{ lb/bolt}$$

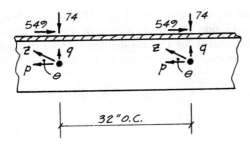

Figure 15.5 Anchor bolt in ledger carrying gravity load and lateral force parallel to wall.

Resultant load on one bolt:

$$z = \sqrt{784^2 + 74^2} = 787 \text{ lb/bolt}$$

Now considering $1.11D + 0.525\rho Q_E + 0.75L_r$:

Perpendicular-to-grain component q:

$$q = 1.11D + 0.75L_r = [1.11(60 \text{ plf}) + 0.75(100 \text{ plf})]\left(\frac{32}{12}\right)$$

$$= (67 + 75)\left(\frac{32}{12}\right) = 379 \text{ lb/bolt}$$

Parallel-to-grain component p:

$$p = 0.55Q_E = 0.55(420 \text{ plf})\left(\frac{32}{12}\right) = 231\left(\frac{32}{12}\right) = 616 \text{ lb/bolt}$$

NOTE: These forces have already been adjusted to an ASD level in order to properly apply load combinations.

Resultant load on one bolt:

$$z = \sqrt{379^2 + 616^2} = 723 \text{ lb/bolt} > 554 \text{ lb/bolt}$$

And finally considering equation $b2$: $1.14D + 0.7\rho Q_E$

Perpendicular-to-grain component q:

$$q = 1.14D = 1.14(60 \text{ plf})\left(\frac{32}{12}\right) = (68)\left(\frac{32}{12}\right) = 182 \text{ lb/bolt}$$

Parallel-to-grain component p:

$$p = 0.7Q_E = 0.7(420 \text{ plf})\left(\frac{32}{12}\right) = 294\left(\frac{32}{12}\right) = 784 \text{ lb/bolt}$$

Resultant load on one bolt:

$$z = \sqrt{182^2 + 784^2} = 805 \text{ lb/bolt} > 723 \text{ lb/bolt} > 554 \text{ lb/bolt}$$

The equation $b2$ load combination will govern for design of the anchor bolt attachment to concrete. In this case the equation $b1$ load combination with the larger perpendicular-to-grain force will likely control for the wood ledger, and will be used for the balance of this example. The reader is encouraged to check the ledger for the other combinations.

Bolt capacity in concrete wall with $f'_c = 2500$ psi is obtained from IBC Table 1912.2.

$$B_v = 2250 \text{ lb} > 805 \qquad OK$$

Bolt capacity in ledger:

Evaluation of the anchor bolt capacity in the ledger requires the use of the yield limit equations since the load acts at an angle to the grain. Computer solutions handle this problem by simply changing θ.

Angle of load to grain of θ:

$$\theta = \tan^{-1} \frac{379}{616} = 31.6 \text{ degrees}$$

$$D = 0.75 \text{ in.}$$

$$F_{yb} = 45 \text{ ksi} \qquad \text{(assumed for A307 bolt)}$$

$$l_s = t_s = 3.5 \text{ in.}$$

$$\theta_s = 31.6 \text{ degrees}$$

$$F_{e\|} = 5600 \text{ psi} \qquad (F_{e\|} \text{ for bolt in DF-L from NDS Table 11.3.2})$$

$$F_{e\perp} = 2600 \text{ psi} \qquad (F_{e\perp} \text{ for } \frac{3}{4}\text{-in. bolt in DF-L from NDS Table 11.3.2})$$

$$F_{es} = F_{e\theta} = \frac{F_{e\|} F_{e\perp}}{F_{e\|} \sin^2 \theta + F_{e\perp} \cos^2 \theta}$$

$$= \frac{5600(2600)}{5600 \sin^2 31.6 + 2600 \cos^2 31.6} = 4250 \text{ psi}$$

For an anchor bolt problem:

$$l_m = 6.0 \text{ in.} \qquad \text{(embedment)}$$

$$F_{em} = 6000 \text{ psi}$$

Coefficients for yield limit equations:

$$R_e = \frac{F_{em}}{F_{es}} = \frac{6000}{4250} = 1.412$$

$$1 + R_e = 1 + 1.412 = 2.412$$

$$1 + 2R_e = 1 + 2(1.412) = 3.824$$

$$2 + R_e = 2 + 1.412 = 3.412$$

$$R_t = \frac{l_m}{l_s} = \frac{6.0}{3.5} = 1.71$$

$$1 + R_t = 1 + 1.71 = 2.71$$

$$1 + R_t + R_t^2 = 2.71 + 1.71^2 = 5.65$$

$$\theta_{max} = 31.6 \text{ degrees}$$

$$K_\theta = 1 + \frac{\theta_{max}}{360} = 1 + \frac{31.6}{360} = 1.088$$

$$k_1 = \frac{\sqrt{R_e + 2R_e^2(1 + R_t + R_t^2) + R_t^2 R_e^3} - R_e(1 + R_t)}{(1 + R_e)}$$

$$= \frac{\sqrt{1.412 + 2(1.412^2)(5.65) + (1.71^2)(1.412^3)} - 1.412(2.71)}{2.412}$$

$$= 0.765$$

$$k_2 = -1 + \sqrt{2(1 + R_e) + \frac{2F_{yb}(1 + 2R_e)D^2}{3F_{em}l_m^2}}$$

$$= -1 + \sqrt{2(2.412) + \frac{2(45,000)(3.824)(0.75^2)}{3(6000)(6.0^2)}}$$

$$= 1.263$$

$$k_3 = -1 + \sqrt{\frac{2(1 + R_e)}{R_e} + \frac{2F_{yb}(2 + R_e)D^2}{3F_{em}l_s^2}}$$

$$= -1 + \sqrt{\frac{2(2.412)}{1.412} + \frac{2(45,000)(3.412)(0.75^2)}{3(6000)(3.5^2)}}$$

$$= 1.049$$

Yield limit equations:

Mode I_m (NDS equation 11.3-1):

$$Z = \frac{Dl_m F_{em}}{4K_\theta} = \frac{0.75(6.0)(6000)}{4(1.088)} = 6204 \text{ lb}$$

Mode I_s (NDS equation 11.3-2):

$$Z = \frac{Dl_s F_{es}}{4K_\theta} = \frac{0.75(3.5)(4250)}{4(1.088)} = 2563 \text{ lb}$$

Mode II (NDS equation 11.3-3):

$$Z = \frac{k_1 D l_s F_{es}}{3.6 K_\theta} = \frac{0.765(0.75)(3.5)(4250)}{3.6(1.088)} = 2178 \text{ lb}$$

Mode III$_m$ (NDS equation 11.3-4):

$$Z = \frac{k_2 D l_m F_{em}}{3.2(1 + 2R_e)K_\theta} = \frac{1.263(0.75)(6.0)(6000)}{3.2(3.824)(1.088)} = 2562 \text{ lb}$$

Mode III$_s$ (NDS equation 11.3-5):

$$Z = \frac{k_3 D l_s F_{em}}{3.2(2 + R_e)K_\theta} = \frac{1.049(0.75)(3.5)(6000)}{3.2(3.412)(1.088)} = 1391 \text{ lb}$$

Mode IV (NDS equation 11.3-6):

$$Z = \frac{D^2}{3.2 K_\theta} \sqrt{\frac{2F_{em}F_{yb}}{3(1 + R_e)}}$$

$$= \frac{0.75^2}{3.2(1.088)} \sqrt{\frac{2(6000)(45,000)}{3(2.412)}} = 1406 \text{ lb}$$

Nominal design value is selected as the smallest value from the yield limit formulas:

$$Z = 1391 \text{ lb}$$

Load duration factor from NDS for load combinations involving seismic loads is $C_D = 1.6$. Other adjustment factors are unity for this problem.

$$Z' = Z(C_D C_M C_t C_g C_\Delta)$$

$$= 1391(1.6)(1.0)(1.0)(1.0)(1.0)$$

$$= 2226 \text{ lb/bolt} > 723 \text{ lb/bolt} \qquad OK$$

> *Use*   ¾-in.-diameter anchor bolts at 32 in. o.c.

d. Provide wall anchors at every other roof framing member (i.e., 4 ft-0 in. o.c.). Perpendicular-to-wall force:

$$P = 0.7(420 \text{ lb/ft}) \times 4 \text{ ft} = 1176 \text{ lb}$$

Choose a wall anchor from manufacturer with Code-recognized load capacity that is greater than 1176 lb.

A series of additional design and detailing requirements for anchorage of concrete and masonry walls was introduced in Sec. 15.4. Of interest for selection of an anchorage device is the requirement that the strength design forces for steel elements of the wall anchorage system, other than the anchor bolts and reinforcing steel, shall be 1.4 times the forces otherwise required (ASCE 7 Sec. 5.5.2.6.3.2). Also of interest when selecting a device is the requirement that the anchorage must address forces towards and away from the diaphragm. An anchorage device capable of resisting both tension and compression forces most directly addresses this issue. Alternately a compression load path that makes use of bearing can be used. Care should be taken, however, to ensure that wood shrinkage and construction tolerances will not leave gaps in the load path. Retrofit of wall anchorage for existing systems of this type is addressed in Ref. 15.8.

The second anchorage detail example is for a diaphragm that connects to the *top* of a concrete or masonry wall. See Example 15.7. The anchor bolts in the wall are used to transfer the lateral forces both parallel and perpendicular to the wall. These forces are not checked concurrently, and the bolt values parallel and perpendicular to grain in the wood plate are used independently. The strength of the bolt in the wall must also be checked.

The problem in this connection centers on the need to avoid cross-grain tension in the wood plate when forces normal to the wall are considered. Two possible connections for preventing cross-grain tension are presented.

Calculations for the anchorage of this type of connection are similar in many respects to those for the ledger connection (Example 15.6), and a numerical example is not provided.

EXAMPLE 15.7   Anchorage at Top of Masonry Wall

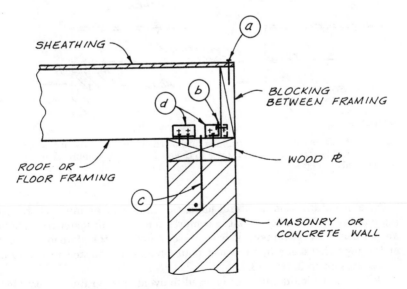

Figure 15.6a Anchorage connections to top of masonry wall.

Parallel-to-Wall Force (Diaphragm Shear)

a. Boundary nailing transfers diaphragm shear from sheathing into blocking.

b. Connection for transferring shear from blocking to plate. Framing anchor is shown. Alternate connection is additional blocking as in Example 9.5 (Sec. 9.4). Toenails are limited to seismic forces not greater than 150 plf in Seismic Design Categories D, E, and F, per IBC Sec. 2305.1.4.

c1. Shear is transferred from plate to wall by anchor bolts. Bolt strength is governed by wall capacity (IBC Table 1912.2 or Sec. 2109.7.3) or by the allowable load parallel to grain of the bolt in the wood plate.

Perpendicular-to-Wall Force (Wind, Seismic, or Code Minimum)

c2. Lateral force normal to wall is transferred from wall to plate by anchor bolts. Bolt strength is governed by wall capacity (IBC Table 1912.2 or Sec. 2109.7.3) or by the allowable load perpendicular to the grain of the bolt in the wood plate.

d. The attachment of the wall plate to the framing presents a problem. The Code requires that anchorage of concrete and masonry walls be accomplished without cross-grain bending or *cross-grain tension.*

The connection in Fig. 15.6*b* is shown with framing anchors on both the right and left sides of the wall plate. In this way cross-grain tension in the plate is not being relied upon for load transfer. The framing anchors shown with solid lines are those that stress the plate in compression perpendicular to grain.

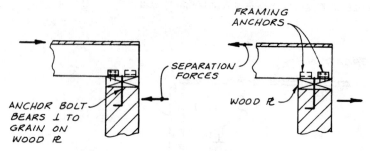

Figure 15.6*b*   For each case, the framing anchor shown with solid lines is capable of transferring the lateral force to the anchor bolt by compression perpendicular to grain.

For a given direction of lateral force, the effective framing anchors cause the wood plate to bear on the anchor bolt in compression. Without fasteners on both sides of the anchor bolt, cross-grain tension would be developed in the plate for one of the directions of loading. The shear in the wood plate should be checked in accordance with the procedures given in Sec. 13.9.

Other connection details can be used to avoid cross-grain tension. One alternative, shown in Fig. 15.6*c,* avoids the problem by tying the framing member directly to the anchor bolt. Use of this type anchorage requires close field coordination. The anchor bolts must be set accurately when the wall is constructed to match the future location of the roof or floor framing.

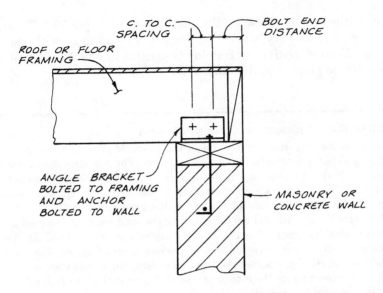

C. TO C. SPACING

BOLT END DISTANCE

ROOF OR FLOOR FRAMING

ANGLE BRACKET BOLTED TO FRAMING AND ANCHOR BOLTED TO WALL

MASONRY OR CONCRETE WALL

PARTIAL CONNECTION DETAIL SHOWING ALT. ANCHORAGE FOR NORMAL TO WALL FORCES

Figure 15.6c Angle bracket is connected directly to framing and anchor bolt. Therefore, normal wall force is transferred directly to the framing without placing the wood plate on top of wall in cross-grain tension.

Subdiaphragm Anchorage

For a discussion of the subdiaphragm anchorage of concrete and masonry walls, see Example 15.5 and Sec. 15.5.

15.5 Subdiaphragm Anchorage of Concrete and Masonry Walls

The anchorage requirements for concrete and masonry walls were outlined in Sec. 15.4. There it was noted that the Code requires *continuous ties* between the diaphragm chords in order to distribute the anchorage forces perpendicular to a wall into the diaphragm (ASCE 7 Sec. 9.5.2.6.3.2). The concept of subdiaphragms (also known as minidiaphragms) was developed to satisfy this Code requirement. The concept is fully developed in Ref. 15.6.

If the framing members in a horizontal diaphragm are continuous members from one side of the building to the other, the diaphragm would naturally have continuous ties between the chords. See Example 15.8. This is possible, however, only in very small buildings or in buildings with closely spaced trusses. Even if continuous framing is present in one direction, it would normally not be present in the other direction.

EXAMPLE 15.8   Diaphragm with Continuous Framing

Continuous framing in the transverse direction would provide continuous ties between diaphragm chords for the transverse lateral force. For this force the chords are parallel to the longitudinal walls. Continuous ties distribute the anchorage force back into the diaphragm. The action of a continuous tie is similar in concept to a collector or drag strut. The tie drags the wall force into the diaphragm.

If the spacing of wall anchors exceeds 4 ft, the wall must be designed to resist bending between the anchors. For the lateral force in the longitudinal direction, no intermediate ties are shown perpendicular to the continuous framing (Fig. 15.7). Ties must be designed and spliced across the transverse framing, or the transverse wall must be designed for bending (at the diaphragm level) between the longitudinal shearwalls (not practical in most buildings).

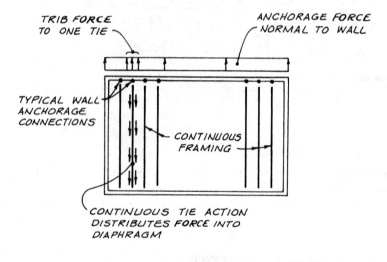

Figure 15.7 Building with continuous cross ties between diaphragm chords in one direction provided by continuous framing.

The majority of buildings do not have continuous framing. In a typical panelized roof system, for example, the subpurlins span between purlins, purlins span between girders, and girders span between columns. See Example 15.9. In addition, the tops of the beams are usually kept at the same elevation, and the diaphragm sheathing is nailed directly to these members. In order for this type of construction to be used, the lighter beams are not continuous across supporting members. Lighter beams are typically suspended from heavier beams with metal hangers (Chap. 14).

If continuous cross ties are to be provided between diaphragm chords in both directions of the building, a larger number of additional connections will be required. In order to meet the continuous-tie requirement literally, each beam would have to be spliced for an axial drag-type force across the members to which it frames.

EXAMPLE 15.9   Diaphragm without Continuous Framing

In a building without continuous framing between the diaphragm chords, a large number of cross-tie splices would be required (Fig. 15.8).

a. Subpurlins spliced across purlins
b. Purlins spliced across girders
c. Girders spliced across hinges.

Splices must be capable of transmitting the wall anchorage force tributary to a given member (Fig. 15.7). With the use of subdiaphragms, the continuous-tie requirement may be satisfied *without* connection splices at every beam crossing.

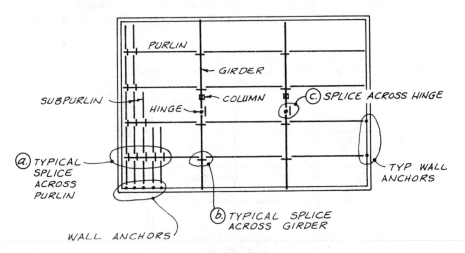

PLAN

Figure 15.8 Continuous cross ties between diaphragm chords by splicing each framing member would require an excessive number of connections.

The idea of a *subdiaphragm* was developed to ensure the proper anchorage of the wall seismic forces without requiring a cross-tie splice at every beam crossing. With this method the designer selects a portion of the total diaphragm (known as a subdiaphragm) which is designed as a separate diaphragm. See Example 15.10. The unit *shear* capacity (wood structural panel thickness and nailing) must be sufficient to carry the tributary wall lateral forces, and the subdiaphragm must have its own *chords* and *continuous cross ties* between the chords. Subdiaphragm chords and cross ties are also referred to as *subchords* and *subcross ties* (or simple *subties*).

The subdiaphragm reactions are in turn carried by the main cross ties which must be continuous across the full horizontal diaphragm. Use of the subdiaphragm avoids the requirement for continuous ties across the full width of the diaphragm except at the subdiaphragm boundaries.

ASCE 7 Sec. 9.5.2.6.3.2 specifically requires continuous ties or struts between diaphragm chords for Seismic Design Category C and higher, and specifically permits use of subdiaphragms to meet this requirement. This section also places an upper limit of 2.5 on the subdiaphragm length-to-width ratio.

EXAMPLE 15.10   Subdiaphragm Analysis

The building from the previous example is redrawn in Fig. 15.9a showing the subdiaphragms which will be used for the wall anchorage. The assumed action of the subdiaphragms for resisting the lateral forces is shown in Figs. 15.9b and 15.9c. The subdiaphragm concept is only a computational device that is used to develop adequate anchorage of concrete and masonry walls. It is recognized that the subdiaphragms are actually part of the total horizontal diaphragm, and they do not deflect independently as shown.

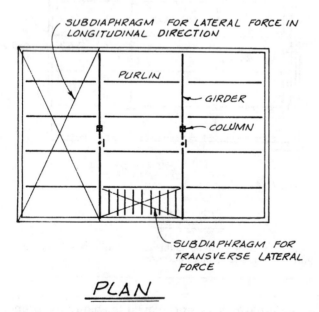

Figure 15.9a Subdiaphragms may be used to provide wall anchorage without requiring the large number of connections described in Fig. 15.8.

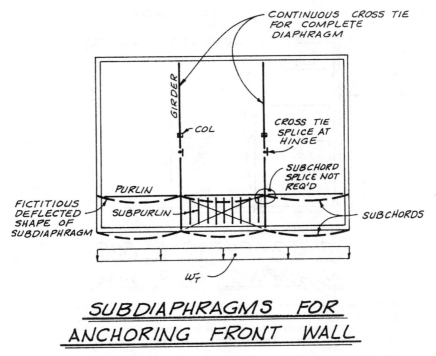

CONTINUOUS CROSS TIE
FOR COMPLETE
DIAPHRAGM

GIRDER

COL

CROSS TIE
SPLICE AT
HINGE

SUBCHORD
SPLICE NOT
REQ'D

PURLIN

FICTITIOUS
DEFLECTED
SHAPE OF
SUBDIAPHRAGM

SUBPURLIN

SUBCHORDS

w_T

SUBDIAPHRAGMS FOR ANCHORING FRONT WALL

Figure 15.9b Subdiaphragms spanning between the continuous cross ties for the complete diaphragm.

Three subdiaphragms for the anchorage of the front longitudinal wall are shown in Fig. 15.9b. Three similar subdiaphragms will also be used for the anchorage of the rear wall. Note that the transverse anchorage force can act in either direction.

The purlin and wall serve as the chords (subchords) for the subdiaphragms. The purlin must be designed for combined bending and axial forces. Because the subdiaphragms are treated as separate diaphragms, the subchord forces are zero at the girder. The purlins, therefore, need not be spliced across the girders.

The subpurlins serve as continuous cross ties (subties) between the subdiaphragm chords and do not have to be spliced across the purlin. The girders act as continuous cross ties for the complete diaphragm, and a cross-tie splice at the hinge is required. The splice must carry the reactions from two subdiaphragms.

The subdiaphragm for anchorage of the left wall is shown in Fig. 15.9c. A similar subdiaphragm is required for anchoring the right wall. The girder and wall serve as chords for the subdiaphragm. The girder must be designed for combined bending from gravity loads plus axial tension or compression caused by lateral forces.

The splice at the girder hinge must be designed for the subdiaphragm chord force. This is an alternate force to the continuous cross-tie force caused by the transverse lateral force in Fig. 15.9b. The purlins serve as continuous ties between subdiaphragm chords and do not have to be spliced across the girder.

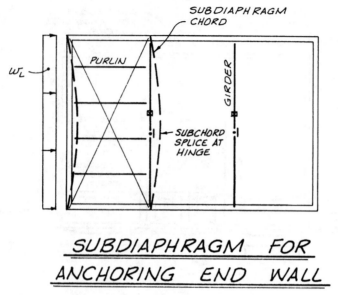

SUBDIAPHRAGM FOR
ANCHORING END WALL

Figure 15.9c A subdiaphragm can have the same span as the complete diaphragm. However, the subdiaphragm is designed for the wall anchorage force, and the complete diaphragm is designed for the total lateral force to the horizontal diaphragm.

The designer has some flexibility in the choice of what portions of the complete diaphragm are used as subdiaphragms. See Example 15.11. Figure 15.10a in this example shows that a number of small subdiaphragms are used to anchor the transverse wall. In Fig. 15.10b, one large subdiaphragm is used to anchor the same wall force between the longitudinal walls only. For additional examples illustrating subdiaphragm choices, see Refs. 15.6 and 15.7.

The flexibility permitted in the choice of subdiaphragm arrangements results from the fact that the subdiaphragm concept is a computational device only. The independent deflection of the subdiaphragms is not possible, and the analysis is used arbitrarily to ensure adequate anchorage connections.

Various types of connections can be used to provide the splices in the cross ties and chords of the diaphragm and subdiaphragms. Because most beams frame into the sides of the supporting members, compression forces are normally assumed to be taken by bearing. Thus the connections are usually designed for tension. This tension tie can be incorporated as part of the basic connection hardware (e.g., the seismic tie splice on the hinge connection in Fig. 14.19), or a separate tension strap may be provided (Fig. 14.22). The connections may be specially designed using the principles of Chaps. 12, 13, and 14, or some form of prefabricated connection hardware may be used.

Where compression forces are to be transferred through bearing, design and detailing attention is needed to make sure that wood shrinkage and construction tolerances do not disrupt the bearing, and that tension and compression cycling of the joint will not lead to premature failure (such as low-cycle fatigue from buckling of cold-formed steel straps). Alternately, connection devices rated for both tension and compression forces may be used.

EXAMPLE 15.11   Alternate Subdiaphragm Configurations

The subdiaphragm arrangement in Fig. 15.10a uses three small subdiaphragms. The two outside subdiaphragms span between the longitudinal shearwalls and the continuous cross ties for the complete diaphragm. The center subdiaphragm spans between the two continuous cross ties. The line of purlins forming the two main cross ties must be spliced at each girder crossing (in this example a column happens to occur at each of these locations) for the tributary wall anchorage force.

The other purlins in the subdiaphragms serve as continuous ties (subties) between the subdiaphragm chords, and they need not be spliced for lateral forces across the girder line.

The transverse wall and first row of girders serve as the subchords. The subdiaphragm chord force is zero at the ends of the girders, and subchord splices are not required at the columns.

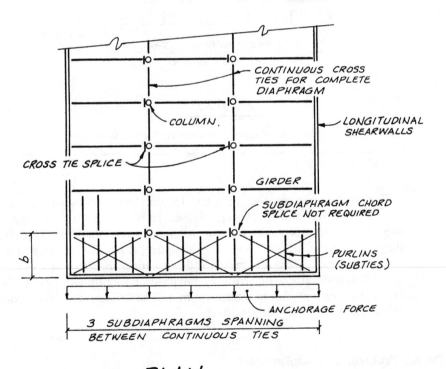

Figure 15.10a Three small subdiaphragms used to anchor transverse end wall.

In the second subdiaphragm layout (Fig. 15.10*b*), one large subdiaphragm spans between the exterior walls. Continuous subtie splices are required for the purlins across the first girder only. The subdiaphragm chord formed by the second line of girders must be tied across the columns for the appropriate subchord force.

The choice of the subdiaphragm configuration in practice will depend on the required number of connections when both the transverse and longitudinal forces are considered.

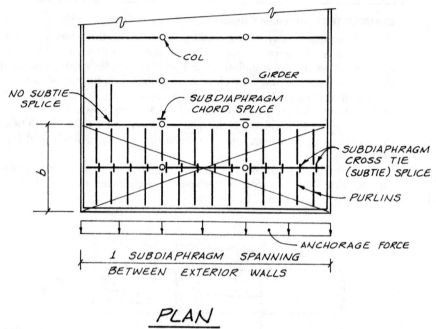

Figure 15.10*b*   One large subdiaphragm used to anchor transverse end wall.

The reader is reminded that the lateral forces shown in the examples in this section are *wall anchorage forces*. In other words, the forces represent one of the four perpendicular-to-wall forces described in Example 15.2, and they are *not* the lateral forces for designing the main horizontal diaphragm.

Subdiaphragms are a significant design concern in structures with concrete and masonry walls. The primary purpose of Examples 15.8 to 15.11 was to define the concept of a subdiaphragm and to illustrate where additional seismic connections may be required.

15.6 Design Problem: Subdiaphragm

In Example 15.12 the wall anchorage requirements for a large one-story building are analyzed. Exterior walls are of reinforced brick masonry construction.

In order to check the unit shear in the subdiaphragms, the construction of the entire roof diaphragm must be known. For this reason the unit shear diagrams for the complete diaphragm are shown along with the nailing and the allowable unit shears. Two different nailing patterns are used for the roof diaphragm. The heavier nailing occurs near the ends of the building in areas of high diaphragm shear. This different nailing may be important when the unit shears in the subdiaphragms are checked.

Along the longitudinal wall a series of small subdiaphragms is used to anchor the wall (Fig. 15.11b). Subdiaphragm 1 spans between girders which serve as continuous cross ties for the entire roof diaphragm in the transverse direction. Because subdiaphragm 1 occurs in both areas of diaphragm nailing, the unit shear in the subdiaphragm is first checked against the lower allowable shear for the complete diaphragm.

The lower shear capacity is found to be adequate for subdiaphragm 1. If the lower shear value had not been adequate, an 8-ft-wide strip of heavier nailing could be added next to the longitudinal wall. An alternative solution would be to increase the size of subdiaphragm 1.

The transverse wall is anchored by a large subdiaphragm that spans between the longitudinal walls (Fig. 15.11c). Continuous cross ties are required only between the subdiaphragm chords. Because a portion of subdiaphragm 2 coincides with the area of lighter nailing for the complete diaphragm, the unit shear is checked against the lower shear capacity of the diaphragm.

The use of the larger-diaphragm shear capacity (425 lb/ft) requires that the heavier boundary nailing be provided at the perimeter of each subdiaphragm. Care must be taken to avoid potential splitting of lumber that might result from the use of closely spaced nails at adjacent subdiaphragm boundaries. See Table 3.1B of the *Wood Structural Panel Shear Wall and Diaphragm Supplement, Allowable Stress Design Manual for Engineered Wood Construction* (Ref. 15.1), footnote c, for possible considerations.

EXAMPLE 15.12   Subdiaphragm Anchorage for Masonry Walls

The complete roof diaphragm for the building in Example 15.11 has been designed for seismic forces in both the transverse and longitudinal directions. The unit shears and plywood nailing are summarized below. A subdiaphragm analysis is to be performed using subdiaphragm 1 (Fig. 15.11b) for the longitudinal walls and subdiaphragm 2 (Fig. 15.11c) for the transverse walls. The calculation of subdiaphragm unit shears and tie forces is demonstrated.

Given Information

Total wall height is 16 ft. Tributary wall height to roof diaphragm:

$$h = \frac{16}{2} = 8 \text{ ft}$$

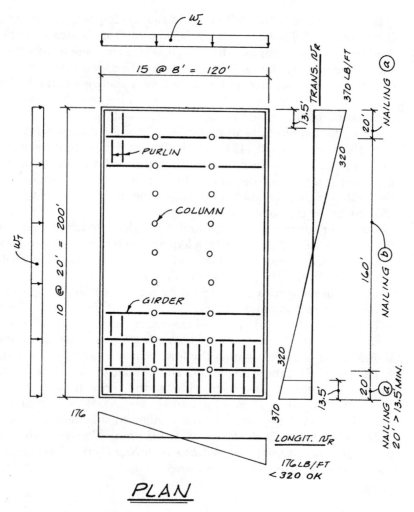

PLAN

Figure 15.11a Lateral forces on entire horizontal diaphragm. Unit shear diagrams caused by transverse and longitudinal lateral forces are also shown.

Dead loads:

Roof dead load D = 11 psf

Wall dead load D = 90 psf

Seismic coefficient: Assume that the strength level seismic coefficient for design of the roof diaphragm is 0.225g, and S_{DS} is 1.0. This is for the main seismic force system. Calculation of the seismic forces for wall anchorage and subdiaphragms will follow.

Entire Roof Diaphragm

Seismic force to roof diaphragm (*strength level*):

$$\text{Transverse } w_u = 0.225[11 \text{ psf}(120 \text{ ft}) + 90 \text{ psf}(8 \text{ ft})(2 \text{ walls})]$$

$$= 621 \text{ lb/ft}$$

$$\text{Longitudinal } w_u = 0.225[11 \text{ psf}(200 \text{ ft}) + 90 \text{ psf}(8 \text{ ft})(2 \text{ walls})]$$

$$= 819 \text{ lb/ft}$$

Roof unit shears (*strength level*):

$$\text{Transverse } V_u = \frac{wL}{2} = \frac{621(200)}{2} = 62,100 \text{ lb}$$

$$v_u = \frac{V}{b} = \frac{62,100}{120} = 518 \text{ lb/ft}$$

$$\text{Longitudinal } V_u = \frac{wL}{2} = \frac{819(120)}{2} = 49,140 \text{ lb}$$

$$v_u = \frac{V}{b} = \frac{49,140}{200} = 246 \text{ lb/ft}$$

Roof unit shears (*allowable stress design level*): Since the shear forces in the roof diaphragm result entirely from seismic forces, the applicable load combinations reduce to 0.7E.

$$\text{Transverse } v = \frac{518}{1.4} = 370 \text{ lb/ft}$$

$$\text{Longitudinal } v = \frac{246}{1.4} = 176 \text{ lb/ft}$$

See unit shear diagrams (Fig. 15.11a).

Plywood sheathing and nailing:

½-in. STR I plywood.

All edges supported and nailed into 2-in. minimum nominal framing (i.e., blocking is used).

Assume plywood load cases 2 and 4 (Table 3.1B of the *Wood Structural Panel Shear Wall and Diaphragm Supplement, Allowable Stress Design Manual for Engineered Wood Construction*) (Ref. 15.1). Note that plywood layout is not shown, but will eventually be included on the plans.

> *Diaphragm nailing a.* 10d common nails at:
>   4-in. o.c. boundary and continuous panel edges
>   6-in. o.c. other panel edges (blocking required)
>   12-in. o.c. field

$$\text{Allow. } v = 425 \text{ lb/ft} > 370 \quad OK$$

> *Diaphragm nailing b.* 10d common nails at:
>   6-in. o.c. all edges (blocking required)
>   12-in. o.c. field

$$\text{Allow. } v = 320 \text{ lb/ft}$$

From similar triangles on the unit shear diagram, a unit shear of 320 lb/ft occurs 13.5 ft from the transverse shearwalls.

∴ Use diaphragm *nailing a* for distance of 20 ft from transverse shearwalls (13.5 ft is minimum), and use diaphragm *nailing b* in central 160 ft of horizontal diaphragm.

NOTE: In the longitudinal direction two allowable shears apply: 425 lb/ft for *nailing a* and 320 lb/ft for *nailing b*. Both of these allowable shears exceed the actual shear of 176 lb/ft. Therefore, unit shear in longitudinal direction is not critical.

Subdiaphragm Anchorage of Longitudinal Masonry Wall

Subdiaphragm 1(8 ft × 20 ft):

The seismic forces for wall anchorage were introduced in Sec. 15.4 and illustrated in Example 15.6, which also used $S_{DS} = 1.0$. From Example 15.6 the critical wall anchorage force was found to be $0.8w_c$, based on ASCE 7 Sec. 9.5.2.6.3.2. This seismic force is at the *strength level* and is used to determine the force normal to the wall.

$$w_u = 0.80W_c = 0.80(90 \text{ psf})$$

$$= 72 \text{ psf}$$

Anchorage force is the "roof reaction" which is obtained as the uniform seismic force times the tributary height to the roof level.

$$w_{T_u} = 72 \text{ psf} \times 8 \text{ ft} = 576 \text{ lb/ft} > \text{Code minimum of } 280$$

Since this force is the result from seismic forces only, the applicable load combinations reduce to 0.7E.

$$w_T = 0.7(576) = 403 \text{ lb/ft} < \text{required wind anchorage (wind analysis not shown)}$$

Subdiaphragm span-to-width ratio:

$$\frac{L}{b} = \frac{20}{8} = 2.5 = 2.5 \quad OK$$

Subdiaphragm unit shear:

A typical 8 ft × 20 ft subdiaphragm (subdiaphragm 1) in the central 100 ft of the building will have the critical unit shear.

$$\text{Max. } V = \frac{wL}{2} = \frac{403(20)}{2} = 4030 \text{ lb}$$

$$v = \frac{V}{b} = \frac{4030}{8} = 504 \text{ lb/ft} > 320 \text{ lb/ft} \quad NG$$

The nailing for the entire roof diaphragm in the center portion of the building (i.e., diaphragm nailing b) is not adequate for subdiaphragm 1. There are several possible options. For example, the 8-ft width of subdiaphragm 1 could be increased to 16 ft to reduce the unit shear. In other words, subdiaphragm 1 could be changed from an 8 × 20 ft subdiaphragm to a 16 × 20 ft subdiaphragm. Although this would reduce the unit shear in the subdiaphragm, it would require the addition of subtie splice connections across the first line of purlins next to the longitudinal wall. An alternative solution could be to provide increased nailing for the original 8 × 20 ft subdiaphragm 1. This could be done by extending nailing a for the entire roof diaphragm, to include an 8-ft-wide strip along the longitudinal walls. The second solution with increased nailing will be assumed for the following demonstration of subdiaphragm design.

For the following anchorage force calculations, refer to Fig. 15.11b.

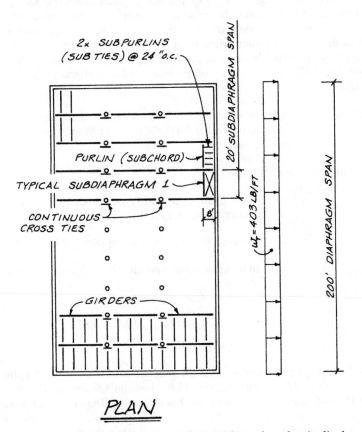

PLAN

Figure 15.11b Subdiaphragm 1 (8 ft × 20 ft) anchors longitudinal masonry wall for seismic force w_T in transverse direction.

Girder—continuous cross-tie force at columns:

$$T = (403 \text{ lb/ft})(20 \text{ ft}) = 8060 \text{ lb}$$

The continuous-cross-tie connections along the girders at the columns must be designed for a force of 8060 lb.

NOTE: The purpose of this example is to define the subdiaphragm problem and to determine the associated design forces. Detailed connection designs for these forces are beyond the scope of the example. See Chaps. 12 to 14 for design principles.

Subpurlin—anchor wall to subpurlin at 4 ft-0 in. o.c.

Every other subpurlin will serve as a continuous cross tie (subtie) for subdiaphragm 1. The masonry wall need not be designed for bending between wall anchors that are spaced 4 ft-0 in. o.c. or less. As with the girder splices above, subpurlin anchors must be designed for a force:

$$T = (403 \text{ lb/ft})(4 \text{ ft}) = 1612 \text{ lb}$$

An anchor bracket with a Code-recognized load capacity in excess of this value should be provided between the masonry wall and the $2\times$ subpurlins (subties).

The anchorage brackets may be attached to subpurlins that occur at the edge of a plywood panel. However, unless the designer is assured of this, it should be assumed that the anchors occur at subpurlins located at the center of a plywood panel. In this case, the capacity of the diaphragm *field nailing* should be checked against the wall anchorage force. The capacity of a 10d common nail in a plywood diaphragm nailed to DF-L framing is determined to be

$Z = 87 \text{ lb/nail}$ (Obtained from NDS Table 11Q for $\frac{7}{16}$-in. plywood. As an alternative to the NDS table, yield limit equations may be evaluated.)

The following adjustments apply:

$C_D = 1.6$ for seismic force (verify local code acceptance before using in practice)

$C_{di} = 1.1$ plywood diaphragm nailing (Sec. 12.6)

Nail penetration is adequate by inspection, and all other adjustment factors are unity.

$$Z' = Z(C_D C_M C_t C_{eg} C_{di} C_{tn})$$

$$= 87(1.6)(1.0)(1.0)(1.0)(1.1)(1.0)$$

$$= 153 \text{ lb/nail}$$

NOTE: Reference 15.9 recommends a 10 percent reduction in nail value when diaphragm nailing is driven into narrow (2-in. nominal) framing. The 0.9 multiplying factor in the next step is to account for this reduction. Depending on materials, other adjustments may apply. Contact APA for the latest information regarding allowable nail design values for diaphragms based on yield limit theory.

$$\text{Modified } Z' = 0.9(153) = 138 \text{ lb/nail}$$

The usual spacing of nails in the field (i.e., along intermediate framing) of a plywood panel is 12 in. o.c. Subpurlins are 8 ft long, and a minimum of 8 (likely 9) nails is available to develop the anchorage force.

$$P = 8(138) = 1104 \text{ lb}$$

This sheathing nailing capacity is less than the subpurlin anchorage force of 1612 lb, so additional fastening is required. It is not unusual to specify a row of diaphragm edge nailing (i.e., nailing at the sheathing edge spacing) at each subpurlin anchor. Vigilance is required, however, to ensure that this requirement is understood by the contractor and installed.

Subdiaphragm chord force purlins (refer again to Fig. 15.11b):

The first line of purlins parallel to the masonry wall serves as the subchord for subdiaphragm 1.

$$T_{\max} = C_{\max} = \frac{M}{b} = \frac{wL^2}{8b} = \frac{403(20)^2}{8(8)} = 2520 \text{ lb}$$

A subchord splice across the girders is not required. However, the subchord must be checked for combined axial stress (from the subchord force) plus bending stress (from gravity loads).

Subdiaphragm Anchorage of Transverse Masonry Wall

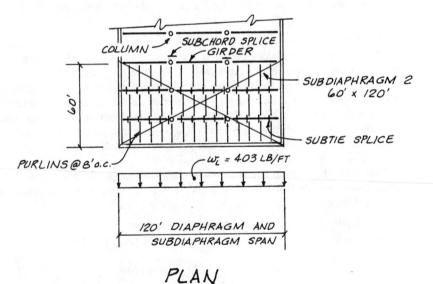

PLAN

Figure 15.11c Subdiaphragm 2 (60 ft $\times$ 120 ft) anchors transverse masonry wall for seismic force w_L in longitudinal direction.

Subdiaphragm 2 (60 ft × 120 ft):

Wall anchorage force (from subdiaphragm 1):

$$w_L = 403 \text{ lb/ft}$$

Span-to-width ratio:

$$\frac{L}{b} = \frac{120}{60} = 2 < 2.5 \qquad OK$$

Subdiaphragm shear:

$$V = \frac{(403)(120)}{2} = 24{,}180 \text{ lb}$$

$$v = \frac{V}{b} = \frac{24{,}180 \text{ lb}}{60} = 403 \text{ lb/ft} > 320 \text{ lb/ft} \qquad NG$$

Diaphragm nailing b is not adequate for subdiaphragm 2, so to qualify for an allowable shear of 425 lb/ft (nailing a), the boundary nailing of 10d common nails at 4 in. o.c. will need to be provided along the second girder line in addition to the usual boundary nailing locations for nailing a.

Purlin—subtie splice force across the first girder line:

$$T = (403 \text{ lb/ft})(8 \text{ ft}) = 3224 \text{ lb}$$

All the subtie splices across the first and second girder lines must be designed for this force.

Purlin—anchor wall to purlins at 8 ft-0 in. o.c.:

The connection hardware between the masonry wall and the purlins has the same design forces as the subtie splices just determined: $T = 3224$ lb.

The nailing between the plywood sheathing and the cross tie for subdiaphragm 2 is not as critical as subdiaphragm 1. The maximum subtie force has doubled ($2 \times 1612 = 3224$ lb), but the cross-tie length has far more than doubled (the length is 60 ft instead of 8 ft). In addition, plywood edge nailing (not just the field nailing) will occur along the subties (purlins) due to the plywood layout. Therefore, a less critical subtie nailing situation exists for subdiaphragm 2.

Note that the masonry wall must be designed for bending because the 8-ft spacing of anchors exceeds the 4-ft Code limit. This is a masonry design problem and is not included here.

Subdiaphragm chord force in second girder line:

The maximum chord force in subdiaphragm 2 (Fig. 15.11d) is

$$T = C = \frac{M}{b} = \frac{wL^2}{8b} = \frac{403(120)^2}{8(60)} = 12{,}090 \text{ lb}$$

The subdiaphragm chord along the second girder should be spliced at the columns for the subchord force of 12,090 lb, instead of the continuous-cross-tie force of 8060 lb (see calculations for subdiaphragm 1).

An alternate for subdiaphragm 2 is to use several smaller subdiaphragms similar to the first arrangement in Example 15.11.

The design of anchorage for concrete and masonry walls to wood roof systems using the concept of subdiaphragms has been utilized since the early 1970s.

15.7 References

[15.1] American Forest and Paper Association (AF&PA). 2001. *Allowable Stress Design Manual for Engineered Wood Construction and Supplements and Guidelines*, 2001 ed., AF&PA, Washington, DC.

[15.2] American Forest and Paper Association (AF&PA). 1999. *Commentary on the National Design Specification for Wood Construction*, 1997 ed., AF&PA, Washington DC.

[15.3] American Forest and Paper Association (AF&PA). 2001. *National Design Specification for Wood Construction and Supplement*, 2001 ed., AF&PA, Washington DC.

[15.4] Hamburger, Ronald O., and McCormick, David L. 1994. "Implications of the January 17, 1994 Northridge Earthquake on Tiltup and Masonry Buildings with Wood Roofs," Proceedings, 1994 Spring Seminar, Structural Engineers Association of Northern California, San Francisco, CA.

[15.5] International Codes Council (ICC). 2003. *International Building Code*, 2003 ed., ICC, Falls Church, VA.

[15.6] Sheedy, Paul. 1983. "Anchorage of Concrete and Masonry Walls," *Building Standards*, September/October, 1983, ICBO, Whittier, CA (reprint available with comments from APA, Tacoma, WA).

[15.7] Sheedy, Paul, and Sheedy, Christine. 1992. "Concrete and Masonry Wall Anchorage," *Buildings Standards*, July/August, 1992, ICBO, Whittier, CA.

[15.8] Structural Engineers Association of Northern California (SEAONC). 2001. *Guidelines for Seismic Evaluation and Rehabilitation of Tilt-up Buildings and Other Rigid Wall / Flexible Diaphragm Structures*. Structural Engineers Association of Northern California, San Francisco, CA.

[15.9] Tissel, John R., and Elliott, James R. 1990. *Plywood Diaphragms, Research Report 138*, APA—The Engineered Wood Association, Engineered Wood Systems, Tacoma, WA.

15.8 Problems

15.1 *Given:* The roof framing of Fig. 9.A (Chap. 9) and the typical section of the same building in Fig. 15.A. The lateral wind pressure is 20 psf over the entire vertical projection of the building surface. Roof dead load = 15 psf and roof live load = 20 psf.

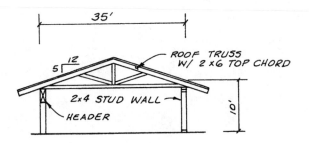

Figure 15.A

SECTION

Find: Design and sketch the roof-to-wall anchorage details for:
  a. The shear longitudinal wall at the 25- and 20-ft headers.
  b. The longitudinal bearing walls.
  c. The right transverse wall. Consider one case with full- height wall
    studs and a second case with 10-ft studs to a double plate and a
    filler truss over.

15.2   *Given:*   The roof framing plan of Fig. 9.A (Chap. 9) and the typical section
    of the same building in Fig. 15.B. Roof dead load = 20 psf; roof live
    load = 20 psf; wall dead load = 75 psf. Assume $f_c' = 2500$ psi, Seismic
    Design Category D with $S_{DS} = 1.0$, $I = 1.0$, $\rho = 1.0$, and $R = 5.0$.

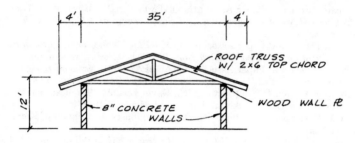

Figure 15.B

SECTION

Find: Design and sketch the roof-to-wall anchorage details for:
  a. The longitudinal bearing walls.
  b. The transverse walls include cross ties at 4 ft-0 in. o.c.

Advanced Topics in Lateral Force Design

16.1 Introduction

A number of important issues have been incorporated into recent editions of the Code. One significant issue is consideration of whether a structure is *regular* or *irregular*. In addition, specific overturning requirements for earthquake forces have been modified. It is important for the designer to understand the similarities and differences in the overturning requirements for wind and seismic forces.

This chapter provides an introduction to seismic irregularity considerations, gives a more detailed review of the overturning requirements for both wind and seismic forces, and takes a detailed look at the diaphragm flexibility issues introduced in Sec. 9.11.

The reader is reminded of the general convention in this book to use the term *load* to refer to a *gravity* effect such as dead load, live load, snow load, and so on. On the other hand, *force* generally refers to a *lateral* effect such as wind force and seismic force. Therefore, the phrase *combination of loads and forces* implies that both gravity and lateral effects are involved. Although there are not hard-and-fast rules regarding the use of these terms, the pattern of addressing vertical (gravity) loads *and* lateral (wind and seismic) forces seems to be emerging.

16.2 Seismic Forces—Regular Structures

Experience has proven that regular structures perform much better in earthquakes than irregular structures. The subject of irregularity has added to the length and complexity of the seismic code. Recall how the SEAOC *Blue Book* and its commentary (Ref. 16.11) have grown in size, and to a large extent this increase is the result of the need to consider structural irregularity.

To demonstrate how involved this subject has become, the seismic code identifies 10 types of structural irregularity (ASCE 7 Sec. 9.5.2.3, as referenced in IBC Sec. 1616.5). In order to organize these into a workable system, the Code summarizes these irregularities in tables, and the tables then reference appropriate sections of the Code which give the detailed requirements for each type of irregularity. The commentary to the Blue Book expands on each item. Obviously a comprehensive treatment of structural irregularity is beyond the scope of this introductory text. However, it is essential that the designer of a wood-frame building have some knowledge of this subject.

Several examples of *regular* and *irregular* structures are given in this chapter to illustrate the general concepts. The wood-frame buildings that were considered in previous chapters meet the basic definition of a regular structure. Dealing with these basic types of structures allowed a number of important subjects to be covered without becoming overly complicated. These subjects included horizontal diaphragms, shearwalls, chords, collectors (drag struts), and anchorage forces.

However, regular structures are rather limited in plan and elevation, and many buildings do not qualify as regular structures. It is important for the designer to be able to recognize what constitutes a structural irregularity and to know what, if any, additional requirements are imposed because of the irregularity. A number of the features that can cause a structure to be classified as irregular are addressed in Sec. 16.3.

Regular structures are those which have no significant physical or structural irregularities in either plan view or elevation. For example, a simple rectangular wood-frame building with a uniform distribution of mass can fit this description. However, in order to be classified as structurally regular, an additional requirement for multistory buildings is that the shearwalls in each upper story must be located directly over the shearwalls in each lower story. See Example 16.1.

EXAMPLE 16.1   Regular Structure

Regular structures have a history of good performance in resisting earthquake forces. These structures have no *plan irregularities* or *vertical irregularities.*

A building will be classified as structurally regular if

1. The plan is essentially *rectangular.*
2. The *mass* is reasonably uniform over the height.
3. The *shearwalls* in a lower story are directly below the shearwalls in each upper story.

Both transverse and longitudinal lateral seismic forces are shown in the sketch. Normal practice is to consider the force in each direction separately.

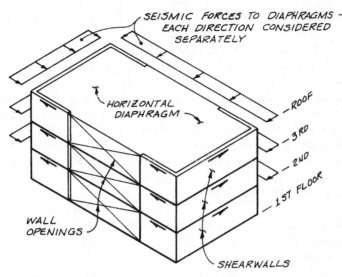

Figure 16.1 Regular structure. Recall from Chaps. 2 and 3 that two different distributions of seismic forces are provided by the Code: F_x for shearwall design, and F_{px} for diaphragm design. Although the sketch shows lateral forces in both directions, each principal direction of the structure is considered separately.

For regular structures the seismic forces are computed and distributed as illustrated in Chaps. 2 and 3. The design of the lateral-force-resisting systems (LFRSs) for these buildings follows the procedures given in Chaps. 9 and 10, and the anchorage details are handled as described in Chap. 15.

The IBC load combinations, load combination factors, and load duration factors were covered in Chap. 2. Before proceeding with a discussion of seismic irregularity, several topics from Chap. 2 will be revisited. Recall that the IBC has two sets of load combinations for allowable stress design (ASD). The newer and preferred combination of loads is known as the ASD *basic load combinations*. The older ASD load combinations from the 1994 (and earlier) Codes may still be used and are now known as the ASD *alternate basic load combinations*. All of the examples in this book make use of the basic load combinations (not the alternate basic load combinations).

Several adjustment factors for load combinations and adjustment factors for allowable stresses in wood and other materials were introduced in Sec. 2.8. It is important for the designer to have a basic understanding of these terms and adjustments.

One of the basic load combinations includes a *load combination factor* (LCF) of 0.75. The LCF is applied to the load combination that involves two or more transient (variable) loads in addition to dead load. One of the transient loads in this combination is a lateral force (wind or seismic). See Sec. 2.16. It should be clear that it is not likely that both transient loads will be at their maximum values simultaneously. The load combination factor reflects this reduced probability of loading.

Although this book does not deal with them at length, it is worth noting that the IBC alternate basic load combinations have a similar adjustment for load combinations that involves wind and seismic forces. Here the adjustment is in the form of an *allowable stress increase* (ASI) of one third (ASI multiplier of 1.33) for any loading involving wind or seismic. Note that the inverse of the LCF of 0.75 is the same as the ASI of 1.33 (i.e., 1/1.33 = 0.75). Because the LCF and the ASI have similar effects, they are not used together. In other words, the LCF is used only in the *basic* load combination involving multiple transient loads, and the ASI is used only in the *alternate basic* load combinations that include wind or seismic.

Because the inverse of the LCF is numerically equal to the ASI of 1.33, it would appear that the two adjustment factors really accomplish the same effect. However, the ASI can be applied to load combinations involving only a single transient load:

$$D + W$$

$$D + E$$

Dead load is obviously a permanent load, and these two cases each involve a *single* transient force (i.e., the lateral force). Further differences exist between the *basic* and *alternate basic* load combinations.

The point of mentioning the two conflicting loading criteria is not to point out the differences in design loads and forces, but to note the long-term national practice of either increasing allowable stresses *or* reducing the loads and forces when the combination includes either wind or seismic. The 2001 NDS (NDS Appendix B.4) notes that the reduction with combined loads and forces is essentially related to the *reduced probability* that the *maximum* loads and forces will act *simultaneously,* and it is independent of the type of material being used. The NDS distinguishes between this universal adjustment, applicable to all materials and systems, and the material property in wood design known as the load duration factor C_D.

The 2003 IBC has clarified use of and combining of these factors. Per IBC Sec. 1605.3.1, when using the ASD *basic load combinations,* use of the factor of 0.75 is permitted with two or more variable (transient) loads. In addition the duration of load factor C_D is specifically permitted, in accordance with the wood chapter (IBC Chap. 23). When using the ASD *alternate load combinations* (IBC Sec. 1605.3.2), load increases specified in the materials chapters (IBC Chaps. 18 to 25) are permitted, including the wood duration of load factor C_D. There is no longer any mention of a one-third increase in the load combination sections.

The *Uniform Building Code* (UBC) for several editions modified the NDS duration of load factors for seismic forces and some wind force conditions. While in the 1990s the NDS had increased C_D to 1.6, the UBC maintained the historically used duration of load increase of 1.33. The IBC does not modify the NDS C_D factors. For designers that followed UBC requirements in the past, use of IBC provisions will result in increased allowable values in many cases. Allowable values for shearwalls and diaphragms remain unchanged because of other compensating changes.

The examples in this book will follow the ASD basic load combinations, applying the load combination factor of 0.75 where appropriate and using applicable C_D factors.

These subjects have been discussed previously in this book, and the reader may find it helpful to review this material (Secs. 2.8, 2.16, and 4.15) before continuing with this chapter. The reason for developing a thorough understanding of these problems is that the Code assigns additional design requirements to buildings classified as irregular structures.

16.3 Seismic Forces—Irregular Structures

An *irregular structure* has one or more significant discontinuities in its configuration or in its lateral-force-resisting system. Several examples of irregular structures are given in this section to illustrate the general concept of structural irregularities.

The 10 general types of irregularity identified in the Code are classified as either *vertical irregularities* (ASCE 7 Table 9.5.2.3.3) or *plan irregularities* (ASCE 7 Table 9.5.2.3.2). Detailed descriptions of the irregular features are given in these tables. As mentioned previously, the commentary in the SEAOC *Blue Book* expands these descriptions and helps to clarify the intent of the Code.

The discussion in this section wil first introduce the irregularities and then discuss the design penalties applied to structures with these irregularities. The first form of irregularity considered in this introduction has an effect on the dynamic properties of a building that is easy to visualize. It is a type of vertical irregularity known as *weight (mass) irregularity* (ASCE 7 Table 9.5.2.3.3, Type 2). This is best described by considering the lumped-mass model of the three-story building in Fig. 2.14b (Sec. 2.12). In this figure the F_x story forces are shown to follow a triangular distribution with zero force at the bottom and maximum force at the roof level. This triangular distribution is obtained if the masses at the story levels are all equal. Obviously some other distribution will result (using the Code formula for F_x) if the masses are not all the same magnitude.

Now, if one of the masses is significantly larger or smaller than the mass at an adjacent level, the dynamic behavior of the system will be altered. Thus, *mass irregularity* is said to exist if the mass of any story is more than 150 percent of the mass of an adjacent story. However, a roof which is lighter than the floor below need not be considered as being mass-irregular. The fundamental mode (first mode) of vibration may not be the critical mode for a structure that has mass irregularity.

Recall that the Code recognizes two basic types of seismic analysis, *static procedures* and *dynamic procedures*. Static procedures (e.g., *equivalent lateral force analysis*) are generally applied to smaller buildings, while dynamic procedures are applied to larger building. The Code now requires that dynamic procedures be used for more structures than in the past. In Seismic Design Categories D, E, and F, ASCE 7 Table 9.5.2.5.1 requires a dynamic analysis procedure for structures with plan irregularity Type 1 and vertical irregularity Types 1, 2, and 3. Buildings of light-frame construction not exceeding three stories in height are exempted as long as they are used for standard occupancies (Seismic Use Group 1). Some irregular wood light-frame buildings greater than three stories, and those used for essential or hazardous facilities, will now require a dynamic analysis procedure. This will be problematic for some time, because dynamic analysis tools for wood-frame structures are just now being developed at a research level and still require much further development and calibration before they are appropriate for design application. Because of the exemption, however, the majority of wood light-frame buildings can still use equivalent lateral force analysis.

The example of mass irregularity is given here as a simple illustration of a vertical irregularity. One can easily visualize how the mode shape in Fig. 2.14*b* could be significantly altered by having one of the lumped masses in the system much larger or smaller than the others. The designer of any structure should generally be aware of mass irregularity and should pay particular attention to tying the structure together for lateral force transfer and continuity.

The question of diaphragm flexibility was also discussed in Chap. 9, and it was noted that wood diaphragms have usually been classified as *flexible* diaphragms. If a building uses a flexible diaphragm in its lateral-force-resisting system, the Code does not require that *torsional irregularity* (Plan Irregularity Type 1) be considered. Chapter 9 suggests, however, that a rigid diaphragm analysis should be considered for buildings with a notable separation between the center of mass and center of rigidity. This separation of the center of mass and center of rigidity is the same characteristic that causes a building to become torsionally irregular, resulting in one side of the building drifting significantly more than the average story drift. It should be noted that very few wood buildings will have a torsional irregularity significant enough to qualify as torsionally irregular under the Code criteria discussed in Sec. 16.9. Recent research (Refs. 16.7 and 16.10) has suggested that current Code design procedures are not adequately identifying the most torsionally irregular wood-frame structures. Until design procedures can be revised, the designer is encouraged to consider irregularity requirements in any building that can be visually identified as having significant torsional eccentricity and low torsional redundancy.

Reentrant corners (ASCE 7 Table 9.5.2.3.2, Type 2) occur frequently in wood-frame structures. Typical L-, T-, and U-plan shaped buildings are common examples of structures with reentrant corners. When *both projections* beyond a reentrant corner are greater than 15 percent of the plan dimension of the structure in a given direction, the building is classified as plan-irregular. See Fig. 16.2. A method of calculating the design forces for collectors and related connections for a typical L-shaped building is demonstrated in Sec. 16.8.

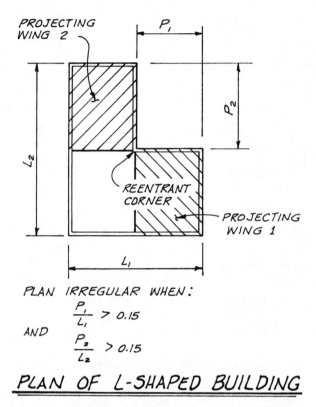

PLAN IRREGULAR WHEN:

$$\frac{P_1}{L_1} > 0.15$$

AND

$$\frac{P_2}{L_2} > 0.15$$

PLAN OF L-SHAPED BUILDING

Figure 16.2 A building with a *reentrant corner* is classified as plan irregularity Type 2 when both projections exceed 15 percent of the corresponding side lengths.

Another type of plan irregularity is described as a *diaphragm discontinuity* (ASCE 7 Table 9.5.2.3.2, Type 3). An example of a diaphragm discontinuity is a large opening in a floor or roof diaphragm. See Fig. 16.3.

Diaphragms with openings have additional diaphragm chords and collectors (drag struts). These are framing members at the boundary of the diaphragm opening. If a free-body diagram is cut through the diaphragm at the opening, the chords at opening plus the chords at the boundary of the diaphragm carry the moment in the diaphragm. A method of analyzing the diaphragm chord and drag strut forces in a horizontal diaphragm with an opening is given in Ref. 16.14 and additional information is available in Refs. 16.5 and 16.6. The evaluation of forces described in these references applies whether or not the opening in the diaphragm is large enough to be classified as a Type 3 plan irregularity.

Tests of diaphragms with openings are also reported in Ref. 16.4. These indicate that large forces can develop at the corners of the openings in the diaphragm and that the magnitude of the forces increases with increased opening size. Design of reinforcing around the opening is required regardless of whether or not design penalties are involved. Limiting the size of the open-

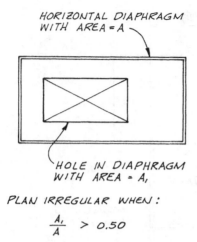

HORIZONTAL DIAPHRAGM
WITH AREA = A

HOLE IN DIAPHRAGM
WITH AREA = A_1

PLAN IRREGULAR WHEN:

$$\frac{A_1}{A} > 0.50$$

PLAN - DIAPHRAGM WITH OPENING

Figure 16.3 A horizontal diaphragm with a large opening is a *diaphragm discontinuity* and is known as plan irregularity Type 3. The discontinuity is classified as plan-irregular if the size of the opening A_1 is greater than 50 percent of the gross area A.

ing will reduce the required reinforcing and improve the performance of the structure.

On the other hand, if an opening is small, Ref. 16.4 offers the following general guide to designers: ". . . When openings are relatively small, chord forces do not increase significantly and it is usually sufficient to simply reinforce perimeter framing and assure that it is continuous. Continuous framing should extend from each corner of the opening both directions into the diaphragm a distance equal to the largest dimension of the opening."

The fourth type of plan irregularity is known as an *out-of-plane offset* (ASCE 7 Table 9.5.2.3.2, Type 4). This is a particularly serious type of discontinuity because it interrupts the flow in the normal path of lateral forces down through the structure and into the foundation. Because of this, the use of the *special seismic load combinations* and the *special seismic (overstrength) force* E_m is required for supporting members.

To describe this type of plan irregularity, consider a two-story building with a horizontal diaphragm at the second floor and roof levels. See Example 16.2. The supporting walls include an interior shearwall as well as the usual exterior shearwalls. This example focuses on the interior shearwall, and the exterior walls are not shown in detail.

Initially the building is considered *without* a plan irregularity (i.e., an out-of-plane offset does not exist) so that the *normal flow* of forces down through the structure can first be understood. See Fig. 16.4a. Note that the interior shearwall on the second story is vertically in line with the shearwall on the first-story level.

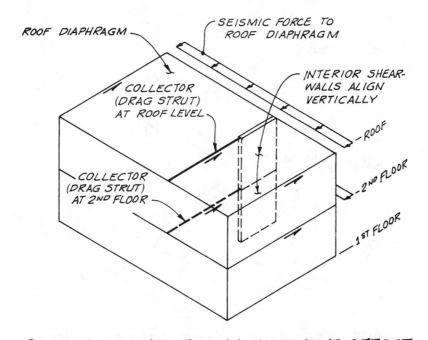

ROOF DIAPHRAGM

SEISMIC FORCE TO
ROOF DIAPHRAGM

INTERIOR SHEAR-
WALLS ALIGN
VERTICALLY

COLLECTOR
(DRAG STRUT)
AT ROOF LEVEL

ROOF

COLLECTOR
(DRAG STRUT)
AT 2ND FLOOR

2ND FLOOR

1ST FLOOR

2-STORY SHEARWALL WITHOUT OFFSET

Figure 16.4a Structurally *regular* two-story building. The alignment of the shear-walls provides a *direct path* for the continuous flow of the lateral force from the roof level down to the foundation. Drag struts function as described in Sec. 9.10.

Because one shearwall is immediately above the other, the lateral force from the upper wall can be transferred *directly* to the lower wall. Note that the shearwall on the first story carries a lateral force from the second-floor dia-phragm in addition to the force from the second-story shearwall. The collectors (drag struts) transfer the shears in the horizontal diaphragms to the interior shearwall. See Sec. 9.10 for a review of collectors.

In Fig. 16.4b the interior shearwalls have been revised so that there is a horizontal offset between the shearwall on the first- and second-story levels. Two problems are created by the *out-of-plane offset:*

1. The force from the second-story wall must travel horizontally through the diaphragm at the second-floor level before it can be transferred into the shearwalls on the first story.

2. The resistance to the overturning moment at the base of the second-story interior shearwall requires special consideration.

These are described in greater detail in Fig. 16.4c and d.

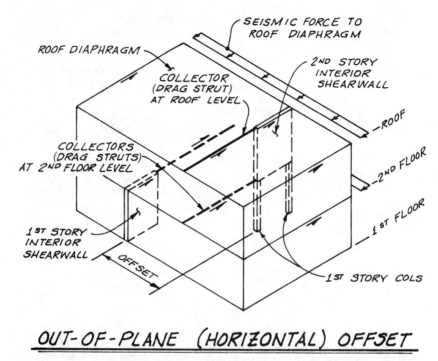

OUT-OF-PLANE (HORIZONTAL) OFFSET

Figure 16.4b Interior shearwall with *out-of-plane offset*. This is classified in ASCE 7 Table 9.5.2.3.2 as plan irregularity Type 4.

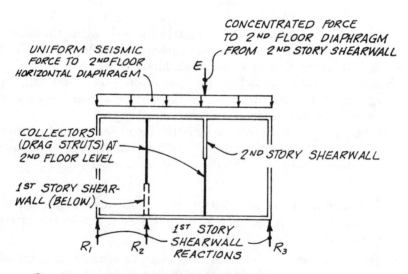

PLAN - 2ND FLOOR DIAPHRAGM

Figure 16.4c *Shear* transfer as a result of out-of-plane offset. The shear from the second-story shearwall is transferred as a concentrated force E to the second-floor diaphragm.

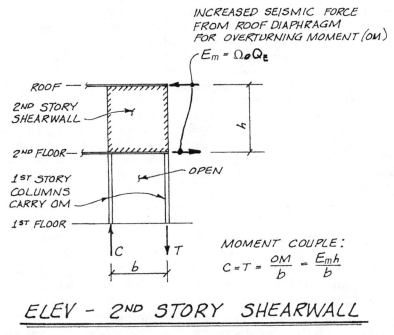

ELEV – 2ND STORY SHEARWALL

Figure 16.4d *Overturning moment* at base of second-story shearwall. Note that the OM is at the *base* of the shearwall and is computed using height h. The overturning at the first-floor level is taken into account by an overturning analysis of the first-story shearwalls. The special seismic forces and load combinations apply only to the structure supporting the shearwall (i.e., first-story columns) and not to the shearwall itself.

EXAMPLE 16.2   Out-of-Plane Offset Added to a Regular Structure

The interior shearwalls in the two-story building in Fig. 16.4a do not involve an out-of-plane offset. The shearwalls on the first and second floors are aligned vertically, one over the other. The revised layout of this building in Fig. 16.4b has the interior shearwall at the first-story level relocated to the left of its original position. A much more critical situation has been created because the shear from the second-story shearwall must now be transferred *horizontally* through the second-floor diaphragm before it can be picked up by the shearwalls on the first floor. The second problem involves the overturning moment at the base of the second-story shearwall. Neither of these problems occurs in the original configuration (Fig. 16.4a) which was structurally regular.

The horizontal seismic force from the roof is carried by the second-story shearwall. At the second-floor diaphragm this force must be distributed horizontally to the shearwalls at the first-story level. See Fig. 16.4c. Although it is possible to develop a continuous path for the transfer of lateral forces down through the structure, the forces do not follow a direct route. The discontinuity introduced by the out-of-plane offset causes the shear to travel horizontally in its path to the foundation. This places a concentrated force on the second-floor diaphragm that is in addition to the usual distributed force.

The concentrated force in Fig. 16.4c accounts for the transfer of the *shear* at the base of the second-story shearwall. The force to the second-floor horizontal diaphragm includes the usual uniformly distributed force plus a concentrated earthquake force E from the second-story shearwall. The second-floor horizontal diaphragm is supported by the first-story shearwall reactions R_1, R_2, and R_3.

The *moment* at the base of the second-story shearwall must also be transferred. This overturning moment is resolved into a couple, and the moment is replaced by a pair of equal and opposite forces (one tension and one compression) separated by a lever arm at the ends of the second-story interior shearwall.

One approach to handling this moment is to provide columns in the story directly below these forces for the purpose of carrying the overturning moment to the foundation. See Fig. 16.4d. These columns may carry gravity loads in addition to the forces due to overturning. Note that these are isolated *columns* in the first story, and there is *no shearwall* between them. Other solutions to the overturning moment problem are possible.

The overturning moment below the second-story interior shearwall may be carried by columns in the first story. The forces T and C are determined by dividing the overturning moment OM by the lever arm b. An out-of-plane offset requires that the columns carrying the overturning moment be designed for the special seismic load combinations using E_m. The increased seismic force to be considered for overturning is significantly larger than the usual earthquake force E. Detailed discussion of design penalties for irregular structures will follow.

In summary, the discontinuity resulting from an out-of-plane offset causes the lateral *shear force* and the *overturning moment* from the second-story shearwall to follow two different paths. The shear is transferred horizontally through the second-floor diaphragm to the first-story shearwalls, and the overturning moment is transferred vertically to the columns directly below the second-story shearwall. As another example, if a second story shearwall were to be supported on a beam, the beam would require design for load combinations using the magnified E_m force. The design penalty assigned by the Code indicates the level of concern about Type 4 discontinuities.

The final type of irregularity to be covered in this section is a form of *vertical irregularity* (Type 4 in ASCE 7 Table 9.5.2.3.3). Although this type of irregularity is classified as a vertical irregularity, it is very similar in concept to the out-of-plane offset (plan irregularity) that was just reviewed. An in-plane-offset of a lateral-force-resisting element also interrupts the normal lateral-force path of the system.

To illustrate an in-plane discontinuity, again consider the regular structure in Fig. 16.4a. The building has shearwalls on the second story that are located directly over the shearwalls on the first story. In Fig. 16.5 this building is reconfigured to have a Type 4 vertical irregularity. The interior shearwalls on the first and second stories are offset by a distance greater than the length of the shearwall elements.

Essentially the same overturning problem occurs with the in-plane offset in Fig. 16.5 and the out-of-plane offset in Fig. 16.4d. For this reason the Code

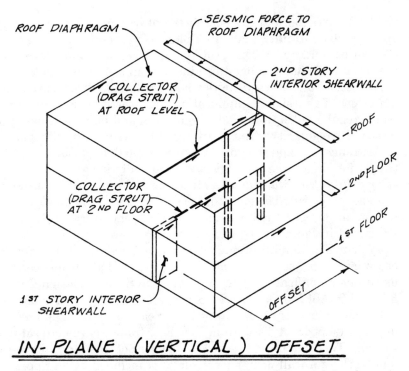

ROOF DIAPHRAGM

SEISMIC FORCE TO
ROOF DIAPHRAGM

COLLECTOR
(DRAG STRUT)
AT ROOF LEVEL

2ND STORY
INTERIOR SHEARWALL

ROOF

2ND FLOOR

COLLECTOR
(DRAG STRUT)
AT 2ND FLOOR

1ST FLOOR

1ST STORY INTERIOR
SHEARWALL

OFFSET

IN- PLANE (VERTICAL) OFFSET

Figure 16.5 An *in-plane-discontinuity in vertical lateral-force-resisting element* is classified as vertical irregularity Type 4 in ASCE 7 Table 9.5.2.3.3.

requires that columns or other members supporting the second-story shear-wall also be designed for the overturning effects caused by the special seismic (overstrength) force. (More detailed discussion of design penalties for irregular structures follows.)

The discussion of design penalties for irregular structures has been separated from the introduction to the irregularities because multiple design penalties will often apply for a given irregularity. The following discussion incorporates several errata to the first printing of ASCE 7-02. The reader is encouraged to consult ASCE for complete errata information.

The five design penalties triggered by plan structural irregularities will be presented first. The first penalty for irregularity takes the form of an increased force used for shear transfer from diaphragm to vertical element, from diaphragm to collector (drag strut), and from collector to vertical element. In addition, unless the collector is already being designed for the special seismic load combinations using E_m forces (with Ω_o as a multiplier), the collector member force must also be multiplied by 1.25. This penalty is found in ASCE 7 Sec. 9.5.2.6.4.2 and requires that the force multiplier be applied to forces determined from Sec. 9.5.2.5, meaning that the multiplier applies to design forces from any of the possible analysis procedures. The multiplier is applicable to plan structural irregularities 1a and 1b through 4 in ASCE 7 Table

9.5.2.3.2 (torsional and extreme torsional irregularities, reentrant corner, diaphragm discontinuity, and out-of-plane offset) and applies in Seismic Design Categories D, E, and F. This provision replaces a similar provision in the UBC that prohibited use of a one-third increase in collectors and their connections.

The second and third penalties address torsional response. The second requires use of a torsional amplification factor for structures determined to have plan irregularity 1a (torsional irregularity) and plan irregularity 1b (extreme torsional irregularity). The third prohibits structures with plan irregularity 1b (extreme torsional irregularity) in Seismic Design Categories E and F. The trigger for identifying a building as torsionally irregular involves calculating story drift at the two ends of a structure and comparison of the maximum to average story drift. The structure is said to have a torsional irregularity if the maximum is more than 1.2 times the average story drift, and an extreme structural irregularity if the maximum is more than 1.4 times the average story drift. The amplification factor can be found in ASCE 7 Sec. 9.5.5.5.2. An exception to this section exempts structures of light-frame construction from use of the amplification. The prohibition in Seismic Design Categories E and F, found in ASCE 7 Sec. 9.5.2.6.5.1, applies regardless of construction type.

The fourth penalty, found in ASCE 7 Sec. 9.5.2.6.2.11, addresses plan irregularity 4 (support of discontinued walls or frames of the lateral-force-resisting system). This is applicable when an element of the lateral-force-resisting system (such as a wall or moment frame) is included at an upper floor but not continued to the foundation, and applies in Seismic Design Categories B and higher. It is required that columns, beams, trusses, or slabs supporting the discontinued elements be designed using the axial (vertical) loads that can develop in accordance with the special seismic load combinations using E_m. This is required for all materials, including wood light-frame structures. The overstrength force is intended to apply to elements below the discontinued element but not to the discontinued element itself. The intent is to protect the system that is providing vertical support against failure, in the event that the discontinued element experiences E_m level forces. Note that it is counterproductive to use the E_m forces for design of the shearwall tie-downs because this could potentially further strengthen the shearwall system, putting additional force into the beam or other support that is being protected. This penalty is not intended to address the design of the diaphragm supporting the discontinued element. This penalty is found in ASCE 7 Sec. 9.5.2.6.2.11, but it is further modified in IBC Sec. 1620.1 by the addition of two exceptions. The first exception provides that the force E_m need not exceed the force that can be developed by the system. The second exception exempts concrete slabs supporting light-frame walls. The second exemption came from the observed good performance of these systems to date, combined with the realization that design using the E_m forces would likely make these systems prohibitively expensive to construct.

The fifth and final penalty applicable to plan irregularities is triggered for nonparallel systems. A system is nonparallel when the shearwall or frames are not at right angles to each other. This is of concern because forces acting along one primary axis will induce forces in all of the resisting elements, increasing interaction of forces. For this type of system, ASCE 7 Sec. 9.5.2.6.3.1 requires that collector elements and their connections to resisting elements be designed for E_m forces. This applies in Seismic Design Categories C and higher; however, an exception is provided for structures braced entirely of light-frame shearwalls. Because many new light-frame shearwall buildings have other elements included, such as steel moment frames, this penalty will apply in a lot of instances.

The five design penalties triggered by vertical structural irregularities will now be presented. The first penalty, applicable for vertical irregularities 1a and 1b through 3 [soft story and extreme soft story, weight (mass) irregularity, and vertical geometric irregularity], places limitations on acceptable analyses for Seismic Design Categories D and higher. This penalty was discussed earlier in this section, where it was noted that there is an exemption for light-frame structures of three stories or less. For light-frame structures of four stories or more, this penalty requires use of analysis that is currently at or beyond the state of the art.

The second penalty is the prohibition of vertical irregularities 1b (extreme soft story) and 5 (weak story) in Seismic Design Categories E and F.

The third and fourth penalties apply to discontinuous systems and were covered in the plan irregularity discussion.

The fifth penalty, applicable to vertical irregularity type 5 (weak story), requires that detailing be provided at openings in shearwalls and diaphragms to distribute the stress concentrations that could occur at the corners. This is applicable in Seismic Design Category B and higher.

When design using the special seismic load combinations is required, the increase in seismic force is Ω_o *times* the usual seismic force, where Ω_o is determined from ASCE 7 Table 9.5.2.2. The large order of magnitude of this force indicates the severe nature of the triggering of discontinuity in the lateral-force path. The Code specifies the increased seismic force for overturning as

$$E_m = \Omega_o Q_E \pm 0.2 S_{DS} D$$

In the case of columns supporting discontinued shear walls, the added force requirement is to guard against the collapse of the lateral-force-resisting system that would be triggered by a failure of the columns carrying the overturning moment. This force level was selected because the columns at such a discontinuity must be capable of supporting the lateral-force-resisting system (LFRS) as it is forced well into the inelastic range. Thus, the performance of these columns is essentially a *strength* design requirement.

Obviously these force levels represent an extreme loading condition. Thus, when a working stress design method (ASD) is used, the allowable stresses for this loading condition may be factored up to an ultimate strength level. When following the provisions of IBC Sec. 1617.1.1.2, this is accomplished by multiplying the allowable stresses for use in the design for overturning by 1.7. Thus, the *net increase* in seismic force for overturning is the load multiplier divided by the allowable-stress increase:

$$\frac{\Omega_o}{1.7}$$

The structural system quality factor R (Sec. 2.13) reflects the ability of the lateral-force-resisting system to sustain cyclic deformations in the inelastic range without collapse. Recall that for the types of buildings covered in this book the values of R range from 5.0 to 6.5. The corresponding value of Ω_o for wood structural panel shearwalls is 3.0. Therefore, the net increase for this common situation is

$$\frac{3.0}{1.7} = 1.75$$

The Code specifically permits use of the load duration factor for wood design C_D in combination with the 1.7 conversion factor.

The value of 1.75 is given here to demonstrate the relative order of magnitude of the increase for overturning brought about by the out-of-plane offset. However, for consistency it is recommended that, in practice, the individual *force* multipliers be used in design calculations. This discussion also applies when special seismic forces are used for design of collectors, shear transfer, etc.

The designer should recall that Sec. 9.7 discussed design of collectors and their connections to shearwalls. Use of the special seismic load combinations and magnified seismic forces E_m is required irrespective of the existence of any structural irregularity. Buildings braced entirely of light-frame wood walls are exempted from the magnified force level; buildings with concrete or masonry walls or mixed systems are not exempt. Where applicable, the design requirements resulting from the E_m force are likely to be more critical than those resulting from the typical design force.

A full understanding of the design requirements for irregular structures requires the reader to first understand the simple regular buildings that are covered in Chaps. 1 to 15. The intent of Sec. 16.3 is to summarize some of the common discontinuities that may be encountered in a typical wood-frame building. The next section continues the discussion of overturning under lateral forces and provides a comparison of the practices for wind and seismic forces.

16.4   Overturning—Background

The general combinations of loads and forces that are required by the IBC were summarized briefly in Sec. 2.16. These combinations essentially define which loads and forces must be applied simultaneously in the design of a building. To a certain extent, a combination reflects the probability that the various loads and forces will occur concurrently. Some of these probabilities were discussed previously and are not repeated here.

As the Code requirements for wind and seismic forces have become increasingly complex, the considerations for overturning (moment) stability have also become more involved. The overturning analysis for a shearwall under wind forces was introduced in Example 2.9 (Sec. 2.10), and additional overturning moment problems were covered in Chap. 10. The reader may find it useful to review these sections in preparation for the detailed summary given here.

A comprehensive study of overturning is introduced by first reviewing a simple statics problem. The principles of statics are then expanded to cover moment stability at the base of a wood-frame shearwall.

It is important that the designer understand how to evaluate overturning effects. Initially this problem is treated purely from a statics approach, without concern for the Code-required factor of safety (or load and force combinations). This essentially defines the moment stability problem, and it introduces the analysis necessary to evaluate soil bearing pressures under combined gravity loads and lateral forces. While shearwalls are often modeled as standalone segments for the purpose of analysis, foundations supporting shearwalls are seldom designed this way. This is because the foundation generally provides considerable strength and stiffness, and it serves to interconnect the many segments of shearwall and other structural and nonstructural portions of the system. As a result, the foundation can be viewed as a continuous beam with many applied loads and forces. For this reason, foundation design for overturning is beyond the scope of this text. The principles of overturning design described in this section, however, can be used in design of foundations.

After the statics of this problem is understood, the Code-required factor safety (or required load and force combinations) is finally introduced. The need for the detailed review given here is the result of the seismic code requirements.

16.5   Overturning—Review

Most designers are first exposed to the subject of *moment stability* in a course in statics or engineering mechanics. See Example 16.3. In this type of problem, the dimensions and weight of an object are given. In addition, the coefficient of static friction μ of the surface upon which the block rests is known.

A typical requirement in such a problem is to determine whether the block *moves* or *remains in static equilibrium* when subjected to a lateral force. If the block moves, the problem is to indicate whether it *slides* along the supporting surface or whether it *tips* (overturns). If the block remains at rest, the *factors of safety* against *sliding* and *overturning* are to be calculated.

EXAMPLE 16.3   Typical Moment Stability Problem from a Course in Statics

GIVEN

The dimensions and weight of a block are known. See Fig. 16.6. The coefficient of static friction μ of the surface upon which the block rests is given.

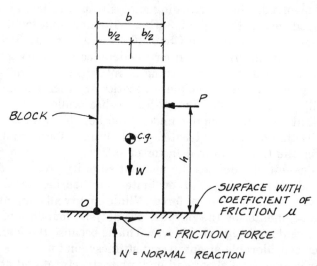

BLOCK

Figure 16.6 Typical statics problem involving moment stability of a block.

FIND

If a lateral force P is applied at height h above the base, determine whether or not the block remains in static equilibrium. If the block moves, indicate whether it slides or tips. If the block remains static, determine the factors of safety against sliding and overturning.

SOLUTION

Sliding

By summing forces in the y direction, the normal reaction at the base equals the weight of the block:

$$\Sigma F_y = 0$$

$$N = W$$

By Σ forces in the x direction, the friction force at the base equals the applied lateral force P:

$$\Sigma F_x = 0$$

$$F = P$$

The maximum available friction force for resisting the lateral force is the normal reaction times the coefficient of static friction:

$$\text{Max. } F = \mu N$$

If the lateral force exceeds the maximum available resisting force

$$P > \text{max. } F$$

sliding will occur.
   On the other hand, if

$$P < \text{max. } F$$

the object does not slide, and the *factor of safety against sliding* can be computed as

$$\text{FS}_{\text{sliding}} = \frac{\text{max. } F}{P} = \frac{\mu N}{P} = \frac{\mu W}{P}$$

Moment Stability

If the block tips, the normal force N shifts to the pivot point at O. By summing moments about point O, N drops out, and the maximum overturning moment OM is the lateral force times the height above the base:

$$\Sigma M_O$$

$$\text{OM} = Ph$$

Also by summing moments about point O, the resisting moment RM is the weight of the block times the lever arm to the line of action of this force:

$$\text{RM} = W \left(\frac{b}{2} \right)$$

If the maximum overturning moment exceeds the available resisting moment

$$\text{OM} > \text{RM}$$

tipping will occur.
   On the other hand, if

$$\text{OM} < \text{RM}$$

the object remains in static equilibrium, and the *factor of safety against overturning* can be computed as

$$\text{FS}_{\text{overturning}} = \frac{\text{RM}}{\text{OM}}$$

Although the example of a block sliding or tipping is an elementary problem, it is helpful to begin a review of overturning for a wood-frame shearwall with these basic principles in mind.

Most wood-frame buildings use a combination of horizontal diaphragms and shearwalls to resist lateral forces. In checking overturning for these types of structures, the usual practice is to examine the moment stability of the vertical elements (i.e., the shearwalls) in the primary LFRS of the structure. If the individual shearwalls are stable, the overall moment stability of the structure is ensured.

It has been noted throughout this book that the key to successful lateral force design is to adequately tie the elements together so that a continuous path is provided for the transfer of forces from the roof level down into the foundation. The simple statics problem is expanded to a shearwall overturning analysis by considering the forces acting on a shearwall from a one-story building. See Example 16.4.

In this example the stability of a free-body diagram (FBD) taken at the base of a shearwall is analyzed. Similar problems were considered in Chap. 10. The approach taken here is purely from a statics point of view, and consideration is not given to whether the lateral force is wind or seismic, and the FS required by the Code is not defined. The expressions necessary to determine the forces in the *shearwall chords* and *anchorage connections* for overturning are again reviewed.

EXAMPLE 16.4   Overturning at Base of Shearwall

If the lateral forces to a shearwall are relatively small, the dead load of the wall plus any roof or floor dead load supported by the wall may be sufficient to resist overturning (see Fig. 16.7). On the other hand, if the lateral forces are large or if the dead loads above the foundation are small, it may be necessary to attach the shearwall to the foundation with anchorage hardware so that the weight of the foundation (and other structure sitting on the foundation) can be included in the moment stability analysis.

The gross overturning moment at the *base of the shearwall* is the lateral force (the reaction from the roof diaphragm) times the height from the top of the foundation to the roof level:

$$\text{Gross OM} = Rh$$

The resisting moment is the sum of any dead loads supported by the wall (including the weight of the wall and any other dead loads framing into the wall) times the lever arm from the pivot point O to the line of action of the dead loads. For a uniformly distributed dead load, the lever arm is simply one-half of the shearwall length:

$$\text{RM} = W_D \left(\frac{b}{2}\right)$$

In order for the shearwall to be stable, the maximum available resisting moment must be greater than or equal to the gross overturning moment:

$$\text{RM} \geq \text{OM}$$

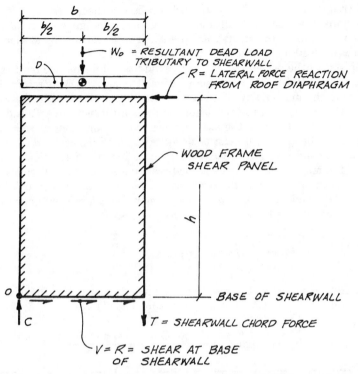

Figure 16.7 Moment stability of a one-story shearwall. FBD cut at base of wall (top of foundation).

Alternatively, this could be expressed by requiring the factor of safety against overturning to be 1.0 or larger:

$$\text{FS} = \frac{\text{RM}}{\text{OM}} \geq +1.0$$

For specific lateral force combinations the IBC requires the RM to be based on 60 percent of the dead load. Thus the Code effectively requires the FS to be $\frac{5}{3} = 1.67$.

Because of the relatively small dead loads involved in wood-frame buildings, it is not uncommon for the OM at the base of a shearwall to exceed the RM. The typical solution in such a case is to attach the shearwall to the supporting foundation with hardware of adequate strength so that the *design overturning moment* is transferred to the foundation. According to statics, the design OM is the difference between the gross OM and the dead load RM based on 60 percent of the dead load:

$$\text{Design OM} = \text{gross OM} - 0.6\,\text{RM}$$

$$= Rh - 0.6\,\text{RM}$$

Obviously if the sign of the design OM is negative, the shearwall is stable without special hold-down attachments to the foundation. In this case the only force to be transferred to the foundation with hardware is the shear force $V = R$. Generally the friction force between the wall and the foundation is ignored, and *anchor bolts* are designed to transfer the shear from the shearwall into the foundation (see Example 10.8 in Sec. 10.10 for additional information).

If the sign of the design OM is positive, additional anchorage hardware is required to transfer the design OM to the foundation. The design OM is resolved into a couple by dividing the moment by the width of the shearwall (distance between shearwall chords). One force from the resulting couple is a tension force T, and the other is a compression force C. Since the force can originate from either direction, anchorage at either end of the shearwall must be designed for both tension and compression forces. See Sec. 10.6 and Fig. 10.11 (Sec. 10.11) for additional information.

When the magnitude of the dead-load-resisting moment is relatively small, the designer may choose to ignore the resisting dead load. In this case the design OM equals the gross OM, and the tension and compression chord forces obtained by resolving the OM into a couple are known as *gross chord forces:*

$$\text{Gross } T = \text{gross } C = \frac{\text{gross OM}}{b}$$

On the other hand, when the resisting dead load is considered, the tension chord force is based on the design OM. The *design tension chord force* is

$$\text{Design } T = \frac{\text{design OM}}{b} = \frac{\text{gross OM} - \text{RM}}{b}$$

It is important to realize that only 60 percent of the dead load of the structure is used to resist overturning. This again reflects the conservative approach used in structural design. There may or may not be gravity *live* loads acting on the shearwall system when lateral forces are applied. There is a greater chance of having floor live loads present during a major windstorm or earthquake than roof live loads. However, the presence of neither type of live load can be ensured, and consequently only gravity dead loads are considered in checking uplift due to overturning.

The specific provisions for overturning under wind and seismic forces will be given in the next sections. To do this, three general classes of practical moment stability problems are identified, and a summary of the important design criteria for each type is given. The intent is to provide the information in a form in which the similarities and differences are clear for each situation. The three classes of overturning problems include one wind and two earthquake problems:

1. Wind overturning—general wind force requirements (Example 16.5)
2. Seismic overturning
    a. Regular structures (Example 16.6)
    b. Irregular structures (Example 16.7)

Each of these cases produces an overturning situation which, according to the Code, is unique in some way. For each overturning problem there are certain basic requirements, including the following:

1. Check the moment stability using the appropriate combination of loads and forces at the base of the shearwall (excluding the weight of the foundation)
2. Determine shearwall chord forces.
   a. Net tension chord force (known also as the *uplift* or *anchorage* force)
   b. Maximum compression chord force (including tributary gravity load)

16.6   Overturning—Wind

The general wind force requirements from the IBC were introduced in Secs. 2.9 to 2.11. The specific requirements for checking overturning under wind forces are summarized in this portion of Chap. 16.

It was previously noted that only 60 percent of the dead load D of the structure is used to resist overturning. When the dead load of a structure is estimated, designers often take an approach that is conservative for vertical load analysis. This may be done by rounding the summation of gravity loads upward and by slightly overestimating the weight of the members so that it will not be necessary to revise design calculations once member sizes have been determined. The practice of overestimating loads is generally considered to be on the side of greater safety.

However, in the moment stability analysis of a structure, an overestimation of dead loads is *unconservative* because the larger loads will produce a calculated resisting moment that is larger than is actually available. For this and other reasons, a factor of safety of 1.67 is provided by the IBC for checking overturning at the base of a structure. This can be stated mathematically in several different forms. The factor of safety for moment stability can be written as

$$FS = \frac{RM}{OM} \geq 1.67$$

Alternatively the moment stability requirement could be stated as a load case. For moment stability under wind, this combination would be

$$0.6D \pm W$$

The consideration of this combination for the moment stability of the complete shearwall-foundation system *and* the corresponding shearwall tension chord force leads to an interesting problem. It is one that has been debated by designers for many years.

The *allowable stresses* for use in the design of shearwall chords and their anchorage connections have a factor of safety already incorporated into the design values. Some designers logically reason that it is not necessary, then, to incorporate the 1.67 factor of safety required for overall moment stability into the design forces for the shearwall chords. It is argued that the full dead-load-resisting moment above the foundation should be used to compute the tension chord force (instead of 60 percent of the dead load-resisting moment). Again, the reasoning is that the wood tension member and the tie-down connection hardware already have appropriate factors of safety included their allowable design values.

The anomaly with this logic occurs when the dead load-resisting moment just equals the overturning moment (i.e., when RM = OM). In this situation, the calculated tension chord force is zero. Accordingly, it would, theoretically, be unnecessary to anchor the shear panel to the concrete footing, because the resisting dead load moment exactly counters the overturning effect. However, *without a tie-down connection to the foundation,* it is clearly not possible to develop the required factor of safety at the base of the foundation. In other words, the dead load of the foundation will not be effective unless there is an adequate connection between the shearwall tension chord and the foundation.

Consequently, the FS of 1.67 used for base overturning is also used to determine the tension chord force at the connection between the shearwall and the foundation. This is done by computing the anchorage requirements for the shearwall using 60 percent of the dead load-resisting moment. A similar analysis may be used to calculate the forces for connecting a shear panel in the second story of a structure to a shear panel in the first story. Thus, the combination of 0.6D ± W is used for all tension chord problems involving general wind force design. See Example 16.5.

The anomaly just described applies to the *tension chord* of a shearwall and its connection to the foundation. However, this problem does not occur with the *compression chord*. The combination of (0.6D ± W) applies to base overturning and to the tension chord and *not* to the compression chord. The critical combination of gravity loads and wind forces for the compression chord will be another of the basic load combinations given in Sec. 2.16. Therefore, the design force for the compression chord is the gross chord force caused by the wind overturning moment plus the tributary gravity loads.

It was noted in Chap. 2 that the IBC wind design provisions require wind uplift, if any, to be considered *simultaneously* with horizontal wind forces. Whether or not wind uplift acts on a given shearwall depends on the roof framing system and whether the shearwall is a bearing or nonbearing wall for vertical loads. If the wall supports gravity loads in addition to its own weight, it is classified as a bearing wall. The same framing members that cause additional gravity loads to be supported by the wall may also cause wind uplift to be transferred to the shearwall.

EXAMPLE 16.5   Overturning Requirements for Wind—General

General IBC requirements are that horizontal wind forces and uplift wind are to be considered simultaneously. See Fig. 16.8a. Base overturning requirements are

Implied FS for base overturning: $FS = \dfrac{1}{0.6} = 1.67$

Combination of loads and forces incorporating FS = 1.67:   0.6D + horizontal wind + uplift wind

Load duration factor for wood members and fasteners, C_D: 1.6

To develop an FS of 1.67 for base overturning, the shearwall tension chord force T is determined by using 60 percent of the resisting moment. Dead load only is considered in computing the resisting moment for the tension chord. See Fig. 16.8b. The critical shearwall compression chord force C will be governed by a combination of loads and forces that uses the full dead load D (not 0.6D).

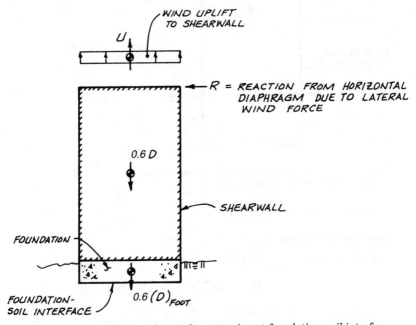

Figure 16.8a   Loads and forces for wind overturning at foundation–soil interface.

The gross overturning moment is

$$\text{Gross OM} = Rh + U\left(\frac{b}{2}\right)$$

The expression developed in the statics analysis of the overturning problem in Example 16.4 are revised as follows to reflect the appropriate factors of safety:

Tension Chord

Load combination: 0.6D + horizontal wind + uplift wind
Design overturning moment (ΣM_O): design OM = gross OM − 0.6RM
Design tension chord force:

$$\text{Design } T = \frac{\text{design OM}}{b}$$

$$= \frac{\text{gross OM} - 0.6\text{RM}}{b}$$

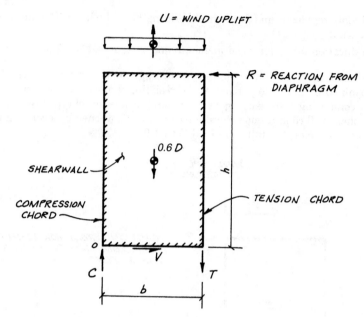

Figure 16.8b Loads and forces for wind overturning at shearwall connection to foundation for *tension chord*.

Load duration factor C_D: 1.6

Compression Chord

Load combinations: D + W

$$D + 0.75[L + (L_r \text{ or } S) + W]$$

Compression chord force:

$$C = \frac{\text{gross OM} + \text{RM}}{b}$$

Load duration factor C_D: 1.6

Note that the free body diagram (FBD) in Fig. 16.8b shows the load and force combination for the tension chord analysis. The DL in the sketch represents all of the dead load to the shearwall above the foundation.

For designing the compression chord, the FBD would have to be revised to reflect the two required load cases. The calculation of the shearwall chord forces may be carried out using the expressions given above. A similar approach with a slightly different format was described in Chap. 10.

Example 16.5 deals with the *general* wind overturning requirements, which involve the addition of overturning due to horizontal wind and overturning due to uplift wind. If the uplift force U is large, the requirement to consider overturning due to the two wind effects simultaneously can be a significant force increase over previous codes.

16.7 Overturning—Seismic

The seismic force requirements from the Code were introduced in Secs. 2.12 to 2.15. The specific requirements for checking overturning under seismic forces are summarized in this section.

The current seismic code contains detailed design provisions for *regular* and *irregular* structures. Because irregular structures are inherently more susceptible to earthquake damage, the Code assigns additional design requirements for these structures. As previously noted, these penalties may take the form of higher design force levels, or other requirements may be imposed. For comparison, the overturning requirements for regular structures will be reviewed first, followed by a summary of the design provisions for irregular structures.

Even in a regular structure the forces generated during an earthquake are erratic, and the magnitude of the force level changes rapidly with time. This is due in part to the changing ground accelerations and in part to the dynamic characteristics of the structure. The transient and reversing nature of seismic forces as compared to wind forces, and observed behavior of buildings in past earthquakes, indicates that the accumulated effects at the base (foundation–soil interface) of a structure are less critical than would be obtained by a direct addition of the seismic overturning forces acting on the structure above. This behavior needs to be balanced against the realization that the seismic forces used for design are not the real forces that will be experienced by the structure. Further study is needed to establish a true understanding of requirements for adequate overturning resistance.

It is interesting to note that very early seismic codes included a coefficient that was used to reduce the seismic overturning moment. Although the previous reduction factor has been deleted for some time, for the equivalent static force procedure, ASCE 7 now reflects this behavior by permitting foundations to be designed for 75 percent of the overturning moment at the foundation-to-soil interface. This can be found in ASCE 7 Sec. 9.5.5.6. The following example will look at the structure-to-foundation interface rather than the foundation-to-soil interface, so the reduction in overturning will not be included.

It is necessary to look at the required load combinations in order to discuss the implied factor of safety against overturning at design level forces. This will be addressed in Example 16.6. Again, because the seismic forces used for design are not those that will be experienced by the structure, the overturning design and the implied factor of safety, together, are merely a design tool and not a predictor of building behavior.

EXAMPLE 16.6   Seismic Overturning Requirements—Regular Structures

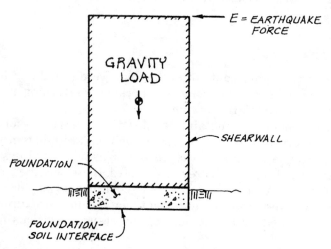

Figure 16.9   Loads and forces for seismic overturning at foundation–soil interface.

Seismic overturning effects are less critical at the foundation–soil interface (Fig. 16.9)
because of the transient and reversing nature of the seismic forces on the structure
above. As an indication of this less critical situation, ASCE 7 reflects this behavior for
the equivalent lateral force design procedure by permitting foundations to be designed
for 75 percent of the overturning moment at the foundation–soil interface.

In order to evaluate overturning for the shearwall in Fig. 16.9, the Code-required
load combinations must be used. Applicable load combinations from IBC Sec. 1605.3.1
are:

$$D + (W \text{ or } 0.7E) + L + (L_r \text{ or } S \text{ or } R)$$

$$0.6D + 0.7E$$

These equations need to be combined with the equations for E from ASCE 7 Sec.
9.5.2.7:

$$E = \rho Q_E + 0.2S_{DS}D$$

$$E = \rho QE - 0.2S_{DS}D$$

Combining these equations and considering D, L, and L_r only provide worse-case equations for upward and downward forces:

$$D + 0.7(\rho Q_E + 0.2S_{DS}D) + L + L_r$$

$$0.6D + 0.7(\rho QE - 0.2S_{DS}D)$$

The reader is reminded that IBC Sec. 1605.3.1.1 permits a factor of 0.75 to be used with multiple transient loads, as was illustrated in Sec. 15.4. If this factor is used, however, alternate combinations without the factor and with only one transient load or force must be explored. In this case, the first equation will be left in its most conservative form, not using the 0.75 factor.

If S_{DS} is set as 1.0 and ρ is assumed to be 1.0, the equations can be simplified to:

$$D + 0.7Q_E + 0.14D + L + L_r = 1.14D + 0.7Q_E + L + L_r$$

$$0.6D + 0.7Q_E - 0.14D = 0.46D + 0.7Q_E$$

The first equation will be critical for downward loading, while the second will be critical for uplift. For comparison to ASD wind loading, a factor of safety against uplift of $0.7/(1.4)(0.46) = 1.14$ could be inferred; however, this is a largely fictitious number.

The gross overturning moment is

$$\text{Gross OM} = 0.7Q_E h$$

The expressions developed in the statics analysis of the overturning problem in Example 16.4 are revised as follows:

Tension Chord

Load and force combination: $0.46D + 0.7Q_E$
Resisting moment:

$$RM = (0.46D)(b/2) = 0.23Db$$

Net overturning moment:

$$\text{Net OM} = \text{gross OM} - RM$$

Net tension chord force:

$$\text{Net } T = \frac{\text{net OM}}{b}$$

$$= \frac{\text{gross OM} - RM}{b}$$

$$= \frac{0.7Q_E h - 0.23Db}{b}$$

Load duration factor C_D: 1.6

Compression Chord

Load and force combinations: $1.14D + 0.7Q_E + L + L_r$
Gross overturning moment:

$$\text{Gross OM} = 0.7Q_E h$$

Resisting moment:

$$RM = (1.14D + L + L_r)(b/2) = 0.57Db + 0.5Lb + 0.5L_rb$$

Net overturning moment:

$$\text{Net OM} = \text{Gross OM} + RM$$

Compression chord force:

$$\text{Net } C = \frac{\text{net OM}}{b}$$

$$= \frac{\text{gross OM} + RM}{b}$$

$$= \frac{0.7Q_E h + 0.57Db + 0.5Lb + 0.5L_rb}{b}$$

Load duration factor $C_D = 1.6$

Note that the free-body diagram in Fig. 16.10 shows the load and force combination for the tension chord analysis. For designing the compression chord, the FBD is shown in Fig. 16.11.

The calculation of the shearwall chord forces may be carried out by using the expressions given above. A similar but slightly different approach was described in Chap. 10.

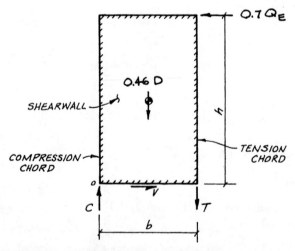

Figure 16.10 Loads and forces for seismic overturning at shearwall tension chord connection to foundation—regular structure.

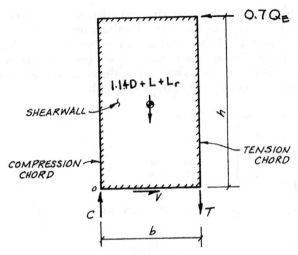

Figure 16.11 Loads and forces for seismic overturning at shearwall compression chord—regular structure.

The seismic overturning requirements for a *regular structure* are summarized in Example 16.6. The reader is cautioned that the sketches in Examples 16.5 and 16.6 show the dead load D of the structure multiplied by a factor. These FBDs are given to illustrate the concept of the load and force combinations covered in the discussion.

However, it is advisable to work, as consistently as possible, in terms of the true dead load of the structure (i.e., the dead load without the multipliers). Then, only when needed to make the final check, should the multiplier be introduced. This practice will avoid mistakenly using a *reduced dead load* in a load combination that requires the *full dead load*. Again, the drawings show the reduced dead load D to illustrate the concept of load and force combinations, rather than to recommend an approach to be followed in practice.

The design penalties that are assigned to *irregular* structures in areas of high seismic risk were outlined in Sec. 16.3. It will be recalled that two particularly critical types of discontinuities are the out-of-plane offset (Fig. 16.4b) and the in-plane-discontinuity (Fig. 16.5). These are *plan irregularity Type 4* and *vertical irregularity Type 4,* respectively. For buildings in Seismic Design Categories B and higher with either Type 4 irregularity, the Code requires that members supporting the shearwall system be designed for greatly increased earthquake force levels. The required load and force combinations for the design of these elements are given in Example 16.7.

EXAMPLE 16.7   Seismic Overturning Requirements—Irregular Structures

The overturning of a second-story shear panel with a Type 4 vertical structural irregularity was introduced in Example 16.2. A special seismic force that approximates the peak capacity of the system (rather than the lesser design forces) is to be considered. Figure 16.12 illustrates the loads and forces resulting from the IBC special seismic load combinations. Figure 16.12 shows the special seismic loads and forces applied to the shearwall in order to illustrate the overturning and resisting moments. The special seismic load combinations, however, apply only to the structure (beams and columns) supporting the shearwall and not to the shearwall itself.

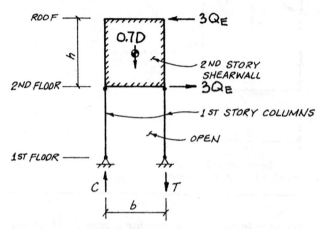

Figure 16.12 Overturning of second-story shearwall with Type 4 vertical or plan irregularity. Loads and forces shown are for seismic overturning for tension chord. The special seismic load combinations, however, apply only to the structure supporting the shearwall (i.e., first-story columns) and not to the shearwall itself. Note that the shear from wall is transferred to second-floor horizontal diaphragm, as shown in Fig. 16.4c.

Load Combinations

In order to evaluate overturning for the shearwall in Fig. 16.12, the Code-required load combinations must be used. For this irregular structure, the IBC special seismic load combinations from Sec. 1605.3.1 will be used:

$$1.2D + f_1L + E_m$$

$$0.9D + E_m$$

The variable f_1 will be taken as 0.5 for this example. The E_m special seismic forces are found in ASCE 7 Sec. 9.5.2.7.1:

$$E_m = \Omega_o Q_E + 0.2S_{DS}D$$

$$E_m = \Omega_o Q_E - 0.2S_{DS}D$$

When combined to reflect the two critical overturning conditions, the following results:

$$1.2D + 0.5L + \Omega_o Q_E + 0.2S_{DS}D$$

$$0.9D + \Omega_o Q_E - 0.2S_{DS}D$$

These are special *strength level* load combinations. When using these combinations with allowable stress design methods, IBC Sec. 1617.1.1.2 permits use of both a multiplier of 1.7 to convert to a strength level, and a duration of load factor C_D. Note that these are to be used to factor up the allowable loads, rather than to decrease the forces, so the load and force diagram will reflect special strength level forces.

For this example, for purposes of illustration, S_{DS} will be set to 1.0 and Ω_o to 3, permitting the equations to be simplified to:

$$1.2D + 0.5L + 3Q_E + 0.2D = 1.4D + 0.5L + 3Q_E$$

$$0.9D + 3Q_E - 0.2D = 0.7D + 3Q_E$$

Tension Column at First Floor

Gross overturning moment:

$$\text{Gross OM} = 3Q_E(h)$$

Resisting moment:

$$RM = 0.7D(b/2) = 0.35Db$$

Net overturning moment:

$$\text{Net OM} = \text{gross OM} - RM$$

Net tension force:

$$\text{Net } T = \frac{\text{net OM}}{b}$$

$$= \frac{\text{gross OM} - RM}{b}$$

$$= \frac{3Q_E h - 0.35Db}{b}$$

This is a strength level force that can be compared to an ASD allowable multiplied by 1.7 to convert to strength level and further multiplied by the duration of load factor C_D for wood members and connections.

NOTE: The free-body diagram in Fig. 16.12 shows the load and force combination for the tension chord analysis. For the compression chord, the FBD would have to be revised to reflect the combination for the compression chord.

For a discussion of the load duration factor C_D the reader is referred to Secs. 2.8 and 16.2. See Sec. 16.3 for background on the 1.7 allowable stress multiplier.

It is hoped that the fairly extensive treatment of the moment stability problem will give the reader a better appreciation of the problem and provide a convenient summary of the design criteria.

16.8 Lateral Analysis of Nonrectangular Buildings

Many of the buildings that have been considered thus far have been rectangular in plan view with exterior and possibly interior shearwalls. These types of buildings have been used to illustrate basic concepts about horizontal diaphragm design.

In practice, however, buildings are often not rectangular. Obviously, an unlimited number of plan configurations are possible. However, a simple L-shaped building can be used to illustrate how the lateral forces on a nonrectangular building can be resisted. See Example 16.8.

For seismic design, buildings of this nature are said to have a plan irregularity known as a *reentrant corner* (ASCE 7 Table 9.5.2.3.2, Type 2). This type of discontinuity was described earlier (Fig. 16.2).

For buildings that are not braced entirely by light-frame wood shearwalls, the design of collectors and their connections to shearwalls will be governed by a requirement for use of the *special seismic load combinations* using the *special seismic (overstrength)* E_m *force level*. This requirement is applied to all collectors of other than all-light-frame buildings in Seismic Design Categories C and higher and is required irrespective of the existence of any structural irregularities. This requirement was discussed in Sec. 9.10.

EXAMPLE 16.8   Lateral Analysis of Nonrectangular Buildings

A building with a nonrectangular plan may use rectangular horizontal diaphragms. Reactions to the shearwalls are calculated on a tributary width basis. The nonuniform seismic force results from the distribution of the dead load (Example 2.15 in Sec. 2.14). A similar analysis is used for lateral forces in the other direction.

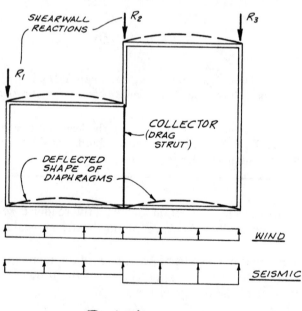

Figure 16.13   Wind and seismic forces to diaphragms in nonrectangular building.

In the approach illustrated, the building is divided into rectangles which are treated as separate horizontal diaphragms. The shearwall that is common to both diaphragms receives a portion of its force from the collector (drag strut). The drag strut collects the roof diaphragm unit shear from the unsupported segments of the two horizontal diaphragms. A numerical example using this same building illustrates the method of calculating the design forces in both directions. See Example 16.9.

The simple span diaphragm approach illustrated in Example 16.9 has been used successfully by a number of designers to analyze buildings of nonrectangular configuration. However, Ref. 16.6 recommends that the continuity of the diaphragm at the interior shearwall be taken into account. In this case the chord in narrower diaphragms will have to be extended (see dashed line along line B in Fig. 16.14b and line 2 in Fig. 16.14c). See Ref. 16.6 for additional information.

EXAMPLE 16.9   L-Shaped Building

Determine the design unit roof shears, chord forces, and maximum collector connection forces for the one-story building in Fig. 16.14a. Consider lateral forces acting in both the N-S and E-W direction, and assume simple span diaphragms.

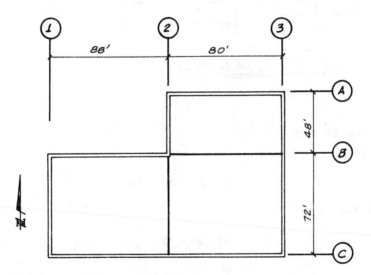

Figure 16.14a   L-shaped building.

Known Information

Roof D = 20 psf
Wall D = 75 psf
Wind ≈ 20 psf
Ω_o = 3.0

Basic Seismic = 0.275W (Sec. 3.4)

Special Seismic = $0.275\Omega_oW = 0.275(3.0)W = 0.825W$ strength level

Trib. wall height to roof level = 12 ft

Lateral Forces in N-S Direction

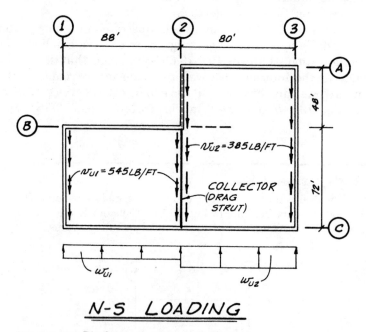

Figure 16.14b Diaphragm forces in N-S direction.

Diaphragm forces (Fig. 16.14b):

Wind

$$w = 20 \times 12 = 240 \text{ lb/ft}$$

Basic Seismic

$$w_{u1} = .275[(20 \times 72) + 2(75 \times 12)]$$

$$= 891 \text{ lb/ft}$$

$$w_{u2} = .275[(20 \times 120) + 2(75 \times 12)]$$

$$= 1155 \text{ lb/ft} \quad \text{seismic governs}$$

Roof shears:

Line 1 and west of line 2:

$$v_{u1} = \frac{V_u}{b} = \frac{w_{u1}L}{2b} = \frac{891(88)}{2(72)} = 545 \text{ lb/ft}$$

Line 3 and east of line 2:

$$v_{u2} = \frac{w_{u2}L}{2b} = \frac{1155(80)}{2(120)} = 385 \text{ lb/ft}$$

Chord forces:

Lines B and C between lines 1 and 2:

$$T_u = C_u = \frac{M_u}{b} = \frac{w_{u1}L^2}{8b} = \frac{.891(88)^2}{8(72)} = 12.00 \text{ k}$$

Lines A and C between lines 2 and 3:

$$T_u = C_u = \frac{w_{u2}L^2}{8b} = \frac{1.155(80)^2}{8(120)} = 7.70 \text{ k}$$

Collector (drag strut) force:

Connection on line 2 at line B: Collector resists unsupported roof shear from both diaphragms. The special seismic load combinations are to be used. From Example 16.7, the load combinations are:

$$1.2D + 0.5L + \Omega_o Q_E + 0.2S_{DS}D$$

$$0.9D + \Omega_o Q_E - 0.2S_{DS}D$$

In some instances collectors will also carry significant vertical loads (such as when a gravity beam is used as a collector), and in other cases the collector only has to support its own weight. For this example it is assumed that the collector supports essentially no vertical load, in which case, both equations are simplified to $\Omega_o Q_E$. For this system Ω_o is taken as 3.0.

$$\text{special seismic } T_u = [(v_{u1} + v_{u2})72]\Omega_o$$

$$\text{special seismic } T_u = [(0.545 + 0.385)72]3.0 = 201 \text{ k}$$

This is a strength level force which can be compared to an allowable stress multiplied by 1.7. See Sec. 16.3. The load duration factor C_D may be used for allowable stress design of this member.

Lateral Forces in E-W Direction

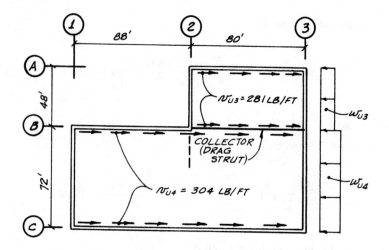

E-W LOADING

Figure 16.14c Diaphragm forces in E-W direction.

Diaphragm forces (Fig. 16.14c):

Wind does not govern.

Basic Seismic

$$w_{u3} = .275[(20 \times 80) + 2(75 \times 12)]$$

$$= 935 \text{ lb/ft}$$

$$w_{u4} = .275[(20 \times 168) + 2(75 \times 12)]$$

$$= 1420 \text{ lb/ft}$$

Roof shears:

Line A and north of line B:

$$v_{u3} = \frac{V_u}{b} = \frac{w_{u3}L}{2b} = \frac{935(48)}{2(80)} = 281 \text{ lb/ft}$$

Line C and south of line B:

$$v_{u4} = \frac{w_{u4}L}{2b} = \frac{1420(72)}{2(168)} = 304 \text{ lb/ft}$$

Chord forces:

Lines 2 and 3 between lines A and B:

$$T_u = C_u = \frac{M_u}{b} = \frac{w_{u3}L^2}{8b} = \frac{.935(48)^2}{8(80)} = 3.37 \text{ k}$$

Lines 1 and 3 between lines B and C:

$$T_u = C_u = \frac{w_{u4}L^2}{8b} = \frac{1.420(72)^2}{8(168)} = 5.48 \text{ k}$$

Collector (drag strut) force:

Connection on line B at line 2. Collector resists unsupported roof shear from both diaphragms. The *special seismic load combinations* are to be used with $\Omega_o = 3.0$.

$$\text{special seismic } T_u = [(v_{u3} + v_{u4})80]\Omega_o$$

$$\text{special seismic } T_u = [(0.281 + 0.304)80]3.0 = 140 \text{ k}$$

This is a strength level force which can be compared to an allowable stress multiplied by 1.7. See Sec. 16.3. The load duration factor C_D may be used for allowable stress design of this member.

16.9 Rigid Diaphragm Analysis

In Sec. 9.11 diaphragm flexibility was discussed in some detail and the classification of diaphragms as *rigid* or *flexible* was introduced. It was noted that the most prevalent design practice is to classify diaphragms as flexible without consideration of the Code criteria for diaphragm classification.

It should be noted that a rigid diaphragm analysis can be used for both wind and seismic forces. As long as the diaphragm is rigid with respect to the shearwalls and the center of the lateral force does not coincide with the center of rigidity, the diaphragm will have the ability to rotate. As a result of rotation the lateral forces will be redistributed between the shearwalls. Example 16.10 is provided to demonstrate the Code criteria for classifying a diaphragm as rigid or flexible. The example uses seismic forces at a strength level. For wind design ASD level forces would be used.

EXAMPLE 16.10   Diaphragm Classification

To demonstrate the use of the Code classification criteria, the building from Example 9.18 (Sec. 9.11) will be used to compare the diaphragm deflection Δ_D with the shearwall story drift Δ_S (see Fig. 16.15a). It will be assumed that both the diaphragm and the shearwalls are sheathed with wood structural panels.

Figure 16.15b provides dimensions for the diaphragm and shearwalls and gives the diaphragm loading, reactions and unit shears. In accordance with the Code procedure for the calculation of the diaphragm deflection, it will be assumed that the shearwalls are infinitely stiff, and the diaphragm spans as a simple beam between shearwalls.

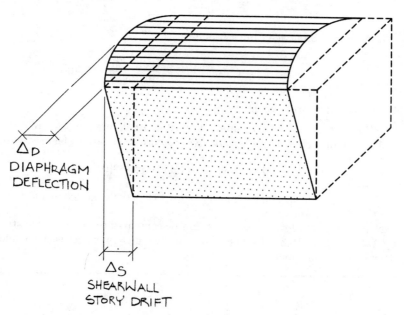

Figure 16.15a Shearwall and diaphragm deflections.

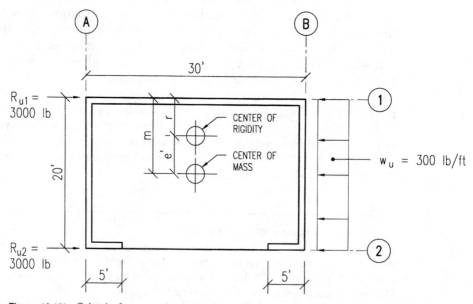

Figure 16.15b Seismic forces and reactions in the longitudinal direction used for diaphragm classification.

The equations for the diaphragm and shearwall deflections come from the *Wood Structural Panel Shear Wall and Diaphragm Supplement, Allowable Stress Design Manual for Engineered Wood Construction* (Ref. 16.2). These equations were developed by APA and further information regarding their use can be found in Refs. 16.4 and 16.5. Both Δ_S and Δ_D are calculated using IBC strength level forces.

Diaphragm and Shear Wall Unit Stresses

Before starting the diaphragm deflection calculations, it is useful to calculate the unit shears in the diaphragm and shearwalls.

$$R_{u1} = R_{u2} = w_u(L)/2 = 300(20)/2 = 3000 \text{ lb}$$

For the diaphragm:

$$v_{u1} = v_{u2} = R_{u1}/b = 3000/30 = 100 \text{ lb/ft}$$

For the shearwalls:

$$v_{u1} = R_{u1}/l_1 = 3000/30 = 100 \text{ lb/ft}$$

$$v_{u2} = R_{u2}/\Sigma l_2 = 3000/(5 + 5) = 300 \text{ lb/ft}$$

Diaphragm Deflection

The first step is to calculate the diaphragm deflection, but first the construction of the diaphragm needs to be defined:

Assume: Diaphragm and shearwall sheathing is 15/32 span-rated wood structural panels with exterior glue; span rating 32/16.

The diaphragm and shear walls are blocked and nailed with 8d common nails at 6 in. on center at panel edges and diaphragm boundaries.

The diaphragm chord is 2-2×4 top plates of No. 2 DF-L

Top plate splices occur at ten feet on center and the slip at each splice is $\frac{1}{16}$ in.

The diaphragm deflection equation can be expressed as:

$$\Delta_D = \Delta_b + \Delta_v + \Delta_n + \Delta_a$$

These four terms represent the same deflection sources that are discussed in Sec. 10.12 for shearwalls. Conceptually they are:

$$\Delta_b = \text{bending deflection of the diaphragm}$$

$$\Delta_v = \text{shear deflection of the diaphragm}$$

$$\Delta_n = \text{deflection due to nail slip}$$

$$\Delta_a = \text{deflection due to chord member slip}$$

The four terms can be evaluated individually and then summed:

Diaphragm Term 1—Bending Deflection

$$\Delta_b = \frac{5vL^3}{8Eab}$$

The variables are as follows:

v = 100 lb/ft maximum diaphragm unit shear

L = 20 feet (Fig. 16.15b)

  = diaphragm span

E = 1,700,000 psi (NDS Supplement Table 4A)

  = diaphragm chord modulus of elasticity

a = 2(3.5)1.5 = 10.5 in.^{2} for 2-2×4

  = diaphragm chord area

b = 30 feet (Fig. 16.15b) = diaphragm width

This results in:

$$\Delta_b = \frac{5(100)(20)^3}{8(1,700,000)(10.5)(30)} = 0.0009 \text{ in.}$$

Diaphragm Term 2—Panel Shear Deformation

$$\Delta_v = \frac{vL}{4Gt}$$

With:

v = 100 lb/ft = maximum diaphragm unit shear

L = 20 feet = diaphragm span

Gt = 27,000 lb/in. (Ref. 16.2, Panel Supplement, Table 3.4)

  = rigidity through the thickness

This results in:

$$\Delta_v = \frac{100(20)}{4(27,000)} = 0.0185 \text{ in.}$$

Diaphragm Term 3—Panel Nail Slip

$$\Delta_n = 0.188 L e_n$$

The variable e_n represents the slip of the sheathing with respect to the supporting framing as a function of the load per nail at design forces. In order to calculate e_n, the load per nail in pounds needs to be calculated. This is the unit shear v, divided by the number of nails per foot. This results in 100 lb/ft ÷ 2 nails per foot which equals 50 lb/nail. Values for e_n are found in Table 3.2 of the *Wood Structural Panel Shear Wall and Diaphragm Supplement* to the *Allowable Stress Design Manual for Engineered Wood Construction*. Table 3.2 provides e_n equations for framing green at time of construction and dry in use (green/dry) and dry at the time of construction and dry in use (dry/dry). The availability of dry framing for new construction varies widely, and local availability should be checked. For this example the green/dry condition will be used. In addition, footnote b to ASD Supplement Table 3.2 notes that the equations apply to framing members with a specific gravity of 0.50 and higher, and that an increase of 1.2 is needed where the sheathing is not STRUCTURAL I. Douglas fir-Larch framing will be assumed with a specific gravity of 0.5, and the sheathing is not STRUCTURAL I, so the multiplier of 1.2 will need to be used. For 8-penny (8d) common nails:

$$e_n = (V_n/857)^{1.869}(1.2)$$

where V_n is the calculated load per nail.

$$e_n = (50/857)^{1.869}(1.2) = 0.0059 \text{ in.} = \text{nail slip coefficient}$$

This results in:

$$\Delta_n = 0.188(20)(0.0059) = 0.0222 \text{ in.}$$

Diaphragm Term 4—Chord Splice Slip

$$\Delta_a = \sum (\Delta_c X)/2b$$

With:

$\Delta_c = 0.0625$ in (assumed slip) diaphragm chord slip at a single chord splice.

$X = 5$ ft and 5 ft (assumed)

= distance from diaphragm chord splice to nearest supporting shearwall.

$b = 30$ ft = diaphragm width

$$\sum (\Delta_c X) = 0.0625 (5) + 0.0625 (5) = 0.625 \text{ in-ft}$$

$$\Delta_a = 0.625/2 (30) = 0.0104 \text{ in.}$$

Total Diaphragm Deflection

The Total diaphragm deflection can be calculated as:

$$\Delta_D = \Delta_b + \Delta_v + \Delta_n + \Delta_a$$

$$= 0.0009 + 0.0185 + 0.0222 + 0.0104 = 0.0520 \text{ in.}$$

Story Drift Calculation

The Code criteria requires that the diaphragm deflection Δ_D be compared to the average story drift Δ_S, based on the tributary lateral load to the shearwalls. In order to determine the average story drift, the drift at each of the two shearwall lines will need to be calculated. One shearwall line has 30 feet of shearwall length and the other has 10 feet. Like the diaphragm deflection equation, the shearwall deflection equation can be divided into four parts. The calculation of shearwall deflections is treated in more detail in Sec. 10.11.

Shearwall Term 1—Chord Flexural Deflection

$$\Delta_b = \frac{8vh^3}{Eab}$$

The variables are as follows:

$v = 100$ and 300 lb/ft = shearwall unit shears

$h = 10$ feet = shearwall height

$E = 1,700,000$ psi (NDS Supplement Table 4A)

= shearwall chord modulus of elasticity

$a = 3.5\ (3.5) = 12.25$ in.^{2} for 4×4

= shearwall chord area

$b = 30$ and 5 feet (Fig. 16.15b)

= shearwall widths

This results in:

$\Delta_b = 8\ (100)(10)^3/1700000\ (12.25)(30) = 0.0013$ ft $= 0.016$ in. for the 30 ft wall

$\Delta_b = 8\ (300)(10)^3/1700000\ (12.25)(5) = 0.0230$ ft $= 0.28$ in. for the two 5 ft walls

Shearwall Term 2—Panel Shear Deformation

$$\Delta_v = \frac{vh}{Gt}$$

With:

$$v = 100/300 \text{ lb/ft} = \text{shearwall unit shears}$$

$$h = 10 \text{ feet} = \text{shearwall height}$$

$$Gt = 27,000 \text{ lb/in. (Ref 16.2, Panel Supplement, Table 3.4)}$$

$$= \text{rigidity through the thickness}$$

This results in:

$$\Delta_v = 100(10)/27,000 = 0.0370 \text{ in. for the 30 ft wall}$$

$$\Delta_v = 300(10)/27,000 = 0.1111 \text{ in. for the two 5 ft walls}$$

Shearwall Term 3—Panel Nail Slip

$$\Delta_n = 0.75he_n$$

The variable e_n is the same variable used to calculate the third term of the diaphragm deflection. In order to calculate e_n, the load per nail in pounds needs to be calculated. This is the unit shear v, divided by the number of nails per foot. This results in 100 lb/ft ÷ 2 nails per foot which equals 50 lb/nail for the 30 foot long wall. This value will be used. For the two 5 foot long walls, 300/2 = 150 lb/ft.

Values for e_n are again found in Table 3.2 of the *Wood Structural Panel Shear Wall and Diaphragm Supplement* to the *Allowable Stress Design Manual for Engineered Wood Construction* (the shearwall deflection formula in Sec. 4.3 refers to Shear Wall and Diaphragm Supplement Table 3.2 for shearwall nail slip). For this example the green/dry condition will again be used. In addition, the footnote b multiplier will again be used since the sheathing is not STRUCTURAL I. For 8-penny (8d) common nails:

$$e_n = (V_n/857)^{1.869}(1.2)$$

where V_n is the calculated load per nail.

$$e_n = (50/857)^{1.869}(1.2) = 0.0059 \text{ in.} = 30 \text{ ft wall nail slip coefficient}$$

$$e_n = (150/857)^{1.869}(1.2) = 0.0462 \text{ in.} = 5 \text{ ft wall nail slip coefficient}$$

With $h = 10$ ft, this results in:

$$\Delta_n = 0.75(10)(0.0059) = 0.0443 \text{ in. for the 30 foot wall}$$

$$\Delta_n = 0.75(10)(0.0462) = 0.3465 \text{ in. for the 5 foot walls}$$

Shearwall Term 4—Chord Anchorage Slip

$$\Delta_a = d_a h/b$$

With:

$d_a = 0.125$ in. based on manufacturer's data = chord anchorage slip

$h = 10$ ft = shearwall height

$b = 30$ and 5 ft = shearwall widths

$\Delta_a = 0.125 \ (10)/30 = 0.0417$ in. for the 30 ft wall

$\Delta_a = 0.125 \ (10)/5 = 0.2500$ in. for the two 5 ft walls

Total Shearwall Deflection and Average Story Drift

The total shearwall deflection can be calculated as:

$$\Delta_S = \Delta_b + \Delta_s + \Delta_n + \Delta_a$$

$$= 0.0013 + 0.0370 + 0.0443 + 0.0417 = 0.1243 \text{ in. for the 30 ft wall}$$

$$= 0.0230 + 0.1111 + 0.3465 + 0.2500 = 0.7306 \text{ in. for two 5 ft walls}$$

The average story drift is the average of these two shearwall defections:

$$\Delta_S = (0.1243 + 0.7306)/2 = 0.4275 \text{ in.}$$

The diaphragm deflection was calculated to be 0.0520 in. The Code criteria defines the diaphragm as flexible if $\Delta_D > 2(\Delta_S)$. In our case, 0.0520 in. $<<$ 2 (0.4275 in.) so the diaphragm is classified as rigid, not flexible. This example uses strength level forces because seismic loading is being considered. The process would be identical for ASD wind forces.

Example 16.10 merely demonstrates the Code criteria for determining the applicability of a rigid diaphragm analysis. Example 16.10 concludes that there is justification for use of a rigid diaphragm analysis. Example 16.11 demonstrates the rigid diaphragm analysis method. A simple rigid diaphragm analysis with a very straight forward solution was demonstrated in Example 9.19. Example 16.11 will demonstrate the more generalized rigid diaphragm analysis approach using the building from Examples 16.10. In Example 16.10, it was already demonstrated that the roof diaphragm can be considered a rigid diaphragm. Now the rigid diaphragm force distribution to the shear walls will be demonstrated.

EXAMPLE 16.11   General Rigid Diaphragm Analysis

This example will again look at the building used in Examples 9.18 and 16.10. This building is assumed to have a wood structural panel roof diaphragm and wood structural panel shearwalls. The configuration shown in Fig. 16.16a will be assumed. The seismic forces shown in Fig. 16.16b will be assumed to include all seismic forces tributary to the roof diaphragm including the weight contribution of exterior walls in both directions, as well as any interior partitions and nonstructural components. For this one story building with wood shearwalls, this means:

$$V = F_x = F_{px} = 300 \text{ lb/ft } (20 \text{ ft}) = 6000 \text{ lb}$$

This example will consider seismic forces in the longitudinal direction.

   In Example 16.10, for the purposes of evaluating diaphragm flexibility, a flexible diaphragm with equal reactions at Lines 1 and 2 was assumed. This distribution will now be re-evaluated using a rigid diaphragm model.

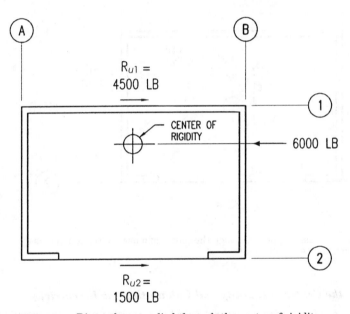

Figure 16.16a Direct shear applied through the center of rigidity.

Locating the Center of Mass

The mass of the building in Fig. 16.15b is shown as uniformly distributed and consequently the seismic force of 300 lb/ft in the longitudinal direction is uniformly distributed. Therefore the center of mass is at the center of the building. The assumption of uniform seismic force is a slight simplification because it ignores lower wall weight at openings. When this type of assumption results in slightly conservative requirements, it is usually acceptable for design purposes. Because not every building will be this simple and symmetrical, the designer should be prepared to calculate the location of the center of mass. This can be thought of as the location of the resultant of the seismic forces. In this case the distance m from Line 1 to the center of mass can be calculated as:

$$m = \frac{\Sigma \text{ Moments of lateral force (lb-ft)}}{\text{Total lateral force (lb)}}$$

$$m = w(\text{SPAN})(\text{SPAN}/2)/W$$

$$= (300 \text{ lb/ft}) (20 \text{ ft}) (10 \text{ ft})/6000 \text{ lb}$$

$$= 10 \text{ ft}$$

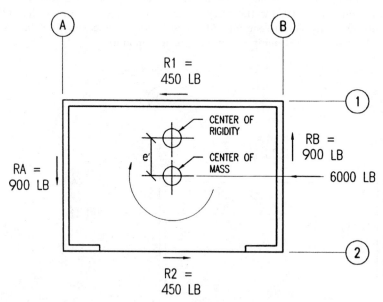

Figure 16.16b Moment applied through the center of mass. The center of mass rotates about the center of rigidity.

Locating the Center of Rigidity and Calculating the Eccentricity

The stiffness of the wood structural panel shearwalls and resulting rigidity is affected by the construction of the walls. In this case, it will be assumed that the construction (including sheathing nail spacing) of all the walls is the same. This will allow an assumption that the wall stiffness is substantially proportional to the wall length. The distance r from Line 1 to the center of rigidity can be calculated as:

$$r = \frac{l_1(0) + \Sigma l_2(\text{SPAN})}{l_1 + \Sigma l_2}$$

$$= \frac{30(0) + (5 + 5)20}{30 + (5 + 5)}$$

$$= \frac{200}{40} = 5.0 \text{ ft}$$

The eccentricity e can then be calculated as:

$$e = m - r = 10.0 - 5.0 = 5.0 \text{ ft}$$

This is called the *actual eccentricity*. For rigid diaphragm design the Code requires that, in addition to the *actual eccentricity*, the center of mass be displaced 5% of the building dimension perpendicular to the force direction. This results in additional eccentricity, referred to as the *accidental eccentricity*. The *design eccentricity e'* is then the *actual* and *accidental eccentricities*:

$$e' = e \pm 0.05 (L) = 5.0 \pm 0.05 (20) = 4.0 \text{ and } 6.0 \text{ ft}$$

The design eccentricity of 4.0 ft need not be investigated because 6.0 ft will control. In the perpendicular direction, by observation, the center of rigidity is at the center of the building.

Resulting Longitudinal Forces

As was discussed in Example 9.18, the forces generated in a rigid diaphragm analysis are generally considered in a two-step process. The first step is the distribution of the F_x force to the walls in proportion to the wall stiffness. Graphically this can be thought of as applying the seismic force to the structure at the center of rigidity. This is illustrated in Fig. 16.16a. Because wall stiffness is being assumed equal to the wall length, the unit shear in all the longitudinal walls will be the same:

$$V = F_x/(l_1 + \Sigma\, l_2) = 6000 \text{ lb}/(30 + 5 + 5 \text{ ft}) = 150 \text{ lb/ft}$$

The wall line reactions from this first of two steps can then be calculated as:

$$R_{u1} = 150 \text{ lb/ft } (30 \text{ ft}) = 4500 \text{ lb}$$

$$R_{u2} = 150 \text{ lb/ft } (5 + 5 \text{ ft}) = 1500 \text{ lb}$$

The second step in distributing seismic forces involves consideration of the moment that results from the force F_x times the design eccentricity e'. It should be noted that this moment is resisted by the shearwalls on all four sides of the building, not just Lines 1 and 2. The distribution of the moment $F_x (e')$ requires that the torsional stiffness of the shearwalls on all four sides of the building be calculated. The following table shows a common approach to distributing this moment.

The second column is the wall stiffness which, as in Example 16.10, will be taken as proportional to the wall length. The third column gives the distance from the wall to the center of rigidity. The fourth column calculates k(r), which is just a convenient intermediate value to be used in the last column. In the fifth column k(r)² is the torsional stiffness of each wall and the sum for the group of walls.

The last column calculates the shear to individual walls due to the moment. This shear is obtained by multiplying the portion of the moment to each wall k(r)²/Σk(r)² times the moment $F_x (e')$ divided by the moment arm r. This can be accomplished by multiplying the moment by the fourth column and dividing it by the sum of the fifth column. The table shows a common approach to distributing this moment to the resisting shearwalls.

Rigid Diaphragm Moment Distribution

Line	stiffness (ft) k	distance (ft) (r)	k(r)	k(r²)	force (lb) $\dfrac{F_x(e')k(r)}{\Sigma k(r^2)}$
1	30	5	150	750	450
2	10	15	150	2250	450
A	20	15	300	4500	900
B	20	15	300	4500	900
				12,000	

The shearwall forces resulting from this moment appear in the last column of the table and are depicted in Fig. 16.16b. It is important to keep track of the direction of the forces due to the moment. The center of applied lateral force can be visualized as rotating about the center of rigidity. In this case, a clockwise rotation of the building will result. The forces due to the moment will add to the direct shear forces on Line 2 and subtract from the direct shear forces on Line 1. The sum of forces on each line is shown in Fig. 16.16c.

The design forces used for Lines A and B need to be the most critical resulting from either the seismic forces in the transverse direction or due to rotation of the diaphragm resulting from forces in the longitudinal direction. In the example building, the force from the longitudinal seismic is 750 lb, which is significantly less than the 3000 lb expected for transverse seismic forces.

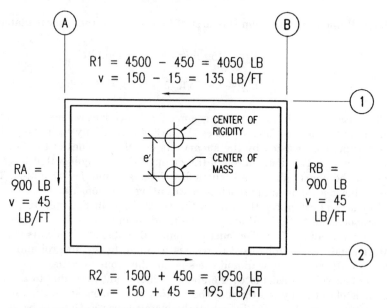

R1 = 4500 − 450 = 4050 LB
v = 150 − 15 = 135 LB/FT

CENTER OF RIGIDITY
CENTER OF MASS

RA = 900 LB
v = 45 LB/FT

RB = 900 LB
v = 45 LB/FT

R2 = 1500 + 450 = 1950 LB
v = 150 + 45 = 195 LB/FT

Figure 16.16c Result of direct shear and moment.

Both the roof diaphragm and the shearwalls in the Example 16.11 building would be designed for the forces resulting from the rigid diaphragm analysis.

A flexible diaphragm analysis would currently be a lot more common in practice. It is recommended that the reader analyze several example buildings using both the flexible and rigid diaphragm analyses to develop a feeling for how much the distribution of shear changes between the two approaches.

While the force distribution to shearwalls will be different in each building studied, the reader will be able to develop an understanding of what type of plan layout results in significant redistribution of forces to shearwalls between the two methods. The layouts resulting in significant differences generally have a substantial distance between their center of mass and center of rigidity. Where this occurs, it is recommended that the diaphragms and shearwalls be examined using a rigid diaphragm analysis. Another approach is to perform both flexible and rigid diaphragm analyses and design each element for the larger of the resulting forces.

16.10   Additional Topics in Horizontal Diaphragm Design

The lateral design forces required by the Code in areas of high seismic risk have increased substantially in recent years. In order to provide design values for more heavily loaded diaphragms, the American Plywood Association conducted a series of tests on diaphragms that were designed to develop high allowable unit shears. These diaphragms are referred to as *high-load* diaphragms, and test results are published in Ref. 16.4. High-load diaphragms are recognized in ICBO Evaluation Service, Inc., Report Number 1952.

One type of high-load diaphragm uses two layers of plywood sheathing. In the test the first layer was attached to the framing members with conventional nailing. The top layer was attached to the first layer with 14-gauge × 1¾-in. staples. The panel edges for the second layer were offset from the edges of the first layer, and the staples were purposely *not* driven into the framing members (i.e., staples penetrated the first layer of plywood only).

Another method of obtaining high-load diaphragms was with the use of relatively thick plywood (⅝ and ¾ in.) and closely spaced fasteners. Tests indicated that closely spaced 10d and 16d nails often caused the framing lumber to split. However, very closely spaced (1-in. o.c.) pneumatically driven wire staples were found to not cause splitting. High unit shears were obtained with these diaphragms.

Among other objectives, these diaphragm tests were designed to measure the effects of openings in horizontal diaphragms. See Refs. 16.4 and 16.6 for information on openings in diaphragms and other advanced topics in diaphragm design.

16.11   References

[16.1]   American Forest and Paper Association (AF&PA). 2001. *National Design Specification for Wood Construction* and Supplement, 2001 ed., AF&PA, Washington, DC.

[16.2]   American Forest and Paper Association (AF&PA). 2002. *Allowable Stress Design Manual for Engineered Wood Construction* and Supplements and Guidelines, 2002 ed., AF&PA, Washington, DC.

[16.3] American Society of Civil Engineers (ASCE). 2002. *Minimum Design Loads for Buildings and Other Structures (ASCE 7-02)*, ASCE, New York, NY.

[16.4] APA—The Engineered Wood Association. 1990. *Research Report 138—Plywood Diaphragms*, APA, Tacoma, WA.

[16.5] APA—The Engineered Wood Association. 2001. *Design / Construction Guide—Diaphragms and Shear Walls*, APA, Tacoma, WA.

[16.6] Applied Technology Council. 1981. *Guidelines for the Design of Horizontal Wood Diaphragms (ATC-7)*, ATC, Redwood City, CA.

[16.7] Cobeen, K., Russell, J. E., and Dolan, J. D. 2003. *Recommendations for Earthquake Resistance in the Design and Construction of Woodframe Buildings* (CUREE Publication W-30), Consortium of Universities for Research in Earthquake Engineering, Richmond, CA.

[16.8] Faherty, Keith F., and Williamson, T. G. (eds.). 1999. *Wood Engineering and Construction Handbook*, 3rd ed., McGraw-Hill, New York, NY.

[16.9] International Codes Council (ICC). 2003. *International Building Code*, 2003 Ed., ICC, Falls Church, VA.

[16.10] Mosalam, K. 2003. *Seismic Evaluation of an Asymmetric Three-Story Woodframe Building* (CUREE Publication W-19), Consortium of Universities for Research in Earthquake Engineering, Richmond, CA.

[16.11] Structural Engineers Association of California (SEAOC). 1999. *Recommended Lateral Force Requirements and Commentary*, 7th ed., SEAOC, Sacramento, CA.

16.12   Problems

16.1   Discuss the differences between a structure that is seismicly regular and one that is irregular.

16.2   Give a brief description and sketch of the following seismic irregularities:
    a. Vertical mass irregularity
    b. Diaphragm discontinuity
    c. Reentrant corner
    d. Out-of-plane offset
    e. In-plane discontinuity

16.3   Classify each of the irregularities in Prob. 16.2 as either a vertical irregularity or a plan irregularity. List the section number in 1997 UBC that gives the design requirements for each type of irregularity.

16.4   Briefly describe the seismic design requirements for the irregularities in Prob. 16.2. Only the requirements that are *in addition* to those for a regular structure need to be discussed.

16.5   Give the critical combinations of loads and forces required for designing the shearwall tension and compression chords for a typical single-story building. The structure has $S_{DS} = 1.00$, and the local snow loads govern over roof live loads. Consider the following cases:
    a. Seismic force for an irregular structure with an out-of-plane offset
    b. Wind force
    c. Seismic force for a regular structure. In addition to the shearwall chords, give the loading condition for use in analyzing the overturning effects at the foundation-soil interface.

16.6 *Given:*   The plan of the U-shaped building in Fig. 16.A. Exterior walls serve as shearwalls. Roof D = 20 psf, and wall D = 60 psf. Tributary wall height to roof diaphragm is 8 ft. Seismic coefficient = 0.275.

   Find: a. Unit shear in the horizontal diaphragms and shearwalls
   b. The magnitude of the drag strut force at the point where these members connect to the shearwalls

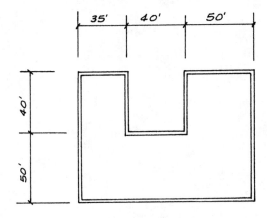

PLAN

Figure 16.A

PLAN

Figure 14.

Equivalent Uniform Weights of Wood Framing

Weights are for Douglas Fir-Larch lumber (S4S) used at an equilibrium moisture content of 15 percent (the maximum found in most covered structures). The unit weight of Douglas Fir-Larch equals or exceeds the unit weight of most softwood species, and the dead loads given below are conservative for most designs.

| | Spacing | | | | | |
| | 12 in. o.c. | | 16 in. o.c. | | 24 in. o.c. | |
Nominal size	Weight, psf	Board feet* per ft^2	Weight, psf	Board feet* per ft^2	Weight, psf	Board feet* per ft^2
2 × 3	0.9	0.50	0.7	0.38	0.4	0.25
2 × 4	1.2	0.67	0.9	0.50	0.6	0.34
2 × 6	1.9	1.00	1.4	0.75	1.0	0.5
2 × 8	2.5	1.33	1.9	1.00	1.3	0.67
2 × 10	3.2	1.67	2.4	1.25	1.6	0.84
2 × 12	3.9	2.00	2.9	1.50	2.0	1.00
3 × 6	3.2	1.50	2.4	1.13	1.6	0.75
3 × 8	4.2	2.00	3.1	1.50	2.1	1.00
3 × 10	5.4	2.50	4.0	1.88	2.7	1.25
3 × 12	6.6	3.00	4.9	2.25	3.3	1.50
3 × 14	7.7	3.50	5.8	2.63	3.9	1.75
4 × 8	5.9	2.67	4.4	2.00	3.0	1.34
4 × 10	7.5	3.33	5.7	2.50	3.8	1.67
4 × 12	9.2	4.00	6.9	3.00	4.6	2.00
4 × 14	11.0	4.67	8.1	3.50	5.4	2.34
4 × 16	12.4	5.33	9.3	4.00	6.2	2.67

*Lumber is ordered and priced by the board foot. A board foot is a volume of lumber corresponding to 1 in. thick by 12 in. wide by 1 ft. long. Nominal dimensions are used to calculate board measure.
SOURCE: *Western Woods Use Book,* Western Wood Products Association, 4th ed., 1996.

Weights of Building Materials

Loads given in Appendix B are typical values. Specific products may have weights which differ considerably from those shown, and manufacturer's catalogs should be consulted for actual loads.

Roof dead loads			
Material	Weight, psf		
Lumber sheathing, 1 in. nominal	2.5		
Plywood, per inch of thickness	3.0		
Timber decking (MC = 15%):	2 in. nom.	3 in. nom.	4 in. nom.
DF-Larch	4.2	7.0	9.8
DF (South)	3.9	6.5	9.1
Hem-Fir	3.7	6.1	8.5
Spruce-Pine-Fir	3.7	6.1	8.5
Western Woods	3.5	5.8	8.1
Western Cedars	3.5	5.8	8.1

Roof dead loads		
Material	Weight, psf	
Aluminum (including laps):	Flat	Corrugated (1½ and 2½ in.)
12 American or B&S gage	1.2	...
14	0.9	1.1
16	0.7	0.9
18	0.6	0.7
20	0.5	0.6
22	...	0.4
Galvanized steel (including laps):	Flat	Corrugated (2½ and 3 in.)
12 U.S. std. gage	4.5	4.9
14	3.3	3.6
16	2.7	2.9
18	2.2	2.4
20	1.7	1.8
22	1.4	1.5
24	1.2	1.3
26	0.9	1.0
Other types of decking (per inch of thickness):		
Concrete plank	6.5	
Insulrock	2.7	
Petrical	2.7	
Porex	2.7	
Poured gypsum	6.5	
Tectum	2.0	
Vermiculite concrete	2.6	
Corrugated asbestos (¼ in.)	3.0	
Felt:		
3-ply	1.5	
3-ply with gravel	5.5	
5-ply	2.5	
5-ply with gravel	6.5	
Insulation (per inch of thickness):		
Expanded polystyrene	0.2	
Fiber glass, rigid	1.5	
Loose	0.5	
Roll roofing	1.0	

Roof dead loads

Material	Weight, psf
Shingles:	
Asphalt (¼ in. approx.)	2.0
Book tile (2 in.)	12.0
Book tile (3 in.)	20.0
Cement asbestos (⅜ in. approx.)	4.0
Clay tile (for mortar add 10 psf)	9.0 to 14.0
Ludowici	10.0
Roman	12.0
Slate (¼ in.)	10.0
Spanish	19.0

Ceiling dead loads

Material	Weight, psf
Acoustical fiber tile	1.0
Channel-suspended system	1.0
For gypsum wallboard and plaster, see *Wall and partition dead loads*	

Floor dead loads

Material	Weight, psf
Hardwood (1 in. nominal)	4.0
Plywood (per inch of thickness)	3.0
Asphalt mastic (per inch of thickness)	12.0
Cement finish (per inch of thickness)	12.0
Ceramic and quarry tile (¾ in.)	10.0
Concrete (per inch of thickness)	
Lightweight	6.0 to 10.0
Reinforced (normal weight)	12.5
Stone	12.0
Cork tile (1⁄16 in.)	0.5
Flexicore (6-in. slab)	46.0
Linoleum (¼ in.)	1.0
Terrazo finish (1½ in.)	19.0
Vinyl tile (⅛ in.)	1.4

Wall and partition dead loads

Material	Weight, psf
Wood paneling (1 in.)	2.5
Wood studs (2 × 4 @ 15% mc DF-Larch):	
12 in. o.c.	1.2
16 in. o.c.	0.9
24 in. o.c.	0.6
Glass block (4 in.)	18.0
Glass (¼-in. plate)	3.3
Glazed tile	18.0
Marble or marble wainscoting	15.0

Wall and partition dead loads	
Material	Weight, psf
Masonry (per 4 in. of thickness):	
Brick	38.0
Concrete block	30.0
Cinder concrete block	20.0
Hollow clay tile, load bearing	23.0
Hollow clay tile, non-load-bearing	18.0
Hollow gypsum block	13.0
Limestone	55.0
Terra-cotta tile	25.0
Stone	55.0
(The average weights of completed reinforced and grouted concrete block and brick walls can be found in Ref. 12.)	
Plaster (1 in.)	8.0
Plaster (1 in.) on wood lath	10.0
Plaster (1 in.) on metal lath	8.5
Gypsum wallboard (1 in.)	~~4.4~~ 5.0
Porcelain-enameled steel	3.0
Stucco (⅞ in.)	10.0
Windows (glass, frame, and sash)	8.0

SOURCE: Weights from *Western Woods Use Book,* Western Wood Products Association, 4th ed., 1996.

Selected Tables and Figures from The International Building Code, 2003 Edition*

Tables and figures for use in determining loads and forces	Table or Figure No.
Net Deisgn Wind Pressure (Component and Cladding) p_{net30} (Exposure B At h = 30 feet with I_w = 1.0)	Table 1609.6.2.1(2)
Roof Overhang Net Design Wind Pressure (Component and Cladding), P_{net30} (Exposure B at h = 30 feet with I_w = 1.0)	Table 1609.6.2.1(3)
Adjustment Factor for Building Height and Exposure, (λ)	Table 1609.6.2.1(4)
Maximum Considered Earthquake Ground Motion Maps[1]	Figures 1615(1)–(10)
Site Class Definitions	Table 1615.1.1
Values of Site Coefficient F_a as a Function of Site Class and Mapped Spectral Response Acceleration at Short Periods (S_S)	Table 1615.1.2(1)
Values of Site Coefficient F_V as a Function of Site Class and Mapped Spectral Response Acceleration at 1-Second Period (S_1)	Table 1615.1.2(2)
Design Response Spectrum	Figure 1615.1.4
Site Classification	Table 1615.1.5
Seismic Design Category Based on Short-Period Response Accelerations	Table 1616.3(1)
Seismic Design Category Based on 1-Second Period Response Acceleration	Table 1616.3(2)
Plan Structural Irregularities	Table 1616.5.1.1
Vertical Structural Irregularities	Table 1616.5.1.2
Allowable Service Load on Embedded Bolts	Table 1912.2
Fastening Schedule	Table 2304.9.1
Maximum Diaphragm Dimension Ratios Horizontal and Sloped Diaphragm	Table 2305.2.3
Maximum Shear Wall Aspect Ratios	Table 2305.3.3
Recommended Shear (Pounds Per Foot) For Wood Structural Panel Diaphragms With Framing of Douglas-Fir-Larch, of Southern Pine for Wind or Seismic Loading	Table 2306.3.1
Allowable Shear (Pounds Per Foot) For Wood Structural Panel Shear Walls with Framing of Douglas-Fir-Larch, or Southern Pine, For Wind Or Seismic Loading	Table 2306.4.1
Allowable Shear For Particleboard Shear Wall Sheathing	Table 2306.4.3
Allowable Shear for Wind or Seismic Forces For Shear Walls of Lath and Plaster or Gypsum Board Wood Frames Wall Assemblies	Table 2306.4.5

---

[1]A compact disk of the seismic hazard maps is available from ICC, if requested when purchasing the IBC. The compact disk permits search for mapped spectral acceleration values based on area code or latitude and longitude.

TABLE 1604.3
DEFLECTION LIMITS[a, b, c, h, i]

CONSTRUCTION	L	S or W[f]	$D + L$[d,g]
Roof members:[e]			
Supporting plaster ceiling	$l/360$	$l/360$	$l/240$
Supporting nonplaster ceiling	$l/240$	$l/240$	$l/180$
Not supporting ceiling	$l/180$	$l/180$	$l/120$
Floor members	$l/360$	—	$l/240$
Exterior walls and interior partitions:			
With brittle finishes	—	$l/240$	—
With flexible finishes	—	$l/120$	—
Farm buildings	—	—	$l/180$
Greenhouses	—	—	$l/120$

For SI: 1 foot = 304.8 mm.

a. For structural roofing and siding made of formed metal sheets, the total load deflection shall not exceed $l/60$. For secondary roof structural members supporting formed metal roofing, the live load deflection shall not exceed $l/150$. For secondary wall members supporting formed metal siding, the design wind load deflection shall not exceed $l/90$. For roofs, this exception only applies when the metal sheets have no roof covering.

b. Interior partitions not exceeding 6 feet in height and flexible, folding and portable partitions are not governed by the provisions of this section. The deflection criterion for interior partitions is based on the horizontal load defined in Section 1607.13.

c. See Section 2403 for glass supports.

d. For wood structural members having a moisture content of less than 16 percent at time of installation and used under dry conditions, the deflection resulting from $L + 0.5D$ is permitted to be substituted for the deflection resulting from $L + D$.

e. The above deflections do not ensure against ponding. Roofs that do not have sufficient slope or camber to assure adequate drainage shall be investigated for ponding. See Section 1611 for rain and ponding requirements and Section 1503.4 for roof drainage requirements.

f. The wind load is permitted to be taken as 0.7 times the "component and cladding" loads for the purpose of determining deflection limits herein.

g. For steel structural members, the dead load shall be taken as zero.

h. For aluminum structural members or aluminum panels used in roofs or walls of sunroom additions or patio covers, not supporting edge of glass or aluminum sandwich panels, the total load deflection shall not exceed $l/60$. For aluminum sandwich panels used in roofs or walls of sunroom additions or patio covers, the total load deflection shall not exceed $l/120$.

i. For cantilever members, l shall be taken as twice the length of the cantilever.

TABLE 1604.5
CLASSIFICATION OF BUILDINGS AND OTHER STRUCTURES FOR IMPORTANCE FACTORS

CATEGORY[a]	NATURE OF OCCUPANCY	SEISMIC FACTOR I_E	SNOW FACTOR I_S	WIND FACTOR I_W
I	Buildings and other structures that represent a low hazard to human life in the event of failure including, but not limited to: • Agricultural facilities • Certain temporary facilities • Minor storage facilities	1.00	0.8	0.87[b]
II	Buildings and other structures except those listed in Categories I, III and IV	1.00	1.0	1.00
III	Buildings and other structures that represent a substantial hazard to human life in the event of failure including, but not limited to: • Buildings and other structures where more than 300 people congregate in one area • Buildings and other structures with elementary school, secondary school or day care facilities with an occupant load greater than 250 • Buildings and other structures with an occupant load greater than 500 for colleges or adult education facilities • Health care facilities with an occupant load of 50 or more resident patients but not having surgery or emergency treatment facilities • Jails and detention facilities • Any other occupancy with an occupant load greater than 5,000 • Power-generating stations, water treatment for potable water, waste water treatment facilities and other public utility facilities not included in Category IV • Buildings and other structures not included in Category IV containing sufficient quantities of toxic or explosive substances to be dangerous to the public if released	1.25	1.1	1.15
IV	Buildings and other structures designed as essential facilities including, but not limited to: • Hospitals and other health care facilities having surgery or emergency treatment facilities • Fire, rescue and police stations and emergency vehicle garages • Designated earthquake, hurricane or other emergency shelters • Designated emergency preparedness, communication, and operation centers and other facilities required for emergency response • Power-generating stations and other public utility facilities required as emergency backup facilities for Category IV structures • Structures containing highly toxic materials as defined by Section 307 where the quantity of the material exceeds the maximum allowable quantities of Table 307.7(2) • Aviation control towers, air traffic control centers and emergency aircraft hangars • Buildings and other structures having critical national defense functions • Water treatment facilities required to maintain water pressure for fire suppression	1.50	1.2	1.15

a. For the purpose of Section 1616.2, Categories I and II are considered Seismic Use Group I, Category III is considered Seismic Use Group II and Category IV is equivalent to Seismic Use Group III.

b. In hurricane-prone regions with $V > 100$ miles per hour, I_w shall be 0.77.

TABLE 1607.1
MINIMUM UNIFORMLY DISTRIBUTED LIVE LOADS AND MINIMUM CONCENTRATED LIVE LOADS[g]

OCCUPANCY OR USE	UNIFORM (psf)	CONCENTRATED (lbs.)
1. Apartments (see residential)	—	—
2. Access floor systems		
Office use	50	2,000
Computer use	100	2,000
3. Armories and drill rooms	150	—
4. Assembly areas and theaters		
Fixed seats (fastened to floor)	60	
Lobbies	100	
Movable seats	100	
Stages and platforms	125	—
Follow spot, projections and control rooms	50	
Catwalks	40	
5. Balconies (exterior)	100	
On one- and two-family residences only, and not exceeding 100 ft.²	60	—
6. Decks	Same as occupancy served[h]	—
7. Bowling alleys	75	—
8. Cornices	60	—
9. Corridors, except as otherwise indicated	100	—
10. Dance halls and ballrooms	100	—
11. Dining rooms and restaurants	100	—
12. Dwellings (see residential)	—	—
13. Elevator machine room grating (on area of 4 in.²)	—	300
14. Finish light floor plate construction (on area of 1 in.²)	—	200
15. Fire escapes	100	
On single-family dwellings only	40	—
16. Garages (passenger vehicles only)	40	Note a
Trucks and buses	See Section 1607.6	
17. Grandstands (see stadium and arena bleachers)	—	—
18. Gymnasiums, main floors and balconies	100	—
19. Handrails, guards and grab bars	See Section 1607.7	
20. Hospitals		
Operating rooms, laboratories	60	1,000
Private rooms	40	1,000
Wards	40	1,000
Corridors above first floor	80	1,000
21. Hotels (see residential)	—	—
22. Libraries		
Reading rooms	60	1,000
Stack rooms	150[b]	1,000
Corridors above first floor	80	1,000
23. Manufacturing		
Light	125	2,000
Heavy	250	3,000
24. Marquees	75	

OCCUPANCY OR USE	UNIFORM (psf)	CONCENTRATED (lbs.)
25. Office buildings		
File and computer rooms shall be designed for heavier loads based on anticipated occupancy		
Lobbies and first-floor corridors	100	2,000
Offices	50	2,000
Corridors above first floor	80	2,000
26. Penal institutions		
Cell blocks	40	—
Corridors	100	
27. Residential		
One- and two-family dwellings		
Uninhabitable attics without storage	10	
Uninhabitable attics with storage	20	
Habitable attics and sleeping areas	30	
All other areas except balconies and decks	40	—
Hotels and multifamily dwellings		
Private rooms and corridors serving them	40	
Public rooms and corridors serving them	100	
28. Reviewing stands, grandstands and bleachers	Note c	—
29. Roofs	See Section 1607.11	
30. Schools		
Classrooms	40	1,000
Corridors above first floor	80	1,000
First-floor corridors	100	1,000
31. Scuttles, skylight ribs and accessible ceilings	—	200
32. Sidewalks, vehicular driveways and yards, subject to trucking	250[d]	8,000[e]
33. Skating rinks	100	—
34. Stadiums and arenas		
Bleachers	100[c]	—
Fixed seats (fastened to floor)	60[c]	
35. Stairs and exits	100	Note f
One- and two-family dwellings	40	
All other	100	
36. Storage warehouses (shall be designed for heavier loads if required for anticipated storage)		—
Light	125	
Heavy	250	
37. Stores		
Retail		
First floor	100	1,000
Upper floors	75	1,000
Wholesale, all floors	125	1,000
38. Vehicle barriers	See Section 1607.7	
39. Walkways and elevated platforms (other than exitways)	60	—
40. Yards and terraces, pedestrians	100	—

(continued)

Notes to Table 1607.1

For SI: 1 inch = 25.4 mm, 1 square inch = 645.16 mm², 1 pound per
square foot = 0.0479 kN/m², 1 pound = 0.004448 kN.
1 pound per cubic foot = 16 kg/m³

a. Floors in garages or portions of buildings used for the storage of motor vehicles shall be designed for the uniformly distributed live loads of Table
   1607.1 or the following concentrated loads: (1) for garages restricted to vehicles accommodating not more than nine passengers, 3,000 pounds acting
   on an area of 4.5 inches by 4.5 inches; (2) for mechanical parking structures
   without slab or deck which are used for storing passenger vehicles only,
   2,250 pounds per wheel.

b. The loading applies to stack room floors that support nonmobile, double-faced library bookstacks, subject to the following limitations:
   1. The nominal bookstack unit height shall not exceed 90 inches;
   2. The nominal shelf depth shall not exceed 12 inches for each face; and

3. Parallel rows of double-faced bookstacks shall be separated by aisles
   not less than 36 inches wide.

c. Design in accordance with the ICC *Standard on Bleachers, Folding and
   Telescopic Seating and Grandstands*.

d. Other uniform loads in accordance with an approved method which contains
   provisions for truck loadings shall also be considered where appropriate.

e. The concentrated wheel load shall be applied on an area of 20 square inches.

f. Minimum concentrated load on stair treads (on area of 4 square inches) is
   300 pounds.

g. Where snow loads occur that are in excess of the design conditions, the
   structure shall be designed to support the loads due to the increased loads
   caused by drift buildup or a greater snow design determined by the building
   official (see Section 1608). For special-purpose roofs, see Section
   1607.11.2.2.

h. See Section 1604.8.3 for decks attached to exterior walls.

TABLE 1607.9.1
LIVE LOAD ELEMENT FACTOR, K_{LL}

ELEMENT	K_{LL}
Interior columns	4
Exterior columns without cantilever slabs	4
Edge columns with cantilever slabs	3
Corner columns with cantilever slabs	2
Edge beams without cantilever slabs	2
Interior beams	2
All other members not identified above including: Edge beams with cantilever slabs Cantilever beams Two-way slabs Members without provisions for continuous shear transfer normal to their span	1

TABLE 1608.2
GROUND SNOW LOADS, p_g, FOR ALASKAN LOCATIONS

LOCATION	POUNDS PER SQUARE FOOT	LOCATION	POUNDS PER SQUARE FOOT	LOCATION	POUNDS PER SQUARE FOOT
Adak	30	Galena	60	Petersburg	150
Anchorage	50	Gulkana	70	St. Paul Islands	40
Angoon	70	Homer	40	Seward	50
Barrow	25	Juneau	60	Shemya	25
Barter Island	35	Kenai	70	Sitka	50
Bethel	40	Kodiak	30	Talkeetna	120
Big Delta	50	Kotzebue	60	Unalakleet	50
Cold Bay	25	McGrath	70	Valdez	160
Cordova	100	Nenana	80	Whittier	300
Fairbanks	60	Nome	70	Wrangell	60
Fort Yukon	60	Palmer	50	Yakutat	150

For SI: 1 pound per square foot = 0.0479 kN/m^2.

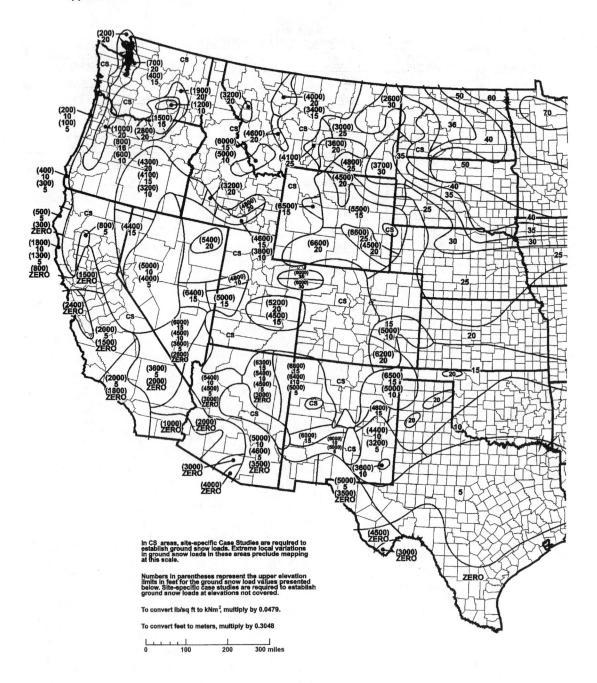

FIGURE 1608.2
GROUND SNOW LOADS, p_g, FOR THE UNITED STATES (psf)

FIGURE 1608.2–continued
GROUND SNOW LOADS, p_g, FOR THE UNITED STATES (psf)

TABLE 1608.3.1
SNOW EXPOSURE FACTOR, C_e

TERRAIN CATEGORY[a]	EXPOSURE OF ROOF[a,b]		
	Fully exposed[c]	Partially exposed	Sheltered
A (see Section 1609.4)	N/A	1.1	1.3
B (see Section 1609.4)	0.9	1.0	1.2
C (see Section 1609.4)	0.9	1.0	1.1
D (see Section 1609.4)	0.8	0.9	1.0
Above the treeline in windswept mountainous areas	0.7	0.8	N/A
In Alaska, in areas where trees do not exist within a 2-mile radius of the site	0.7	0.8	N/A

For SI: 1 mile = 1609 m.

a. The terrain category and roof exposure condition chosen shall be representative of the anticipated conditions during the life of the structure. An exposure factor shall be determined for each roof of a structure.

b. Definitions of roof exposure are as follows:

1. Fully exposed shall mean roofs exposed on all sides with no shelter afforded by terrain, higher structures or trees. Roofs that contain several large pieces of mechanical equipment, parapets which extend above the height of the balanced snow load, h_b, or other obstructions are not in this category.

2. Partially exposed shall include all roofs except those designated as "fully exposed" or "sheltered."

3. Sheltered roofs shall mean those roofs located tight in among conifers that qualify as "obstructions."

c. Obstructions within a distance of $10 h_o$ provide "shelter," where h_o is the height of the obstruction above the roof level. If the only obstructions are a few deciduous trees that are leafless in winter, the "fully exposed" category shall be used except for terrain category "A." Note that these are heights above the roof. Heights used to establish the terrain category in Section 1609.4 are heights above the ground.

TABLE 1608.3.2
THERMAL FACTOR, C_t

THERMAL CONDITION[a]	C_t
All structures except as indicated below	1.0
Structures kept just above freezing and others with cold, ventilated roofs in which the thermal resistance (R-value) between the ventilated space and the heated space exceeds $25h \cdot ft^2 \cdot °F/Btu$	1.1
Unheated structures	1.2
Continuously heated greenhouses[b] with a roof having a thermal resistance (R-value) less than $2.0h \cdot ft^2 \cdot °F/Btu$	0.85

For SI: $°C = [(°F)-32]/1.8$, 1 British thermal unit per hour = 0.2931W.

a. The thermal condition shall be representative of the anticipated conditions during winters for the life of the structure.

b. A continuously heated greenhouse shall mean a greenhouse with a constantly maintained interior temperature of 50°F or more during winter months. Such greenhouse shall also have a maintenance attendant on duty at all times or a temperature alarm system to provide warning in the event of a heating system failure.

TABLE 1609.1.4
WIND-BORNE DEBRIS PROTECTION FASTENING
SCHEDULE FOR WOOD STRUCTURAL PANELS[a,b,c]

FASTENER TYPE	FASTENER SPACING (Inches)			
	Panel span ≤ 2 feet	2 feet < Panel span ≤ 4 feet	4 feet < Panel span ≤ 6 feet	6 feet < Panel span ≤ 8 feet
$2^{1}/_{2}$ No. 6 Wood screws	16	16	12	9
$2^{1}/_{2}$ No. 8 Wood screws	16	16	16	12

For SI: 1 inch = 25.4 mm, 1 foot = 304.8 mm, 1 pound = 4.4 N,
1 mile per hour = 0.44 m/s.

a. This table is based on a maximum wind speed (3-second gust) of 130 mph and mean roof height of 33 feet or less.

b. Fasteners shall be installed at opposing ends of the wood structural panel.

c. Where screws are attached to masonry or masonry/stucco, they shall be attached utilizing vibration-resistant anchors having a minimum withdrawal capacity of 490 pounds.

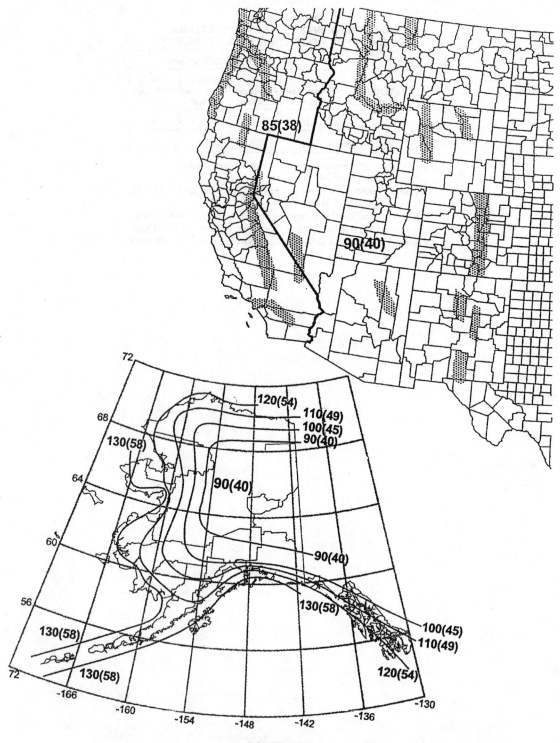

FIGURE 1609
BASIC WIND SPEED (3-SECOND GUST)

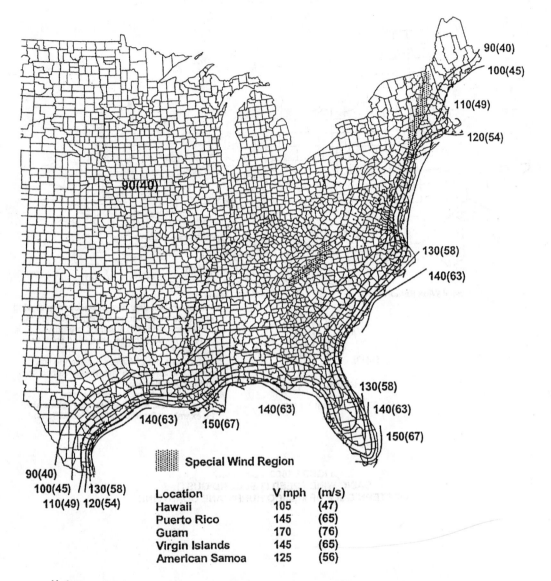

Location	V mph	(m/s)
Hawaii	105	(47)
Puerto Rico	145	(65)
Guam	170	(76)
Virgin Islands	145	(65)
American Samoa	125	(56)

Notes:

1. Values are nominal design 3-second gust wind speeds in miles per hour (m/s) at 33 ft (10 m) above ground for Exposure C category.
2. Linear interpolation between wind contours is permitted.
3. Islands and coastal areas outside the last contour shall use the last wind speed contour of the coastal area.
4. Mountainous terrain, gorges, ocean promontories, and special wind regions shall be examined for unusual wind conditions.

FIGURE 1609—continued
BASIC WIND SPEED (3-SECOND GUST)

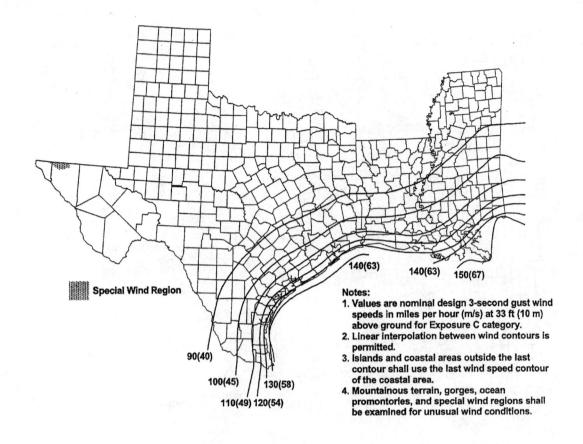

Special Wind Region

140(63)   140(63)   150(67)

Notes:
1. Values are nominal design 3-second gust wind speeds in miles per hour (m/s) at 33 ft (10 m) above ground for Exposure C category.
2. Linear interpolation between wind contours is permitted.
3. Islands and coastal areas outside the last contour shall use the last wind speed contour of the coastal area.
4. Mountainous terrain, gorges, ocean promontories, and special wind regions shall be examined for unusual wind conditions.

90(40)

100(45)   130(58)

110(49) 120(54)

FIGURE 1609—continued
BASIC WIND SPEED (3-SECOND GUST)
WESTERN GULF OF MEXICO HURRICANE COASTLINE

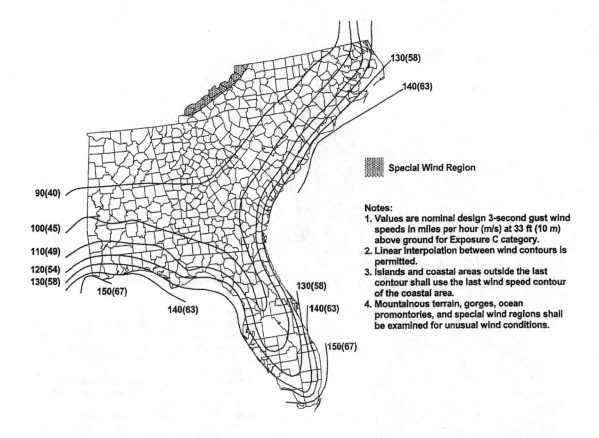

130(58)

140(63)

Special Wind Region

90(40)

100(45)

110(49)

120(54)
130(58)

150(67)

130(58)

140(63)

140(63)

150(67)

Notes:
1. Values are nominal design 3-second gust wind
   speeds in miles per hour (m/s) at 33 ft (10 m)
   above ground for Exposure C category.
2. Linear interpolation between wind contours is
   permitted.
3. Islands and coastal areas outside the last
   contour shall use the last wind speed contour
   of the coastal area.
4. Mountainous terrain, gorges, ocean
   promontories, and special wind regions shall
   be examined for unusual wind conditions.

FIGURE 1609–continued
BASIC WIND SPEED (3-SECOND GUST)
EASTERN GULF OF MEXICO AND SOUTHEASTERN U.S. HURRICANE COASTLINE

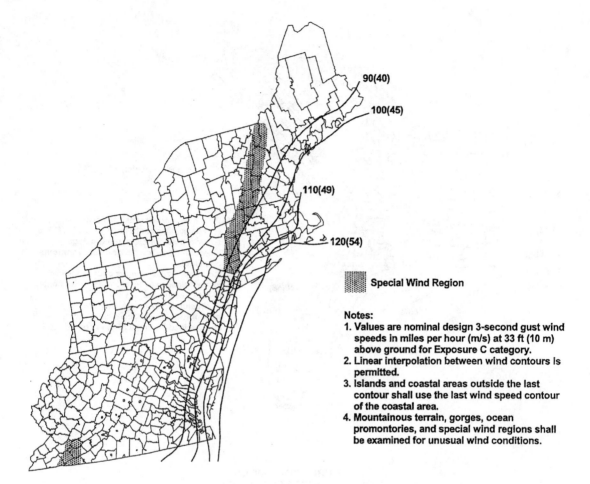

Special Wind Region

Notes:
1. Values are nominal design 3-second gust wind speeds in miles per hour (m/s) at 33 ft (10 m) above ground for Exposure C category.
2. Linear interpolation between wind contours is permitted.
3. Islands and coastal areas outside the last contour shall use the last wind speed contour of the coastal area.
4. Mountainous terrain, gorges, ocean promontories, and special wind regions shall be examined for unusual wind conditions.

FIGURE 1609—continued
BASIC WIND SPEED (3-SECOND GUST)
MID AND NORTHERN ATLANTIC HURRICANE COASTLINE

TABLE 1609.3.1
EQUIVALENT BASIC WIND SPEEDS[a,b,c]

V_{3S}	85	90	100	105	110	120	125	130	140	145	150	160	170
V_{fm}	70	75	80	85	90	100	105	110	120	125	130	140	150

For SI: 1 mile per hour = 0.44 m/s.
a. Linear interpolation is permitted.
b. V_{3S} is the 3-second gust wind speed (mph).
c. V_{fm} is the fastest mile wind speed (mph).

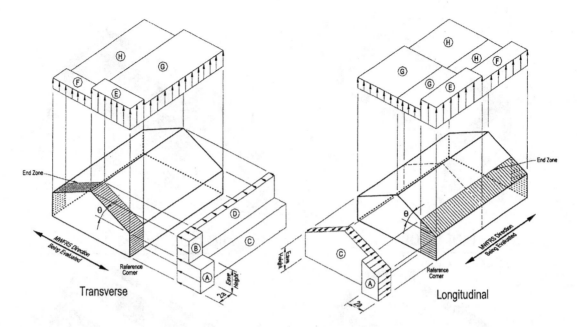

FIGURE 1609.6.2.1
MAIN WINDFORCE LOADING DIAGRAM

For SI: 1 foot = 304.8 mm, 1 degree = 0.0174 rad.

Notes:

1. Pressures are applied to the horizontal and vertical projections for Exposure B, at $h = 30$ feet, for $I_w = 1.0$. Adjust to other exposures and heights with adjustment factor λ.

2. The load patterns shown shall be applied to each corner of the building in turn as the reference corner.

3. For the design of the longitudinal MWFRS, use $\theta = 0°$, and locate the Zone E/F, G/H boundary at the mid-length of the building.

4. Load Cases 1 and 2 must be checked for $25° < \theta \leq 45°$. Load Case 2 at $25°$ is provided only for interpolation between $25°$ to $30°$.

5. Plus and minus signs signify pressures acting toward and away from the projected surfaces, respectively.

6. For roof slopes other than those shown, linear interpolation is permitted.

7. The total horizontal load shall not be less than that determined by assuming $p_S = 0$ in Zones B and D.

8. The zone pressures represent the following:
   Horizontal pressure zones — Sum of the windward and leeward net (sum of internal and external) pressures on vertical projection of:

A – End zone of wall	C – Interior zone of wall
B – End zone of roof	D – Interior zone of roof

   Vertical pressure zones — Net (sum of internal and external) pressures on horizontal projection of:

E – End zone of windward roof	G – Interior zone of windward roof
F – End zone of leeward roof	H – Interior zone of leeward roof

9. Where Zone E or G falls on a roof overhang on the windward side of the building, use E_{OH} and G_{OH} for the pressure on the horizontal projection of the overhang. Overhangs on the leeward and side edges shall have the basic zone pressure applied.

10. Notation:
   a: 10 percent of least horizontal dimension or 0.4h, whichever is smaller, but not less than either 4 percent of least horizontal dimension or 3 feet.
   h: Mean roof height, in feet (meters), except that eave height shall be used for roof angles $<10°$.
   θ: Angle of plane of roof from horizontal, in degrees.

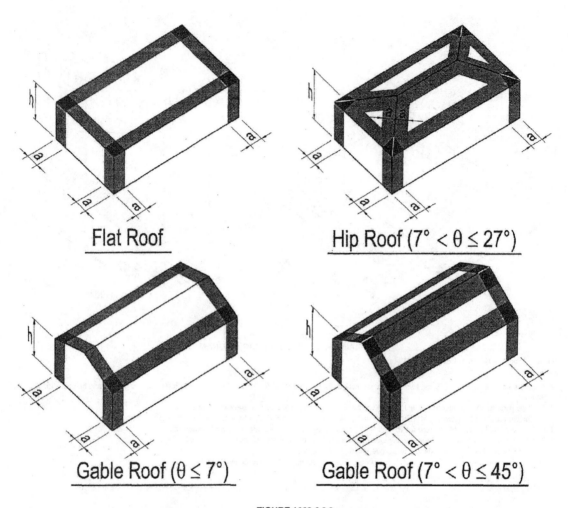

FIGURE 1609.6.2.2
COMPONENT AND CLADDING PRESSURE

For SI: 1 foot = 304.8 mm, 1 degree = 0.0174 rad.

Notes:

1. Pressures are applied normal to the surface for Exposure B, at h = 30 feet, for I_w = 1.0. Adjust to other exposures and heights with adjustment factor λ.
2. Plus and minus signs signify pressures acting toward and away from the surfaces, respectively.
3. For hip roofs with $\theta \le 25°$, Zone 3 shall be treated as Zone 2.
4. For effective areas between those given, the value is permitted to be interpolated, otherwise use the value associated with the lower effective area.
5. Notation:

    a: 10 percent of least horizontal dimension or 0.4h, whichever is smaller, but not less than either 4 percent of least horizontal dimension or 3 feet.

    h: Mean roof height, in feet (meters), except that eave height shall be used for roof angles <10°.

    θ: Angle of plane of roof from horizontal, in degrees.

TABLE 1609.6.2.1(1)
SIMPLIFIED DESIGN WIND PRESSURE (MAIN WINDFORCE-RESISTING SYSTEM), p_{s30} (Exposure B at h = 30 feet with I_w = 1.0) (psf)

BASIC WIND SPEED (mph)	ROOF ANGLE (degrees)	ROOF RISE IN 12"	LOAD CASE	ZONES									
				Horizontal Pressures				Vertical Pressures				Overhangs	
				A	B	C	D	E	F	G	H	E_{OH}	G_{OH}
85	0 to 5°	Flat	1	11.5	-5.9	7.6	-3.5	-13.8	-7.8	-9.6	-6.1	-19.3	-15.1
	10°	2	1	12.9	-5.4	8.6	-3.1	-13.8	-8.4	-9.6	-6.5	-19.3	-15.1
	15°	3	1	14.4	-4.8	9.6	-2.7	-13.8	-9.0	-9.6	-6.9	-19.3	-15.1
	20°	4	1	15.9	-4.2	10.6	-2.3	-13.8	-9.6	-9.6	-7.3	-19.3	-15.1
	25°	6	1	14.4	2.3	10.4	2.4	-6.4	-8.7	-4.6	-7.0	-11.9	-10.1
			2	—	—	—	—	-2.4	-4.7	-0.7	-3.0	—	—
	30° to 45°	7 to 12	1	12.9	8.8	10.2	7.0	1.0	-7.8	0.3	-6.7	-4.5	-5.2
			2	12.9	8.8	10.2	7.0	5.0	-3.9	4.3	-2.8	-4.5	-5.2
90	0 to 5°	Flat	1	12.8	-6.7	8.5	-4.0	-15.4	-8.8	-10.7	-6.8	-21.6	-16.9
	10°	2	1	14.5	-6.0	9.6	-3.5	-15.4	-9.4	-10.7	-7.2	-21.6	-16.9
	15°	3	1	16.1	-5.4	10.7	-3.0	-15.4	-10.1	-10.7	-7.7	-21.6	-16.9
	20°	4	1	17.8	-4.7	11.9	-2.6	-15.4	-10.7	-10.7	-8.1	-21.6	-16.9
	25°	6	1	16.1	2.6	11.7	2.7	-7.2	-9.8	-5.2	-7.8	-13.3	-11.4
			2	—	—	—	—	-2.7	-5.3	-0.7	-3.4	—	—
	30° to 45°	7 to 12	1	14.4	9.9	11.5	7.9	1.1	-8.8	0.4	-7.5	-5.1	-5.8
			2	14.4	9.9	11.5	7.9	5.6	-4.3	4.8	-3.1	-5.1	-5.8
100	0 to 5°	Flat	1	15.9	-8.2	10.5	-4.9	-19.1	-10.8	-13.3	-8.4	-26.7	-20.9
	10°	2	1	17.9	-7.4	11.9	-4.3	-19.1	-11.6	-13.3	-8.9	-26.7	-20.9
	15°	3	1	19.9	-6.6	13.3	-3.8	-19.1	-12.4	-13.3	-9.5	-26.7	-20.9
	20°	4	1	22.0	-5.8	14.6	-3.2	-19.1	-13.3	-13.3	-10.1	-26.7	-20.9
	25°	6	1	19.9	3.2	14.4	3.3	-8.8	-12.0	-6.4	-9.7	-16.5	-14.0
			2	—	—	—	—	-3.4	-6.6	-0.9	-4.2	—	—
	30° to 45°	7 to 12	1	17.8	12.2	14.2	9.8	1.4	-10.8	0.5	-9.3	-6.3	-7.2
			2	17.8	12.2	14.2	9.8	6.9	-5.3	5.9	-3.8	-6.3	-7.2
110	0 to 5°	Flat	1	19.2	-10.0	12.7	-5.9	-23.1	-13.1	-16.0	-10.1	-32.3	-25.3
	10°	2	1	21.6	-9.0	14.4	-5.2	-23.1	-14.1	-16.0	-10.8	-32.3	-25.3
	15°	3	1	24.1	-8.0	16.0	-4.6	-23.1	-15.1	-16.0	-11.5	-32.3	-25.3
	20°	4	1	26.6	-7.0	17.7	-3.9	-23.1	-16.0	-16.0	-12.2	-32.3	-25.3
	25°	6	1	24.1	3.9	17.4	4.0	-10.7	-14.6	-7.7	-11.7	-19.9	-17.0
			2	—	—	—	—	-4.1	-7.9	-1.1	-5.1	—	—
	30° to 45°	7 to 12	1	21.6	14.8	17.2	11.8	1.7	-13.1	0.6	-11.3	-7.6	-8.7
			2	21.6	14.8	17.2	11.8	8.3	-6.5	7.2	-4.6	-7.6	-8.7
120	0 to 5°	Flat	1	22.8	-11.9	15.1	-7.0	-27.4	-15.6	-19.1	-12.1	-38.4	-30.1
	10°	2	1	25.8	-10.7	17.1	-6.2	-27.4	-16.8	-19.1	-12.9	-38.4	-30.1
	15°	3	1	28.7	-9.5	19.1	-5.4	-27.4	-17.9	-19.1	-13.7	-38.4	-30.1
	20°	4	1	31.6	-8.3	21.1	-4.6	-27.4	-19.1	-19.1	-14.5	-38.4	-30.1
	25°	6	1	28.6	4.6	20.7	4.7	-12.7	-17.3	-9.2	-13.9	-23.7	-20.2
			2	—	—	—	—	-4.8	-9.4	-1.3	-6.0	—	—
	30° to 45°	7 to 12	1	25.7	17.6	20.4	14.0	2.0	-15.6	0.7	-13.4	-9.0	-10.3
			2	25.7	17.6	20.4	14.0	9.9	-7.7	8.6	-5.5	-9.0	-10.3
130	0 to 5°	Flat	1	26.8	-13.9	17.8	-8.2	-32.2	-18.3	-22.4	-14.2	-45.1	-35.3
	10°	2	1	30.2	-12.5	20.1	-7.3	-32.2	-19.7	-22.4	-15.1	-45.1	-35.3
	15°	3	1	33.7	-11.2	22.4	-6.4	-32.2	-21.0	-22.4	-16.1	-45.1	-35.3
	20°	4	1	37.1	-9.8	24.7	-5.4	-32.2	-22.4	-22.4	-17.0	-45.1	-35.3
	25°	6	1	33.6	5.4	24.3	5.5	-14.9	-20.4	-10.8	-16.4	-27.8	-23.7
			2	—	—	—	—	-5.7	-11.1	-1.5	-7.1	—	—
	30° to 45°	7 to 12	1	30.1	20.6	24.0	16.5	2.3	-18.3	0.8	-15.7	-10.6	-12.1
			2	30.1	20.6	24.0	16.5	11.6	-9.0	10.0	-6.4	-10.6	-12.1

continued

TABLE 1609.6.2.1(1)-continued
SIMPLIFIED DESIGN WIND PRESSURE (MAIN WINDFORCE-RESISTING SYSTEM), p_{s30} (Exposure B at h = 30 feet with I_w = 1.0) (psf)

BASIC WIND SPEED (mph)	ROOF ANGLE (degrees)	ROOF RISE IN 12"	LOAD CASE	ZONES									
				Horizontal Pressures				Vertical Pressures				Overhangs	
				A	B	C	D	E	F	G	H	E_{OH}	G_{OH}
140	0 to 5°	Flat	1	31.1	-16.1	20.6	-9.6	-37.3	-21.2	-26.0	-16.4	-52.3	-40.9
	10°	2	1	35.1	-14.5	23.3	-8.5	-37.3	-22.8	-26.0	-17.5	-52.3	-40.9
	15°	3	1	39.0	-12.9	26.0	-7.4	-37.3	-24.4	-26.0	-18.6	-52.3	-40.9
	20°	4	1	43.0	-11.4	28.7	-6.3	-37.3	-26.0	-26.0	-19.7	-52.3	-40.9
	25°	6	1	39.0	6.3	28.2	6.4	-17.3	-23.6	-12.5	-19.0	-32.3	-27.5
			2	—	—	—	—	-6.6	-12.8	-1.8	-8.2	—	—
	30° to 45°	7 to 12	1	35.0	23.9	27.8	19.1	2.7	-21.2	0.9	-18.2	-12.3	-14.0
			2	35.0	23.9	27.8	19.1	13.4	-10.5	11.7	-7.5	-12.3	-14.0
150	0 to 5°	Flat	1	35.7	-18.5	23.7	-11.0	-42.9	-24.4	-29.8	-18.9	-60.0	-47.0
	10°	2	1	40.2	-16.7	26.8	-9.7	-42.9	-26.2	-29.8	-20.1	-60.0	-47.0
	15°	3	1	44.8	-14.9	29.8	-8.5	-42.9	-28.0	-29.8	-21.4	-60.0	-47.0
	20°	4	1	49.4	-13.0	32.9	-7.2	-42.9	-29.8	-29.8	-22.6	-60.0	-47.0
	25°	6	1	44.8	7.2	32.4	7.4	-19.9	-27.1	-14.4	-21.8	-37.0	-31.6
			2	—	—	—	—	-7.5	-14.7	-2.1	-9.4	—	—
	30° to 45°	7 to 12	1	40.1	27.4	31.9	22.0	3.1	-24.4	1.0	-20.9	-14.1	-16.1
			2	40.1	27.4	31.9	22.0	15.4	-12.0	13.4	-8.6	-14.1	-16.1
170	0 to 5°	Flat	1	45.8	-23.8	30.4	-14.1	-55.1	-31.3	-38.3	-24.2	-77.1	-60.4
	10°	2	1	51.7	-21.4	34.4	-12.5	-55.1	-33.6	-38.3	-25.8	-77.1	-60.4
	15°	3	1	57.6	-19.1	38.3	-10.9	-55.1	-36.0	-38.3	-27.5	-77.1	-60.4
	20°	4	1	63.4	-16.7	42.3	-9.3	-55.1	-38.3	-38.3	-29.1	-77.1	-60.4
	25°	6	1	57.5	9.3	41.6	9.5	-25.6	-34.8	-18.5	-28.0	-47.6	-40.5
			2	—	—	—	—	-9.7	-18.9	-2.6	-12.1	—	—
	30° to 45°	7 to 12	1	51.5	35.2	41.0	28.2	4.0	-31.3	1.3	-26.9	-18.1	-20.7
			2	51.5	35.2	41.0	28.2	19.8	-15.4	17.2	-11.0	-18.1	-20.7

For SI: 1 inch = 25.4 mm, 1 foot = 304.8 mm, 1 degree = 0.0174 rad, 1 mile per hour = 0.44 m/s, 1 pound per square foot = 47.9 N/m².

TABLE 1609.6.2.1(2)
NET DESIGN WIND PRESSURE (COMPONENT AND CLADDING), p_{net30} (Exposure B at h = 30 feet with I_w = 1.0)

	ZONE	EFFECTIVE WIND AREA	BASIC WIND SPEED V (mph—3-second gust)																
			85		90		100		110		120		130		140		150		170	
Roof 0 to 7 degrees	1	10	5.3	-13.0	5.9	-14.6	7.3	-18.0	8.9	-21.8	10.5	-25.9	12.4	-30.4	14.3	-35.3	16.5	-40.5	21.1	-52.0
	1	20	5.0	-12.7	5.6	-14.2	6.9	-17.5	8.3	-21.2	9.9	-25.2	11.6	-29.6	13.4	-34.4	15.4	-39.4	19.8	-50.7
	1	50	4.5	-12.2	5.1	-13.7	6.3	-16.9	7.6	-20.5	9.0	-24.4	10.6	-28.6	12.3	-33.2	14.1	-38.1	18.1	-48.9
	1	100	4.2	-11.9	4.7	-13.3	5.8	-16.5	7.0	-19.9	8.3	-23.7	9.8	-27.8	11.4	-32.3	13.0	-37.0	16.7	-47.6
	2	10	5.3	-21.8	5.9	-24.4	7.3	-30.2	8.9	-36.5	10.5	-43.5	12.4	-51.0	14.3	-59.2	16.5	-67.9	21.1	-87.2
	2	20	5.0	-19.5	5.6	-21.8	6.9	-27.0	8.3	-32.6	9.9	-38.8	11.6	-45.6	13.4	-52.9	15.4	-60.7	19.8	-78.0
	2	50	4.5	-16.4	5.1	-18.4	6.3	-22.7	7.6	-27.5	9.0	-32.7	10.6	-38.4	12.3	-44.5	14.1	-51.1	18.1	-65.7
	2	100	4.2	-14.1	4.7	-15.8	5.8	-19.5	7.0	-23.6	8.3	-28.1	9.8	-33.0	11.4	-38.2	13.0	-43.9	16.7	-56.4
	3	10	5.3	-32.8	5.9	-36.8	7.3	-45.4	8.9	-55.0	10.5	-65.4	12.4	-76.8	14.3	-89.0	16.5	-102.2	21.1	-131.3
	3	20	5.0	-27.2	5.6	-30.5	6.9	-37.6	8.3	-45.5	9.9	-54.2	11.6	-63.6	13.4	-73.8	15.4	-84.7	19.8	-108.7
	3	50	4.5	-19.7	5.1	-22.1	6.3	-27.3	7.6	-33.1	9.0	-39.3	10.6	-46.2	12.3	-53.5	14.1	-61.5	18.1	-78.9
	3	100	4.2	-14.1	4.7	-15.8	5.8	-19.5	7.0	-23.6	8.3	-28.1	9.8	-33.0	11.4	-38.2	13.0	-43.9	16.7	-56.4
Roof > 7 to 27 degrees	1	10	7.5	-11.9	8.4	-13.3	10.4	-16.5	12.5	-19.9	14.9	-23.7	17.5	-27.8	20.3	-32.3	23.3	-37.0	30.0	-47.6
	1	20	6.8	-11.6	7.7	-13.0	9.4	-16.0	11.4	-19.4	13.6	-23.0	16.0	-27.0	18.5	-31.4	21.3	-36.0	27.3	-46.3
	1	50	6.0	-11.1	6.7	-12.5	8.2	-15.4	10.0	-18.6	11.9	-22.2	13.9	-26.0	16.1	-30.2	18.5	-34.6	23.8	-44.5
	1	100	5.3	-10.8	5.9	-12.1	7.3	-14.9	8.9	-18.1	10.5	-21.5	12.4	-25.2	14.3	-29.3	16.5	-33.6	21.1	-43.2
	2	10	7.5	-20.7	8.4	-23.2	10.4	-28.7	12.5	-34.7	14.9	-41.3	17.5	-48.4	20.3	-56.2	23.3	-64.5	30.0	-82.8
	2	20	6.8	-19.0	7.7	-21.4	9.4	-26.4	11.4	-31.9	13.6	-38.0	16.0	-44.6	18.5	-51.7	21.3	-59.3	27.3	-76.2
	2	50	6.0	-16.9	6.7	-18.9	8.2	-23.3	10.0	-28.2	11.9	-33.6	13.9	-39.4	16.1	-45.7	18.5	-52.5	23.8	-67.4
	2	100	5.3	-15.2	5.9	-17.0	7.3	-21.0	8.9	-25.5	10.5	-30.3	12.4	-35.6	14.3	-41.2	16.5	-47.3	21.1	-60.8
	3	10	7.5	-30.6	8.4	-34.3	10.4	-42.4	12.5	-51.3	14.9	-61.0	17.5	-71.6	20.3	-83.1	23.3	-95.4	30.0	-122.5
	3	20	6.8	-28.6	7.7	-32.1	9.4	-39.6	11.4	-47.9	13.6	-57.1	16.0	-67.0	18.5	-77.7	21.3	-89.2	27.3	-114.5
	3	50	6.0	-26.0	6.7	-29.1	8.2	-36.0	10.0	-43.5	11.9	-51.8	13.9	-60.8	16.1	-70.5	18.5	-81.0	23.8	-104.0
	3	100	5.3	-24.0	5.9	-26.9	7.3	-33.2	8.9	-40.2	10.5	-47.9	12.4	-56.2	14.3	-65.1	16.5	-74.8	21.1	-96.0
Roof > 27 to 45 degrees	1	10	11.9	-13.0	13.3	-14.6	16.5	-18.0	19.9	-21.8	23.7	-25.9	27.8	-30.4	32.3	-35.3	37.0	-40.5	47.6	-52.0
	1	20	11.6	-12.3	13.0	-13.8	16.0	-17.1	19.4	-20.7	23.0	-24.6	27.0	-28.9	31.4	-33.5	36.0	-38.4	46.3	-49.3
	1	50	11.1	-11.5	12.5	-12.8	15.4	-15.9	18.6	-19.2	22.2	-22.8	26.0	-26.8	30.2	-31.1	34.6	-35.7	44.5	-45.8
	1	100	10.8	-10.8	12.1	-12.1	14.9	-14.9	18.1	-18.1	21.5	-21.5	25.2	-25.2	29.3	-29.3	33.6	-33.6	43.2	-43.2
	2	10	11.9	-15.2	13.3	-17.0	16.5	-21.0	19.9	-25.5	23.7	-30.3	27.8	-35.6	32.3	-41.2	37.0	-47.3	47.6	-60.8
	2	20	11.6	-14.5	13.0	-16.3	16.0	-20.1	19.4	-24.3	23.0	-29.0	27.0	-34.0	31.4	-39.4	36.0	-45.3	46.3	-58.1
	2	50	11.1	-13.7	12.5	-15.3	15.4	-18.9	18.6	-22.9	22.2	-27.2	26.0	-32.0	30.2	-37.1	34.6	-42.5	44.5	-54.6
	2	100	10.8	-13.0	12.1	-14.6	14.9	-18.0	18.1	-21.8	21.5	-25.9	25.2	-30.4	29.3	-35.3	33.6	-40.5	43.2	-52.0
	3	10	11.9	-15.2	13.3	-17.0	16.5	-21.0	19.9	-25.5	23.7	-30.3	27.8	-35.6	32.3	-41.2	37.0	-47.3	47.6	-60.8
	3	20	11.6	-14.5	13.0	-16.3	16.0	-20.1	19.4	-24.3	23.0	-29.0	27.0	-34.0	31.4	-39.4	36.0	-45.3	46.3	-58.1
	3	50	11.1	-13.7	12.5	-15.3	15.4	-18.9	18.6	-22.9	22.2	-27.2	26.0	-32.0	30.2	-37.1	34.6	-42.5	44.5	-54.6
	3	100	10.8	-13.0	12.1	-14.6	14.9	-18.0	18.1	-21.8	21.5	-25.9	25.2	-30.4	29.3	-35.3	33.6	-40.5	43.2	-52.0
Wall	4	10	13.0	-14.1	14.6	-15.8	18.0	-19.5	21.8	-23.6	25.9	-28.1	30.4	-33.0	35.3	-38.2	40.5	-43.9	52.0	-56.4
	4	20	12.4	-13.5	13.9	-15.1	17.2	-18.7	20.8	-22.6	24.7	-26.9	29.0	-31.6	33.7	-36.7	38.7	-42.1	49.6	-54.1
	4	50	11.6	-12.7	13.0	-14.3	16.1	-17.6	19.5	-21.3	23.2	-25.4	27.2	-29.8	31.6	-34.6	36.2	-39.7	46.6	-51.0
	4	100	11.1	-12.2	12.4	-13.6	15.3	-16.8	18.5	-20.4	22.0	-24.2	25.9	-28.4	30.0	-33.0	34.4	-37.8	44.2	-48.6
	4	500	9.7	-10.8	10.9	-12.1	13.4	-14.9	16.2	-18.1	19.3	-21.5	22.7	-25.2	26.3	-29.3	30.2	-33.6	38.8	-43.2
	5	10	13.0	-17.4	14.6	-19.5	18.0	-24.1	21.8	-29.1	25.9	-34.7	30.4	-40.7	35.3	-47.2	40.5	-54.2	52.0	-69.6
	5	20	12.4	-16.2	13.9	-18.2	17.2	-22.5	20.8	-27.2	24.7	-32.4	29.0	-38.0	33.7	-44.0	38.7	-50.5	49.6	-64.9
	5	50	11.6	-14.7	13.0	-16.5	16.1	-20.3	19.5	-24.6	23.2	-29.3	27.2	-34.3	31.6	-39.8	36.2	-45.7	46.6	-58.7
	5	100	11.1	-13.5	12.4	-15.1	15.3	-18.7	18.5	-22.6	22.0	-26.9	25.9	-31.6	30.0	-36.7	34.4	-42.1	44.2	-54.1
	5	500	9.7	-10.8	10.9	-12.1	13.4	-14.9	16.2	-18.1	19.3	-21.5	22.7	-25.2	26.3	-29.3	30.2	-33.6	38.8	-43.2

For SI: 1 foot = 304.8 mm, 1 degree = 0.0174 rad, 1 mile per hour = 0.44 m/s, 1 pound per square foot = 47.9 N/m².
Note: For effective areas between those given above, the load is permitted to be interpolated, otherwise use the load associated with the lower effective area.

TABLE 1609.6.2.1(3)
ROOF OVERHANG NET DESIGN WIND PRESSURE (COMPONENT AND CLADDING), p_{net30} (Exposure B at h = 30 feet with I_w = 1.0) (psf)

	ZONE	EFFECTIVE WIND AREA (sq. ft.)	BASIC WIND SPEED V (mph—3-second gust)							
			90	100	110	120	130	140	150	170
Roof 0 to 7 degrees	2	10	-21.0	-25.9	-31.4	-37.3	-43.8	-50.8	-58.3	-74.9
	2	20	-20.6	-25.5	-30.8	-36.7	-43.0	-49.9	-57.3	-73.6
	2	50	-20.1	-24.9	-30.1	-35.8	-42.0	-48.7	-55.9	-71.8
	2	100	-19.8	-24.4	-29.5	-35.1	-41.2	-47.8	-54.9	-70.5
	3	10	-34.6	-42.7	-51.6	-61.5	-72.1	-83.7	-96.0	-123.4
	3	20	-27.1	-33.5	-40.5	-48.3	-56.6	-65.7	-75.4	-96.8
	3	50	-17.3	-21.4	-25.9	-30.8	-36.1	-41.9	-48.1	-61.8
	3	100	-10.0	-12.2	-14.8	-17.6	-20.6	-23.9	-27.4	-35.2
Roof > 7 to 27 degrees	2	10	-27.2	-33.5	-40.6	-48.3	-56.7	-65.7	-75.5	-96.9
	2	20	-27.2	-33.5	-40.6	-48.3	-56.7	-65.7	-75.5	-96.9
	2	50	-27.2	-33.5	-40.6	-48.3	-56.7	-65.7	-75.5	-96.9
	2	100	-27.2	-33.5	-40.6	-48.3	-56.7	-65.7	-75.5	-96.9
	3	10	-45.7	-56.4	-68.3	-81.2	-95.3	-110.6	-126.9	-163.0
	3	20	-41.2	-50.9	-61.6	-73.3	-86.0	-99.8	-114.5	-147.1
	3	50	-35.3	-43.6	-52.8	-62.8	-73.7	-85.5	-98.1	-126.1
	3	100	-30.9	-38.1	-46.1	-54.9	-64.4	-74.7	-85.8	-110.1
Roof > 27 to 45 degrees	2	10	-24.7	-30.5	-36.9	-43.9	-51.5	-59.8	-68.6	-88.1
	2	20	-24.0	-29.6	-35.8	-42.6	-50.0	-58.0	-66.5	-85.5
	2	50	-23.0	-28.4	-34.3	-40.8	-47.9	-55.6	-63.8	-82.0
	2	100	-22.2	-27.4	-33.2	-39.5	-46.4	-53.8	-61.7	-79.3
	3	10	-24.7	-30.5	-36.9	-43.9	-51.5	-59.8	-68.6	-88.1
	3	20	-24.0	-29.6	-35.8	-42.6	-50.0	-58.0	-66.5	-85.5
	3	50	-23.0	-28.4	-34.3	-40.8	-47.9	-55.5	-63.8	-82.2
	3	100	-22.2	-27.4	-33.2	-39.5	-46.4	-53.8	-61.7	-79.3

For SI: 1 foot = 304.8 mm, 1 degree = 0.0174 rad, 1 mile per hour = 0.45 m/s, 1 pound per square foot = 47.9 N/m^2.

Note: For effective areas between those given above, the load is permitted to be interpolated, otherwise use the load associated with the lower effective area.

TABLE 1609.6.2.1(4)
ADJUSTMENT FACTOR FOR BUILDING HEIGHT AND EXPOSURE, (λ)

MEAN ROOF HEIGHT (feet)	EXPOSURE		
	B	C	D
15	1.00	1.21	1.47
20	1.00	1.29	1.55
25	1.00	1.35	1.61
30	1.00	1.40	1.66
35	1.05	1.45	1.70
40	1.09	1.49	1.74
45	1.12	1.53	1.78
50	1.16	1.56	1.81
55	1.19	1.59	1.84
60	1.22	1.62	1.87

For SI: 1 foot = 304.8 mm.

a. All table values shall be adjusted for other exposures and heights by multiplying by the above coefficients.

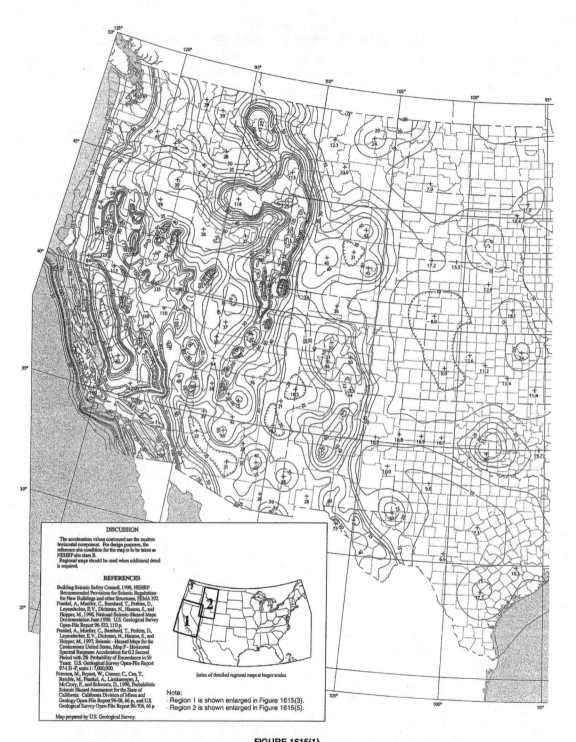

FIGURE 1615(1)
MAXIMUM CONSIDERED EARTHQUAKE GROUND MOTION FOR THE CONTERMINOUS UNITED STATES
OF 0.2 SEC SPECTRAL RESPONSE ACCELERATION (5 PERCENT OF CRITICAL DAMPING), SITE CLASS B

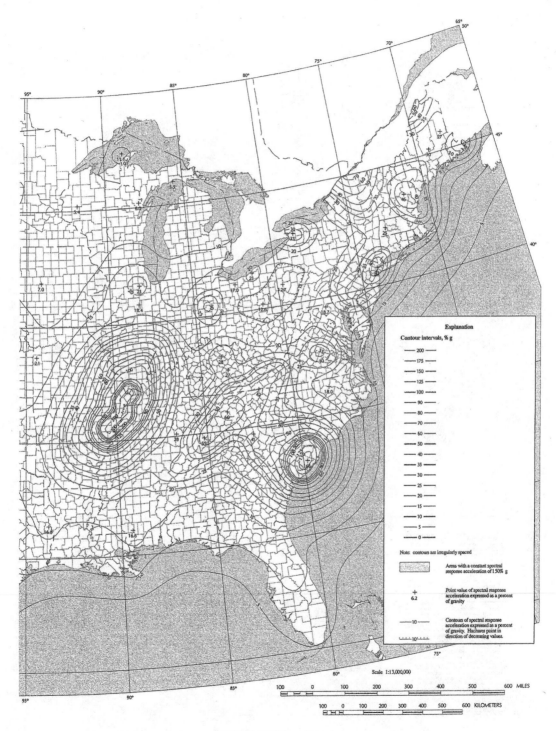

FIGURE 1615(1)–continued
MAXIMUM CONSIDERED EARTHQUAKE GROUND MOTION FOR THE CONTERMINOUS UNITED STATES
OF 0.2 SEC SPECTRAL RESPONSE ACCELERATION (5 PERCENT OF CRITICAL DAMPING), SITE CLASS B

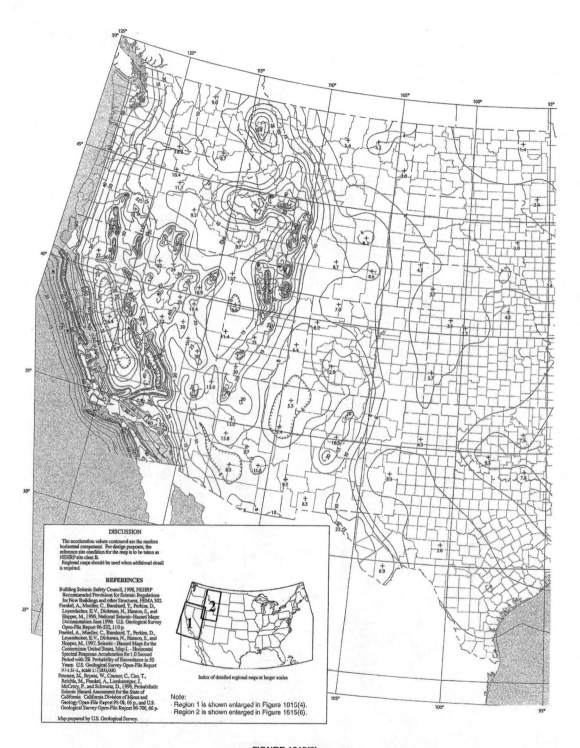

FIGURE 1615(2)
MAXIMUM CONSIDERED EARTHQUAKE GROUND MOTION FOR THE CONTERMINOUS UNITED STATES
OF 1.0 SEC SPECTRAL RESPONSE ACCELERATION (5 PERCENT OF CRITICAL DAMPING), SITE CLASS B

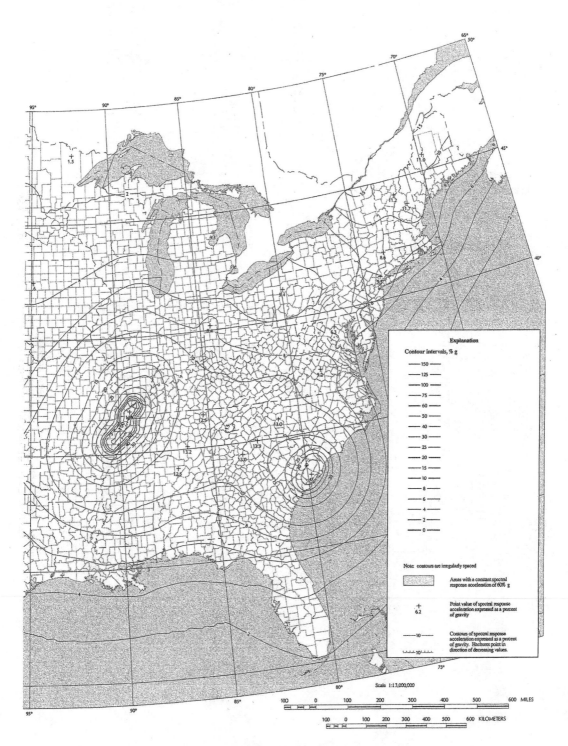

FIGURE 1615(2)–continued
MAXIMUM CONSIDERED EARTHQUAKE GROUND MOTION FOR THE CONTERMINOUS UNITED STATES
OF 1.0 SEC SPECTRAL RESPONSE ACCELERATION (5 PERCENT OF CRITICAL DAMPING), SITE CLASS B

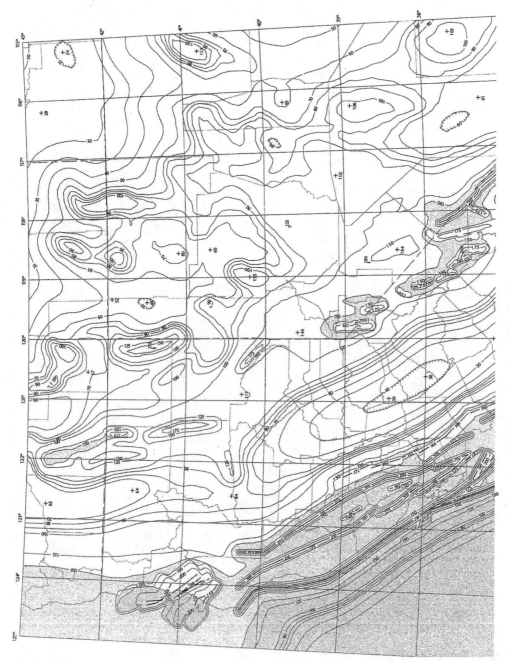

FIGURE 1615(3)
MAXIMUM CONSIDERED EARTHQUAKE GROUND MOTION FOR REGION 1 OF 0.2 SEC SPECTRAL RESPONSE ACCELERATION (5 PERCENT OF CRITICAL DAMPING), SITE CLASS B

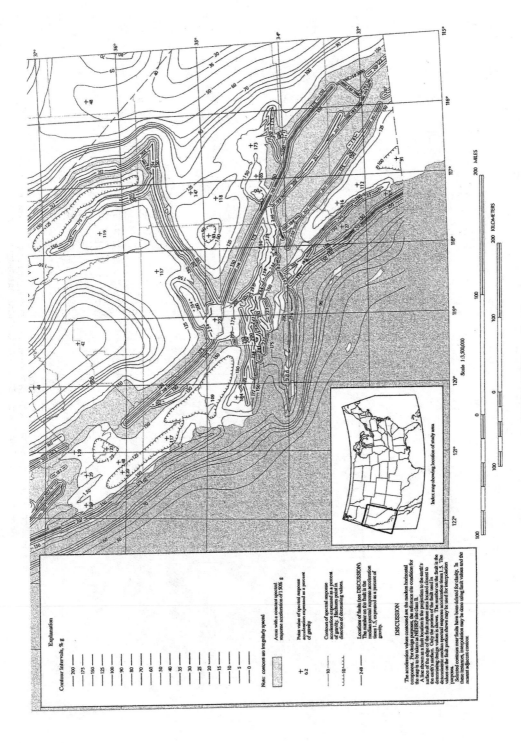

FIGURE 1615(3)—continued
**MAXIMUM CONSIDERED EARTHQUAKE GROUND MOTION FOR REGION 1 OF 0.2 SEC SPECTRAL
RESPONSE ACCELERATION (5 PERCENT OF CRITICAL DAMPING), SITE CLASS B**

FIGURE 1615(4)
MAXIMUM CONSIDERED EARTHQUAKE GROUND MOTION FOR REGION 1 OF 1.0 SEC SPECTRAL
RESPONSE ACCELERATION (5 PERCENT OF CRITICAL DAMPING), SITE CLASS B

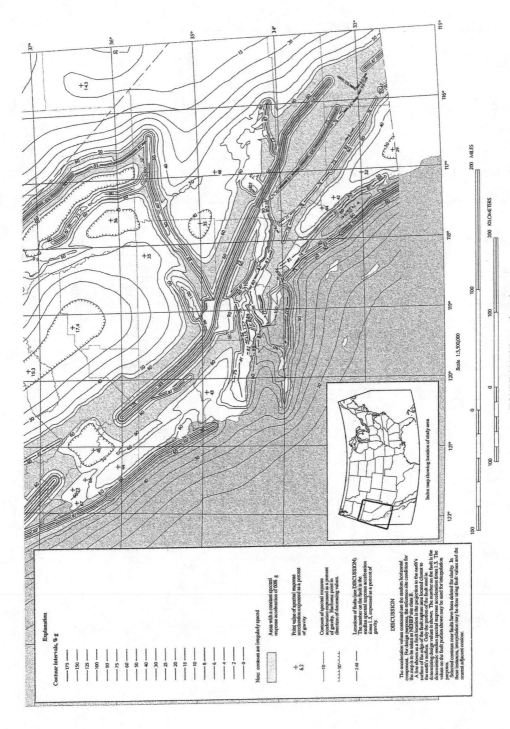

FIGURE 1615(4)—continued
**MAXIMUM CONSIDERED EARTHQUAKE GROUND MOTION FOR REGION 1 OF 1.0 SEC SPECTRAL
RESPONSE ACCELERATION (5 PERCENT OF CRITICAL DAMPING), SITE CLASS B**

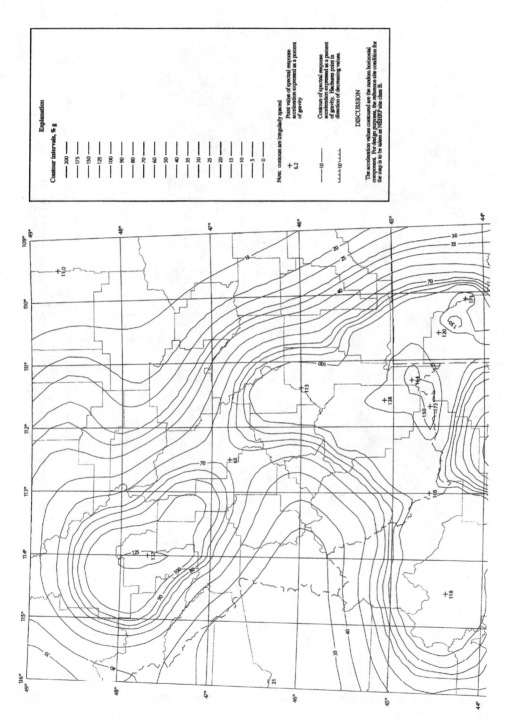

FIGURE 1615(5)
MAXIMUM CONSIDERED EARTHQUAKE GROUND MOTION FOR REGION 2 OF 0.2 SEC SPECTRAL
RESPONSE ACCELERATION (5 PERCENT OF CRITICAL DAMPING), SITE CLASS B

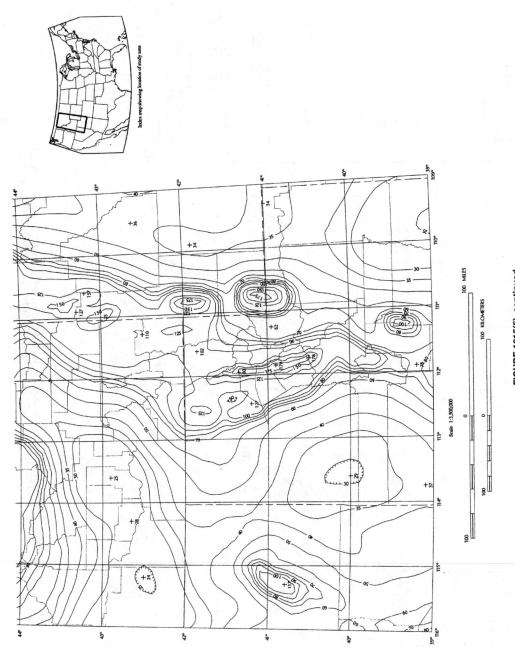

FIGURE 1615(5)—continued
MAXIMUM CONSIDERED EARTHQUAKE GROUND MOTION FOR REGION 2 OF 0.2 SEC SPECTRAL
RESPONSE ACCELERATION (5 PERCENT OF CRITICAL DAMPING), SITE CLASS B

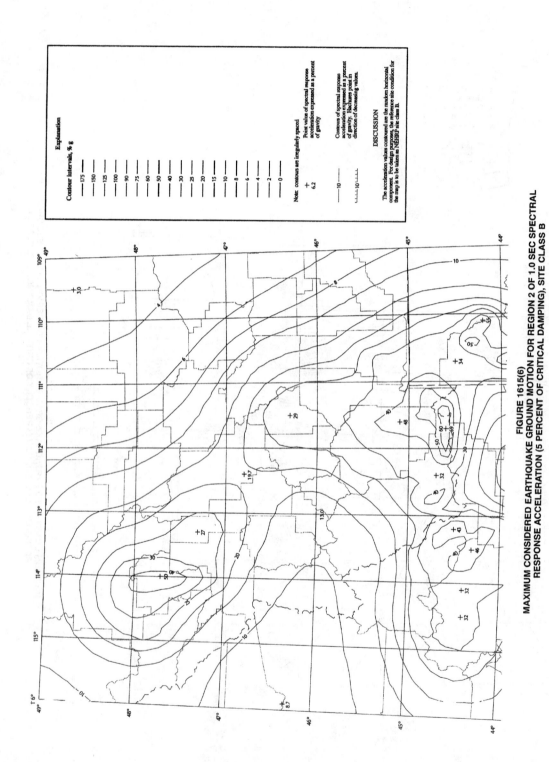

FIGURE 1615(6)

MAXIMUM CONSIDERED EARTHQUAKE GROUND MOTION FOR REGION 2 OF 1.0 SEC SPECTRAL
RESPONSE ACCELERATION (5 PERCENT OF CRITICAL DAMPING), SITE CLASS B

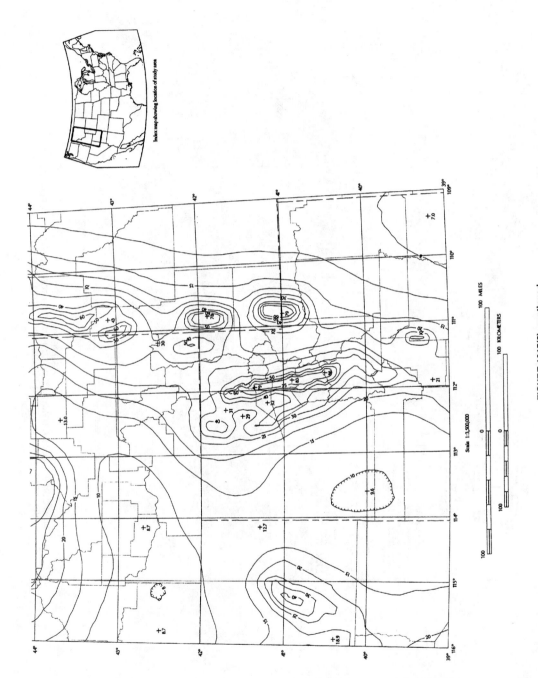

**FIGURE 1615(6)—continued
MAXIMUM CONSIDERED EARTHQUAKE GROUND MOTION FOR REGION 2 OF 1.0 SEC SPECTRAL
RESPONSE ACCELERATION (5 PERCENT OF CRITICAL DAMPING), SITE CLASS B**

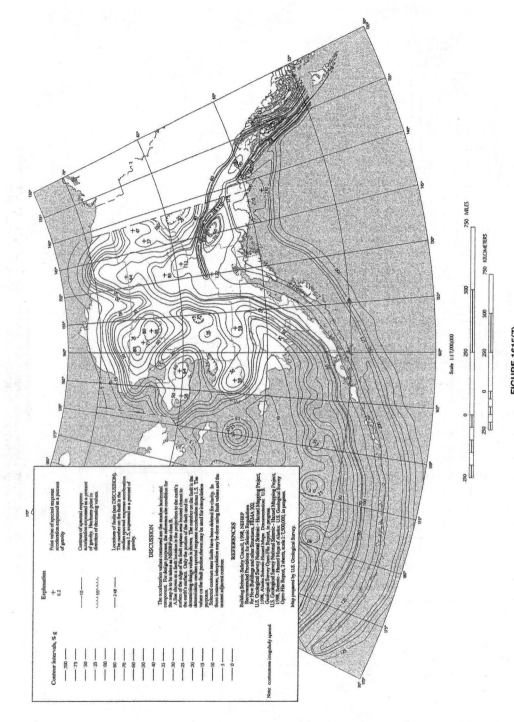

FIGURE 1615(7)

MAXIMUM CONSIDERED EARTHQUAKE GROUND MOTION FOR ALASKA OF 0.2 SEC SPECTRAL RESPONSE ACCELERATION (5 PERCENT OF CRITICAL DAMPING), SITE CLASS B

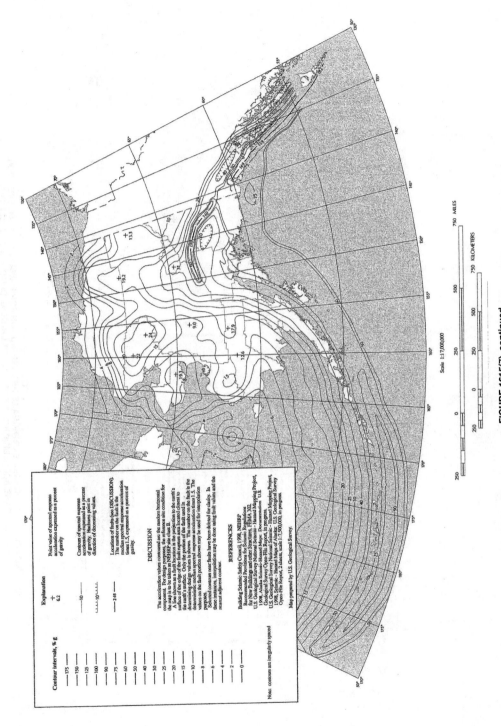

FIGURE 1615(7)—continued
MAXIMUM CONSIDERED EARTHQUAKE GROUND MOTION FOR ALASKA OF 0.2 SEC SPECTRAL
RESPONSE ACCELERATION (5 PERCENT OF CRITICAL DAMPING), SITE CLASS B

Scale 1:17,000,000

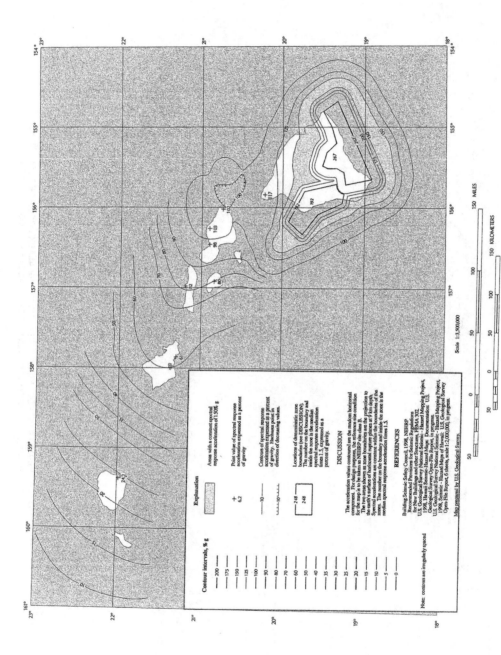

FIGURE 1615(8)

MAXIMUM CONSIDERED EARTHQUAKE GROUND MOTION FOR HAWAII OF 0.2 SEC SPECTRAL RESPONSE ACCELERATION (5 PERCENT OF CRITICAL DAMPING), SITE CLASS B

C.38

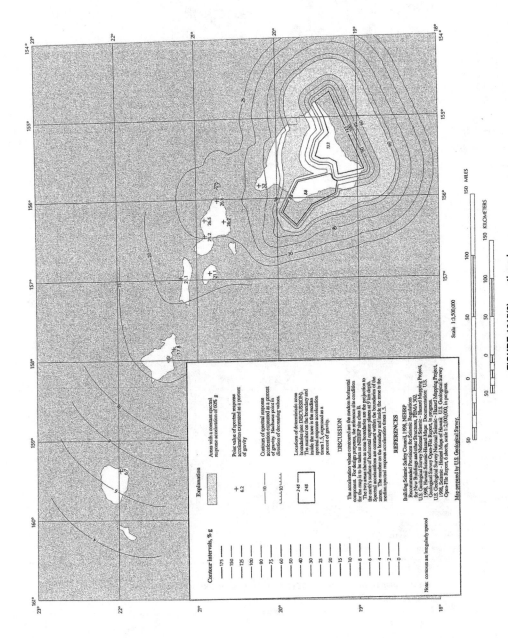

FIGURE 1615(8)—continued
MAXIMUM CONSIDERED EARTHQUAKE GROUND MOTION FOR HAWAII OF 0.2 SEC SPECTRAL
RESPONSE ACCELERATION (5 PERCENT OF CRITICAL DAMPING), SITE CLASS B

C.39

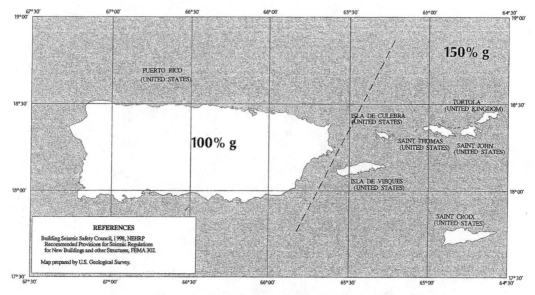

0.2 SEC SPECTRAL RESPONSE ACCELERATION (5% OF CRITICAL DAMPING)

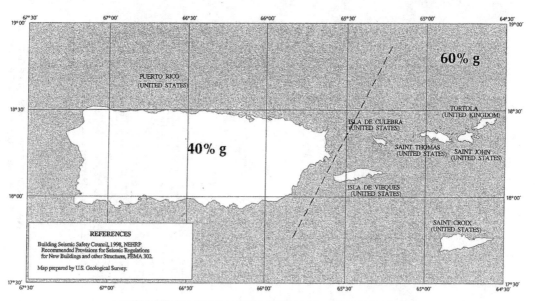

1.0 SEC SPECTRAL RESPONSE ACCELERATION (5% OF CRITICAL DAMPING)

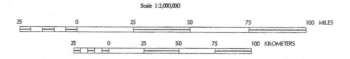

FIGURE 1615(9)
MAXIMUM CONSIDERED EARTHQUAKE GROUND MOTION FOR PUERTO RICO,
CULEBRA, VIEQUES, ST. THOMAS, ST. JOHN, AND ST. CROIX OF 0.2 AND 1.0 SEC SPECTRAL
RESPONSE ACCELERATION (5 PERCENT OF CRITICAL DAMPING), SITE CLASS B

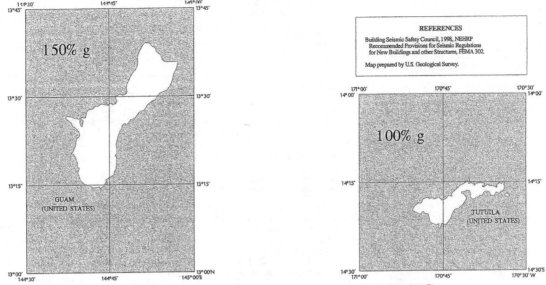

0.2 SEC SPECTRAL RESPONSE ACCELERATION (5% OF CRITICAL DAMPING)

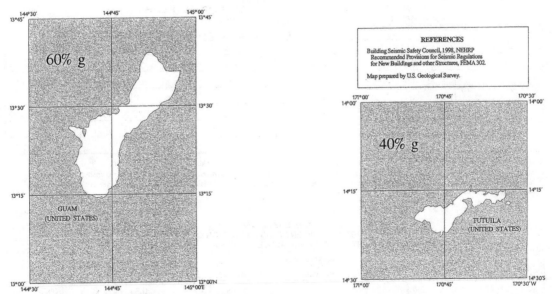

1.0 SEC SPECTRAL RESPONSE ACCELERATION (5% OF CRITICAL DAMPING)

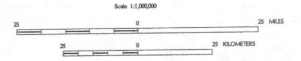

FIGURE 1615(10)
MAXIMUM CONSIDERED EARTHQUAKE GROUND MOTION FOR GUAM AND TUTUILLA
OF 0.2 AND 1.0 SEC SPECTRAL RESPONSE ACCELERATION
(5 PERCENT OF CRITICAL DAMPING), SITE CLASS B

TABLE 1615.1.1
SITE CLASS DEFINITIONS

SITE CLASS	SOIL PROFILE NAME	AVERAGE PROPERTIES IN TOP 100 feet, AS PER SECTION 1615.1.5		
		Soil shear wave velocity, $\bar{v}_s$, (ft/s)	Standard penetration resistance, $\bar{N}$	Soil undrained shear strength, $\bar{s}_u$, (psf)
A	Hard rock	$\bar{v}_s > 5{,}000$	N/A	N/A
B	Rock	$2{,}500 < \bar{v}_s \le 5{,}000$	N/A	N/A
C	Very dense soil and soft rock	$1{,}200 < \bar{v}_s \le 2{,}500$	$\bar{N} > 50$	$\bar{s}_u \ge 2{,}000$
D	Stiff soil profile	$600 \le \bar{v}_s \le 1{,}200$	$15 \le \bar{N} \le 50$	$1{,}000 \le \bar{s}_u \le 2{,}000$
E	Soft soil profile	$\bar{v}_s < 600$	$\bar{N} < 15$	$\bar{s}_u < 1{,}000$
E	—	Any profile with more than 10 feet of soil having the following characteristics: 1. Plasticity index $PI > 20$, 2. Moisture content $w \ge 40\%$, and 3. Undrained shear strength $\bar{s}_u < 500$ psf		
F	—	Any profile containing soils having one or more of the following characteristics: 1. Soils vulnerable to potential failure or collapse under seismic loading such as liquefiable soils, quick and highly sensitive clays, collapsible weakly cemented soils. 2. Peats and/or highly organic clays ($H > 10$ feet of peat and/or highly organic clay where H = thickness of soil) 3. Very high plasticity clays ($H > 25$ feet with plasticity index $PI > 75$) 4. Very thick soft/medium stiff clays ($H > 120$ feet)		

For SI: 1 foot = 304.8 mm, 1 square foot = 0.0929 m², 1 pound per square foot = 0.0479 kPa. N/A = Not applicable

TABLE 1615.1.2(1)
VALUES OF SITE COEFFICIENT F_a AS A FUNCTION OF SITE CLASS
AND MAPPED SPECTRAL RESPONSE ACCELERATION AT SHORT PERIODS (S_s)[a]

SITE CLASS	MAPPED SPECTRAL RESPONSE ACCELERATION AT SHORT PERIODS				
	$S_s \le 0.25$	$S_s = 0.50$	$S_s = 0.75$	$S_s = 1.00$	$S_s \ge 1.25$
A	0.8	0.8	0.8	0.8	0.8
B	1.0	1.0	1.0	1.0	1.0
C	1.2	1.2	1.1	1.0	1.0
D	1.6	1.4	1.2	1.1	1.0
E	2.5	1.7	1.2	0.9	0.9
F	Note b	Note b	Note b	Note b	Note b

a. Use straight-line interpolation for intermediate values of mapped spectral response acceleration at short period, S_s.
b. Site-specific geotechnical investigation and dynamic site response analyses shall be performed to determine appropriate values, except that for structures with periods of vibration equal to or less than 0.5 second, values of F_a for liquefiable soils are permitted to be taken equal to the values for the site class determined without regard to liquefaction in Section 1615.1.5.1.

TABLE 1615.1.2(2)
VALUES OF SITE COEFFICIENT F_v AS A FUNCTION OF SITE CLASS
AND MAPPED SPECTRAL RESPONSE ACCELERATION AT 1-SECOND PERIOD (S_1)[a]

SITE CLASS	MAPPED SPECTRAL RESPONSE ACCELERATION AT SHORT PERIODS				
	$S_1 \le 0.1$	$S_1 = 0.2$	$S_1 = 0.3$	$S_1 = 0.4$	$S_1 \ge 0.5$
A	0.8	0.8	0.8	0.8	0.8
B	1.0	1.0	1.0	1.0	1.0
C	1.7	1.6	1.5	1.4	1.3
D	2.4	2.0	1.8	1.6	1.5
E	3.5	3.2	2.8	2.4	2.4
F	Note b	Note b	Note b	Note b	Note b

a. Use straight-line interpolation for intermediate values of mapped spectral response acceleration at 1-second period, S_1.
b. Site-specific geotechnical investigation and dynamic site response analyses shall be performed to determine appropriate values, except that for structures with periods of vibration equal to or less than 0.5 second, values of F_v for liquefiable soils are permitted to be taken equal to the values for the site class determined without regard to liquefaction in Section 1615.1.5.1.

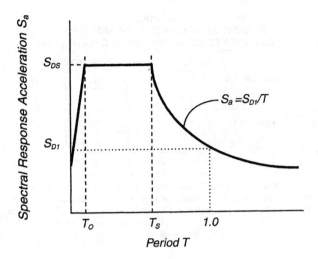

FIGURE 1615.1.4
DESIGN RESPONSE SPECTRUM

TABLE 1615.1.5
SITE CLASSIFICATION[a]

SITE CLASS	$\bar{v}_s$	$\bar{N}$ or $\bar{N}_{ch}$	$\bar{s}_u$
E	< 600 ft/s	< 15	< 1,000 psf
D	600 to 1,200 ft/s	15 to 50	1,000 to 2,000 psf
C	1,200 to 2,500 ft/s	> 50	> 2,000

For SI: 1 foot per second = 304.8 mm per second, 1 pound per square foot = 0.0479 kN/m^2.
a. If the $\bar{s}_u$ method is used and the $\bar{N}_{ch}$ and $\bar{s}_u$ criteria differ, select the category with the softer soils (for example, use Site Class E instead of D).

TABLE 1616.3(1)
SEISMIC DESIGN CATEGORY BASED ON
SHORT-PERIOD RESPONSE ACCELERATIONS

VALUE OF S_{DS}	SEISMIC USE GROUP		
	I	II	III
$S_{DS} < 0.167g$	A	A	A
$0.167g \leq S_{DS} < 0.33g$	B	B	C
$0.33g \leq S_{DS} < 0.50g$	C	C	D
$0.50g \leq S_{DS}$	D[a]	D[a]	D[a]

a. Seismic Use Group I and II structures located on sites with mapped maximum considered earthquake spectral response acceleration at 1-second period, S_1, equal to or greater than 0.75g, shall be assigned to Seismic Design Category E, and Seismic Use Group III structures located on such sites shall be assigned to Seismic Design Category F.

TABLE 1616.3(2)
SEISMIC DESIGN CATEGORY BASED ON
1-SECOND PERIOD RESPONSE ACCELERATION

VALUE OF S_{DI}	SEISMIC USE GROUP		
	I	II	III
$S_{DI} < 0.067g$	A	A	A
$0.067g \leq S_{DI} < 0.133g$	B	B	C
$0.133g \leq S_{DI} < 0.20g$	C	C	D
$0.20g \leq S_{DI}$	D[a]	D[a]	D[a]

a. Seismic Use Group I and II structures located on sites with mapped maximum considered earthquake spectral response acceleration at 1-second period, S_1, equal to or greater than 0.75g, shall be assigned to Seismic Design Category E, and Seismic Use Group III structures located on such sites shall be assigned to Seismic Design Category F.

TABLE 1616.5.1.1
PLAN STRUCTURAL IRREGULARITIES

	IRREGULARITY TYPE AND DESCRIPTION	REFERENCE SECTION	SEISMIC DESIGN CATEGORY[a] APPLICATION
1a	Torsional Irregularity—to be considered when diaphragms are not flexible as determined in Section 1602.1.1 Torsional irregularity shall be considered to exist when the maximum story drift, computed including accidental torsion, at one end of the structure transverse to an axis is more than 1.2 times the average of the story drifts at the two ends of the structure.	9.5.5.5.2 of ASCE 7 1620.4.1 9.5.2.5.1 of ASCE 7 9.5.5.7.1 of ASCE 7	C, D, E and F D, E and F D, E and F C, D, E and F
1b	Extreme Torsional Irregularity—to be considered when diaphragms are not flexible as determined in Section 1602.1. Extreme torsional irregularity shall be considered to exist when the maximum story drift, computed and including accidental torsion, at one end of the structure transverse to an axis is more than 1.4 times the average of the story drifts at the two ends of the structure.	9.5.5.5.2 of ASCE 7 1620.4.1 1620.5.1 9.5.2.5.1 of ASCE 7 9.5.5.7.1 of ASCE 7	C, D, E and F D E and F D, E and F C, D, E and F
2	Reentrant Corners Plan configurations of a structure and its lateral-force-resisting system contain reentrant corners where both projections of the structure beyond a reentrant corner are greater than 15 percent of the plan dimension of the structure in the given direction.	1620.4.1	D, E and F
3	Diaphragm Discontinuity Diaphragms with abrupt discontinuities or variations in stiffness, including those having cutout or open areas greater than 50 percent of the gross enclosed diaphragm area, or changes in effective diaphragm stiffness of more than 50 percent from one story to the next.	1620.4.1	D, E and F
4	Out-of-Plane Offsets Discontinuities in a lateral-force-resistance path, such as out-of-plane offsets of the vertical elements.	1620.4.1 9.5.2.5.1 of ASCE 7 1620.2.9	D, E and F D, E and F B, C, D, E and F
5	Nonparallel Systems The vertical lateral-force-resisting elements are not parallel to or symmetric about the major orthogonal axes of the lateral-force-resisting system.	1620.3.2	C, D, E and F

a. Seismic design category is determined in accordance with Section 1616.

TABLE 1616.5.1.2
VERTICAL STRUCTURAL IRREGULARITIES

	IRREGULARITY TYPE AND DESCRIPTION	REFERENCE SECTION	SEISMIC DESIGN CATEGORY[a] APPLICATION
1a	Stiffness Irregularity—Soft Story A soft story is one in which the lateral stiffness is less than 70 percent of that in the story above or less than 80 percent of the average stiffness of the three stories above.	9.5.2.5.1 of ASCE 7	D, E, and F
1b	Stiffness Irregularity—Extreme Soft Story An extreme soft story is one in which the lateral stiffness is less than 60 percent of that in the story above or less than 70 percent of the average stiffness of the three stories above.	1620.5.1 9.5.2.5.1 of ASCE 7	E and F D, E and F
2	Weight (Mass) Irregularity Mass irregularity shall be considered to exist where the effective mass of any story is more than 150 percent of the effective mass of an adjacent story. A roof that is lighter than the floor below need not be considered.	9.5.2.5.1 of ASCE 7	D, E and F
3	Vertical Geometric Irregularity Vertical geometric irregularity shall be considered to exist where the horizontal dimension of the lateral-force-resisting system in any story is more than 130 percent of that in an adjacent story.	9.5.2.5.1 of ASCE 7	D, E and F
4	In-plane Discontinuity in Vertical Lateral-Force-Resisting Elements An in-plane offset of the lateral-force-resisting elements greater than the length of those elements or a reduction in stiffness of the resisting element in the story below.	1620.4.1 9.5.2.5.1 of ASCE 7 1620.2.9	D, E and F D, E and F B, C, D, E and F
5	Discontinuity in Capacity—Weak Story A weak story is one in which the story lateral strength is less than 80 percent of that in the story above. The story strength is the total strength of seismic-resisting elements sharing the story shear for the direction under consideration.	1620.2.3 9.5.2.5.1 of ASCE 7 1620.5.1	B, C, D, E and F D, E and F E and F

a. Seismic design category is determined in accordance with Section 1616.

TABLE 1912.2
ALLOWABLE SERVICE LOAD ON EMBEDDED BOLTS (pounds)

BOLT DIAMETER (inches)	MINIMUM EMBEDMENT (inches)	EDGE DISTANCE (inches)	SPACING (inches)	MINIMUM CONCRETE STRENGTH (psi)					
				$f'_c = 2,500$		$f'_c = 3,000$		$f'_c = 4,000$	
				Tension	Shear	Tension	Shear	Tension	Shear
$1/4$	$2^1/_2$	$1^1/_2$	3	200	500	200	500	200	500
$3/_8$	3	$2^1/_4$	$4^1/_2$	500	1,100	500	1,100	500	1,100
$1/_2$	4 4	3 5	6 5	950 1,450	1,250 1,600	950 1,500	1,250 1,650	950 1,550	1,250 1,750
$5/_8$	$4^1/_2$ $4^1/_2$	$3^3/_4$ $6^1/_4$	$7^1/_2$ $7^1/_2$	1,500 2,125	2,750 2,950	1,500 2,200	2,750 3,000	1,500 2,400	2,750 3,050
$3/_4$	5 5	$4^1/_2$ $7^1/_2$	9 9	2,250 2,825	3,250 4,275	2,250 2,950	3,560 4,300	2,250 3,200	3,560 4,400
$7/_8$	6	$5^1/_4$	$10^1/_2$	2,550	3,700	2,550	4,050	2,550	4,050
1	7	6	12	3,050	4,125	3,250	4,500	3,650	5,300
$1^1/_8$	8	$6^3/_4$	$13^1/_2$	3,400	4,750	3,400	4,750	3,400	4,750
$1^1/_4$	9	$7^1/_2$	15	4,000	5,800	4,000	5,800	4,000	5,800

For SI: 1 inch = 25.4 mm, 1 pound per square inch = 0.00689 MPa, 1 pound = 4.45 N.

TABLE 2304.9.1
FASTENING SCHEDULE

CONNECTION	FASTENING[a,m]	LOCATION
1. Joist to sill or girder	3 - 8d common 3 - 3″ × 0.131″ nails 3 - 3″ 14 gage staples	toenail
2. Bridging to joist	2 - 8d common 2 - 3″ × 0.131″ nails 2 - 3″14 gage staples	toenail each end
3. 1″ × 6″ subfloor or less to each joist	2 - 8d common	face nail
4. Wider than 1″ × 6″ subfloor to each joist	3 - 8d common	face nail
5. 2″ subfloor to joist or girder	2 - 16d common	blind and face nail
6. Sole plate to joist or blocking	16d at 16″ o.c. 3″ × 0.131″ nails at 8″ o.c. 3″ 14 gage staples at 12″ o.c.	typical face nail
Sole plate to joist or blocking at braced wall panel	3 - 16d at 16″ 4 - 3″ × 0.131″ nails at 16″ 4 - 3″ 14 gage staples per 16″	braced wall panels
7. Top plate to stud	2 - 16d common 3 - 3″ × 0.131″ nails 3 - 3″ 14 gage staples	end nail
8. Stud to sole plate	4 - 8d common 4 - 3″ × 0.131″ nails 3 - 3″ 14 gage staples	toenail
	2 - 16d common 3 - 3″ × 0.131″ nails 3 - 3″ 14 gage staples	end nail
9. Double studs	16d at 24″ o.c. 3″ × 0.131″ nail at 8″ o.c. 3″ 14 gage staple at 8″ o.c.	face nail
10. Double top plates	16d at 16″ o.c. 3″ × 0.131″ nail at 12″ o.c. 3″ 14 gage staple at 12″ o.c.	typical face nail
Double top plates	8-16d common 12 - 3″ × 0.131″ nails 12 - 3″ 14 gage staples typical face nail	lap splice
11. Blocking between joists or rafters to top plate	3 - 8d common 3 - 3″ × 0.131″ nails 3 - 3″ 14 gage staples	toenail
12. Rim joist to top plate	8d at 6″ (152 mm) o.c. 3″ × 0.131″ nail at 6″ o.c. 3″ 14 gage staple at 6″ o.c.	toenail
13. Top plates, laps and intersections	2 - 16d common 3 - 3″ × 0.131″ nails 3 - 3″ 14 gage staples	face nail
14. Continuous header, two pieces	16d common	16″ o.c. along edge
15. Ceiling joists to plate	3 - 8d common 5 - 3″ × 0.131″ nails 5 - 3″ 14 gage staples	toenail
16. Continuous header to stud	4 - 8d common	toenail

(continued)

TABLE 2304.9.1—continued
FASTENING SCHEDULE

CONNECTION	FASTENING[a,m]	LOCATION
17. Ceiling joists, laps over partitions (see Section 2308.10.4.1, Table 2308.10.4.1)	3 - 16d common minimum, Table 2308.10.4.1 4 - 3″ × 0.131″ nails 4 - 3″ 14 gage staples	face nail
18. Ceiling joists to parallel rafters (see Section 2308.10.4.1, Table 2308.10.4.1)	3 - 16d common minimum, Table 2308.10.4.1 4 - 3″ × 0.131″ nails 4 - 3″ 14 gage staples	face nail
19. Rafter to plate (see Section 2308.10.1, Table 2308.10.1)	3 - 8d common 3 - 3″ × 0.131″ nails 3 - 3″ 14 gage staples	toenail
20. 1″ diagonal brace to each stud and plate	2 - 8d common 2 - 3″ × 0.131″ nails 2 - 3″ 14 gage staples face nail	face nail
21. 1″ × 8″ sheathing to each bearing wall	2 - 8d common	face nail
22. Wider than 1″× 8″ sheathing to each bearing	3 - 8d common	face nail
23. Built-up corner studs	16d common 3″ × 0.131″ nails 3″ 14 gage staples	24″ o.c. 16″ o.c. 16″ o.c.
24. Built-up girder and beams	20d common 32″ o.c. 3″ × 0.131″ nail at 24″ o.c. 3″ 14 gage staple at 24″ o.c.	face nail at top and bottom staggered on opposite sides
	2 - 20d common 3 - 3″ × 0.131″ nails 3 - 3″ 14 gage staples	face nail at ends and at each splice
25. 2″ planks	16d common	at each bearing
26. Collar tie to rafter	3 - 10d common 4 - 3″ × 0.131″ nails 4 - 3″ 14 gage staples face nail	face nail
27. Jack rafter to hip	3 - 10d common 4 - 3″ × 0.131″nails 4 - 3″ 14 gage staples	toenail
	2 - 16d common 3 - 3″ × 0.131″ nails 3 - 3″ 14 gage staples	face nail
28. Roof rafter to 2-by ridge beam	2 - 16d common 3 - 3″ × 0.131″ nails 3 - 3″ 14 gage staples	toenail
	2 - 16d common 3 - 3″ × 0.131″ nails 3 - 3″ 14 gage staples	face nail
29. Joist to band joist	3 - 16d common 5 - 3″ × 0.131″ nails 5 - 3″ 14 gage staples	face nail

(continued)

TABLE 2304.9.1—continued
FASTENING SCHEDULE

CONNECTION	FASTENING[a,m]	LOCATION
30. Ledger strip	3 - 16d common 4 - 3″ × 0.131″ nails 4 - 3″ 14 gage staples	face nail

31. Wood structural panels and particleboard:[b]
Subfloor, roof and wall sheathing (to framing):
	$^1/_2$″ and less 6d[e,l]
	                    2 $^3/_8$″ × 0.113″ nail[n]
	                    1 $^3/_4$″ 16 gage[o]
	$^{19}/_{32}$″ to $^3/_4$″ 8d[d] or 6d[e]
	                    2 $^3/_8$″ × 0.113″ nail[p]
	                    2″ 16 gage[p]
	$^7/_8$″ to 1″ 8d[e]	

Single Floor (combination subfloor-underlayment to framing):	1 $^1/_8$″ to 1 $^1/_4$″ 10d[d] or 8d[e] $^3/_4$″ and less 6d[e] $^7/_8$″ to 1″ 8d[e] 1 $^1/_8$″ to 1 $^1/_4$″ 10d[d] or 8d[e]	
32. Panel siding (to framing)	$^1/_2$″ or less 6d[f] $^5/_8$″ 8d[f]	

33. Fiberboard sheathing:[g]	$^1/_2$″ No. 11 gage roofing nail[h]
	                 6d common nail
	                 No. 16 gage staple[i]
	$^{25}/_{32}$″ No. 11 gage roofing nail[h]
	                 8d common nail
	                 No. 16 gage staple[i]	

34. Interior paneling	$^1/_4$″	4d[j]
	$^3/_8$″	6d[k]

For SI: 1 inch = 25.4 mm.

a. Common or box nails are permitted to be used except where otherwise stated.

b. Nails spaced at 6 inches on center at edges, 12 inches at intermediate supports except 6 inches at supports where spans are 48 inches or more. For nailing of wood structural panel and particleboard diaphragms and shear walls, refer to Section 2305. Nails for wall sheathing are permitted to be common, box or casing.

c. Common or deformed shank.

d. Common.

e. Deformed shank.

f. Corrosion-resistant siding or casing nail.

g. Fasteners spaced 3 inches on center at exterior edges and 6 inches on center at intermediate supports.

h. Corrosion-resistant roofing nails with $^7/_{16}$-inch-diameter head and 1$^1/_2$-inch length for $^1/_2$-inch sheathing and 1 $^3/_4$-inch length for $^{25}/_{32}$-inch sheathing.

i. Corrosion-resistant staples with nominal $^7/_{16}$-inch crown and 1 $^1/_8$-inch length for $^1/_2$-inch sheathing and 1 $^1/_2$-inch length for $^{25}/_{32}$-inch sheathing. Panel supports at 16 inches (20 inches if strength axis in the long direction of the panel, unless otherwise marked).

j. Casing or finish nails spaced 6 inches on panel edges, 12 inches at intermediate supports.

k. Panel supports at 24 inches. Casing or finish nails spaced 6 inches on panel edges, 12 inches at intermediate supports.

l. For roof sheathing applications, 8d nails are the minimum required for wood structural panels.

m. Staples shall have a minimum crown width of $^7/_{16}$ inch.

n. For roof sheathing applications, fasteners spaced 4 inches on center at edges, 8 inches at intermediate supports.

o. Fasteners spaced 4 inches on center at edges, 8 inches at intermediate supports for subfloor and wall sheathing and 3 inches on center at edges, 6 inches at intermediate supports for roof sheathing.

p. Fasteners spaced 4 inches on center at edges, 8 inches at intermediate supports.

TABLE 2305.3.3
MAXIMUM SHEAR WALL ASPECT RATIOS

TYPE	MAXIMUM HEIGHT-WIDTH RATIO
Wood structural panels or particleboard, nailed edges	For other than seismic: $3^1/_2$:1 For seismic: 2:1[a]
Diagonal sheathing, single	2:1
Fiberboard	$1^1/_2$:1
Gypsum board, gypsum lath, cement plaster	$1^1/_2$:1[b]

a. For design to resist seismic forces, shear wall aspect ratios greater than 2:1, but not exceeding $3^1/_2$:1, are permitted provided the factored shear resistance values in Table 2306.4.1 are multiplied by $2w/h$.

b. Ratio shown is for unblocked construction. Aspect ratio is permitted to be 2:1 where the wall is installed as blocked construction in accordance with Section 2306.4.5.1.2.

TABLE 2306.3.1
RECOMMENDED SHEAR (POUNDS PER FOOT) FOR WOOD STRUCTURAL PANEL DIAPHRAGMS WITH FRAMING OF DOUGLAS-FIR-LARCH, OR SOUTHERN PINE[a] FOR WIND OR SEISMIC LOADING

PANEL GRADE	COMMON NAIL SIZE OR STAPLE[f] LENGTH AND GAGE	MINIMUM FASTENER PENETRATION IN FRAMING (inches)	MINIMUM NOMINAL PANEL THICKNESS (inch)	MINIMUM NOMINAL WIDTH OF FRAMING MEMBER (inches)	BLOCKED DIAPHRAGMS — Fastener spacing (inches) at diaphragm boundaries (all cases) at continuous panel edges parallel to load (Cases 3, 4), and at all panel edges (Cases 5 and 6)[b]				UNBLOCKED DIAPHRAGMS — Fasteners spaced 6" max. At supported edges[b]	
					6	4	$2\frac{1}{2}$[c]	2[c]	Case 1 (No unblocked edges or continuous joints parallel to load)	All other configurations (Cases 2, 3, 4, 5 and 6)
					Fastener spacing (inches) at other panel edges (Cases 1, 2, 3 and 4)[b]					
					6	6	4	3		
Structural I Grades	6d[e]	$1\frac{1}{4}$	$\frac{5}{16}$	2	185	250	375	420	165	125
				3	210	280	420	475	185	140
	$1\frac{1}{2}$ 16 Gage	1		2	155	205	310	350	135	105
				3	175	230	345	390	155	115
	8d	$1\frac{3}{8}$	$\frac{3}{8}$	2	270	360	530	600	240	180
				3	300	400	600	675	265	200
	$1\frac{1}{2}$ 16 Gage	1		2	175	235	350	400	155	115
				3	200	265	395	450	175	130
	10d[d]	$1\frac{1}{2}$	$\frac{15}{32}$	2	320	425	640	730	285	215
				3	360	480	720	820	320	240
	$1\frac{1}{2}$ 16 Gage	1		2	175	235	350	400	155	120
				3	200	265	395	450	175	130
Sheathing, single floor and other grades covered in DOC PS 1 and PS 2	6d[e]	$1\frac{1}{4}$	$\frac{5}{16}$	2	170	225	335	380	150	110
				3	190	250	380	430	170	125
	$1\frac{1}{2}$ 16 Gage	1		2	140	185	275	315	125	90
				3	155	205	310	350	140	105
	6d[e]	$1\frac{1}{4}$	$\frac{3}{8}$	2	185	250	375	420	165	125
				3	210	280	420	475	185	140
	8d	$1\frac{3}{8}$		2	240	320	480	545	215	160
				3	270	360	540	610	240	180

(continued)

TABLE 2306.3.1—continued
RECOMMENDED SHEAR (POUNDS PER FOOT) FOR WOOD STRUCTURAL PANEL DIAPHRAGMS WITH FRAMING OF DOUGLAS-FIR-LARCH, OR SOUTHERN PINE[a] FOR WIND OR SEISMIC LOADING

PANEL GRADE	COMMON NAIL SIZE OR STAPLE[f] LENGTH AND GAGE	MINIMUM FASTENER PENETRATION IN FRAMING (inches)	MINIMUM NOMINAL PANEL THICKNESS (inch)	MINIMUM NOMINAL WIDTH OF FRAMING MEMBER (inches)	BLOCKED DIAPHRAGMS				UNBLOCKED DIAPHRAGMS	
					Fastener spacing (inches) at diaphragm boundaries (all cases), at continuous panel edges parallel to load (Cases 3, 4), and at all panel edges (Cases 5 and 6)[b]				Fasteners spaced 6" max. at supported edges[b]	
					6	4	2½[c]	2[c]	Case 1 (No unblocked edges or continuous joints parallel to load)	All other configurations (Cases 2, 3, 4, 5 and 6)
					___ Fastener spacing (inches) at other panel edges (Cases 1, 2, 3 and 4)[b] ___					
					6	6	4	3		
Sheathing, single floor and other grades covered in DOC PS 1 and PS 2 (continued)	1½ 16 Gage	1	3/8	2	160	210	315	360	140	105
				3	180	235	355	400	160	120
	8d	1 3/8	7/16	2	255	340	505	575	230	170
				3	285	380	570	645	255	190
	1½ 16 Gage	1	15/32	2	165	225	335	380	150	110
				3	190	250	375	425	165	125
	8d	1 3/8	15/32	2	270	360	530	600	240	180
				3	300	400	600	675	265	200
	10d[d]	1½	15/32	2	290	385	575	655	255	190
				3	325	430	650	735	290	215
	1½ 16 Gage	1	19/32	2	160	210	315	360	140	105
				3	180	235	355	405	160	120
	10d[d]	1½	19/32	2	320	425	640	730	285	215
				3	360	480	720	820	320	240
	1¾ 16 Gage	1	19/32	2	175	235	350	400	155	115
				3	200	265	395	450	175	130

(continued)

TABLE 2306.3.1—continued
RECOMMENDED SHEAR (POUNDS PER FOOT) FOR WOOD STRUCTURAL
PANEL DIAPHRAGMS WITH FRAMING OF DOUGLAS-FIR-LARCH,
OR SOUTHERN PINE[a] FOR WIND OR SEISMIC LOADING

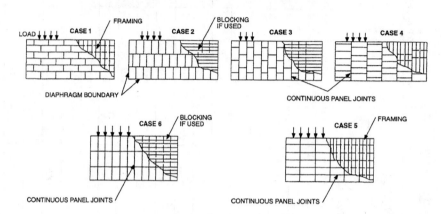

For SI: 1 inch = 25.4 mm, 1 pound per foot = 14.5939 N/m.

a. For framing of other species: (1) Find specific gravity for species of lumber in AFPA National Design Specification. (2) For staples find shear value from table above for Structural I panels (regardless of actual grade) and multiply value by 0.82 for species with specific gravity of 0.42 or greater, or 0.65 for all other species. (3) For nails find shear value from table above for nail size for actual grade and multiply value by the following adjustment factor: Specific Gravity Adjustment Factor = [1-(0.5 - SG)], where SG = Specific Gravity of the framing lumber. This adjustment factor shall not be greater than 1.

b. Space fasteners maximum 12 inches o.c. along intermediate framing members (6 inches o.c. where supports are spaced 48 inches o.c.).

c. Framing at adjoining panel edges shall be 3 inches nominal or wider, and nails shall be staggered where nails are spaced 2 inches o.c. or 2 $^1/_2$ inches o.c.

d. Framing at adjoining panel edges shall be 3 inches nominal or wider, and nails shall be staggered where both of the following conditions are met: (1) 10d nails having penetration into framing of more than 1$^1/_2$ inches and (2) nails are spaced 3 inches o.c. or less.

e. 8d is recommended minimum for roofs due to negative pressures of high winds.

f. Staples shall have a minimum crown width of $^7/_{16}$ inch.

TABLE 2306.4.1
ALLOWABLE SHEAR (POUNDS PER FOOT) FOR WOOD STRUCTURAL PANEL SHEAR WALLS WITH FRAMING OF DOUGLAS-FIR-LARCH, OR SOUTHERN PINE[a] FOR WIND OR SEISMIC LOADING[b, h, i, j]

PANEL GRADE	MINIMUM NOMINAL PANEL THICKNESS (inch)	MINIMUM FASTENER PENETRATION IN FRAMING (inches)	PANELS APPLIED DIRECT TO FRAMING					PANELS APPLIED OVER 1/2" OR 5/8" GYPSUM SHEATHING				
			NAIL (common or galvanized box) or staple size[k]	Fastener spacing at panel edges (inches)				NAIL (common or galvanized box) or staple size[k]	Fastener spacing at panel edges (inches)			
				6	4	3	2*		6	4	3	2*
Structural I Sheathing	5/16	1 1/4	6d	200	300	390	510	8d	200	300	390	510
		1	1 1/2 16 Gage	165	245	325	415	2 16 Gage	125	185	245	315
	3/8	1 3/8	8d	230d	360d	460d	610d	10d	280	430	550f	730
		1	1 1/2 16 Gage	155	235	315	400	2 16 Gage	155	235	310	400
	7/16	1 3/8	8d	255d	395d	505d	670d	10d	280	430	550f	730
		1	1 1/2 16 Gage	170	260	345	440	2 16 Gage	155	235	310	400
	15/32	1 3/8	8d	280	430	550	730	10d	280	430	550f	730
		1	1 1/2 16 Gage	185	280	375	475	2 16 Gage	155	235	300	400
		1 1/2	10d	340	510	665f	870	10d	—	—	—	—
Sheathing, plywood siding[g] except Group 5 Species	5/16 or 1/4 c	1 1/4	6d	180	270	350	450	8d	180	270	350	450
		1	1 1/2 16 Gage	145	220	295	375	2 16 Gage	110	165	220	285
	3/8	1 1/4	6d	200	300	390	510	8d	200	300	390	510
		1 3/8	8d	220d	320d	410d	530d	10d	260	380	490f	640
		1	1 1/2 16 Gage	140	210	280	360	2 16 Gage	140	210	280	360
	7/16	1 3/8	8d	240d	350d	450d	585d	10d	260	380	490f	640
		1	1 1/2 16 Gage	155	230	310	395	2 16 Gage	140	210	280	360
	15/32	1 3/8	8d	260	380	490	640	10d	260	380	490f	640
		1 1/2	10d	310	460	600f	770	—	—	—	—	—
		1	1 1/2 16 Gage	170	255	335	430	2 16 Gage	140	210	280	360
	19/32	1 1/2	10d	340	510	665f	870	—	—	—	—	—
		1	1 3/4 16 Gage	185	280	375	475	—	—	—	—	—
			Nail Size (galvanized casing)					Nail Size (galvanized casing)				
	5/16 c	1 1/4	6d	140	210	275	360	8d	140	210	275	360
	3/8	1 3/8	8d	160	240	310	410	10d	160	240	310f	410

For SI: 1 inch = 25.4 mm, 1 pound per foot = 14.5939 N/m.

a. For framing of other species: (1) Find specific gravity for species of lumber in AF&PA National Design Specification. (2) For staples find shear value from table above for Structural I panels (regardless of actual grade) and multiply value by 0.82 for species with specific gravity of 0.42 or greater, or 0.65 for all other species. (3) For nails find shear value from table above for nail size for actual grade and multiply value by the following adjustment factor: Specific Gravity Adjustment Factor = [1−(0.5 − SG)], where SG = Specific Gravity of the framing lumber. This adjustment factor shall not be greater than 1.

b. Panel edges backed with 2-inch nominal or wider framing. Install panels either horizontally or vertically. Space fasteners maximum 6 inches on center along intermediate framing members for 3/8-inch and 7/16-inch panels installed on studs spaced 24 inches on center. For other conditions and panel thickness, space fasteners maximum 12 inches on center on intermediate supports.

c. 3/8-inch panel thickness or siding with a span rating of 16 inches on center is the minimum recommended where applied direct to framing as exterior siding.

d. Shears are permitted to be increased to values shown for 15/32-inch sheathing with same nailing provided (a) studs are spaced a maximum of 16 inches on center, or (b) if panels are applied with long dimension across studs.

e. Framing at adjoining panel edges shall be 3 inches nominal or wider, and nails shall be staggered where nails are spaced 2 inches on center.

f. Framing at adjoining panel edges shall be 3 inches nominal or wider, and nails shall be staggered where both of the following conditions are met: (1) 10d nails having penetration into framing of more than 1 1/2 inches and (2) nails are spaced 3 inches on center.

g. Values apply to all-veneer plywood. Thickness at point of fastening on panel edges governs shear values.

h. Where panels are applied on both faces of a wall and nail spacing is less than 6 inches o.c. on either side, panel joints shall be offset to fall on different framing members. Or framing shall be 3 inch nominal or thicker and nails on each side shall be staggered.

i. In Seismic Design Category D, E or F, where shear design values exceed 490 pounds per lineal foot (LRFD) or 350 pounds per lineal foot (ASD) all framing members receiving edge nailing from abutting panels shall not be less than a single 3-inch nominal member. Plywood joint and sill plate nailing shall be staggered in all cases. See Section 2305.3.10 for sill plate side and anchorage requirements.

j. Galvanized nails shall be hot dipped or tumbled.

k. Staples shall have a minimum crown width of 7/16 inch.

C.53

TABLE 2306.4.3
ALLOWABLE SHEAR FOR PARTICLEBOARD SHEAR WALL SHEATHING

PANEL GRADE	MINIMUM NOMINAL PANEL THICKNESS (inch)	MINIMUM NAIL PENETRATION IN FRAMING (inches)	PANELS APPLIED DIRECT TO FRAMING				
			Nail size (common or galvanized box)	Allowable shear (pounds per foot) nail spacing at panel edges (inches)[a]			
				6	4	3	2
M-S "Exterior Glue" and M-2 "Exterior Glue"	$^3/_8$	$1\,^1/_2$	6d	120	180	230	300
	$^3/_8$	$1\,^1/_2$	8d	130	190	240	315
	$^1/_2$			140	210	270	350
	$^1/_2$	$1\,^5/_8$	10d	185	275	360	460
	$^5/_8$			200	305	395	520

For SI: 1 inch = 25.4 mm, 1 pound per foot = 14.5939 N/m.

a. Values are not permitted in Seismic Design Category D, E or F.

TABLE 2306.4.5
ALLOWABLE SHEAR FOR WIND OR SEISMIC FORCES FOR SHEAR WALLS OF LATH
AND PLASTER OR GYPSUM BOARD WOOD FRAMED WALL ASSEMBLIES

TYPE OF MATERIAL	THICKNESS OF MATERIAL	WALL CONSTRUCTION	FASTENER SPACING[b] MAXIMUM (inches)	SHEAR VALUE[a,e] (plf)	MINIMUM FASTENER SIZE[c,d,j,k]
1. Expanded metal or woven wire lath and portland cement plaster	$7/8''$	Unblocked	6	180	No. 11 gage $1^1/_2''$ long, $7/_{16}''$ head 16 Ga. Galv. Staple, $7/_8''$ legs
2. Gypsum lath, plain or perforated	$3/_8''$ lath and $1/_2''$ plaster	Unblocked	5	100	No. 13 gage, $1^1/_8''$ long, $19/_{64}''$ head, plasterboard nail 16 Ga. Galv. Staple, $1^1/_8''$ long $0.120''$ Nail, min. $3/_8''$ head, $1^1/_4''$ long
3. Gypsum sheathing	$1/_2'' \times 2' \times 8'$	Unblocked	4	75	No. 11 gage, $1^3/_4''$ long, $7/_{16}''$ head, diamond-point, galvanized
	$1/_2'' \times 4'$	Blocked[f]	4	175	16 Ga. Galv. Staple, $1^3/_4''$ long
		Unblocked	7	100	
	$5/_8'' \times 4'$	Blocked	4'' edge/7'' field	200	6d galvanized $0.120''$ Nail, min. $3/_8''$ head, $1^3/_4''$ long
4. Gypsum board, gypsum veneer base, or water-resistant gypsum backing board	$1/_2''$	Unblocked[f]	7	75	5d cooler or wallboard $0.120''$ Nail, min. $3/_8''$ head, $1^1/_2''$ long 16 Gage Staple, $1^1/_2''$ long
		Unblocked[f]	4	110	
		Unblocked	7	100	
		Unblocked	4	125	
		Blocked[g]	7	125	
		Blocked[g]	4	150	
		Unblocked	8/12[h]	60	No. 6-$1^1/_4''$ screws[i]
		Blocked[g]	4/16[h]	160	
		Blocked[g]	4/12[h]	155	
		Blocked[f, g]	8/12[h]	70	
		Blocked[g]	6/12[h]	90	
	$5/_8''$	Unblocked[f]	7	115	6d cooler or wallboard $0.120''$ Nail, min. $3/_8''$ head, $1^3/_4''$ long 16 Gage Staple, $1^1/_2''$ legs, $1^5/_8''$ long
		Unblocked[f]	4	145	
		Blocked[g]	7	145	
		Blocked[g]	4	175	
		Blocked[g] Two-ply	Base ply: 9 Face ply: 7	250	Base ply—6d cooler or wallboard $1^3/_4'' \times 0.120''$ Nail, min. $3/_8''$ head $1^5/_8''$ 16 Ga. Galv. Staple Face ply—8d cooler or wallboard $0.120''$ Nail, min. $3/_8''$ head, $2^3/_8''$ long 15 Ga. Galv. Staple, $2^1/_4''$ long
		Unblocked	8/12[h]	70	No. 6-$1^1/_4''$ screws[i]
		Blocked[g]	8/12[h]	90	

For SI: 1 inch = 25.4 mm, 1 foot = 304.8 mm, 1 pound per foot = 14.5939 N/m.

a. These shear walls shall not be used to resist loads imposed by masonry or concrete construction (see Section 2305.1.5). Values shown are for short-term loading due to wind or seismic loading in Seismic Design Categories A, B and C. Walls resisting seismic loads shall be subject to the limitations in Section 1617.6. Values shown shall be reduced 25 percent for normal loading.

b. Applies to nailing at studs, top and bottom plates and blocking.

c. Alternate nails are permitted to be used if their dimensions are not less than the specified dimensions. Drywall screws are permitted to be substituted for the 5d, 6d (cooler) nails listed above. $1^1/_4$ inches Type S or W, No. 6 for 6d (cooler) nails.

d. For properties of cooler nails, see ASTM C 514.

e. Except as noted, shear values are based on a maximum framing spacing of 16 inches on center.

f. Maximum framing spacing of 24 inches on center.

g. All edges are blocked, and edge nailing is provided at all supports and all panel edges.

h. First number denotes fastener spacing at the edges; second number denotes fastener spacing in the field.

i. Screws are Type W or S.

j. Staples shall have a minimum crown width of $7/_{16}$ inch, measured outside the legs.

k. Staples for the attachment of gypsum lath and woven-wire lath shall have a minimum crown width of $3/_4$ inch, measured outside the legs.

Selected Tables and Figures from Minimum Design Loads for Buildings and Other Structures, ASCE 7-02*

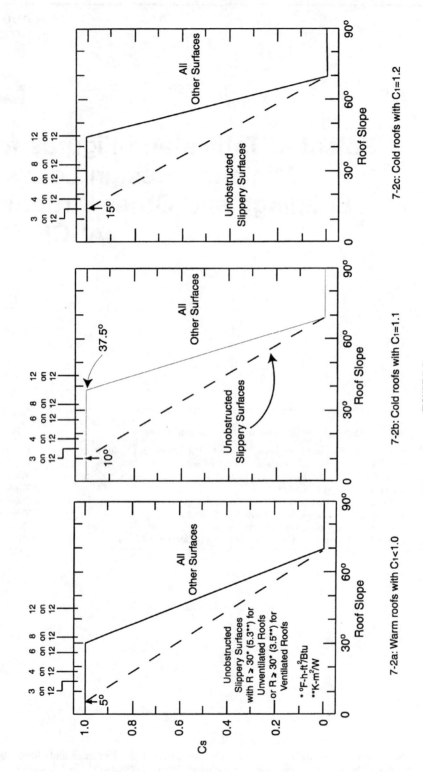

7-2a: Warm roofs with $C_1 < 1.0$

7-2b: Cold roofs with $C_1 = 1.1$

7-2c: Cold roofs with $C_1 = 1.2$

FIGURE 7-2

GRAPHS FOR DETERMINING ROOF SLOPE FACTOR C_s FOR WARM AND COLD ROOFS (SEE TABLE 7-3 FOR C_1 DEFINITIONS)

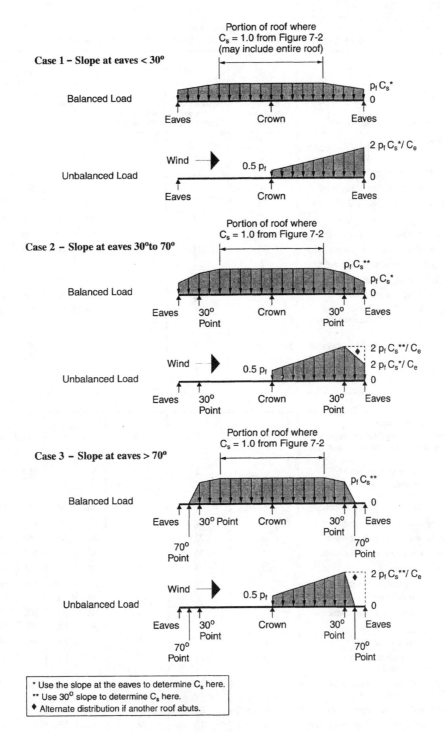

FIGURE 7-3
BALANCED AND UNBALANCED LOADS FOR CURVED ROOFS

TABLE 9.5.2.2
DESIGN COEFFICIENTS AND FACTORS FOR BASIC SEISMIC FORCE-RESISTING SYSTEMS

Basic Seismic Force-Resisting System	Response Modification Coefficient, R^a	System Over-strength Factor, $W_0{}^g$	Deflection Amplification Factor, $C_d{}^b$	Structural System Limitations and Building Height (ft) Limitations[c] Seismic Design Category				
				A&B	C	D[d]	E[e]	F[e]
Bearing Wall Systems								
Ordinary steel concentrically braced frames	4	2	$3\frac{1}{2}$	NL	NL	35[k]	35[k]	NP[k]
Special reinforced concrete shear walls	5	$2\frac{1}{2}$	5	NL	NL	160	160	100
Ordinary reinforced concrete shear walls	4	$2\frac{1}{2}$	4	NL	NL	NP	NP	NP
Detailed plain concrete shear walls	$2\frac{1}{2}$	$2\frac{1}{2}$	2	NL	NP	NP	NP	NP
Ordinary plain concrete shear walls	$1\frac{1}{2}$	$2\frac{1}{2}$	$1\frac{1}{2}$	NL	NP	NP	NP	NP
Special reinforced masonry shear walls	5	$2\frac{1}{2}$	$3\frac{1}{2}$	NL	NL	160	160	100
Intermediate reinforced masonry shear walls	$3\frac{1}{2}$	$2\frac{1}{2}$	$2\frac{1}{4}$	NL	NL	NP	NP	NP
Ordinary reinforced masonry shear walls	2	$2\frac{1}{2}$	$1\frac{3}{4}$	NL	160	NP	NP	NP
Detailed plain masonry shear walls	2	$2\frac{1}{2}$	$1\frac{3}{4}$	NL	NP	NP	NP	NP
Ordinary plain masonry shear walls	$1\frac{1}{2}$	$2\frac{1}{2}$	$1\frac{1}{4}$	NL	NP	NP	NP	NP
Light-framed walls sheathed with wood structural panels rated for shear resistance or steel sheets	6	3	4	NL	NL	65	65	65
Light-framed walls with shear panels of all other materials	2	$2\frac{1}{2}$	2	NL	NL	35	NP	NP
Light-framed wall systems using flat strap bracing	4	2	$3\frac{1}{2}$	NL	NL	65	65	65
Building Frame Systems								
Steel eccentrically braced frames, moment resisting, connections at columns away from links	8	2	4	NL	NL	160	160	100
Steel eccentrically braced frames, non-moment resisting, connections at columns away from links	7	2	4	NL	NL	160	160	100
Special steel concentrically braced frames	6	2	5	NL	NL	160	160	100
Ordinary steel concentrically braced frames	5	2	$4\frac{1}{2}$	NL	NL	35[k]	35[k]	NP[k]
Special reinforced concrete shear walls	6	$2\frac{1}{2}$	5	NL	NL	160	160	100
Ordinary reinforced concrete shear walls	5	$2\frac{1}{2}$	$4\frac{1}{2}$	NL	NL	NP	NP	NP
Detailed plain concrete shear walls	3	$2\frac{1}{2}$	$2\frac{1}{2}$	NL	NP	NP	NP	NP
Ordinary plain concrete shear walls	2	$2\frac{1}{2}$	2	NL	NP	NP	NP	NP
Composite eccentrically braced frames	8	2	4	NL	NL	160	160	100
Composite concentrically braced frames	5	2	$4\frac{1}{2}$	NL	NL	160	160	100
Ordinary composite braced frames	3	2	3	NL	NL	NP	NP	NP
Composite steel plate shear walls	$6\frac{1}{2}$	$2\frac{1}{2}$	$5\frac{1}{2}$	NL	NL	160	160	100
Special composite reinforced concrete shear walls with steel elements	6	$2\frac{1}{2}$	5	NL	NL	160	160	100
Ordinary composite reinforced concrete shear walls with steel elements	5	$2\frac{1}{2}$	$4\frac{1}{4}$	NL	NL	NP	NP	NP
Special reinforced masonry shear walls	$5\frac{1}{2}$	$2\frac{1}{2}$	4	NL	NL	160	160	100
Intermediate reinforced masonry shear walls	4	$2\frac{1}{2}$	4	NL	NL	NP	NP	NP
Ordinary reinforced masonry shear walls	$2\frac{1}{2}$	$2\frac{1}{2}$	$2\frac{1}{4}$	NL	160	NP	NP	NP

TABLE 9.5.2.2 — continued
DESIGN COEFFICIENTS AND FACTORS FOR BASIC SEISMIC FORCE-RESISTING SYSTEMS

Basic Seismic Force-Resisting System	Response Modification Coefficient, R^a	System Over-strength Factor, $W_0{}^g$	Deflection Amplification Factor, $C_d{}^b$	Structural System Limitations and Building Height (ft) Limitations[c]				
				Seismic Design Category				
				A&B	C	D[d]	E[e]	F[e]
Detailed plain masonry shear walls	$2\frac{1}{2}$	$2\frac{1}{2}$	$2\frac{1}{4}$	NL	160	NP	NP	NP
Ordinary plain masonry shear walls	$1\frac{1}{2}$	$2\frac{1}{2}$	$1\frac{1}{4}$	NL	NP	NP	NP 65	NP
Light-framed walls sheathed with wood structural panels rated for shear resistance or steel sheets	$6\frac{1}{2}$	$2\frac{1}{2}$	$4\frac{1}{4}$	NL	NL	65	65	65
Light-framed walls with shear panels of all other materials	$2\frac{1}{2}$	$2\frac{1}{2}$	$2\frac{1}{2}$	NL	NL	35	NP	NP
Moment Resisting Frame Systems								
Special steel moment frames	8	3	$5\frac{1}{2}$	NL	NL	NL	NL	NL
Special steel truss moment frames	7	3	$5\frac{1}{2}$	NL	NL	160	100	NP
Intermediate steel moment frames	4.5	3	4	NL	NL	35[h]	NP[h, i]	NP[h, i]
Ordinary steel moment frames	3.5	3	3	NL	NL	NP[h, i]	NP[h, i]	NP[h, i]
Special reinforced concrete moment frames	8	3	$5\frac{1}{2}$	NL	NL	NL	NL	NL
Intermediate reinforced concrete moment frames	5	3	$4\frac{1}{2}$	NL	NL	NP	NP	NP
Ordinary reinforced concrete moment frames	3	3	$2\frac{1}{2}$	NL	NP	NP	NP	NP
Special composite moment frames	8	3	$5\frac{1}{2}$	NL	NL	NL	NL	NL
Intermediate composite moment frames	5	3	$4\frac{1}{2}$	NL	NL	NP	NP	NP
Composite partially restrained moment frames	6	3	$5\frac{1}{2}$	160	160	100	NP	NP
Ordinary composite moment frames	3	3	$2\frac{1}{2}$	NL	NP	NP	NP	NP
Special masonry moment frames	$5\frac{1}{2}$	3	5	NL	NL	160	160	100
Dual Systems with Special Moment Frames Capable of Resisting at Least 25% of Prescribed Seismic Forces								
Steel eccentrically braced frames, moment resisting connections, at columns away from links	8	$2\frac{1}{2}$	4	NL	NL	NL	NL	NL
Steel eccentrically braced frames, non-moment resisting connections, at columns away from links	7	$2\frac{1}{2}$	4	NL	NL	NL	NL	NL
Special steel concentrically braced frames	8	$2\frac{1}{2}$	$6\frac{1}{2}$	NL	NL	NL	NL	NL
Special reinforced concrete shear walls	8	$2\frac{1}{2}$	$6\frac{1}{2}$	NL	NL	NL	NL	NL
Ordinary reinforced concrete shear walls	7	$2\frac{1}{2}$	6	NL	NL	NP	NP	NP
Composite eccentrically braced frames	8	$2\frac{1}{2}$	4	NL	NL	NL	NL	NL
Composite concentrically braced frames	6	$2\frac{1}{2}$	5	NL	NL	NL	NL	NL
Composite steel plate shear walls	8	$2\frac{1}{2}$	$6\frac{1}{2}$	NL	NL	NL	NL	NL
Special composite reinforced concrete shear walls with steel elements	8	$2\frac{1}{2}$	$6\frac{1}{2}$	NL	NL	NL	NL	NL

TABLE 9.5.2.2 – continued
DESIGN COEFFICIENTS AND FACTORS FOR BASIC SEISMIC FORCE-RESISTING SYSTEMS

Basic Seismic Force-Resisting System	Response Modification Coefficient, R^a	System Over-strength Factor, $W_0{}^g$	Deflection Amplification Factor, $C_d{}^b$	Structural System Limitations and Building Height (ft) Limitations^c Seismic Design Category				
				A&B	C	D^d	E^e	F^e
Ordinary composite reinforced concrete shear walls with steel elements	7	$2\frac{1}{2}$	6	NL	NL	NP	NP	NP
Special reinforced masonry shear walls	7	3	$6\frac{1}{2}$	NL	NL	NL	NL	NL
Intermediate reinforced masonry shear walls	6	$2\frac{1}{2}$	5	NL	NL	NL	NL	NL
Ordinary steel concentrically braced frames	6	$2\frac{1}{2}$	5	NL	NL	NL	NL	NL
Dual Systems with Intermediate Moment Frames Capable of Resisting at Least 25% of Prescribed Seismic Forces								
Special steel concentrically braced frames^f	$4\frac{1}{2}$	$2\frac{1}{2}$	$4\frac{1}{2}$	NL	NL	35	NP	NP$^{h,\,i}$
Special reinforced concrete shear walls	6	$2\frac{1}{2}$	5	NL	NL	160	100	100
Ordinary reinforced masonry shear walls	3	3	$2\frac{1}{2}$	NL	160	NP	NP	NP
Intermediate reinforced masonry shear walls	5	3	$4\frac{1}{2}$	NL	NL	NP	NP	NP
Composite concentrically braced frames	5	$2\frac{1}{2}$	$4\frac{1}{2}$	NL	NL	160	100	NP
Ordinary composite braced frames	4	$2\frac{1}{2}$	3	NL	NL	NP	NP	NP
Ordinary composite reinforced concrete shear walls with steel elements	5	3	$4\frac{1}{2}$	NL	NL	NP	NP	NP
Ordinary steel concentrically braced frames	5	$2\frac{1}{2}$	$4\frac{1}{2}$	NL	NL	160	100	NP
Ordinary reinforced concrete shear walls	$5\frac{1}{2}$	$2\frac{1}{2}$	$4\frac{1}{2}$	NL	NL	NP	NP	NP
Inverted Pendulum Systems and Cantilevered Column Systems								
Special steel moment frames	$2\frac{1}{2}$	2	$2\frac{1}{2}$	NL	NL	NL	NL	NL
Ordinary steel moment frames	$1\frac{1}{4}$	2	$2\frac{1}{2}$	NL	NL	NP	NP	NP
Special reinforced concrete moment frames	$2\frac{1}{2}$	2	$1\frac{1}{4}$	NL	NL	NL	NL	NL
Structural Steel Systems Not Specifically Detailed for Seismic Resistance	3	3	3	NL	NL	NP	NP	NP

^{a} Response modification coefficient, R, for use throughout the standard. Note R reduces forces to a strength level, not an allowable stress level.

^{b} Deflection amplification factor, C_d, for use in Sections 9.5.3.7.1 and 9.5.3.7.2

^{c} NL = Not Limited and NP = Not Permitted. For metric units use 30 m for 100 ft and use 50 m for 160 ft. Heights are measured from the base of the structure as defined in Section 9.2.1.

^{d} See Section 9.5.2.2.4.1 for a description of building systems limited to buildings with a height of 240 ft (75 m) or less.

^{e} See Sections 9.5.2.2.4 and 9.5.2.2.4.5 for building systems limited to buildings with a height of 160 ft (50 m) or less.

^{f} Ordinary moment frame is permitted to be used in lieu of intermediate moment frame in Seismic Design Categories B and C.

^{g} The tabulated value of the overstrength factor, W_0, may be reduced by subtracting $\frac{1}{2}$ for structures with flexible diaphragms but shall not be taken as less than 2.0 for any structure.

^{h} Steel ordinary moment frames and intermediate moment frames are permitted in single-story buildings up to a height of 60 ft, when the moment joints of field connections are constructed of bolted end plates and the dead load of the roof does not exceed 15 psf.

^{i} Steel ordinary moment frames are permitted in buildings up to a height of 35 ft where the dead load of the walls, floors, and roofs does not exceed 15 psf.

^{k} Steel ordinary concentrically braced frames are permitted in single-story buildings up to a height of 60 ft when the dead load of the roof does not exceed 15 psf and in penthouse structures.

TABLE 9.5.2.5.1
PERMITTED ANALYTICAL PROCEDURES

Seismic Design Category	Structural Characteristics	Index Force Analysis Section 9.5.3	Simplified Analysis Section 9.5.4	Equivalent Lateral Force Analysis Section 9.5.5	Modal Response Spectrum Analysis Section 9.5.6	Linear Response History Analysis Section 9.5.7	Nonlinear Response History Analysis Section 9.5.8
A	All structures	P	P	P	P	P	P
B, C	SUG-1 buildings of light-framed construction not exceeding three stories in height	NP	P	P	P	P	P
	Other SUG-1 buildings not exceeding two stories in height	NP	P	P	P	P	P
	All other structures	NP	NP	P	P	P	P
D, E, F	SUG-1 buildings of light-framed construction not exceeding three stories in height	NP	P	P	P	P	
	Other SUG-1 buildings not exceeding two stories in height	NP	P	P	P	P	P
	Regular structures with $T < 3.5\ T_s$ and all structures of light-frame construction	NP	NP	P	P	P	P
	Irregular structures with $T < 3.5\ T_s$ and having only plan irregularities type 2, 3, 4, or 5 of Table 9.5.2.3.2 or vertical irregularities type 4 or 5 of Table 9.5.2.3.3	NP	NP	P	P	P	P
	All other structures	NP	NP	NP	P	P	P

Notes: P—indicates permitted, NP—indicates not permitted

TABLE 9.5.2.8
ALLOWABLE STORY DRIFT, Δ_a[a]

Structure	Seismic Use Group		
	I	II	III
Structures, other than masonry shear wall or masonry wall frame structures, four stories or less with interior walls, partitions, ceilings and exterior wall systems that have been designed to accommodate the story drifts.	$0.025h_{sx}$[b]	$0.020h_{sx}$	$0.015h_{sx}$
Masonry cantilever shear wall structures[c]	$0.010h_{sx}$	$0.010h_{sx}$	$0.010h_{sx}$
Other masonry shear wall structures	$0.007h_{sx}$	$0.007h_{sx}$	$0.007h_{sx}$
Masonry wall frame structures	$0.013h_{sx}$	$0.013h_{sx}$	$0.010h_{sx}$
All other structures	$0.020h_{sx}$	$0.015h_{sx}$	$0.010h_{sx}$

[a] h_{sx} is the story height below Level x.
[b] There shall be no drift limit for single-story structures with interior walls, partitions, ceilings, and exterior wall systems that have been designed to accommodate the story drifts. The structure separation requirement of Section 9.5.2.8 is not waived.
[c] Structures in which the basic structural system consists of masonry shear walls designed as vertical elements cantilevered from their base or foundation support which are so constructed that moment transfer between shear walls (coupling) is negligible.

TABLE 9.5.5.3.2
VALUES OF APPROXIMATE PERIOD PARAMETERS C_t AND x

Structure Type	C_t	x
Moment resisting frame systems of steel in which the frames resist 100% of the required seismic force and are not enclosed or adjoined by more rigid components that will prevent the frames from deflecting when subjected to seismic forces	$0.028(0.068)$[a]	0.8
Moment resisting frame systems of reinforced concrete in which the frames resist 100% of the required seismic force and are not enclosed or adjoined by more rigid components that will prevent the frame from deflecting when subjected to seismic forces	$0.016(0.044)$[a]	0.9
Eccentrically braced steel frames	$0.03(0.07)$[a]	0.75
All other structural systems	$0.02(0.055)$	0.75

[a] — metric equivalents are shown in parentheses

TABLE 9.6.2.2
ARCHITECTURAL COMPONENT COEFFICIENTS

Architectural Component or Element	a_p [a]	R_p [b]
Interior Nonstructural Walls and Partitions		
Plain (unreinforced) masonry walls	1	1.5
All other walls and partitions	1	2.5
Cantilever Elements (Unbraced or Braced to Structural Frame Below Its Center of Mass)		
Parapets and cantilever interior nonstructural walls	2.5	2.5
Chimneys and stacks when laterally braced or supported by the structural frame	2.5	2.5
Cantilever Elements (Braced to Structural Frame Above Its Center of Mass)		
Parapets	1.0	2.5
Chimneys and stacks	1.0	2.5
Exterior nonstructural walls	1.0^b	2.5
Exterior Nonstructural Wall Elements and Connections		
Wall element	1	2.5
Body of wall panel connections	1	2.5
Fasteners of the connecting system	1.25	1
Veneer		
Limited deformability elements and attachments	1	2.5
Low deformability elements and attachments	1	2.5
Penthouses (Except when Framed by an Extension of the Building Frame)	2.5	3.5
Ceilings		
All	1	2.5
Cabinets		
Storage cabinets and laboratory equipment	1	2.5
Access Floors		
Special access floors (designed in accordance with Section 9.6.2.7.2)	1	2.5
All other	1	1.5
Appendages and Ornamentations	2.5	2.5
Signs and Billboards	2.5	2.5
Other Rigid Components		
High deformability elements and attachments	1	3.5
Limited deformability elements and attachments	1	2.5
Low deformability materials and attachments	1	1.5
Other Flexible Components	2.5	3.5
High deformability elements and attachments	2.5	2.5
Limited deformability elements and attachments	2.5	1.5
Low deformability materials and attachments		

[a] A lower value for a_p shall not be used unless justified by detailed dynamic analysis. The value for a_p shall not be less than 1.00. The value of $a_p = 1$ is for equipment generally regarded as rigid and rigidly attached. The value of $a_p = 2.5$ is for equipment generally regarded as flexible or flexibly attached. See Section 9.2.1 for definitions of rigid and flexible.
[b] Where flexible diaphragms provide lateral support for walls and partitions, the design forces for anchorage to the diaphragm shall be as specified in Section 9.5.2.6.

TABLE 9.6.3.2
MECHANICAL AND ELECTRICAL COMPONENTS SEISMIC
COEFFICIENTS

Mechanical and Electrical Component or Element[b]	a_p[a]	R_p
General Mechanical Equipment		
Boilers and furnaces	1.0	2.5
Pressure vessels on skirts and free-standing	2.5	2.5
stacks	2.5	2.5
Cantilevered chimneys	2.5	2.5
Other	1.0	2.5
Manufacturing and Process Machinery		
General	1.0	2.5
Conveyors (non-personnel)	2.5	2.5
Piping Systems		
High deformability elements and attachments	1.0	3.5
Limited deformability elements and attachments	1.0	2.5
Low deformability elements and attachments	1.0	1.5
HVAC Systems		
Vibration isolated	2.5	2.5
Nonvibration isolated	1.0	2.5
Mounted in-line with ductwork	1.0	2.5
Other	1.0	2.5
Elevator Components	1.0	2.5
Escalator Components	1.0	2.5
Trussed Towers (free-standing or guyed)	2.5	2.5
General Electrical		
Distribution systems (bus ducts, conduit, cable tray)	2.5	5.0
Equipment	1.0	2.5
Lighting Fixtures	1.0	1.5

[a] A lower value for a_p shall not be used unless justified by detailed dynamic analyses. The value for a_p shall not be less than 1.00. The value of $a_p = 1$ is for equipment generally regarded as rigid or rigidly attached. The value of $a_p = 2.5$ is for equipment generally regarded as flexible or flexibly attached. See Section 9.2.2 for definitions of rigid and flexible.
[b] Components mounted on vibration isolation systems shall have a bumper restraint or snubber in each horizontal direction. The design force shall be taken as $2F_p$ if the maximum clearance (air gap) between the equipment support frame and restraint is greater than 1/4 in. If the maximum clearance is specified on the construction documents to be not greater than 1/4 in., the design force may be taken as F_p.

E

SI Metric Units

Introduction

In 1960 the Eleventh General Conference of Weights and Measures adopted the name International System of Units (accepted abbreviation SI, from Système International d'Unites) for a practical and consistent set of units of measure. Rules for common usage, notation, and abbreviation were adopted. Most industrial nations in the world have adopted and converted to SI. SI is not the old metric system (cgs or MKS).

A large number of code and industry design tables are required to carry out a timber structural design. Many of these tables now have dual units which includes both U.S. Customary units and SI metric units. However, the primary units are still U.S. Customary units, and the SI equivalent dimensions are generally provided for information purposes ("soft" conversion). Most wood products are still produced in traditional U.S. Customary unit sizes, and a "hard" conversion to metric sizes has not taken place. Consequently most practical structural design in the U.S. is still done in U.S. Customary units.

However, the future trend will be toward the use of SI units. The following brief introduction to SI is included to aid the structural engineer in conversion between U.S. Customary (USC) units and SI units. Additional information can be found in ASTM E380 "Metric Practice Excerpts."

Notation

SI is made up of seven base units, two supplementary units, and many consistent derived units. The following units are pertinent to structural design.

	Quantity	Unit	SI symbol
Base unit	Length	meter	m
	Mass	kilogram	kg
	Time	second	s
Supplementary unit	Plane angle	radian	rad
Derived unit	Density (weight)		N/m^3
	Density (mass)		kg/m^3
	Energy	joule	$J (N \cdot m)$
	Force	newton	$N (kg \cdot m/s^2)$
	Frequency	hertz	$Hz (1/s)$
	Moment of force		$N \cdot m (kg \cdot m^2/s^2)$
	Moment of inertia		m^4
	Stress, pressure	pascal	$Pa (N/m^2)$

Prefixes

The following are SI-approved prefixes and their abbreviations used to denote very small or very large quantities.

Multiplication factor	Name	Symbol
$0.000\ 000\ 001 = 10^{-9}$	nano	n
$0.000\ 001 = 10^{-6}$	micro	μ
$0.001 = 10^{-3}$	milli	m
$1000 = 10^3$	kilo	k
$1\ 000\ 000 = 10^6$	mega	M
$1000\ 000\ 000 = 10^9$	giga	G

Conversion Factors

The following is a list of structurally pertinent conversions between U.S. Customary units and SI units.

Area

$$1 \text{ in.}^2 = 645.2 \text{ mm}^2$$

$$1 \text{ ft}^2 = 92.90 \times 10^{-3} \text{ m}^2$$

Bending Moment or Torque

$$1 \text{ in.-lb} = 0.1130 \text{ N} \cdot \text{m}$$

$$1 \text{ ft-lb} = 1.356 \text{ N} \cdot \text{m}$$

$$1 \text{ ft-k} = 1.356 \text{ kN} \cdot \text{m}$$

Lengths or Displacements

$$1 \text{ in.} = 25.40 \text{ mm}$$

$$1 \text{ ft} = 0.3048 \text{ m}$$

Loads

$$1 \text{ lb} = 4.448 \text{ N}$$

$$1 \text{ k} = 4.448 \text{ kN}$$

$$1 \text{ lb/ft} = 14.59 \text{ N/m}$$

$$1 \text{ k/ft} = 14.59 \text{ kN/m}$$

Moment of Inertia

$$1 \text{ in.}^4 = 0.4162 \times 10^6 \text{ mm}^4$$

$$1 \text{ ft}^4 = 8.631 \times 10^{-3} \text{ m}^4$$

Section Modulus or Volume

$$1 \text{ in.}^3 = 16.39 \times 10^3 \text{ mm}^3$$

$$1 \text{ ft}^3 = 28.32 \times 10^{-3} \text{ m}^3$$

Stress and Modulus of Elasticity

$$1 \text{ psi} = 6.895 \text{ kPa}$$

$$1 \text{ ksi} = 6.895 \text{ MPa}$$

$$1 \text{ psf} = 47.88 \text{ Pa}$$

Unit Weight, Density

$$1 \text{ lb/ft}^3 = 0.157 \text{ kN/m}^3$$

Index

ABOUT THE AUTHORS

DONALD E. BREYER, P.E., is a Professor Emeritus in the Department of Engineering Technology at California State Polytechnic University, Pomona, California.

KENNETH J. FRIDLEY, Ph.D., is a Professor and Head of the Department of Civil and Environmental Engineering at the University of Alabama, Tuscaloosa, Alabama.

DAVID G. POLLOCK, P.E., Ph.D., is an Associate Professor in the Department of Civil and Environmental Engineering at Washington State University, Pullman, Washington.

KELLY E. COBEEN, S.E., is the Principal of Cobeen & Associates Structural Engineering, Lafayette, California.

The authors all actively serve as members of the Wood Design Standards Committee of the American Forest and Paper Association, which is responsible for maintaining and revising the design provisions of the National Design Specification for Wood Construction. Ms. Cobeen is also a member of Technical Subcommittee 7, Wood Structures, of the Provisions Update Committee for the NEHRP Recommended Provisions for Seismic Regulations for New Buildings and Other Structures.